Handbook of Spatial Statistics

Chapman & Hall/CRC
Handbooks of Modern Statistical Methods

Series Editor

Garrett Fitzmaurice
Department of Biostatistics
Harvard School of Public Health
Boston, MA, U.S.A.

Aims and Scope

The objective of the series is to provide high-quality volumes covering the state-of-the-art in the theory and applications of statistical methodology. The books in the series are thoroughly edited and present comprehensive, coherent, and unified summaries of specific methodological topics from statistics. The chapters are written by the leading researchers in the field, and present a good balance of theory and application through a synthesis of the key methodological developments and examples and case studies using real data.

The scope of the series is wide, covering topics of statistical methodology that are well developed and find application in a range of scientific disciplines. The volumes are primarily of interest to researchers and graduate students from statistics and biostatistics, but also appeal to scientists from fields where the methodology is applied to real problems, including medical research, epidemiology and public health, engineering, biological science, environmental science, and the social sciences.

Published Titles

Longitudinal Data Analysis
Edited by Garrett Fitzmaurice, Marie Davidian, Geert Verbeke, and Geert Molenberghs

Handbook of Spatial Statistics
Edited by Alan E. Gelfand, Peter J. Diggle, Montserrat Fuentes, and Peter Guttorp

Chapman & Hall/CRC
Handbooks of Modern
Statistical Methods

Handbook of Spatial Statistics

Edited by

Alan E. Gelfand
Peter J. Diggle
Montserrat Fuentes
Peter Guttorp

CRC Press
Taylor & Francis Group
Boca Raton London New York

CRC Press is an imprint of the
Taylor & Francis Group, an **informa** business
A CHAPMAN & HALL BOOK

CRC Press
Taylor & Francis Group
6000 Broken Sound Parkway NW, Suite 300
Boca Raton, FL 33487-2742

Printed in the United States of America on acid-free paper
10 9 8 7 6 5 4 3 2

International Standard Book Number: 978-1-4200-7287-7 (Hardback)

Library of Congress Cataloging-in-Publication Data

Handbook of spatial statistics / [edited by] Alan E. Gelfand ... [et al.].
p. cm. -- (Chapman & Hall/CRC handbooks of modern statistical methods)
Includes bibliographical references and index.
ISBN 978-1-4200-7287-7 (hardcover : alk. paper)
1. Spatial analysis (Statistics) 2. Mathematical statistics--Methodology. I. Gelfand, Alan E., 1945-
II. Title. III. Series.

QA278.2.H374 2010
519.5--dc22 2010006620

Visit the Taylor & Francis Web site at
http://www.taylorandfrancis.com

and the CRC Press Web site at
http://www.crcpress.com

Contents

Part III Discrete Spatial Variation

Part IV Spatial Point Patterns

Part V Spatio-Temporal Processes

Preface

Spatial statistics has an unusual history as a field within the discipline of statistics. The stochastic process theory that underlies much of the field was developed within the mathematical sciences by probabilists, whereas, early on, much of the statistical methodology was developed quite independently. In fact, this methodology grew primarily from the different areas of application, including mining engineering, which led to the development of geostatistics by Matheron and colleagues; agriculture with spatial considerations informed by the thinking of Fisher on randomization and blocking; and forestry, which was the setting for the seminal PhD thesis of Matérn. As a result, for many years spatial statistics labored on the fringe of mainstream statistics. However, the past 20 years has seen an explosion of interest in space and space–time problems. This has been largely fueled by the increased availability of inexpensive, high-speed computing (as has been the case for many other areas). Such availability has enabled the collection of large spatial and spatio-temporal datasets across many fields, has facilitated the widespread usage of sophisticated geographic information systems (GIS) software to create attractive displays, and has endowed the ability to investigate (fit and infer under) challenging, evermore appropriate and realistic models.

In the process, spatial statistics has been brought into the mainstream of statistical research with a proliferation of books (including this one), conferences and workshops, as well as courses and short courses. Moreover, while there has been a body of strong theoretical work developed since the 1950s (Whittle, Bartlett, Besag, etc.), it is safe to say that, broadly, spatial statistics has been changed from a somewhat ad hoc field to a more model-driven one. Though the entire field continues to be in flux, we, as editors taking on this project, feel that it is now mature enough to warrant a handbook. In this regard, we hope that this volume will serve as a worthy successor to Noel Cressie's (1993) encyclopedic effort, which has served as *the* "handbook" since its publication. However, this observation further argues the need for a new handbook. In addition to a dramatic increase in size since Cressie's book appeared, the literature has become exceptionally diverse, in part due to the diversity of applications. Collecting a review of the major portion of this work in one place, with an extensive bibliography, should certainly assist future research in the field.

Spatial statistics is generally viewed as being comprised of three major branches: (1) continuous spatial variation, i.e., point referenced data; (2) discrete spatial variation, including lattice and areal unit data; and (3) spatial point patterns. However, this handbook consists of 31 chapters spread over six sections. Our rationale is as follows. The three main areas form the subject matter of Parts II, III and IV, respectively, and span the majority of the book. However, we felt that a historical introduction detailing the aforementioned evolution of the field would be valuable. In addition, with so much space–time work being available, it was clear that a section on this material was needed. Finally, as anticipated with such a project, we needed an "Additional Topics" section to collect some important topics that build upon material presented in the earlier sections, but, in some sense, seemed beyond those earlier ones. We acknowledge that some readers will feel that certain topics are underrepresented or inadequately discussed. We cannot disagree, noting only that no selection would make all readers happy.

The list of contributors to this volume is outstanding, a collection of very prominent researchers in the field. Moreover, we have attempted to give the handbook a more unified, integrated feel. We specifically precluded the notion of stand-alone contributions. Rather, for each chapter, the editors provided the contributor(s) with a core set of material to be covered as well as chapter reviews, which reinforced this. In addition, though there is inherent variability in technical levels across the chapters, we have targeted a background/preparation at the level of a master's in statistics. We hope that this will make the book accessible to researchers in many other fields, researchers seeking an easy entry to particular topics. In this regard, we have tried to balance theory and application while targeting a strong emphasis on modeling and introducing as many real data analysis examples as feasible.

We have focused a bit less on computation (though there are some exceptions) and not at all on GIS displays and software (arguing that this is primarily descriptive). In this spirit, we have also avoided discussion of what is often referred to as *spatial analysis* because this work resides primarily in the geography literature and does not tend to be stochastic. However, with regard to computing issues, typically, there is some discussion in most of the nontheoretical chapters. We also decided not to say much about available software packages. In our view, the material we have included is perceived as durable, while software for spatial and spatio-temporal analysis is evolving rapidly and is perhaps best pursued through appropriate references cited in the volume.

In summary, we have enjoyed assembling this handbook, finding collegial ease and consistent agreement across the editorial team. We greatly appreciate the efforts of our contributors who provided their chapters and their revisions in a timely manner to enable this book to come together with minimal delay. And, we appreciate the encouragement of Rob Calver from Taylor & Francis to undertake this project and Sarah Morris and Marsha Pronin, also from T&F, who helped put the entire package together.

Alan E. Gelfand

Peter J. Diggle

Montserrat Fuentes

Peter Guttorp

List of Contributors

Jeffrey L. Anderson
National Center for Atmospheric Research
Boulder, Colorado

Adrian Baddeley
CSIRO Mathematical and Information Sciences
Perth, Western Australia
and
School of Mathematics and Statistics
University of Western Australia
Crawley, Western Australia

Sudipto Banerjee
School of Public Health
University of Minnesota
Minneapolis, Minnesota

David R. Brillinger
Department of Statistics
University of California
Berkeley, California

Brad Carlin
Division of Biostatistics
School of Public Health
University of Minnesota
Minneapolis, Minnesota

Peter J. Diggle
Division of Medicine
Lancaster University
Lancaster, United Kingdom

Montserrat Fuentes
Department of Statistics
North Carolina State University
Raleigh, North Carolina

Edith Gabriel
IUT, Département STID
University of Avignon
Avignon, France

Dani Gamerman
Instituto de Matemática – UFRJ
Rio de Janeiro, Brazil

Alan E. Gelfand
Department of Statistical Science
Duke University
Durham, North Carolina

Tilmann Gneiting
Institut fur Angewandte Mathematik
Universitat Heidelberg
Heidelberg, Germany

Peter Guttorp
Department of Statistics
University of Washington
Seattle, Washington

Leonhard Held
Biostatistics Unit
Institute of Social and Preventive Medicine
University of Zurich
Zurich, Switzerland

Valerie Isham
Department of Statistical Science
University College London
London, United Kingdom

James Lesage
Department of Finance and Economics
McCoy College of Business Administration
Texas State University – San Marcos
San Marcos, Texas

Hilary Lyons
Department of Statistics
University of Washington
Seattle, Washington

Jesper Møller
Department of Mathematical Sciences
Aalborg University
Aalborg, Denmark

Douglas W. Nychka
National Center for Atmospheric Research
Boulder, Colorado

Kelley Pace
Department of Finance
E.J. Ourso College of Business
Louisiana State University
Baton Rouge, Louisiana

Hävard Rue
Department of Mathematical Sciences
Norwegian University of Science and Technology
Trondheim, Norway

Paul D. Sampson
Department of Statistics
University of Washington
Seattle, Washington

Mark Steel
Department of Statistics
University of Warwick
Coventry, United Kingdom

Michael Stein
Department of Statistics
University of Chicago
Chicago, Illinois

Marie-Colette van Lieshout
Probability and Stochastic Networks (PNA2)
Amsterdam, the Netherlands

Jonathan Wakefield
Departments of Statistics and Biostatistics
University of Washington
Seattle, Washington

Lance Waller
Rollins School of Public Health
Emory University
Atlanta, Georgia

Chris Wikle
Department of Statistics
University of Missouri
Columbia, Missouri

Jim Zidek
Department of Statistics
University of British Columbia
Vancouver, British Columbia, Canada

Dale Zimmerman
Department of Statistics and Actuarial Science
University of Iowa
Iowa City, Iowa

Part I

1

Historical Introduction

Peter J. Diggle

CONTENTS

In this chapter, I aim to set the scene for what follows by summarizing the historical development of spatial statistics as a recognizable sub-branch of the statistics discipline. As with many other areas of statistical methodology, the important developments came not from the study of mathematics for its own sake, but from the needs of substantive applications, including in this case areas as diverse as astronomy, agriculture, ecology and mineral exploration.

1.1 Antecedents

The fundamental feature of spatial statistics is its concern with phenomena whose spatial location is either of intrinsic interest or contributes directly to a stochastic model for the phenomenon in question. Perhaps the earliest examples are problems in geometrical probability.

Buffon's needle, named after the Comte de Buffon (1707–1788), is perhaps the most famous problem of this kind. The question is easily posed. Suppose that a needle of length x is thrown at random onto a table marked by parallel lines, a distance $d > x$ apart. What is the probability that the needle crosses one of the parallel lines? To solve the problem, we need to say what we mean by "at random." A natural definition is that the distance of the needle's center from the closest parallel line should be uniform on $(0, d/2)$ and that the acute angle that the needle makes relative to the parallel lines should be uniform on $(0, \pi/2)$. However, the first part of this definition begs the slightly subtle question of what happens near the edges of the table or, if the table is deemed to be of infinite size so as to avoid the question, what physical meaning can we attach to throwing the needle at random onto the table? The mathematical resolution of this difficulty lies in the theory of point processes. Spatial point processes (meaning point processes in Euclidean space of dimension two or three) first occurred (in the form of Poisson processes, though without that name) in physics and astronomy. To my knowledge, the first introduction of a Poisson process was in 1858 when Rudolf Clausius needed to calculate the mean free path of a molecule in a volume of gas in order to defend from criticism the then new molecular theory of heat (Peter Guttorp, personal communication). Properties of the Poisson process were derived and used as benchmarks relative to which the empirical properties of astronomical data could be judged, a mode of inquiry that we might now call significance testing with the Poisson process acting as a *dividing hypothesis*, using this term in the same sense as used by Cox (1977). In the current context, the Poisson process serves as a demarcation line between spatial point processes whose realizations can be described as being more regular, or less regular, than those of the Poisson process.

As an early example, Hertz (1909) derived the probability distribution of the distance from an arbitrary point of a three-dimensional Poisson process to the nearest other point. Comparison of this theoretical distribution with the observed distribution of these so-called nearest neighbor distances could be used to test whether an observed pattern is compatible with the Poisson process model, an idea that resurfaced many years later in ecology (Skellam, 1952; Clark and Evans, 1954).

Other early work considered the properties of geometrical figures made up of specified numbers of points. Between 1859 and 1860, Professor Simon Newcomb wrote a series of papers in the then newly established *Mathematical Monthly* that included, for example, a calculation of the probability that six stars would lie in a given degree square, on the assumption that stars were scattered at random over the sky, i.e., formed a homogeneous Poisson process (Newcomb, 1859–1860). The study of geometrical probability problems of this kind eventually developed into the field of stochastic geometry (Harding and Kendall, 1974; Stoyan, Kendall and Mecke, 1987).

Unsurprisingly, the homogeneous Poisson process was soon found wanting as a model for naturally occurring patterns of points in space. Two widely occurring phenomena that the Poisson model fails to capture were on the one hand, spatial clustering of points into local aggregations and, on the other, spatial regularity typically arising from the fact that "points" are usually abstractions of nonoverlapping objects of finite size; e.g., cells in a biological tissue section or trees in a forest.

The earliest attempt to build a class of models for spatial clustering was by Neyman (1939), who wanted a model to describe the positions of insect larvae after they had hatched in a field of view from a Poisson process of egg clusters. Although Neyman set up his model to include the spatial dispersal of individual larvae from their point of hatching, this was lost in the subsequent approximations that he made in deriving the probablity distribution of the number of larvae located within a field of view. Hence, Neyman's Type A distribution is simply that of the sum of a Poisson number of Poisson counts, which is strictly appropriate only if larvae from the same egg cluster remain co-located. Neyman and Scott (1958) revisited this construction in three spatial dimensions, again motivated by astronomy, to give what we should probably recognize as the first genuine model for spatially dispersed clusters of points emanating from randomly distributed parent locations.

The natural starting point for building models that recognize points as reference locations for finite-sized objects is to ask what would be meant by a random spatial distribution of nonintersecting circles. The generally accepted definition is a homogeneous Poisson process conditioned by the requirement that, within a designated finite region D, no two points of the process are separated by a distance less than a prespecified value $\delta > 0$. Ripley and Kelly (1977) embedded this construction within their definition of Markov point processes, which remain the most widely used family of models to describe point patterns whose realizations exhibit some degree of spatial regularity.

1.2 Agricultural Field Trials

Spatial considerations were implicit in R. A. Fisher's (1966) seminal work on the development of design-based inference for data from agricultural field trials of the kind that he encountered at the Rothamsted Experimental Station in Hertforshire, England. Fisher was employed at Rothamsted between 1919 and 1933, devoting much of his time to the development of a coherent methodology for the analysis of data from agricutural field trials.

Figure 1.1 is a representation of data from a uniformity trial conducted at Rothamsted and reported in Mercer and Hall (1911). Each small rectangle in the diagram corresponds to a rectangular plot of dimension 3.30 by 2.59 meters, and the plots are color-coded to show the yield of wheat grain from each. Because only a single cultivar was sown, the variation in yield among the 500 plots presumably arises substantially from spatial variation in the microenvironment (e.g., soil fertility, slope, aspect, etc.). It may be reasonable to think of this variation as being stochastic in nature, in which case, a first, albeit naive, model for the data might be

$$Y_{ij} = \mu + Z_{ij} : i = 1, \ldots, 20; \quad j = 1, \ldots, 25 \tag{1.1}$$

where μ is the (hypothetical) population mean yield per plot, i and j denote rows and columns respectively, and the Z_{ij} are mutually independent, zero-mean random perturbations. A modern approach to inference from Equation (1.1) might be to add the assumption that the Z_{ij} are normally distributed and use the likelihood of the resulting stochastic model as the basis for inference. This model is naive because it is obvious from Figure 1.1 that near-neighboring plots tend to give similar yields, thus violating the asumption of mutually independent Z_{ij}. Fisher recognized this difficulty, commenting, for example, on "the widely verified fact that patches in close proximity are commonly more alike, as judged by the yield of crops, than those which are farther apart" (Fisher, 1966, p. 66). Rather than adopt a model-based solution, Fisher advocated the use of blocking as a fundamental design principle to combat the malevolent influence of extraneous, in this case spatial, variation on the precision of the experiment. To illustrate, suppose that we consider the individual columns of Figure 1.2 as blocks. A less naive model than Equation (1.1)

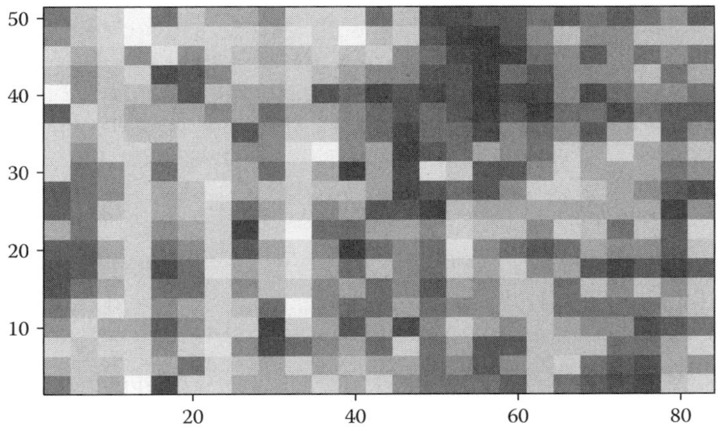

FIGURE 1.1
Data from a wheat uniformity trial reported in Mercer and Hall (1911). Each square represents a plot of dimension 3.30 × 2.59 meters (10.82 × 8.50 feet). Squares are coded to show the yield of wheat grain, in pounds, from each plot, from 2.73 (black) to 5.16 (white).

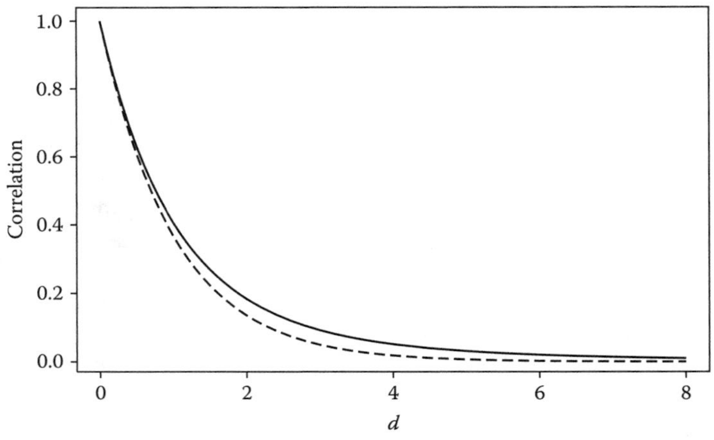

FIGURE 1.2
Correlation between spatial averages over two contiguous squares of side-length d (solid line) and correlation function $\rho(d) = \exp(-d)$ of the underlying spatially continuous process (dashed line).

would now be

$$Y_{ij} = \mu + \beta_j + Z_{ij} : i = 1, ..., 20; \quad j = 1, ..., 25 \tag{1.2}$$

where now the β_j, constrained to sum to zero, represent the amounts by which the expected yield within each block differs from the overall expectation, μ. In effect, Equation (1.2) is a primitive spatial model in which any systematic spatial variation is assumed to operate only in the horizontal direction, but is otherwise arbitrary. The beneficial effects of blocking are well known. In this case, the estimated variance of the Z_{ij} is 0.210 under model (1.1), but reduces to 0.150 under Equation (1.2) and to 0.135 if, perhaps more naturally in the absence of additional information, we define blocks as contiguous 5 by 5 arrays of plots. Fisher himself expressed the view that "it is therefore a safe rule to make the blocks as compact as possible" (Fisher, 1966, p. 66).

Blocking can be thought of as a form of covariate adjustment under the implicit assumption that systematic spatial variation, if it exists at all, is piecewise constant within blocks. An alternative strategy is to adjust plot yields to take account of the average yield in neighboring plots. This form of covariate adjustment was suggested by Papadakis (1937) and turns out to be closely connected with Markov random field models for discrete spatial variation (see Section 1.4.1 below, and Part II of this book). A modern interpretation of the Papadakis adjustment is as a conditional model for the distribution of each plot yield, Y_{ij}, given the average yield, $\bar{Y}_{ij}$ say, over plots deemed to be neighbors of ij, namely,

$$Y_{ij} | \{Y_{kl} : (k, l) \neq (i, j)\} \sim \mathrm{N}(\mu + \beta(\bar{Y}_{ij} - \mu), \tau^2). \tag{1.3}$$

The model (1.3) reduces to Equation (1.1) when $\beta = 0$, whereas the general case is an example of what would now be called a Gaussian Markov random field (Rue and Held, 2005). The connection between Papadakis adjustment and Markov random fields was pointed out by Sir David Cox in the discussion of Besag (1974), and made explicit in Bartlett (1978).

W. F. Gosset (Student) also thought carefully about spatial correlation in this context, but sought what we would now call a model-based solution. In a letter to Karl Pearson in December 1910, he wrote: "Now in general the correlation weakens as the unit of time or space grows larger and I can't help thinking that it would be a great thing to work out the law according to which the correlation is likely to weaken with increase of unit" (Pearson, 1990).

One way to formalize this request for a "law" of spatial correlation is to consider a stationary, spatially continuous stochastic process $S(x)$ with covariance function $\mathrm{Cov}\{S(x), S(y)\} = \sigma^2\rho(u)$ where u is the distance between the locations x and y. The covariance between spatial averages of $S(\cdot)$ over two regions A and B is

$$\gamma(A, B) = (|A| \times |B|)^{-1}\sigma^2 \int_A \int_B \rho(||x - y||)dxdy,$$

where $|\cdot|$ denotes area and $||\cdot||$ distance. Figure 1.2 shows the resulting correlation when A and B are contiguous squares each of side-length d, and $\rho(u) = \exp(-u)$, showing how the correlation decays more slowly than does the underlying exponential correlation function of the process $S(\cdot)$.

Gosset's use of "law" in the singular here is interesting, implying as it does a belief that there is a natural law of spatial variation, akin to a physical law, that could account quite generally for the phenomenon in question. In a similar spirit, Fairfield Smith (1938) proposed that the correlation should diminish according to a power law, a proposition explored further by Whittle (1956, 1962), and more recently by McCullagh and Clifford (2006). The McCullagh and Clifford paper also reviews recent literature on spatial stochastic models for the analysis of agricultural field trials, amongst which important contributions include Wilkinson et al. (1983), Besag and Kempton (1986), and Besag and Higdon (1999).

1.3 Modeling Continuous Spatial Variation

A less ambitious and, in the author's view, more realistic goal than a search for a universal law is to develop a parsimonious class of models that can succeed collectively in capturing the empirical behavior of extraneous spatial variation in a range of scientific settings. From this point of view, a convenient working assumption is that the spatial phenomenon under investigation can be modeled as a Gaussian spatial process, $S(x)$ say, whose mean can be described by a suitable linear model. To complete the specification of the model, we then need only to specify the covariance between $S(x)$ and $S(x')$ for any two locations, x and x'. This task is considerably simplified if we are prepared to assume spatial stationarity, in which case the covariance specification reduces to a scalar parameter $\sigma^2 = \mathrm{Var}\{S(x)\}$ and a correlation function $\rho(u) = \mathrm{Corr}\{S(x), S(x')\}$ where u is the distance between x and x'.

It took a while for this idea to bear fruit, and when it did, it did so independently in at least two very different fields: forestry (Matérn, 1960) and mining engineering (Krige, 1951).

Bertil Matérn spent the majority of his career in the Swedish Royal College of Forestry, now part of the Swedish University of Agricultural Sciences. His 1960 Stockholm University doctoral dissertation (Matérn, 1960, reprinted as Matérn, 1978) set out the class of models for the correlation structure of real-valued, spatially continuous stationary processes that now bear his name. The Matérn correlation function takes the form

$$\rho(u) = \{2^{\kappa-1}\Gamma(\kappa)\}^{-1}(u/\phi)^{\kappa}K_{\kappa}(u/\phi), \tag{1.4}$$

where ϕ is a scale parameter and $K_\kappa(\cdot)$ denotes a modified Bessel function of order κ. One of the attractions of this class of models is that the integer part of κ determines the mean square differentiability of the underlying process. This matters because the differentiability affects the behavior of predictions made using the model, but κ is poorly identified in typically sized datasets. For this reason, the fact that the numerical value of κ has a tangible interpretation in terms of the smoothness of the underlying spatial process is very helpful in informing its choice or, in a Bayesian context, in choosing a suitably informative prior. Guttorp and Gneiting (2006) give a scholarly account of the multiple origins of the parametric family (1.4) and its many areas of application.

D. G. Krige was for many years a professor at the University of the Witwatersrand, South Africa. He promoted the use of statistical methods in mineral exploration and, in Krige (1951), set the seeds for the later development, by Georges Mathéron and colleagues at L'École des Mines in Fontainbleau, France, of the branch of spatial statistics known as geostatistics (see Section 1.3.1 below and Part II). The spatial prediction method known as kriging is named in his honor. In a different scientific setting, the objective analysis of Gandin (1960), essentially the same thing as kriging, was for a long time the standard tool for constructing spatially continuous weather maps from spatially discrete observations on the ground and in the air.

1.3.1 Geostatistics

As noted above, geostatistics has its origins in the South African mining industry, but was developed into a self-contained methodology for spatial prediction at L'École des Mines, Fontainebleau, France (Mathéron, 1955, 1963, 1965). The original practical problem for which geostatistical methods were developed is to predict the likely yield of a mining operation over a spatial region D, given the results of samples of ore extracted from a finite set of locations. Put more formally, suppose that $S(x)$ is a spatially continuous random process and that data $Y_i : i = 1, ..., n$ are obtained as, possibly noise-corrupted versions of, the realized values of $S(x)$ at sampling locations $x_i : i = 1, ..., n$, hence $Y_i = S(x_i) + Z_i$ where the Z_i are independent with mean zero and variance τ^2. Let $T = \int_D S(x)dx$. Then, the canonical geostatistical problem is to use the data $Y_i : i = 1, ..., n$ to make predictive inference about T. In its simplest manifestation (called, not unreasonably, *simple* kriging), the process $S(\cdot)$ is assumed to have constant mean, estimated by the sample mean of the data, and known covariance structure. The point predictor of T is then the integral of the best (in mean square error sense) linear predictor of $S(x)$. A somewhat better method, known as the *ordinary kriging*, replaces the sample mean by the generalized least squares estimate of μ, thereby taking account of the estimated covariance structure of $S(\cdot)$. A further extension, called *universal kriging*, replaces the constant mean μ by a regression model, $\mu(x) = d(x)'\beta$, where $d(x)$ is a vector of spatially referenced explanatory variables.

For some time, the work of the Fontainebleau School remained relatively unconnected to the mainstream of spatial statistical methodology, and vice versa. Watson (1972) drew attention to the close connections between Fontainebleau-style geostatistical methods and more theoretically oriented work on stochastic process prediction (see, for example, Whittle, 1963). Ripley (1981, Sec. 4.4) made this connection explicit by giving an elegantly concise but essentially complete rederivation of kriging methods using the language of stochastic process prediction.

1.3.2 Forestry

Bertil Matérn's doctoral dissertation is a remarkable work that remains highly relevant almost 50 years after its first appearance. As a young research student, I remember being puzzled as to how a doctoral dissertation came to be so widely cited. I recently learned from Sir David Cox that Matérn had visited London in the late 1960s to give what turned out to be a very influential series of lectures, after which copies of the dissertation began to circulate within the fledgling community of spatial statistics researchers in the United Kingdom.

Matérn's parametric family of correlation functions has become the family of choice for many geostatistical applications, although some may feel this choice to be slightly less straightforward than is expressed by Stein (1999). His summary of practical suggestions when interpolating spatial data begins with the blunt statement: "Use the Matérn model."

Matérn's dissertation also makes important contributions to spatial point processes, random sets and spatial sampling theory. In the first of these areas, he considers models for spatial point processes that impose a minimum distance between any two points of the process; an obvious case in point being when the points represent the locations of mature trees in a dense forest is clear. In so doing, he lays down the elements of what would now be called Markov point processes. Matérn also gives early examples of models for random sets. One such, consisting of the union of copies of a circular disk translated by the points of a Poisson process, would later be rediscovered as an example of a Boolean scheme (Serra, 1982), in which the circular disk is replaced by a more or less arbitrary geometrical shape. Finally, the dissertation gives an extensive discussion of the efficiency of spatial sampling schemes showing, in particular, the advantages of systematic over random sampling for estimating spatial averages. The key theoretical idea here was to recognize, and formulate precisely, the distinction between estimating a mean and predicting a spatial average. This distinction is crucial to an understanding of spatial asyptotic theory, and in particular to the choice between in-fill and expanding domain asymptotics (Stein, 1999). With in-fill asymptotics, the values of a spatial process $S(x)$ in a fixed spatial region D are sampled at the points of a regular lattice with progressively finer spacing. This leads to consistent prediction of the spatial average, $T = |D|^{-1} \int_D S(x)dx$ where $|\cdot|$ denotes area, but not to consistent estimation of the (assumed constant) expectation, $\mu = \mathrm{E}[S(x)]$. Conversely, with expanding domain asymptotics, the process is sampled at the points of a lattice with fixed spacing, but covering an increasing sequence of spatial regions. This leads to consistent estimation of μ (and of other model parameters), but not to consistent prediction of T because the number of sample points relevant to predicting the values of $S(x)$ for $x \in D$ do not increase with the total sample size.

1.4 Two Methodological Breakthroughs

The overall structure of this handbook recognizes three major branches of spatial statistics: (1) continuous spatial variation, (2) discrete spatial variation, and (3) spatial point processes. We have seen how the first of these grew out of a series of initially unconnected strands of work that were separated both by geography and by the area of scientific motivation. In the author's opinion, the seminal contributions to the systematic development of the second and third branches of spatial statistics are to be found in two papers read at meetings of the Royal Statistical Society during the 1970s. Besag (1974) proposed models and associated methods of inference for analyzing spatially discrete, or "lattice," data, while Ripley (1977) set out a systematic approach for analyzing spatial point process data. At the time of writing, Google Scholar lists 2690 and 783 citations for these two papers, respectively.

Interestingly, both papers exemplify the connections between statistical modeling of spatial data and ideas in statistical physics. As pointed out by M. S. Bartlett in the discussion of Besag (1974), the autologistic model for binary lattice data is related to the Ising model of ferromagnetism, while the pairwise interaction point process models described in Ripley (1977) are related to idealized models of liquids (Bernal, 1960).

1.4.1 Spatial Interaction and the Statistical Analysis of Lattice Systems

The title of Besag (1974), with its reference to "*lattice* systems" (my italics) reveals its genesis in seeking to analyze data such as those shown in Figure 1.1. Julian Besag began his academic career (immediately after completing his undergraduate education) as a research assistant to Maurice Bartlett, whose work in spatial statistics was, as already noted, motivated in part by early work on the analysis of agricultural field trials. Besag's automodels were constructed as regression models with spatially lagged values of the response variable treated as explanatory variables. Crucially, *regression* here refers to the specification of a set of conditional distributions for each of a set of spatially referenced random variables $Y_i : i = 1, ..., n$ given all other members of the set, with no necessary resriction to linearity. Hence, for example, Besag's autologistic model for binary Y_i (Besag, 1972) specified the logit of the conditional probability of $Y_i = 1$ given all other $Y_j : j \neq i$ to be $\alpha + \beta N_i$ where N_i is the number of values $Y_j = 1$ amongst lattice points j deemed to be neighbors of i (for example, horiziontal and vertical nearest neighbors in a square lattice arrangement).

Even in the linear Gaussian setting, the choice between so-called conditional and simultaneous spatially autoregressive models is important in a rather surprising way. In time series analysis, it is easy to show that the conventional *simultaneous* specification of a first-order autoregressive process as $Y_t = \alpha Y_{t-1} + Z_t$, where the Z_t are mutually independent, $N(0, \sigma^2)$ is equivalent to the *conditional* specification $Y_t|\{Y_s : s < t\} \sim N(\alpha Y_{t-1}, \sigma^2)$. However, the analogous specifications in one spatial dimension,

$$Y_i = \alpha(Y_{i-1} + Y_{i+1}) + Z_i : Z_i \sim N(0, \sigma^2) \tag{1.5}$$

and

$$Y_i|\{Y_j : j \neq i\} \sim N(\alpha(Y_{i-1} + Y_{i+1}), \sigma^2) \tag{1.6}$$

are not equivalent. In the conditional version of (1.5), the conditional expectation of Y_i given all other Y_i depends not only on Y_{i-1} and Y_{i+1}, but also on Y_{i-2} and Y_{i+2}.

In general, any mutually compatible set of conditional distributions $P(Y_i|\{Y_j : j \neq i\})$ (the so-called *full conditionals*) exactly determines the joint distribution of $(Y_1, ..., Y_n)$. The ramifications of this extend far beyond spatial statistics. In the discussion of Besag (1974), A. J. Hawkes commented that he was "extremely grateful to Mr. Besag for presenting this paper with his elegant general treatment of distributions on lattices—or, indeed, *for any multivariate distribution at all*" (my italics). This prescient remark predated by six years the first explicit reference to graphical modeling of multivariate data (Darroch, Lauritzen and Speed, 1980).

1.4.2 Modeling Spatial Patterns

Brian Ripley's paper "Modeling spatial patterns" (Ripley, 1977) brought together what had previously been apparently unrelated strands of work on modeling and analyzing spatial point pattern data including point process models specified through local interactions, functional summary statistics as model-building tools, and Monte Carlo methods for assessing goodness-of-fit.

As previously noted, Matérn (1960) had described models for patterns that are, in some sense, completely random apart from the constraint that no two points can be separated by less than some distance, δ say. Strauss (1975) proposed a more general class of models in which the joint density for a configuration of n points in D is proportional to θ^t, where t is the number of pairs of points separated by less than δ. Matérn's construction is the special case $\theta = 0$, while $\theta = 1$ gives the homogeneous Poisson process. Strauss envisaged the case $\theta > 1$ as a model for clustering. However, Kelly and Ripley (1976) subsequently showed that this cannot correspond to any well-defined process in the plane, essentially because as D and n increase in proportion, the limiting density cannot be normalized when $\theta > 1$ because the term θ^t can grow too quickly. The case $0 < \theta < 1$ provides a useful class of models for patterns that are spatially more regular than realizations of a homogeneous Poisson process, but without a strict minimum distance constraint. Ripley further generalized the

Strauss process to the class of pairwise interaction processes, in which the density of a configuration of points $x_i : i = 1, ..., n$ in a spatial region d is proportional to the product of terms $h(||x_i - x_j||)$ over all distinct pairs of points. A sufficient condition for this to be well-defined is that $0 \leq h(u) \leq 1$ for all u. For further discussion, see Chapters 16 and 17.

The essence of Ripley's approach to model fitting was to compare functional summary statistics of the data with those of simulated realisations of a proposed model. Perhaps the simplest example to illustrate is the Strauss process with $\theta = 0$. At least in principle, this could be simulated by rejection sampling, i.e., drawing random samples of size n from the uniform distibution on D and rejecting all such samples that violate the constraint that no two points can be separated by less than a distance δ. In practice, this is hopelessly inefficient for realistically large numbers of points. Instead, Ripley used a sequential algorithm that exploits the equivalence between a pairwise interaction point process and the equilibrium distribution of a spatial birth-and-death process (Preston, 1977). This algorithm, subsequently published as Ripley (1979), is to the best of my knowledge the first explicit example within the statistical literature of a Markov chain Monte Carlo algorithm being used as a tool for inference (albeit non-Bayesian).

The data analytic tool most strongly associated with Ripley (1977), although it first appeared in Ripley (1976), is a nonparametric estimate of the reduced second moment measure, or K-function. This is one of several mathematically equivalent descriptions of the second moment properties of a stationary spatial point process, others including the spectrum (Bartlett, 1964), and the curiously named pair correlation function ("curiously" because it is not a correlation, but rather a scaled second-order intensity, see Chapter 16). From a data analytic perspective, the key feature of the K-function is that it is cumulative in nature and can be defined as a scaled conditional expectation: If λ denotes the intensity, or mean number of points per unit area, of the process, then $\lambda K(s)$ is the expected number of additional points of the process within distance s of the origin, conditional on there being a point of the process at the origin. This immediately suggests how $K(s)$ can be estimated from the empirical distribution of pairwise interpoint distances without the need for any arbitrary smoothing or histogram-like binning, an important consideration for the sparse datasets that, typically, were all that were available at the time.

The sampling distributions of functional summary statistics, such as the estimated K-function, are intractable, even for the simplest of models. Ripley's 1976 and 1977 papers were the first to make use of the now ubiquitous device of using envelopes from functional summaries of simulated realizations of a proposed model to assess goodness-of-fit.

1.5 Maturity: Spatial Statistics as Generic Statistical Modeling

Arguably the most important post-1960s development in statistical methodology for independently replicated data is Nelder and Wedderburn's embedding of previously separate regression-like methods within the encompassing framework of generalized linear modeling (Nelder and Wedderburn, 1972). By the same token, statistical methodology for dependent data has been transformed by the development of hierarchically specified random effects models, sometimes called latent graphical models, coupled with the ability to make inferences for such models using Monte Carlo methods.

The core idea in graphical modeling is to build complex patterns of interdependence in a high-dimensional random vector by the combination of relatively simple local dependence. In hierarchical, or latent graphical modeling, the stochastic process of interest is not observed directly, but only indirectly through extant random variables whose distributions are specified conditionally on the underlying, latent process. An early, nonspatial example is the Kalman filter (Kalman, 1960; Kalman and Bucy, 1961). Modern spatial statistics uses models of this kind in areas of application as diverse as disease mapping (Clayton and Kaldor, 1987) and image restoration (Besag, York and Mollié, 1991).

The basic geostatistical model was described in Section 1.3.1 in the form $Y_i = S(x_i) + Z_i$: $i = 1, ..., n$, where $S(\cdot)$ is a stationary Gaussian process with mean μ, variance σ^2 and correlation function $\rho(\cdot)$, and the Z_i are mutually independent $\mathrm{N}(0, \tau^2)$ random variables. However, this model can equally be represented in a hierarchical form as

$$\begin{aligned} S(\cdot) &\sim \mathrm{SGP}(\mu, \sigma^2, \rho(\cdot)) \\ Y_i | S(\cdot) &\sim \mathrm{N}(0, \tau^2). \end{aligned} \tag{1.7}$$

Diggle, Moyeed and Tawn (1998) later used the hierarchical form (1.7) to model discrete responses Y_i, using a generalized linear model in place of the normal conditional for Y_i given $S(\cdot)$.

The challenges posed by the need to analyze spatially referenced data have led to their playing a prominent role in the development of modern statistical methodology. As I hope this short historical introduction has illustrated, many statistical models and methods proposed originally in a spatial setting now sit firmly within the statistical mainstream of models and methods for dependent data, whether that dependence arises in space, time or, in the guise of graphical modeling, as independently replicated multivariate data.

References

Bartlett, M.S. (1964). Spectral analysis of two-dimensional point processes. *Biometrika*, **51**, 299–311.

Bartlett, M.S. (1978). Nearest neighbour models in the analysis of field experiments. *Journal of the Royal Statistical Society*, B **40**, 147–58.

Bernal, J.D. (1960). Geometry of the structure of monatomic liquids. *Nature*, **185**, 68–70.

Besag, J.E. (1972). Nearest-neighbour systems and the auto-logistic model for binary data. *Journal of the Royal Statistical Society*, B **34**, 75–83.

Besag, J.E. (1974). Spatial interaction and the statistical analysis of lattice systems (with discussion). *Journal of the Royal Statistical Society*, B **36**, 192–225.

Besag, J.E. and Higdon, D. (1999). Bayesian analysis of agricultural field experiments (with discussion). *Journal of the Royal Statistical Society*, B **61**, 691–746.

Besag, J.E. and Kempton, R. (1986). Statistical analysis of field experiments using neighbouring plots. *Biometrics*, **42**, 231–251.

Besag, J.E., York, J. and Mollié, A. (1991). Bayesian image restoration, with two applications in spatial statistics (with discussion). *Annals of the Institute of Statistical Mathematics*, **43**, 1–59.

Clark, P.J. and Evans, F.C. (1954). Distance to nearest neighbour as a measure of spatial relationships in populations. *Ecology*, **35**, 445–53.

Clayton, D. and Kaldor, J. (1987). Empirical Bayes estimates of age-standardised relative risks for use in disease mapping. *Biometrics*, **43**, 671–81.

Cox, D.R. (1977). The role of significance tests. *Scandinavian Journal of Statistics*, **4**, 49–71.

Darroch, J.N., Lauritzen, S.L. and Speed, T.P. (1980). Markov fields and log-linear interaction models for contingency tables. *Annals of Statistics*, **8**, 522–539.

Diggle, P.J., Moyeed, R.A. and Tawn, J.A. (1998). Model-based geostatistics (with discussion). *Applied Statistics*, **47**, 299–350.

Fairfield Smith, H. (1938). An empirical law describing heterogeneity in the yields of agricultural crops. *Journal of Agricultural Science*, **28**, 1–23.

Fisher, R.A. (1966). *The Design of Experiments*, 8th ed. London: Oliver and Boyd (first edition, 1935).

Gandin, L.S., 1960. On optimal interpolation and extrapolation of meteorological fields. *Trudy Main Geophys. Obs.* **114**, 75–89.
Guttorp, P. and Gneiting, T. (2006). Studies in the history of probability and statistics XLIX. On the Matérn correlation family. *Biometrika*, **93**, 989–995.
Harding, E.F. and Kendall, D.G. (1974). *Stochastic Geometry*. Chichester, U.K.: John Wiley & Sons.
Hertz, P. (1909). Uber die gegenseitigen durchshnittlichen Abstand von Punkten, die mit bekannter mittlerer Dichte im Raum angeordnet sind. *Mathematische Annalen*, **67**, 387–398.
Kalman, R.E. (1960). A new approach to linear filtering and prediction problems. *Journal of Basic Engineering*, **82**, 35–45.
Kalman, R.E. and Bucy, R.S. (1961). New results in linear filtering and prediction problems. *Journal of Basic Engineering*, **83**, 95–108.
Kelly, F.P. and Ripley, B.D. (1976). A note on Strauss' model for clustering. *Biometrika*, **63**, 357–360.
Krige, D.G. (1951). A statistical approach to some basic mine valuation problems on the Witwatersrand. *Journal of the Chemical, Metallurgical and Mining Society of South Africa*, **52**, 119–139.
Matérn, B. (1960). *Spatial variation*. Meddelanden fran Statens Skogsforsknings Institut, Stockholm. Band 49, No. 5.
Matérn, B. (1986). *Spatial Variation*, 2nd ed. Berlin: Springer-Verlag.
Mathéron, G. (1955). Applications des méthodes stastistiques à l'évaluation des gisements. *Annales des Mines*, **12**, 50–75.
Mathéron, G. (1963). Principles of geostatistics. *Economic Geology*, **58**, 1246–1266.
Mathéron, G. (1965). *Les Variables Régionalisées et leur Estimation. Une Application de la Théorie des Fonctions Aléatoires aux Sciences de la Nature*. Paris: Masson.
McCullagh, P. and Clifford, D. (2006). Evidence for conformal invariance of crop yields. *Proceedings of the Royal Society*, A **462**, 2119–2143.
Mercer, W.B. and Hall, A.D. (1911). The experimental error of field trials. *Journal of Agricultural Science*, **4**, 107–132.
Nelder, J.A. and Wedderburn, R.W.M. (1972). Generalized linear models. *Journal of the Royal Statistical Society*, A **135**, 370–384.
Newcomb, S. (1859–60). Notes on the theory of probabilities. *Mathematical Monthly*, **1**, 136–139, 233–235, 331–335, 349–350, and **2**, 134–140, 272–275.
Neyman, J. (1939). On a new class of contagious distributions, applicable in entomology and bacteriology. *Annals of Mathematical Statistics*, **10**, 35–57.
Neyman, J. and Scott, E.L. (1958). Statistical approach to problems of cosmology. *Journal of the Royal Statistical Society*, Series B, **20**, 1–43.
Papadakis, J.S. (1937). Méthode statistique pour des expériences sur champ. *Bull. Inst. Amél. Plantes á Salonique*, (Greece) **23**.
Pearson, E.S. (1990). *Student: A Statistical Biography of William Sealy Gosset*, R.L. Plackett and G.A. Barnard, Eds. Oxford: Oxford University Press.
Preston, C.J. (1977). Spatial birth-and-death processes. *Bulletin of the International Statistical Institute*, **46**(2), 371–391.
Ripley, B.D. (1976). The second-order analysis of stationary point processes. *Journal of Applied Probability*, **13**, 255–266.
Ripley, B.D. (1977). Modelling spatial patterns (with discussion). *Journal of the Royal Statistical Society*, B **39**, 172–212.
Ripley, B.D. (1979). Simulating spatial patterns: Dependent samples from a multivariate density. *Applied Statistics*, **28**, 109–112.
Ripley, B.D. (1981). *Spatial Statistics*. Chichester, U.K.: John Wiley & Sons.
Ripley, B.D. and Kelly, F.P. (1977). Markov point processes. *Journal of the London Mathematical Society*, **15**, 188–192.
Rue, H. and Held, L. (2005). *Gaussian Markov Random Fields: Theory and Applications*. Boca Raton, FL: CRC Press.
Serra, J. (1982). *Image Analysis and Mathematical Morphology*. London: Academic Press.
Skellam, J.G. (1952). Studies in statistical ecology I, spatial pattern. *Biometrika*, **39**, 346–362.
Stein, M. (1999). *Interpolation of Spatial Data*. New York: Springer.
Stoyan, D., Kendall, W.S. and Mecke, J. (1987). *Stochastic Geometry and Its Applications*. Chichester, U.K.: John Wiley & Sons.

Strauss, D.J. (1975). A model for clustering. *Biometrika*, **62**, 467–475.

Watson, G.S. (1972). Trend surface analysis and spatial correlation. *Geological Society of America Special Paper*, **146**, 39–46.

Whittle, P. (1956). On the variation of yield variance with plot size. *Biometrika*, **43**, 337–343.

Whittle, P. (1962). Topographic correlation, power law covariance functions, and diffusion. *Biometrika*, **49**, 305–314.

Whittle, P. (1963). *Prediction and Regulation*, 2nd ed. Minneapolis: University of Minnesota.

Wilkinson, G.N., Eckert, S.R., Hancock, T.W. and Mayo, O. (1983). Nearest neighbour (NN) analysis with field experiments (with discussion). *Journal of the Royal Statistical Society*, B **45**, 151–178.

Part II

Continuous Spatial Variation

The objective of Part II is to provide a coherent and complete coverage of the traditional, modern, and state-of-art models, methods, theory and approaches for continuous spatial processes. The focus is just on spatial processes (rather than spatial-temporal processes; there is Part V for that). In Part II, we start with some theoretical background for continuous-parameter stochastic processes. Then, we introduce geostatistical modeling and inference, likelihood-based approaches for spatial data, spectral methods, asymptotic results relevant for the modeling and inference of spatial data, hierarchical modeling (with a case study), and spatial design (with a case study). Last, we review the methods, models and approaches for nonstationarity, and also present nonparametric methods for spatial data. Some of the chapters within this part introduce the topic in the context of a very well-developed application.

2

Continuous Parameter Stochastic Process Theory

Tilmann Gneiting and Peter Guttorp

CONTENTS

2.1 Spatial Stochastic Processes

In this chapter, we consider probability models for a spatial variable that varies over a continuous domain of interest, $\mathcal{D} \subseteq \mathbb{R}^d$, where the spatial dimension is typically $d = 2$ or $d = 3$. Our approach relies on the notion of a spatial stochastic process $\{Y(s) : s \in \mathcal{D} \subseteq \mathbb{R}^d\}$, in the sense that

$$Y(s) = Y(s, \omega) \tag{2.1}$$

$$\begin{array}{cc} \text{spatial location} & \text{chance} \\ s \in \mathcal{D} \subseteq \mathbb{R}^d & \omega \in \Omega \end{array}$$

is a collection of random variables with a well-defined joint distribution. At any single spatial location $s \in \mathcal{D}$, we think of $Y(s)$ as a random variable that can more fully be written as $Y(s; \omega)$, where the elementary event ω lies in some abstract sample space, Ω. If we restrict attention to any fixed, finite set of spatial locations $\{s_1, \ldots, s_n\} \subset \mathcal{D}$, then

$$(Y(s_1), \ldots, Y(s_n))^T \tag{2.2}$$

is a random vector, whose multivariate distribution reflects the spatial dependencies in the variable of interest. Each component corresponds to a spatial site. Conversely, if we fix any elementary event $\omega \in \Omega$, then

$$\{Y(s, \omega) : s \in \mathcal{D} \subseteq \mathbb{R}^d\} \qquad \text{and} \qquad (y_1, \ldots, y_n)^T = (Y(s_1, \omega), \ldots, Y(s_n, \omega))^T$$

are realizations of the spatial stochastic process (2.1) and the induced random vector (2.2), respectively. The observed data are considered but one such realization. A generalization to be discussed in Chapter 28 is that of a multivariate spatial stochastic process, for which $Y(s)$ is a random vector rather than just a random variable.

In applications, the sample space remains abstract, and the dependency on elementary events is suppressed in the notation. However, it is important to ensure a valid mathematical specification of the spatial stochastic process. Specifically, the distribution of the process $\{Y(s) : s \in \mathcal{D} \subseteq \mathbb{R}^d\}$ is given by the associated collection of the finite-dimensional joint distributions

$$F(y_1, \ldots, y_n; s_1, \ldots, s_n) = \mathbb{P}(Y(s_1) \leq y_1, \ldots, Y(s_n) \leq y_n) \tag{2.3}$$

of the random vector (2.2) for every n and every collection $s_1, \ldots, s_n$ of sites in $\mathcal{D}$. The celebrated Kolmogorov existence theorem states that the stochastic process model is valid if the family of the finite-dimensional joint distributions is consistent under reordering of the sites and marginalization. Intuitively, this is unsurprising; however, the details are cumbersome and we refer to Billingsley (1986) for a technical exposition.

An important special case is that of a *Gaussian* process where the finite-dimensional distributions (2.3) are multivariate normal and, therefore, characterized by their mean vectors and covariance matrices. In this particular case, the consistency conditions of the Kolmogorov existence theorem reduce to the usual requirement that covariance matrices are nonnegative definite. The case of non-Gaussian spatial stochastic processes is considerably more complex, and we refer the reader to Chapter 11.

2.2 Stationary and Intrinsically Stationary Processes

A spatial stochastic process is called *strictly stationary* if the finite dimensional joint distributions are invariant under spatial shifts. Essentially, this means that for all vectors $\boldsymbol{h} \in \mathbb{R}^d$ we have

$$F(y_1, \ldots, y_n; s_1 + \boldsymbol{h}, \ldots, s_n + \boldsymbol{h}) = F(y_1, \ldots, y_n; s_1, \ldots, s_n).$$

In the case of a Gaussian process $\{Y(s) : s \in \mathbb{R}^d\}$, where the finite dimensional distributions are determined by their second-order properties, we get, in particular, that

$$\mathbb{E}(Y(s)) = \mathbb{E}(Y(s + \boldsymbol{h})) = \mu$$

and

$$\text{Cov}(Y(s), Y(s + \boldsymbol{h})) = \text{Cov}(Y(\mathbf{0}), Y(\boldsymbol{h})) = C(\boldsymbol{h}),$$

where the function $C(\boldsymbol{h}), \boldsymbol{h} \in \mathbb{R}^d$, is called the *covariance function*. A process, be it Gaussian or not, which satisfies these two conditions is called *weakly stationary* or *second-order stationary*. It follows that a Gaussian process, which is second-order stationary, is also strictly stationary. This is a very special property that depends critically on the Gaussian assumption.

Matheron (1971) proposed the use of the *semivariogram* or *variogram*, γ, which he defined as

$$\gamma(\boldsymbol{h}) = \frac{1}{2}\,\text{var}(Y(s + \boldsymbol{h}) - Y(s)),$$

as an alternative to the covariance function. Elementary calculations show that if Y is a second-order stationary spatial stochastic process with covariance function $C(\boldsymbol{h})$, then

$$\gamma(\boldsymbol{h}) = C(\mathbf{0}) - C(\boldsymbol{h}).$$

The variogram can be used in some cases where a covariance function does not exist. More generally, Matheron (1973) introduced the class of the *intrinsically stationary* processes, which are such that certain spatial increments are second-order stationary, so that a *generalized covariance function* can be defined. For details, we refer to Chapter 11, Cressie (1993, Sec. 5.4), and Chilès and Delfiner (1999, Chap. 4).

Classical *Brownian motion* in one dimension provides an example of a Gaussian process that is intrinsically stationary, but not stationary. This process has variogram $\gamma(h) = |h|$, stationary and independent increments, and continuous sample paths. Interesting generalizations include Brownian motion and the *Brownian sheet* on spatial domains (Khoshnevisan, 2002). The Brownian motion process in one dimension allows for a series expansion in terms of a certain orthonormal system, which is often referred to as the *reproducing kernel Hilbert space* for the nonstationary Brownian motion covariance, $\text{Cov}(Y(s), Y(t)) = \min(s, t)$ for $s, t > 0$. The coefficients in the expansion are independent, identically distributed, Gaussian random variables. This representation is commonly called a *Karhunen–Loève expansion* and relates closely to the Fourier representation discussed in Chapter 5. It can be generalized to much broader classes of spatial processes, under regularity conditions that generally correspond to the continuity of sample paths. See Breiman (1968, Chap. 12.7) and Wahba (1990) for details.

2.3 Nugget Effect

The classical geostatistical model of Chapter 3 decomposes a spatial stochastic process as

$$Y(\boldsymbol{s}) = \mu(\boldsymbol{s}) + \eta(\boldsymbol{s}) + \epsilon(\boldsymbol{s})$$

where $\mu(\boldsymbol{s}) = \mathbb{E}(Y(\boldsymbol{s}))$, the mean function, is deterministic and smooth, the process $\eta(\boldsymbol{s})$ has mean zero and continuous realizations, $\epsilon(\boldsymbol{s})$ is a field of spatially uncorrelated mean zero errors, and the processes η and ϵ are independent. The error process ϵ has covariance function

$$\text{Cov}(\epsilon(\boldsymbol{s}), \epsilon(\boldsymbol{s} + \boldsymbol{h})) = \begin{cases} \sigma^2 \geq 0, & \boldsymbol{h} = \boldsymbol{0}, \\ 0, & \boldsymbol{h} \neq \boldsymbol{0}, \end{cases} \tag{2.4}$$

and is often referred to as a *nugget effect*. The nugget effect describes the observational error in (potentially) repeated measurements at any single site, or to microscale variability, occurring at such small scales that is cannot be distinguished from the effect of measurement error. The terminology stems from mining applications, where the occurrence of gold nuggets shows substantial microscale variability. Figure 2.1 shows a realization from a second-order stationary spatial stochastic process in the Euclidean plane with and without a nugget component. Without the nugget effect, the sample path is smooth; with the nugget effect, it is irregular and nondifferentiable.

In the following, we restrict our attention to the continuous part, $\eta(\boldsymbol{s})$, and to second-order stationary processes. Hence, from now on we consider spatial stochastic processes, which admit a continuous covariance function. The nugget effect (2.4) is the only discontinuous covariance function of practical interest (Gneiting, Sasvári and Schlather, 2001).

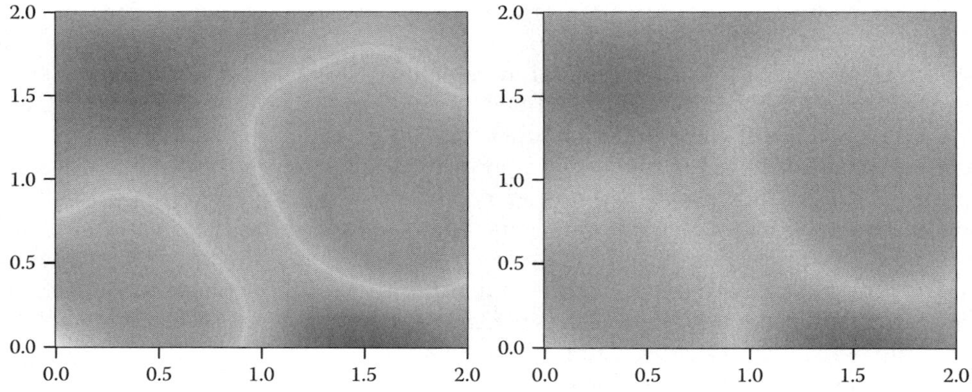

FIGURE 2.1
Realizations of a planar Gaussian process with isotropic Matérn covariance function (2.13) with smoothness parameter $\nu = \frac{3}{2}$ and scale parameter $\theta = 1$, without (left) and with a (very small) nugget effect (right). The sample paths were generated using the R package RandomFields (Schlather, 2001).

2.4 Bochner's Theorem

Suppose that $\{Y(s) : s \in \mathbb{R}^d\}$ is a second-order stationary spatial stochastic process with covariance function C. Given any finite set of spatial locations $s_1, \ldots, s_n \in \mathbb{R}^d$, the covariance matrix of the finite dimensional joint distribution (2.3) is

$$\begin{pmatrix} C(\mathbf{0}) & C(s_1 - s_2) & \cdots & C(s_1 - s_n) \\ C(s_2 - s_1) & C(\mathbf{0}) & \cdots & C(s_2 - s_n) \\ \vdots & \vdots & \ddots & \vdots \\ C(s_n - s_1) & C(s_n - s_2) & \cdots & C(\mathbf{0}) \end{pmatrix}, \tag{2.5}$$

which needs to be a valid (that is, a nonnegative definite) covariance matrix. Another way of expressing this fact is to state that the covariance function is *positive definite*.* Conversely, given any positive definite function C that generates valid covariance matrices via (2.5) there exists a spatial Gaussian process with covariance function C.

By a classical theorem of Bochner (1933, 1955), a real-valued continuous function C is positive definite if and only if it is the Fourier transform of a symmetric, nonnegative measure F on $\mathbb{R}^d$, that is, if and only if

$$C(\mathbf{h}) = \int_{\mathbb{R}^d} \exp(i\, \mathbf{h}^T \mathbf{x})\, \mathrm{d}F(\mathbf{x}) = \int_{\mathbb{R}^d} \cos(\mathbf{h}^T \mathbf{x})\, \mathrm{d}F(\mathbf{x}). \tag{2.6}$$

We refer to (2.6) as the *spectral representation* of the covariance function. In particular, any real-valued characteristic function can serve as a valid correlation function. In most cases of practical interest, the *spectral measure*, F, has a Lebesgue density, f, called the *spectral density*, so that

$$C(\mathbf{h}) = \int_{\mathbb{R}^d} \exp(i\, \mathbf{h}^T \mathbf{x})\, f(\mathbf{x})\, \mathrm{d}\mathbf{x} = \int_{\mathbb{R}^d} \cos(\mathbf{h}^T \mathbf{x})\, f(\mathbf{x})\, \mathrm{d}\mathbf{x}. \tag{2.7}$$

* A function C on $\mathbb{R}^d$ is *positive definite* if the matrix (2.5) is nonnegative definite for all finite collections of sites $s_1, \ldots, s_n \in \mathbb{R}^d$. It is *strictly positive definite* if the matrix (2.5) is positive definite for all collections of distinct sites $s_1, \ldots, s_n \in \mathbb{R}^d$. The terminology used here, which differs for matrices and functions, stems from an unfortunate, yet well established, tradition in the literature.

If the covariance function, C, is integrable over $\mathbb{R}^d$, the spectral density and the covariance are related via the standard Fourier inversion formula,

$$f(\mathbf{x}) = \frac{1}{(2\pi)^d} \int_{\mathbb{R}^d} \cos(\mathbf{h}^T \mathbf{x})\, C(\mathbf{h})\, \mathrm{d}\mathbf{h}.$$

Chapter 5 provides a much more detailed discussion of spectral representations for stochastic processes and spectral methods in spatial statistics. Spectral densities also play a critical role in asymptotic theory for inference in spatial Gaussian processes (see Chapter 6 and Stein (1999)).

2.5 Isotropic Covariance Functions

A particular case of second-order stationary processes is when the covariance function, $C(\mathbf{h})$, depends on the spatial separation vector, $\mathbf{h}$, only through its Euclidean length, $\|\mathbf{h}\|$. We then call both the process and the covariance function *isotropic*. Note that the isocovariance curves for an isotropic process are circles or spheres around the point with which we are computing the covariance. A slight extension is that to *geometrically anisotropic* processes for which the isocovariance curves are ellipsoids rather than circles or spheres. See Chapter 3 for details. In applications, the isotropy assumption is frequently violated; however, isotropic processes remain fundamental, in that they form the basic building blocks of more complex, anisotropic and nonstationary spatial stochastic process models. Guan, Sherman and Calvin (2004) and Fuentes (2005), among others, propose tests for stationarity and isotropy in spatial data.

Without loss of generality, we may assume a standardized process, so that the covariance function satisfies $C(\mathbf{0}) = 1$. For an isotropic covariance, we can then write

$$C(\mathbf{h}) = \varphi(\|\mathbf{h}\|), \qquad \mathbf{h} \in \mathbb{R}^d, \tag{2.8}$$

for some continuous function $\varphi : [0, \infty) \to \mathbb{R}$ with $\varphi(0) = 1$. Let Φ_d denote the class of the continuous functions φ that generate a valid isotropic covariance function in $\mathbb{R}^d$ via the relationship (2.8). It is then obvious that

$$\Phi_1 \supseteq \Phi_2 \supseteq \cdots \qquad \text{and} \qquad \Phi_d \downarrow \Phi_\infty = \textstyle\bigcap_{d \geq 1} \Phi_d,$$

because we can restrict an isotropic process in $\mathbb{R}^d$ to any lower-dimensional subspace. However, an element of the class Φ_d need not belong to $\Phi_{d'}$ if $d' > d$. Armstrong and Jabin (1981) give a simple, striking example, in that an isotropic, triangular-shaped correlation function is valid in one dimension, but not in higher dimensions. The members of the class Φ_∞ are valid in all dimensions.

Schoenberg (1938) studied Bochner's representation in the special case of an isotropic or spherically symmetric function. He showed that a function $\varphi : [0, \infty) \to \mathbb{R}$ belongs to the class Φ_d if and only if it is of the form

$$\varphi(t) = \int_{[0,\infty)} \Omega_d(rt)\, \mathrm{d}F_0(r), \tag{2.9}$$

where F_0 is a probability measure on the positive half-axis, often referred to as the *radial spectral measure*, and

$$\Omega_d(t) = \Gamma(d/2) \left(\frac{2}{t}\right)^{(d-2)/2} J_{(d-2)/2}(t), \tag{2.10}$$

TABLE 2.1

Generator Ω_d of the Class Φ_d

Dimension d	1	2	3	···	∞
$\Omega_d(t)$	$\cos t$	$J_0(t)$	$t^{-1}\sin t$	···	$\exp(-t^2)$
Lower bound for $\Omega_d(t)$	-1	-0.403	-0.218	···	0

where Γ is Euler's gamma function and J is a Bessel function. In other words, the members of the class Φ_d are scale mixtures of a *generator*, Ω_d, and as such they have lower bound $\inf_{t\geq 0} \Omega_d(t)$. Table 2.1 and Figure 2.2 provide closed-form expressions, numerical values for the bound, and a graphical illustration. For example, in dimension $d = 2$, the generator is $\Omega_2(t) = J_0(t)$ and the lower bound is $\inf_{t\geq 0} J_0(t) = -0.403$.

In most cases of practical interest, the radial spectral measure, F_0, is absolutely continuous with *radial spectral density*, f_0, which relates to the spectral density, f, in Bochner's representation (2.7) through the relationship

$$f_0(\|x\|) = 2\,\frac{\pi^{d/2}}{\Gamma(d/2)}\,\|x\|^{d-1}\,f(x), \qquad x \in \mathbb{R}^d.$$

Matheron (1965, 1973) and Gneiting (2002) describe operators, such as the *turning bands operator*, the *descente* and the *montée*, that build on spectral representations to map elements of the class Φ_d to some other class $\Phi_{d'}$.

Schoenberg (1938) furthermore showed that a function $\varphi : [0, \infty) \to \mathbb{R}$ belongs to the class Φ_∞ if and only if it is of the form

$$\varphi(t) = \int_{[0,\infty)} \exp(-r^2 t^2)\, \mathrm{d}F(r), \tag{2.11}$$

where F is a probability measure on $[0, \infty)$. One way of seeing this is by noting that $\lim_{d\to\infty} \Omega_d((2d)^{1/2}\, t) = \exp(-t^2)$ uniformly in $t \geq 0$. The members of the class Φ_∞ therefore are scale mixtures of a squared exponential generator, $\Omega_\infty(t) = \exp(-t^2)$. In particular, they are strictly positive and strictly decreasing functions, and have infinitely many derivatives away from the origin.

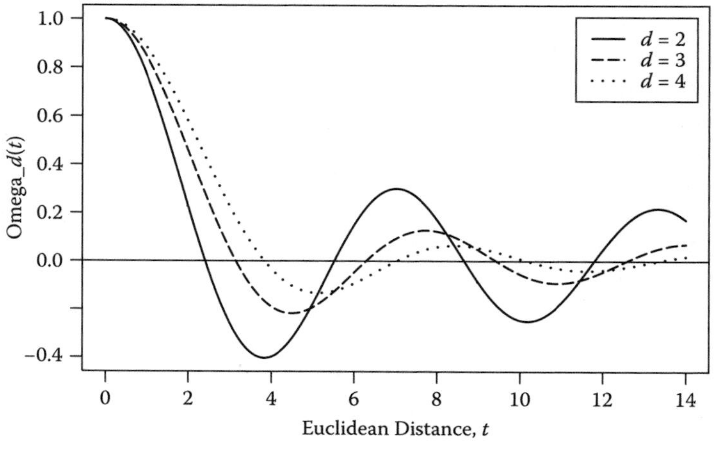

FIGURE 2.2
Generator Ω_d of the class Φ_d.

In general, it can be difficult to check whether or not a function φ belongs to the class Φ_d and generates a valid isotropic covariance function. Hence, it is advisable to work with a parametric family that is known to be valid. From these known examples, many others can be constructed because sums, convex mixtures, products, and convolutions of valid covariance functions remain valid, whether or not the components are isotropic. In dimension $d = 1$, Pólya's criterion gives a simple sufficient condition: If φ is continuous and convex with $\varphi(0) = 1$ and $\lim_{t\to\infty} \varphi(t) = 0$, then $\varphi \in \Phi_1$. For example, the criterion applies if $\varphi(t) = \exp(-t)$ or $\varphi(t) = (1+t)^{-1}$. Gneiting (2001) describes similar criteria in dimension $d \geq 2$.

2.6 Smoothness Properties

A spatial stochastic process $\{Y(\mathbf{s}) : \mathbf{s} \in D \subseteq \mathbb{R}^d\}$ is called *mean square continuous* if $\mathbb{E}(Y(\mathbf{s}) - Y(\mathbf{s}+\mathbf{h}))^2 \to 0$ as $\|\mathbf{h}\| \to 0$. For a second-order stationary process,

$$\mathbb{E}(Y(\mathbf{s}) - Y(\mathbf{s}+\mathbf{h}))^2 = 2(C(\mathbf{0}) - C(\mathbf{h})),$$

and, thus, mean square continuity is equivalent to the covariance function being continuous at the origin (and therefore everywhere). However, a process that is mean square continuous need not have continuous sample paths and vice versa (Banerjee and Gelfand, 2003). Adler (1981, Sec. 3.4) gives a covariance condition that guarantees the existence of a version with continuous sample paths for a stationary Gaussian process, and Kent (1989) has similar criteria that apply to any, Gaussian or non-Gaussian, stationary random field.

Turning now to isotropic processes, the properties of the member φ of the class Φ_d translate into properties of the associated Gaussian spatial process on $\mathbb{R}^d$ (Cramér and Leadbetter, 1967; Adler, 1981; Banerjee and Gelfand, 2003). In particular, the behavior of $\varphi(t)$ at the origin determines the smoothness of the sample paths, which is of great importance in spatial prediction problems (Stein, 1999). Specifically, suppose that there exists an $\alpha \in (0, 2]$ such that

$$1 - \varphi(t) \sim t^\alpha \quad \text{as} \quad t \downarrow 0, \tag{2.12}$$

then the realizations of the associated Gaussian spatial process have fractal or Hausdorff dimension $D = d + 1 - \frac{\alpha}{2}$. The larger α, the smaller the dimension D, and the smoother the realizations. If $\alpha = 2$, then the symmetrically continued function $c(u) = \varphi(|u|)$, $u \in \mathbb{R}$, is at least twice differentiable and the following holds for all positive integers m: The function $c(u)$ is $2m$ times differentiable at the origin if and only if the sample, paths of the associated Gaussian spatial process admit m derivatives. For instance, Figure 2.3 shows sample paths of Gaussian stochastic processes where $\varphi(t) = \exp(-t)$ and $\varphi(t) = (1+t)\exp(-t)$, that is, with the Matérn covariance (2.13) with smoothness parameter $\nu = \frac{1}{2}$ and $\nu = \frac{3}{2}$. In the former case, $\alpha = 1$, so the realization is not differentiable; in the latter $\alpha = 2$ and the sample path admits $m = 1$ derivative. Note that these are properties of the stochastic process model; when plotting sample paths, visual appearance depends on the resolution, which needs to be chosen judiciously.

2.7 Examples of Isotropic Covariance Functions

We now give examples of parametric families within the class Φ_d that generate valid isotropic covariance functions via the relationship (2.8).

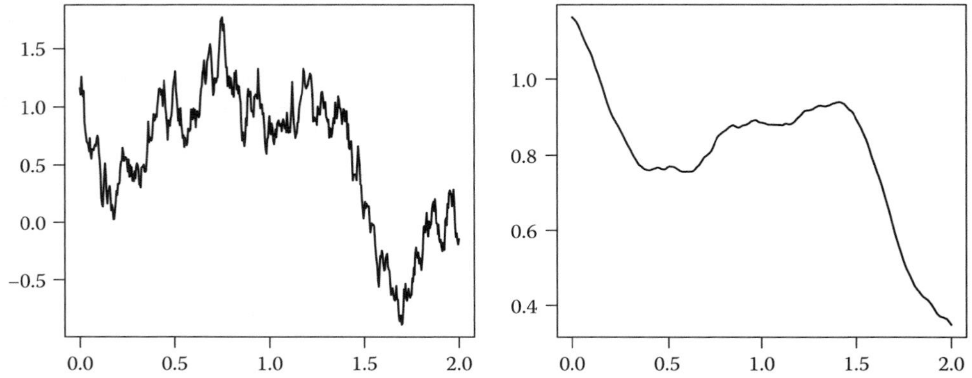

FIGURE 2.3
Sample paths of the one-dimensional Gaussian process with Matérn correlation function (2.13) where $\theta = 1$. The smoothness parameter is $\nu = \frac{1}{2}$ (left) and $\nu = \frac{3}{2}$ (right). For both panels, the resolution on the horizontal axis is in increments of 0.004.

The most popular and most often used family is the *Matérn class* (Matérn, 1960; Handcock and Stein, 1993; Guttorp and Gneiting, 2006), for which

$$\varphi(t) = \frac{2^{1-\nu}}{\Gamma(\nu)} \left(\frac{t}{\theta}\right)^{\nu} K_\nu\left(\frac{t}{\theta}\right) \qquad (\nu > 0,\ \theta > 0), \tag{2.13}$$

where K_ν is a modified Bessel function, and $\nu > 0$ and $\theta > 0$ are smoothness and scale parameters, respectively. The Matérn correlation function (2.13) admits the relationship (2.12) where $\alpha = 2\min(\nu, 1)$. The associated Gaussian sample paths are m times differentiable if and only if $m < \nu$, as illustrated in Figure 2.3.

The members of the Matérn family belong to the class Φ_∞, so they are valid in all dimensions $d \geq 1$, with a spectral density function that is proportional to $(\theta^2\|x\|^2 + 1)^{-\nu-\frac{d}{2}}$. Some special cases are noted in Table 2.2 and illustrated in Figure 2.4. If $\nu = \frac{1}{2}$, we obtain the exponential correlation function, $\varphi(t) = \exp(-t)$. The nugget effect (2.4) arises in the limit as $\nu \to 0$. The squared exponential correlation function, $\varphi(t) = \exp(-t^2)$, is also a limiting member, arising in the limit as $\nu \to \infty$ with the scale parameter set at $\theta = 1/(2\sqrt{\nu})$.

Whittle (1954, 1963) studied the stochastic fractional differential equation

$$\left(\frac{\partial^2}{\partial s_1^2} + \cdots + \frac{\partial^2}{\partial s_d^2} - \frac{1}{\theta^2}\right)^{(2\nu+d)/4} Y(s) = \epsilon(s),$$

TABLE 2.2
Special Cases of the Matérn Correlation Function (2.13) with Scale Parameter $\theta = 1$

Parameter	Correlation Function
$\nu = \frac{1}{2}$	$\varphi(t) = \exp(-t)$
$\nu = 1$	$\varphi(t) = t\,K_1(t)$
$\nu = \frac{3}{2}$	$\varphi(t) = (1 + t)\exp(-t)$
$\nu = \frac{5}{2}$	$\varphi(t) = (1 + t + \frac{t^2}{3})\exp(-t)$

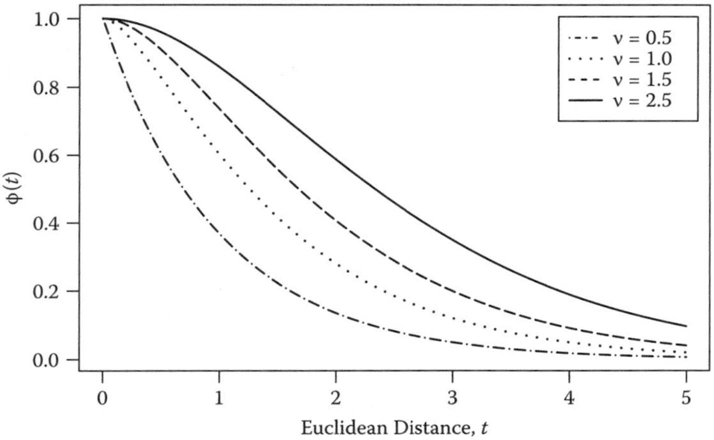

FIGURE 2.4
Special cases of the Matérn correlation function (2.13) with scale parameter $\theta = 1$.

where $s = (s_1, \ldots, s_d)'$ and ϵ is a spatial white noise process. Essentially, the latter is a Gaussian process such that integrals of it become a spatial Brownian motion. Whittle (1963) showed that the solution to the stochastic differential equation is a Gaussian process with isotropic Matérn correlation function of the form (2.13). This construction can be thought of as an analog of the autoregressive moving average (ARMA) approach to time series modeling (Box and Jenkins, 1970) in continuous space. For a similar approach to lattice processes, see Chapter 13.

Another important class of correlation functions is the *powered exponential family*, which was employed by Diggle, Tawn and Moyeed (1998), among others. Its members are of the form

$$\varphi(t) = \exp\left(-\left(\frac{t}{\theta}\right)^{\alpha}\right) \qquad (0 < \alpha \le 2,\ \theta > 0), \tag{2.14}$$

belong to the class Φ_∞, and admit the relationship (2.12).[†] Hence, the smoothness parameter $\alpha \in (0, 2]$ determines the fractal dimension of the Gaussian sample paths, while $\theta > 0$ is a scale parameter. However, the sample paths are either infinitely differentiable (when $\alpha = 2$) or not differentiable at all (when $\alpha < 2$), so the powered exponential family provides a less flexible parametrization than the Matérn class. The exponential correlation function arises when $\alpha = 1$.

Gneiting and Schlather (2004) introduced the *Cauchy family*, whose members are of the form

$$\varphi(t) = \left(1 + \left(\frac{t}{\theta}\right)^{\alpha}\right)^{-\beta/\alpha} \qquad (0 < \alpha \le 2,\ \beta > 0,\ \theta > 0) \tag{2.15}$$

and belong to the class Φ_∞. The smoothness parameter $\alpha \in (0, 2]$ and the scale parameter $\theta > 0$ are interpreted in the same way as for the exponential power family. However, there is an additional long-memory parameter, $\beta > 0$, that determines the asymptotic power law, $\varphi(t) \sim t^{-\beta}$, with which the correlation decays as $t \to \infty$. The smaller the β, the stronger the long-range dependence (Beran, 1994). Long-memory dependence is unlikely to be relevant in spatial interpolation, but can be of critical importance in problems of estimation and inference for spatial data (Whittle, 1962; Haslett and Raftery, 1989).

[†] If $\alpha > 2$, the powered exponential model (2.14) does not belong to any of the classes Φ_d. Similarly, if $\alpha > 2$ the Cauchy model (2.15) does not belong to any of the classes Φ_d.

The Matérn, powered exponential and Cauchy families belong to the class Φ_∞, that is, they generate valid isotropic covariance functions in all spatial dimensions. Consequently, they admit the representation (2.11) and so they are strictly positive and strictly decreasing.

In applications, these assumptions might be too restrictive. For example, moderate negative correlations are occasionally observed and referred to as a *hole effect*. Figure 2.2 shows that the generator Ω_d of the class Φ_d exhibits this type of behavior. Another option is to fit an *exponentially damped cosine* function,

$$\varphi(t) = \exp\left(-\tau \frac{t}{\theta}\right) \cos\left(\frac{t}{\theta}\right) \qquad \left(\tau \geq \frac{1}{\tan \frac{\pi}{2d}}, \ \theta > 0\right), \tag{2.16}$$

where the restriction on the decay parameter, τ, is necessary and sufficient for membership in the class Φ_d (Zastavnyi, 1993).

Covariance functions with *compact support* allow for computationally efficient estimation, prediction, and simulation (Gneiting, 2002; Furrer, Genton and Nychka, 2006). The most popular, compactly supported model is the *spherical* correlation function with scale parameter $\theta > 0$, for which

$$\varphi(t) = 1 - \frac{3}{2}\frac{t}{\theta} + \frac{1}{2}\left(\frac{t}{\theta}\right)^3 \tag{2.17}$$

if $t < \theta$ and $\varphi(t) = 0$ otherwise. This function belongs to the class Φ_d if $d \leq 3$, but is not valid in higher dimensions. Many other models with compact support are available (Gneiting, 2002). For example, the compactly supported function

$$\varphi(t) = \left(1 + \frac{80}{47}\frac{t}{\theta} + \frac{2500}{2209}\frac{t^2}{\theta^2} + \frac{32000}{103823}\frac{t^3}{\theta^3}\right)\left(1 - \frac{10}{47}\frac{t}{\theta}\right)^8 \tag{2.18}$$

if $t < \frac{47}{10}\theta$ and $\varphi(t) = 0$ otherwise, where $\theta > 0$ is a scale parameter, is valid in dimension $d \leq 3$, is smooth, and approximates the squared exponential correlation function, in that $\sup_{t \geq 0} |\varphi(t) - \exp(-t^2/(2\theta^2))| = 0.0056$.

2.8 Prediction Theory for Second-Order Stationary Processes

The most common problem in spatial statistics is to predict, or interpolate, the value of the process at a location $s \in \mathbb{R}^d$ where no observation has been made. If we want to do least squares prediction, that is, if we wish to minimize the expected squared prediction error, it is well known (Ferguson, 1967) that the optimal predictor $\hat{Y}(s)$ is the conditional expectation given the observations, that is,

$$\hat{Y}(s) = \mathbb{E}(Y(s) \,|\, Y(s_1) = y_1, \ldots, Y(s_n) = y_n). \tag{2.19}$$

Generally speaking, this predictor is difficult to evaluate in closed form. However, the Gaussian case is an exception, for if

$$\begin{pmatrix} U \\ V \end{pmatrix} \sim \mathcal{N}\left(\begin{pmatrix} \mu_U \\ \mu_V \end{pmatrix}, \begin{pmatrix} \Sigma_{UU} & \Sigma_{UV} \\ \Sigma_{VU} & \Sigma_{VV} \end{pmatrix}\right),$$

then the conditional distribution of V given U is normal,

$$(V|U) \sim \mathcal{N}(\mu_V + \Sigma_{VU}\Sigma_{UU}^{-1}(U - \mu_U), \ \Sigma_{VV} - \Sigma_{VU}\Sigma_{UU}^{-1}\Sigma_{UV}). \tag{2.20}$$

Let $V = Y(s)$, and let $U = (Y(s_1), \ldots, Y(s_n))^T$ denote the observed data. Assuming that we have a second-order stationary process with mean μ and covariance function C, so that $\mu_U = \mu \mathbf{1}^T \in \mathbb{R}^n$, $\mu_V = \mu$, $\Sigma_{UU} = [C(s_i - s_j)]_{i,j=1}^n \in \mathbb{R}^{n \times n}$, $\Sigma_{UV} = [C(s - s_i)]_{i=1}^n \in \mathbb{R}^n$ and $\Sigma_{VV} = C(\mathbf{0})$, Equation (2.19) and Equation (2.20) give

$$\hat{Y}(s) = \mu + (C(s - s_1), \ldots, C(s - s_n))[C(s_i - s_j)]^{-1} \begin{pmatrix} Y(s_1) - \mu \\ \vdots \\ Y(s_n) - \mu \end{pmatrix}. \qquad (2.21)$$

We see that for a Gaussian process with a known mean and a known covariance function the conditional expectation is linear in the observed data. Moreover, the *ordinary kriging predictor* defined by (2.21) is the best linear predictor in the least squares sense even when the process is not necessarily Gaussian (Whittle, 1963, Chap. 5; Ripley, 1981, Sec. 4.4). If one estimates the covariance, the predictor is no longer linear and, thus, not optimal. For additional discussion, see Chapter 3.

By elementary matrix manipulations, the ordinary kriging predictor (2.21) can be rewritten in *dual kriging* form, that is,

$$\hat{Y}(s) = \mu + (Y(s_1) - \mu, \ldots, Y(s_n) - \mu) \left[C(s_i - s_j)\right]^{-1} \begin{pmatrix} C(s - s_1) \\ \vdots \\ C(s - s_n) \end{pmatrix}. \qquad (2.22)$$

From this we see that, when viewed as a function of $s \in \mathbb{R}^d$, the ordinary kriging predictor is a linear combination of a constant and terms of the form $w_i\, C(s - s_i)$, where the coefficients, w_i, depend on the observed data and the (known) second-order structure, akin to interpolation with radial basis functions (Wendland, 2005; Fasshauer, 2007). In other words, the predicted surface interpolates the observations (unless there is a nugget effect) and inherits its appearance from the covariance function, in that locally, in the neighborhood of an observation site, s_i, it behaves like the covariance function at the origin.

It is important to note that the kriging predictor may no longer be optimal if we abandon the quadratic loss function that underlies least squares prediction. Under other loss functions, such as linear or piecewise linear loss, other predictors might be optimal. For example, the conditional median is optimal if we wish to minimize the expected absolute error, rather than the expected squared error (Ferguson, 1967), a distinction that can become important in non-Gaussian settings.

References

Adler, R.J. (1981). *The Geometry of Random Fields*, Chichester, U.K.: John Wiley & Sons.

Armstrong, M. and Jabin, R. (1981). Variogram models must be positive definite. *Mathematical Geology* 13, 455–459.

Banerjee, S. and Gelfand, A.E. (2003). On smoothness properties of spatial processes. *Journal of Multivariate Analysis* 84, 85–100.

Beran, J. (1994). *Statistics for Long-Memory Processes*, London: Chapman & Hall.

Billingsley, P. (1986). *Probability and Measure*, 2nd ed., Hoboken, NJ: John Wiley & Sons.

Bochner, S. (1933). Monotone Funktionen, Stieltjessche Integrale und Harmonische Analyse. *Mathematische Annalen* 108, 378–410.

Bochner, S. (1955). *Harmonic Analysis and the Theory of Probability*, Berkeley, CA: University of California Press.

Box, G.E.P. and Jenkins, G.M. (1970). *Time Series Analysis: Forecasting and Control*, San Francisco: Holden-Day.

Breiman, L. (1968). *Probability*, Reading, MA: Addison-Wesley.

Chilès, J.P. and Delfiner, P. (1999). *Geostatistics: Modeling Spatial Uncertainty*, New York: John Wiley & Sons.

Cramér, H. and Leadbetter, M.R. (1967). *Stationary and Related Random Processes*, New York: John Wiley & Sons.

Cressie, N.A.C. (1993). *Statistics for Spatial Data*, revised ed., New York: John Wiley & Sons.

Diggle, P.J., Tawn, J.A. and Moyeed, R.A. (1998). Model-based geostatistics (with discussion), *Applied Statistics* 47, 299–350.

Fasshauer, G.E. (2007). *Meshfree Approximation Methods with MATLAB*, Singapore: World Scientific.

Ferguson, T.S. (1967). *Mathematical Statistics: A Decision Theoretic Approach*, New York: Academic Press.

Fuentes, M. (2005). A formal test for nonstationarity of spatial stochastic processes, *Journal of Multivariate Analysis* 96, 30–54.

Furrer, R., Genton, M.G. and Nychka, D. (2006). Covariance tapering for interpolation of large spatial datasets, *Journal of Computational and Graphical Statistics* 15, 502–523.

Gneiting, T. (2001). Criteria of Pólya type for radial positive definite functions, *Proceedings of the American Mathematical Society* 129, 2309–2318.

Gneiting, T. (2002). Compactly supported correlation functions, *Journal of Multivariate Analysis* 83, 493–508.

Gneiting, T. and Schlather, M. (2004). Stochastic models that separate fractal dimension and the Hurst effect, *SIAM Review* 46, 269–282.

Gneiting, T., Sasvári, Z. and Schlather, M. (2001). Analogies and correspondences between variograms and covariance functions, *Advances in Applied Probability* 33, 617–630.

Guan, Y., Sherman, M. and Calvin, J.A. (2004). A nonparametric test for spatial isotropy using subsampling, *Journal of the American Statistical Association* 99, 810–821.

Guttorp, P. and Gneiting, T. (2006). Studies in the history of probability and statistics XLIX: On the Matérn correlation family, *Biometrika* 93, 989–995.

Handcock, M.S. and Stein, M.L. (1993). A Bayesian analysis of kriging, *Technometrics* 35, 403–410.

Haslett, J. and Raftery, A.E. (1989). Space-time modelling with long-memory dependence: Assessing Ireland's wind-power resource (with discussion), *Applied Statistics* 38, 1–50.

Kent, J.T. (1989). Continuity properties for random fields, *Annals of Probability* 17, 1432–1440.

Khoshnevisan, D. (2002). *Multiparameter Processes: An Introduction to Random Fields*, New York: Springer.

Matérn, B. (1960). *Spatial Variation: Stochastic Models and Their Application to Some Problems in Forest Surveys and Other Sampling Investigations*, Medd. Statens Skogsforskningsinstitut, Stockholm, 49.

Matheron, G. (1965). *Les variables régionalisées et leur estimation*, Paris, Masson.

Matheron, G. (1971). *The Theory of Regionalized Variables and Its Applications*, Les cahiers du Centre de Morphologie Mathématique de Fontainebleau, Fascicule 5, Ecole des Mines de Paris.

Matheron, G. (1973). The intrinsic random functions and their applications, *Advances in Applied Probability* 5, 439–468.

Ripley, B.D. (1981). *Spatial Statistics*, New York: John Wiley & Sons.

Schlather, M. (2001). Simulation and analysis of random fields, *R News* 1 (2), 18–20.

Schoenberg, I.J. (1938). Metric spaces and completely monotone functions, *Annals of Mathematics* 39, 811–841.

Stein, M.L. (1999). *Interpolation of Spatial Data: Some Theory for Kriging*, New York: Springer.

Wahba, G. (1990). *Spline Models for Observational Data*, Philadelphia: SIAM.

Wendland, H. (2005). *Scattered Data Approximation*, Cambridge, U.K.: Cambridge University Press.

Whittle, P. (1954). On stationary processes in the plane. *Biometrika* 41, 434–449.

Whittle, P. (1962). Topographic correlation, power-law covariance functions, and diffusion, *Biometrika* 49, 305–314.

Whittle, P. (1963). Stochastic processes in several dimensions, *Bulletin of the International Statistical Institute* 40, 974–994.

Zastavnyi, V.P. (1993). Positive definite functions depending on a norm, *Russian Academy of Sciences Doklady Mathematics* 46, 112–114.

3

Classical Geostatistical Methods

Dale L. Zimmerman and Michael Stein

CONTENTS

3.1 Overview

Suppose that a spatially distributed variable is of interest, which in theory is defined at every point over a bounded study region of interest, $D \subset R^d$, where $d = 2$ or 3. We suppose further that this variable has been observed (possibly with error) at each of n distinct points in D, and that from these observations we wish to make inferences about the process that governs how this variable is distributed spatially and about values of the variable at locations where it was not observed. The geostatistical approach for achieving these objectives is to assume that the observed data are a sample (at the n data locations) of one realization of a continuously indexed spatial stochastic process (random field) $Y(\cdot) \equiv \{Y(\mathbf{s}) : \mathbf{s} \in D\}$. Chapter 2 reviewed some probabilistic theory for such processes. In this chapter, we are concerned with how to use the sampled realization to make statistical inferences about the process. In particular, we discuss a body of spatial statistical methodology that has come to be known as "classical geostatistics." Classical geostatistical methods focus on estimating the first-order (large-scale or global trend) structure and especially the second-order (small-scale or local) structure of $Y(\cdot)$, and on predicting or interpolating (kriging) values of $Y(\cdot)$ at unsampled locations using linear combinations of the observations and evaluating the performance of these predictions by their (unconditional) mean squared errors. However, if the process Y is sufficiently non-Gaussian, methods based on considering just the first two moments of Y may be misleading. Furthermore, some common practices in classical geostatistics are problematic even for Gaussian processes, as we shall note herein.

Because good prediction of $Y(\cdot)$ at unsampled locations requires that we have at our disposal estimates of the structure of the process, the estimation components of a geostatistical analysis necessarily precede the prediction component. It is not clear, however, which structure, first-order or second-order, should be estimated first. In fact, an inherent circularity exists—to properly estimate either structure, it appears we must know the other. We note that likelihood-based methods (see Chapter 4) quite neatly avoid this circularity problem, although they generally require a fully specified joint distribution and a parametric model for the covariance structure (however, see Im, Stein, and Zhu, 2007). The classical solution to this problem is to provisionally estimate the first-order structure by a method that ignores the second-order structure. Next, use the residuals from the provisional first-order fit to estimate the second-order structure, and then finally reestimate the first-order structure by a method that accounts for the second-order structure. This chapter considers each of these stages of a classical geostatistical analysis in turn, plus the kriging stage. We begin, however, with a description of the geostatistical model upon which all of these analyses are based.

3.2 Geostatistical Model

Because only one realization of $Y(\cdot)$ is available, and the observed data are merely an incomplete sample from that single realization, considerable structure must be imposed upon the process for inference to be possible. The classical geostatistical model imposes structure by specifying that

$$Y(\mathbf{s}) = \mu(\mathbf{s}) + e(\mathbf{s}), \tag{3.1}$$

where $\mu(\mathbf{s}) \equiv E[Y(\mathbf{s})]$, the mean function, is assumed to be deterministic and continuous, and $e(\cdot) \equiv \{e(\mathbf{s}) : \mathbf{s} \in D\}$ is a zero-mean random "error" process satisfying a stationarity assumption. One common stationarity assumption is that of second-order stationarity, which specifies that

$$\mathrm{Cov}[e(\mathbf{s}), e(\mathbf{t})] = C(\mathbf{s} - \mathbf{t}), \text{ for all } \mathbf{s}, \mathbf{t} \in D. \tag{3.2}$$

In other words, this asserts that the covariance between values of $Y(\cdot)$ at any two locations depends on only their *relative* locations or, equivalently, on their spatial lag vector. The function $C(\cdot)$ defined in (3.2) is called the covariance function. Observe that nothing is assumed about higher-order moments of $e(\cdot)$ or about its joint distribution. Intrinsic stationarity, another popular stationary assumption, specifies that

$$\frac{1}{2}\mathrm{var}[e(\mathbf{s}) - e(\mathbf{t})] = \gamma(\mathbf{s} - \mathbf{t}), \text{ for all } \mathbf{s}, \mathbf{t} \in D. \tag{3.3}$$

The function $\gamma(\cdot)$ defined by (3.3) is called the semivariogram (and the quantity $2\gamma(\cdot)$ is known as the variogram). A second-order stationary random process with covariance function $C(\cdot)$ is intrinsically stationary, with semivariogram given by

$$\gamma(\mathbf{h}) = C(\mathbf{0}) - C(\mathbf{h}), \tag{3.4}$$

but the converse is not true in general. In fact, intrinsically stationary processes exist for which $\mathrm{var}[Y(\mathbf{s})]$ is not even finite at any $\mathbf{s} \in D$. An even weaker stationarity assumption is that satisfied by an intrinsic random field of order k (IRF-k), which postulates that certain linear combinations of the observations known as kth-order generalized increments have mean zero and a (generalized) covariance function that depends only on the spatial lag vector. IRF-ks were introduced in Chapter 2, to which we refer the reader for more details.

Model (3.1) purports to account for large-scale spatial variation (trend) through the mean function $\mu(\cdot)$, and for small-scale spatial variation (spatial dependence) through the process $e(\cdot)$. In practice, however, it is usually not possible to unambiguously identify and separate these two components using the available data. Quoting from Cressie (1991, p. 114), "One person's deterministic mean structure may be another person's correlated error structure."

Consequently, the analyst will have to settle for a plausible, but admittedly nonunique, decomposition of spatial variation into large-scale and small-scale components.

In addition to capturing the small-scale spatial variation, the error process $e(\cdot)$ in (3.1) accounts for measurement error that may occur in the data collection process. This measurement error component typically has no spatial structure; hence, for some purposes it may be desirable to explicitly separate it from the spatially dependent component. That is, we may write

$$e(\mathbf{s}) = \eta(\mathbf{s}) + \epsilon(\mathbf{s}), \tag{3.5}$$

where $\eta(\cdot)$ is the spatially dependent component and $\epsilon(\cdot)$ is the measurement error. Such a decomposition is discussed in more detail in Section 3.5.

The stationarity assumptions introduced above specify that the covariance or semivariogram depends on locations $\mathbf{s}$ and $\mathbf{t}$ only through their lag vector $\mathbf{h} = \mathbf{s} - \mathbf{t}$. A stronger property, not needed for making inference from a single sampled realization but important nonetheless, is that of isotropy. Here we describe just intrinsic isotropy (and anisotropy); second-order isotropy differs only by imposing an analogous condition on the covariance function rather than the semivariogram. An intrinsically stationary random process with semivariogram $\gamma(\cdot)$ is said to be (intrinsically) isotropic if $\gamma(\mathbf{h}) = \gamma(h)$, where $h = (\mathbf{h}'\mathbf{h})^{1/2}$; that is, the semivariogram is a function of the locations only through the (Euclidean) distance between them. If the process is not isotropic, it is said to be anisotropic. Perhaps the most tractable form of anisotropy is geometric anisotropy, for which $\gamma(\mathbf{h}) = \gamma((\mathbf{h}'\mathbf{A}\mathbf{h})^{1/2})$ where $\mathbf{A}$ is a positive definite matrix. Isotropy can be regarded as a special case of geometric anisotropy in which $\mathbf{A}$ is an identity matrix. Contours along which the semivariogram is constant (so-called isocorrelation contours when $Y(\cdot)$ is second-order stationary) are d-dimensional spheres in the case of isotropy and d-dimensional ellipsoids in the more general case of geometric anisotropy.

The objectives of a geostatistical analysis, which were noted in general terms in Section 3.1, can now be expressed more specifically in terms of model (3.1). Characterization of the spatial structure is tantamount to the estimation of $\mu(\cdot)$ and either $C(\cdot)$ or $\gamma(\cdot)$. The prediction objective can be reexpressed as seeking to predict the value of $Y(\mathbf{s}_0) = \mu(\mathbf{s}_0) + e(\mathbf{s}_0)$ at an arbitrary site $\mathbf{s}_0$.

3.3 Provisional Estimation of the Mean Function

The first stage of a classical geostatistical analysis is to specify a parametric model, $\mu(\mathbf{s}; \boldsymbol{\beta})$, for the mean function of the spatial process, and then provisionally estimate this model by a method that requires no knowledge of the second-order dependence structure of $Y(\cdot)$. The most commonly used parametric mean model is a *linear* function, given by

$$\mu(\mathbf{s}; \boldsymbol{\beta}) = \mathbf{X}(\mathbf{s})^T \boldsymbol{\beta}, \tag{3.6}$$

where $\mathbf{X}(\mathbf{s})$ is a vector of covariates (explanatory variables) observed at $\mathbf{s}$, and $\boldsymbol{\beta}$ is an unrestricted parameter vector. Alternative choices include nonlinear mean functions, such as sines/cosines (with unknown phase, amplitude, and period) or even semiparametric or nonparametric (locally smooth) mean functions, but these appear to be used very rarely.

One possible approach to spatial interpolation is to place all of the continuous variation of the process into the mean function, i.e., assume that the observations equal a true but unknown continuous mean function plus independent and identically distributed errors, and use nonparametric regression methods, such as kernel smoothers, local polynomials, or splines. Although nonparametric regression methods provide a viable approach to spatial interpolation, we prefer for the following reasons the geostatistical approach when $\mathbf{s}$ refers to a location in physical space. First, the geostatistical approach allows us to take advantage of properties, such as stationarity and isotropy, that do not usually arise in nonparametric regression. Second, the geostatistical approach naturally generates uncertainty estimates for interpolated values even when the underlying process is continuous and is observed with little or no measurement error. Uncertainty estimation is problematic with nonparametric regression methods, especially if the standard deviation of the error term is not large compared to the changes in the underlying function between neighboring observations. It should be pointed out that smoothing splines, which can be used for nonparametric regression, yield spatial interpolants that can be interpreted as kriging predictors (Wahba, 1990). The main difference, then, between smoothing splines and kriging is in how one goes about estimating the degree of smoothing and in how one provides uncertainty estimates for the interpolants.

The covariates associated with a point $\mathbf{s}$ invariably include an overall intercept term, equal to one for all data locations. Note that if this is the only covariate and the error process $e(\cdot)$ in (3.1) is second-order (or intrinsically) stationary, then $Y(\cdot)$ itself is second-order (or intrinsically) stationary. The covariates may also include the geographic coordinates (e.g., latitude and longitude) of $\mathbf{s}$, mathematical functions (such as polynomials) of those coordinates, and attribute variables. For example, in modeling the mean structure of April 1 snow water equivalent (a measure of how much water is contained in the snowpack) over the western United States in a given year, one might consider, in addition to an overall intercept, latitude and longitude, such covariates as elevation, slope, aspect, average wind speed, etc., to the extent that data on these attribute variables are available. If data on potentially useful attribute variables are not readily available, the mean function often is taken to be a polynomial function of the geographic coordinates only. Such models are called *trend surface models*. For example, the first-order (planar) and second-order (quadratic) polynomial trend surface models for the mean of a two-dimensional process are respectively as follows, where $\mathbf{s} = (s_1, s_2)$:

$$\mu(\mathbf{s}; \boldsymbol{\beta}) = \beta_0 + \beta_1 s_1 + \beta_2 s_2,$$
$$\mu(\mathbf{s}; \boldsymbol{\beta}) = \beta_0 + \beta_1 s_1 + \beta_2 s_2 + \beta_{11} s_1^2 + \beta_{12} s_1 s_2 + \beta_{22} s_2^2.$$

Using a "full" qth-order polynomial, i.e., a polynomial that includes all pure and mixed monomials of degree $\leq q$, is recommended because this will ensure that the fitted surface is invariant to the choice of origin and orientation of the (Euclidean) coordinate system.

It is worth noting that realizations of a process with constant mean, but strong spatial correlation, frequently appear to have trends; therefore, it is generally recommended that one refrain from using trend surfaces that cannot be justified apart from examining the data.

The standard method for fitting a provisional linear mean function to geostatistical data is ordinary least squares (OLS). This method yields the OLS estimator $\hat{\boldsymbol{\beta}}_{OLS}$ of $\boldsymbol{\beta}$, given by

$$\hat{\boldsymbol{\beta}}_{OLS} = \text{argmin} \sum_{i=1}^{n} [Y(\mathbf{s}_i) - \mathbf{X}(\mathbf{s}_i)^T \boldsymbol{\beta}]^2.$$

Equivalently, $\hat{\boldsymbol{\beta}}_{OLS} = (\mathbf{X}^T\mathbf{X})^{-1}\mathbf{X}^T\mathbf{Y}$ where $\mathbf{X} = [\mathbf{X}(\mathbf{s}_1), \mathbf{X}(\mathbf{s}_2), \ldots, \mathbf{X}(\mathbf{s}_n)]^T$ and $\mathbf{Y} = [Y(\mathbf{s}_1), Y(\mathbf{s}_2), \ldots, Y(\mathbf{s}_n)]^T$, it being assumed without loss of generality that $\mathbf{X}$ has full column rank. Fitted values and fitted residuals at data locations are given by $\hat{\mathbf{Y}} = \mathbf{X}^T \hat{\boldsymbol{\beta}}_{OLS}$ and $\hat{\mathbf{e}} = \mathbf{Y} - \hat{\mathbf{Y}}$, respectively. The latter are passed to the second stage of the geostatistical analysis, to be described in the next section. While still at this first stage, however, the results of the OLS fit should be evaluated and used to suggest possible alternative mean functions. For this purpose, the standard arsenal of multiple regression methodology, such as transformations of the response, model selection, and outlier identification, may be used, but in an exploratory rather than confirmatory fashion since the independent errors assumption upon which this OLS methodology is based is likely not satisfied by the data.

As a result of the wide availability of software for fitting linear regression models, OLS fitting of a linear mean function to geostatistical data is straightforward. However, there are some practical limitations worth noting, as well as some techniques/guidelines for overcoming these limitations. First, and in particular for polynomial trend surface models, the covariates can be highly multicollinear, which causes the OLS estimators to have large variances. This is mainly a numerical problem, not a statistical one, unless the actual value of the regression coefficients is of interest and it can be solved by centering the covariates (i.e., subtracting their mean values) or, if needed, by orthogonalizing the terms in some manner prior to fitting. Second, the fitted surface in portions of the spatial domain of interest where no observations are taken may be distorted so as to better fit the observed data. This problem is avoided, however, if the sampling design has good spatial coverage. Finally, as with least squares estimation in any context, the OLS estimators are sensitive to outliers and thus one may instead wish to fit the mean function using one of many available general procedures for robust and resistant regression. If the data locations form a (possibly partially incomplete) rectangular grid, one robust alternative to OLS estimation is median polish (Cressie, 1986), which iteratively sweeps out row and column medians from the observed data (and thus is implicitly based on an assumed row–column effects model for the first-order structure). However, the notion of what constitutes an outlier can be tricky with spatially dependent data, so robust methods should be used with care.

3.4 Nonparametric Estimation of the Semivariogram

The second stage of a geostatistical analysis is to estimate the second-order dependence structure of the random process $Y(\cdot)$ from the residuals of the fitted provisional mean function. To describe this in more detail, we assume that $e(\cdot)$ is intrinsically stationary, in which case the semivariogram is the appropriate mode of description of the second-order dependence. We also assume that $d = 2$, though extensions to $d = 3$ are straightforward.

Consider first a situation in which the data locations form a regular rectangular grid. Let $\mathbf{h}_1 = \begin{pmatrix} h_{11} \\ h_{12} \end{pmatrix}, \ldots, \mathbf{h}_k = \begin{pmatrix} h_{k1} \\ h_{k2} \end{pmatrix}$ represent the distinct lags between data locations (in units of the grid spacings), with displacement angles $\phi_u = \tan^{-1}(h_{u2}/h_{u1}) \in [0, \pi)$ $(u = 1, \ldots, k)$. Attention may be restricted to only those lags with displacement angles in $[0, \pi)$ without any loss of information because $\gamma(\mathbf{h})$ is an even function. For $u = 1, \ldots, k$, let $N(\mathbf{h}_u)$ represent the number of times that lag $\mathbf{h}_u$ occurs among the data locations. Then the *empirical semivariogram* is defined as follows:

$$\hat{\gamma}(\mathbf{h}_u) = \frac{1}{2N(\mathbf{h}_u)} \sum_{\mathbf{s}_i - \mathbf{s}_j = \mathbf{h}_u} \{\hat{e}(\mathbf{s}_i) - \hat{e}(\mathbf{s}_j)\}^2 \qquad (u = 1, \ldots, k),$$

where $\hat{e}(\mathbf{s}_i)$ is the residual from the fitted provisional mean function at the ith data location and is thus the ith element of the vector $\hat{\mathbf{e}}$ defined in the previous section. We call $\hat{\gamma}(\mathbf{h}_u)$ the uth ordinate of the empirical semivariogram. Observe that $\hat{\gamma}(\mathbf{h}_u)$ is a method-of-moments type of estimator of $\gamma(\mathbf{h}_u)$. Under model (3.1) with constant mean, this estimator is unbiased; if the mean is not constant in model (3.1), the estimator is biased as a consequence of

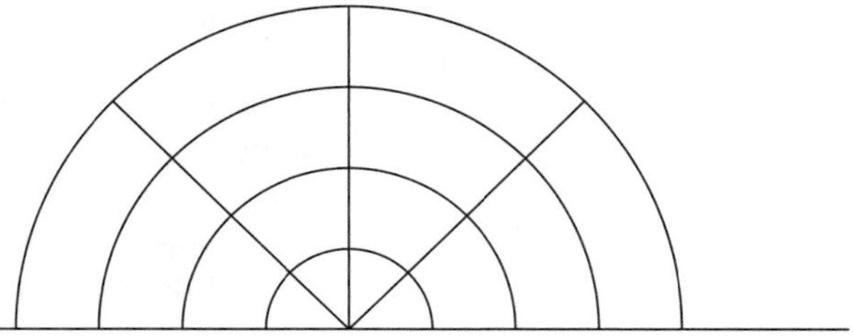

FIGURE 3.1
A polar partition of the lag space.

estimating the mean structure, but the bias is not large in practice (provided that the mean structure that is estimated is correctly specified).

When data locations are irregularly spaced, there is generally little to no replication of lags among the data locations. To obtain quasireplication of lags, we first partition the lag space $H = \{\mathbf{s} - \mathbf{t}: \mathbf{s}, \mathbf{t} \in D\}$ into lag classes or "bins" $H_1, \ldots, H_k$, say, and assign each lag with displacement angle in $[0, \pi)$ that occurs among the data locations to one of the bins. Then, we use a similar estimator:

$$\hat{\gamma}(\mathbf{h}_u) = \frac{1}{2N(H_u)} \sum_{\mathbf{s}_i - \mathbf{s}_j \in H_u} \{\hat{e}(\mathbf{s}_i) - \hat{e}(\mathbf{s}_j)\}^2 \qquad (u = 1, \ldots, k). \tag{3.7}$$

Here $\mathbf{h}_u$ is a representative lag for the entire bin H_u, and $N(H_u)$ is the number of lags that fall into H_u. The bin representative, $\mathbf{h}_u$, is sometimes taken to be the centroid of H_u, but a much better choice is the average of all the lags that fall into H_u. The most common partition of the lag space is a "polar" partition, i.e., a partitioning into angle and distance classes, as depicted in Figure 3.1. A polar partition naturally allows for the construction and plotting of a directional empirical semivariogram, i.e., a set of empirical semivariogram ordinates corresponding to the same angle class, but different distance classes, in each of several directions. It also allows for lags to be combined over all angle classes to yield the ordinates of an omnidirectional empirical semivariogram. The polar partition of the lag space is not the only possible partition; however, some popular software for estimating semivariograms use a rectangular partition instead.

Each empirical semivariogram ordinate in the case of irregularly spaced data locations is approximately unbiased for its corresponding true semivariogram ordinate, as it is when the data locations form a regular grid, but there is an additional level of approximation or blurring in the irregularly spaced case due to the grouping of unequal lags into bins.

How many bins should be used to obtain the empirical semivariogram, and how large should they be? Clearly, there is a trade-off involved: The more bins that are used, the smaller they are and the better the lags in H_u are approximated by $\mathbf{h}_u$, but the fewer the number of observed lags belonging to H_u (with the consequence that the sampling variation of the empirical semivariogram ordinate corresponding to that lag is larger). One popular rule of thumb is to require $N(\mathbf{h}_u)$ to be at least 30 and to require the length of $\mathbf{h}_u$ to be less than half the maximum lag length among data locations. But, there may be many partitions that meet these criteria, and so the empirical semivariogram is not actually uniquely defined when data locations are irregularly spaced. Furthermore, as we shall see in the simulation below, at lags that are a substantial fraction of the dimensions of the observation domain, $\hat{\gamma}(\mathbf{h}_u)$ may be highly variable even when $N(\mathbf{h}_u)$ is much larger than 30. The problem is that the various terms making up the sum in (3.7) are not independent and the dependence can be particularly strong at larger lags.

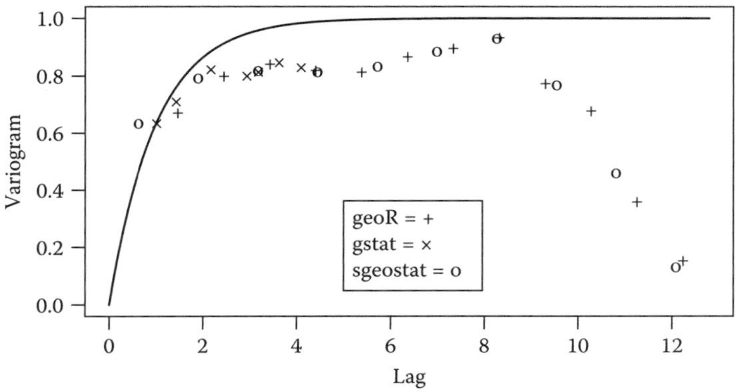

FIGURE 3.2
Empirical semivariograms of simulated data obtained via three R programs.

One undesirable feature of the empirical semivariogram is its sensitivity to outliers, a consequence of each of its ordinates being a scaled sum of squares. An alternative and more robust estimator, due to Cressie and Hawkins (1980), is

$$\bar{\gamma}(\mathbf{h}_u) = \frac{\{\frac{1}{N(H_u)}\sum_{\mathbf{s}_i-\mathbf{s}_j\in H_u}|\hat{e}(\mathbf{s}_i)-\hat{e}(\mathbf{s}_j)|^{1/2}\}^4}{.914+[.988/N(H_u)]} \qquad (u=1,\ldots,k).$$

As an example, let us consider empirical semivariograms obtained from three programs available in R with all arguments left at their default values. Specifically, we simulate an isotropic Gaussian process Y with constant mean and exponential semivariogram with sill and range parameters equal to 1 on a 10×10 square grid with distance 1 between neighboring observations. (See Section 3.5 for definitions of the exponential semivariogram and its sill and range parameters.) Figure 3.2 shows the resulting empirical semivariograms using the command `variog` from geoR, the command `est.variogram` from sgeostat, and the command `variogram` from gstat. The first two programs do not automatically impose an upper bound on the distance lags and we can see that the estimates of γ at the longer lags are very poor in this instance, even though, for example, for `est.variogram` from sgeostat, the estimate for the second longest lag (around 10.8) is based on 80 pairs of observations and the estimate for the third longest lag (aound 9.5) is based on 326 pairs. For `variogram` in gstat, the default gives a largest lag of around 4.08. Another important difference between the gstat program and the other two is that gstat, as we recommend, uses the mean distance within the bin rather than the center of the bin as the ordinate on the horizontal axis. For haphazardly sited data, the differences between the two may often be small, but here we find that for regular data, the differences can be dramatic. In particular, gstat and sgeostat give the same value for $\hat{\gamma}$ at the shortest lag (0.6361), but gstat gives the corresponding distance as 1, whereas sgeostat gives this distance as 0.6364. In fact, with either program, every pair of points used in the estimator is exactly distance 1 apart, so the sgeostat result is quite misleading. It would appear that, in this particular setting, the default empirical variogram in gstat is superior to those in geoR and sgeostat. However, even with the best of programs, one should be very careful about using default parameter values for empirical semivariograms. Furthermore, even with well-chosen bins, it is important to recognize that empirical semivariograms do not necessarily contain all of the information in the data about the true semivariogram, especially, as noted by Stein (1999, Sec. 6.2), for differentiable processes.

3.5 Modeling the Semivariogram

Next, it is standard practice to smooth the empirical semivariogram by fitting a parametric model to it. Why smooth the empirical semivariogram? There are several reasons. First, it is often quite bumpy; a smoothed version may be more reliable (have smaller variance) and therefore may increase our understanding of the nature of the spatial dependence. Second, the empirical semivariogram will often fail to be conditionally nonpositive definite, a property which must be satisfied to ensure that at the prediction stage to come, the prediction error variance is nonnegative at every point in D. Finally, prediction at arbitrary locations requires estimates of the semivariogram at lags not included among the bin representatives $\mathbf{h}_1, \ldots, \mathbf{h}_k$ nor existing among the lags between data locations, and smoothing can provide these needed estimates.

To smooth the empirical semivariogram, a valid parametric model for the semivariogram and a method for fitting that model must be chosen. The choice of model among the collection of valid semivariogram models is informed by an examination of the empirical semivariogram, of course, but other considerations (prior knowledge, computational simplicity, sufficient flexibility) may be involved as well. The following three conditions are necessary and sufficient for a semivariogram model to be valid (provided that they hold for all $\boldsymbol{\theta} \in \Theta$, where Θ is the parameter space for $\boldsymbol{\theta}$):

1. Vanishing at 0, i.e., $\gamma(\mathbf{0}; \boldsymbol{\theta}) = 0$
2. Evenness, i.e., $\gamma(-\mathbf{h}; \boldsymbol{\theta}) = \gamma(\mathbf{h}; \boldsymbol{\theta})$ for all $\mathbf{h}$
3. Conditional negative definiteness, i.e., $\sum_{i=1}^{n} \sum_{j=1}^{n} a_i a_j \gamma(\mathbf{s}_i - \mathbf{s}_j; \boldsymbol{\theta}) \leq 0$ for all n, all $\mathbf{s}_1, \ldots, \mathbf{s}_n$, and all $a_1, \ldots, a_n$ such that $\sum_{i=1}^{n} a_i = 0$

Often, the empirical semivariogram tends to increase roughly with distance in any given direction, up to some point at least, indicating that the spatial dependence decays with distance. In other words, values of $Y(\cdot)$ at distant locations tend to be less alike than values at locations in close proximity. This leads us to consider primarily those semivariogram models that are monotone increasing functions of the intersite distance (in any given direction). Note that this is not a requirement for validity, however. Moreover, the modeling of the semivariogram is made easier if isotropy can be assumed. The degree to which this assumption is tenable has sometimes been assessed informally via "rose diagrams" (Isaaks and Srivastava, 1989) or by comparing directional empirical semivariograms. It is necessary to make comparisons in at least three, and preferably more, directions so that geometric anisotropy can be distinguished from isotropy. Moreover, without some effort to attach uncertainty estimates to semivariogram ordinates, we consider it dangerous to assess isotropy based on visual comparisons of directional empirical semivariograms. Specifically, directional empirical semivariograms for data simulated from an isotropic model can appear to show clear anisotropies (e.g., the semivariogram in one direction being consistently higher than in another direction) that are due merely to random variation and the strong correlations that occur between estimated semivariogram ordinates at different lags. More formal tests for isotropy have recently been developed; see Guan, Sherman, and Calvin (2004).

A large variety of models satisfy the three aforementioned validity requirements (in R^2 and R^3), plus monotonicity and isotropy, but the following five appear to be the most commonly used:

- *Spherical*

$$\gamma(h; \boldsymbol{\theta}) = \begin{cases} \theta_1 \left(\frac{3h}{2\theta_2} - \frac{h^3}{2\theta_2^3} \right) & \text{for } 0 \leq h \leq \theta_2 \\ \theta_1 & \text{for } h > \theta_2 \end{cases}$$

- *Exponential*

$$\gamma(h;\boldsymbol{\theta}) = \theta_1\{1 - \exp(-h/\theta_2)\}$$

- *Gaussian*

$$\gamma(h;\boldsymbol{\theta}) = \theta_1\left\{1 - \exp\left(-h^2/\theta_2^2\right)\right\}$$

- *Matérn*

$$\gamma(h;\boldsymbol{\theta}) = \theta_1\left(1 - \frac{(h/\theta_2)^\nu \mathcal{K}_\nu(h/\theta_2)}{2^{\nu-1}\Gamma(\nu)}\right)$$

where $\mathcal{K}_\nu(\cdot)$ is the modified Bessel function of the second kind of order ν

- *Power*

$$\gamma(h;\boldsymbol{\theta}) = \theta_1 h^{\theta_2}$$

These models are displayed in Figure 3.3. For each model, θ_1 is positive; similarly, θ_2 is positive in each model except the power model, for which it must satisfy $0 \leq \theta_2 < 2$. In the Matérn model, $\nu > 0$. It can be shown that the Matérn models with $\nu = 0.5$ and $\nu \to \infty$ coincide with the exponential and Gaussian models, respectively.

Several attributes of an isotropic semivariogram model are sufficiently important to single out. The *sill* of $\gamma(h;\boldsymbol{\theta})$ is defined as $\lim_{h\to\infty}\gamma(h;\boldsymbol{\theta})$ provided that the limit exists. If this limit exists, then the process is not only intrinsically stationary, but also second-order stationary, and $C(\mathbf{0};\boldsymbol{\theta})$ coincides with the sill. Note that the spherical, exponential, Gaussian, and Matérn models have sills (equal to θ_1 in each of the parameterizations given above), but the power model does not. Furthermore, if the sill exists, then the *range* of $\gamma(h;\boldsymbol{\theta})$ is the smallest value of h for which $\gamma(h;\boldsymbol{\theta})$ equals its sill, if such a value exists. If the range does not exist, there is a related notion of an *effective range*, defined as the smallest value of h for which $\gamma(h;\boldsymbol{\theta})$ is equal to 95% of its sill; in this case, the effective range is often a function of a single parameter called the *range parameter*. Of those models listed above that have a sill, only the spherical has a range (equal to θ_2); however, the exponential and Gaussian models have effective ranges of approximately $3\theta_2$ and $\sqrt{3}\theta_2$, respectively, with θ_2 then being the range parameter. Range parameters can be difficult to estimate even with quite large datasets, in particular when, as is often the case, the range is not much smaller than the dimensions of the observation region (see Chapter 6). This difficulty is perhaps an argument for using the power class of variograms, which is essentially the Matérn class for $\nu < 1$ with the range set to infinity, thus, avoiding the need to estimate a range.

The Matérn model has an additional parameter ν known as the *smoothness parameter*, as the process $Y(\cdot)$ is m times mean square differentiable if and only if $\nu > m$. The smoothness of the semivariogram near the origin (i.e., at small lags) is a key attribute for efficient spatial prediction (Stein, 1988; Stein and Handcock, 1989). Finally, the *nugget effect* of $\gamma(h;\boldsymbol{\theta})$ is defined as $\lim_{h\to 0}\gamma(h;\boldsymbol{\theta})$. The nugget effect is zero for all the models listed above, but a nonzero nugget effect can be added to any of them. For example, the exponential model with nugget effect θ_3 is given by

$$\gamma(h;\boldsymbol{\theta}) = \begin{cases} 0 & \text{if } h = 0 \\ \theta_3 + \theta_1\{1 - \exp(-h/\theta_2)\} & \text{if } h > 0. \end{cases} \tag{3.8}$$

One rationale for the nugget effect can be given in terms of the measurement error model (3.5). If $\eta(\cdot)$ in that model is intrinsically stationary and mean square continuous with a nuggetless exponential semivariogram, if $\epsilon(\cdot)$ is an iid (white noise) measurement error process with variance θ_3, and if $\eta(\cdot)$ and $\epsilon(\cdot)$ are independent, then the semivariogram of $e(\cdot)$ will coincide with (3.8).

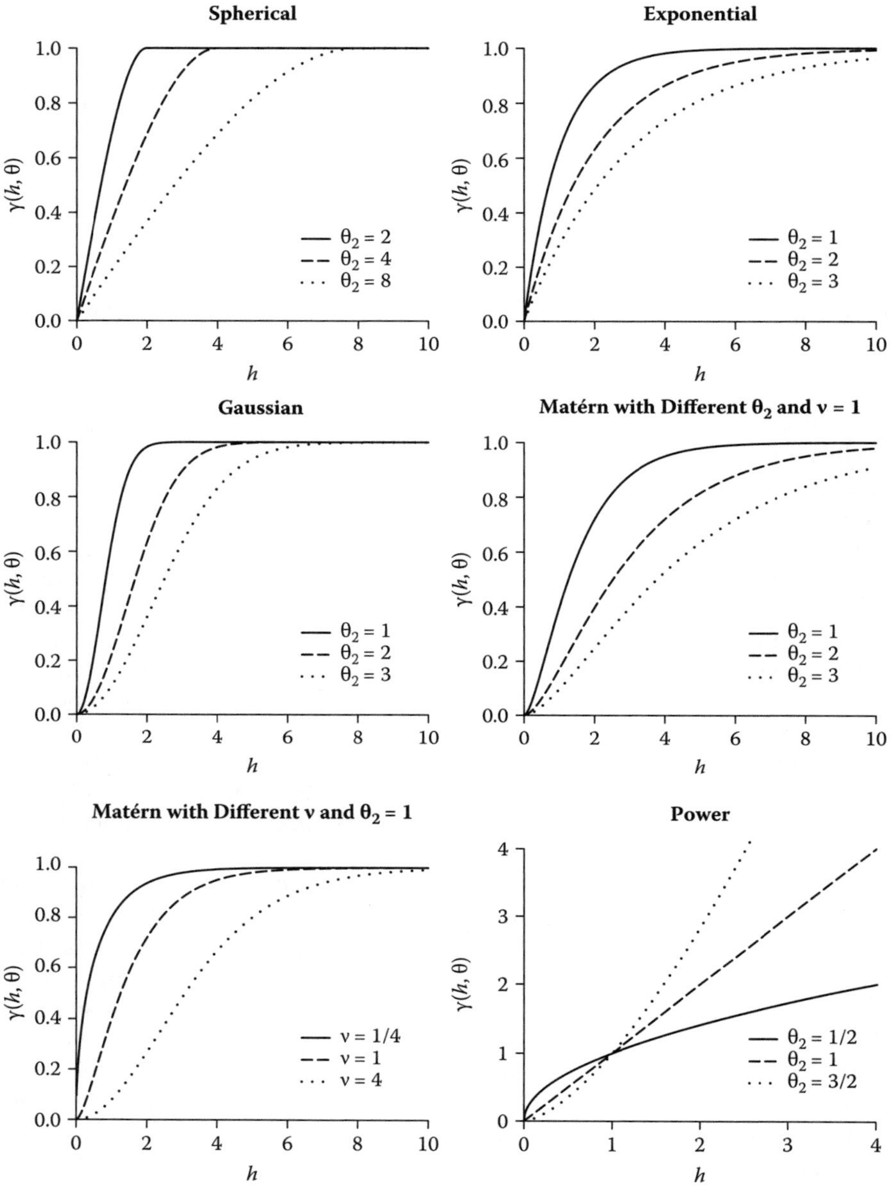

FIGURE 3.3
Semivariogram models.

Gaussian semivariograms correspond to processes that are extremely smooth—too much so to generally serve as good models for natural processes. For differentiable spatial processes, a Matérn model with $\nu > 1$, but not very large, is generally preferable. However, if one has an underlying smooth process with a sufficiently large nugget effect, it may sometimes not matter much whether one uses a Gaussian or Matérn model. Spherical semivariograms are very popular in the geostatistical community, but less so among statisticians, in part because the semivariogram is only once differentiable in θ_2 at $\theta_2 = h$, which leads to rather odd looking likelihood functions for the unknown parameters. There can be computational advantages to using semivariograms with a finite range if this range is substantially smaller than the dimensions of the observation domain, but even if one wants to use a semivariogram with finite range for computational reasons, there may be better alternatives than the spherical semivariogram (Furrer, Genton, and Nychka, 2006).

Any valid isotropic semivariogram model can be generalized to make it geometrically anisotropic, simply by replacing the argument h with $(\mathbf{h}'\mathbf{A}\mathbf{h})^{1/2}$, where $\mathbf{A}$ is a $d \times d$ positive definite matrix of parameters. For example, a geometrically anisotropic exponential semivariogram in R^2 is given by

$$\gamma(\mathbf{h};\boldsymbol{\theta}) = \theta_1 \left\{1 - \exp\left[-\left(h_1^2 + 2\theta_3 h_1 h_2 + \theta_4 h_2^2\right)^{1/2} / \theta_2^2\right]\right\}.$$

Thus, for example, if $\theta_3 = 0$ and $\theta_4 = 4$, the effective range of the spatial correlation is twice as large in the E–W direction as in the N–S direction, and the effective range in all other directions is intermediate between these two. The isotropic exponential semivariogram corresponds to the special case in which $\theta_3 = 0$, $\theta_4 = 1$. Anisotropic models that are not geometrically anisotropic—so-called zonally anisotropic models—have sometimes been used, but they are problematic, both theoretically and practically (see Zimmerman (1993)).

Two main procedures for estimating the parameters of a chosen semivariogram model have emerged: weighted least squares (WLS) and maximum likelihood (ML) or its variant, restricted (or residual) maximum likelihood (REML). The WLS approach is very popular among practitioners due to its relative simplicity, but, because it is not based on an underlying probabilistic model for the spatial process, it is suboptimal and does not rest on as firm a theoretical footing as the likelihood-based approaches (though it is known to yield consistent and asymptotically normal estimators under certain regularity conditions and certain asymptotic frameworks) (see Lahiri, Lee, and Cressie (2002)). Nevertheless, at least for nondifferentiable processes, its performance is not greatly inferior to those that are likelihood-based (Zimmerman and Zimmerman, 1991; Lark, 2000). The remainder of this section describes the WLS approach only; likelihood-based approaches are the topic of the next chapter.

The WLS estimator of $\boldsymbol{\theta}$ in the parametric model $\gamma(\mathbf{h};\boldsymbol{\theta})$ is given by

$$\hat{\boldsymbol{\theta}} = \operatorname{argmin} \sum_{u \in U} \frac{N(\mathbf{h}_u)}{[\gamma(\mathbf{h}_u;\boldsymbol{\theta})]^2} [\hat{\gamma}(\mathbf{h}_u) - \gamma(\mathbf{h}_u;\boldsymbol{\theta})]^2 \tag{3.9}$$

where all quantities are defined as in the previous section. Observe that the weights, $N(\mathbf{h}_u)/[\gamma(\mathbf{h}_u;\boldsymbol{\theta})]^2$, are small if either $N(\mathbf{h}_u)$ is small or $\gamma(\mathbf{h}_u;\boldsymbol{\theta})$ is large. This has the effect, for the most commonly used semivariogram models (which are monotone increasing) and for typical spatial configurations of observations, of assigning relatively less weight to ordinates of the empirical semivariogram corresponding to large lags. For further details on the rationale for these weights, see Cressie (1985), although the argument is based on an assumption of independence between the terms in the sum (3.9), so it may tend to give too much weight to larger lags. Since the weights depend on $\boldsymbol{\theta}$, the WLS estimator must be obtained iteratively, updating the weights on each iteration until convergence is deemed to have occurred.

Comparisons of two or more fitted semivariogram models are usually made rather informally. If the models are non-nested and have the same number of parameters (e.g., the spherical and exponential models, with nuggets), the minimized weighted residual sum of squares (the quantity minimized in (3.9)) might be used to choose from among the competing models. However, we are unaware of any good statistical arguments for such a procedure and, indeed, Stein (1999) argues that an overly strong emphasis on making parametric estimates of semivariograms match the empirical semivariogram represents a serious flaw in classical geostatistics.

3.6 Reestimation of the Mean Function

Having estimated the second-order dependence structure of the random process, there are two tacks the geostatistical analysis may take next. If the analyst has no particular interest in estimating the effects of covariates on $Y(\cdot)$, then he/she may proceed directly to kriging, as described in the next section. If the analyst has such an interest, however, the next stage is to estimate the mean function again, but this time accounting for the second-order dependence structure. The estimation approach of choice in classical geostatistics is estimated generalized least squares (EGLS), which is essentially the same as generalized least squares (GLS) except that the variances and covariances of the elements of $\mathbf{Y}$, which are assumed known for GLS, are replaced by estimates. Note that second-order stationarity, not merely intrinsic stationarity, of $e(\cdot)$ must be assumed here to ensure that these variances and covariances exist and are functions of lag only.

A sensible method for estimating the variances and covariances, and one which yields a positive definite estimated covariance matrix, is as follows. First, estimate the common variance of the $Y(\mathbf{s}_i)$s by the sill of the fitted semivariogram model, $\gamma(\mathbf{h};\hat{\boldsymbol{\theta}})$, obtained at the previous stage; denote this estimated variance by $\hat{C}(\mathbf{0})$. Then, motivated by (3.4), estimate the covariance between $Y(\mathbf{s}_i)$ and $Y(\mathbf{s}_j)$ for $i \neq j$ as $\hat{C}(\mathbf{s}_i - \mathbf{s}_j) = \hat{C}(\mathbf{0}) - \gamma(\mathbf{s}_i - \mathbf{s}_j;\hat{\boldsymbol{\theta}})$. These estimated variances and covariances may then be arranged appropriately to form an estimated variance–covariance matrix

$$\hat{\Sigma} = (\hat{C}(\mathbf{s}_i - \mathbf{s}_j)).$$

The EGLS estimator of $\boldsymbol{\beta}$, $\hat{\boldsymbol{\beta}}_{EGLS}$ is then given by

$$\hat{\boldsymbol{\beta}}_{EGLS} = (\mathbf{X}^T\hat{\Sigma}^{-1}\mathbf{X})^{-1}\mathbf{X}^T\hat{\Sigma}^{-1}\mathbf{Y}.$$

The sampling distribution of $\hat{\boldsymbol{\beta}}_{EGLS}$ is much more complicated than that of the OLS or GLS estimator. It is known, however, that $\hat{\boldsymbol{\beta}}_{EGLS}$ is unbiased under very mild conditions, and that, if the process is Gaussian, the variance of $\hat{\boldsymbol{\beta}}_{EGLS}$ is larger than that of the GLS estimator were $\boldsymbol{\theta}$ to be known, i.e., larger than $(\mathbf{X}^T\Sigma^{-1}\mathbf{X})^{-1}$ (Harville, 1985). (Here, by "larger," we mean that the difference, $\mathrm{var}(\hat{\boldsymbol{\beta}}_{EGLS}) - (\mathbf{X}^T\Sigma^{-1}\mathbf{X})^{-1}$, is nonnegative definite.) Nevertheless, for lack of a simple satisfactory alternative, the variance of $\hat{\boldsymbol{\beta}}_{EGLS}$ is usually estimated by the plug-in estimator, $(\mathbf{X}^T\hat{\Sigma}^{-1}\mathbf{X})^{-1}$.

If desired, the EGLS residuals, $\mathbf{Y} - \mathbf{X}\hat{\boldsymbol{\beta}}_{EGLS}$, may be computed and the semivariogram reestimated from them. One may even iterate between mean estimation and semivariogram estimation several times, but, in practice, this procedure usually stops with the first EGLS fit. REML, described in Chapter 4, avoids this problem by estimating $\boldsymbol{\theta}$ using only linear combinations of the observations whose distributions do not depend on $\boldsymbol{\beta}$.

3.7 Kriging

The final stage of a classical geostatistical analysis is to predict the values of $Y(\cdot)$ at desired locations, perhaps even at all points, in D. Methods dedicated to this purpose are called *kriging*, after the South African mining engineer D. G. Krige, who was the first to develop and apply them. Krige's original method, now called ordinary kriging, was based on the special case of model (3.1) in which the mean is assumed to be constant. Here, we describe the more general method of universal kriging, which is identical to best linear unbiased prediction under model (3.1) with mean function assumed to be of the linear form (3.6).

Let $\mathbf{s}_0$ denote an arbitrary location in D; usually this will be an unsampled location, but it need not be. Consider the prediction of $Y(\mathbf{s}_0)$ by a predictor, $\hat{Y}(\mathbf{s}_0)$, that minimizes the prediction error variance, $\text{var}[\hat{Y}(\mathbf{s}_0) - Y(\mathbf{s}_0)]$, among all predictors satisfying the following two properties:

1. Linearity, i.e., $\hat{Y}(\mathbf{s}_0) = \boldsymbol{\lambda}^T\mathbf{Y}$, where $\boldsymbol{\lambda}$ is a vector of fixed constants
2. Unbiasedness, i.e., $E[\hat{Y}(\mathbf{s}_0)] = E[Y(\mathbf{s}_0)]$, or equivalently $\boldsymbol{\lambda}^T\mathbf{X} = \mathbf{X}(\mathbf{s}_0)$

Suppose for the moment that the semivariogram of $Y(\cdot)$ is known. Then the solution to this constrained minimization problem, known as the universal kriging predictor of $Y(\mathbf{s}_0)$, is given by

$$\hat{Y}(\mathbf{s}_0) = [\boldsymbol{\gamma} + \mathbf{X}(\mathbf{X}^T\boldsymbol{\Gamma}^{-1}\mathbf{X})^{-1}(\mathbf{x}_0 - \mathbf{X}^T\boldsymbol{\Gamma}^{-1}\boldsymbol{\gamma})]^T\boldsymbol{\Gamma}^{-1}\mathbf{Y}, \tag{3.10}$$

where $\boldsymbol{\gamma} = [\gamma(\mathbf{s}_1 - \mathbf{s}_0), \ldots, \gamma(\mathbf{s}_n - \mathbf{s}_0)]^T$, $\boldsymbol{\Gamma}$ is the $n \times n$ symmetric matrix with ijth element $\gamma(\mathbf{s}_i - \mathbf{s}_j)$ and $\mathbf{x}_0 = \mathbf{X}(\mathbf{s}_0)$. This result may be obtained using differential calculus and the method of Lagrange multipliers. However, a geometric proof is more instructive and following is an example.

Let us assume that the first component of $\mathbf{x}(\mathbf{s})$ is identically 1, which guarantees that the error of any linear predictor of $Y(\mathbf{s}_0)$ that satisfies the unbiasedness constraint is a contrast, so that its variance can be obtained from the semivariogram of $Y(\cdot)$. Let us also assume that there exists a linear predictor satisfying the unbiasedness constraint. Suppose $\boldsymbol{\lambda}^T\mathbf{Y}$ is such a predictor. Consider any other such predictor $\boldsymbol{\nu}^T\mathbf{Y}$ and set $\boldsymbol{\mu} = \boldsymbol{\nu} - \boldsymbol{\lambda}$. Since $E(\boldsymbol{\lambda}^T\mathbf{Y}) = E(\boldsymbol{\nu}^T\mathbf{Y})$ for all $\boldsymbol{\beta}$, we must have $\mathbf{X}^T\boldsymbol{\mu} = \mathbf{0}$. And,

$$\begin{aligned}
\text{var}\{\boldsymbol{\nu}^T\mathbf{Y} - Y(\mathbf{s}_0)\} &= \text{var}[\boldsymbol{\mu}^T\mathbf{Y} + \{\boldsymbol{\lambda}^T\mathbf{Y} - Y(\mathbf{s}_0)\}] \\
&= \text{var}(\boldsymbol{\mu}^T\mathbf{Y}) + \text{var}\{\boldsymbol{\lambda}^T\mathbf{Y} - Y(\mathbf{s}_0)\} + 2\,\text{Cov}\{\boldsymbol{\mu}^T\mathbf{Y}, \boldsymbol{\lambda}^T\mathbf{Y} - Y(\mathbf{s}_0)\} \\
&\geq \text{var}\{\boldsymbol{\lambda}^T\mathbf{Y} - Y(\mathbf{s}_0)\} + 2\,\text{Cov}\{\boldsymbol{\mu}^T\mathbf{Y}, \boldsymbol{\lambda}^T\mathbf{Y} - Y(\mathbf{s}_0)\} \\
&= \text{var}\{\boldsymbol{\lambda}^T\mathbf{Y} - Y(\mathbf{s}_0)\} + 2\boldsymbol{\mu}^T(-\boldsymbol{\Gamma}\boldsymbol{\lambda} + \boldsymbol{\gamma}).
\end{aligned}$$

If we can choose $\boldsymbol{\lambda}$ such that $\boldsymbol{\mu}^T(-\boldsymbol{\Gamma}\boldsymbol{\lambda} + \boldsymbol{\gamma}) = 0$ for all $\boldsymbol{\mu}$ satisfying $\mathbf{X}^T\boldsymbol{\mu} = \mathbf{0}$, then $\boldsymbol{\lambda}$ is the solution we seek, since we then have $\text{var}\{\boldsymbol{\nu}^T\mathbf{Y} - Y(\mathbf{s}_0)\} \geq \text{var}\{\boldsymbol{\lambda}^T\mathbf{Y} - Y(\mathbf{s}_0)\}$ for any predictor $\boldsymbol{\nu}^T\mathbf{Y}$ satisfying the unbiasedness constraint. But, since the column space of $\mathbf{X}$ is the orthogonal complement of its left null space, this condition holds if and only if $-\boldsymbol{\Gamma}\boldsymbol{\lambda} + \boldsymbol{\gamma}$ is in the column space of $\mathbf{X}$, which is equivalent to the existence of a vector $\boldsymbol{\alpha}$ satisfying $-\boldsymbol{\Gamma}\boldsymbol{\lambda} + \mathbf{X}\boldsymbol{\alpha} = -\boldsymbol{\gamma}$. Putting this condition together with the unbiasedness constraint yields the system of linear equations for $\boldsymbol{\lambda}$ and $\boldsymbol{\alpha}$

$$\begin{pmatrix} -\boldsymbol{\Gamma} & \mathbf{X} \\ \mathbf{X}^T & \mathbf{O} \end{pmatrix} \begin{pmatrix} \boldsymbol{\lambda} \\ \boldsymbol{\alpha} \end{pmatrix} = \begin{pmatrix} -\boldsymbol{\gamma} \\ \mathbf{0} \end{pmatrix},$$

where $\mathbf{0}$ and $\mathbf{O}$ indicate a vector and a matrix of zeroes, respectively. If $\boldsymbol{\Gamma}$ is invertible and $\mathbf{X}$ is of full rank, then simple row reductions yields $\boldsymbol{\lambda}$ as in (3.10).

The minimized value of the prediction error variance is called the (universal) kriging variance and is given by

$$\sigma^2(\mathbf{s}_0) = \boldsymbol{\gamma}^T \boldsymbol{\Gamma}^{-1}\boldsymbol{\gamma} - (\mathbf{X}^T\boldsymbol{\Gamma}^{-1}\boldsymbol{\gamma} - \mathbf{x}_0)^T(\mathbf{X}^T\boldsymbol{\Gamma}^{-1}\mathbf{X})^{-1}(\mathbf{X}^T\boldsymbol{\Gamma}^{-1}\boldsymbol{\gamma} - \mathbf{x}_0). \tag{3.11}$$

The universal kriging predictor is an example of the best linear unbiased predictor, or BLUP, as it is generally abbreviated. If $Y(\cdot)$ is Gaussian, the kriging variance can be used to construct a nominal $100(1-\alpha)\%$ prediction interval for $Y(\mathbf{s}_0)$, which is given by

$$\hat{Y}(\mathbf{s}_0) \pm z_{\alpha/2}\sigma(\mathbf{s}_0),$$

where $0 < \alpha < 1$ and $z_{\alpha/2}$ is the upper $\alpha/2$ percentage point of a standard normal distribution. If $Y(\cdot)$ is Gaussian and $\gamma(\cdot)$ is known, then $\hat{Y}(\mathbf{s}_0) - Y(\mathbf{s}_0)$ is normally distributed and the coverage probability of this interval is exactly $1-\alpha$.

If the covariance function for Y exists and $\boldsymbol{\sigma} = [C(\mathbf{s}_1 - \mathbf{s}_0), \ldots, C(\mathbf{s}_n - \mathbf{s}_0)]^T$, then the formula for the universal kriging predictor (3.10) holds with $\boldsymbol{\gamma}$ replaced by $\boldsymbol{\sigma}$ and $\boldsymbol{\Gamma}$ by $\boldsymbol{\Sigma}$. It is worthwhile to compare this formula to that for the best (minimum mean squared error) linear predictor when $\boldsymbol{\beta}$ is known: $\mathbf{x}_0^T\boldsymbol{\beta} + \boldsymbol{\sigma}^T\boldsymbol{\Sigma}^{-1}(\mathbf{Y} - \mathbf{X}\boldsymbol{\beta})$. A straightforward calculation shows that the universal kriging predictor is of this form with $\boldsymbol{\beta}$ replaced by $\hat{\boldsymbol{\beta}}_{GLS}$. Furthermore, the expression (3.11) for the kriging variance is replaced by

$$C(\mathbf{0}) - \boldsymbol{\sigma}^T\boldsymbol{\Sigma}^{-1}\boldsymbol{\sigma} + (\mathbf{x}_0 - \mathbf{X}^T\boldsymbol{\Sigma}^{-1}\boldsymbol{\sigma})^T(\mathbf{X}^T\boldsymbol{\Sigma}^{-1}\mathbf{X})^{-1}(\mathbf{x}_0 - \mathbf{X}^T\boldsymbol{\Sigma}^{-1}\boldsymbol{\sigma}).$$

The first two terms, $C(\mathbf{0}) - \boldsymbol{\sigma}^T\boldsymbol{\Sigma}^{-1}\boldsymbol{\sigma}$, correspond to the mean squared error of the best linear predictor, so that the last term, which is always nonnegative, is the penalty for having to estimate $\boldsymbol{\beta}$.

In practice, two modifications are usually made to the universal kriging procedure just described. First, to reduce the amount of computation required, the prediction of $Y(\mathbf{s}_0)$ may be based not on the entire data vector $\mathbf{Y}$, but on only those observations that lie in a specified neighborhood around $\mathbf{s}_0$. The range, the nugget-to-sill ratio, and the spatial configuration of data locations are important factors in choosing this neighborhood (for further details, see Cressie (1991, Sec. 3.2.1)). Generally speaking, larger nuggets require larger neighborhoods to obtain nearly optimal predictors. However, there is no simple relationship between the range and the neighborhood size. For example, Brownian motion is a process with no finite range for which the kriging predictor is based on just the two nearest neighbors. Conversely, there are processes with finite ranges for which observations beyond the range play a nontrivial role in the kriging predictor (Stein, 1999, p. 67). When a spatial neighborhood is used, the formulas for the universal kriging predictor and its associated kriging variance are of the same form as (3.10) and (3.11), but with $\boldsymbol{\gamma}$ and $\mathbf{Y}$ replaced by the subvectors, and $\boldsymbol{\Gamma}$ and $\mathbf{X}$ replaced by the submatrices, corresponding to the neighborhood.

The second modification reckons with the fact that the semivariogram that appears in (3.10) and (3.11) is in reality unknown. It is common practice to substitute $\hat{\boldsymbol{\gamma}} = \boldsymbol{\gamma}(\hat{\boldsymbol{\theta}})$ and $\hat{\boldsymbol{\Gamma}} = \boldsymbol{\Gamma}(\hat{\boldsymbol{\theta}})$ for $\boldsymbol{\gamma}$ and $\boldsymbol{\Gamma}$ in (3.10) and (3.11), where $\hat{\boldsymbol{\theta}}$ is an estimate of $\boldsymbol{\theta}$ obtained by, say, WLS. The resulting *empirical* universal kriging predictor is no longer a linear function of the data, but remarkably it remains unbiased under quite mild conditions (Kackar and Harville, 1981). The empirical kriging variance tends to underestimate the actual prediction error variance of the empirical universal kriging predictor because it does not account for the additional error incurred by estimating $\boldsymbol{\theta}$. Zimmerman and Cressie (1992) give a modified estimator of the prediction error variance of the empirical universal kriging predictor, which performs well when the spatial dependence is not too strong. However, Bayesian methods are arguably a more satisfactory approach for dealing with the uncertainty of spatial dependence parameters in prediction (see Handcock and Stein (1993)). Another possibility is to estimate the prediction error variance via a parametric bootstrap (Sjöstedt-de Luna and Young, 2003).

Universal kriging yields a predictor that is a "location estimator" of the conditional distribution of $Y(\mathbf{s}_0)$ given $\mathbf{Y}$; indeed, if the error process $e(\cdot)$ is Gaussian, the universal kriging predictor coincides with the conditional mean, $E(Y(\mathbf{s}_0)|\mathbf{Y})$ (assuming $\gamma(\cdot)$ is known and putting a flat improper prior on any mean parameters). If the error process is non-Gaussian, then generally the optimal predictor, the conditional mean, is a nonlinear function of the observed data. Variants, such as disjunctive kriging and indicator kriging, have been developed for spatial prediction of conditional means or conditional probabilities for non-Gaussian processes (see Cressie, 1991, pp. 278–283), but we are not keen about them, as the first is based upon strong, difficult to verify assumptions and the second tends to yield unstable estimates of conditional probabilities. In our view, if the process appears to be badly non-Gaussian and a transformation doesn't make it sufficiently Gaussian, then the analyst should "bite the bullet" and develop a decent non-Gaussian model for the data.

The foregoing has considered *point kriging*, i.e., prediction at a single point. Sometimes a block kriging predictor, i.e., a predictor of the average value $Y(B) \equiv \int_B Y(\mathbf{s})d\mathbf{s}/|B|$ over a region (block) $B \subset D$ of positive d-dimensional volume $|B|$ is desired, rather than predictors of $Y(\cdot)$ at individual points. Historically, for example, mining engineers were interested in this because the economics of mining required the extraction of material in relatively large blocks. Expressions for the universal block kriging predictor of $Y(B)$ and its associated kriging variance are identical to (3.10) and (3.11), respectively, but with $\boldsymbol{\gamma} = [\gamma(B, \mathbf{s}_1), \ldots, \gamma(B, \mathbf{s}_n)]^T$, $\mathbf{x}_0 = [X_1(B), \ldots, X_p(B)]^T$ (where p is the number of covariates in the linear mean function), $\gamma(B, \mathbf{s}_i) = |B|^{-1} \int_B \gamma(\mathbf{u} - \mathbf{s}_i)\, d\mathbf{u}$ and $X_j(B) = |B|^{-1} \int_B X_j(\mathbf{u})\, d\mathbf{u}$.

Throughout this chapter, it was assumed that a single spatially distributed variable, namely $Y(\cdot)$, was of interest. In some situations, however, there may be two or more variables of interest, and the analyst may wish to study how these variables co-vary across the spatial domain and/or predict their values at unsampled locations. These problems can be handled by a multivariate generalization of the univariate geostatistical approach we have described. In this multivariate approach, $\{\mathbf{Y}(\mathbf{s}) \equiv [Y_1(\mathbf{s}), \ldots, Y_m(\mathbf{s})]^T : \mathbf{s} \in D\}$ represents the m-variate spatial process of interest and a model $\mathbf{Y}(\mathbf{s}) = \boldsymbol{\mu}(\mathbf{s}) + \mathbf{e}(\mathbf{s})$ analogous to (3.1) is adopted in which the second-order variation is characterized by either a set of m semivariograms and $m(m-1)/2$ cross-semivariograms $\gamma_{ij}(\mathbf{h}) = \frac{1}{2}\text{var}[Y_i(\mathbf{s}) - Y_j(\mathbf{s}+\mathbf{h})]$, or a set of m covariance functions and $m(m-1)/2$ cross-covariance functions $C_{ij}(\mathbf{h}) = \text{Cov}[Y_i(\mathbf{s}), Y_j(\mathbf{s}+\mathbf{h})]$, depending on whether intrinsic or second-order stationarity is assumed. These functions can be estimated and fitted in a manner analogous to what we described for univariate geostatistics; likewise, the best (in a certain sense) linear unbiased predictor of $\mathbf{Y}(\mathbf{s}_0)$ at an arbitrary location $\mathbf{s}_0 \in D$, based on observed values $\mathbf{Y}(\mathbf{s}_1), \ldots, \mathbf{Y}(\mathbf{s}_n)$, can be obtained by an extension of kriging known as cokriging. Good sources for further details are Ver Hoef and Cressie (1993) and Chapter 27 in this book.

While we are strong supporters of the general geostatistical framework to analyzing spatial data, we have, as we have indicated, a number of concerns about common geostatistical practices. For a presentation of geostatistics from the perspective of "geostatisticians" (that is, researchers who can trace their lineage to Georges Matheron and the French School of Geostatistics), we recommend the book by Chilès and Delfiner (1999).

References

Chilès, J.-P. and Delfiner, P. (1999). *Geostatistics: Modeling Spatial Uncertainty.* New York: John Wiley & Sons.

Cressie, N. (1985). Fitting variogram models by weighted least squares. *Journal of the International Association for Mathematical Geology*, **17**, 563–586.

Cressie, N. (1986). Kriging nonstationary data. *Journal of the American Statistical Association*, 81, 625–634.

Cressie, N. (1991). *Statistics for Spatial Data.* New York: John Wiley & Sons.

Cressie, N. and Hawkins, D.M. (1980). Robust estimation of the variogram, I. *Journal of the International Association for Mathematical Geology*, **12**, 115–125.

Furrer, R., Genton, M.G., and Nychka, D. (2006). Covariance tapering for interpolation of large spatial datasets. *Journal of Computational and Graphical Statistics*, **15**, 502–523.

Guan, Y., Sherman, M., and Calvin, J.A. (2004). A nonparametric test for spatial isotropy using subsampling. *Journal of the American Statistical Association*, **99**, 810–821.

Handcock, M.S. and Stein, M.L. (1993). A Bayesian analysis of kriging. *Technometrics*, **35**, 403–410.

Harville, D.A. (1985). Decomposition of prediction error. *Journal of the American Statistical Association*, **80**, 132–138.

Im, H.K., Stein, M.L., and Zhu, Z. (2007). Semiparametric estimation of spectral density with irregular observations. *Journal of the American Statistical Association*, **102**, 726–735.

Isaaks, E.H. and Srivastava, R.M. (1989). *An Introduction to Applied Geostatistics*. London: Academic Press.

Kackar, R.N. and Harville, D.A. (1981). Unbiasedness of two-stage estimation and prediction procedures for mixed linear models. *Communications in Statistics—Theory and Methods*, **10**, 1249–1261.

Lahiri, S.N., Lee, Y., and Cressie, N. (2002). On asymptotic distribution and asymptotic efficiency of least squares estimators of spatial variogram parameters. *Journal of Statistical Planning and Inference*, **103**, 65–85.

Lark, R.M. (2000). Estimating variograms of soil properties by the method-of-moments and maximum likelihood. *European Journal of Soil Science*, **51**, 717–728.

Sjöstedt-de Luna, S. and Young, A. (2003). The bootstrap and kriging prediction intervals. *Scandinavian Journal of Statistics*, **30**, 175–192.

Stein, M.L. (1988). Asymptotically efficient prediction of a random field with a misspecified covariance function. *Annals of Statistics*, **16**, 55–63.

Stein, M.L. (1999). *Interpolation of Spatial Data: Some Theory for Kriging*. New York: Springer.

Stein, M.L. and Handcock, M.S. (1989). Some asymptotic properties of kriging when the covariance function is misspecified. *Mathematical Geology*, **21**, 171–190.

Ver Hoef, J.M. and Cressie, N. (1993). Multivariable spatial prediction. *Mathematical Geology*, **25**, 219–240.

Wahba, G. (1990). *Spline Models for Observational Data.* Philadelphia: SIAM.

Zimmerman, D.L. (1993). Another look at anisotropy in geostatistics. *Mathematical Geology*, **25**, 453–470.

Zimmerman, D.L. and Cressie, N. (1992). Mean squared prediction error in the spatial linear model with estimated covariance parameters. *Annals of the Institute of Statistical Mathematics*, **44**, 27–43.

Zimmerman, D.L. and Zimmerman, M.B. (1991). A comparison of spatial semivariogram estimators and corresponding ordinary kriging predictors. *Technometrics*, **33**, 77–91.

4

Likelihood-Based Methods

Dale L. Zimmerman

CONTENTS

4.1 Overview

The previous chapter considered estimation of the parameters of a geostatistical model by a combination of method-of-moments and least squares methods. Those methods, collectively known as "classical geostatistics," are relatively simple and do not explicitly require any distributional assumptions, but they are not optimal in any known sense. In this chapter, we present the estimation of parametric geostatistical models by likelihood-based approaches, which adhere to the likelihood principle (of course!) and, under appropriate regularity conditions, may be expected to have certain optimality properties.

As in Chapter 3, we consider here a single spatial variable, Y, defined (in principle) at every point over a bounded study region of interest, $D \subset R^d$, where $d = 2$ or 3, and modeled as a random process $Y(\cdot) \equiv \{Y(\mathbf{s}) : \mathbf{s} \in D\}$. We suppose that $Y(\cdot)$ has been observed (possibly with error) at each of n distinct points in D, and that from these observations we wish to make inferences about the process. Likelihood-based approaches to making these inferences are most fully developed for Gaussian random fields, which are random processes whose finite-dimensional distributions are all multivariate normal. Accordingly, most of this chapter considers inference for only such processes. In Section 4.8, we briefly consider likelihood-based inference for some non-Gaussian processes. Throughout, we emphasize frequentist methods, though we do mention a few Bayesian procedures. For a more thorough treatment of the subject from a Bayesian perspective, see Chapter 7.

4.2 Maximum Likelihood Estimation

Consider the geostatistical model (3.1), and suppose that the mean function is linear and the error process is Gaussian. That is, suppose that

$$Y(\mathbf{s}) = \mathbf{X}(\mathbf{s})^T\boldsymbol{\beta} + e(\mathbf{s})$$

where $\mathbf{X}(\mathbf{s})$ is a p-vector of observable covariates at $\mathbf{s}$, $\boldsymbol{\beta}$ is a p-vector of unknown unrestricted parameters, and $e(\cdot) \equiv \{e(\mathbf{s}) : \mathbf{s} \in D\}$ is a zero-mean Gaussian process. Suppose further that $e(\cdot)$ has covariance function

$$\mathrm{Cov}[e(\mathbf{s}), e(\mathbf{t})] \equiv C(\mathbf{s}, \mathbf{t}; \boldsymbol{\theta}),$$

where $\boldsymbol{\theta}$ is an m-vector of unknown parameters belonging to a given parameter space, $\Theta \subset R^m$, within which the covariance function is positive definite. The joint parameter space for $\boldsymbol{\beta}$ and $\boldsymbol{\theta}$ is taken to be simply the Cartesian product of R^p and Θ. Observe that in this formulation we do not require that $e(\cdot)$ be stationary in any sense. This is in contrast to the classical geostatistics of Chapter 3, for which some form of stationarity (either second-order or intrinsic) was required.

Denote the n-vector of values of $Y(\cdot)$ and the $n \times p$ matrix of values of $\mathbf{X}(\cdot)$ observed at the data locations $\mathbf{s}_1, \ldots, \mathbf{s}_n$ as $\mathbf{Y} = [Y(\mathbf{s}_1), Y(\mathbf{s}_2), \ldots, Y(\mathbf{s}_n)]^T$, and $\mathbf{X} = [\mathbf{X}(\mathbf{s}_1), \mathbf{X}(\mathbf{s}_2), \ldots, \mathbf{X}(\mathbf{s}_n)]^T$, respectively, and assume that $n > p$ and $\mathrm{rank}(\mathbf{X}) = p$. Finally, let $\boldsymbol{\Sigma} = \boldsymbol{\Sigma}(\boldsymbol{\theta})$ denote the $n \times n$ covariance matrix, $\mathrm{var}(\mathbf{Y})$, whose (i, j)th element is $C(\mathbf{s}_i, \mathbf{s}_j; \boldsymbol{\theta})$.

Because $Y(\cdot)$ is Gaussian, the joint distribution of $\mathbf{Y}$ is multivariate normal, with mean vector $\mathbf{X}\boldsymbol{\beta}$ and covariance matrix $\boldsymbol{\Sigma}(\boldsymbol{\theta})$. Therefore, the likelihood function is given by

$$l(\boldsymbol{\beta}, \boldsymbol{\theta}; \mathbf{Y}) = (2\pi)^{-n/2}|\boldsymbol{\Sigma}(\boldsymbol{\theta})|^{-1/2}\exp\{-(\mathbf{Y} - \mathbf{X}\boldsymbol{\beta})^T\boldsymbol{\Sigma}^{-1}(\boldsymbol{\theta})(\mathbf{Y} - \mathbf{X}\boldsymbol{\beta})/2\}.$$

Maximum likelihood (ML) estimates of $\boldsymbol{\beta}$ and $\boldsymbol{\theta}$ are defined as any values of these parameters (in R^p and Θ, respectively) that maximize $l(\boldsymbol{\beta}, \boldsymbol{\theta}; \mathbf{Y})$. Note that any such values also maximize the log likelihood function,

$$\log l(\boldsymbol{\beta}, \boldsymbol{\theta}; \mathbf{Y}) = -\frac{n}{2}\log(2\pi) - \frac{1}{2}\log|\boldsymbol{\Sigma}(\boldsymbol{\theta})| - \frac{1}{2}(\mathbf{Y} - \mathbf{X}\boldsymbol{\beta})^T\boldsymbol{\Sigma}^{-1}(\boldsymbol{\theta})(\mathbf{Y} - \mathbf{X}\boldsymbol{\beta}).$$

Furthermore, for any fixed $\boldsymbol{\theta} \in \Theta$, say $\boldsymbol{\theta}_0$, the unique value of $\boldsymbol{\beta}$ that maximizes $\log l(\boldsymbol{\beta}, \boldsymbol{\theta}_0; \mathbf{Y})$ is given by

$$\hat{\boldsymbol{\beta}} = (\mathbf{X}^T\boldsymbol{\Sigma}^{-1}(\boldsymbol{\theta}_0)\mathbf{X})^{-1}\mathbf{X}^T\boldsymbol{\Sigma}^{-1}(\boldsymbol{\theta}_0)\mathbf{Y},$$

which would be the generalized least squares estimator of $\boldsymbol{\beta}$ if $\mathrm{var}(\mathbf{Y})$ was equal to $\boldsymbol{\Sigma}(\boldsymbol{\theta}_0)$. Thus, ML estimates of $\boldsymbol{\theta}$ and $\boldsymbol{\beta}$ are $\hat{\boldsymbol{\theta}}$ and $\hat{\boldsymbol{\beta}}$, where $\hat{\boldsymbol{\theta}}$ is any value of $\boldsymbol{\theta} \in \Theta$ that maximizes

$$L(\boldsymbol{\theta}; \mathbf{Y}) = -\frac{1}{2}\log|\boldsymbol{\Sigma}(\boldsymbol{\theta})| - \frac{1}{2}\mathbf{Y}^T\mathbf{P}(\boldsymbol{\theta})\mathbf{Y} \tag{4.1}$$

and $\hat{\boldsymbol{\beta}} = (\mathbf{X}^T\boldsymbol{\Sigma}^{-1}(\hat{\boldsymbol{\theta}})\mathbf{X})^{-1}\mathbf{X}^T\boldsymbol{\Sigma}^{-1}(\hat{\boldsymbol{\theta}})\mathbf{Y}$. Here,

$$\mathbf{P}(\boldsymbol{\theta}) = \boldsymbol{\Sigma}^{-1}(\boldsymbol{\theta}) - \boldsymbol{\Sigma}^{-1}(\boldsymbol{\theta})\mathbf{X}(\mathbf{X}^T\boldsymbol{\Sigma}^{-1}(\boldsymbol{\theta})\mathbf{X})^{-1}\mathbf{X}^T\boldsymbol{\Sigma}^{-1}(\boldsymbol{\theta}). \tag{4.2}$$

Eliminating $\boldsymbol{\beta}$ in this manner from the log likelihood function is called *profiling* and the function $L(\boldsymbol{\theta}; \mathbf{Y})$ is called the profile log likelihood function of $\boldsymbol{\theta}$.

The problem of maximizing $L(\boldsymbol{\theta}; \mathbf{Y})$ for $\boldsymbol{\theta} \in \Theta$ is a constrained (over the subset Θ of R^m) nonlinear optimization problem for which a closed-form solution exists only in very special cases. In general, therefore, ML estimates must be obtained numerically. One possible numerical approach is a brute force grid search, which is often reasonably effective when the parameter space for $\boldsymbol{\theta}$ is low-dimensional. Alternatively, iterative algorithms can be used. One important class of iterative algorithms is gradient algorithms. In a gradient algorithm, the $(k+1)$st iterate $\boldsymbol{\theta}^{(k+1)}$ is computed by updating the kth iterate $\boldsymbol{\theta}^{(k)}$ according to the equation

$$\boldsymbol{\theta}^{(k+1)} = \boldsymbol{\theta}^{(k)} + \rho^{(k)}\mathbf{M}^{(k)}\mathbf{g}^{(k)},$$

where $\rho^{(k)}$ is a scalar, $\mathbf{M}^{(k)}$ is an $m \times m$ matrix, and $\mathbf{g}^{(k)}$ is the gradient of L evaluated at $\boldsymbol{\theta} = \boldsymbol{\theta}^{(k)}$, i.e., $\mathbf{g}^{(k)} = \partial L(\boldsymbol{\theta}; \mathbf{Y})/\partial\boldsymbol{\theta}|_{\boldsymbol{\theta}=\boldsymbol{\theta}^{(k)}}$. The matrix product of $\mathbf{M}^{(k)}$ and $\mathbf{g}^{(k)}$ can be thought of as defining the search direction (relative to the kth iterate $\boldsymbol{\theta}^{(k)}$), while $\rho^{(k)}$ defines the size of the step to be taken in that direction.

Two gradient algorithms commonly used in conjunction with maximizing a log likelihood function are the Newton–Raphson and Fisher scoring procedures. In the Newton–Raphson procedure, $\mathbf{M}^{(k)}$ is the inverse of the $m \times m$ matrix whose (i,j)th element is $-\partial^2 L(\boldsymbol{\theta}; \mathbf{Y})/\partial\theta_i \partial\theta_j|_{\boldsymbol{\theta}=\boldsymbol{\theta}^{(k)}}$. In the Fisher scoring algorithm, $\mathbf{M}^{(k)} = (\mathbf{B}^{(k)})^{-1}$ where $\mathbf{B}^{(k)}$ is the Fisher information matrix associated with $L(\boldsymbol{\theta}; \mathbf{Y})$ evaluated at $\boldsymbol{\theta}^{(k)}$, i.e., $\mathbf{B}^{(k)}$ is the $m \times m$ matrix whose (i, j)th element is $E\{-\partial^2 L(\boldsymbol{\theta}; \mathbf{Y})/\partial\theta_i \partial\theta_j|_{\boldsymbol{\theta}=\boldsymbol{\theta}^{(k)}}\}$. For both algorithms, $\rho^{(k)} = 1$. Thus, Fisher scoring is identical to Newton–Raphson except that the second-order partial derivatives are replaced by their expectations. Expressions for both the second-order partial derivatives and their expectations may be found in Mardia and Marshall (1984).

Not all potentially useful iterative algorithms are gradient algorithms. The Nelder–Mead simplex algorithm (Nelder and Mead, 1965), for example, may also be effective.

Several practical decisions must be made to implement any of these iterative algorithms. These include choices of starting value for $\boldsymbol{\theta}$, convergence criterion, parameterization of the covariance function, and methods for accommodating the constraints on $\boldsymbol{\theta}$, which typically are linear inequality constraints. Some guidance on these and other implementation issues is provided by Harville (1977). It should be noted that one possible choice of starting value is an estimate of $\boldsymbol{\theta}$ obtained via classical geostatistical methods.

In many standard statistical problems, a unique ML estimate exists. In the present context, however, there is no guarantee of existence or uniqueness, nor is there even a guarantee that all local maxima of the likelihood function are global maxima. Indeed, Warnes and Ripley (1987) and Handcock (1989) show that the likelihood function corresponding to gridded observations of a stationary Gaussian process with spherical covariance function often has multiple modes, and that these modes may be well separated. Rasmussen and Williams (2006) display a bimodal likelihood surface for a case of a stationary Gaussian process with a Gaussian covariance function, which is observed at seven irregularly spaced locations on a line. However, the results of Handcock (1989) and Mardia and Watkins (1989), plus the experience of this author, suggest that multiple modes are extremely rare in practice for covariance functions within the Matérn class, such as the exponential function, and for datasets of the size typical of most applications. (See Chapter 2 for a definition of the Matérn class.) In any case, a reasonable practical strategy for determining whether a local maximum obtained by an iterative algorithm is likely to be the unique global maximum is to repeat the algorithm from several widely dispersed starting values.

4.3 REML Estimation

Although ML estimators of the spatial covariance parameters making up $\boldsymbol{\theta}$ have several desirable properties, they have a well-known shortcoming — they are biased as a consequence of the "loss in degrees of freedom" from estimating $\boldsymbol{\beta}$ (Harville, 1977). This bias may be substantial even for moderately sized samples if either the spatial correlation is strong or p (the dimensionality of $\boldsymbol{\beta}$) is large. However, the bias can be reduced substantially and, in some special cases, eliminated completely by employing a variant of maximum likelihood estimation known as restricted (or residual) maximum likelihood (REML) estimation. REML was originally proposed for use in components-of-variance models by Patterson and Thompson (1971, 1974); the first to propose it for use in spatial models was Kitanidis (1983). For both types of models, it is now as popular as maximum likelihood estimation, if not more so. In REML estimation, the likelihood (or equivalently the log likelihood) function associated with $n - p$ linearly independent linear combinations of the observations known as error contrasts, rather than the likelihood function associated with the observations, is maximized. An error contrast is a linear combination of the observations, i.e., $\mathbf{a}^T\mathbf{Y}$, that has expectation zero for all $\boldsymbol{\beta}$ and all $\boldsymbol{\theta} \in \Theta$; furthermore, two error contrasts $\mathbf{a}^T\mathbf{Y}$ and $\mathbf{b}^T\mathbf{Y}$ are said to be linearly independent if $\mathbf{a}$ and $\mathbf{b}$ are linearly independent vectors. Any set of $n - p$ linearly independent elements of $[\mathbf{I} - \mathbf{X}(\mathbf{X}^T\mathbf{X})^{-1}\mathbf{X}^T]\mathbf{Y}$ may serve as the required error contrasts for the restricted likelihood, but it actually makes no difference what set of error contrasts is used as long as they number $n - p$ and are independent, because the log likelihood function associated with any such set differs by at most an additive constant (which does not depend on $\boldsymbol{\beta}$ or $\boldsymbol{\theta}$) from the function

$$L_R(\boldsymbol{\theta};\mathbf{Y}) = -\frac{1}{2}\log|\boldsymbol{\Sigma}(\boldsymbol{\theta})| - \frac{1}{2}\log|\mathbf{X}^T\boldsymbol{\Sigma}^{-1}(\boldsymbol{\theta})\mathbf{X}| - \frac{1}{2}\mathbf{Y}^T\mathbf{P}(\boldsymbol{\theta})\mathbf{Y},$$

where $\mathbf{P}(\boldsymbol{\theta})$ was defined by (4.2). Observe that $L_R(\boldsymbol{\theta};\mathbf{Y})$ differs from the profile log likelihood function $L(\boldsymbol{\theta};\mathbf{Y})$ (given by (4.1)) only additively, by an extra term, $-\frac{1}{2}\log|\mathbf{X}^T\boldsymbol{\Sigma}^{-1}(\boldsymbol{\theta})\mathbf{X}|$. A REML estimate of $\boldsymbol{\theta}$ is any value $\tilde{\boldsymbol{\theta}} \in \Theta$ at which L_R attains its maximum. This estimate generally must be obtained via the same kinds of numerical procedures used to obtain a ML estimate. Once a REML estimate of $\boldsymbol{\theta}$ is obtained, the corresponding estimate of $\boldsymbol{\beta}$ is obtained as its generalized least squares estimator evaluated at $\boldsymbol{\theta} = \tilde{\boldsymbol{\theta}}$, i.e., $\tilde{\boldsymbol{\beta}} = (\mathbf{X}^T\boldsymbol{\Sigma}^{-1}(\tilde{\boldsymbol{\theta}})\mathbf{X})^{-1}\mathbf{X}^T\boldsymbol{\Sigma}^{-1}(\tilde{\boldsymbol{\theta}})\mathbf{Y}$.

Does REML effectively reduce the bias incurred by maximum likelihood estimation? Two published simulation studies have compared the performance of REML and ML estimators in spatial models. An early study by Zimmerman and Zimmerman (1991) for stationary Gaussian random fields with constant mean and exponential covariance function without nugget indicated that in this case the REML estimators of the sill and range parameters were indeed less biased than their ML counterparts, but that the upper tail of the distribution of the sill's REML estimator could be quite heavy. In a recent, more thorough study, Irvine, Gitelman, and Hoeting (2007) compared estimator performance for stationary Gaussian random fields with constant mean and exponential-plus-nugget covariance functions, the latter having several different effective ranges and nugget-to-sill ratios. They obtained results broadly similar to those of Zimmerman and Zimmerman (1991), but even more unfavorable to REML; for example, the upper tail of the empirical distribution of the REML estimator of the effective range was, in many cases, so heavy that this estimator's mean squared error was much larger than that of the ML estimator. The heavy upper tail problem became worse as either the effective range or the nugget-to-sill ratio increased. Thus, while REML does appear to effectively reduce bias, it may also increase estimation variance unless the persistence of spatial dependence (as measured by the effective range) is weak. Whether this would cause one to prefer ML to REML estimation depends on which problem one considers to be more serious: badly underestimating or badly overestimating the effective range and sill, leading respectively to overly optimistic or overly conservative inferences on mean parameters and predictions. On the other hand, it is worth noting that both of the studies mentioned here considered only processes with constant means, i.e., $p = 1$. Zimmerman (1986) demonstrated that the relative performance of REML to ML estimation for the nuggetless exponential covariance function improves considerably as the dimensionality of β increases, which should come as no surprise given REML's *raison d'etre*. Still, it seems to this author that further research is needed before comprehensive guidelines can be provided as to which estimator, REML or ML, is best in which situation(s).

REML estimation does have one distinct advantage, however. It may be used to estimate parameters of certain nonstationary processes known as Gaussian intrinsic random fields of order k (IRF-ks), which were introduced in Chapter 2. Such processes are characterized by the probabilistic structure of certain linear combinations known as generalized increments; more specifically, the kth-order generalized increments of a Gaussian IRF-k are jointly Gaussian with mean zero and a generalized covariance function. The generalized covariance function does not specify a covariance structure for the observations themselves, but only for the generalized increments. An intrinsically stationary random field and an IRF-0 are merely different names for the same process; furthermore, the generalized increments and the generalized covariance function of an IRF-0 are merely $\{Y(\mathbf{s}) - Y(\mathbf{t}) : \mathbf{s}, \mathbf{t} \in D\}$ and the semivariogram, respectively. It turns out that generalized increments for any Gaussian IRF-k coincide with error contrasts for a Gaussian random field with kth-order polynomial mean structure, so that REML estimation may proceed for such processes. Owing to the lack of a fully specified covariance structure for the observations, however, ML cannot. For example, REML, but not ML, may be used to estimate the power semivariogram (described in Chapter 3) of an IRF-0. The REML log likelihood in this case is

$$L_R(\boldsymbol{\theta}; \mathbf{Y}) = -\frac{1}{2}\log|\mathbf{A}^T\mathbf{K}(\boldsymbol{\theta})\mathbf{A}| - \frac{1}{2}\mathbf{Y}^T\mathbf{A}(\mathbf{A}^T\mathbf{K}(\boldsymbol{\theta})\mathbf{A})^{-1}\mathbf{A}^T\mathbf{Y}$$

where $\mathbf{A}$ is an $(n-1) \times n$ matrix whose ith row has ith element 1, $(i+1)$st element -1, and zeros elsewhere, and $\mathbf{K}(\boldsymbol{\theta})$ is the $n \times n$ matrix with (i, j)th element $\gamma(\mathbf{s}_i - \mathbf{s}_j; \boldsymbol{\theta}) = \theta_1 \|\mathbf{s}_i - \mathbf{s}_j\|^{\theta_2}$.

4.4 Asymptotic Results

For purposes of making inferences about the parameters of a Gaussian random field, e.g., estimating standard errors and constructing confidence intervals, knowledge of the asymptotic properties of the likelihood-based estimators is useful. In contrast to most other areas of statistics, however, for statistics of continuous spatial processes there are two (at least) quite different asymptotic frameworks to which one could reasonably appeal: (1) increasing domain asymptotics, in which the minimum distance between data locations is bounded away from zero and, thus, the spatial domain of observation is unbounded, and (2) infill (also called fixed-domain) asymptotics, in which observations are taken ever more densely in a fixed and bounded domain. In Chapter 6, Michael Stein gives, in considerable detail and within both frameworks, some known asymptotic results relevant to spatial prediction and estimation. Below, we briefly summarize some of the results pertaining to estimation.

In the increasing-domain asymptotic framework, it is known that, under certain regularity conditions, the ML and REML estimators of the parameters of Gaussian random fields are consistent and asymptotically normal (Mardia and Marshall, 1984; Cressie and Lahiri, 1996). The available results under infill asymptotics, however, are considerably more limited. Moreover, the few known results indicate that the asymptotic behavior of ML estimators can be quite different in this framework. For example, for a one-dimensional, zero-mean, Gaussian random field model with nuggetless exponential covariance function, the ML estimators of the sill and the range parameter are not consistent under infill asymptotics (although their product is consistent for the product of these two parameters) (Ying, 1991). For the same process and for one with a nugget but otherwise identical, Zhang and Zimmerman (2005) found that for those parameters for which the ML estimator is consistent under both asymptotic frameworks, the approximations to the ML estimator's finite-sample distribution provided by the two frameworks perform about equally well, but for those parameters that cannot be estimated consistently under infill asymptotics, the finite-sample approximation provided by the infill asymptotic framework performs better. My recommendation, therefore, is to base inferences on infill asymptotic results if and when they are available.

4.5 Hypothesis Testing and Model Comparisons

Performing a hypothesis test on the parameters $\boldsymbol{\beta}$ and $\boldsymbol{\theta}$ of the model may be viewed as a comparison of two models — the "full" and "reduced" models, corresponding to the alternative and null hypotheses — and may be carried out via a likelihood ratio test. This amounts to comparing twice the difference in the log likelihood functions evaluated at the ML estimators for the two models to percentiles of the chi-square distribution with q degrees of freedom, where q is the difference in the dimensionality of the parameter space for the two models. The success of this procedure is predicated, of course, on the assumption that twice the negative log of the likelihood ratio statistic is asymptotically distributed, under the null hypothesis, as a chi-squared random variable with q degrees of freedom. This assumption is justified under increasing domain asymptotics and appropriate regularity conditions.

Non-nested models may be compared informally within the likelihood framework using penalized likelihood criteria, such as Akaike's information criterion,

$$AIC = -2\log l(\hat{\boldsymbol{\beta}}, \hat{\boldsymbol{\theta}}) + 2(p+m),$$

or Schwarz's Bayesian information criterion,

$$BIC = -2\log l(\hat{\boldsymbol{\beta}}, \hat{\boldsymbol{\theta}}) + (p+m)\log n.$$

In comparing two models, the model with the smaller value of one of these criteria is judged to be the better of the two models according to that criterion. Both criteria trade model fit and model parsimony off against one another, but in slightly different ways, with the result that *BIC* tends to favor smaller models than *AIC*. Observe that models with different mean functions or different covariance functions, or both, may be compared, though it is not clear that penalizing mean and covariance parameters equally (as these criteria do) is appropriate.

Within the REML framework, hypotheses about $\boldsymbol{\theta}$ may be tested analogously via a comparison of restricted log likelihoods evaluated at REML estimates. To compare non-nested models, penalized restricted likelihood criteria may again be used; in this context

$$AIC = -2L_R(\tilde{\boldsymbol{\theta}}) + 2m$$

and

$$BIC = -2L_R(\tilde{\boldsymbol{\theta}}) + m\log n.$$

It is extremely important, however, that such comparisons be made between models that differ only with respect to their covariance structure. Valid comparisons of models with different mean structure cannot be made within the REML framework, as the error contrasts are different for two such models.

A popular criterion for model comparisons within a Bayesian framework is the deviance information criterion (DIC),

$$DIC = \overline{D} + p_D$$

(Spiegelhalter, Best, Carlin, and van der Linde, 2002). Here, $\overline{D} = E[D(\boldsymbol{\beta}, \boldsymbol{\theta})|\mathbf{Y}]$, where $D(\boldsymbol{\beta}, \boldsymbol{\theta}) = -2\log l(\boldsymbol{\beta}, \boldsymbol{\theta}; \mathbf{Y}) + c$ and c is a constant that can be ignored because it is identical across models, and $p_D = \overline{D} - D(\overline{\boldsymbol{\beta}}, \overline{\boldsymbol{\theta}})$ where $(\overline{\boldsymbol{\beta}}, \overline{\boldsymbol{\theta}}) = E[(\boldsymbol{\beta}, \boldsymbol{\theta})|\mathbf{Y}]$. $\overline{D}$ is a measure of model fit, while p_D, known as the effective number of parameters, is a measure of model complexity, and *DIC* balances these two against each other. *DIC* may be calculated easily from the samples generated by a Markov chain Monte Carlo (MCMC) simulation, simply by averaging $D(\boldsymbol{\beta}, \boldsymbol{\theta})$ and $(\boldsymbol{\beta}, \boldsymbol{\theta})$ over the samples to obtain $\overline{D}$ and $(\overline{\boldsymbol{\beta}}, \overline{\boldsymbol{\theta}})$, respectively, and evaluating D at the latter quantity.

4.6 Computational Issues

The likelihood-based estimation procedures we have described here generally require extensive computations. This is due primarily to the necessity of repeatedly forming and inverting the covariance matrix and evaluating its determinant, either as parameter estimates are updated within an iterative algorithm or as a grid search proceeds. This computational burden can be a serious obstacle to the implementation of likelihood-based approaches in practice. Fortunately, however, in some cases it is possible to substantially reduce the amount of computation required for exact likelihood-based inference.

One situation in which the computational burden can be reduced occurs when one of the covariance function's parameters is a scale parameter, in which case it is possible to express the covariance matrix as a scalar multiple of another matrix that is free of this parameter, i.e., as $\boldsymbol{\Sigma}(\boldsymbol{\theta}) = \theta_1 \mathbf{W}(\boldsymbol{\theta}_{-1})$ for some matrix $\mathbf{W}$, where $\boldsymbol{\theta}_{-1} = (\theta_2, \theta_3, \ldots, \theta_m)^T \in \Theta_{-1}$. Here, Θ_{-1} is defined such that $\Theta = \{\theta_1 > 0, \boldsymbol{\theta}_{-1} \in \Theta_{-1}\}$. One case of this occurs, for example, when $Y(\cdot)$ is stationary, for then the elements of $\mathbf{Y}$ are homoscedastic and $\boldsymbol{\Sigma}(\boldsymbol{\theta})$ can be expressed in the aforementioned way, with θ_1 being the common variance (or semivariogram sill) of the observations and $\mathbf{W}(\boldsymbol{\theta}_{-1})$ being a correlation matrix. In any such case, it is easily shown that $L(\boldsymbol{\theta}; \mathbf{Y})$ assumes its maximum value at $(\hat{\theta}_1, \hat{\boldsymbol{\theta}}_{-1}^T)^T$, where

$$\hat{\theta}_1 = \mathbf{Y}^T \mathbf{Q}(\hat{\boldsymbol{\theta}}_{-1})\mathbf{Y}/n$$

and $\hat{\boldsymbol{\theta}}_{-1}$ is any value of $\boldsymbol{\theta}_{-1} \in \Theta_{-1}$ that maximizes

$$L_{-1}(\boldsymbol{\theta}_{-1}; \mathbf{Y}) = -\frac{1}{2}\log|\mathbf{W}(\boldsymbol{\theta}_{-1})| - \frac{n}{2}\log(\mathbf{Y}^T\mathbf{Q}(\boldsymbol{\theta}_{-1})\mathbf{Y}).$$

Here,

$$\mathbf{Q}(\boldsymbol{\theta}_{-1}) = \mathbf{W}^{-1}(\boldsymbol{\theta}_{-1}) - \mathbf{W}^{-1}(\boldsymbol{\theta}_{-1})\mathbf{X}(\mathbf{X}^T\mathbf{W}^{-1}(\boldsymbol{\theta}_{-1})\mathbf{X})^{-1}\mathbf{X}^T\mathbf{W}^{-1}(\boldsymbol{\theta}_{-1}).$$

This is another instance of profiling, applied in this case to the scale parameter θ_1 (rather than to the mean parameters $\boldsymbol{\beta}$). The advantage of maximizing $L_{-1}(\boldsymbol{\theta}_{-1};\mathbf{Y})$ rather than $L(\boldsymbol{\theta};\mathbf{Y})$ is that the dimensionality of the maximization problem is reduced from m to $m-1$. This can yield substantial savings in computation. The same approach works equally well with the REML log likelihood function.

Another situation in which the computations associated with likelihood-based inference may be reduced occurs when $\boldsymbol{\Sigma}(\boldsymbol{\theta})$ (or $\mathbf{W}(\boldsymbol{\theta}_{-1})$) has a patterned structure that can be exploited to speed up the computation of its determinant and the quadratic form(s) involving its inverse. When $\boldsymbol{\Sigma}(\boldsymbol{\theta})$ is $n \times n$, the required computations are $O(n^3)$ in general; however, for certain models and spatial configurations of data locations, the structure within $\boldsymbol{\Sigma}(\boldsymbol{\theta})$ is sufficiently specialized that the amount of computation can be reduced significantly. For example, if the covariance function has a range that is small relative to the spatial domain of observation, then the covariance matrix will be "sparse" (i.e., it will have many elements equal to zero), and methods for sparse matrices may be used to good effect (Barry and Pace, 1997). Finally, another situation in which major reductions in computation are possible occurs when the covariance function (or generalized covariance function) is either isotropic or separable and the data locations form a regular rectangular grid (Zimmerman, 1989a, 1989b).

4.7 Approximate and Composite Likelihood

The previous section described some methods for reducing the computational burden of exact likelihood-based estimation. A completely different strategy for reducing computations is to approximate the likelihood function by a function that is more easily maximized. This strategy is described below.

Vecchia (1988) proposed an approximation to the likelihood for geostatistical data, which is based on a partitioning of the observation vector $\mathbf{Y}$ into subvectors $\mathbf{Y}_1, \ldots, \mathbf{Y}_b$. Letting $\mathbf{Y}_{(j)} = (\mathbf{Y}_1^T, \ldots, \mathbf{Y}_j^T)^T$, letting $p(\mathbf{Y};\boldsymbol{\beta},\boldsymbol{\theta})$ denote the joint density of $\mathbf{Y}$, and letting $p(\mathbf{Y}_j|\mathbf{Y}_{(j-1)};\boldsymbol{\beta},\boldsymbol{\theta})$ denote the conditional density of $\mathbf{Y}_j$ given $\mathbf{Y}_{(j-1)}$, the (exact) likelihood may be written as

$$p(\mathbf{Y};\boldsymbol{\beta},\boldsymbol{\theta}) = p(\mathbf{Y}_1;\boldsymbol{\beta},\boldsymbol{\theta}) \prod_{j=2}^{b} p(\mathbf{Y}_j|\mathbf{Y}_{(j-1)};\boldsymbol{\beta},\boldsymbol{\theta}). \tag{4.3}$$

Vecchia's (1988) proposal was to approximate (4.3) by replacing each complete conditioning vector $\mathbf{Y}_{(j-1)}$ with a subvector $\mathbf{S}_{(j-1)}$ of $\mathbf{Y}_{(j-1)}$ so that the matrices whose inverses and determinants must be evaluated to compute the conditional densities are much smaller, thereby reducing the computations required to maximize (4.3). Since there is no unique way to partition $\mathbf{Y}$ in this scheme, a natural question is: How should $\mathbf{Y}_j$ and $\mathbf{S}_{(j-1)}$ be chosen? Vecchia only considered vectors $\mathbf{Y}_j$ of length one, ordered by the values of either of the two coordinate axes of data locations, and he recommended choosing $\mathbf{S}_{(j-1)}$ to consist of the q most proximate observations to $\mathbf{Y}_j$, where q is much smaller than n. The smaller the value of q, of course, the more efficient the computation, but the cruder the approximation to the true likelihood. For datasets of size $n = 100$ simulated from several Gaussian random fields, Vecchia showed that taking $q = 10$ resulted in good approximations; in practice, of course, an appropriate choice of q will depend on the range of spatial dependence relative to the distance between observations.

Stein, Chi, and Welty (2004) extended Vecchia's original proposal by allowing the $\mathbf{Y}_j$s to be of nonuniform (and nonunity) lengths and by applying it to the restricted likelihood function rather than the ordinary likelihood function. For simulated Gaussian data of size

$n = 1000$, they found that taking $q = 16$ was sufficient for the (statistical) efficiencies of the approximate REML estimators, relative to the exact REML estimators, to be in the range 80 to 95%. In contrast to Vecchia, however, they found that it was sometimes advantageous to include some observations in $\mathbf{S}_{(j-1)}$ that were rather distant from $\mathbf{Y}_j$. This was particularly true when the spatial dependence was strong relative to the spatial domain of observation, a situation that was not represented in any of Vecchia's examples.

A similar but slightly different approximate likelihood estimation strategy is that of pseudolikelihood or composite likelihood (Besag, 1975; Lindsay, 1988). Such a likelihood is formed by multiplying individual component likelihood functions, each of which is a valid conditional or marginal likelihood, as if these component likelihoods correspond to independent subvectors of the data. In this way, one avoids having to evaluate the determinant and inverse of the $n \times n$ covariance matrix, replacing these evaluations with ones for much smaller matrices. The cost of this computational simplicity, as with the Vecchia/Stein et al. strategy, is a loss of statistical efficiency as a result of ignoring information about the components' covariance structure.

Curriero and Lele (1999) consider a particular composite likelihood estimation scheme. For an intrinsically stationary Gaussian random field, they form a composite log likelihood function from the marginal densities of all pairwise differences among the observations. Ignoring constant terms, this function is

$$CL(\boldsymbol{\theta}; \mathbf{Y}) = -\frac{1}{2}\sum_{i=1}^{n-1}\sum_{j>i}\left\{\log(\gamma(\mathbf{s}_i - \mathbf{s}_j; \boldsymbol{\theta}) + \frac{(Y(\mathbf{s}_i) - Y(\mathbf{s}_j))^2}{2\gamma(\mathbf{s}_i - \mathbf{s}_j; \boldsymbol{\theta})}\right\}.$$

Maximization of $CL(\boldsymbol{\theta}; \mathbf{Y})$ with respect to $\boldsymbol{\theta}$ yields a composite maximum likelihood estimator (MLE). Note that by considering only pairwise differences, the unknown constant mean parameter is eliminated from this composite log likelihood, in keeping with the spirit of REML. The composite MLE is consistent and asymptotically normal under certain regularity conditions within an increasing domain asymptotic framework, and it turns out that this consistency holds even if the random field is not Gaussian. Moreover, it can be shown (Gotway and Schabenberger, 2005, pp. 171–172) that this particular composite likelihood estimator is equivalent to the estimator obtained by applying (nonlinear) weighted least squares in the model

$$[Y(\mathbf{s}_i) - Y(\mathbf{s}_j)]^2 = 2\gamma(\mathbf{s}_i - \mathbf{s}_j; \boldsymbol{\theta}) + \epsilon_{ij}, \quad E(\epsilon_{ij}) = 0, \ \mathrm{var}(\epsilon_{ij}) = 8[\gamma(\mathbf{s}_i - \mathbf{s}_j; \boldsymbol{\theta})]^2.$$

A third strategy for approximating the likelihood is that of covariance tapering (Kaufman, Schervish, and Nychka, 2009). In this approach, elements of the covariance matrix corresponding to spatially distant pairs of observations are set to zero, so that the algorithms for sparse matrix inversion and determinant evaluation mentioned in the previous section may be used. Suppose that $Y(\cdot)$ is an isotropic Gaussian random field with linear mean function and covariance function $C_0(h; \boldsymbol{\theta})$, and let $C_T(h; \gamma)$, where γ is known, be a "tapering function," i.e., an isotropic, continuous correlation function that is identically zero whenever $h \geq \gamma$. Then, let

$$C_1(h; \boldsymbol{\theta}, \gamma) = C_0(h; \boldsymbol{\theta}) C_T(h; \gamma), \quad h \geq 0.$$

Note that $C_1(h; \boldsymbol{\theta}, \gamma)$ is a valid (positive definite) covariance function that differs from C_0 for distant locations, but retains the same variance as C_0. Furthermore, Kaufman et al. show that if C_0 belongs to the Matérn family of covariance functions, then a tapering function C_T may be found for which the resulting covariance function C_1 has the same behavior near the origin as C_0 and corresponds, in fact, to a mean-zero Gaussian measure equivalent to that corresponding to C_0. Now, if C_1 rather than C_0 was actually the covariance function of

$Y(\cdot)$, then the log likelihood function corresponding to the vector of observations $\mathbf{Y}$ would be

$$L_T(\boldsymbol{\theta}, \beta; \mathbf{Y}) = -\frac{1}{2}\log|\boldsymbol{\Sigma}(\boldsymbol{\theta}) \circ \mathbf{T}(\gamma)| - \frac{1}{2}(\mathbf{Y} - \mathbf{X}\beta)^T[\boldsymbol{\Sigma}(\boldsymbol{\theta}) \circ \mathbf{T}(\gamma)]^{-1}(\mathbf{Y} - \mathbf{X}\beta)$$

where $\mathbf{T}(\gamma)$ is the matrix with (i, j)th element $C_T(\|\mathbf{s}_i - \mathbf{s}_j\|; \gamma)$ and "$\circ$" refers to the elementwise, or Schur, product of two matrices. Kaufman et al. consider the "one-taper estimator" obtained by maximizing this approximation to the log likelihood, and another, the "two-taper estimator," which is based on the theory of unbiased estimating equations and, therefore, is less biased. They show via simulation that both estimators, but especially the two-taper estimator, perform quite well, even when the degree of tapering is severe.

A final strategy is to use spectral methods to approximate the likelihood. For a stationary Gaussian random field observed on a regular grid of points, Whittle (1954) developed the following approximation:

$$L(\boldsymbol{\theta}; \mathbf{Y}) \doteq L_W(\boldsymbol{\theta}; \mathbf{Y}) = -\frac{n}{(2\pi)^2}\sum_{\omega \in F}\{\log f(\omega, \boldsymbol{\theta}) + I_n(\omega; \mathbf{Y})[f(\omega, \boldsymbol{\theta})]^{-1}\}$$

where F is the set of Fourier frequencies, f is the spectral density of the random field, and I_n is the periodogram. Using the fast Fourier transform, $L_W(\boldsymbol{\theta}; \mathbf{Y})$ can be calculated very efficiently, with only $O(n \log_2 n)$ operations. Fuentes (2007) extended this methodology for use with irregularly spaced data by integrating the random field $Y(\cdot)$ over grid cells. Using simulation, she found that this approach was as efficient statistically as the approach of Stein et al. (2004) except for capturing the behavior of $Y(\cdot)$ at very short distances, and that it was considerably more efficient computationally.

4.8 Methods for Non-Gaussian Data

To this point in the chapter we have assumed that the observations are Gaussian. What can be done to estimate parameters of continuous spatial processes that yield non-normal observations? For observations that are continuous but skewed or have bounded support (such as proportions), one possibility is to transform the data to a scale on which the data are more nearly normal (see de Oliveira, Kedem, and Short (1997) for a Bayesian implementation of this idea). For discrete observations, such as counts or presence/absence data, one may base an analysis upon a class of models called spatial generalized linear mixed models (GLMMs). We conclude this chapter with a brief discussion of GLMMs and of likelihood-based estimation of their parameters. Some other models and methods for non-Gaussian data are presented in Chapter 11.

In a classical generalized linear model (GLM) for a vector of independent observations with mean $\boldsymbol{\mu}$, some function of the mean, $g(\boldsymbol{\mu})$, called the link function, is assumed to be a linear combination of fixed effects, i.e., $g(\boldsymbol{\mu}) = \mathbf{X}\beta$. These effects and any dispersion parameters of the model may be estimated by either maximum likelihood or quasilikelihood (Wedderburn, 1974), depending on whether one assumes a distribution (from the exponential family) for the data or only its first and second moments. Liang and Zeger (1986) and Zeger and Liang (1986) extended the quasilikelihood methodology to GLMs with serially correlated observations from longitudinal studies, and Albert and McShane (1995) and Gotway and Stroup (1997) adapted this extension to spatial data. However, a spatial GLM does not, by itself, provide a way to model and estimate spatial correlation. Spatial GLMMs, introduced by Diggle, Tawn, and Moyeed (1998), offer such an opportunity by adding spatially correlated random effects to the model. In a spatial GLMM, a latent zero-mean, stationary Gaussian random field $\{b(\mathbf{s}) : \mathbf{s} \in D\}$, with covariance function $C(\mathbf{h}; \boldsymbol{\theta})$, is assumed to exist. Conditionally on $b(\cdot)$, $Y(\cdot)$ is taken to be an independent process with distribution specified by the conditional mean $E\{Y(\mathbf{s})|b(\mathbf{s})\}$, and for some link function g,

$$g[E\{Y(\mathbf{s})|b(\mathbf{s})\}] = \mathbf{x}_j^T(\mathbf{s})\boldsymbol{\beta} + b(\mathbf{s}).$$

The likelihood function for the parameters of a spatial GLMM is given by

$$L(\boldsymbol{\beta}, \boldsymbol{\theta}; \mathbf{Y}) = \int_{R^n} \left\{ \prod_{i=1}^{n} f_i(Y(\mathbf{s}_i)|\mathbf{b}, \boldsymbol{\beta}) \right\} f_{\mathbf{b}}(\mathbf{b}|\boldsymbol{\theta})\, d\mathbf{b}$$

where $\mathbf{b} = (b(\mathbf{s}_1, \ldots, b(\mathbf{s}_n))^T$, f_i is the conditional density of $Y(\mathbf{s}_i)$ given $\mathbf{b}$, and $f_{\mathbf{b}}$ is the multivariate Gaussian density of $\mathbf{b}$. Owing to the high dimensionality of this integral, obtaining the MLE by direct maximization of L is not feasible, so a number of alternatives have been put forward. Zhang (2002) presented an expectation-maximization (EM) gradient algorithm for maximizing L. Varin, Host, and Skare (2005) and Apanasovich, Ruppert, Lupton, Popovic (2007) developed composite (pairwise) likelihood estimation approaches. Diggle et al. (1998) and Christensen and Waagepetersen (2002) obtained parameter estimates using Bayesian MCMC methods.

References

Albert, P.A. and McShane, L.M. (1995). A generalized estimating equations approach for spatially correlated data: Applications to the analysis of neuroimaging data. *Biometrics*, **51**, 627–638.

Apanasovich, T.V., Ruppert, D., Lupton, J.R., Popovic, N., Turner, N.D., Chapkin, R.S., and Carroll, R.J. (2007). Aberrant crypt foci and semiparametric modeling of correlated binary data. *Biometrics*, DOI: 10.1111/j.1541-0420.2007.00892.x

Barry, R.P. and Pace, R.K. (1997). Kriging with large data sets using sparse matrix techniques. *Communications in Statistics — Simulation and Computation*, **26**, 619–629.

Besag, J. (1975). Statistical analysis of non-lattice data. *The Statistician*, **24**, 179–195.

Christensen, O.F. and Waagepetersen, R. (2002). Bayesian prediction of spatial count data using generalized mixed models. *Biometrics*, **58**, 280–286.

Cressie, N. and Lahiri, S.N. (1996). Asymptotics for REML estimation of spatial covariance parameters. *Journal of Statistical Planning and Inference*, **50**, 327–341.

Curriero, F.C. and Lele, S. (1999). A composite likelihood approach to semivariogram estimation. *Journal of Agricultural, Biological, and Environmental Statistics*, **4**, 9–28.

de Oliveira, V., Kedem, B., and Short, D.A. (1997). Bayesian prediction of transformed Gaussian random fields. *Journal of the American Statistical Association*, **92**, 1422–1433.

Diggle, P.J., Tawn, J.A., and Moyeed, R.A. (1998). Model-based geostatistics (with discussion). *Applied Statistics*, **47**, 299–350.

Fuentes, M. (2007). Approximate likelihood for large irregularly spaced spatial data. *Journal of the American Statistical Association*, **102**, 321–331.

Gotway, C.A. and Schabenberger, O. (2005). *Statistical Methods for Spatial Data Analysis*. Boca Raton, FL: Chapman & Hall.

Gotway, C.A. and Stroup, W.W. (1997). A generalized linear model approach to spatial data analysis and prediction. *Journal of Agricultural, Biological, and Environmental Statistics*, **2**, 157–178.

Handcock, M.S. (1989). Inference for spatial Gaussian random fields when the objective is prediction. PhD dis., Department of Statistics, University of Chicago.

Harville, D.A. (1977). Maximum likelihood approaches to variance component estimation and to related problems. *Journal of the American Statistical Association*, **72**, 320–340.

Irvine, K.M., Gitelman, A.I., and Hoeting, J.A. (2007). Spatial designs and properties of spatial correlation: Effects on covariance estimation. *Journal of Agricultural, Biological, and Environmental Statistics*, **12**, 450–469.

Kaufman, C.G., Schervish, M.J., and Nychka, D.W. (2008). Covariance tapering for likelihood-based estimation in large spatial datasets. Forthcoming. *Journal of the American Statistical Association*, **103**, 1545–1555.

Kitanidis, P.K. (1983). Statistical estimation of polynomial generalized covariance functions and hydrologic applications. *Water Resources Research*, **19**, 909–921.

Liang, K.-Y. and Zeger, S.L. (1986). Longitudinal data analysis using generalized linear models. *Biometrika* **73**, 13–22.

Lindsay, B.G. (1988). Composite likelihood methods. *Contemporary Mathematics*, **80**, 221–239.

Mardia, K.V. and Marshall, R.J. (1984). Maximum likelihood estimation of models for residual covariance in spatial regression. *Biometrika*, **71**, 135–146.

Mardia, K.V. and Watkins, A.J. (1989). On multimodality of the likelihood in the spatial linear model. *Biometrika*, **76**, 289–295.

Nelder, J.A. and Mead, R. (1965). A simplex method for function minimization. *The Computer Journal*, **7**, 308–313.

Patterson, H.D. and Thompson, R. (1971). Recovery of inter-block information when block sizes are unequal. *Biometrika*, **58**, 545–554.

Patterson, H.D. and Thompson, R. (1974). Maximum likelihood estimation of components of variance. In *Proceedings of the 8th International Biometrics Conference*, 1974, pp. 197–207. Alexandria, VA: The Biometric Society.

Rasmussen, C.E. and Williams, C.K.I. (2006). *Gaussian Processes for Machine Learning*. Cambridge, MA: MIT Press.

Spiegelhalter, D.J., Best, N.G., Carlin, B.P., and van der Linde, A. (2002). Bayesian measures of model complexity and fit (with discussion). *Journal of the Royal Statistical Society Series B*, **64**, 583–639.

Stein, M.L., Chi, Z., and Welty, L.J. (2004). Approximating likelihoods for large spatial data sets. *Journal of the Royal Statistical Society, Series B*, **66**, 275–296.

Varin, C., Host, G., and Skare, O. (2005). Pairwise likelihood inference in spatial generalized linear mixed models. *Computational Statistics and Data Analysis*, **49**, 1173–1191.

Vecchia, A.V. (1988). Estimation and model identification for continuous spatial processes. *Journal of the Royal Statistical Society Series B*, **50**, 297–312.

Warnes, J.J. and Ripley, B.D. (1987). Problems with likelihood estimation of covariance functions of spatial Gaussian processes. *Biometrika*, **74**, 640–642.

Wedderburn, R.W.M. (1974). Quasi-likelihood functions, generalized linear models and the Gauss–Newton method. *Biometrika*, **61**, 439–447.

Whittle, P. (1954). On stationary processes in the plane. *Biometrika*, **41**, 434–449.

Ying, Z. (1991). Asymptotic properties of a maximum likelihood estimator with data from a Gaussian process. *Journal of Multivariate Analysis*, **36**, 280–296.

Zeger, S.L. and Liang, K.-Y. (1986). Longitudinal data analysis for discrete and continuous outcomes. *Biometrics*, **42**, 121–130.

Zhang, H. (2002). On estimation and prediction for spatial generalized linear mixed models. *Biometrics*, **58**, 129–136.

Zhang, H. and Zimmerman, D.L. (2005). Toward reconciling two asymptotic frameworks in spatial statistics. *Biometrika*, **92**, 921–936.

Zimmerman, D.L. (1986). A random field approach to spatial experiments. PhD dis., Department of Statistics, Iowa State University.

Zimmerman, D.L. (1989a). Computationally exploitable structure of covariance matrices and generalized covariance matrices in spatial models. *Journal of Statistical Computation and Simulation*, **32**, 1–15.

Zimmerman, D.L. (1989b). Computationally efficient restricted maximum likelihood estimation of generalized covariance functions. *Mathematical Geology*, **21**, 655–672.

Zimmerman, D.L. and Zimmerman, M.B. (1991). A comparison of spatial semivariogram estimators and corresponding ordinary kriging predictors. *Technometrics*, **33**, 77–91.

5

Spectral Domain

Montserrat Fuentes and Brian Reich

CONTENTS

Spectral methods are a powerful tool for studying the spatial structure of spatial continuous processes and sometimes offer significant computational benefits. Using the spectral representation of a spatial process we can easily construct valid (*positive definite*) covariance functions and introduce new models for spatial fields. Likelihood approaches for large spatial datasets are often very difficult, if not infeasible, to implement due to computational limitations. Even when we can assume normality, exact calculations of the likelihood for a Gaussian spatial process observed at n locations requires $O(n^3)$ operations. The spectral version of the Gaussian log likelihood for gridded data requires $O(nlog_2n)$ operations and does not involve calculating determinants.

In Section 5.1 of this chapter we offer a review of the Fourier transform and introduce the spectral representation of a stationary spatial process, we also present Bochner's theorem to obtain the spectral representation of a covariance function and, in particular, of a d-dimensional isotropic covariance function (for any value of d). In Section 5.2 we describe some commonly used classes of spectral densities. In Section 5.3, we introduce the periodogram, a nonparametric estimate of the spectrum, and we study its properties. In Section 5.4, we present an approximation to the Gaussian likelihood using spectral methods. In Section 5.5 we apply all of these methods in a case study to illustrate the potential of the spectral methods presented in this chapter for modeling, estimation, and prediction of spatial processes.

5.1 Spectral Representation

In this section, we start with some background material, and we present a review of the Fourier transform and its properties; we also discuss the aliasing phenomenon in the spectral domain. This aliasing effect is a result of the loss of information when we take a discrete set of observations on a continuous process.

5.1.1 Continuous Fourier Transform

A Fourier analysis of a spatial process, also called a harmonic analysis, is a decomposition of the process into sinusoidal components (sines and cosines waves). The coefficients of these sinusoidal components are the Fourier transform of the process.

Suppose that g is a real or complex-valued function that is integrable over $\mathcal{R}^d$. Define

$$G(\boldsymbol{\omega}) = \int_{\mathcal{R}^d} g(\mathbf{s})\exp(i\boldsymbol{\omega}^t\mathbf{s})d\mathbf{s} \tag{5.1}$$

for $\boldsymbol{\omega} \in \mathcal{R}^d$. The function G in Equation (5.1) is said to be the Fourier transform of g. Then, if G is integrable over $\mathcal{R}^d$, g has the representation

$$g(\mathbf{s}) = \frac{1}{(2\pi)^d}\int_{\mathcal{R}^d} G(\boldsymbol{\omega})\exp(-i\boldsymbol{\omega}^t\mathbf{s})d\boldsymbol{\omega}, \tag{5.2}$$

so that $G(\boldsymbol{\omega})$ represents the amplitude associated with the complex exponential with frequency $\boldsymbol{\omega}$. When $d = 2$, we call $\boldsymbol{\omega}$ a spatial frequency. A spatial frequency is also called a *wavenumber*, as the measurement of the number of repeating units of a propagating wave (the number of times a wave has the same phase) per unit of space. The norm of G is called the Fourier spectrum of g. The right-hand side of Equation (5.2) is called the Fourier integral representation of g. The functions g and G are said to be a Fourier transform pair. If both g and G are integrable, there is no difficulty in defining (5.1) and (5.2). However, the integrals in (5.1) and (5.2) are also defined in other more general settings (see, e.g., Sogge, 1993).

It is often useful to think of functions and their transforms as occupying two domains. These domains are referred to as the upper and the lower domains in older texts, "as if functions circulated at ground level and their transforms in the underworld" (Bracewell, 2000). They are also referred to as the function and transform domains, but in most physics applications they are called the time (or space) and frequency domains, respectively. Operations performed in one domain have corresponding operations in the other. For example, the convolution operation in the time (space) domain becomes a multiplication operation in the frequency domain. The reverse is also true. Such results allow one to move between domains so that operations can be performed where they are easiest or most advantageous.

5.1.2 Aliasing

If we decompose a continuous process $Z(\mathbf{s})$ with $\mathbf{s} \in \mathcal{R}^d$, into a discrete superposition of harmonic oscillations, it is easy to see that such a decomposition cannot be uniquely restored from observations of Z in $\Delta\mathcal{Z}^d = \delta_1\mathcal{Z} \times \cdots \times \delta_d\mathcal{Z}$, where the vector $\Delta = (\delta_1, \ldots, \delta_d)$ is the distance between neighboring observations, and $\mathcal{Z}^d$ the integer d-dimensional lattice. The equal spacing in the space domain of the observations introduces an *aliasing* effect for the frequencies. Indeed, for any $\mathbf{z}_1 = (z_{11}, \ldots, z_{1d})$ and $\mathbf{z}_2 = (z_{21}, \ldots, z_{2d})$ in $\mathcal{Z}^d$, we have

$$\exp\left(i\boldsymbol{\omega}^t\mathbf{z}_1\Delta\right) = \exp\left\{i\left(\boldsymbol{\omega} + \mathbf{z}_2 2\pi/\Delta\right)^t \mathbf{z}_1\Delta)\right\} = \exp\left(i\boldsymbol{\omega}^t\mathbf{z}_1\Delta\right)\exp\left(i2\pi\mathbf{z}_2^t\mathbf{z}_1\right),$$

where $\mathbf{z}_1\Delta = (z_{11}\delta_1, \ldots, z_{1d}\delta_d)$ and $\mathbf{z}_1/\Delta = (z_{11}/\delta_1, \ldots, z_{1d}/\delta_d)$. We simply cannot distinguish an oscillation with a spatial frequency $\boldsymbol{\omega}$ from all the oscillations with frequencies $\boldsymbol{\omega} + 2\pi\mathbf{z}_2/\Delta$. The frequencies $\boldsymbol{\omega}$ and $\boldsymbol{\omega}' = \boldsymbol{\omega} + 2\pi\mathbf{z}_2/\Delta$ are indistinguishable and, hence, are aliases of each other. The impossibility of distinguishing the harmonic components with frequencies differing by an integer multiple of $2\pi/\Delta$ by observations in the d-dimensional integer lattice with spacing Δ is called the *aliasing* effect.

Then, if observation of a continuous process Z is carried out only at uniformly spaced spatial locations Δ units apart, the spectrum of observations of the sample sequence $Z(\Delta\mathbf{z}_i)$, for $\mathbf{z}_i \in \mathcal{Z}^d$, is concentrated within the finite frequency d-dimensional interval $-\boldsymbol{\pi}/\Delta \leq \boldsymbol{\omega} < \boldsymbol{\pi}/\Delta$, where $\boldsymbol{\pi}$ is a d-dimensional vector with all components equal to π, and $\mathbf{z}_1 < \mathbf{z}_2$ for $\mathbf{z}_i \in \mathcal{Z}^d$ denotes $z_{1i} < z_{2i}$ for all $i = 1, \ldots, d$. Every frequency not in that d-dimensional interval has an alias in the interval, which is termed its principal alias. The whole frequency spectrum is partitioned into intervals of length $2\pi/\Delta$ by *fold points* $(2\mathbf{z}_i + \mathbf{1})\pi/\Delta$, where $\mathbf{z}_i \in \mathcal{Z}^d$ and $\mathbf{1}$ is a d-dimensional vector with all components equal to 1. Then, the power distribution within each of the intervals distinct from the principal interval $-\boldsymbol{\pi}/\Delta \leq \boldsymbol{\omega} < \boldsymbol{\pi}/\Delta$, is superimposed on the power distribution within the principal interval. Thus, if we wish that the spectral characteristics of the process Z to be determined accurately enough from the observed sample, then the *Nyquist frequency* $\boldsymbol{\pi}/\Delta$ must necessarily be so high that still higher frequencies $\boldsymbol{\omega}$ make only a negligible contribution to the total power of the process. This means that we observe a dense sample of Z (small Δ). The Nyquist frequency is also called the *folding frequency*, since higher frequencies are effectively folded down into the d-dimensional interval $-\boldsymbol{\pi}/\Delta \leq \boldsymbol{\omega} < \boldsymbol{\pi}/\Delta$.

It should be noted that aliasing is a relatively simple phenomenon. In general, when one takes a discrete set of observations on a continuous function, information is lost. It is an advantage of the trigonometric functions that this loss of information is manifest in the easily understood form of aliasing.

5.1.3 Spectral Representation of a Continuous Spatial Process

In this section, we first define the concept of mean square continuity of a spatial process, then we introduce the spectral representation of a stationary mean square continuous spatial process using sine and cosine waves. We also present Bochner's theorem to obtain a spectral representation for the covariance function.

5.1.3.1 Mean Square Continuity

A spatial process $Z(\mathbf{s})$ is mean square continuous at $\mathbf{s}$ if $\lim_{\mathbf{y}\to\mathbf{s}} E\{Z(\mathbf{y}) - Z(\mathbf{s})\}^2 = 0$. If Z is a weakly stationary process with covariance function C, then $Z(\mathbf{s})$ is mean square continuous at $\mathbf{s}$, if and only if C is continuous at the origin. Because a weakly stationary process is either mean square continuous everywhere or nowhere, we can say that Z is mean square continuous if and only if C is continuous at the origin. However, the mean square continuity of Z does not imply that its realizations are continuous.

5.1.3.2 The Spectral Representation Theorem

Suppose $Y_1, \ldots, Y_m$ are mean zero complex random variables with $E(Y_i Y_j) = 0$ for $i \neq j$, and $E|Y_j|^2 = f_j$ for each j, and $\omega_1, \ldots, \omega_m \in \mathcal{R}^d$. Consider

$$Z(\mathbf{s}) = \sum_{j=1}^{m} \exp(i\omega_j{}^t\mathbf{s})Y_j, \tag{5.3}$$

then Z is a weakly stationary process in $\mathcal{R}^d$ with covariance $C(\mathbf{s}) = \sum_{j=1}^{m} \exp\left(i\omega_j{}^t\mathbf{s}\right) f_j$. Equation (5.3) provides an example of a spectral representation of a complex spatial process Z. In Figure 5.1a, following the representation in Equation 5.3, we present the real part of a realization of a stationary process Z in one dimension as a sum of 100 sinusoidal functions with random amplitudes; in Figure 5.1b, we show the 100 values of f_i, the variance of the random amplitude functions. The values of f_i correspond to what we will later call a squared exponential spectral density function. In Figure 5.1c, we present another realization of a stationary process Z, but in this case the spectrum (shown in 5.1d) has higher values (more power) at higher frequencies, so this makes the process Z less smooth. The values of f_i in 5.1(d) correspond to what we will later call the spectral density function of an exponential covariance function. By taking L^2 limits of the sums in Equation (5.3), we obtain spectral representations of all mean square continuous weakly stationary processes. That is, to every mean square continuous weakly stationary process $Z(\mathbf{s})$, with mean 0 and covariance C, there can be assigned a process $Y(\omega)$ with orthogonal increments, such that we have for

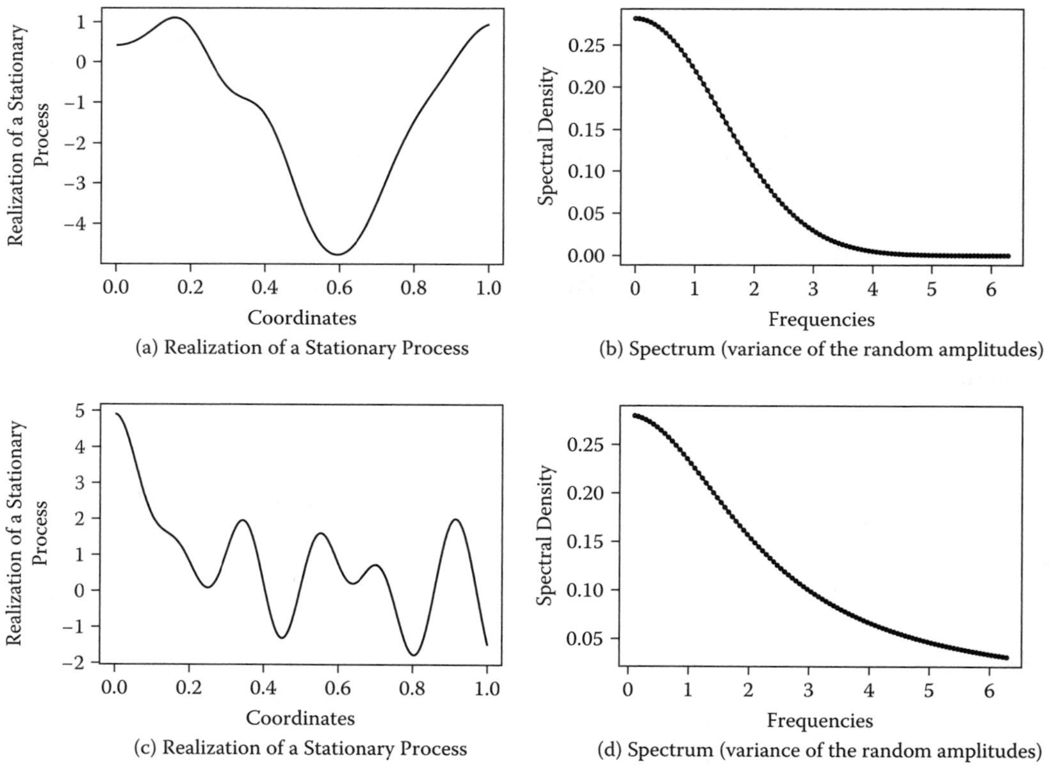

(a) Realization of a Stationary Process
(b) Spectrum (variance of the random amplitudes)
(c) Realization of a Stationary Process
(d) Spectrum (variance of the random amplitudes)

FIGURE 5.1
Realization of two stationary processes in one dimension using a sum of 100 sinusoidal functions with different random amplitudes.

each fixed $\mathbf{s}$ the following stochastic integral that gives the spectral representation (e.g., Yaglom, 1987):

$$Z(\mathbf{s}) = \int_{\mathcal{R}^d} \exp(i\omega^t \mathbf{s}) dY(\omega). \tag{5.4}$$

This integral can be interpreted as a limit in L^2 of a Fourier series. $Y(\omega)$ is defined up to an additive random variable. The Y process is called the spectral process associated with a stationary process Z. The random spectral process Y has the following properties:

$$E(Y(\omega)) = 0,$$

because the mean of Z is 0. The process Y has orthogonal increments:

$$E[(Y(\omega_3) - Y(\omega_2))(Y(\omega_1) - Y(\omega_0))] = 0,$$

when $\omega_3 < \omega_2 < \omega_1 < \omega_0$.

If we define F as

$$E[|dY(\omega)|^2] = F(d\omega),$$

where $|F(d\omega)| < \infty$ for all ω. F is a positive finite measure.

The spectral representation theorem may be proved by various methods: using Hilbert space theory or by means of trigonometric integrals. A good reference is Cramér and Leadbetter (1967).

5.1.3.3 Bochner's Theorem

We introduce the spectral representation of the autocovariance function C:

$$C(\mathbf{s}) = \int_{\mathcal{R}^d} \exp(i\mathbf{s}^t \omega) F(d\omega).$$

Bochner's theorem states that a continuous function C is nonnegative definite if and only if it can be represented in the form above where F is a positive finite measure. Thus, the spatial structure of Z could be analyzed with a spectral approach or equivalently by estimating the autocovariance function (Cramér and Leadbetter, 1967).

If we compare the spectral representation of $C(\mathbf{s})$ and $Z(\mathbf{s})$ (e.g., Loéve, 1955),

$$C(\mathbf{s}) = \int_{\mathcal{R}^d} \exp(i\omega^t \mathbf{s}) F(d\omega),$$

$$Z(\mathbf{s}) = \int_{\mathcal{R}^d} \exp(i\omega^t \mathbf{s}) dY(\omega),$$

it will be seen that the elementary harmonic oscillations are respectively $\exp(i\omega^t \mathbf{s}) F(d\omega)$, $\exp(i\omega^t \mathbf{s}) dY(\omega)$.

If we think of $Y(\omega)$ as representing the spatial development of some concrete physical systems, the spectral representation gives the decomposition of the total fluctuation in its elementary harmonic components. The spectral d.f. $F(\{\omega' : \omega' < \omega\})$ determines the distribution of the total average power in the $Z(\mathbf{s})$ fluctuation over the range of spatial frequency ω. The average power assigned to the frequency d-dimensional interval $A = [\omega_1, \omega_2]$ is $F(A)$, which for the whole infinite ω range becomes

$$E|Z(\mathbf{s})|^2 = C(\mathbf{0}) = F(\mathcal{R}^d).$$

Thus, F determines the power spectrum of the Z process. We may think of this as a distribution of a spectral mass of total amount $C(\mathbf{0})$ over the ω axis. F only differs by a multiplicative constant from an ordinary d.f.

If F has a density with respect to Lebesgue measure, this density is the spectral density, f. When the spectral density exists, if the covariance function C is a continuous function, we have the Fourier inversion formula

$$f(\omega) = \frac{1}{(2\pi)^d} \int_{\mathcal{R}^d} \exp(-i\omega^t \mathbf{x}) C(\mathbf{x}) d\mathbf{x}. \tag{5.5}$$

Throughout this chapter, we will assume that the spectral density exists. To get the most general result, we have seen that one needs to replace $f(\omega)d\omega$ with $F(d\omega)$, but this extension is not of much practical value in spatial settings for which one rarely gets positive mass at individual frequencies.

If Z is observed only at N uniformly spaced spatial locations Δ units apart, the spectrum of observations of the sample sequence $Z(\Delta\mathbf{x})$, for $\mathbf{x} \in \mathcal{Z}^2$, is concentrated within the finite frequency interval $-\pi/\Delta \leq \omega < \pi/\Delta$ (aliasing phenomenon). The spectral density f_Δ of the process on the lattice can be written in terms of the spectral density f of the continuous process Z as

$$f_\Delta(\omega) = \sum_{Q \in \mathcal{Z}^2} f\left(\omega + \frac{2\pi Q}{\Delta}\right) \tag{5.6}$$

for $\omega \in \Pi_\Delta^2 = [-\pi/\Delta, \pi/\Delta]^2$.

5.1.4 Spectral Representation of Isotropic Covariance Functions

If the d-dimensional process Z is isotropic with continuous covariance C and spectral density f, then for $\mathbf{h} = (h_1, \ldots, h_d)$, we have $C(\mathbf{h}) = C_0(\|\mathbf{h}\|)$, where $\|\mathbf{h}\| = [h_1^2 + \ldots + h_d^2]^{1/2}$, for some function C_0 of a univariate argument. We denote $\|\mathbf{h}\|$ as h. For $d = 1$, the class of all (one-dimensional) isotropic covariance functions coincides with the class of all real covariance functions of stationary processes. If, however, $d > 1$, then not every real positive definite function can be an isotropic covariance function.

Let $\mathcal{K}$ represent all nonnegative real numbers. Since any d-dimensional isotropic covariance function C is necessarily a covariance function of a stationary process, by using Bochner's theorem, C can be written in spherical coordinates as

$$C(\mathbf{h}) = \int_{\mathcal{R}^d} \exp(i\mathrm{h}^t\omega) f(\omega) d\omega = \int_{\mathcal{K}^d} \exp(i\mathrm{h}^t\omega \cos(\psi_{\mathrm{h}^t\omega})) f(\omega) d\omega$$

where $\psi_{\mathrm{h}^t\omega}$ is the angle between vectors $\mathbf{h}$ and ω. This representation of C simplifies to $C_0(h)$,

$$C_0(h) = \int_{\mathcal{K}^d} Y_d(\omega h) f(\omega) d\omega = \int_{(0,\infty)} Y_d(\omega h)\, \Phi(d\omega), \tag{5.7}$$

where

$$\Phi(\omega) = \int \cdots \int_{\|\omega\| < \omega} f(\omega) d\omega$$

is nondecreasing on $[0, \infty)$ with $\int \Phi(d\omega) < \infty$ (this implies C_0 falls off rapidly enough at infinity), and

$$Y_d(h) = \left(\frac{2}{h}\right)^{(d-2)/2} \Gamma\left(\frac{d}{2}\right) J_{(d-2)/2}(h),$$

where $J_\nu(\cdot)$ denotes the Bessel function of the first kind of order ν, which can be defined as

$$J_\nu(x) = \frac{(x/2)^\nu}{\Gamma(\nu + 1/2)\Gamma(1/2)} \int_0^\pi \exp(\pm i x \cos(\theta)) \sin^{2\nu}\theta d\theta,$$

when $\nu + 1/2 > 0$. The Bessel function of the first kind can be more generally defined by a contour integral,

$$J_\nu(x) = \frac{1}{2\pi i} \oint e^{[x/2][t-1/t]} t^{-\nu-1} dt,$$

where the contour encloses the origin and is traversed in a counterclockwise direction (Arfken and Weber, 1995, p. 416).

The representation of the covariance in the form (5.7) is called the spectral representation of a d-dimensional isotropic covariance function.

This illustrates the most general strategy for constructing an isotropic stationary covariance function, we use Equation (5.7) with an arbitrary nondecreasing Φ. Conversely, any conjectured covariance that cannot be written in this form cannot be positive definite and, hence, is not the covariance of a valid stationary process.

If the random field Z has a spectral density $f(\boldsymbol{\omega})$, then f can be determined from the known covariance function C using Bochner's theorem. If we have isotropy,

$$f(\boldsymbol{\omega}) = \frac{1}{(2\pi)^{d/2}} \int_0^\infty \frac{J_{(d-2)/2}(\omega h)}{(\omega h)^{(d-2)/2}} h^{d-1} C_0(h) dh \tag{5.8}$$

where $\omega = \|\boldsymbol{\omega}\|$. In the particular cases where $d = 2$ or $d = 3$, Equation (5.8) is of the form

$$f(\boldsymbol{\omega}) = \frac{1}{2\pi} \int_0^\infty J_0(\omega h) h C_0(h) dh,$$

for $d = 2$, and

$$f(\boldsymbol{\omega}) = \frac{1}{2\pi^2} \int_0^\infty \frac{\sin(\omega h)}{\omega h} h^2 C_0(h) dh,$$

for $d = 3$. The function f is usually called the d-dimensional spectral density of the isotropic random field Z in $\mathcal{R}^d$.

A d-dimensional isotropic covariance function with $d > 1$ is also a covariance function of some real stationary random process. Therefore, in looking for examples of isotropic covariance functions, we can examine only the real functions $C(\mathbf{s})$ that are stationary covariance functions. To check whether or not the given function C (which falls off rapidly enough at infinity) is a d-dimensional isotropic covariance function, one only needs to obtain, using Equation (5.8), the corresponding spectral density $f(\boldsymbol{\omega})$ and examine whether or not this function $f(\boldsymbol{\omega})$ is everywhere nonnegative. This is under the assumption f does exist, if not, one would obtain F and examine whether or not F is a positive finite measure function.

5.1.5 Principal Irregular Term

Before introducing different classes of isotropic spectral densities, let us study the behavior of an isotropic covariance functions C in a neighborhood of 0, since that is related to the high frequency behavior. This high frequency behavior is the most critical factor for the *kriging* prediction (see Chapter 6). The concept of principal irregular term (PIT) is a characterization of the covariance function at the origin. It is natural to describe this behavior at short distances using a series expansion in $\|\mathbf{h}\| = h$ about 0. For an isotropic covariance function C, let us informally define its PIT as the first term in the series expansion about 0 for C as a function of h that is not proportional to h raised to an even power (Matheron, 1971, p. 58). If $g(h) = \alpha h^\beta$ is a PIT of C, we call β the power and α the coefficient of the PIT.

If $C(\mathbf{h}) = C_0(h)$ is an exponential function, then

$$C_0(h) = \pi\phi\alpha^{-1}\exp(-\alpha h) = \pi\phi\alpha^{-1} - \pi\phi h + O(h^2),$$

as $h \downarrow 0$, and the nonzero coefficient multiplying h in the series expansion indicates that C is not differentiable at 0. For $C_0(h) = \pi\phi\alpha^{-1}\exp(-\alpha h)$, the PIT is $-\pi\phi h$, so the coefficient of the PIT does not depend on α. This suggests that the local behavior of the corresponding process is not much affected by α. This observation is more clear in the spectral domain. The spectral density corresponding to C_0 is $f_0(\omega) = \phi(\alpha^2 + \omega^2)^{-1}$, which is approximately $\phi\omega^{-2}$ for high frequencies (as $|\omega| \uparrow \infty$), so the high-frequency behavior of the spectral densities also does not depend on α. For the squared exponential covariance model in the series expansion, the coefficient multiplying h raised to an odd number is always zero indicating that that process has mean square derivatives of all orders. It is not easy to give a formal definition of the PIT, because it need not be of the form αh^β. Stein (1999) generalizes Matheron's definition of the PIT, and suggests that for most of the common covariance models, the PIT is of the form αh^β, $\alpha \in \mathcal{R}$, $\beta > 0$ and not an even integer, or $\alpha \log(h) h^\beta$, $\alpha \in \mathcal{R}$, β an even integer.

Chapter 6 explores this connection between the high-frequency behavior of the spectral density and the coefficient of the PIT, as well as its impact on the kriging prediction.

5.2 Some Spectral Densities

We describe in this section some commonly used classes of spectral densities. We consider a real process, thus the spectral density is an even function. We also assume that the covariance is isotropic, so that the spectral density is a function of a single frequency.

5.2.1 Triangular Model

For a spatial process with a triangular isotropic covariance:

$$C(\mathbf{h}) = C_0(h) = \sigma(a - h)^+,$$

for σ and a positive, where $(a)^+ = a$ if $a > 0$, otherwise $(a)^+ = 0$, and h denotes the Euclidean norm of the d-dimensional vector $\mathbf{h}$. This model is only valid for $d = 1$. The corresponding spectral density (for $d = 1$) is

$$f(\omega) = f_0(\omega) = \sigma\pi^{-1}\{1 - \cos(\alpha\omega)\}/\omega^2,$$

for $\|\omega\| = \omega$, $\omega > 0$, and $f(\mathbf{0}) = \sigma\pi^{-1}\alpha^2/2$. The oscillating behavior of the spectral density probably would be quite unrealistic for many physical processes. There is usually no reason to assume the spectrum has much more mass near the frequency $(2n+1)\pi$ than near $2n\pi$ for n large, which is the case for the spectral density $\{1 - \cos(\alpha\omega)\}/\omega^2$. Some kriging predictors under this model have strange properties as a consequence of the oscillations of the spectral density at high frequencies.

5.2.2 Spherical Model

One of the most commonly used models for isotropic covariance functions in geological and hydrological applications is the spherical

$$C_0(h) = \begin{cases} \sigma\left(1 - \frac{3}{2\rho}h + \frac{1}{2\rho^3}h^3\right) & h \le \rho \\ 0 & h > \rho \end{cases} \tag{5.9}$$

for positive constants σ and ρ. This function is not a valid covariance in higher dimensions than 3. The parameter ρ is called the range and is the distance at which correlations become exactly 0. This function is only once differentiable at $h = \rho$ and this can lead to problems when using likelihood methods for estimating the parameters of this model. In three dimensions, the corresponding isotropic spectral density has oscillations at high frequencies similar to the triangular covariance function in one dimension. This could cause some abnormal behavior in the spatial predicted fields when using this spherical model in three dimensions (Stein and Handcock, 1989).

5.2.3 Squared Exponential Model

The density of a spatial process with an isotropic squared exponential covariance:

$$C_0(h) = \sigma e^{-\alpha h^2}$$

is

$$f_0(\omega) = \frac{1}{2}\sigma(\pi\alpha)^{-1/2}e^{-\omega^2/(4\alpha)}.$$

Note that C_0 and f_0 both are the same type of exponential functions when $\gamma = 2$. The parameter σ is the variance of the process and α^{-1} is a parameter that explains how fast the correlation decays.

5.2.4 Matérn Class

A class of practical variograms and autocovariance functions for a process Z can be obtained from the Matérn (1960) class of spectral densities

$$f(\omega) = f_0(\omega) = \phi(\alpha^2 + \omega^2)^{(-\nu-\frac{d}{2})} \tag{5.10}$$

with parameters $\nu > 0$, $\alpha > 0$, and $\phi > 0$ (the value d is the dimension of the spatial process Z). Here, the vector of covariance parameters is $\boldsymbol{\theta} = (\phi, \nu, \alpha)$. The parameter α^{-1} can be interpreted as the autocorrelation range. The parameter ν measures the degree of smoothness of the process Z, the higher the value of ν the smoother Z would be, in the sense that the degree of differentiability of Z would increase. The parameter ϕ is proportional to the ratio of the variance σ and the range (α^{-1}) to the $2\nu^{\text{th}}$ power, $\phi \propto \sigma\alpha^{2\nu}$.

The corresponding covariance for the Matérn class is

$$C(\mathbf{h}) = C_0(h) = \frac{\pi^{d/2}\phi}{2^{\nu-1}\Gamma(\nu + d/2)\alpha^{2\nu}}(\alpha h)^{\nu}\mathcal{K}_{\nu}(\alpha h), \tag{5.11}$$

where $\mathcal{K}_\nu$ is a modified Bessel function of the third kind,

$$\mathcal{K}_\nu(h) = \frac{\pi}{2}\left(\frac{I_{-\nu}(h) - I_\nu(h)}{\sin(\pi\nu)}\right),$$

with I_ν the modified Bessel function of the first kind, which can be defined by a contour integral,

$$I_\nu(x) = \oint e^{[x/2][t+1/t]}t^{-\nu-1}dt,$$

where the contour encloses the origin and is traversed in a counterclockwise direction (Arfken and Weber, 1995, p. 416). In the Matérn class, when $\nu = \frac{1}{2}$, we get the exponential covariance function

$$C_0(h) = \pi\phi\alpha^{-1}\exp(-\alpha h).$$

When ν is of the form $m + \frac{1}{2}$ with m a nonnegative integer, the Matérn covariance function is of the form $e^{-\alpha h}$ times a polynomial in h of degree m, (Abramowitz and Stegun, 1965; Stein, 1999, p. 31).

Handcock and Wallis (1994) suggested the following parametrization of the Matérn covariance that does not depend on d:

$$C_0(h) = \frac{\sigma}{2^{\nu-1}\Gamma(\nu)}(2\nu^{1/2}h/\rho)^{\nu}\mathcal{K}_{\nu}(2\nu^{1/2}h/\rho), \tag{5.12}$$

but the corresponding spectral density then depends on d:

$$f_0(\omega) = \frac{\sigma g(\nu, \rho)}{(4\nu/\rho^2 + \omega^2)^{\nu+d/2}},$$

where

$$g(\nu, \rho) = \frac{\Gamma(\nu + d/2)(4\nu)^{\nu}}{\pi^{d/2}\rho^{2\nu}\Gamma(\nu)}$$

with $\sigma = \text{var}(Z(\mathbf{s}))$, the parameter ρ measures how the correlation decays with distance, and generally this parameter is called the *range*. The parameter α^{-1} has a very similar interpretation to ρ, but ρ is approximately independent of ν, while α^{-1} is not. ρ and α^{-1} have also different asymptotic properties under an infill asymptotic model described in Chapter 6. If we consider the limit as $\nu \to \infty$, we get the squared exponential covariance

$$C_0(h) = \sigma e^{-h^2/\rho^2}.$$

The smoothness of a random field, the parameter ν in the Matérn class, plays a critical role in interpolation problems. This parameter is difficult to estimate accurately from data. A number of the commonly used models for the covariance structure, including exponential and squared exponential structures assume that the smoothness parameter is known a priori.

As an alternative to the Matérn covariance, sometimes the powered exponential model could be used:

$$C_0(h) = \sigma e^{-\alpha h^{\gamma}}$$

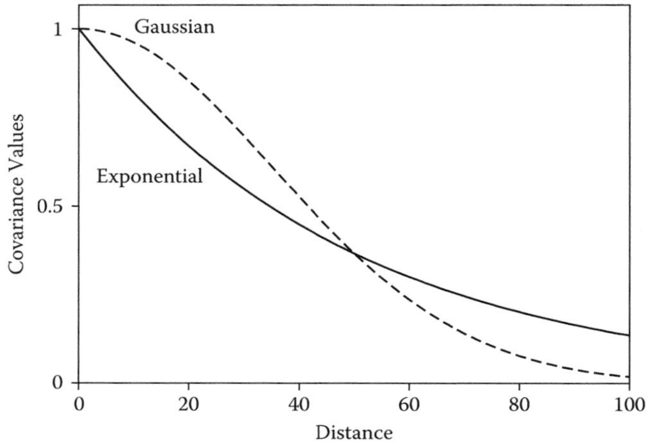

FIGURE 5.2
Covariance models: Exponential (solid line) and squared exponential also called Gaussian covariance (dotted line).

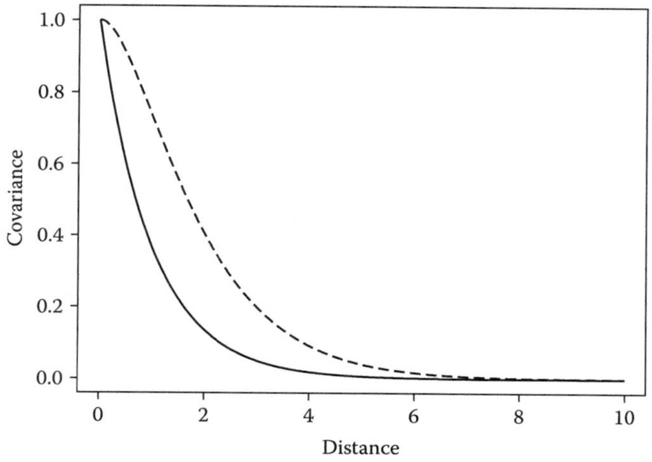

FIGURE 5.3
Matérn covariance functions. The solid line represents a Matérn covariance with $\nu = 1/2$ (exponential covariance) and the dashed line a Matérn with $\nu = 3/2$.

with $\alpha > 0$ and $\gamma \in (0, 2]$. The parameter γ (when $\gamma < 2$) plays the same role as 2ν in the Matérn, and for $\gamma = 2$ it corresponds to $\nu = \infty$. However, for values of $1 \leq \nu < \infty$, the powered exponential has no elements providing similar local behavior as the Matérn.

Figure 5.2 shows two Matérn covariances: a squared exponential covariance (also known as Gaussian) and an exponential covariance. The squared exponential is more flat at the origin, this indicates that the spatial process is very smooth. On the other hand, the exponential is almost linear at the origin, indicating that the corresponding spatial process is not very smooth; in fact, this process is not even once mean square differentiable.

Figure 5.3 shows another two Matérn covariances with $\nu = 1/2$ (exponential) and with $\nu = 3/2$, which corresponds to a process that is once differentiable. Figure 5.4 shows the corresponding spectral densities.

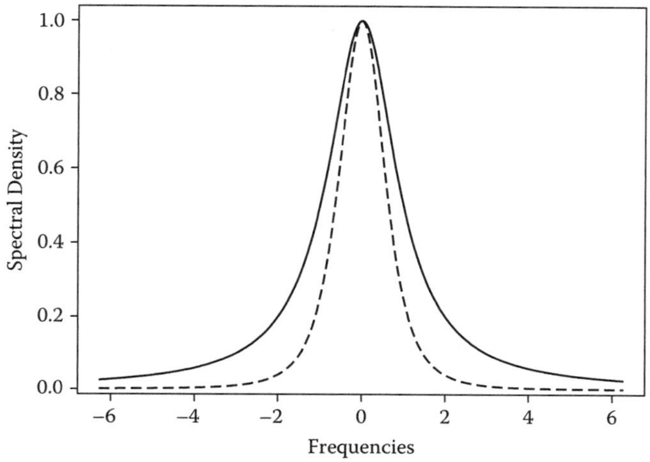

FIGURE 5.4
Spectral densities. The solid line represents a Matérn spectral density with $\nu = 1/2$ (the corresponding covariance function is exponential), and the dashed line a Matérn spectral density with $\nu = 3/2$.

5.3 Estimating the Spectral Density

The periodogram, a nonparametric estimate of the spectral density, is a powerful tool for studying the properties of stationary processes observed on a d-dimensional lattice. Use and properties of spatial periodograms for stationary processes have been investigated by and Whittle (1954), Ripley (1981), Guyon (1982, 1992), Rosenblatt (1985), Stein (1995, 1999), and Fuentes (2001, 2002, 2005), among others.

This section is organized as follows. First, we introduce the periodogram, then, by using least squares, we present a parametric fitting algorithm for the spectral density. In this section, to simplify the notation, we assume we observe the process in a two-dimensional space.

5.3.1 Periodogram

Consider a spatial stationary process Z with a covariance function C. We observe the process at N equally spaced locations in a two-dimensional regular grid $D(n_1 \times n_2)$, where $N = n_1 n_2$. The vector distance between neighboring observations is $\Delta = (\delta_2, \delta_2)$. The periodogram is a nonparametric estimate of the spectral density, which is the Fourier transform of the covariance function. We define $I_N(\omega_0)$ to be the periodogram at a frequency ω_0,

$$I_N(\omega_0) = \delta_1\delta_2(2\pi)^{-2}(n_1 n_2)^{-1}\left|\sum_{s_1=1}^{n_1}\sum_{s_2=1}^{n_2} Z(\Delta\mathbf{s})\exp(-i\Delta\mathbf{s}^t\omega)\right|^2. \tag{5.13}$$

If the spectral representation of Z is

$$Z(\mathbf{s}) = \int_{\mathcal{R}^2} \exp(i\omega^t\mathbf{s})dY(\omega),$$

we define $J(\omega)$, a discrete version of the spectral process $Y(\omega)$, which is the Fourier transform of Z (see, e.g., Priestley, 1981),

$$J(\omega) = (\delta_1\delta_2)^{-1/2}(2\pi)^{-1}(n_1 n_2)^{-1/2}\sum_{s_1=1}^{n_1}\sum_{s_2=1}^{n_2} Z(\Delta\mathbf{s})\exp(-i\Delta\mathbf{s}^t\omega).$$

Using the spectral representation of Z and proceeding formally,

$$C(\mathbf{x}) = \int_{\mathcal{R}^2} \exp(i\omega^t\mathbf{x})F(\mathbf{d}\omega) \tag{5.14}$$

where the function F is called the spectral measure or spectrum for Z. F is a positive finite measure, defined by

$$E\{d|Y(\omega)|^2\} = F(d\omega). \tag{5.15}$$

Thus, we get

$$I_N(\omega) = |J(\omega)|^2; \tag{5.16}$$

this expression for I_N is consistent with the definition of the spectral measure F in Equation (5.15), as a function of the spectral processes Y. The periodogram (5.13) is simply the discrete Fourier transform of the sample covariance, defined as $c_N(\Delta\mathbf{h}) = N^{-1}\sum_{s_1=1}^{n_1}\sum_{s_2=1}^{n_2} Z(\Delta\mathbf{s})Z(\Delta(\mathbf{s}+\mathbf{h}))$. Since, the periodogram can be rewritten in terms of c_N as follows,

$$I_N(\omega_0) = \delta_1\delta_2(2\pi)^{-2}\sum_{h_1=1}^{n_1}\sum_{h_2=1}^{n_2} c_N(\Delta\mathbf{h})\exp(-i\Delta\mathbf{h}^t\omega), \tag{5.17}$$

where $\mathbf{h} = (h_1, h_2)$.

In practice, the periodogram estimate for $\boldsymbol{\omega}$ is computed over the set of Fourier frequencies $2\pi(\frac{\mathbf{j}/}{\mathbf{n}\Delta})$ where $\mathbf{j}/\mathbf{n} = \left(\frac{j_1}{n_1}, \frac{j_2}{n_2}\right)$, and $\mathbf{j} \in J_N$, for

$$J_N = \{\lfloor -(n_1 - 1)/2 \rfloor, \ldots, n_1 - \lfloor n_1/2 \rfloor\} \times \{\lfloor -(n_2 - 1)/2 \rfloor, \ldots, n_2 - \lfloor n_2/2 \rfloor\} \tag{5.18}$$

where $\lfloor u \rfloor$ denotes the largest integer less than or equal to u.

5.3.1.1 Theoretical Properties of the Periodogram

The expected value of the periodogram at $\boldsymbol{\omega}_0$ is given by (see, e.g., Fuentes, 2002)

$$E(I_N(\boldsymbol{\omega}_0)) = (2\pi)^{-2}(n_1 n_2)^{-1} \int_{\Pi_\Delta^2} f_\Delta(\boldsymbol{\omega}) W(\boldsymbol{\omega} - \boldsymbol{\omega}_0) \mathbf{d}\boldsymbol{\omega},$$

where $\Pi_\Delta^2 = (-\pi/\delta_1, \pi/\delta_1) \times (-\pi/\delta_2, \pi/\delta_2)$, and

$$W_N(\boldsymbol{\omega}) = \prod_{j=1}^{2} \frac{\sin^2\left(\frac{n_j \omega_j}{2}\right)}{\sin^2\left(\frac{\omega_j}{2}\right)}$$

for $\boldsymbol{\omega} = (\omega_1, \omega_2) = 2\pi(\frac{\mathbf{j}/}{\mathbf{n}\Delta})$ and $\mathbf{j} \in J_N \backslash \{0\}$, and $f_\Delta(\boldsymbol{\omega})$ is the spectral density of the process Z on the lattice with spacing Δ. The side lobes (subsidiary peaks) of the function W_N can lead to substantial bias in $I_N(\boldsymbol{\omega}_0)$ as an estimator of $f_\Delta(\boldsymbol{\omega}_0)$ because they allow the value of f_Δ at frequencies far from $\boldsymbol{\omega}_0$ to contribute to the expected value. Figure 5.5 shows a graph of W_N along the vertical axis ($n_2 = 500$). As $n_1 \to \infty$ and $n_2 \to \infty$, then $W_N(\boldsymbol{\omega})$ becomes the Dirac delta function, which has mass 1 at zero or zero otherwise.

For a fixed N, if the side lobes of W_N were substantially smaller, we could reduce this source of bias for the periodogram considerably. Tapering is a technique that effectively

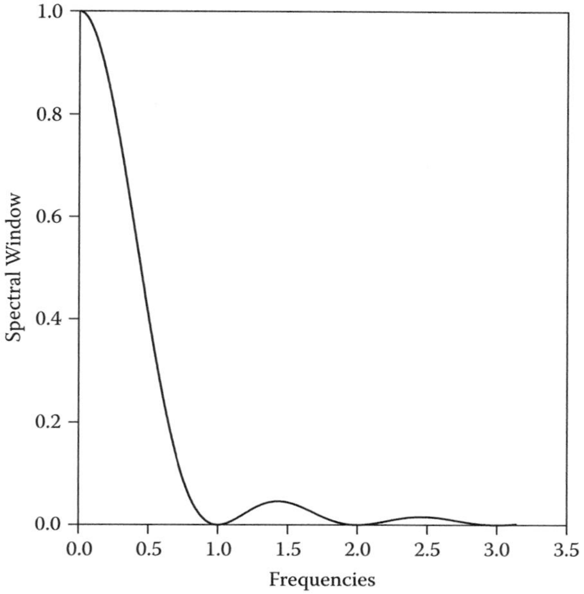

FIGURE 5.5
Spectral window along the vertical axis. The x-axis (horizontal line) in the graph is the spectral frequencies projected on the 1-dimensional interval $(0, \pi/\delta_2)$ with $\delta_2 = 1$, while the y-axis shows the spectral window along the vertical axis for the periodogram (without tapering), for $n_2 = 500$.

reduces the side lobes associated with the spectral window W. We form the product $h(\mathbf{s}) Z(\mathbf{s})$ for each value of $\mathbf{s}$, where $\{h(\mathbf{s})\}_{\mathbf{s}}$ is a suitable sequence of real-valued constants called a *data taper*, and then we compute the periodogram for the tapered data.

The periodogram values are approximately independent, and this facilitates the use of techniques, such as nonlinear least squares (NLS) to fit a theoretical spectral model to the periodogram values.

5.3.1.1.1 *Asymptotic Properties of the Periodogram*

Theorem 1 (Brillinger, 1981): Consider a Gaussian stationary process Z with spectral density $f(\omega)$ on a lattice D. We assume Z is observed at N equally spaced locations in D ($n_1 \times n_2$), where $N = n_1 n_2$, and the spacing between observations is Δ. We define the periodogram function, $I_N(\omega)$, as in (5.13).

Assume $n_1 \to \infty$, $n_2 \to \infty$, $n_1/n_2 \to \lambda$, for a constant $\lambda > 0$.

Then, we get:

(i) The expected value of the periodogram, $I_N(\omega)$, is asymptotically $f_\Delta(\omega)$.

(ii) The asymptotic variance of $I_N(\omega)$ is $f_\Delta^2(\omega)$.

(iii) The periodogram values $I_N(\omega)$, and $I_N(\omega')$ for $\omega \neq \omega'$, are asymptotically independent.

By Theorem 1(i) the periodogram I_N is asymptotically an unbiased estimate of the spectral density, f_Δ on the lattice. Note, that if f is the continuous process Z and f_Δ the spectral density on the lattice, then using *increasing-domain* asymptotics, I_N is not asymptotically an unbiased estimate of f, but of f_Δ, the spectral density of the sampled sequence $Z(\Delta\mathbf{x})$.

By Theorem 1(ii) the variance of the periodogram at ω is asymptotically $f_\Delta^2(\omega)$. The traditional approach to this inconsistency problem is to smooth the periodogram across frequencies.

By Theorem 1(iii), the periodogram values at any two fixed Fourier frequencies are approximately independent. However, in the space domain, the empirical covariance or variogram values are correlated, which thwarts the use of least squares to fit a parametric covariance model to the sample covariance/variogram function.

Dahlhaus and Künsch (1987) show that the periodogram of tapered data is efficient in only one dimension. However, the periodogram of tapered data is an efficient estimate of the spectral density in two and three dimensions.

5.3.1.1.2 *Asymptotic Distribution of the Periodogram*

If the process Z is stationary, such that the absolute value of the joint cumulants of order k are integrable (for all k), then the periodogram has asymptotically a distribution that is a multiple of a χ_2^2. More specifically, the periodogram $I_N(\omega_j)$, where ω_j is a Fourier frequency, has asymptotically a $f(\omega_j)\chi_2^2/2$ distribution (Brillinger, 1981).

5.3.2 Lattice Data with Missing Values

In this section, we introduce a version of the periodogram by Fuentes (2007) for lattices with missing values. Consider Z a lattice process with spectral density f. We assume Z is a weakly stationary real-valued Gaussian process having mean zero and finite moments. The process Z is defined on a rectangle $P_N = \{1, \dots, n_1\} \times \{1, \dots, n_2\}$ of sample size $N = n_1 n_2$. The covariance C of the process Z satisfies the following condition: $\Sigma_{\mathbf{s}}[1 + \|\mathbf{s}\|]|C(\mathbf{s})| < \infty$, where $C(\mathbf{s}) = \text{Cov}\{Z(\mathbf{s}+\mathbf{y}), Z(\mathbf{y})\}$. This condition implies that the spectral density of Z exits and has uniformly bounded first derivatives (Brillinger, 1981, Sec. 2.6).

The process Z is not directly observed. Rather, we observe Y, an amplitude modulated version of Z for the observations on the grid, given by

$$Y(\mathbf{s}) = g(\mathbf{s}/\mathbf{n})Z(\mathbf{s}), \tag{5.19}$$

where $\mathbf{s}/\mathbf{n} = (s_1/n_1, s_2/n_2)$. The weight function g in practice could take zero values at the locations at which we do not have observations and one everywhere else. We call this function a zero-filling taper.

We then introduce a version of the periodogram for the incomplete lattice:

$$\tilde{I}_N(\boldsymbol{\omega}) = \frac{1}{H_2(\mathbf{0})}\left|\sum_{k=1}^{N}(Y(\mathbf{s}_k) - g(\mathbf{s}_k/\mathbf{n})\tilde{Z})\exp\{-i\boldsymbol{\omega}^t\mathbf{s}_k\}\right|^2 \tag{5.20}$$

where $H_j(\boldsymbol{\lambda}) = (2\pi)^2\sum_{k=1}^{N} g^j(\mathbf{s}_k/\mathbf{n})\exp\{i\boldsymbol{\lambda}^t\mathbf{s}_k\}$, then $H_2(\mathbf{0}) = (2\pi)^2\sum_{k=1}^{N} g(\mathbf{s}_k/\mathbf{n})^2$, and

$$\tilde{Z} = \left(\sum_{k=1}^{N} Y(\mathbf{s}_k)\right)\bigg/\left(\sum_{k=1}^{N} g(\mathbf{s}_k/\mathbf{n})\right).$$

If $g(\mathbf{s}_k/\mathbf{n}) = 1$ for all $\mathbf{s}_k$ in P_N, then $Y \equiv Z$ on the lattice, and $\tilde{I}_N$ reduces to the standard definition of the periodogram. When g takes some zero values (due to missing data), the difference between the traditional periodogram for Z, I_N, and the new definition given here, $\tilde{I}_N$, is a multiplicative factor, $(n_1n_2)/H_2(\mathbf{0})$. This is the bias adjustment that needs to be made to the periodogram function due to the missing values.

Fuentes (2007) studied the asymptotic properties of this estimate of f, under some weak assumptions for g (g is bounded and of bounded variation), as $N \to \infty$. The expected value of $\tilde{I}_N$ is

$$E[\tilde{I}_N(\boldsymbol{\omega})] = \int_{-\pi}^{\pi}\int_{-\pi}^{\pi} f(\boldsymbol{\omega} - \boldsymbol{\phi})|H_1(\boldsymbol{\phi})|^2 d\boldsymbol{\phi}. \tag{5.21}$$

Thus, $E[\tilde{I}_N(\boldsymbol{\omega})]$ is a weighted integral of $f(\boldsymbol{\omega})$. Asymptotically,

$$E[\tilde{I}_N(\boldsymbol{\omega})] = f(\boldsymbol{\omega}) + O(N^{-1}). \tag{5.22}$$

Sharp changes in g make its Fourier transform and the squared modulus of its Fourier transform exhibit side lobes. The scatter associated with a large number of missing values creates very large side lobes in Equation (5.21). Even if asymptotically the bias is negligible by Equation (5.22), it could have some impact for small samples. Fuentes (2007) obtains the asymptotic variance for $\tilde{I}_Z$,

$$\mathrm{var}\{\tilde{I}_N(\boldsymbol{\omega})\} = |H_2(\mathbf{0})|^{-2}\left\{H_2(\mathbf{0})^2 + H_2(2\boldsymbol{\omega})^2\right\}f(\boldsymbol{\omega})^2 + O(N^{-1}). \tag{5.23}$$

The quantity multiplying f in the expression (5.23) for the asymptotic variance is greater than 1 when we have missing values, and it is 1 when there are no missing values. Thus, a large number of missing values would increase the variance of the estimated spectrum.

Expression 5.20 provides then a version of the periodogram for lattices with missing values; that it is an asymptotically unbiased estimate of the spectral density of the process. A strategy here is to fill with zeros the values of the process at the locations with missing values. It has also been extended to model data not observed on a grid (Fuentes, 2007).

5.3.3 Least Squares Estimation in the Spectral Domain

Consider modeling the spatial structure of Z by fitting a spectral density f to the periodogram values. We could use a weighted nonlinear least squares (WNLS) procedure that gives more weight to frequencies with smaller variance for the periodogram (this generally corresponds to higher frequency values).

Thus, we propose using a parametric model for f with weights $f(\omega)^{-1}$ (based on the asymptotic results for the periodogram variance). For large N, the approximate standard deviation of the periodogram I_N is $f(\omega)$. Thus, the proposed weights $f(\omega)^{-1}$ stabilize the variance of the periodogram values. This is similar to the weighted least squares method used in the space domain to fit a variogram model (Cressie, 1985). We recommend using weighted least squares in the spectral domain rather than in the space domain because periodogram values are approximately independent while sample variogram values are not. Therefore, in the spectral domain, we do not need a generalized least squares approach to be able to make proper inference about the estimated parameters and their variances, while in the special domain we would.

For spatial prediction, the behavior of the process at high frequencies is more relevant (Stein, 1999). We can obtain asymptotically (as $N \to \infty$) optimal prediction when the spectral density at short frequencies is misspecified. An approximate expression for the spectral density of the Matérn class for high frequency values is obtained from Equation (5.10) by letting $\|\omega\|$ go to ∞:

$$f(\omega) = \phi \|\omega\|^{(-2\nu-d)} \tag{5.24}$$

Thus, the degree of smoothness, ν, and ϕ are the critical parameters (and not the range α^{-1}). This is consistent with the description of the principal irregular term given in Section 5.1.5. Then, an alternative approach to the WNLS for the high-frequencies model in Equation (5.24) is to fit in the log scale a linear model using OLS (ordinary least squares):

$$\log(f(\omega)) = \beta_0 + \beta_1 X \tag{5.25}$$

where $X = \log(\omega)$, $\beta_0 = \log(\phi)$, and $\beta_1 = 2\left(-\nu - \frac{d}{2}\right)$.

5.4 Likelihood Estimation in the Spectral Domain

For large datasets, calculating the determinants that we have in the likelihood function can be often infeasible. Spectral methods could be used to approximate the likelihood and obtain the maximum likelihood estimates (MLEs) of the covariance parameters: $\boldsymbol{\theta} = (\theta_1, \ldots, \theta_r)$.

Spectral methods to approximate the spatial likelihood have been used by Whittle (1954), Guyon (1982), Dahlhaus and Künsch, (1987), Stein (1995, 1999), and Fuentes (2007), among others. These spectral methods are based on Whittle's approximation to the Gaussian negative log likelihood:

$$\frac{N}{(2\pi)^2} \int_{\mathcal{R}^2} \left\{ \log f(\omega) + I_N(\omega) f(\omega)^{-1} \right\} d\omega \tag{5.26}$$

where the integral is approximated with a sum evaluated at the discrete Fourier frequencies, I_N is the periodogram, and f is the spectral density of the lattice process. The approximated likelihood can be calculated very efficiently by using the fast Fourier transform. This approximation requires only $O(N \log_2 N)$ operations, assuming n_1 and n_2 are highly composite. Simulation studies conducted by the authors seem to indicate that N needs to be at least 100 to get good, estimated MLE parameters using Whittle's approximation.

The asymptotic covariance matrix of the MLE estimates of $\theta_1, \ldots, \theta_r$ is

$$\frac{2}{N} \left[\left\{ \frac{1}{4\pi^2} \int_{[-\pi,\pi]} \int_{[-\pi,\pi]} \frac{\partial \log f(\omega_1)}{\partial \theta_j} \frac{\partial \log f(\omega_2)}{\partial \theta_k} d\omega_1 d\omega_2 \right\}_{jk} \right]^{-1}; \tag{5.27}$$

the exponent -1 in Equation (5.27) denotes the inverse of a matrix. The expression in Equation (5.27) is much easier to compute than the inverse of the Fisher information matrix.

Guyon (1982) proved that when the periodogram is used to approximate the spectral density in the Whittle likelihood function, the periodogram bias contributes a non-negligible component to the mean squared error (mse) of the parameter estimates for two-dimensional processes, and for three dimensions, this bias dominates the mse. Thus, the MLE parameters of the covariance function based on the Whittle likelihood are only efficient in one dimension, but not in two- and higher-dimensional problems, although they are consistent. Guyon demonstrated that this problem can be solved by using a different version of the periodogram, an "unbiased periodogram," which is the discrete Fourier transform of an unbiased version of the sample covariance. Dahlhaus and Künsch (1987) demonstrated that tapering also solves this problem.

The Whittle approximation to the likelihood assumes complete lattices. Fuentes (2007) introduces an extension of the Whittle approximated likelihood that can be applied to lattice data with missing values to obtain MLEs and the variance of the MLEs, by using the version of the periodogram in Equation (5.20). Fuentes also introduces a more general version of the periodogram for irregularly spaced datasets that can be used to obtain an approximated likelihood for irregularly spaced spatial data.

5.5 Case Study: Analysis of Sea Surface Temperature

In this section, we apply spectral methods to estimate the spatial structure of sea surface temperature fields using the Tropical Rainfall Measuring Mission (TRMM) microwave imager (TMI) satellite data for the Pacific Ocean. Global sea surface temperature (SST) fields are very useful for monitoring climate change, as an oceanic boundary condition for numerical atmospheric models, and as a diagnostic tool for comparison with the SSTs produced by ocean numerical models. SSTs can be estimated from satellites, for example, using the TRMM TMI. The spatial scales and structure of SST fields are the main factor to identify phenomena, such as El Niño and La Niña, that occur in the equatorial Pacific and influence weather in the Western Hemisphere. Spatial patterns of SST in the Pacific are also being used as one of the main climate factors to identify tropical cyclones (hurricanes) that form in the south of Mexico and strike Central America and Mexico from June to October. Studying the spatial structure of SST in the Pacific is also important to understanding the exchange of water between the north and south equatorial currents. A good understanding of the SST's spatial variability is crucial for guiding future research on the variability and predictability of the world ocean SST and the climate that it influences.

We analyze TMI data to estimate the spatial structure of SST over the northeast Pacific Ocean. Currently, most of the operational approaches to estimate the covariance parameters of the TMI SST fields, in particular the mesoscale and zone scale parameters (ranges of correlation) (Reynolds and Smith, 1994), are empirical methods and there is not a reliable measure of the uncertainty associated to the estimated parameters. Likelihood methods are difficult to implement because the satellite datasets are very large.

The satellite observations in this application are obtained from a radiometer onboard the TRMM satellite. This radiometer, the TRMM TMI, is well calibrated and contains lower-frequency channels required for SST retrievals. The measurement of SST through clouds by satellite microwave radiometers has been an elusive goal for many years. The TMI data have roughly 25 km $\times$ 25 km spatial resolution and are available on the Internet (www.remss.com/tmi/tmi_browse.html). Results gathered for March 1998.

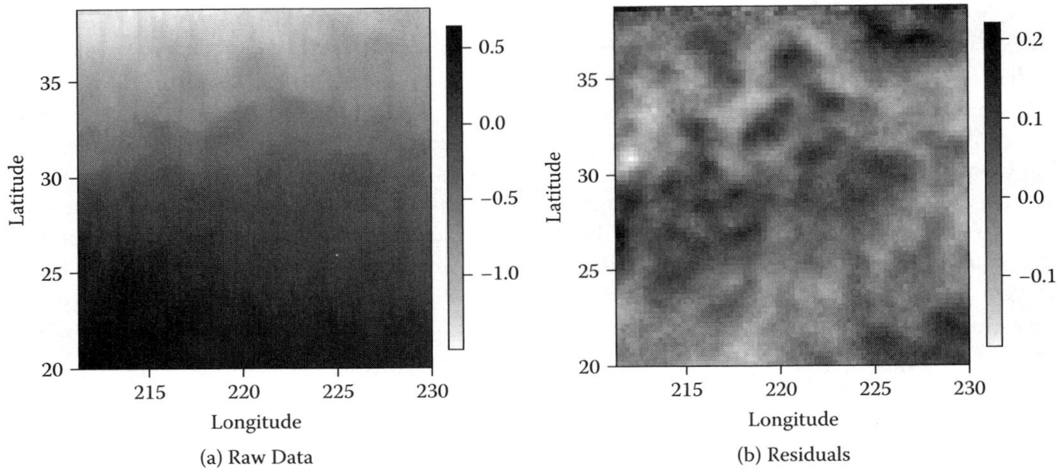

FIGURE 5.6
Sea surface temperature data (degrees Celsius) and residuals after removing a second-order polynomial trend.

5.5.1 Exploratory Analysis

The data in Figure 5.6a exhibit a strong north/south gradient. Therefore, we remove the second-order polynomial trend

$$\beta_0 + \beta_1 * LONG + \beta_2 * LAT + \beta_3 * LONG^2 + \beta_4 * LAT^2 + \beta_5 * LONG * LAT,$$

where the regression parameters are estimated using ordinary least squares. The residuals in Figure 5.6b show no large-scale spatial trend.

Figure 5.7a plots the sample semivariogram for the residuals with bin width 2 km. In the semivariogram, we use the actual latitude and longitude for each location, which do not precisely lie on a grid due to satellite projections. The semivariogram is fairly constant after 500 km, indicating there is little spatial correlation in the residuals past this distance. Because north/south (different distance from the equator) and east/west (different distance from the coast) neighbors share different geographic features, we may question the isotropy assumption. We inspect the isotropy assumption using the sample semivariogram and periodogram in Figure 5.7c along different directions. The semivariogram shows slightly more variation in the north/south direction (0°) than in the east/west direction (90°). The periodogram shows a similar pattern. In this plot, $\omega_1 = 0$ corresponds to north/south variation (0° in the semivariogram), $\omega_2 = 0$ corresponds to east/west variation (90°). As with the semivariogram, the periodogram shows the most variation in the north/south direction. Although these plots suggest there may be moderate anisotropy, we proceed with the isotropic model for ease of presentation. For a discussion of anisotropic model fitting, see Chapter 3.

5.5.2 Parameter Estimation

The data are on a 75 × 75 rectangular grid defined by latitude and longitude for a total of 5,625 observations. Due to the large amount of data, a full maximum likelihood analysis for a continuous Gaussian process is infeasible. Therefore, we consider two computationally efficient methods (both run in a few seconds on an ordinary PC) for estimating the spatial covariance parameters: maximum likelihood using the Whittle approximation and weighted least squares using the semivariogram. For variogram analyses, we

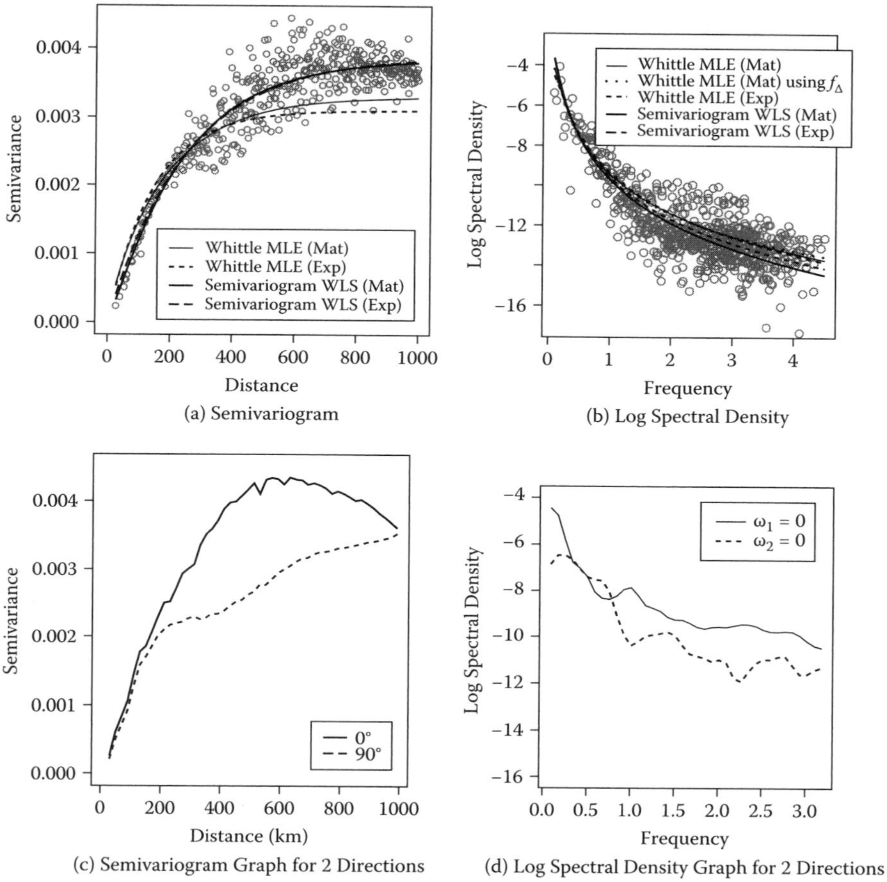

FIGURE 5.7
Sample (points) and fitted semivariograms and periodograms (lines) for the sea surface temperature data.

use geodesic distance between points. For spectral analyses, we approximate distance between points using a rectangular grid. This flat-Earth approximation is reasonable in this small subregrion of the Pacific Ocean; variogram analyses are similar with and without this approximation.

The semivariogram and periodogram in Figure 5.7a suggest that the Matérn covariance is appropriate for these data. Since these satellite data are smoothed during preprocessing, we do not include a nugget effect. Table 5.1 gives the Matérn covariance parameter estimates and their standard errors for both approaches using the Handcock–Wallis parameterization, the MLE estandard errors are obtained using Equation (5.27). The estimates are fairly similar, but there are some important differences. Since most of the information for the smoothness parameter ν comes from short-range differences and the semivariogram bins all short-range differences into only a few observations, the standard error for ν is larger for the weighted least squares (WLS) estimate (0.050) than for the MLE (0.007). Also, since the standard errors for the WLS regression of the semivariogram do not account for correlation between bins, the standard errors for the sill σ and the range ρ are smaller than the MLE standard errors (5.27) and may be too optimistic. Smith (2001, pp. 54–61) proposes a method for calculating the standard errors for the WLS regression of the semivariogram that account for correlation between bins; however, we do no implement this method here.

TABLE 5.1

SST Covariance Parameter Estimates (Standard Errors)

Method	Sill (σ)	Range (ρ)	Smoothness (ν)
MLE – Whittle (Matérn)	0.0033 (0.00012)	264.6 (13.5)	0.425 (0.007)
MLE – Whittle (Matérn) using f_Δ	0.0032 (0.00012)	252.5 (11.5)	0.486 (0.006)
WLS – Semivariogram (Matérn)	0.0038 (0.00003)	331.9 (12.6)	0.588 (0.050)
MLE – Whittle (Exponential)	0.0031 (0.00010)	206.1 (6.9)	0.5
MLE – Whittle (Exponential) using f_Δ	0.0032 (0.00012)	238.4 (8.9)	0.5
WLS – Semivariogram (Exponential)	0.0039 (0.00004)	352.8 (15.4)	0.5

The fitted Matérn semivariograms are plotted in Figure 5.7a. The spectral estimate's semivariogram increases more rapidly near the origin than the WLS estimate's semivariogram because the spectral estimate of the smoothness parameter is smaller than the WLS estimate. Because the spectral estimate of the range is smaller than the WLS estimate, the spectral estimate's semivariogram plateaus sooner than the WLS estimate's semivariogram. The spectral densities in Figure 5.7b show similar relationships between the two estimates. The MLE estimate is smaller than the WLS estimate for small frequencies that measure large-scale effects, and the MLE estimate is larger than the WLS estimate for large frequencies that measure small-scale effects.

Both estimates of the smoothness parameter ν are close to 0.5, indicating that an exponential covariance may also fit the data well. Table 5.1 presents the covariance parameter estimates for the exponential model. Setting ν to 0.5 illustrates that despite the asymptotic independence under the Handcock–Wallis parameterization, for finite sample, there may still be a negative relationship between the smoothness parameter and the range. For the MLE, increasing the smoothness from 0.425 to 0.5 causes the range to decrease from 264.6 to 206.1. For the WLS method, decreasing the smoothness from 0.588 to 0.5 causes the range to increase from 331.9 to 352.8. Figures 5.7a,b show that the Matérn and exponential models give similar semivariograms and spectral densities.

In the above spectral analysis, we have used the spectral density f_0 in Equation (5.5) evaluated at the Fourier frequencies. However, to be more precise, the spectral density for these gridded data is the infinite sum f_Δ in Equation (5.6). The parameter estimates using f_Δ ("MLE – Whittle (Matérn) using f_Δ") are given in Table 5.1. To calculate f_Δ, we use only the first 100 terms of the infinite sum. The estimates are slightly different using f_Δ rather than f_0. However, the fitted spectral density for the Matérn covariance in Figure 5.7b is virtually identical using either f_Δ or f, so it appears that using f as the spectral density is an adequate approximation.

This section demonstrates that spectral methods are useful tools for exploratory analysis and covariance estimation, especially for large datasets. Compared with variogram approaches, spectral methods produce similar exploratory analysis and parameter estimates. However, because spectral methods approximate the true likelihood, the standard errors used for inference are more reliable.

References

Abramowitz, M. and Stegun, I.A. (1965). *Handbook of Mathematical Functions*. New York: Dover Publications.

Arfken, G. and Weber, H.J. (1995). *Mathematical Methods for Physicists*, 4th ed. San Diego, CA: Academic Press.

Bracewell, R. (2000). *The Fourier Transform and Its Applications*, 3rd ed. New York: McGraw-Hill.

Brillinger, D.R. (1981). *Time Series Data Analysis and Theory.* San Francisco: Holden-Day.
Cramér, H. and Leadbetter, M.R. (1967). *Stationary and Related Stochastic Processes. Sample Function Properties and Their Applications.* New York: John Wiley & Sons.
Cressie, N. (1985). Fitting models by weighted least squares. *Journal of Mathematical Geology*, **17**, 563–586.
Dahlhaus, R. and Künsch, H. (1987). Edge effects and efficient parameter estimation for stationary random fields. *Biometrika*, **74**, 877–882.
Fuentes, M. (2001). A new high frequency kriging approach for nonstationary environmental processes. *Environmetrics*, **12**, 469–483.
Fuentes, M. (2002). Spectral methods for nonstationary spatial processes. *Biometrika*, **89**, 197–210.
Fuentes, M. (2005). A formal test for nonstationarity of spatial stochastic processes. *Journal of Multivariate Analysis*, **96**, 30–54.
Fuentes, M. (2007). Approximate likelihood for large irregularly spaced spatial data. *Journal of the American Statistical Association, Theory and Methods*, **102**, 321–331.
Guyon, X. (1982). Parameter estimation for a stationary process on a d-dimensional lattice. *Biometrika*, **69**, 95–105.
Guyon, X. (1992). *Champs aléatoires sur un réseau*. Paris: Masson.
Handcock, M. and Wallis, J. (1994). An approach to statistical spatial-temporal modeling of meteorological fields. *Journal of the American Statistical Association*, **89**, 368–378.
Loeve, M. (1955). *Probability Theory*. Princeton, NJ: Van Nostrand.
Matérn, B. (1960). *Spatial Variation.* Meddelanden från Statens Skogsforskningsinstitut, **49**, No. 5. Almaenna Foerlaget, Stockholm. Second edition (1986), Berlin: Springer-Verlag.
Matheron, G. (1971). *The Theory of Regionalized Variables and Its Applications.* Fontainbleau, France: Ecole des Mines.
Priestley, M. (1981). *Spectral Analysis and Time Series*. San Diego, CA: Academic Press.
Reynolds, R.W. and Smith, T.M. (1994). Improved global sea surface temperature analysis. *Journal of Climate*, **7**, 929–948.
Ripley, B.D. (1981). *Spatial Statistics*, New York: John Wiley & Sons.
Rosenblatt, M.R. (1985). *Stationary Sequences and Random Fields*, Boston: Birkhäuser.
Smith, R.L. (2001). Environmental statistics (course notes). University of North Carolina, Chapel Hill. Website: http://www.stat.unc.edu/postscript/rs/envnotes.pdf
Sogge, C.D. (1993). *Fourier Integrals in Classical Analysis*. Cambridge, U.K.: Cambridge University Press.
Stein, M.L. (1995). Fixed-domain asymptotics for spatial periodograms. *Journal of the American Statistical Association*, **90**, 1277–1288.
Stein, M.L. (1999). *Interpolation of Spatial Data: Some Theory for Kriging.* New York: Springer-Verlag.
Stein, M.L. and Handcock, M.S. (1989). Some asymptotic properties of kirging when the covariance function is misspecified. *Mathematical Geology*, **21**, 171–190.
Whittle, P. (1954). On stationary processes in the plane, *Biometrika*, **41**, 434–449.
Yaglom, A.M. (1987). *Correlation Theory of Stationary and Related Random Functions.* New York: Springer-Verlag.

6

Asymptotics for Spatial Processes

Michael Stein

CONTENTS

6.1 Asymptotics

Yakowitz and Szidarovszky (1985) posed what appeared to be a fundamental challenge to the use of kriging for spatial interpolation. Specifically, they noted that it is not generally possible to estimate consistently the variogram of a spatial process based on observations in a bounded domain and, hence, it was not clear that one could obtain good kriging predictors and good assessments of kriging variance based on estimated variograms even if one had a very large number of observations in some bounded domain of interest. It is still not possible to give a definitive response to this challenge, but there is now at least some good reason to believe that the objection raised by Yakowitz and Szidarovszky does not expose a fundamental flaw in the use of kriging for spatial interpolation. However, it is still very much the case that what might be considered very basic results in the large sample properties of kriging are unproven and, furthermore, the prospects of rigorous proofs appearing in the foreseeable future are not great.

Before proceeding to review the present state of knowledge in the area, it is worth asking if Yakowitz and Szidarovszky had somehow "rigged" the game. Specifically, their approach to asymptotics is to consider what happens as observations are allowed to get increasingly dense in a bounded domain, what we will call fixed-domain asymptotics, but which is sometimes also called infill asymptotics (Cressie, 1993). Because the point of their paper was to compare kriging to nonparametric regression approaches (such as kernel estimates) as methods of spatial interpolation, their choice of fixed-domain asymptotics was natural, since asymptotic results for nonparametric regression are inevitably based on allowing observations to get dense in some bounded region of covariate space. One could instead let the area of the observation domain grow proportionally with the number of observations, which occurs if, for example, observations are taken at all $\mathbf{s} = (s_1, s_2)$ with s_1 and s_2 integers between 1 and n and then letting $n \to \infty$. In this case, one might expect to be able to estimate the variogram consistently under some appropriate stationarity and ergodicity assumptions on the underlying spatial process, although one still has to be careful as even this result is not universally true (e.g., if a process is observed on a regular lattice, it will not be possible without a parametric model to estimate the variogram consistently at all distance lags). Some works have considered asymptotics in which the observations become dense throughout $\mathbb{R}^d$ as their number increases (Hall and Patil, 1994; Lahiri, et al., 1999), although the idea has been used in the time series context at least as far back as Härdle and Tuan (1986). Taking limits in this manner makes it possible, for example, to obtain consistent estimates of the spectral density of a stationary Gaussian process at all frequencies (Hall and Patil, 1994). However, existing asymptotic results that allow the observation domain to increase with the number of observations generally assume that, at sufficiently large distances, the process will be very nearly independent, so they would appear not to be helpful when, as is fairly common, the effective range of a variogram is comparable or even larger than the dimensions of the observation region. Still, this "mixed-domain" approach to asymptotics may turn out to provide the most fruitful way to obtain rigorous theoretical results in spatial statistics that are of relevance in practice.

Although there are some good arguments for taking a fixed-domain perspective, it is not possible to say that one asymptotic approach is right and another is wrong. Asymptotic methods have two important functions in statistics. One is to provide useful approximations in settings where exact methods are unavailable. To the extent that a Bayesian approach to kriging provides an adequate practical solution to the problem of carrying out spatial interpolation with appropriate assessments of uncertainty, the need for asymptotics as a source of approximations is arguably less than it was in the past. The other important function of asymptotics is to provide insight into the properties of statistical methods, and this need is as great as ever for understanding kriging. In particular, results on the behavior of kriging predictors under fixed-domain asymptotics provide at least a partial response to the challenge of Yakowitz and Szidarovszky (1985).

To understand these results, it is necessary to say something about a somewhat advanced topic in probability theory: equivalence and orthogonality of probability measures. Suppose we get to observe some random object X (e.g., a random vector or a stochastic process) and that one of two probability measures (a fancy word for probability laws), P_0 or P_1, is the correct one for X. Roughly speaking, P_0 and P_1 are equivalent if, no matter what value of X is observed, it is impossible to know for sure which of the two measures is correct. The measures are orthogonal if, no matter what value of X is observed, it is always possible to determine which of the two measures is correct. For X a finite-dimensional random vector, it is quite easy to determine if two possible probability measures are equivalent, orthogonal, or neither. Figure 6.1 considers X a scalar random variable and gives examples of pairs of probability densities for each of these three cases. The equivalent pair is on top; note that equivalent densities are not identical and there is *some* information in an observation $X = x$ about which density is correct, there is just not enough information to say for sure. In the middle plot, the supports of the two densities are disjoint, so if we observe an x in $(0, 1)$, we know for sure that P_0 is correct and if we observe an x in $(2, 3)$, we know for sure P_1 is correct. The bottom plot displays a case in which the two laws are neither equivalent nor orthogonal: if x is in $(0, 1)$, we know P_0 is correct, if x is in $(2, 3)$, we know P_1 is correct, and if x is in $(1, 2)$, we do not know which measure is correct. A general theorem (Kuo, 1975) says that Gaussian probability measures are, in great generality in finite or infinite dimensions, either equivalent or orthogonal; that is, cases like the bottom plot in Figure 6.1 cannot happen for Gaussian processes.

Let us give an example of a class of infinite-dimensional Gaussian measures in which these issues arise. Suppose $Y(t)$ is a zero mean Gaussian process on the interval $[0, 1]$ with autocovariance function $K(t) = \theta e^{-\phi|t|}$. Let us assume θ and ϕ are both positive, in which case, K is a positive definite function. For $j = 0, 1$, let P_j be the Gaussian process law with $(\theta, \phi) = (\theta_j, \phi_j)$. It is possible to show that if $\theta_0\phi_0 = \theta_1\phi_1$, then P_0 and P_1 are equivalent Gaussian measures and if $\theta_0\phi_0 \neq \theta_1\phi_1$, they are orthogonal. Thus, for example, writing K_j for the autocovariance function under P_j, if $K_0(t) = e^{-|t|}$ and $K_1(t) = 2e^{-|t|/2}$, then P_0 and P_1 are equivalent. That is, if we know that either P_0 or P_1 is true, then, despite the fact

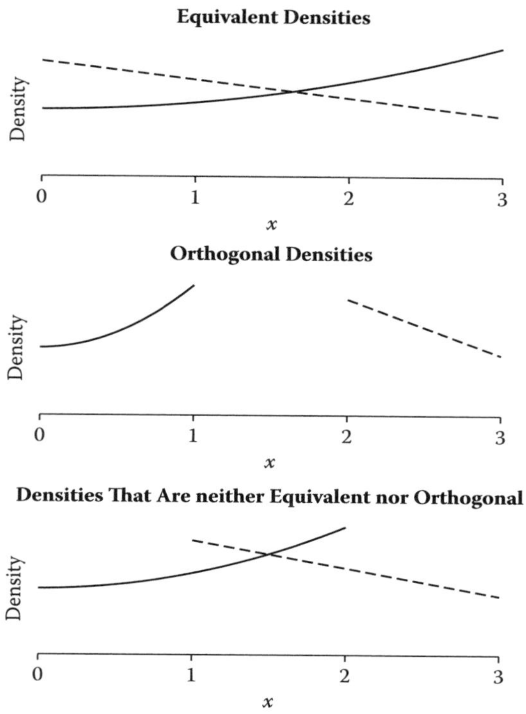

FIGURE 6.1
Plots showing equivalent probability measures (top), orthogonal probability measures (middle), and measures that are neither equivalent nor orthogonal (bottom).

that we observe $Y(t)$ for all t in $[0, 1]$, we cannot say for sure which of the two measures is correct. On the other hand, if $K_0(t) = e^{-|t|}$ and $K_1(t) = e^{-|t|/2}$, then P_0 and P_1 are orthogonal and we can say for sure (that is, with probability 1) which measure is correct. For example, consider the sequence of statistics

$$S_n = \sum_{j=1}^{2^n} \{Y(j2^{-n}) - Y((j-1)2^{-n})\}^2$$

for $n = 1, 2, \ldots$. Under P_0, it is possible to show that $S_n \to 2$ (with probability 1) as $n \to \infty$ and under P_1, $S_n \to 1$ (with probability 1) as $n \to \infty$. Thus, in this case, looking at the infinite sequence of S_ns provides us with a way of saying for sure which of P_0 or P_1 is correct.

We can gain some insight into equivalence and orthogonality by looking at the behavior of the autocovariance functions near the origin. For $K(t) = \theta e^{-\phi|t|}$, we have $K(t) = \theta - \theta\phi|t| + O(t^2)$ for t in a neighborhood of the origin. Thus, $-\theta\phi|t|$ is the principal irregular term for the model (defined in Chapter 2) and we see that, in this case, at least, P_0 and P_1 are equivalent if and only if their corresponding autocovariance functions have the same principal irregular term. Alternatively, we can look in the spectral domain. The spectral density corresponding to $K(t) = \theta e^{-\phi|t|}$ is $f(\omega) = \pi^{-1}\phi\theta/(\theta^2 + \omega^2)$, which satisfies $f(\omega) = \pi^{-1}\phi\theta\omega^{-2} + O(\omega^{-4})$ as $\omega \to \infty$. We see that P_0 and P_1 are equivalent if and only if their spectral densities have the same asymptotic behavior at high frequencies.

In fact, it turns out that the high frequency behavior of the spectral densities is a better guide for determining equivalence and orthogonality of Gaussian measures than the autocovariance functions, since behavior of autocovariance functions away from the origin can affect equivalence (Stein, 1999), whereas, under mild conditions, the behavior of the spectral density in any bounded set cannot. Suppose Y is a stationary Gaussian process on $\mathbb{R}^d$ and the spectral density f_0 satisfies

$$f_0(\boldsymbol{\omega})|\boldsymbol{\omega}|^{\alpha} \text{ is bounded away from 0 and } \infty \text{ as } |\boldsymbol{\omega}| \to \infty. \tag{6.1}$$

This condition is stronger than necessary for the result below, but it is satisfied by a wide range of covariance functions, including all Matérn covariance functions. Recall that specifying the first two moments of a Gaussian process defines its probability law. Define $G_D(m, K)$ to be the probability measure of a Gaussian process on a domain D with mean function m and covariance function K. If f_0 satisfies (6.1) and

$$\int_{|\boldsymbol{\omega}|>C} \left\{ \frac{f_1(\boldsymbol{\omega}) - f_0(\boldsymbol{\omega})}{f_0(\boldsymbol{\omega})} \right\}^2 \mathbf{d}\boldsymbol{\omega} < \infty \tag{6.2}$$

for some $C < \infty$, then $G_D(0, K_0)$ and $G_D(0, K_1)$ are equivalent on all bounded domains D. In one dimension, if (6.1) holds, then (6.2) is close to being necessary for equivalence of the corresponding Gaussian measures. For example, Ibragimov and Rozanov (1978, p. 107) show that, under (6.1), $\omega^{1/2}\{f_1(\omega) - f_0(\omega)\}/f_0(\omega) \to \infty$ as $\omega \to \infty$ implies $G_D(0, K_0)$ and $G_D(0, K_1)$ are orthogonal on any interval D. Conditions for orthogonality of Gaussian measures in more than one dimension are not so simple; some are given in Skorokhod and Yadrenko (1973).

So what does this have to do with the challenge of Yakowitz and Szidarovszky (1985) to kriging? Suppose P_0 was the correct model for some Gaussian process on a bounded domain D, but we instead used the equivalent Gaussian measure P_1 to compute both kriging predictors and to evaluate the mean squared errors of these predictors. If either the resulting predictors or their presumed mean squared errors were not very good when one had a large number of observations in D, then that would indicate a problem for kriging because we would not be able to distinguish reliably between the correct P_0 and the incorrect P_1, no matter how many observations we had in D. Fortunately, it turns out this is not the case. Specifically, Stein (1988) showed that as a sequence of observations gets dense in D, it makes no asymptotic difference whether we use the correct P_0 or the incorrect (but equivalent) P_1 to carry out the kriging.

To explain this result, we need some new notation. Suppose we wish to predict $Y(\mathbf{s}_0)$ for some $\mathbf{s}_0$ and $\mathbf{s}_1, \mathbf{s}_2, \ldots$ is a sequence of observation locations. Suppose $EY(\mathbf{s}) = \boldsymbol{\beta}'\mathbf{m}(\mathbf{s})$, where $\mathbf{m}$ is a known vector-valued function and $\boldsymbol{\beta}$ a vector of unknown coefficients. The correct covariance structure for Y is given by $K_0(\mathbf{s}, \mathbf{s}') = \mathrm{Cov}\{Y(\mathbf{s}), Y(\mathbf{s}')\}$ and the presumed covariance structure by $K_1(\mathbf{s}, \mathbf{s}')$. Let $e_j(\mathbf{s}_0, n)$ be the error of the BLUP (best linear unbiased predictor, or the universal kriging predictor) of $Y(\mathbf{s}_0)$ based on $Y(\mathbf{s}_1), \ldots, Y(\mathbf{s}_n)$ under K_j. Note that we are assuming the BLUP exists, which it will under very mild conditions on $\mathbf{m}$. To avoid confusion, let us call the BLUP of $Y(\mathbf{s}_0)$ under K_1 the pseudo-BLUP. Furthermore, let E_j indicate expected values under K_j. Because errors of BLUPs are contrasts (linear combinations of Y with mean 0 for all $\boldsymbol{\beta}$), specifying K_j is sufficient to determine quantities like $E_j e_k(\mathbf{s}_0, n)^2$ for $j, k = 0, 1$, which is all we need here. Then $E_0 e_0(\mathbf{s}_0, n)^2$ is the mean squared error of the BLUP of $Y(\mathbf{s}_0)$ based on $Y(\mathbf{s}_1), \ldots, Y(\mathbf{s}_n)$. Furthermore, $E_0 e_1(\mathbf{s}_0, n)^2$ is the actual mean squared error of the pseudo-BLUP. We necessarily have $E_0 e_1(\mathbf{s}_0, n)^2 / E_0 e_0(\mathbf{s}_0, n)^2 \geq 1$ and a value near 1 indicates that the pseudo-BLUP is nearly optimal relative to the BLUP. When we presume K_1 is true, we evaluate the mean squared prediction error for the pseudo-BLUP by $E_1 e_1(\mathbf{s}_0, n)^2$, whereas its actual mean squared error is $E_0 e_1(\mathbf{s}_0, n)^2$. Thus, a value of $E_0 e_1(\mathbf{s}_0, n)^2 / E_1 e_1(\mathbf{s}_0, n)^2$ near 1 indicates that the presumed mean squared prediction error for the pseudo-BLUP is close to its actual mean squared prediction error.

Note that the preceding paragraph says nothing about Y being Gaussian and in fact the theorem below does not require Y to be Gaussian. However, it does include a condition about the equivalence of the Gaussian measures $G_D(0, K_0)$ and $G_D(0, K_1)$ and, thus, its practical relevance for non-Gaussian processes is unclear.

Theorem 1: Consider a process Y defined on a domain D with finite second moments and $EY(\mathbf{s}) = \boldsymbol{\beta}'\mathbf{m}(\mathbf{s})$. Suppose $\mathbf{s}_0, \mathbf{s}_1, \ldots$ are in D, $G_D(0, K_0)$ and $G_D(0, K_1)$ are equivalent probability measures, and $e_j(\mathbf{s}_0, n)$ is the error of the BLUP of $Y(\mathbf{s}_0)$ under K_j based on $Y(\mathbf{s}_1), \ldots, Y(\mathbf{s}_n)$. If $E_0 e_0(\mathbf{s}_0, n)^2 \to 0$ as $n \to \infty$, but $E_0 e_0(\mathbf{s}_0, n)^2 > 0$ for all n, then

$$\lim_{n\to\infty} \frac{E_0 e_1(\mathbf{s}_0, n)^2}{E_0 e_0(\mathbf{s}_0, n)^2} = 1, \tag{6.3}$$

$$\lim_{n\to\infty} \frac{E_0 e_1(\mathbf{s}_0, n)^2}{E_1 e_1(\mathbf{s}_0, n)^2} = 1. \tag{6.4}$$

See Stein (1988) for a proof. It is not necessary to assume separately that $E_1 e_1(\mathbf{s}_0, n)^2 > 0$, since $E_0 e_0(\mathbf{s}_0, n)^2 > 0$ and the equivalence of the two measures in fact implies that $E_1 e_1(\mathbf{s}_0, n)^2 > 0$. Indeed, Theorem 1 holds without the restriction $E_0 e_0(\mathbf{s}_0, n)^2 > 0$ for all n as long as one defines $0/0 = 1$. However, $E_0 e_0(\mathbf{s}_0, n)^2 \to 0$ as $n \to \infty$ is required and is satisfied if, for example, Y is mean square continuous on D and $\mathbf{s}_0$ is a limit point of $\mathbf{s}_1, \mathbf{s}_2, \ldots$. Equation (6.3) says that the relative impact of using the incorrect K_1 rather than the correct K_0 on mean squared prediction errors is asymptotically negligible; i.e., predictions under K_1 are asymptotically optimal. Equation (6.4) says that the presumed mean squared prediction error under the incorrect K_1, given by $E_1 e_1(\mathbf{s}_0, n)^2$, is, on a relative basis, close to the actual mean squared error of this prediction, $E_0 e_1(\mathbf{s}_0, n)^2$. Thus, for purposes of both point prediction and evaluating mean squared prediction error, asymptotically, there is no harm in using K_1 rather than K_0.

If $\mathbf{s}_1, \mathbf{s}_2, \ldots$ is dense in D, Stein (1990) shows that the convergence in these limits is uniform over all possible predictands obtainable by taking linear combinations of $Y(\mathbf{s})$ for $\mathbf{s}$ in D and their mean squared limits (i.e., including predictands such as weighted integrals over D). Theorem 1 also holds if the observations include uncorrelated, equal variance measurement errors, as long as the variance of these errors is taken to be the same under both P_0 and P_1. In the case when the mean of Y is known, then these results follow from a much more general result due to Blackwell and Dubins (1962) that shows that the predictions of two Bayesians must, in great generality, be asymptotically identical if their priors are equivalent probability measures.

For stationary processes with spectral densities, it is possible to give a substantially weaker and simpler condition than the equivalence of corresponding Gaussian measures. Specifically, if f_j is the spectral density associated with the autocovariance function K_j and f_0 satisfies Equation (6.1), then $f_1(\omega)/f_0(\omega) \to 1$ as $|\omega| \to \infty$ is sufficient for Theorem 1 to hold. This condition is, in practice, substantially weaker than Equation (6.2) and is, in fact, not sufficient to imply that $G_D(0, K_0)$ and $G_D(0, K_1)$ are equivalent. Stein (1998, 1999) gives many more detailed results along these lines, including rates of convergence in Equation (6.3) and Equation (6.4) in a number of circumstances, although mainly for processes in $\mathbb{R}^1$.

Let us consider a numerical illustration. Suppose we observe a stationary process Y at 25 randomly located points in the unit square $D = [0, 1]^2$ (Figure 6.2) and wish to predict Y at (0, 0) and (0.5, 0.5), a corner and the middle of the observation domain, respectively. Assume the mean of Y is an unknown constant, so that we will be using ordinary kriging. The true autocovariance function of Y is $K_0(\mathbf{s}) = e^{-|\mathbf{s}|}$ and the presumed autocovariance function is $K_1(\mathbf{s}) = 2e^{-|\mathbf{s}|/2}$. Using Equation (6.2), it is possible to show that $G_D(0, K_0)$

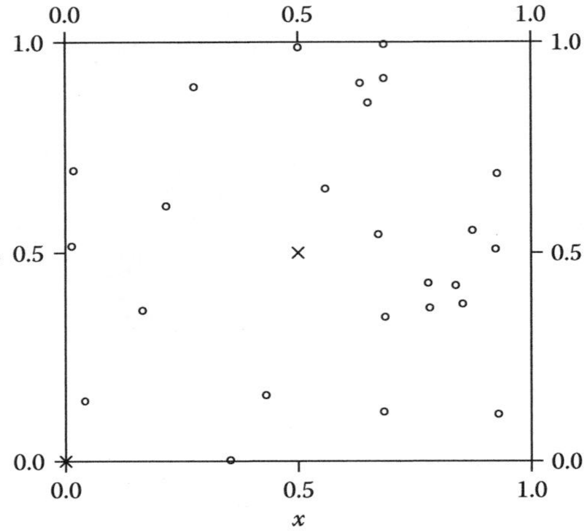

FIGURE 6.2
Shown is a plot of observation locations (○) and prediction locations (×) for numerical illustration.

and $G_D(0, K_1)$ are equivalent and, hence, Theorem 1 applies. Thus, for a sufficiently large number of randomly located observations in D and any $\mathbf{s}_0$ in D, we should expect both $E_0 e_1(\mathbf{s}_0, n)^2 / E_0 e_0(\mathbf{s}_0, n)^2$ and $E_0 e_1(\mathbf{s}_0, n)^2 / E_1 e_1(\mathbf{s}_0, n)^2$ to be near 1. Table 6.1 gives the values for these ratios and, indeed, all of them are somewhat close to 1, but some are much closer than others. In particular, for both prediction locations, the first ratio is much closer to 1 than the second, indicating that the effect of using K_1 instead of the correct K_0 is much greater on the assessment of mean squared error than on the actual mean squared error of the pseudo-BLUP. The phenomenon that getting the covariance function wrong tends to have a greater impact on the evaluation of mean squared error than on the efficiency of the point predictions has long been observed in the geostatistical community. Stein (1999) provides further numerical results and some theoretical support for this observation.

The other important pattern displayed in Table 6.1 is that both ratios are much closer to 1 when predicting at the center of the square rather than at the corner of the square. We might reasonably call predicting at the center of the square an interpolation and predicting at the corner an extrapolation. Stein (1999) gives extensive numerical and theoretical evidence that this distinction between interpolation and extrapolation holds quite generally. That is, using an incorrect covariance structure K_1 instead of the correct K_0 generally matters more when extrapolating than interpolating. It is perhaps not surprising that model misspecifications should matter more when extrapolating than interpolating; one only has to think about regression problems to realize that statistical statements about what happens outside of the range of available data are more sensitive to the choice of model than statements within that range. In the present context, we obtain the conclusion that there is very little need to distinguish between K_0 and K_1 with $G_D(0, K_0)$ and $G_D(0, K_1)$ equivalent if we wish to

TABLE 6.1
Properties of Pseudo-BLUPs

	Error Variance = 0		Error Variance = 0.4	
	Corner	**Center**	**Corner**	**Center**
$E_0 e_0^2$	0.2452	0.1689	0.3920	0.2340
$E_0 e_1^2 / E_0 e_0^2$	1.0046	1.0001	1.0083	1.0003
$E_0 e_1^2 / E_1 e_1^2$	0.9623	0.9931	0.9383	0.9861

interpolate and a somewhat greater need if we wish to make predictions near a boundary of the observation domain.

The frequency domain provides a helpful perspective for understanding these results. Specifically, when interpolating, it turns out that the low-frequency behavior of the process has very little impact. Thus, models with similar high-frequency behavior produce highly similar interpolations. As the prediction becomes more of an extrapolation, the low-frequency behavior affects predictions more. Thus, if we considered predicting at a location even farther from the observations than the corner, we would tend to find the ratios in Table 6.1 even further from 1. Table 6.1 also gives results when there is a measurement error with variance 0.4 under either model. Of course, $E_0 e_0^2$ goes up when there is measurement error, but more interestingly, all of the ratios in Table 6.1 are now further from 1. When there is measurement error, BLUPs make greater use of more distant observations, so that the low frequency behavior matters more.

In spatial statistics, one is usually interested in interpolation rather than extrapolation, whereas in time series, extrapolation is usually of much greater interest. This difference provides one possible argument for using fixed-domain asymptotics in spatial statistics, but increasing-domain asymptotics in time series. Specifically, the insight that fixed-domain asymptotics brings to prediction problems on the near irrelevance of low-frequency behavior is much more pertinent when interpolating than extrapolating. It is interesting to note, though, that in recent years, a great deal of fixed-domain asymptotics has been done in the finance literature, spurred by the interest in the fluctuations of various financial instruments on very short time scales (see, e.g., Zhang, Mykland and Aït-Sahalia, 2005).

Although Theorem 1 and its various extensions are theoretically interesting, they do not come close to proving that if one plugs an estimated covariance function into the standard kriging formulas (plug-in prediction), then the resulting predictors are asymptotically optimal or the presumed mean squared prediction errors are asymptotically correct. I would guess that such a result is true under fixed-domain asymptotics in broad generality for Gaussian processes with homogeneous covariance functions, but, except for a rather limited result due to Putter and Young (2001) to be discussed below, no substantial progress has been made on this problem.

6.2 Estimation

The availability of asymptotic results for estimation of spatial covariance structures depends greatly on the question one asks. Generally speaking, there are many more results under increasing-domain asymptotics than under fixed-domain asymptotics, more for observations on a regular grid than irregularly sited observations, and more for explicitly defined estimators, such as those based on empirical variograms, than for implicitly defined estimators, such as restricted maximum likelihood (REML). For example, Guyon (1995) gives many asymptotic results for estimators of spatial covariance based on gridded data and increasing-domain asymptotics for both parametric and nonparametric cases, Gaussian and non-Gaussian, and even some for maximum likelihood (ML) estimators under certain special models. However, if one's goal is to estimate the spatial covariance structure of a process on $\mathbb{R}^2$, which is what is needed for kriging, it may not be a good idea to place all observations on a lattice (Pettitt and McBratney, 1993). In contrast, it is difficult to point to a single meaningful result for irregularly sited observations under fixed-domain asymptotics. Indeed, even for gridded data under fixed-domain asymptotics, results for ML or REML estimates only exist in some very simple cases, largely for processes in one dimension or for separable processes in more than one dimension for which one can obtain explicit expressions for the likelihood function (Ying, 1991, 1993; Chen, Simpson and Ying, 2000; Loh, 2005).

Under increasing-domain asymptotics, one might generally expect that under parametric models for the covariance structure, all parameters can be consistently estimated and that ML and REML estimators should obey the "usual" asymptotics. Specifically, we might expect that these estimators are asymptotically normal with asymptotic mean given by the true value of the parameters and asymptotic covariance matrix given by the inverse of the Fisher information matrix. Mardia and Marshall (1984) give a general result to this effect about ML estimators, but despite a strong effort by the authors to give results that can be verified in practice, it is not an easy matter to verify the conditions of their results for observations that are not on a regular lattice. Cressie and Lahiri (1993, 1996) give similar results for REML estimators. It does not appear that any of these results apply to intrinsic random functions that are not stationary.

One of the difficulties in obtaining fixed-domain asymptotic results even for Gaussian processes observed on a grid is that, under any model including something like a range parameter, there will generally be at least one function of the parameters that cannot be consistently estimated as the number of observations increases. For example, if D is a bounded infinite subset of $\mathbb{R}^d$ for $d \leq 3$ and K_0 and K_1 are two Matérn covariance functions, then the ratio of the corresponding spectral densities tending to 1 as the frequency tends to infinity is necessary and sufficient for the corresponding Gaussian measures to be equivalent (see Zhang, 2004, Theorem 2). As a special case of this result, consider $K(\mathbf{s}; \theta, \phi) = \theta \exp(-\phi|\mathbf{s}|)$, with θ and ϕ unknown positive parameters. Then the corresponding Gaussian measures are equivalent on any bounded infinite subset D in 3 or fewer dimensions if and only if $\theta_0\phi_0 = \theta_1\phi_1$. It immediately follows that it is not possible to estimate either θ or ϕ consistently based on observations in D, but it may be possible to estimate their product, $\theta\phi$, consistently. Ying (1991) shows this is, in fact, the case for the ML estimator for (θ, ϕ) when $d = 1$. Because so much of the asymptotic theory of estimation is based on the existence of consistent estimators of the unknown parameter vector, much of the mathematical machinery that has been built so carefully over the years must be abandoned. Even with models for which all parameters can be consistently estimated based on observations in a bounded domain (e.g., a nugget effect plus a linear variogram), one generally does not get the usual $n^{-1/2}$ convergence rate (Stein, 1987), which also leads to considerable mathematical difficulties.

Zhang and Zimmerman (2005) make an intriguing effort to compare the approximations given by fixed and increasing domain asymptotics. Part of the difficulty is to find a setting in which fixed-domain asymptotics provides a useful explicit approximation for all components of a parameter vector. They provide some evidence in the limited setting of a Gaussian process on the line with exponential variogram observed with or without Gaussian errors, the fixed-domain approach provides greater qualitative insight and sometimes better approximations than the increasing domain approach. However, without some further advances in deriving and computing the limiting distribution of parameter estimates under fixed-domain asymptotics, it is not clear how to extend these results to more realistic settings.

One of the difficulties of studying the asymptotic properties of ML and REML estimates in spatial settings is that one rarely has explicit expressions for the estimates, which causes particular difficulties with fixed-domain asymptotics for which standard asymptotic results do not generally hold. For spatial data on a regular grid, there are some more general fixed-domain asymptotic results for statistical methods that are not based on the likelihood function. These results go back to Lévy (1940), who showed that the sum of squared increments for Brownian motion observed at n evenly spaced locations on the unit interval converges to the variance parameter of the Brownian motion as $n \to \infty$. In the ensuing years, there have been many extensions to this basic result that the parameters controlling the local behavior of a Gaussian process can be estimated consistently under fixed-domain asymptotics using increment-based estimators. Davies and Hall (1999) provide a good entry into this literature, with a focus on estimating the "fractal index" of stationary Gaussian processes, which, for isotropic processes, is directly related to the power in the principal irregular term of the variogram when that power is less than 2 (see Chapter 2) Chan and Wood (2004) extend some of these results to processes that are a pointwise transformation of a stationary Gaussian random process. Anderes and Chatterjee (2008) show how increment-based estimators can be used to estimate the local anisotropy of nonstationary Gaussian processes. Stein (1995) and Lim and Stein (2008) consider the spatial periodogram of a Gaussian random field and show that results similar to those that hold for the periodogram under increasing-domain asymptotics sometimes hold in the fixed-domain setting as long as one first applies an appropriate prewhitening filter to the data.

6.2.1 Prediction and Estimation

Putter and Young (2001) are essentially the only ones whose work provides some theory about kriging with estimated covariance functions that applies in the fixed-domain setting. They give a general-looking result based on contiguity of sequences of measures (a topic related to equivalence of two measures, see Roussas, 1972) that indicates a version of Theorem 1 holds when P_1 is replaced by an estimated model. However, they only explicitly verify their conditions for some very simple models in one dimension. Furthermore, their proof actually requires a slightly stronger condition than they give in their paper (the contiguity condition has to apply not just to the measures generated by the observations, but also to the measures generated by the observations together with the predictand).

Despite this lack of direct theoretical evidence, for Gaussian processes with correctly specified parametric models for the covariance function, I would guess that Theorem 1 does generally hold when P_1 is replaced by an estimated model, as long as a "good" method of parameter estimation, such as REML, is used. This result can be expected to hold to a greater degree of accuracy when predicting at locations not near a boundary of D. Even if it is proven some day, given its dependence on such strong model assumptions, practitioners would still need to use great care in selecting and fitting spatial covariance functions when the goal is spatial prediction.

References

Anderes, E. and Chatterjee, S. (2008). Consistent estimates of deformed isotropic Gaussian random fields on the plane. *Annals of Statistics*, 37, 2324–2350.

Blackwell, D. and Dubins, L.E. (1962). Merging of opinions with increasing information. *Annals of Mathematical Statistics*, 33, 882–886.

Chan, G. and Wood, A.T.A. (2004). Estimation of fractal dimension for a class of non-Gaussian stationary processes and fields. *Annals of Statistics*, 32, 1222–1260.

Chen, H.S., Simpson, D.G., and Ying, Z. (2000). Infill asymptotics for a stochastic process model with measurement error. *Statistica Sinica*, 10, 141–156.

Cressie, N. (1993). *Statistics for Spatial Data*, revised ed. John Wiley & Sons, New York.

Cressie, N. and Lahiri, S.N. (1993). The asymptotic distribution of REML estimators. *Journal of Multivariate Analysis*, 45, 217–233.

Cressie, N. and Lahiri, S.N. (1996). Asymptotics for REML estimation of spatial covariance parameters. *Journal of Statistical Planning and Inference*, 50, 327–341.

Davies, S. and Hall, P. (1999). Fractal analysis of surface roughness by using spatial data (with discussion). *Journal of the Royal Statistical Society, Series B*, 61, 3–37.

Guyon, X. (1995). *Random Fields on a Network: Modeling, Statistics and Applications*. Springer-Verlag, New York.

Hall, P., Patil, P. (1994). Properties of nonparametric estimators of autocovariance for stationary random-fields. *Probability Theory and Related Fields*, 99, 399–424.

Härdle, W., Tuan, P.D. (1986). Some theory on *M* smoothing of time series. *Journal of Time Series Analysis*, 7, 191–204.

Ibragimov, I.A., Rozanov, Y.A. (1978). *Gaussian Random Processes*, Trans. AB Aries. Springer-Verlag, New York.

Kuo, H. (1975). *Gaussian Measures in Banach Spaces*, Lecture Notes in Mathematics No. 463. Springer-Verlag, New York.

Lahiri, S.N., Kaiser, M.S., Cressie, N., Hsu, N.J. (1999). Prediction of spatial cumulative distribution functions using subsampling. *Journal of the American Statistical Association*, 94, 86–97.

Lévy, P. (1940). Le mouvement Brownien plan. *American Journal of Mathematics*, 62, 487–550.

Lim, C., Stein, M.L. (2008). Asymptotic properties of spatial cross-periodograms using fixed-domain asymptotics. *Journal of Multivariate Analysis*, 99, 1962–1984.

Loh, W.L. (2005). Fixed-domain asymptotics for a subclass of Matérn-type Gaussian random fields. *Annals of Statistics*, 33, 2344–2394.

Mardia, K.V., Marshall, R.J. (1984). Maximum likelihood estimation of models for residual covariance in spatial regression. *Biometrika*, 73, 135–146.

Pettitt, A.N., McBratney, A.B. (1993). Sampling designs for estimating spatial variance components. *Applied Statistics*, 42, 185–209.

Putter, H., Young, G.A. (2001). On the effect of covariance function estimation on the accuracy of kriging predictors. *Bernoulli*, 7, 421–438.

Roussas, G.G. (1972). *Contiguity of Probable Measures: Some Applications in Statistics*. Cambridge, U.K.: Cambridge University Press.

Skorokhod, A.V., Yadrenko, M.I. (1973). On absolute continuity of measures corresponding to homogeneous Gaussian fields. *Theory of Probability and Its Applications*, 18, 27–40.

Stein, M.L. (1987). Minimum norm quadratic estimation of spatial variograms. *Journal of the American Statistical Association*, 82, 765–772.

Stein, M.L. (1988). Asymptotically efficient prediction of a random field with a misspecified covariance function. *Annals of Statistics*, 16, 55–63.

Stein, M.L. (1990). Uniform asymptotic optimality of linear predictions of a random field using an incorrect second-order structure. *Annals of Statistics*, 18, 850–872.

Stein, M.L. (1995). Fixed domain asymptotics for spatial periodograms. *Journal of the American Statistical Association*, 90, 1277–1288.

Stein, M.L. (1998). Predicting random fields with increasingly dense observations. *Annals of Applied Probability*, 9, 242–273.

Stein, M.L. (1999). *Interpolation of Spatial Data: Some Theory for Kriging*. Springer, New York.

Yakowitz, S., Szidarovszky, F. (1985). A comparison of kriging with nonparametric regression methods. *Journal of Multivariate Analysis*, 16, 21–53.

Ying, Z. (1991). Asymptotic properties of a maximum likelihood estimator with data from a Gaussian process. *Journal of Multivariate Analysis*, 36, 280–296.

Ying, Z. (1993). Maximum likelihood estimation of parameters under a spatial sampling scheme. *Annals of Statistics*, 21, 1567–1590.

Zhang, H. (2004). Inconsistent estimation and asymptotically equal interpolations in model-based geostatistics. *Journal of the American Statistical Association*, 99, 250–261.

Zhang, L., Mykland, P.A., Aït-Sahalia, Y. (2005). A tale of two time scales: Determining integrated volatility with noisy high-frequency data. *Journal of the American Statistical Association*, 100, 1394–1411.

Zhang, H., Zimmerman, D.L. (2005). Towards reconciling two asymptotic frameworks in spatial statistics. *Biometrika*, 92, 921–936.

7

Hierarchical Modeling with Spatial Data

Christopher K. Wikle

CONTENTS

7.1 Introduction

In spatial statistics, one often must develop statistical models in the presence of complicated processes, multiple sources of data, uncertainty in parameterizations, and various degrees of scientific knowledge. One can approach such complex problems from either a joint or conditional viewpoint. Although it may be intuitive to consider processes from a joint perspective, such an approach can present serious challenges to statistical modeling. For example, it can be very difficult to specify joint multivariate dependence structures for related spatial datasets. It may be much easier to factor such joint distributions into a series of conditional models. For example, it is simpler (and a reasonable scientific assumption) to consider a near-surface ozone process conditional upon the near-surface ambient air temperature (especially in the summer), rather than consider the ozone and temperature processes jointly. Indeed, it is often possible to simplify modeling specifications, account for uncertainties, and use scientific knowledge in a series of conditional models, coherently linked together by simple probability rules. This is the essence of hierarchical modeling.

In principle, hierarchical models can be considered from either a classical or Bayesian perspective. However, as the level of complexity increases, the Bayesian paradigm becomes a necessity. Indeed, the increased use of such approaches has coincided with the revolution in Bayesian computation exemplified by the adoption and further development of Markov chain Monte Carlo (MCMC) simulation approaches (Gelfand and Smith, 1990). The fact that such simulation approaches are used requires that, in some cases, modeling choices have to be made to facilitate computation. Of course, this is true of more traditional modeling/estimation approaches as well, but the hierarchical modeler must explicitly account for these challenges when building models, and such choices are often problem (if not dataset) specific.

The remainder of this chapter will consider hierarchical models for spatial data. The discussion is not intended to be a comprehensive review of the entire literature on hierarchical models in spatial statistics. Rather, it is intended to provide an introduction and glimpse of how hierarchical models have been and can be used to facilitate modeling of spatial processes. The basic principles of hierarchical modeling are well established and the cited references can provide additional perspective as well as increased technical detail. Section 7.2 will describe the hierarchical approach from a simple schematic perspective. This will then be followed in Sections 7.3 and 7.4 with specific discussion of hierarchical models as they pertain to Gaussian and non-Gaussian geostatistical spatial processes, respectively. Given that this part of the handbook is concerned with continuous spatial variation, we will limit the discussion of hierarchical models to such models, although the ideas generalize to other types of spatial support and processes. Finally, Section 7.5 contains a brief overview and discussion.

7.2 An Overview of Hierarchical Modeling

The basic ideas of hierarchical modeling arise from simple probability rules. Although the concept is not inherently Bayesian, over time most of the literature has been developed in that context, and the best pedagogical descriptions are most often found in the Bayesian literature (e.g., Gelman, Carlin, Stern, and Rubin, 2004). Our purpose here is not to reproduce such general descriptions, but rather to describe hierarchical models in the context of a framework that is relevant to spatial modeling. These ideas follow those in Berliner (1996) and Wikle (2003).

Hierarchical modeling is based on the basic fact from probability theory that the joint distribution of a collection of random variables can be decomposed into a series of conditional distributions and a marginal distribution. That is, if A, B, C are random variables, then we can write the joint distribution in terms of factorizations, such as $[A, B, C] = [A|B, C][B|C][C]$, where the bracket notation $[C]$ refers to a probability distribution for C, and $[B|C]$ refers to the conditional probability distribution of B given C, etc. For a spatial process, the joint distribution describes the stochastic behavior of the spatially referenced data, true (latent) spatial process, and parameters. This can be difficult (if not impossible) to specify for many problems. It is often much easier to specify the distribution of the relevant conditional models (e.g., conditioning the observed data on the true process and parameters, etc.). In this case, the product of a series of relatively simple conditional models leads to a joint distribution that can be quite complex.

For complicated processes in the presence of data, it is useful to approach the problem by (conceptually, if not actually) breaking it into three primary stages (e.g., Berliner, 1996):

Stage 1. Data Model: [*data*|*process, parameters*]

Stage 2. Process Model: [*process*|*parameters*]

Stage 3. Parameter Model: [*parameters*]

The first stage is concerned with the observational process or "data model," which specifies the distribution of the data *given* the process of interest as well as the parameters that describe the data model. The second stage then describes a distribution (i.e., model) for the process, conditional on other parameters. The last stage accounts for the uncertainty in the parameters by endowing them with distributions. In general, each of these stages may have multiple substages. For example, if the process is multivariate and spatial, it might be modeled as a product of physically motivated distributions for one process given the others, as suggested by some scientific relationship (e.g., ozone conditioned on temperature). Similar decompositions are possible in the data and parameter stages.

Ultimately, we are interested in the distribution of the process and parameters updated by the data, also known as the "posterior" distribution. This is obtained by Bayes' rule in which the posterior distribution is proportional to the product of the data, process, and parameter distributions:

$$[\mathit{process}, \mathit{parameters}|\mathit{data}] \propto [\mathit{data}|\mathit{process}, \mathit{parameters}] \\ \times [\mathit{process}|\mathit{parameters}][\mathit{parameters}],$$

where the normalizing constant represents the integral of the right-hand side with respect to the process and parameters. This formula serves as the basis for hierarchical Bayesian analysis. Before we apply this directly to spatial data, we discuss in more detail the three primary components of the general hierarchical model.

7.2.1 Data Models

The primary advantage of modeling the conditional distribution of the data given the true process is that substantial simplifications in model form are possible. For example, let Y be data observed for some (spatial) process η (which cannot be observed without error), and let θ_y represent parameters. The data model distribution is written: $[Y|\eta, \theta_y]$. Usually, this conditional distribution is much simpler than the unconditional distribution of $[Y]$ due to the fact that most of the complicated dependence structure comes from the process η. Often, the error structure of this model simply represents measurement error (and/or small-scale spatial variability in the case of spatial processes). In this general framework, the error need not be additive. Furthermore, this framework can also accommodate data that is at a different support and/or alignment than the process, η (e.g., see Gelfand, Zhu, and Carlin, 2001; Wikle, Milliff, Nychka, and Berliner, 2001; Wikle and Berliner, 2005).

The hierarchical data model also provides a natural way to combine datasets. For example, assume that Y_a and Y_b represent data from two different sources. Again, let η be the process of interest and $\theta_{y_a}, \theta_{y_b}$ parameters. In this case, the data model is often written

$$[Y_a, Y_b|\eta, \theta_{y_a}, \theta_{y_b}] = [Y_a|\eta, \theta_{y_a}][Y_b|\eta, \theta_{y_b}]. \tag{7.1}$$

That is, conditioned on the true process, the data are assumed to be independent. This does not suggest that the two datasets are unconditionally independent, but rather, the dependence among the datasets is largely due to the process, η. Such an assumption of independence must be assessed and/or justified for each problem.

The conditional independence assumption can also be applied to multivariate models. That is, if the processes of interest are denoted η_a and η_b, with associated observations Y_a and Y_b, respectively, then one might assume:

$$[Y_a, Y_b|\eta_a, \eta_b, \theta_{y_a}, \theta_{y_c}] = [Y_a|\eta_a, \theta_{y_a}][Y_b|\eta_b, \theta_{y_b}]. \tag{7.2}$$

Again, this assumption must be evaluated for validity in specific problems. When appropriate, this can lead to dramatic simplifications in estimation.

7.2.2 Process Models

It is often the case that developing the process model distribution is the most critical step in constructing the hierarchical model. As mentioned previously, this distribution can sometimes be further factored hierarchically into a series of submodels. For example, assume the process of interest is composed of two subprocesses, η_a and η_b (e.g., η_a might represent lower tropospheric ozone and η_b might represent surface temperature over the same region, as in Royle and Berliner (1999)). In addition, let the parameters $\theta_\eta = \{\theta_{\eta_a}, \theta_{\eta_b}\}$ be associated with these two processes. Then, one might consider the decomposition,

$$[\eta_a, \eta_b|\theta_\eta] = [\eta_a|\eta_b, \theta_\eta][\eta_b|\theta_\eta]. \tag{7.3}$$

Further assumptions on the parameters lead to simplifications so that the right-hand side of Equation (7.3) can often be written as $[\eta_a|\eta_b, \theta_{\eta_a}][\eta_b|\theta_{\eta_b}]$. The modeling challenge is the specification of these component distributions. The utility and implementation of such a model is facilitated by scientific understanding of the processes of interest, but the validity of the hierarchical decomposition is not dependent upon such understanding.

7.2.3 Parameter Models

As is the case with the data and process model distributions, the parameter distribution is often partitioned into a series of distributions. For example, given the data model (7.2) and process model (7.3), one would need to specify the parameter distribution $[\theta_{y_a}, \theta_{y_b}, \theta_{\eta_a}, \theta_{\eta_b}]$. Often, one can make reasonable independence assumptions regarding these distributions, such as $[\theta_{y_a}, \theta_{y_b}, \theta_{\eta_a}, \theta_{\eta_b}] = [\theta_{y_a}][\theta_{y_b}][\theta_{\eta_a}][\theta_{\eta_b}]$. Of course, such assumptions are problem specific.

There are often appropriate submodels for parameters as well, leading to additional levels of the model hierarchy. For complicated processes, there may be scientific insight available that can go into developing the parameter models similar to the development of the process models (e.g., Wikle et al., 2001). In other cases, one may not know much about the parameter distributions, suggesting "noninformative priors" or the use of data-based estimates for hyperparameters. In many cases, the parameter distributions are chosen mainly (or partly) to facilitate computation.

Historically, the specification of parameter distributions has sometimes been the focus of objections due to its inherent "subjectiveness." The formulation of the data and process models is quite subjective as well, but those choices have not generated as much concern. Indeed, subjectivity at these stages is a part of classical inference as well. A strength of the hierarchical (Bayesian) approach is the quantification of such subjective judgment. Hierarchical models provide a coherent framework in which to incorporate explicitly in the model the uncertainty related to judgment, scientific reasoning, pragmatism, and experience. Yet, if one chooses to eliminate as much subjectivity as possible, there are various "objective" choices for prior distributions that may be considered. One must be careful in the hierarchical modeling framework that such choices lead to a proper posterior distribution. For a more philosophical discussion of the issues surrounding subjective and objective prior development in Bayesian statistics, see Berger (2006) and Goldstein (2006) as well as the accompanying discussion.

7.2.4 Hierarchical Spatial Models

The hierarchical framework can be applied to most of the standard spatial statistical modeling approaches described in this volume. Rather than develop a single general framework to encompass all of these, we illustrate hierarchical spatial models for Gaussian and non-Gaussian geostatistical problems in the following sections. The common theme throughout is the general three-stage hierarchical structure described above, the reliance on a latent Gaussian spatial process within this framework, and the associated Bayesian implementation. These examples are not intended to be exhaustive. More detailed and complete descriptions can be found in texts, such as Banerjee, Carlin, and Gelfand (2004), Diggle and Ribeiro (2007), and the references therein.

7.3 Hierarchical Gaussian Geostatistical Model

Assume there are m observations of a spatial process, denoted by $\mathbf{Y} = (Y(\tilde{\mathbf{s}}_1), \ldots, Y(\tilde{\mathbf{s}}_m))'$ and we define a latent spatial vector, $\boldsymbol{\eta} = (\eta(\mathbf{s}_1), \ldots, \eta(\mathbf{s}_n))'$, where $\eta(\mathbf{s}_i)$ is from a Gaussian spatial process where $\mathbf{s}_i$ is a spatial location (assumed here to be in some subset of two-dimensional real space, although generalization to higher dimensions is not difficult in principle). In general, the locations $\{\mathbf{s}_1, \ldots, \mathbf{s}_n\}$ corresponding to the latent vector $\boldsymbol{\eta}$ may not coincide with the observation locations $\{\tilde{\mathbf{s}}_1, \ldots, \tilde{\mathbf{s}}_m\}$. Furthermore, most generally, the supports of the observation and prediction locations need not be the same (e.g., see Chapter 29 of this volume), although here we will assume both the observations and the latent spatial process have point-level support.

In terms of our general hierarchical framework, we must specify a data model, process model, and parameter models. For example, a hierarchical spatial model might be given by

$$\text{Data Model:} \quad \mathbf{Y} \mid \boldsymbol{\beta}, \boldsymbol{\eta}, \sigma_\epsilon^2 \sim Gau(\mathbf{X}\boldsymbol{\beta} + \mathbf{H}\boldsymbol{\eta}, \sigma_\epsilon^2 \mathbf{I}) \tag{7.4}$$

$$\text{Process Model:} \quad \boldsymbol{\eta} \mid \boldsymbol{\theta} \sim Gau(\mathbf{0}, \boldsymbol{\Sigma}(\boldsymbol{\theta})) \tag{7.5}$$

$$\text{Parameter Model:} \quad [\boldsymbol{\beta}, \sigma_\epsilon^2, \boldsymbol{\theta}] \tag{7.6}$$

where $\mathbf{X}$ is an $m \times p$ matrix of covariates, often associated with the large-scale spatial mean (or "trend" as discussed in Chapter 3 in this volume), $\boldsymbol{\beta}$ is the associated $p \times 1$ vector of parameters, $\mathbf{H}$ is an $m \times n$ matrix that associates the observations $\mathbf{Y}$ with the latent process $\boldsymbol{\eta}$, σ_ϵ^2 corresponds to an independent, small-scale (nugget) spatial effect and/or measurement error process ε, and $\boldsymbol{\theta}$ is a vector of parameters used for the spatial covariance function associated with the spatial latent process (e.g., a variance, spatial range parameter, smoothness parameter, etc.). Note, the choice of the specific parameter distribution(s) is often dependent on the estimation approach and/or the specific process being considered. It is worth noting that the "mapping matrix" $\mathbf{H}$ is quite flexible. For example, in its simplest form, $\mathbf{H}$ is simply the identity matrix if the observation and latent process locations coincide. Alternatively, $\mathbf{H}$ might be an incidence matrix of ones and zeros if some (but not all) of the observation locations correspond directly to latent process locations. Or, $\mathbf{H}$ might be a weighting or averaging matrix in certain circumstances (e.g., see Wikle, Berliner, and Cressie, 1998; Wikle et al., 2001; Wikle and Berliner, 2005).

One might consider an alternative hierarchical parameterization in which the spatial mean (trend) is moved down a level of the hierarchy. In this case, the data model is $\mathbf{Y}|\boldsymbol{\eta}, \sigma_\epsilon^2 \sim Gau(\mathbf{H}\boldsymbol{\eta}, \sigma_\epsilon^2\mathbf{I})$, the process model is given by $\boldsymbol{\eta}|\boldsymbol{\beta}, \boldsymbol{\theta} \sim Gau(\mathbf{X}\boldsymbol{\beta}, \boldsymbol{\Sigma}(\boldsymbol{\theta}))$, and the parameter distribution is unchanged as in Equation (7.6). In principle, such a "hierarchical centering" can improve Bayesian estimation (Gelfand, Sahu, and Carlin, 1995).

Yet another representation arises if one integrates out the latent spatial process, essentially combining the data and process stages into one. Specifically,

$$[\mathbf{Y}|\boldsymbol{\beta}, \sigma_\epsilon^2, \boldsymbol{\theta}] = \int [\mathbf{Y}|\boldsymbol{\eta}, \boldsymbol{\beta}, \sigma_\epsilon^2][\boldsymbol{\eta}|\boldsymbol{\theta}]d\boldsymbol{\eta} \tag{7.7}$$

for the models in Equation (7.4) and Equation (7.5), with the integration over the support of η, gives

$$\text{Data Model:} \quad \mathbf{Y}|\beta, \sigma_\epsilon^2, \boldsymbol{\theta} \sim Gau(\mathbf{X}\beta, \boldsymbol{\Sigma}(\boldsymbol{\theta}) + \sigma_\epsilon^2\mathbf{I}). \tag{7.8}$$

$$\text{Parameter Model:} \quad [\beta, \sigma_\epsilon^2, \boldsymbol{\theta}]. \tag{7.9}$$

The connection between the fully hierarchical formulation suggested by Equations (7.4), (7.5), and (7.6) and this marginal formulation in Equations (7.8) and (7.9) is exactly the same as for traditional linear mixed model analysis (e.g., longitudinal analysis, Verbeke and Molenberghs (2000)). In such cases, when inference is concerned with the parameters β (such as with a "spatial regression analysis"), it is often convenient to proceed in terms of the marginal model, without the need for specific prediction of the random effects (i.e., the spatial latent process in the spatial case). For most traditional geostatistical applications, one is interested in predictions of the spatial process, and the hierarchical formulation would seem to be more appropriate. However, as discussed below, given the posterior distribution of the parameters $[\beta, \sigma_\epsilon^2, \boldsymbol{\theta}|\mathbf{Y}]$, one can easily obtain the posterior distribution $[\eta|\mathbf{Y}]$. In addition, there are cases in both classical and Bayesian estimation, regardless of the ultimate prediction goal, where one uses the marginal formulation to facilitate estimation of parameters.

7.3.1 Posterior Analysis

Given suitable choices for the prior distributions of the parameters (7.6), we seek the posterior distribution, which is proportional to the product of Equations (7.4), (7.5), and (7.6):

$$[\eta, \beta, \boldsymbol{\theta}, \sigma_\epsilon^2|\mathbf{Y}] \propto [\mathbf{Y}|\eta, \beta, \sigma_\epsilon^2][\eta|\boldsymbol{\theta}][\beta, \sigma_\epsilon^2, \boldsymbol{\theta}]. \tag{7.10}$$

One cannot obtain the analytical representation of the normalizing constant in this case and, thus, Monte Carlo approaches are utilized for estimation, inference, and prediction (see Robert and Casella (2004) for an overview).

Alternatively, one can use the marginal model (7.8) directly to obtain the posterior distribution for the parameters:

$$[\beta, \sigma_\epsilon^2, \boldsymbol{\theta}|\mathbf{Y}] \propto [\mathbf{Y}|\beta, \sigma_\epsilon^2, \boldsymbol{\theta}][\beta, \sigma_\epsilon^2, \boldsymbol{\theta}]. \tag{7.11}$$

Again, the normalizing constant for (7.11) is not available analytically and Monte Carlo methods must be used to obtain samples from the posterior distribution. In addition, note that since

$$[\eta|\mathbf{Y}] = \int [\eta|\boldsymbol{\theta}][\boldsymbol{\theta}|\mathbf{Y}]d\boldsymbol{\theta}, \tag{7.12}$$

if we have samples from the marginal posterior distribution $[\boldsymbol{\theta}|\mathbf{Y}]$, we can obtain samples from $[\eta|\mathbf{Y}]$ by direct Monte Carlo sampling as long as we know $[\eta|\boldsymbol{\theta}]$ in closed form (as we do in this case). Thus, in the hierarchical Gaussian geostatistical model, there really is no need to consider the full joint posterior (7.10) directly. In fact, as discussed in Banerjee et al. (2004), the covariance structure in the marginal model (7.8) is more computationally stable (due to the additive constant on the diagonal of the marginal covariance matrix) and, in the case of MCMC implementations, estimation is more efficient the more that one can marginalize the hierarchical model analytically.

It is typical in geostatistical analyses to make a distinction between the elements of the latent spatial process vector η that correspond to the observed data, say η_d (an $m \times 1$ vector), and those nonobservation locations for which prediction is desired, say η_0 (an $(n-m) \times 1$ vector). In this case, $\eta = (\eta_d', \eta_0')'$ and the process model (7.5) can be factored as

$$[\eta_0, \eta_d|\boldsymbol{\theta}] = [\eta_0|\eta_d, \boldsymbol{\theta}][\eta_d|\boldsymbol{\theta}]. \tag{7.13}$$

One can then obtain the posterior predictive distribution $[\boldsymbol{\eta}_0|\mathbf{Y}, \boldsymbol{\theta}]$ from

$$[\boldsymbol{\eta}_0|\mathbf{Y}, \boldsymbol{\theta}] = \int [\boldsymbol{\eta}_0|\boldsymbol{\eta}_d, \boldsymbol{\theta}][\boldsymbol{\eta}_d|\mathbf{Y}]d\boldsymbol{\eta}_d, \tag{7.14}$$

where $[\boldsymbol{\eta}_d|\mathbf{Y}]$ is the posterior distribution for those latent process locations corresponding to the observation locations and, in this case, $[\boldsymbol{\eta}_0|\boldsymbol{\eta}_d, \boldsymbol{\theta}]$ is a multivariate Gaussian distribution easily obtained (analytically) from basic properties of conditional multivariate normal distributions. Note, however, that this posterior predictive distribution is equivalent to the posterior distribution of $\mathbf{H}\boldsymbol{\eta}$, where $\mathbf{H} = [\mathbf{I}_{m\times m}\ \mathbf{0}_{m\times(n-m)}]$. That is, one can predict the full spatial vector $\boldsymbol{\eta}$ and then simply "pick off" the elements corresponding to prediction locations. The advantage of considering the decomposition in Equations (7.13) and (7.14) is computational. Specifically, for inference associated with the basic process model (7.5), one will have to consider the inverse of the full $n \times n$ covariance matrix $\boldsymbol{\Sigma}(\boldsymbol{\theta})$. However, by decomposing the latent process, one only needs to compute the inverse of an $m \times m$ covariance matrix. This is true for both classical and Bayesian estimation. It should be noted that there can be a computational advantage when predicting the full $\boldsymbol{\eta}$ as well, particularly in very high dimensional problems where $\boldsymbol{\eta}$ can be defined on a grid and/or a reduced rank representation is considered for $\boldsymbol{\eta}$ (e.g., see Chapter 8 in this volume).

7.3.2 Parameter Model Considerations

As is generally the case with hierarchical models, specific choices for parameter distributions and alternative parameterizations for variance and covariance parameters are often chosen to facilitate computation. It is usually the case that the parameters are considered to be independent, $[\boldsymbol{\beta}, \sigma_\epsilon^2, \boldsymbol{\theta}] = [\boldsymbol{\beta}][\sigma_\epsilon^2][\boldsymbol{\theta}]$. Clearly, other choices are available and one should verify that the posterior inference is not overly sensitive to the choice of the parameter distributions. Such sensitivity analyses are critical when conducting Bayesian hierarchical analyses.

The spatial covariance function that comprises $\boldsymbol{\Sigma}(\boldsymbol{\theta})$ can be chosen from any of the valid spatial covariance functions that have been developed for geostatistical modeling. However, for hierarchical models, the spatial process is often considered to be stationary, homogeneous, and isotropic, typically following one of the well-known classes of spatial covariance functions, such as the Matérn or power exponential class. Thus, we might write $\boldsymbol{\Sigma}(\boldsymbol{\theta}) = \sigma_\eta^2 \mathbf{R}(\phi, \kappa)$, where $\mathbf{R}(\phi, \kappa)$ is the spatial correlation matrix, σ_η^2 is the variance, and ϕ and κ are spatial range and smoothness parameters, respectively. The choice of priors for these parameters is again up to the modeler. In some cases, the smoothness parameter is assumed to be known (e.g., the exponential correlation model). Inverse gamma or uniform priors for the variance and informative gamma priors for the spatial correlation parameters are sometimes used. In addition, for computational simplicity, discrete (uniform) priors are often specified for the spatial correlation parameters. This simplicity arises since it is always possible to generate random samples from the associated discrete distribution in MCMC analyses (e.g., see Robert and Casella, 2004, Chap. 2). In practice, one may have to adjust the discretization interval in the prior to a fine enough resolution to allow the Markov chain to move. For more detail, see the aforementioned references as well as the examples in Sections 7.3.4 and 7.4.2 and the discussion in Section 7.5. In addition, note that the choice of noninformative priors for variance components can be difficult in some cases, as discussed in Browne and Draper (2006) and Gelman (2006).

Computationally, it is often helpful to reparameterize the marginal model variances. For example, Diggle and Ribeiro (2007) suggest the marginal model parameterization:

$$\mathbf{Y}|\boldsymbol{\beta}, \phi, \kappa, \sigma_\eta^2, \tau^2 \sim Gau(\mathbf{X}\boldsymbol{\beta}, \sigma_\eta^2[\mathbf{R}(\phi, \kappa) + \tau^2\mathbf{I}]), \tag{7.15}$$

where $\tau^2 \equiv \sigma_\epsilon^2/\sigma_\eta^2$ is the ratio of the nugget variation to the process variation and has the advantage of being scale free. Alternatively, Yan, Cowles, Wang, and Armstrong (2007) propose a different marginal reparameterization:

$$\mathbf{Y}|\boldsymbol{\beta}, \phi, \kappa, \sigma^2, \xi \sim Gau(\mathbf{X}\boldsymbol{\beta}, \sigma^2[(1-\xi)\mathbf{R}(\phi, \kappa) + \xi\mathbf{I}]), \tag{7.16}$$

where $\sigma^2 = \sigma_\eta^2 + \sigma_\epsilon^2$ and $\xi = \sigma_\epsilon^2/\sigma^2$. In this case, ξ has a nice interpretation as the fraction of the total variation in $\mathbf{Y}$ contributed by the nugget effect. The parameter ξ has bounded support, (0, 1), which facilitates some types of MCMC implementations (e.g., slice sampling).

7.3.3 Bayesian Computation

Bayesian estimation, prediction, and inference for hierarchical spatial models must proceed via Monte Carlo methods. The specific choice of Monte Carlo algorithm depends on the choice of parameter models. In general, MCMC algorithms can be used as described in Banerjee et al. (2004). In certain cases, direct Monte Carlo simulation can be used as described in Diggle and Ribeiro (2007). In problems with large numbers of data and/or prediction locations, careful attention must be given to computational considerations. It is often more efficient in such cases to consider various dimension reduction or decorrelation approaches for modeling the spatial process. Chapter 8 in this volume presents some such approaches.

Increasingly, specialized computational software is being made available to accommodate ever more complex hierarchical spatial models. The interested reader is referred to the aforementioned texts as a place to start.

7.3.4 Gaussian Geostatistical Example: Midwest U.S. Temperatures

For illustration, consider the monthly average maximum temperature observations shown in Figure 7.1 for an area of the U.S. central plains, centered on Iowa. These data are from 131 stations in the U.S. Historical Climate Network (USHCN) for the month of November, 1941.

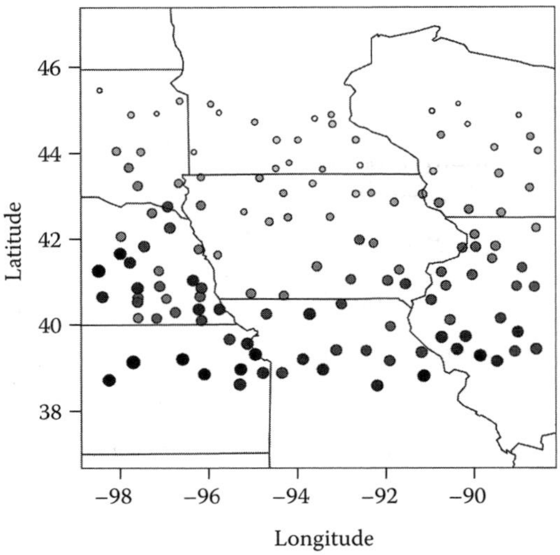

FIGURE 7.1
Monthly average maximum temperatures (in °F) for stations in the U.S. Historical Climate Network for November, 1941. Note that the size (bigger is warmer) and color (dark is warm, light is cool) of the circles are proportional to the temperature at each station with a maximum of 58.6°F and a minimum of 41.5°F.

In this example, say we are interested in predicting (interpolating) the true (without noise) monthly average maximum temperature at a grid of points over the region. Gridded datasets facilitate climatological analysis and are necessary when using observational data as input in numerical weather and climate models. In both cases, the uncertainty associated with the gridding is as important as the gridded field itself.

We assume the marginal form of the model shown in (7.15) with $\mathbf{X}$ having three columns corresponding to an intercept, trend in longitude, and trend in latitude, respectively. Thus, it is reasonable over such a relatively homogeneous and small geographical region to assume the spatial correlation model is stationary and isotropic. In particular, we assume an exponential correlation function here, with spatial dependence parameter, ϕ (i.e., $r(d, \phi) = \exp(-d/\phi)$ where d is the distance between two spatial locations). We select a relatively noninformative prior for $\boldsymbol{\beta}$, $\boldsymbol{\beta} \sim N(\mathbf{0}, \boldsymbol{\Sigma}_\beta)$, where $\boldsymbol{\Sigma}_\beta$ is diagonal, with each variance equal to 1000. We also considered a noninformative (uniform) prior on the β parameters and the posteriors were not sensitive to this choice. We chose a scaled-inverse chi-square prior for σ_η^2 with prior mean 2.7 and degrees of freedom = 2, corresponding to a fairly vague prior. The choice of the prior mean was based on a preliminary regression analysis in which the spatial trend, with no spatially correlated error term, accounted for about 83% of the variation in the temperatures. Because the overall variance in the temperature data is about 16 deg^2, we assumed there was about 2.7 deg^2 available to be explained by the spatial error process. It should be noted that we did consider several other values for the prior mean, as well as the nonimformative prior $p(\sigma_\eta^2) \propto 1/\sigma_\eta^2$, and the posteriors were not very sensitive to these choices. The prior for ϕ is assumed to be discrete uniform from 0.1 to 4.0 with increments of 0.1. This choice was based on the fact that the practical range (the smallest distance between points that gives negligible spatial correlation) for the exponential model is defined to be 3ϕ (see Schabenberger and Pierce, 2002, p. 583). Based on a practical range varying between 0.3 (deg) and 12 (deg), which covers the extent of our prediction domain, this corresponds to choices of ϕ between 0.1 and 4.0. Similarly, the prior for τ^2 is given by a discrete uniform distribution from 0.05 to 2.0 with increments of 0.05. This allows the spatial variation to be between 20 times larger to 2 times smaller than the nugget. With temperature data on this spatial scale, it would be scientifically implausible that the nugget variance would be much bigger than the spatial variation. The posterior distribution is not sensitive to this upper bound. Although for large geographical areas, one needs to factor in the curvature of the Earth when calculating distances and choosing valid covariance functions, the domain considered here is small enough that such considerations are not needed. Estimation is carried out using Bayesian direct simulation with the **geoR** package, as described in Diggle and Ribeiro (2007).

Samples from the posterior distributions for the model parameters are summarized by the histograms in Figure 7.2. In addition, the prior distributions over this portion of the parameter space are shown for comparison. The first thing we notice from these histograms is that there is definitely Bayesian learning in these marginal posterior distributions relative to their respective priors. Not surprisingly, given that temperatures over this region generally decrease with increasing latitude in November, the lattitude effect (β_2) in the large scale spatial mean is much more "significant" than the longitude effect (β_1). The latent spatial process is fairly important given a strong spatial dependence (the posterior median (mean) of ϕ is 2.0 (2.2)) and the posterior median (mean) of the process variance, σ_η^2 is 3.0 (3.2). Note, there is evidence of small-scale variability in τ^2 given it has a posterior median (mean) of 0.25 (0.31), suggesting median (mean) nugget variance of 0.75 (1.0). Although the USHCN stations are often supported by volunteer observers, which might lead to measurement errors, it is typically thought that such errors are pretty minimal when considered over a monthly average, further supporting the interpretation that σ_ϵ^2 is reflecting small-scale spatial variability.

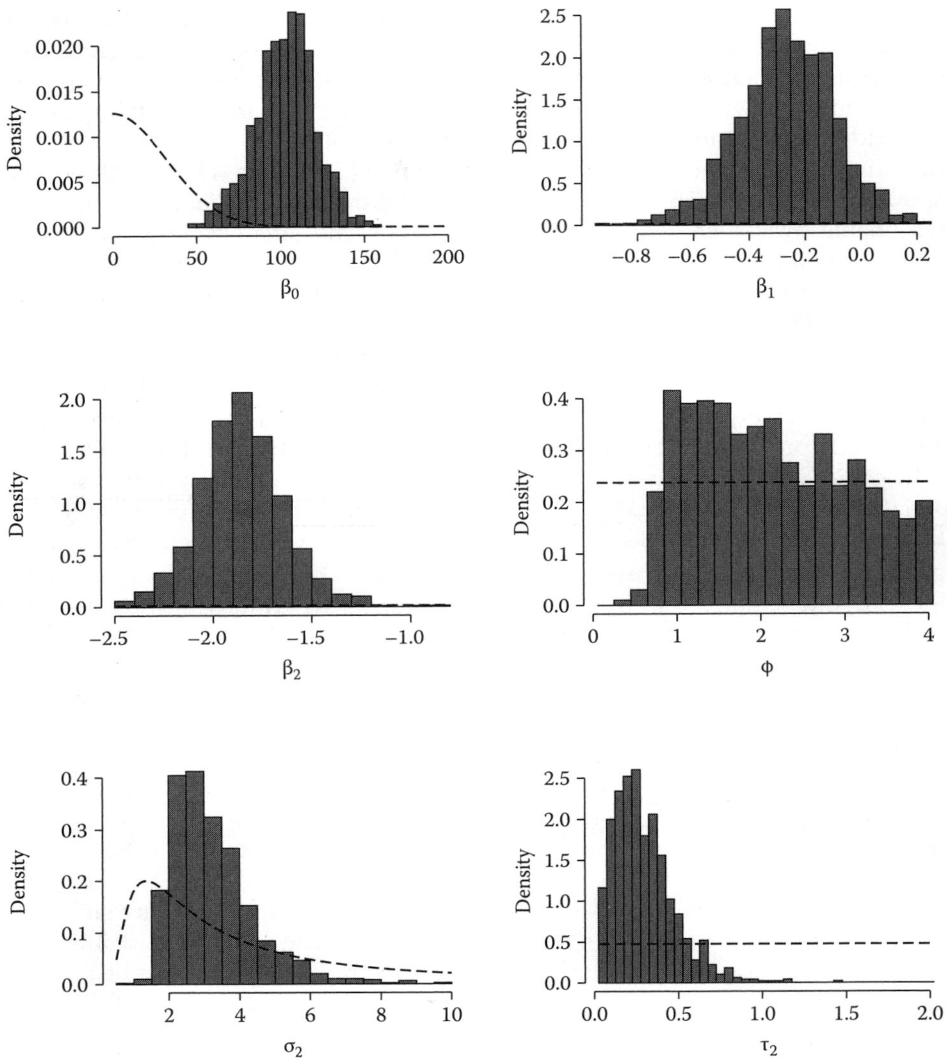

FIGURE 7.2
Posterior histograms of parameters for the USHCN temperature example. Upper left: β_0 (intercept), upper right: β_1 (Longitude), middle left: β_2 (latitude), middle right: ϕ (spatial dependence), lower left: σ_η^2 (process variance), lower right: τ^2 (ratio of nugget to process variance). In each figure, the dashed line represents the prior distribution over the shown portion of the parameter space.

Finally, a plot of the posterior means and variance of the predicted process on a grid of points is shown in Figure 7.3. Note that the prediction location corresponds to the center of each grid box, with the image plot provided to help with visualization only; we are not predicting on a different support than our data. The north-to-south spatial trend is clearly evident in the predictions, yet there is also evidence of spatial dependence in addition to the trend. The posterior predictive variances show the expected pattern of much smaller predictive variance near observation locations and much larger variances in areas outside the data locations (the edges of the map).

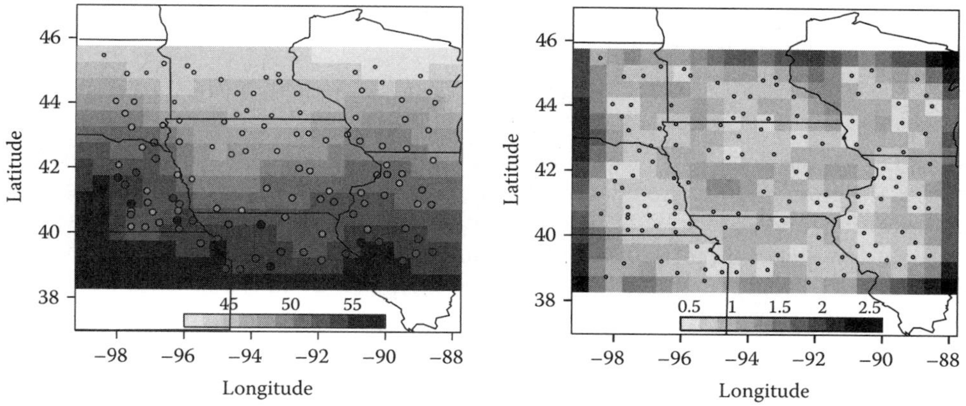

FIGURE 7.3
Posterior prediction means and variances for the USHCN temperature example. The left plot shows the mean from the posterior predictive distribution of temperature on a grid, with the circles showing the data locations. The right plot shows the posterior predictive variance for the prediction grid locations.

7.4 Hierarchical Generalized Linear Geostatistical Models

The advantage of the hierarchical approach in the context of the basic Gaussian geostatistical model presented in Section 7.3 is primarily that it accounts for the uncertainty in parameter estimation, unlike the classical geostatistical approach. For a nice illustration of this with conventional and Bayesian kriging, see the example in Diggle and Ribeiro (2007, Chap. 7.4). This example clearly shows that the Bayesian prediction variance is typically (but not always) larger than the corresponding plug-in predictive variance. Although this is important, the real strength and power of the hierarchical approach becomes evident when one considers non-Gaussian data models. There are many real-world processes in which the data are clearly something other than Gaussian, yet exhibit spatial dependence. For example, count data are common in many biological, ecological, and environmental problems. In such datasets, there are usually discernible mean-variance relationships, and the data are clearly discrete and positive valued. It is natural to consider Poisson and binomial models in these situations, and ignoring this structure and blindly applying the Gaussian geostatistical model will lead to some inefficiency in both estimation and prediction. Just as the Gaussian geostatistical model presented above can be thought of as a special case of a linear mixed model, it is natural that non-Gaussian spatial data problems may be formally analyzed within the context of generalized linear mixed models (GLMMs, e.g., see McCulloch and Searle (2001) for an overview) as proposed by Diggle, Tawn, and Moyeed (1998). This approach is described below from the hierarchical perspective. For more details, see Diggle and Ribeiro (2007), Banerjee et al. (2004), and the references therein.

Formulation of a GLMM requires specification of the likelihood of the random variable $Y(\mathbf{s})$ associated with the spatial data. One assumes that this has a data model that is a member of the exponential family. As in classical generalized linear models, there is a canonical parameter corresponding to this distribution that is nominally a function $g(\;)$ (the *link function*) of the location parameter for the distribution. To incorporate a spatial process, $\boldsymbol{\eta}$ (as above), we assume $[Y(\mathbf{s}_i)|\boldsymbol{\eta}, \boldsymbol{\beta}]$, $i = 1, \ldots, m$, is conditionally independent for any location $\mathbf{s}_i$ in the domain of interest, with conditional mean $E(Y(\mathbf{s}_i)|\boldsymbol{\eta}, \boldsymbol{\beta}) = \mu(\mathbf{s}_i)$, where this mean $\mu(\mathbf{s}_i)$ is a linear function of spatial covariates and a latent spatial random process. Specifically, the spatially correlated latent process is incorporated into the linear predictor through the link function:

$$g(\boldsymbol{\mu}) = \mathbf{X}\boldsymbol{\beta} + \mathbf{H}\boldsymbol{\eta} + \boldsymbol{\varepsilon} \tag{7.17}$$

where, $\boldsymbol{\mu} = (\mu(\mathbf{s}_1), \ldots, \mu(\mathbf{s}_m))'$, and, as with the Gaussian geostatistical model, $\mathbf{X}\boldsymbol{\beta}$ accounts for the spatial covariate effects, $\mathbf{H}\boldsymbol{\eta}$ accounts for the latent spatial process effects, and the additional noise term $\boldsymbol{\varepsilon}$, accounting for site-specific random variation, may or may not be included, depending on the application. For example, this term may be needed to account for the over-dispersion due to small-scale spatial effects. The addition of such a term is somewhat controversial, with advocates and opponents. Although not directly interpretable as a measurement error, the small-scale spatial process interpretation is valid in many circumstances. However, such an effect may not be identifiable without appropriate replication in the observations and/or prior knowledge. As before, the process stage of the hierarchical model requires specifications, such as $\boldsymbol{\eta} \sim \text{Gau}(\mathbf{0}, \Sigma(\boldsymbol{\theta}))$ and $\boldsymbol{\varepsilon} \sim \text{Gau}(\mathbf{0}, \sigma_\epsilon^2\mathbf{I})$, with spatial covariance matrix parameterized by $\boldsymbol{\theta}$. To complete the hierarchy, distributions must be specified for the parameters, $[\boldsymbol{\beta}, \sigma_\epsilon^2, \boldsymbol{\theta}]$. Note, we may also need to specify distributions for a dispersion parameter, say γ, associated with the exponential family data model. Specific choices for parameter distributions will depend on the context and computational considerations, as was the case for the Gaussian data model. Thus, the primary difference between this class of GLMM spatial models and the hierarchical Gaussian geostatistical model presented in Section 7.3 is the data stage. It is critical in this case that the data observations are conditionally independent, given the Gaussian latent spatial process and the covariates.

7.4.1 Computational Considerations

Based on the hierarchical model described above, we are interested in the posterior distribution:

$$[\boldsymbol{\eta}, \boldsymbol{\beta}, \sigma_\epsilon^2, \boldsymbol{\theta}, \gamma|\mathbf{Y}] \propto [\mathbf{Y}|\boldsymbol{\eta}, \boldsymbol{\beta}, \sigma_\epsilon^2, \gamma][\boldsymbol{\eta}|\boldsymbol{\theta}][\boldsymbol{\beta}, \sigma_\epsilon^2, \boldsymbol{\theta}, \gamma]. \tag{7.18}$$

As with the hierarchical Gaussian data geostatistical model, one cannot find the normalizing constant for Equation (7.18) analytically and MCMC methods must be used. However, unlike the Gaussian data case, we are unable to marginalize out the the spatial latent process analytically to facilitate computation. Thus, MCMC sampling procedures must include updates of $\boldsymbol{\eta}$ (or its components) directly.

Careful attention must be focused on specific computational approaches for MCMC sampling of hierarchical generalized linear spatial models. One major concern is the efficient update of the spatial latent process $\boldsymbol{\eta}$, which usually is accomplished by a Metropolis–Hastings update within the Gibbs sampler. The primary concern in this case is the efficiency of the algorithm when sampling dependent processes. A potentially useful approach in this case is the Langevin update scheme as discussed in the spatial context by Diggle and Ribeiro, (2007). Alternatively, when one keeps the small-scale process term $\boldsymbol{\varepsilon}$ in the model, updates of the $\boldsymbol{\eta}$ process can be accomplished through conjugate multivariate normal Gibbs updates (e.g., see Wikle, 2002; Royle and Wikle, 2005), which can be substantially more efficient than Metropolis updates. Of course, in this case, one must estimate the parameters associated with the $\boldsymbol{\varepsilon}$ error process. This can be problematic without the benefit of repeat observations, and/or substantial prior knowledge. In some cases, it is reasonable to treat these parameters as "nuisance parameters" (e.g., assume they are known or have very tight prior distributions) simply to facilitate computation and prediction of $\boldsymbol{\eta}$.

A second concern is the update of the latent spatial process covariance parameters, $\boldsymbol{\theta}$. Reparameterizations are often necessary as described, for example, by Christensen, Roberts and Sköld (2006), Palacios and Steel (2006), and Zhang (2004). This is an active area of current research and the state of the art is evolving rapidly. Experience suggests that successful implementation of these models can be difficult and is certainly model and dataset specific.

7.4.2 Non-Gaussian Data Example: Mapping Bird Counts

As an illustrative example, consider data collected from the annual North American Breeding Bird Survey (BBS, see Robbins, Bystrak, and Geissler 1986). In this survey, conducted in May and June of each year, volunteer observers traverse roadside sampling routes that are 39.2 km in length, each containing 50 stops. At each stop, the observer records the number of birds (by species) seen and heard over a three-minute time span. There are several thousand BBS routes in North America. We focus on mourning dove (*Zenaida macroura*) counts from 45 routes in the state of Missouri as observed in 2007 and shown in Figure 7.4. Given the total count (aggregated over the 50 stops) of doves on the BBS route assigned to a spatial location at the route centroid, our goal is to produce a map of dove relative abundance within the state, as well as characterize the uncertainty associated with these predictions of relative abundance. Such maps are used to study bird/habitat relationships as well as species range.

Following the hierarchical generalized linear spatial modeling framework described previously, we first specify a Poisson data model for dove relative abundance (count) at centroid location $\mathbf{s}_i$, $Y(\mathbf{s}_i)$, conditional on the Poisson mean $\lambda(\mathbf{s}_i)$:

$$Y(\mathbf{s}_i)|\lambda(\mathbf{s}_i) \sim ind.\ \text{Poisson}(\lambda(\mathbf{s}_i)),\ i = 1, \ldots, m. \tag{7.19}$$

Note, we assume that conditional upon the spatially dependent mean process (λ), the BBS counts are independent. Although it is technically possible that this independence assumption could be violated if the same bird was counted at multiple routes, it is not likely given the nature of the BBS sampling protocol (e.g., Robbins et al., 1986). Given that we expect dependence in the counts due to the spatially dependent nature of the underlying

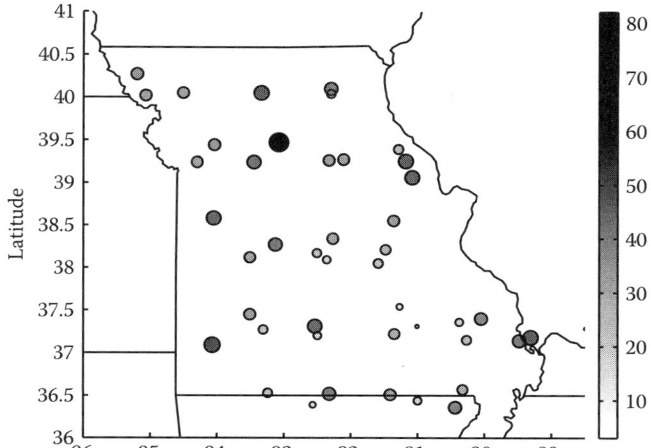

FIGURE 7.4
BBS total mourning dove counts at the centroid of sampling routes in Missouri from 2007. The color and size of the circle are proportional to the observed counts at each station.

habitat preferences of the birds, we employ a Gaussian spatial process model to describe spatial variation in $\lambda(\mathbf{s}_i)$ and use the canonical log-link function:

$$\log(\lambda(\mathbf{s}_i)) = \beta + \eta(\mathbf{s}_i), \tag{7.20}$$

where β is the mean, assumed to be constant over space, which may be generalized to accommodate interesting covariates, and $\eta(\mathbf{s}_i)$ is the latent spatial process. Specifically, $\eta(\mathbf{s}_i)$ is Gaussian with mean 0, variance σ_η^2, and correlation function $r(\mathbf{s}_i, \mathbf{s}_j; \phi)$. As usual, one may consider many different correlation models. Given we have a relatively small geographical area of concern, we specify $r_\eta(\mathbf{s}_i, \mathbf{s}_j; \phi)$ to be the single parameter isotropic exponential model: $r_\eta(\mathbf{s}_i, \mathbf{s}_j; \phi) = \exp(-||\mathbf{s}_i - \mathbf{s}_j||/\phi)$ where ϕ controls the rate of decay in correlation as distance between sites increases. Note, in this case, we do not include the additional small-scale variability process, ε. For an example of a model for BBS counts that does include such a term, see Wikle (2002) and Royle and Wikle (2005). The primary reason why they included the term was that they were predicting on a grid for which there were multiple BBS counts associated with specific grid locations, thus providing enough replication on small scales to estimate the parameters associated with ε. In addition, they required a very efficient estimation and prediction algorithm as they were considering prediction over the continental United States and could exploit the benefits of having the extra error term in the model as described previously.

Estimation of the parameters β, σ_η^2, and ϕ, prediction of $\eta(\mathbf{s})$ at a grid of locations, and prediction of $Y(\mathbf{s})$ at those locations is carried out within a Bayesian framework using MCMC techniques (e.g., see Diggle et al., 1998; Wikle, 2002; Banerjee et al., 2004). In particular, we used the **geoRglm** package, as described in Diggle and Ribeiro (2007). In this case, we assumed the program's default flat prior for β and default uniform prior for σ_η^2. For the spatial dependence parameter, we considered a discrete uniform prior between 0 and 1.7, with increments of 0.005. The upper bound of 1.7 was motivated by the practical range, which for the exponential model is 3ϕ, and we assumed that over our limited domain, a reasonable upper bound for this range would be around 5 deg, implying an upper bound for ϕ near 1.7. Histograms of samples from the posterior distribution and the associated priors for the model parameters are shown in Figure 7.5. Note that in each case there appears to be Bayesian learning.

A map showing the predicted median BBS relative abundance (number of birds) over a grid of points is shown in the left panel of Figure 7.6). In this case, the prediction location is actually a point specified by the center of the grid box. The grid box is filled in, in this case, for illustration purposes only, as we are not predicting on a different support than is (assumed to be) given by our data. Note the relatively large-scale spatial dependence that is evident in these predictions. In addition, we note that the lower predicted relative abundance in the southeast portion of the state corresponds to the higher elevations of the Missouri Ozark Mountains. Uncertainty is quantified by a robust method based on one fourth of the length of the 95% prediction interval. This is shown in the right panel of Figure 7.6. The dominant feature apparent in these maps is the strong relationship between the predicted relative abundance (bird count), and its uncertainty—a consequence of modeling within the Poisson framework. In addition, note that there are quite a few prediction locations outside of Missouri, and these are clearly extrapolation locations. Were we actually interested in these areas for the current analysis, we would use BBS counts from routes in the states bordering Missouri. Clearly, without such observations, we expect a great deal of uncertainty for the predictions at these locations, and this is verified by the predictive uncertainty map.

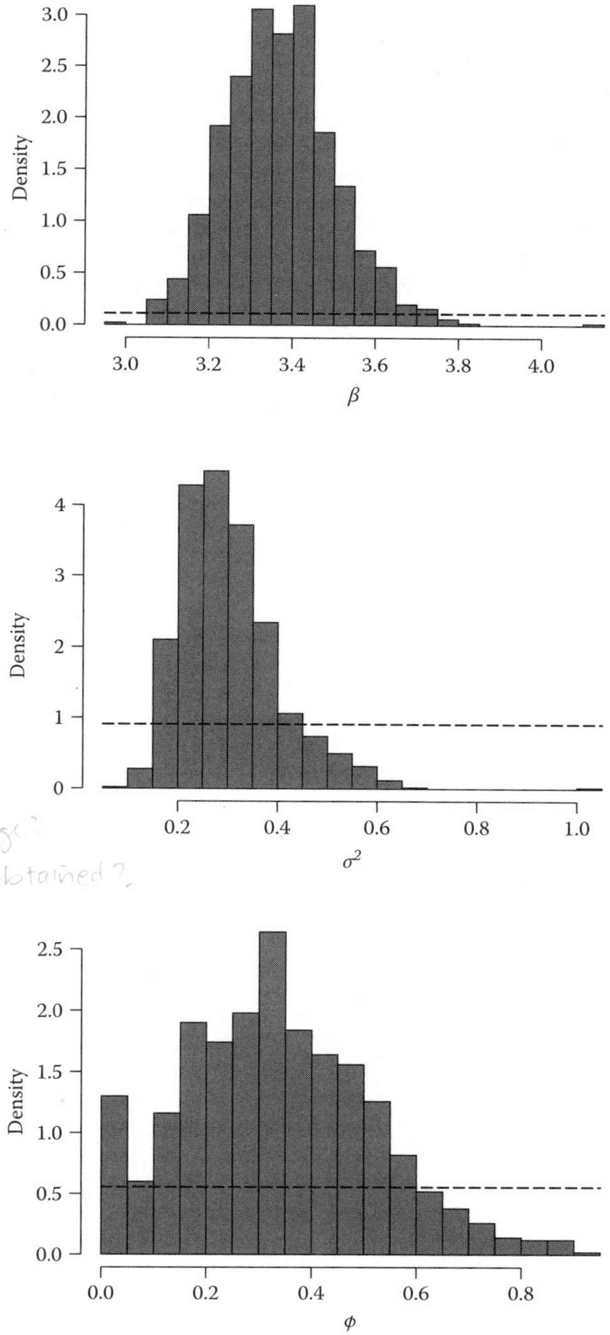

FIGURE 7.5
Posterior histograms of parameters for the Breeding Bird Survey mourning dove example. Top: β (intercept), middle: σ_η^2 (variance of random effect), and bottom: ϕ (spatial dependence). In each figure, the dashed line represents the prior distribution over the shown portion of the parameter space.

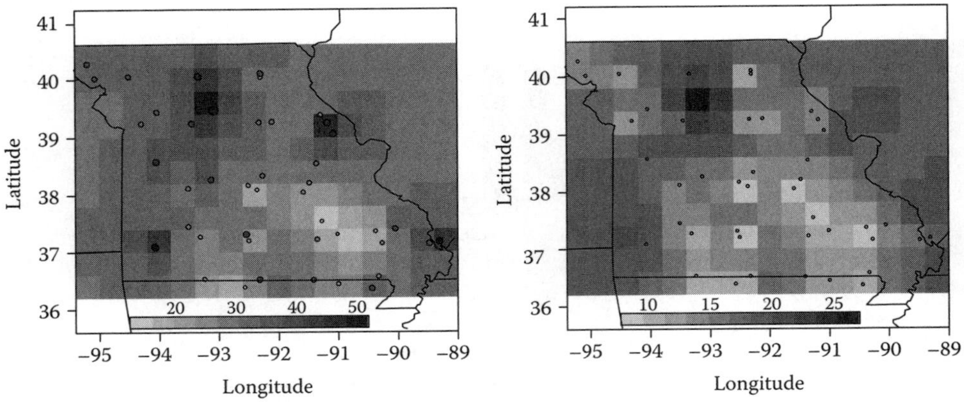

FIGURE 7.6
Posterior prediction medians and uncertainty for the BBS mourning dove example. The left plot shows the median from the posterior predictive distribution of relative abundance on a grid, with the circles showing the data locations. The right plot shows the associated prediction uncertainty at the prediction grid locations. "Uncertainty" is defined as the length of the 95% prediction interval divided by 4.

7.5 Discussion

This chapter outlines a general hierarchical framework that can facilitate modeling of complicated spatial processes. The key advantage of this framework is that it allows the decomposition of a complicated dataset into three primary components that are all linked by simple rules of probability. That is, one has a model for the data conditioned on the true processes and associated parameters, a model for the processes conditioned on different parameters, and finally, models for the parameters. This hierarchical partitioning can lead to relatively simpler models at each stage, yet when combined, describes a very complex joint data, process, and parameter distribution. In addition to accounting for uncertainty in each stage, this approach also facilitates the accommodation of multiple datasets, and multiple processes and complicated parameter structure as well as the inclusion of scientific information. Hierarchical models are typically framed in the Bayesian paradigm and estimation and prediction is typically based on Monte Carlo approaches.

In the context of spatial models, the hierarchical approach was demonstrated in a traditional Gaussian data, geostatistical setting, as well as a non-Gaussian data, generalized linear mixed spatial model framework. The primary differences in these hierarchical models is the data stage, as they both rely on the conditioning on a spatial random process to induce spatial dependence in the observations. In addition, there are differences corresponding to computational concerns. For example, estimation and prediction for the Gaussian data model can be facilitated by analytical marginalization, but the non-Gaussian data model does not allow such marginalization. For both types of models, computational concerns can suggest specific parameterizations of variance components and spatial dependence parameters.

Although several approaches for parameterization and prior selection were discussed briefly in Sections 7.3 and 7.4, these choices are not always so easy to make for hierarchical spatial models. For example, even in simple (nonspatial) hierarchical linear mixed models, the choice of prior distributions for variance parameters can be quite tricky (e.g., see Browne and Draper, 2006, and the discussion by Gelman, 2006, and others). Experience shows that this issue is compounded by the addition of a spatially correlated process. The bottom line is that one must be very careful when making these choices, whether one is using relatively common statistical software, or whether writing problem specific programs. Although there has been some work done searching for "objective" priors for spatial processes (e.g., Berger, De Oliveira, and Sanso, 2001), these are not always appropriate for subcomponents in a larger hierarchical framework.

Another school of thought suggests that one typically has *some* notion about whether there is spatial dependence or not (at least for processes for which there are observations), and thus it is reasonable in those cases to use more informative priors. Of course, one has to be careful that the priors selected (whether they are "informative" or not) have meaning relative to the process and data of interest. For example, simple simulation of the hierarchical model used for the mourning dove counts suggests that the simulated counts quickly become unrealistic when β (corresponding to the log of the mean) is much outside of the range of about 3 to 4. Similarly, the simulated counts are quite unrealistic when the process variance σ_η^2 is much outside the range of about 0.05 to 2. By contrast, the simulated counts are not nearly as sensitive to the spatial dependence parameter, ϕ. A natural question then is whether the so-called "vague" priors chosen for these parameters (recall, we chose the prior $\beta \sim N(0, 100)$) are reasonable when they assign mass to portions of the parameter space that could never give a meaningful model? The use of such a predictive outcome-driven, science-based approach to prior elicitation is not generally accepted by all Bayesian statisticians as optimal Bayesian procedure, yet it provides a good "reality check." For more interesting discussion about priors in Bayesian models (see Berger (2006) and Goldstein (2006) and the associated discussion).

Finally, the true power of hierarchical modeling is more apparent (and the prior elicitation often more critical) when one is considering very complicated processes. In such cases, the hierarchical spatial models discussed here might be relatively minor components of some larger hierarchical structure. For example, in a multivariate spatial-temporal modeling problem, it could be the case that the evolution parameters for a spatial process could be represented by a set of spatially correlated parameters, and these parameters might be related to spatial parameters for some other process, etc. The possibilities are endless, and the algorithmic development and prior selection issues are numerous. The ability to answer scientific questions, and to make predictions while accounting for uncertainty in such cases, is increasing the utility of advanced statistical modeling across a wide range of disciplines.

References

Banerjee, S., Carlin, B.P., and A.E. Gelfand (2004). *Hierarchical Modeling and Analysis for Spatial Data*. Boca Raton, FL: Chapman & Hall/CRC. 452 pp.

Berger, J. (2006). The case for objective Bayesian analysis. *Bayesian Analysis*, **1**, 385–402.

Berger, J., De Oliveira, V., and Sanso, B. (2001). Objective Bayesian analysis of spatially correlated data. *Journal of the American Statistical Association*, **96**, 1361–1374.

Berliner, L.M. (1996). Hierarchical Bayesian time series models. In *Maximum Entropy and Bayesian Methods*, K. Hanson and R. Silver, Eds., Kluwer Academic Publishers, Dordrecht, the Netherlands, 15–22.

Browne, W.J. and Draper, D. (2006). A comparison of Bayesian and likelihood-based methods for fitting multilevel models. *Bayesian Analysis*, **1**, 473–514.

Christensen, O.F., Roberts, G.O., and Sköld, M. (2006). Robust Markov chain Monte Carlo methods for spatial generalized linear mixed models. *Journal of Computational and Graphical Statistics*, **15**, 1–17.

Diggle, P.J. and Ribeiro Jr., P.J. (2007). *Model-Based Geostatistics*. Springer Science+Business Media, New York, 228 pp.

Diggle, P.J., Tawn, J.A., and Moyeed, R.A. (1998). Model-based geostatistics (with discussion). *Applied Statistics*, **47**, 299–350.

Gelfand, A.E. and Smith, A.F.M. (1990). Sampling based approaches to calculating marginal densities. *Journal of the American Statistical Association*, **85**, 398–409.

Gelfand, A.E., Sahu, S.K., and Carlin, B.P. (1995). Efficient parameterization for normal linear mixed models. *Biometrika*, **82**, 479–488.

Gelfand, A.E., Zhu, L., and Carlin, B.P. (2001). On the change of support problem for spatial-temporal data. *Biostatistics*, **2**, 31–45.

Gelman, A. (2006). Prior distributions for variance parameters in hierarchical models. *Bayesian Analysis*, **1**, 515–533.

Gelman, A., Carlin, J.B., Stern, H.S., and Rubin, D.B. (2004). *Bayesian Data Analysis*, 2nd ed. Chapman & Hall/CRC. Boca Raton, FL. 668 pp.

Goldstein, M. (2006). Subjective Bayesian analysis: Principles and practice. *Bayesian Analysis*, **1**, 403–420.

McCulloch, C.E. and Searle, S.R. (2001) *Generalized, Linear and Mixed Models.* John Wiley & Sons. New York. 325 pp.

Palacios, M.B. and Steel, M.F.J. (2006). Non-Gaussian Bayesian geostatistical modeling. *Journal of the American Statistical Association*, **101**, 604–618.

Robbins, C.S., Bystrak, D.A., and Geissler, P.H. (1986). *The Breeding Bird Survey: Its First Fifteen Years, 1965–1979.* USDOI, Fish and Wildlife Service Resource Publication 157. Washington, D.C.

Robert, C.P. and Casella, G. (2004). *Monte Carlo Statistical Methods*, 2nd ed. Springer Science+Business Media, New York, 645 pp.

Royle, J.A. and Berliner, L.M. (1999). A hierarchical approach to multivariate spatial modeling and prediction. *Journal of Agricultural, Biological, and Environmental Statistics*, **4**, 29–56.

Royle, J.A. and Wikle, C.K. (2005). Efficient statistical mapping of avian count data. *Environmental and Ecological Statistics*, **12**, 225–243.

Schabenberger, O. and Pierce, F.J. (2002). *Contemporary Statistical Models for the Plant and Soil Sciences.* Taylor & Francis. Boca Raton, FL.

Verbeke, G. and Molenberghs, G. (2000). *Linear Mixed Models for Longitudinal Data*. Springer-Verlag. New York. 568 pp.

Wikle, C.K. (2002). Spatial modeling of count data: A case study in modeling breeding bird survey data on large spatial domains. In *Spatial Cluster Modeling.* A. Lawson and D. Denison, eds. Chapman & Hall/CRC, Boca Raton, FL. 199–209.

Wikle, C.K. (2003). Hierarchical models in environmental science. *International Statistical Review*, **71**, 181–199.

Wikle, C.K. and Berliner, L.M. (2005). Combining information across spatial scales. *Technometrics*, **47**, 80–91.

Wikle, C.K., Berliner, L.M., and Cressie, N. (1998). Hierarchical Bayesian space-time models. *Journal of Environmental and Ecological Statistics*, **5**, 117–154.

Wikle, C.K., Milliff, R.F., Nychka, D., and Berliner, L.M. (2001). Spatiotemporal hierarchical Bayesian modeling: Tropical ocean surface winds. *Journal of the American Statistical Association*, **96**, 382–397.

Yan, J., Cowles, M., Wang, S., and Armstrong, M. (2007). Parallelizing MCMC for Bayesian spatial-temporal geostatistical models. *Statistics and Computing*, **17**, 323–335.

Zhang, H.Z. (2004). Inconsistent estimation and asymptotically equal interpolations in model-based geostatistics. *Journal of the American Statistical Association*, **99**, 250–261.

8

Low-Rank Representations for Spatial Processes

Christopher K. Wikle

CONTENTS

As with many disciplines, spatial statistical applications have transitioned from a "data poor" setting to a "data rich" setting in recent years. For example, when the primary interest is in predicting ore reserves, one might only have a limited number of observations available, due to the expense of ore extraction. In such cases, the standard geostatistical prediction formulas are easy to implement with basic computer programs, requiring little more than a few low-dimensional matrix (i.e., multiplication and inverse) manipulations. As more data have become available through increases in the number and extent of automated observation platforms (e.g., global weather station networks, remote sensing satellites, ground penetrating radar, lidar, medical imagery, etc.), the issues related to practical implementation of spatial prediction algorithms become significant. These issues are present regardless of whether one takes a Bayesian or frequentist perspective when performing inference and prediction.

In order to proceed with spatial statistical prediction in the presence of "very large" or "massive" spatial datasets, one must reduce the dimensionality. A key to this, in the context of spatial prediction, is to represent the process in terms of a lower-dimensional latent Gaussian process, or to carefully choose the structure/representation (e.g., spectral forms) so as to gain efficiency in computation. This chapter is focused on the reduced-dimensional latent process approach, but as will be discussed, this approach is not unrelated to the restructuring/spectral representation approach as well. In this chapter we will show how many of the methodologies commonly used in practice (e.g., discrete kernel convolutions, orthogonal polynomials, empirical orthogonal functions, splines, wavelets) all fit into this general construct.

The first section briefly reviews the traditional geostatistical prediction equations. This is followed by a presentation of an alternative parameterization in which the spatial process is modeled hierarchically by a low-dimensional latent random process. The computational advantages of this so-called "reduced rank" representation are then described for both Markov chain Monte Carlo (MCMC) implementations as well as traditional kriging. The choice of the expansion matrix that maps this latent process to the spatial process of interest is then discussed. This is followed by a brief discussion of extensions to this low rank representation to non-Gaussian data models and spatiotemporal processes.

8.1 Full-Rank Geostatistical Setup

Assume we are interested in a Gaussian spatial process $\{\eta(\mathbf{s})\}$ for any spatial location $\mathbf{s}$ in some specified spatial domain. In this case, $E[\eta(\mathbf{s})] = \mu(\mathbf{s})$ and the process is assumed to be dependent across space, with covariance function $c_\eta(\mathbf{s}, \mathbf{r}) = \text{Cov}[\eta(\mathbf{s}), \eta(\mathbf{r})]$. Assume also that we have observations of this process at n spatial locations. Thus, we can write a simple data model:

$$\mathbf{Y} = \boldsymbol{\eta} + \boldsymbol{\epsilon}, \quad \boldsymbol{\epsilon} \sim Gau(\mathbf{0}, \boldsymbol{\Sigma}_\epsilon), \tag{8.1}$$

where $\mathbf{Y} = (Y(\mathbf{s}_1), \ldots, Y(\mathbf{s}_n))'$ represents a data vector, $\boldsymbol{\eta} = (\eta(\mathbf{s}_1), \ldots, \eta(\mathbf{s}_n))'$ corresponds to the true (latent) spatial process vector of interest, and $\boldsymbol{\epsilon} = (\epsilon(\mathbf{s}_1), \ldots, \epsilon(\mathbf{s}_n))'$ is a Gaussian error vector, assumed to have zero mean and covariance matrix, $\boldsymbol{\Sigma}_\epsilon$. This error vector corresponds to measurement error and/or representativeness error associated with the true process $\eta(\mathbf{s})$, such as occurs when the observations include information at smaller spatial scales than accommodated by the process $\eta(\mathbf{s})$ (i.e., a nugget effect). We also note that the distribution of the underlying latent spatial process at these measurement locations is given by

$$\boldsymbol{\eta} \sim Gau(\boldsymbol{\mu}, \boldsymbol{\Sigma}_\eta), \tag{8.2}$$

where $\boldsymbol{\mu} = (\mu(\mathbf{s}_1), \ldots, \mu(\mathbf{s}_n))'$ represents the spatial mean vector and there is inherent residual spatial dependence as represented by nontrivial structure in $\boldsymbol{\Sigma}_\eta$. That is, the (i, j)th element of $\boldsymbol{\Sigma}_\eta$ is given by $c_\eta(\mathbf{s}_i, \mathbf{s}_j)$. The spatial mean could be parameterized in terms of known covariates, such as $\boldsymbol{\mu} = \mathbf{X}\boldsymbol{\beta}$. However, we will assume $\mu(\mathbf{s}) = 0$ for all $\mathbf{s}$ in the remainder of this chapter, just to simplify notation; analogous expressions that involve implicitly estimating $\boldsymbol{\beta}$ can be similarly derived without this assumption. As discussed above, we assume that underlying Equation (8.2) is a Gaussian process with a given covariance structure. Thus, we can predict the true process at any locations in our domain of interest, say $\{\mathbf{r}_1, \ldots, \mathbf{r}_m\}$. We will denote the process at these prediction locations by $\boldsymbol{\eta}_0 = (\eta(\mathbf{r}_1), \ldots, \eta(\mathbf{r}_m))'$.

Ultimately, we are interested in the predictive distribution, $[\boldsymbol{\eta}_0|\mathbf{Y}]$, given by

$$[\boldsymbol{\eta}_0|\mathbf{Y}] = \int [\boldsymbol{\eta}_0|\boldsymbol{\eta}][\boldsymbol{\eta}|\mathbf{Y}]d\boldsymbol{\eta}, \tag{8.3}$$

where the bracket notation "[ ]" refers generically to a distribution, either joint, conditional, or marginal. Fortunately, given the observations $\mathbf{y}$, and the Gaussian data and the process models (8.1) and (8.2), respectively, (8.3) can be found analytically:

$$\boldsymbol{\eta}_0|\mathbf{y} \sim Gau(\boldsymbol{\mu}_{\eta_0|y}, \boldsymbol{\Sigma}_{\eta_0|y}), \tag{8.4}$$

where

$$\boldsymbol{\mu}_{\eta_0|y} = \boldsymbol{\Sigma}_{\eta_0,\eta}(\boldsymbol{\Sigma}_\eta + \boldsymbol{\Sigma}_\epsilon)^{-1}\mathbf{y}, \tag{8.5}$$

and

$$\Sigma_{\eta_0|y} = \Sigma_{\eta_0} - \Sigma_{\eta_0,\eta}(\Sigma_\eta + \Sigma_\epsilon)^{-1}\Sigma'_{\eta_0,\eta'} \tag{8.6}$$

with $\Sigma_{\eta_0,\eta} \equiv \text{Cov}(\eta_0, \eta)$ and $\Sigma_{\eta_0} \equiv \text{Cov}(\eta_0, \eta_0)$.

The key challenge in implementing these equations when n is large lies in the evaluation of the inverses in Equation (8.5) and Equation (8.6). Typically, Σ_ϵ is assumed to have a simple homogeneous error structure (e.g., $\Sigma_\epsilon = \sigma_\epsilon^2 \mathbf{I}$). However, without strong assumptions about the form of Σ_η, the matrix inversion entails on the order of n^3 operations. Not only is the inverse prohibitively expensive to calculate, but one may have difficulty even storing the matrix Σ_η itself for very large n.

Note that this discussion assumes the covariance matrices are "known." Of course, this is not the case in reality, but the critical computational issues can be illuminated even by the case when they are known. In various places in the following exposition, estimation issues are discussed when relevant.

8.2 Reduced-Rank Random Effects Parameterizations

Consider an alternative *hierarchical* representation of the spatial model:

$$Y = \eta + \epsilon, \quad \epsilon \sim Gau(\mathbf{0}, \Sigma_\epsilon), \tag{8.7}$$

$$\eta = \mathbf{H}\alpha + \xi, \quad \xi \sim Gau(\mathbf{0}, \Sigma_\xi), \tag{8.8}$$

and

$$\alpha \sim Gau(\mathbf{0}, \Sigma_\alpha), \tag{8.9}$$

where $\eta, Y, \epsilon, \Sigma_\epsilon$ are defined as given above. Furthermore, define $\alpha \equiv (\alpha_1, \ldots, \alpha_p)'$ to be a p-dimensional random (effects) vector, such that $p \ll n$ and $\mathbf{H}$ is then an $n \times p$ expansion matrix that maps the low-dimensional latent process, α, to the true spatial process of interest, η. The residual error term, ξ, is assumed to be Gaussian and accounts for differences between η and its low-dimensional representation, $\mathbf{H}\alpha$. This process, ξ, is also assumed to have zero mean and covariance matrix, Σ_ξ.

It is important to note that we can also express a model for the latent spatial process at unobserved locations given α:

$$\eta_0 = \mathbf{H}_0\alpha + \xi_0, \tag{8.10}$$

where $\mathbf{H}_0$ is an $m \times p$ expansion matrix with elements $h_i(j), i = 1, \ldots, m; j = 1, \ldots, p$, such that

$$\eta_0(\boldsymbol{r}_i) = \sum_{j=1}^{p} h_i(j)\alpha_j + \xi_0(\boldsymbol{r}_i), \tag{8.11}$$

and the residual vector is defined by $\xi_0 \equiv (\xi_0(r_1), \ldots, \xi_0(r_m))'$.

As before, we are interested in $[\eta_0|Y]$. This predictive distribution can be obtained by first integrating out α to get the marginal distribution of the spatial vector, η: $\eta \sim Gau(\mathbf{0}, \mathbf{H}\Sigma_\alpha\mathbf{H}' + \Sigma_\xi)$. Then, $[\eta_0|Y] = \int[\eta_0|\eta][\eta|Y]d\eta$, which, given the above model assumptions, is

$$\eta_0|Y \sim Gau(\tilde{\mu}, \tilde{\Sigma}), \tag{8.12}$$

where

$$\tilde{\mu} = (\mathbf{H}_0\Sigma_\alpha\mathbf{H}' + \Sigma_{\xi_0,\xi})(\mathbf{H}\Sigma_\alpha\mathbf{H}' + \Sigma_\xi + \Sigma_\epsilon)^{-1}y, \tag{8.13}$$

and

$$\tilde{\Sigma} = (\mathbf{H}_0\Sigma_\alpha\mathbf{H}_0' + \Sigma_{\xi_0}) - (\mathbf{H}_0\Sigma_\alpha\mathbf{H}' + \Sigma_{\xi_0,\xi})(\mathbf{H}\Sigma_\alpha\mathbf{H}' + \Sigma_\xi + \Sigma_\epsilon)^{-1}(\mathbf{H}\Sigma_\alpha\mathbf{H}_0' + \Sigma_{\xi,\xi_0}), \tag{8.14}$$

with $\boldsymbol{\Sigma}_{\xi,\xi_0} \equiv \text{Cov}(\boldsymbol{\xi}, \boldsymbol{\xi}_0)$, and $\boldsymbol{\Sigma}_{\xi_0} \equiv \text{Cov}(\boldsymbol{\xi}_0, \boldsymbol{\xi}_0)$. A typical assumption is that $\boldsymbol{\Sigma}_\xi$ and $\boldsymbol{\Sigma}_{\xi_0}$ are diagonal with $\boldsymbol{\Sigma}_{\xi,\xi_0} = \mathbf{0}$ (i.e., the ξ are independent), and $\boldsymbol{\Sigma}_\epsilon = \sigma_\epsilon^2 \mathbf{I}$. Under these assumptions, and letting $\boldsymbol{V} \equiv \boldsymbol{\Sigma}_\xi + \sigma_\epsilon^2 \mathbf{I}$, we get

$$\tilde{\boldsymbol{\mu}} = (\mathbf{H}_0 \boldsymbol{\Sigma}_\alpha \mathbf{H}')(\mathbf{H} \boldsymbol{\Sigma}_\alpha \mathbf{H}' + \boldsymbol{V})^{-1} \boldsymbol{y}, \tag{8.15}$$

and

$$\tilde{\boldsymbol{\Sigma}} = (\mathbf{H}_0 \boldsymbol{\Sigma}_\alpha \mathbf{H}_0' + \boldsymbol{\Sigma}_{\xi_0}) - (\mathbf{H}_0 \boldsymbol{\Sigma}_\alpha \mathbf{H}')(\mathbf{H} \boldsymbol{\Sigma}_\alpha \mathbf{H}' + \boldsymbol{V})^{-1} (\mathbf{H} \boldsymbol{\Sigma}_\alpha \mathbf{H}_0'). \tag{8.16}$$

At this point, it is not evident that the hierarchical, reduced-dimension, random effects representation has gained much for us, as we seem to require the inverse of a potentially high dimensional ($n \times n$) matrix to calculate the predictive mean vector and covariance matrix. However, as described in the next section, the hierarchical framework does actually give us a significant computational advantage from two different perspectives.

8.3 Computational Advantages of the Reduced-Rank Approach

There are two ways that one can take advantage of the hierarchical, reduced-rank, random effects parameterization. The first of these is based on a Monte Carlo implementation and the second takes advantage of a well-known matrix identity.

8.3.1 MCMC Approach

Since the reparameterized model is hierarchical, it is natural to consider an MCMC inferential approach. For purposes of this illustration, we will assume the parameters that make up the various covariance matrices are known. Of course, this would not be the case in practice, but estimation and inference associated with unknown parameters are straightforward, as shown for example, in Chapter 7.

Consider the hierarchical model:

$$\mathbf{Y} | \boldsymbol{\eta}, \sigma_\epsilon^2 \sim Gau(\boldsymbol{\eta}, \sigma_\epsilon^2 \mathbf{I}), \tag{8.17}$$

$$\boldsymbol{\eta} | \boldsymbol{\alpha}, \boldsymbol{\Sigma}_\xi \sim Gau(\mathbf{H}\boldsymbol{\alpha}, \boldsymbol{\Sigma}_\xi), \tag{8.18}$$

$$\boldsymbol{\alpha} \sim Gau(\mathbf{0}, \boldsymbol{\Sigma}_\alpha). \tag{8.19}$$

We are interested in the posterior distribution of $\boldsymbol{\eta}$, $\boldsymbol{\alpha}$, and $\boldsymbol{\eta}_0$ given the observed data $\boldsymbol{y}$. A Gibbs sampler will easily give samples from this posterior distribution. To implement this sampler, one requires the following full conditional distributions for $\boldsymbol{\eta}$ and $\boldsymbol{\alpha}$:

$$\boldsymbol{\eta} | \cdot \sim Gau\left(\left(\frac{1}{\sigma_\epsilon^2} \mathbf{I} + \boldsymbol{\Sigma}_\xi^{-1} \right)^{-1} (\boldsymbol{y}/\sigma_\epsilon^2 + \boldsymbol{\Sigma}_\xi^{-1} \mathbf{H}\boldsymbol{\alpha}), \left(\frac{1}{\sigma_\epsilon^2} \mathbf{I} + \boldsymbol{\Sigma}_\xi^{-1} \right)^{-1} \right) \tag{8.20}$$

$$\boldsymbol{\alpha} | \cdot \sim Gau((\mathbf{H}' \boldsymbol{\Sigma}_\xi^{-1} \mathbf{H} + \boldsymbol{\Sigma}_\alpha^{-1})^{-1} \mathbf{H}' \boldsymbol{\Sigma}_\xi^{-1} \boldsymbol{\eta}, (\mathbf{H}' \boldsymbol{\Sigma}_\xi^{-1} \mathbf{H} + \boldsymbol{\Sigma}_\alpha^{-1})^{-1}). \tag{8.21}$$

Using this Gibbs sampler to obtain samples $\boldsymbol{\alpha}^{(i)}$ from $[\boldsymbol{\alpha}|\boldsymbol{y}]$, we then can get the desired posterior predictive distribution for $\boldsymbol{\eta}_0$:

$$[\boldsymbol{\eta}_0 | \boldsymbol{y}] = \int [\boldsymbol{\eta}_0 | \boldsymbol{\alpha}][\boldsymbol{\alpha} | \boldsymbol{y}] d\boldsymbol{\alpha}, \tag{8.22}$$

by direct Monte Carlo integration, given that $\boldsymbol{\eta}_0 | \boldsymbol{\alpha} \sim Gau(\mathbf{H}_0 \boldsymbol{\alpha}, \boldsymbol{\Sigma}_{\xi_0})$.

If Σ_ξ is assumed to be diagonal, the matrix inverse operations in Equation (8.20) are trivial. Critically, the matrix inverses necessary to obtain the sample in Equation (8.21) are all of dimension $p \times p$. Since $p \ll n$, and the matrix inverse is an $O(p^3)$ operation, we have significantly reduced the computational demand by utilizing the hierarchical framework. Critical to this implementation is the presence of the component ξ and the additional assumption that its covariance structure, Σ_ξ, is "simple" (e.g., diagonal) and facilitates computation of the inverse. This assumption may, or may not, be realistic for a given application.

8.3.2 Fixed-Rank Kriging Approach

The key difficulty in using kriging Equation (8.15) and Equation (8.16) when there are many observation locations stems from the matrix inverse $(\mathbf{H}\Sigma_\alpha\mathbf{H}' + V)^{-1}$. Consider the well-known matrix inverse result (see, p. 29 in Rao, 1965) for nonsingular matrices $\mathbf{A}$ and $\mathbf{D}$ of size $a \times a$ and $d \times d$, respectively, and for an $a \times d$ matrix $\mathbf{B}$:

$$(\mathbf{A} + \mathbf{BDB}')^{-1} = \mathbf{A}^{-1} - \mathbf{A}^{-1}\mathbf{B}(\mathbf{B}'\mathbf{A}^{-1}\mathbf{B} + \mathbf{D}^{-1})^{-1}\mathbf{B}'\mathbf{A}^{-1}. \tag{8.23}$$

Thus, for the reduced dimensional random effects Equation (8.15) and Equation (8.16), this matrix identity gives

$$(\mathbf{H}\Sigma_\alpha\mathbf{H}' + V)^{-1} = V^{-1} - V^{-1}\mathbf{H}(\mathbf{H}'V^{-1}\mathbf{H} + \Sigma_\alpha^{-1})^{-1}\mathbf{H}'V^{-1}, \tag{8.24}$$

which only requires the inverses of the simply structured (e.g., diagonal) matrix, V, and low ($p \ll n$)-dimensional matrices Σ_α and $(\mathbf{H}'V^{-1}\mathbf{H} + \Sigma_\alpha^{-1})$. The use of Equation (8.23) to facilitate computation is, of course, not new. For example, it is used in the development of the Kalman filter equations in time series and spatiotemporal statistics (see p. 317 in Shumway and Stoffer, 2000.) In geostatistical analysis of very large datasets, this approach has recently been labeled "fixed rank kriging" (see Cressie and Johannesson, 2008; Shi and Cressie, 2007).

Some discussion of the implications of this "fixed-rank" approach is in order. First, the terminology arises since the rank of the matrix $\mathbf{H}\Sigma_\alpha\mathbf{H}'$ is p, which we have assumed is much smaller than n. Of course, the addition of the matrix V, accounting for the discrepancy between $\boldsymbol{\eta}$ and $\mathbf{H}\boldsymbol{\alpha}$ as well as the "measurement error," gives a full (n) rank matrix. Thus, the two error processes, ξ and ϵ, are very important. Clearly, the assumption that Σ_ξ is not zero and is simply structured (e.g., diagonal) is also critically important here, as otherwise V^{-1} would not be a simple (linear in n) matrix operation. One must be clear that this is, in fact, an assumption, and thus, the "fixed rank" kriging approach is not strictly equivalent to traditional kriging under this assumption. In some cases, one can assume ξ and ϵ represent very small-scale spatial dependence and measurement error processes, respectively. In that case, there is an equivalency between the reduced rank representation and the "nugget" effect in traditional kriging (see Cressie and Johannesson, 2008).

8.4 Choice of Expansion Matrix, H

In Section 8.3, it was shown that the hierarchical representation of a spatial process in terms of a relatively low dimensional random effects vector multiplied by an expansion matrix, plus a residual vector, can greatly facilitate spatial prediction for very large datasets. We did not specify the form of the expansion matrix, $\mathbf{H}$. Many choices for this matrix have appeared in the literature, often under the guise of being a "different" methodology. The choice is somewhat subjective, but there can be advantages and disadvantages to each, depending on the problem at hand. Some of the most common choices are discussed below.

8.4.1 Orthogonal Basis Functions

There is a long tradition in the mathematical sciences of using orthogonal basis functions in expansions. Thus, it is reasonable to use orthogonal basis functions in the expansion matrix $\mathbf{H}$. We can write

$$\mathbf{H} = \begin{pmatrix} \mathbf{h}_1' \\ \mathbf{h}_2' \\ \vdots \\ \mathbf{h}_n' \end{pmatrix} \tag{8.25}$$

where $\mathbf{h}_i' \equiv (h_i(1), \ldots, h_i(p))$ corresponds to the ith spatial location, and we have the orthogonality constraint,

$$\mathbf{h}_i'\mathbf{h}_j = \begin{cases} \delta, & i = j, \\ 0, & i \neq j, \end{cases} \tag{8.26}$$

the vectors being called "orthonormal" if $\delta = 1$. One choice for $\mathbf{H}$ in this case can come from the class of known orthogonal basis functions (e.g., Fourier, orthogonal polynomials, such as Hermite polynomials, certain wavelets, eigenvectors from a specified covariance matrix, etc.). One advantage of such a choice is that $\mathbf{h}_i$ is defined for any spatial location $\boldsymbol{s}_i$ in the domain of interest. Perhaps more importantly, in the case where $\boldsymbol{\Sigma}_\xi = \sigma_\xi^2\mathbf{I}$, we can write $\boldsymbol{V} = (\sigma_\xi^2 + \sigma_\epsilon^2)\mathbf{I} = \sigma_v^2\mathbf{I}$, so that $\mathbf{H}'\boldsymbol{V}^{-1}\mathbf{H} = \sigma_v^2\mathbf{H}'\mathbf{H} = \sigma_v^2\mathbf{I}$. This can greatly simplify the computations in both the MCMC and the "fixed-rank kriging" implementation. Also, for suitable choices of the basis function (e.g., Fourier or wavelet), fast computational algorithms exist (e.g., the fast fourier transform or discrete wavelet transform, respectively) such that operations $\mathbf{H}'\boldsymbol{x}$ (the transform) and $\mathbf{H}\mathbf{a}$ (the inverse transform) are very efficient; they can be performed without actually forming (or storing) the matrix $\mathbf{H}$. For these basis functions, we note that when $\boldsymbol{\Sigma}_\xi = \sigma_\xi^2\mathbf{I}$, there really isn't a great advantage computationally to having $p \ll n$ if $\boldsymbol{\Sigma}_\alpha$ has relatively simple structure (e.g., if it is diagonal). Typically, as p increases, the residual structure $\boldsymbol{\eta} - \mathbf{H}\boldsymbol{\alpha}$ will tend to have smaller scale spatial dependence, so that the assumption $\boldsymbol{\Sigma}_\xi = \sigma_\xi^2\mathbf{I}$ is often reasonable.

Note that the choice of $\boldsymbol{\Sigma}_\alpha$ in this case is important, particularly if p is relatively large. Ideally, one would like this matrix to be unstructured and estimate all of its parameters without restriction. However, dimensionality issues can quickly make estimation (and computation) somewhat difficult in this case. Thus, some fairly simple structure is typically imposed. Often, this is well justified. For example, if the basis vectors that make up $\mathbf{H}$ are from Fourier functions, then given that the spatial process is second-order stationary, one can choose $\boldsymbol{\Sigma}_\alpha$ to correspond to the analytical variance structure associated with the Fourier modes. That is, $\boldsymbol{\alpha}$ is approximately independent and, thus, $\boldsymbol{\Sigma}_\alpha$ is reasonably modeled as a diagonal matrix. Furthermore, if a known class of stationary covariance models is chosen, one knows these diagonal elements up to the parameters that control the spatial dependence. This can facilitate computation and estimation, as shown in Paciorek, 2007; Royle and Wikle, 2005; and Wikle, 2002.

8.4.1.1 Karhunen–Loéve Expansion

Consider the spatial process, $\eta(\boldsymbol{s})$ for $\boldsymbol{s} \in D$ (some spatial domain) where $E[\eta(\boldsymbol{s})] = 0$ and the covariance function is given by $c_\eta(\boldsymbol{s}, \boldsymbol{r}) = E[\eta(\boldsymbol{s})\eta(\boldsymbol{r})]$. The Karhunen–Loéve (K–L) expansion of this covariance function is

$$c_\eta(\boldsymbol{s}, \boldsymbol{r}) = \sum_{k=1}^{\infty} \lambda_k \phi_k(\boldsymbol{s})\phi_k(\boldsymbol{r}), \tag{8.27}$$

where $\{\phi_k(\cdot) : k = 1, \ldots, \infty\}$ are the eigenfunctions and $\{\lambda_k : k = 1, \ldots, \infty\}$ are the associated eigenvalues of the Fredholm integral equation

$$\int_D c_\eta(\mathbf{s}, \mathbf{r})\phi_k(\mathbf{s})d\mathbf{s} = \lambda_k \phi_k(\mathbf{r}), \tag{8.28}$$

where the eigenfunctions are orthonormal. One can then expand the spatial process in terms of these basis functions, $\eta(\mathbf{s}) = \sum_{k=1}^{\infty} a_k \phi_k(\mathbf{s})$. If one truncates this expansion, say $\eta_p(\mathbf{s}) = \sum_{k=1}^{p} a_k \phi_k(\mathbf{s})$, then it can be shown that among all basis sets of order p, this truncated decomposition minimizes the variance of the truncation error and thus is optimal. In practice, one must often solve numerically the Fredholm integral equation to obtain the basis functions. For example, numerical quadrature or Monte Carlo approaches can be used and give estimates for the eigenvalues and eigenfunctions that are weighted according to the spatial distribution of the data locations (see Cohen and Jones 1969; Buell, 1972). Such approaches are limited in that they obtain the eigenfunctions at spatial locations for which there is an observation. One can consider other spline-based approaches for discretizing the K–L integral equation that effectively interpolate the eigenfunctions to locations where data are not available (see Obled and Creutin, 1986).

The continuous K–L representation is typically not used in applications due to the discrete nature of the data and the difficulty of solving the K–L integral equation. In cases where there are repeat obserations (i.e., over time) or some other information (i.e., deterministic model output, historical observations) that allows one to calculate an empirical covariance matrix for the spatial process, then one can simply perform a principal component analysis (PCA). That is, one estimates $\boldsymbol{\Sigma}_\eta$ and then solves the eigensystem $\boldsymbol{\Sigma}_\eta \boldsymbol{\Phi} = \boldsymbol{\Phi}\boldsymbol{\Lambda}$, where $\boldsymbol{\Phi}$ is the matrix of eigenvectors and $\boldsymbol{\Lambda}$ is a diagonal matrix with the eigenvalues on the main diagonal; these are just the eigenvectors and eigenvalues of the symmetric decomposition of the covariance matrix $\boldsymbol{\Sigma}_\eta$. In spatial statistics, the eigenvectors from this PCA decomposition are called "empirical orthogonal functions" or EOFs. If a discretization of the K–L integral equation assumes equal areas of influence for each spatial observation, such a discretization is equivalent to the PCA-derived EOFs. Conversely, an EOF decomposition of irregularly spaced data without consideration of the relative area of influence for each observation can lead to inappropriate weighting of the importance of each element of the covariance matrix $\boldsymbol{\Sigma}_\eta$ (for further discussion, see Buell, 1972, 1975, and Jolliffe, 2002, p. 297).

The use of EOFs as basis vectors in $\mathbf{H}$ can lead to dramatic dimension reduction for spatial fields. Specifically, we let $\mathbf{H}$ correspond to the first p eigenvectors (columns) of $\boldsymbol{\Phi}$. A significant advantage of the EOF basis set is that there is no assumption of spatial stationarity used to obtain the estimated EOFs. The use of EOFs has the additional advantage that the associated random effects covariance matrix, $\boldsymbol{\Sigma}_\alpha$, is diagonal (given the orthogonality of the EOFs). Note that it is generally the case that the EOFs associated with the largest eigenvalues represent larger-scale spatial variation, and conversely, the EOFs associated with the smallest eigenvalues correspond to smaller-scale spatial variation. Thus, if p is sufficiently large so that the truncated eigenvectors account for the large-scale variability, it might be reasonable to assume the residual spatial structure is uncorrelated and possibly homogeneous, that is, $\boldsymbol{\Sigma}_\xi = \sigma_\xi^2 \mathbf{I}$. As discussed above, such an assumption is critical when working with high-dimensional datasets. Note, it is certainly possible that the residual spatial structure after truncation is not sufficiently uncorrelated. In that case, one can model $\boldsymbol{\Sigma}_\xi$ as an expansion in terms of the next $p + 1$ to $p + n_{eof}$ EOF basis functions. This can be done in a computationally efficient way as described in Berliner, Wikle, and Cressie (2000).

The obvious disadvantages of using EOFs as basis functions are seen in Equation (8.1) predicting at offsite locations, and Equation (8.2) obtaining high-quality estimates of the covariance matrix used to obtain the EOF decomposition. To deal with the first issue, one can "interpolate" the eigenvectors in some reasonable manner. A sensible approach to this is to go back to the K–L integral equation and use canonical spline basis functions that give eigenfunctions for any spatial location, as described in Obled and Creutin (1986). Alternatively, one could "shrink" the empirical covariance estimate (at observation locations) to some (typically) stationary covariance matrix derived from a valid covariance model and available for any pair of locations, observed or not, effectively smoothing or regularizing the covariance estimate (see Nott and Dunsmuir, 2002). Alternatively, one might "presmooth" the data (instead of the estimated covariance matrix) using a standard (typically nonparametric) smoothing procedure (see Wikle and Cressie, 1999). As for the second disadvantage, one can imagine that it is problematic to obtain a direct estimate of the covariance matrix for the process η because we can't observe it without error. If there is sufficient replication, one can obtain the covariance estimate of the data $\hat{\boldsymbol{\Sigma}}_y$, and the measurement variance, $\hat{\sigma}_\epsilon^2$, and base the EOF decomposition on $\hat{\boldsymbol{\Sigma}}_\eta = \hat{\boldsymbol{\Sigma}}_y - \hat{\sigma}_\epsilon^2 \mathbf{I}$, taking care that $\hat{\boldsymbol{\Sigma}}_\eta$ is positive definite. Such an approach is typically reasonable since we are just using the estimate to obtain a relatively small set of basis functions for use in the hierarchical formulation.

8.4.2 Nonorthogonal Basis Functions

Although there are some computational advantages to choosing orthogonal basis functions when constructing $\mathbf{H}$, there is no requirement to do so. There are a myriad choices for nonorthogonal basis functions, including wavelets, radial basis functions, spline functions, and kernel functions. Typically, with nonorthogonal basis functions, it can be difficult to specify (a priori) a simple structure for the random effects covariance matrix $\boldsymbol{\Sigma}_\alpha$ and, thus, it is sometimes allowed to be unstructured in this case (see Cressie and Johannesson, 2008). In other cases, restrictions are imposed on this matrix (see Nychka, Wikle, and Royle, 2002).

Rather than consider all of the different choices for nonorthogonal basis functions, we focus on the use of discrete kernel basis functions in this section.

8.4.2.1 Kernel Basis Functions

It has long been known (e.g., see discussion in Matérn, 1986) that correlated stochastic processes can be written in terms of a convolution (or, moving average) of a Brownian motion process. This idea experienced a renaissance in spatial statistics when it became clear that this idea could also be used to build nonstationary spatial models in practice (see Berry and Ver Hoef, 1996; Fuentes, 2002; Higdon, 1998; Higdon, Swall, and Kern, 1999; Paciorek and Schervish, 2006). Although such convolutions are expressed in terms of continuous space, they can also be used to motivate the selection of the basis functions $\mathbf{H}$ relative to certain "support points" that make up $\boldsymbol{\alpha}$.

Consider the spatial process represented in terms of a discrete kernel expansion:

$$\eta(s) = \sum_{j=1}^{p} h(s, \boldsymbol{r}_j; \boldsymbol{\theta}_s)\alpha_j + \xi(s), \tag{8.29}$$

where $h(s, \boldsymbol{r}_j; \boldsymbol{\theta}_s)$ corresponds to some kernel function and its value (weight) for a given spatial location $\boldsymbol{s}$ depends on the location and value of $j = 1, \ldots, p$ support points, $\boldsymbol{\alpha}$. The kernel parameters, $\boldsymbol{\theta}_s$, may vary in space. In the convolution analog, the kernel is defined as a function of the difference in locations between the spatial location $\boldsymbol{s}$ and the support locations and the kernel parameters are not assumed to change with space, $h(\boldsymbol{s} - \boldsymbol{r}_j; \boldsymbol{\theta})$. In this case, if we define $\mathbf{h}_i(\boldsymbol{\theta}) \equiv (h(\boldsymbol{s}_i - \boldsymbol{r}_1; \boldsymbol{\theta}), \ldots, h(\boldsymbol{s}_i - \boldsymbol{r}_p; \boldsymbol{\theta}))'$, then

$$\mathbf{H}(\boldsymbol{\theta}) = \begin{pmatrix} \mathbf{h}_1'(\boldsymbol{\theta}) \\ \vdots \\ \mathbf{h}_n'(\boldsymbol{\theta}) \end{pmatrix}. \tag{8.30}$$

Thus, subject to the estimation of the kernel parameters, $\boldsymbol{\theta}$, the basis function matrix $\mathbf{H}$ is defined and spatial prediction is computationally efficient as long as $p \ll n$. As described previously, the process model random effects parameterization is $\boldsymbol{\eta} = \mathbf{H}\boldsymbol{\alpha} + \boldsymbol{\xi}$ and $var(\boldsymbol{\eta}) = \mathbf{H}\boldsymbol{\Sigma}_\alpha\mathbf{H}' + \boldsymbol{\Sigma}_\xi$. In the discrete kernel approach described above, one may simply specify $\boldsymbol{\alpha} \sim Gau(\mathbf{0}, \sigma_\alpha^2\mathbf{I})$, so that one is effectively building spatial dependence by smoothing a white noise process defined at the p support locations. As mentioned previously, such simple structure for $\boldsymbol{\Sigma}_\alpha$ allows for quite efficient computation. It isn't always clear in this case what should be the structure for the residual covariance, $\boldsymbol{\Sigma}_\xi$, but it is assumed to have simple structure to facilitate computation. Assuming that p is sufficiently large, this is usually a reasonable assumption. We also note that one might allow $\boldsymbol{\Sigma}_\alpha$ to be unstructured and estimate the associated parameters (see Cressie and Johannesson, 2008). Or, since the support points can be thought of as a spatial process as well, it is reasonable to allow $\boldsymbol{\Sigma}_\alpha$ to have spatial dependence (e.g., in terms of a well-known valid spatial covariance function).

The primary advantage of the kernel-based decomposition, other than computational simplicity for large datasets, is the ease in which fairly complicated nonstationary processes can be modeled. For example, the kernel parameters $\boldsymbol{\theta}$ may vary with space (Higdon et al., 1999), in which case the ith row of $\mathbf{H}$ depends on parameters $\boldsymbol{\theta}_i$. Clearly, the number of parameters needed to estimate $\mathbf{H}$ is much larger since we must estimate $\{\boldsymbol{\theta}_1, \ldots, \boldsymbol{\theta}_n\}$. However, if one takes a Bayesian hierarchical modeling perspective, it is likely that these parameters, indexed in space, can be modeled in terms of (relatively low-dimensional) spatial random fields as well. An alternative approach allows the spatial support points to be defined as a mixture of spatial random processes, so that the neighborhood around each support point corresponds to a stationary spatial process with its own set of governing parameters, implying a more complicated structure on $\boldsymbol{\Sigma}_\alpha$, as seen in Fuentes (2002).

8.5 Extensions

The reduced-dimension random effects approach described above for traditional Gaussian data spatial prediction can also be used in other spatial contexts. For example, this approach can facilitate computation in non-Gaussian spatial data analyses, spatiotemporal modeling, multivariate spatial modeling, or as part of a complex multilevel hierarchical model. The non-Gaussian and spatiotemporal cases are discussed briefly below.

8.5.1 Non-Gaussian Data Models

The traditional non-Gaussian spatial model can be written (e.g., see Paciorek, 2007):

$$Y(\mathbf{s}_i) \sim indep.\ \mathcal{F}(\mu(\mathbf{s}_i), \kappa), \tag{8.31}$$

$$g(\mu(\mathbf{s}_i)) = \mathbf{x}_i'\boldsymbol{\beta} + \eta(\mathbf{s}_i), \tag{8.32}$$

where the observations are assumed to be conditionally independent from a distribution $\mathcal{F}$ chosen from the exponential family. This distribution has spatially indexed mean $\mu(\mathbf{s}_i)$, dispersion parameter κ, and link function $g(\cdot)$. Specifically, the link function is the transformation of the mean response that accommodates a linear function of some covariates plus a spatial random process, $\eta(\mathbf{s}_i)$. As before, it is reasonable to add an extra hierarchical decomposition of the spatial process, $\eta(\mathbf{s}_i) = \mathbf{h}_i'\boldsymbol{\alpha} + \xi(\mathbf{s}_i)$, where $\boldsymbol{\alpha} \sim Gau(\mathbf{0}, \boldsymbol{\Sigma}_\alpha)$ and $\xi(\mathbf{s}_i)$ has zero mean, potentially (but not practically) with spatial dependence. Thus, we can write the model:

$$Y(s_i) \sim indep.\ \mathcal{F}(\mu(s_i), \kappa), \tag{8.33}$$

$$g(\mu(s_i)) = x_i'\boldsymbol{\beta} + \mathbf{h}_i'\boldsymbol{\alpha} + \xi(s_i), \quad \boldsymbol{\xi} \sim Gau(\mathbf{0}, \boldsymbol{\Sigma}_\xi), \tag{8.34}$$

$$\boldsymbol{\alpha} \sim Gau(\mathbf{0}, \boldsymbol{\Sigma}_\alpha). \tag{8.35}$$

The same benefits of having $\boldsymbol{\alpha}$ of dimension $p \ll n$ apply in this case as with the Gaussian data model described previously. Estimation and prediction are, however, typically more difficult in the non-Gaussian data case. But, in an MCMC estimation context, one gains an added advantage in the case of non-Gaussian data models by including the residual error process, $\xi(s_i)$, as long as the structure on the covariance associated with this process is simple. The chief benefit is that the spatial process (and "trend" parameters, $\boldsymbol{\beta}$) can now be updated jointly by an efficient conjugate Gibbs step within the overall MCMC. Although there is somewhat of a tradeoff between this efficient update versus the fact that the spatial parameters are "farther away from the data" (leading to slow mixing in the Markov chain), in very high–dimensional problems, it can sometimes be helpful to reparameterize the spatial process in this fashion (see Paciorek, 2007 and Royle and Wikle, 2005).

8.5.2 Spatiotemporal Processes

The hierarchical random effects representation of a spatial process is ideal for modeling spatiotemporal processes. Consider the spatiotemporal data vector $\mathbf{Y}_t = (Y_t(s_1), \ldots, Y_t(s_n))'$. Complicated spatiotemporal dependence often can be adequately modeled by a spatial process that evolves dynamically. That is, we have the state–space model for $t = 1, \ldots, T$:

$$\mathbf{Y}_t = \boldsymbol{\eta}_t + \boldsymbol{\epsilon}_t, \quad \boldsymbol{\epsilon}_t \sim Gau(\mathbf{0}, \mathbf{R}) \tag{8.36}$$

$$\boldsymbol{\eta}_t = \mathbf{M}_\eta \boldsymbol{\eta}_{t-1} + \boldsymbol{\gamma}_t, \quad \boldsymbol{\gamma}_t \sim Gau(\mathbf{0}, \mathbf{Q}), \tag{8.37}$$

where, for simplicity of presentation, all vectors are $n \times 1$, corresponding to spatial locations, with $\mathbf{M}_\eta$ the $n \times n$ state transition matrix and $\mathbf{R}$, $\mathbf{Q}$ the measurement and state process covariance matrices, respectively. One either has to specify a distribution for $\boldsymbol{\eta}_t$, $t = 0$, or assume it is known, to form a complete model. Note that this model can be made more general by allowing for missing data, nonzero means, and time-varying transition and covariance matrices. Clearly, if n is very large, predicting the state process will be challenging. Typically, state prediction is accomplished by means of Kalman filter/smoother algorithms, which rely on inverses of $n \times n$ covariance matrices (e.g., see Shumway and Stoffer, 2000). Thus, it is natural to think about a dimension reduced version of the Kalman filter.

As before, we consider the decomposition:

$$\boldsymbol{\eta}_t = \mathbf{H}\boldsymbol{\alpha}_t + \boldsymbol{\xi}_t, \quad \boldsymbol{\xi}_t \sim Gau(\mathbf{0}, \boldsymbol{\Sigma}_\xi), \tag{8.38}$$

where $\boldsymbol{\alpha}_t = (\alpha_t(1), \ldots, \alpha_t(p))'$ is a spatiotemporal random effects process and the residual process $\boldsymbol{\xi}_t$ is assumed to be independent across time, but possibly dependent across space. It is assumed that the $\boldsymbol{\alpha}_t$ process evolves dynamically as well. Thus, rewriting the state–space model gives:

$$\mathbf{Y}_t = \mathbf{H}\boldsymbol{\alpha}_t + \boldsymbol{\xi}_t + \boldsymbol{\epsilon}_t, \tag{8.39}$$

$$\boldsymbol{\alpha}_t = \mathbf{M}_\alpha \boldsymbol{\alpha}_{t-1} + \boldsymbol{\nu}_t, \quad \boldsymbol{\nu}_t \sim Gau(\mathbf{0}, \mathbf{Q}_\nu), \tag{8.40}$$

for $t = 1, \ldots, T$. If $\boldsymbol{\xi}_t$ is assumed to exhibit spatial dependence, then one can develop prediction equations that take this into account (e.g., see Wikle and Cressie, 1999). If $\boldsymbol{\xi}_t$ and

ϵ_t are assumed to have simple dependence structures, then the Kalman update equations, or the analogous MCMC updates, can be shown to depend only on $p \times p$ matrix inverses.

As with purely spatial modeling, one can choose from a variety of basis vectors to form $\mathbf{H}$. Often there is professed novelty in the choice of a "new" expansion (i.e., choice of $\mathbf{H}$) on which to base a spatiotemporal dynamical process, but the general procedure does not change. Again, state prediction can be based on classical or Bayesian approaches. Examples in the literature include Berliner, Wikle, and Cressie, 2000; Calder, 2007; Stroud, Mueller, and Sanso, 2001; Wikle and Cressie, 1999; Wikle, Milliff, Nychka, and Berliner, 2001; and Xu and Wikle, 2007.

8.6 Discussion

This exposition of low-rank representations of spatial processes did not include much discussion concerning parameter estimation. Certain modeling assumptions (say, for $\boldsymbol{\Sigma}_\alpha$) can make a substantial difference to the number of such parameters that must be estimated. Similarly, one may fix $\mathbf{H}$ (as in the case of orthogonal functions) or parameterize it (as with the kernel matrices). One might be tempted to allow both $\mathbf{H}$ and $\boldsymbol{\Sigma}_\alpha$ to be unstructured and estimate all of the parameters. As might be expected, without prior information or restrictions on at least one of these matrices, this can lead to model nonidentifiability. The same issue applies to traditional factor analysis and the estimation of $\mathbf{H}$ and $\mathbf{M}_\alpha$ in the spatiotemporal case. In general, the more complicated the parameterizations, the more prior information needs to be provided to obtain good estimates. The literature includes cases where parameter estimation is through method of moments, maximum likelihood, expectation-maximization (EM) algorithm, and various degrees of Bayesian analysis. Examination of these methods are beyond the scope of this chapter, but the references include such details.

Given the general nature of the low-rank framework, one might question how to make the various modeling choices. For example, one must choose the reduced rank, p, the basis functions that make up the expansion matrix $\mathbf{H}$, the truncation covariance $\boldsymbol{\Sigma}_\xi$, and the random effects covariance $\boldsymbol{\Sigma}_\alpha$. Unfortunately, with few exceptions (see Shi and Cressie, 2007), little is known about coherent model selection strategies to elucidate these choices. This is especially the case with regard to $\mathbf{H}$, where there seems to be quite a lot of subjectivity rather than objective decision making. There are also largely unsubstantiated claims that orthogonal basis functions are better/worse than nonorthogonal, or implications that the ability to model nonstationary processes is inherently better/worse than stationary process modeling, etc. In general, as is well-known for random effects modeling of dependent processes, it is almost always true that *any* model that includes dependence is going to be helpful, and that the specific form of that dependence is much less critical. Nevertheless, there needs to be substantially more research into the selection of appropriate low-rank models for spatial processes.

References

L.M. Berliner, C.K. Wikle, and N. Cressie, Long-lead prediction of Pacific SSTs via Bayesian dynamic modeling, *J. Climate*, 13, 3953–3968. 2000.

R.P. Berry and J. Ver Hoef, Blackbox kriging: Spatial prediction without specifying variogram models, *J. Ag., Biol., Ecol. Statist.*, 1, 297–322. 1996.

C.E. Buell, Integral equation representation for factor analysis, *J. Atmos. Sci.*, 28, 1502–1505. 1972.

C.E. Buell, The topography of empirical orthogonal functions, *Fourth Conference on Probability and Statistics in Atmospheric Science*, American Meteorological Society, 188–193. 1975.

C.A. Calder, A dynamic process convolution approach to modeling ambient particulate matter concentrations, *Environmetrics*, 19, 39–48. 2007.

A. Cohen and R.H. Jones, Regression on a random field, *J. American Statist. Assoc.*, 64, 1172–1182. 1969.

N. Cressie and G. Johannesson, Fixed rank kriging for very large spatial data sets, *J.R. Statist. Soc. B*, 70, 209–226. 2008.

M. Fuentes, Modeling and prediction of non-stationary spatial processes, *Statistical Modeling*, 2, 281–298. 2002.

D. Higdon, A process-convolution approach to modelling temperatures in the North Atlantic ocean, *Environ. Ecol. Stat.*, 5, 173–190. 1998.

D. Higdon, J. Swall, and J. Kern, Non-stationary spatial modeling, in J. Bernardo, J. Berger, A. Sawid, and A. Smith (Eds.), *Bayesian Statistics 6*, Oxford University Press, Oxford, U.K., 761–768. 1999.

I.T. Jolliffe, *Principal Component Analysis*, 2nd ed., Springer, New York. 2002.

K. Mardia, C. Goodall, E. Refern, and F. Alonso, The kriged Kalman filter, *Test*, 7, 217–285 (with discussion). 1998.

B. Matern, *Spatial Variation*, 2nd ed., Springer-Verlag, Berlin. 1986.

D.J. Nott and W.T.M. Dunsmuir, Estimation of nonstationary spatial covariance structure, *Biometrika*, 89, 819–829. 2002.

D. Nychka, C.K. Wikle, and J.A. Royle, Multiresolution models for nonstationary spatial covariance functions, *Stat. Modeling*, 2, 315–332. 2002.

Ch. Obled and J.D. Creutin, Some developments in the use of empirical orthogonal functions for mapping meteorological fields, *J. Clim. Appl. Meteor.*, 25, 1189–1204. 1986.

C.J. Paciorek and M. Schervish, Spatial modelling using a new class of nonstationary covariance functions, *Environmetrics*, 17, 483–506. 2006.

C.J. Paciorek, Bayesian smoothing with Gaussian processes using Fourier basis functions in the spectral GP package, *J. Stat. Software*, 19, 1–28. 2007.

C.R. Rao, *Linear Statistical Inference and Its Applications*, John Wiley & Sons, New York, 1965.

J.A. Royle and C.K. Wikle, Efficient statistical mapping of avian count data, *Environ. Eco. Stat.*, 12, 225–243. 2005.

T. Shi and N. Cressie, Global statistical analysis of MISR aerosol data: A massive data product from NASA's Terra satellite, *Environmetrics*, 18, 665–680. 2007.

R.H. Shumway and D.S. Stoffer, *Time Series Analysis and Its Applications*, Springer-Verlag, New York. 2000.

J. Stroud, P. Mueller, and B. Sanso, Dynamic models for spatial-temporal data, *J. Royal Statist. Soc. Ser. B*, 63, 673–689, 2001.

C. Wikle, Spatial modeling of count data: A case study in modelling breeding bird survey data on large spatial domains, in A. Lawson and D. Denison (Eds.), *Spatial Cluster Modelling*, Chapman & Hall, Boca Raton, FL, pp. 199–209. 2002.

C.K. Wikle and N. Cressie, A dimension-reduced approach to space-time Kalman filtering, *Biometrika*, 86, 815–829. 1999.

C.K. Wikle, R. Milliff, D. Nychka, and L. Berliner, Spatiotemporal hierarchical Bayesian modeling: Tropical ocean surface winds, *J. Am. Stat. Assoc.*, 96, 382–397, 2001.

K. Xu and C.K. Wikle, Estimation of parameterized spatial-temporal dynamic models, *J. Stat. Plan. Inf.*, 137, 567–588. 2007.

9

Constructions for Nonstationary Spatial Processes

Paul D. Sampson

CONTENTS

9.1 Overview

Modeling of the spatial dependence structure of environmental processes is fundamental to almost all statistical analyses of data that are sampled spatially. The classical geostatistical model for a *spatial* process $\{Y(\mathbf{s}) : \mathbf{s} \in D\}$ defined over the spatial domain $D \subset \mathbb{R}^d$, specifies a decomposition into mean (or trend) and residual fields, $Y(\mathbf{s}) = \mu(\mathbf{s}) + e(\mathbf{s})$. The process is commonly assumed to be second order stationary, meaning that the spatial covariance function can be written $C(\mathbf{s}, \mathbf{s}+\mathbf{h}) = \text{Cov}(Y(\mathbf{s}), Y(\mathbf{s}+\mathbf{h})) = \text{Cov}(e(\mathbf{s}), e(\mathbf{s}+\mathbf{h})) = C(\mathbf{h})$, so that the covariance between any two locations depends only on the spatial lag vector connecting them. There is a long history of modeling the spatial covariance under an assumption of "intrinsic stationarity" in terms of the *semivariogram*, $\gamma(\mathbf{h}) = \frac{1}{2}\text{var}(Y(\mathbf{s}+\mathbf{h}) - Y(\mathbf{s}))$. However, it is now widely recognized that most, if not all, environmental processes manifest spatially nonstationary or heterogeneous covariance structure when considered over sufficiently large spatial scales.

A fundamental notion underlying most of the current modeling approaches is that the spatial correlation structure of environmental processes can be considered to be approximately stationary over relatively small or "local" spatial regions. This local structure is typically anisotropic. The methods can then be considered to describe spatially varying, locally stationary, anisotropic covariance structure. The models should reflect the effects of known explanatory environmental processes (wind/transport, topography, point sources, etc.). Ideally we would like to model these effects directly, but there have been only a few recent approaches aiming at such explicit modeling (see Calder, 2008).

We distinguish our focus on nonstationarity in spatial covariance from nonstationarity in the mean or trend, as commonly addressed by variants of universal kriging, and from nonstationary processes modeled by intrinsic functions of order k (IRF-F) and characterized by generalized covariance functions, including the one-dimensional special cases of fractional and integrated Brownian motions. Filtered versions of these processes, or "spatial increments of order k," are stationary. In some cases, appropriately identified universal kriging and intrinsic random function kriging are essentially equivalent (Christensen, 1990). See also Stein (2001) and Buttafuoco and Castrignanò (2005).

The "early" literature (reaching back only to the 1980s) on modeling of nonstationary spatial covariance structure was primarily in the context of models for space–time random fields. Prior to 1990, the only apparent approach to this feature of environmental monitoring data (outside of local analyses in subregions where the process might be more nearly stationary) derived from an empirical orthogonal function decomposition of the space–time data matrix, a technique common in the atmospheric science literature. Reference to this approach in the statistical literature dates at least back to Cohen and Jones (1969) and Buell (1972, 1978), although perhaps the most useful elaboration of the method for spatial analysis appears in Obled and Creutin (1986). A number of new computational approaches were introduced in the late 1980s and early 1990s, beginning with Guttorp and Sampson's spatial deformation approach, first mentioned in print in a 1989 comment in a paper by Haslett and Raftery (1989). Shortly following was Haas' "moving window" spatial estimation (Haas, 1990a, 1990b, 1995), although this approach estimates covariance structure locally without providing a (global) model; Sampson and Guttorp's elaboration of their first approach to the spatial deformation model based on multidimensional scaling (1992); an empirical Bayes shrinkage approach of Loader and Switzer (1992); and Oehlert's kernel smoothing approach (1993). Guttorp and Sampson (1994) reviewed this literature on methods for estimating heterogeneous spatial covariance functions with comments on further extensions of the spatial deformation method. In this chapter we focus on the developments from the late 1990s to the present, updating the review of methods provided by Sampson (2001). There has been considerable development and application of kernel and process convolution models, beginning with the work of Higdon (1998) and Fuentes (2001). But despite a substantial growth in the literature of methods on nonstationary modeling, there is almost no conveniently available software at this point in time for the various methods reviewed here. This chapter presents no illustrative case studies and we refer the reader to the original sources for applications.

We review the current literature under the headings of: smoothing and kernel methods, basis function models, process convolution models, and spatial deformation models, concluding with brief mention of parametric models and further discussion.

9.2 Smoothing and Kernel-Based Methods

Perhaps the simplest approaches to dealing with nonstationary spatial covariance structure begin either from the perspective of locally stationary models, which are empirically smoothed over space, or from the perspective of the smoothing and/or interpolation of empirical covariances estimated among a finite number of monitoring sites. Neither of these perspectives incorporate any other explicit modeling of the spatial heterogeneity in the spatial covariance structure. Haas' approach to spatial estimation for nonstationary processes (Haas 1990a, 1990b, 1995) simply computes local estimates of the spatial covariance structure, but does not integrate these into a global model. Oehlert's (1993) kernel smoothing approach and Loader and Switzer's (1992) empirical Bayesian shrinkage and interpolation both aim to smoothly interpolate empirical covariances.

Papers by Fuentes (2001, 2002a, 2002b) and by Nott and Dunsmuir (2002) propose conceptually related approaches for representing nonstationary spatial covariance structure in terms of spatially weighted combinations of stationary spatial covariance functions assumed to represent the local covariance structure in different regions. First, consider dividing the spatial domain D into k subregions S_i, each with a sufficient number of points to estimate a (stationary) variogram or spatial covariance function locally. Fuentes (2001) represents the spatial process $Y(\mathbf{s})$, as a weighted average of "orthogonal local stationary processes":

$$Y(\mathbf{s}) = \sum_{i=1}^{k} w_i(\mathbf{s}) Y_i(\mathbf{s}) \tag{9.1}$$

where $w_i(\mathbf{s})$ is a chosen weight function, such as inverse squared distance between $\mathbf{s}$ and the center of subregion S_i. The nonstationary spatial covariance structure is given by

$$\begin{aligned}\mathrm{Cov}(Y(\mathbf{s}), Y(\mathbf{u})) &= \sum\nolimits_{i=1}^{k} w_i(\mathbf{s}) w_i(\mathbf{u}) \mathrm{Cov}(Y_i(\mathbf{s}), Y_i(\mathbf{u})) \\ &= \sum\nolimits_{i=1}^{k} w_i(\mathbf{s}) w_i(\mathbf{u}) C_{\theta_i}(\mathbf{s} - \mathbf{u}) \end{aligned} \tag{9.2}$$

where $C_{\theta_i}(\mathbf{s} - \mathbf{u})$ represents a stationary spatial covariance function. Fuentes chooses the number of subgrids, k, using a Bayesian information criterion (BIC). The stationary processes $Y_i(\mathbf{s})$ are actually "local" only in the sense that their corresponding covariance functions, $C_{\theta_i}(\mathbf{s} - \mathbf{u})$, are estimated locally, and they are "orthogonal" by assumption in order to represent the overall nonstationary covariance simply as a weighted sum of covariances. Fuentes estimates the parameters with a Bayesian approach providing predictive distributions accounting for uncertainty in the parameter estimates without resorting to computationally intensive MCMC methods.

Fuentes and Smith (2001) proposed to extend the finite decomposition of $Y(x)$ of Fuentes (2001) to a continuous convolution of local stationary processes:

$$Y(\mathbf{x}) = \int_D w(\mathbf{x} - \mathbf{s}) Y_{\theta(\mathbf{s})}(\mathbf{x}) d\mathbf{s}. \tag{9.3}$$

Estimation would require that the spatial field of parameter vectors $\theta(s)$, indexing the stationary Gaussian processes, be constrained to vary smoothly. In practice, the integrals of (9.3) and spectral representations of the spatial covariance (Fuentes, 2002a) are approximated with discrete sums involving k independent spatial locations s_i and corresponding processes $Y_{\theta_i}(\mathbf{s})$, as in Equation (9.2) above. (See also Fuentes, 2002b.)

Nott and Dunsmuir's (2002) approach, proposed as a more computationally feasible alternative to something like the spatial deformation model of Sampson and Guttorp (1992), has the stated aim of reproducing an empirical covariance matrix at a set of monitoring sites and then describing the conditional behavior given monitoring site values with a collection of stationary processes. We will use the same notation as that above, although for Nott and Dunsmuir, i will index the monitoring sites rather than a smaller number of subregions, and the $C_{\theta_i}(x - y)$ represent local *residual* covariance structure after conditioning on values at the monitoring sites. These are derived from locally fitted stationary models. In their general case, Nott and Dunsmuir's representation of the spatial covariance structure can be written

$$\mathrm{Cov}(Y(\mathbf{x}), Y(\mathbf{y})) = \Sigma_0(\mathbf{x}, \mathbf{y}) + \sum\nolimits_{i=1}^{k} w_i(\mathbf{x}) w_i(\mathbf{y}) C_{\theta_i}(\mathbf{x} - \mathbf{y})$$

where $\Sigma_0(x, y)$ is a function of the empirical covariance matrix at the monitoring sites, $\mathbf{C} = [c_{ij}]$, and the local stationary models computed so that $\mathrm{Cov}(Y(\mathbf{x}_i), Y(\mathbf{x}_j)) = c_{ij}$. They further propose to replace the empirical covariance matrix $\mathbf{C}$ by the Loader and Switzer (1992) empirical Bayes shrinkage estimator $\hat{\mathbf{C}} = \gamma \mathbf{C} + (1 - \gamma)\mathbf{C}_\theta$, where $\mathbf{C}_\theta$ is a covariance matrix obtained by fitting some parametric covariance function model. In this case, it can be shown that the Nott and Dunsmuir estimate for covariances between monitored and unmonitored sites is the same as that of the proposed extrapolation procedure of Loader and Switzer, but the estimate for covariances among unmonitored sites is different, and in particular, not dependent on the order with which these unmonitored sites are considered, as was the case for Loader and Switzer's proposal.

Guillot et al. (2001) proposed a kernel estimator similar to the one introduced by Oehlert (1993), although they do not reference this earlier work. Let D denote the spatial domain so that the covariance function $C(\mathbf{x}, \mathbf{y})$ is defined on $D \times D$, and suppose an empirical covariance matrix $\mathbf{C} = [c_{ij}]$ computed for sites $\{x_i, i = 1, \ldots, n\}$. Define a nonnegative kernel K integrating to one on $D \times D$ and let $K_\varepsilon(u, v) = \varepsilon^{-4} K(u/\varepsilon, v/\varepsilon)$ for any real positive ε. Then define a partition $\{D_1, \ldots, D_n\}$ of D (such as the Voronoi partition). The nonparametric, nonstationary estimator of C obtained by regularization of $\mathbf{C}$ is

$$\hat{C}_\varepsilon(\mathbf{x}, \mathbf{y}) = \sum_{i,j} c_{ij} \int_{D_i \times D_j} K_\varepsilon(\mathbf{x} - \mathbf{u}, \mathbf{y} - \mathbf{v}) d\mathbf{u} d\mathbf{v}. \tag{9.4}$$

The authors prove positive definiteness of the estimator for positive definite kernels, discuss selection of the bandwidth parameter ε, and demonstrate an application where, surprisingly, kriging with the nonstationary covariance model is outperformed by kriging with a fitted stationary model.

Finally, we note the nonsmooth, piecewise Gaussian model approach of Kim, Mallick and Holmes (2005), which automatically partitions the spatial domain into disjoint regions using Voronoi tessellations. This model structure, specifying stationary processes within regions (tiles of the tessellation) and independence across regions, is fitted within a Bayesian framework. It is applied to a soil permeability problem where this discrete nonstationary structure seems justified.

9.3 Basis Function Models

The earliest modeling strategy in the literature for nonstationary spatial covariance structure in the context of spatial-temporal applications was based on decompositions of spatial processes in terms of empirical orthogonal functions (EOFs). The original methodology in this field has received renewed attention recently in the work of Nychka and colleagues (Nychka and Saltzman, 1998; Holland et al., 1998; Nychka et al., 2002). Briefly, considering the same spatial-temporal notation as above, the $n \times n$ empirical covariance matrix $\mathbf{C}$ may be written with a spectral decomposition as

$$\mathbf{S} = \mathbf{F}^T \Lambda \mathbf{F} = \sum_{k=1}^{n_T} \lambda_k \mathbf{F}_k \mathbf{F}_k^T \tag{9.5}$$

where $n_T = \min(n, T)$. The extension of this finite decomposition to the continuous spatial case represents the spatial covariance function as

$$C(\mathbf{x}, \mathbf{y}) = \sum_{k=1}^{\infty} \lambda_k F_k(\mathbf{x}) F_k(\mathbf{y}) \tag{9.6}$$

where the eigenfunctions $F_k(\mathbf{x})$ represent solutions to the Fredholm integral equation and correspond to the Karhunen–Loève decomposition of the (mean-centered) field as

$$Y(\mathbf{x}, t) = \sum_{k=1}^{\infty} A_k(t) F_k(\mathbf{x}). \tag{9.7}$$

The modeling and computational task here is in computing a numerical approximation to the Fredholm integral equation, or equivalently, choosing a set of generating functions $e_1(\mathbf{x}), \ldots, e_p(\mathbf{x})$ that are the basis for an extension of the finite eigenvectors $\mathbf{F}_k$ to eigenfunctions $F_k(\mathbf{x})$. (See Guttorp and Sampson (1994), Creutin and Obled (1982), Obled and Creutin (1986), and Preisendorfer (1988, Sec. 2d) for further details.)

In Holland et al. (1998), the spatial covariance function is represented as the sum of a conventional stationary isotropic spatial covariance model and a finite decomposition in terms of empirical orthogonal functions. This corresponds to a decomposition of the spatial process as a sum of a stationary isotropic process and a linear combination of M additional basis functions with random coefficients, the latter sum representing the deviation of the spatial structure from stationarity.

Nychka et al. (2002) introduced a multiresolution wavelet basis function decomposition with a computational focus on large problems with observations discretized to the nodes of a (large) $N \times M$ grid. The example application in this chapter is to air quality model output on a modest 48×48 grid. In the current notation, suppressing the temporal index, they write

$$Y(\mathbf{x}) = \sum_{k=1}^{NM} A_k F_k(\mathbf{x}). \tag{9.8}$$

In the discrete case, they write $\mathbf{F} = [F_{ki}]$, where $F_{ki} = F_k(\mathbf{x}_i)$, $\mathbf{x}_i$ being the ith grid point, so that one can write $\mathbf{Z} = \mathbf{FA}$ and $\mathbf{C} = \mathbf{F}\Sigma_A\mathbf{F}^T$. For the basis functions F_k, they use a "W" wavelet basis with parent forms that are piecewise quadratic splines that are *not* orthogonal or compactly supported. These were chosen because they can approximate the shape of common covariance models, such as the exponential, Gaussian and Matérn, depending on the specification (and off-diagonal sparcity) of the matrix Σ_A. Recent work (Matsuo, Nychka, and Paul, 2008) has extended the methodology to accommodate irregularly spaced monitoring data and a Monte Carlo expectation-maximization (EM) estimation procedure practical for large datasets. They analyze an ozone monitoring network dataset with 397 sites discretized (again) to a 48×48 grid.

Pintore, Holmes, and colleagues (Pintore and Holmes, 2004; Stephenson et al., 2005) work with both Karhunen–Loève and Fourier expansions. Nonstationarity is introduced by evolving the stationary spectrum over space in terms of a latent spatial power process. The resulting models are valid in terms of the original covariance function, but with local parameters. A Bayesian framework is used with MCMC estimation.

9.4 Process Convolution Models

Higdon (1998) introduced a process convolution approach for accommodating nonstationary spatial covariance structure. (See also Higdon, Swall, and Kern (1999).) The basic idea is to consider the fact that any stationary Gaussian process $Z(s)$ with correlogram $\rho(d) = \int_{\mathbb{R}^2} k(s)k(s-d)ds$ can be expressed as the convolution of a Gaussian white noise process $\zeta(s)$ with kernel $k(s)$

$$Y(\mathbf{s}) = \int_{\mathbb{R}^2} k(\mathbf{s}-\mathbf{u})\zeta(\mathbf{u})d\mathbf{u}. \tag{9.9}$$

A particular case of interest is the choice of bivariate Gaussian density functions with 2×2 covariance matrix Σ for the kernel, which results in processes with stationary anisotropic Gaussian correlation functions with the principal axes of Σ determining the directions of the anisotropic structure.

To account for nonstationarity, Higdon (1998) and Higdon et al. (1999) let the kernel vary smoothly with spatial location. Letting $k_s(\cdot)$ denote a kernel centered at the point s, with a shape depending on s, the correlation between two points $\mathbf{s}$ and $\mathbf{s}'$ is

$$\rho(\mathbf{s}, \mathbf{s}') = \int_{\mathbb{R}^2} k_{\mathbf{s}}(\mathbf{u}) k_{\mathbf{s}'}(\mathbf{u}) d\mathbf{u}. \tag{9.10}$$

Higdon et al. (1999) demonstrate the particular case where the $k_s(\cdot)$ are bivariate Gaussian densities characterized by the shape of ellipses underlying the 2×2 covariance matrices. The kernels are constrained to evolve smoothly in space by estimating the local ellipses under a Bayesian paradigm that specifies a prior distribution on the parameters of the ellipse (the relative location of the foci) as a Gaussian random field with a smooth (in fact, Gaussian) spatial covariance function. It should be noted that the form of the kernel determines the shape of the local spatial correlation function, with a Gaussian kernel corresponding to a Gaussian covariance function. Other choices of kernels can lead to approximations of other common spatial correlation functions.

Paciorek and Schervish (2006) extend this approach and create a class of closed-form nonstationary covariance functions, including a nonstationary Matérn covariance parameterized by spatially varying covariance parameters in terms of an eigen-decomposition of the kernel covariance matrix $k_s(\cdot)$.

Calder and Cressie (2007) discuss a number of topics associated with convolution-based modeling including the computational challenges of large datasets. Calder (2007, 2008) extends the approach to dynamic process convolutions for multivariate space–time monitoring data.

D'Hondt et al. (2007) apply the process convolution model with Gaussian kernels (which they call a nonstationary anisotropic Gaussian kernel (AGK) model) to the nonstationary anisotropic texture in synthetic aperture radar (SAR) images. The Gaussian kernels are estimated locally, in contrast to the Bayesian smoothing methods of Higdon and Paciorek and Schervish.

9.5 Spatial Deformation Models

The spatial deformation approach to modeling nonstationary or nonhomogeneous spatial covariance structures has been considered by a number of authors since the early work represented in Sampson and Guttorp (1992) and Guttorp and Sampson (1994). We first review the modeling approach, as presented by Meiring et al. (1997). We will then review some of the other work on this methodology, focusing on recently introduced Bayesian methods.

Suppose that temporally independent samples $Y_{it} = Y(\mathbf{x}_i, t)$ are available at N sites $\{\mathbf{x}_i, i = 1, \ldots, N$, typically in $R^2\}$ and at T points in time $\{t = 1, \ldots, T\}$. $\mathbf{X} = \left[\underline{X}_1\ \underline{X}_2\right]$ represents the matrix of geographic locations. We now write the underlying spatial-temporal process as

$$Y(\mathbf{x}, t) = \mu(\mathbf{x}, t) + \nu(\mathbf{x})^{1/2} E_t(\mathbf{x}) + E_\varepsilon(\mathbf{x}, t), \tag{9.11}$$

where $\mu(\mathbf{x}, t)$ is the mean field, and $E_t(\mathbf{x})$ is a zero mean, variance one, continuous second-order spatial Gaussian process, i.e., $\mathrm{Cov}(E_t(\mathbf{x}), E_t(\mathbf{y})) \to [\mathbf{x} \geq \mathbf{y}]1$.

The correlation structure of the spatial process is expressed as a function of Euclidean distances between site locations in a bijective transformation of the geographic coordinate system

$$\text{cor}(E_t(\mathbf{x}), E_t(\mathbf{y})) = \rho_\theta(\|f(\mathbf{x}) - f(\mathbf{y})\|), \tag{9.12}$$

where $f(\cdot)$ is a transformation that expresses the spatial nonstationarity and anisotropy, ρ_θ belongs to a parametric family with unknown parameters θ, $\nu(\mathbf{x})$ is a smooth function representing spatial variance, and $E_\varepsilon(\mathbf{x}, t)$ represents measurement error and/or very short scale spatial structure, assumed Gaussian and independent of E_t. For mappings from R^2 to R^2, the geographic coordinate system has been called the "G-plane" and the space representing the images of these coordinates under the mapping is called the "D-plane," Perrin and Meiring (1999) prove that this spatial deformation model is identifiable for mappings from R^k to R^k assuming only differentiability of the isotropic correlation function $\rho_\theta()$. Perrin and Senoussi (2000) derive analytic forms for the mappings $f(\cdot)$ under differentiability assumptions on the correlation structure for both the model considered here, where $\rho_\theta()$ is considered to be a stationary and isotropic correlation function ("stationary and isotropic reducibility"), and for the case where this correlation function is stationary, but not necessarily isotropic ("stationary reducibility").

Mardia and Goodall (1992) were the first to propose likelihood estimation and an extension to modeling of multivariate spatial fields (multiple air quality parameters) assuming a Kronecker structure for the space $\times$ species covariance structure. Likelihood estimation and an alternative radial basis function approach to representation of spatial deformations was proposed by Richard Smith in an unpublished report in 1996.

Meiring et al. (1997) fit the spatial deformation model to the empirically observed correlations among a set of monitoring sites by numerical optimization of a weighted least squares criterion constrained by a smoothness penalty on the deformation computed as a thin-plate spline. The problem is formulated so that the optimization is with respect to the parameters, θ, of the isotropic correlation model and the coordinates of the monitoring sites, $\xi_i = f(\mathbf{x}_i)$, in the deformation of the coordinate system. This is a large and often difficult optimization problem. It becomes excessively taxing when uncertainty in the estimated model is assessed by resampling methods or cross-validation. However, it is the approach that is implemented in the most conveniently available software for fitting the deformation model. These are the EnviRo.stat R programs that accompany the text by Le and Zidek (2006) on the analysis of environmental space–time processes (http://enviro.stat.ubc.ca/).

Iovleff and Perrin (2004) implemented a simulated annealing algorithm for fitting the spatial deformation model by optimization, with respect to correlation function parameters θ and D-plane coordinates of the monitoring sites, $\xi_i = f(\mathbf{x}_i)$, of a least squares criterion of goodness-of-fit to an empirical sample covariance matrix. Rather than impose an analytic smoothness constraint on the mapping (such as the thin-plate, spline-based, bending energy penalty of Meiring et al. (1997)), they use a Delaunay triangulation of the monitoring sites to impose constraints on the random perturbations of the D-plane coordinates ξ_i that guarantee that the resulting mapping $f(\mathbf{x}_i)$ is indeed bijective, i.e., it does not "fold." Using any of the other methods discussed here, the achievement of bijective mappings has relied on appropriate tuning of a smoothness penalty or prior probability model for the family of deformations.

Damian et al. (2001, 2003) and Schmidt and O'Hagan (2003) independently proposed similar Bayesian modeling approaches for inference concerning this type of spatial deformation model and for subsequent spatial estimation accounting for uncertainty in the estimation of the spatial deformation model underlying the spatial covariance structure. We present here details of the model of Damian et al. (2001, 2003).

For a Gaussian process with constant mean, $\mu(\mathbf{x}, t) \equiv \mu$, and assuming a flat prior for μ, the marginal likelihood for the covariance matrix Σ has the Wishart form

$$f(\{y_{it}|\Sigma\}) = |2\pi\Sigma|^{-(T-1)/2} \exp\left\{-\frac{T}{2} tr\, \Sigma^{-1}\mathbf{C}\right\} \tag{9.13}$$

where $\mathbf{C}$ is the sample covariance with elements,

$$c_{ij} = \frac{1}{T}\sum_{t=1}^{T}(y_{it} - \bar{y}_i)\left(y_{jt} - \bar{y}_j\right), \tag{9.14}$$

and the true covariance matrix is parameterized as $\Sigma = \Sigma(\theta, \nu_i, \xi_i)$, with $\Sigma_{ij} = (\nu_i \nu_j)^{1/2} \rho_\theta(\|\xi_i - \xi_j\|)$, and $\xi_i = f(\mathbf{x}_i)$. The parameters to be estimated are $\{\theta, \nu_i, \xi_i; i = 1, \ldots, N\}$.

The Bayesian approach requires a prior on all of these parameters. The novel and challenging aspect of the problem concerns the prior for the spatial configuration of the ξ_i. Writing the matrix $\Xi = [\xi_1, \ldots, \xi_N]^T = [\,\Xi_1\ \Xi_2\,]$, Damian et al. (2001, 2003) use a prior of the form

$$\pi(\Xi) \propto \exp\left\{-\frac{1}{2\tau^2}\left[\Xi_1^T\mathbf{K}\Xi_1 + \Xi_2^T\mathbf{K}\Xi_2\right]\right\} \tag{9.15}$$

where $\mathbf{K}$ is a function of the geographic coordinates only—the "bending energy matrix" of a thin-plate spline (see Bookstein, 1989)—and τ is a scale parameter penalizing "nonsmoothness" of the transformation f. Mardia, Kent, and Walder (1991) first used a prior of this form in the context of a deformable template problem in image analysis. It should be noted that the bending energy matrix $\mathbf{K}$ is of rank $n - 3$ and the quadratic forms in the exponent of this prior are zero for all affine transformations, so that the prior is flat over the space of all affine deformations and thus is improper.

The parameter space is highly multidimensional and the posterior distributions are not of closed form, therefore, a Metropolis–Hastings algorithm was implemented to sample from the posterior. (See Damian et al. (2001) for details of the MCMC estimation scheme.) Once estimates for the new locations have been obtained, the transformation is extrapolated to the whole area of interest using a pair of thin-plate splines.

Schmidt and O'Hagan (2003) work with the same Gaussian likelihood, but utilize a general Gaussian process prior for the deformation. When considered in terms of the coordinates ξ_i, the effect of this on the form of the prior $\pi(\Xi)$ is to center the coordinate vectors Ξ_j, $j = 1, 2$, at their geographic locations and to replace $\mathbf{K}$ with a full rank covariance matrix of a form to be specified. Utilizing the known interpretation of thin-plate splines as kriging for an intrinsic random function with a particular form of (generalized) covariance matrix, we see that the Damian et al. (2001) approach may be considered similarly to correspond to a prior for the deformation considered as an intrinsic random function. Schmidt and O'Hagan (2003) also differ from Damian et al. (2001) in their choice of parametric isotropic correlation models and in many of the details of the MCMC estimation scheme, but they are otherwise similarly designed methods.

The atmospheric science literature includes a number of papers with deformation models motivated or determined explicitly by physical processes. (See, for example, Riishojgaard (1998) and Fu et al. (2004).) Xiong et al. (2007) implement a nonlinear mapping model for nonstationary covariance-based kriging in a high-dimensional ($p = 19$) metamodeling problem using computer simulation data.

Anderes and Stein (2008) are the first authors to address the application of the deformation model to the case of a single realization of a spatial process obtained as the deformation of an isotropic Gaussian random field. They present a complete mathematical analysis and methodology for observations from a dense network with approximate likelihood computations derived from partitioning the observations into neighborhoods and assuming independence of the process across partitions.

9.6 Discussion

There is clearly much active work on the development and application of models for nonstationary spatial processes in an expanding range of fields beyond the atmospheric science and environmental applications that motivated most of the early work in this field. We have seen novel applications in image analysis (D'Hondt et al., 2007) and "metamodeling in engineering design" (Xiong et al., 2007). It appears unlikely that there will prove to be one "best" approach for all applications from among the major classes reviewed here: kernel smoothing, process convolution models, spectral and basis functions models, and deformation models.

Although this chapter covers substantial literature, the recent methodologies are still not mature in a number of respects. First, most of the approaches reviewed here are not easily applied as the developers of these methods have, for the most part, not made software available for use by other investigators. A number of questions of practical importance remain to be addressed adequately through analysis and application. Most of the literature reviewed above addresses the application of the fitted spatial covariance models to problems of spatial estimation, as in kriging. The Bayesian methods, all propose to account for the uncertainty in the estimation of the spatial covariance structure, but the practical effects of this uncertainty have not yet been demonstrated. There remains a need for further development of diagnostic methods and experience in diagnosing the fit of these alternative models. In particular, the nature of the nonstationarity, or equivalently, the specification or estimation of the appropriate degree of spatial smoothness in these models expressed in prior distributions or regularization parameters, needs further work. For the Bayesian methods, this translates into a need for further understanding and/or calibration of prior distributions.

This chapter has focused on nonparametric approaches to the modeling of nonstationary spatial covariance structure for univariate spatial processes. In some cases one may wish to formally test the hypothesis of nonstationarity (Fuentes, 2005; Corstanje et al., 2008). Mardia and Goodall (1992), Gelfand et al. (2004), and Calder (2007, 2008) address multivariate problems that are addressed in further detail in Chapter 28. Some parametric models have also been introduced. These include parametric approaches to the spatial deformation model, including Perrin and Monestiez' (1998) parametric radial basis function approach to the representation of two-dimensional deformations. Parametric models appropriate for the characterization of certain point source effects have been introduced by Hughes-Oliver et al. (1998, 1999, 2009).

References

Anderes, E.B. and Stein, M.L. (2008). Estimating deformations of isotropic Gaussian random fields on the plane. *Annals of Statistics*, **36**, 719–741.

Bookstein, F.L. (1989). Principal warps – Thin-plate splines and the decomposition of deformations. *IEEE Transactions on Pattern Analysis*, **11**, 567–585.

Buell, E.C. (1972). Integral equation representation for factor analysis. *Journal of Atmospheric Science*, **28**, 1502–1505.

Buell, E.C. (1978). The number of significant proper functions of two-dimensional fields. *Journal of Applied Meteorology*, **17**, 717–722.

Buttafuoco, G. and Castirgnanò, A. (2005). Study of the spatio-temporal variation of soil moisture under forest using intrinsic random functions of order k. *Geoderma*, **128**, 208–220.

Calder, C.A. (2007). Dynamic factor process convolution models for multivariate space-time data with application to air quality assessment. *Environmental and Ecological Statistics*, **14**, 229–247.

Calder, C.A. (2008). A dynamic process convolution approach to modeling and ambient particulate matter concentrations. *Environmetrics*, **19**, 39–48.

Calder, C.A. and Cressie, N. (2007). Some topics in convolution-based spatial modeling. In: *Proceedings of the 56th Session of the International Statistical Institute*, Lisboa. International Statistical Institute. Voorburg, the Netherlands.

Christensen, R. (1990). The equivalence of predictions from universal kriging and intrinsic random function kriging. *Mathematical Geology*, **22**, 655–664.

Cohen, A., and Jones, R.H. (1969). Regression on a random field. *Journal of the American Statistical Association*, **64**, 1172–1182.

Corstanje, R., Grunwald, S., and Lark, R.M. (2008). Inferences from fluctuations in the local variogram about the assumption of stationarity in the variance. *Geoderma*, **143**, 123–132.

Creutin, J.D. and Obled, C. (1982). Objective analyses and mapping techniques for rainfall fields: An objective comparison. *Water Resources Research*, **18**, 413–431.

Damian, D., Sampson, P.D., and Guttorp, P. (2001). Bayesian estimation of semi-parametric non-stationary spatial covariance structures. *Environmetrics*, **12**, 161–178.

Damian, D., Sampson, P.S., and Guttorp, P. (2003). Variance modeling for nonstationary spatial processes with temporal replicates. *Journal of Geophysical Research-Atmospheres*, **108**, D24, 8778.

D'Hondt, O., Lopez-Martiez, C., Ferro-Famil, L., and Pottier, E. (2007). Spatially nonstationary anisotropic texture analysis in SAR images. *IEEE Transactions on Geoscience and Remote Sensing*, **45**, 3905–3918.

Fu, W.W., Zhou, G.Q., and Wang, H.J. (2004). Ocean data assimilation with background error covariance derived from OGCM outputs. *Advances in Atmospheric Sciences*, **21**, 181–192.

Fuentes, M. (2001). A new high frequency kriging approach for nonstationary environmental processes. *Environmetrics*, **12**, 469–483.

Fuentes, M. (2002a). Spectral methods for nonstationary spatial processes. *Biometrika*, **89**, 197–210.

Fuentes, M. (2002b). Interpolation of nonstationary air pollution processes: A spatial spectral approach. *Statistical Modelling*, **2**, 281–298.

Fuentes, M. (2005). A formal test for nonstationarity of spatial stochastic processes. *Journal of Multivariate Analysis*, **96**, 30–54.

Fuentes, M. and Smith, R.L. (2001). A new class of nonstationary spatial models. Tech report, North Carolina State Univ. Institute of Statistics Mimeo Series #2534, Raleigh.

Gelfand, A.E., Schmidt, A.M., Banerjee, S., and Sirmans, C.F. (2004). Nonstationary multivariate process modeling through spatially varying coregionalization. *Test*, **13**, 263–312.

Guillot, G., Senoussi, R., and Monestiez, P. (2001). A positive definite estimator of the non stationary covariance of random fields. In: *GeoENV 2000: Third European Conference on Geostatistics for Environmental Applications*, P. Monestiez, D. Allard, and R. Froidevaux, Eds., Kluwer, Dordrecht, the Netherlands, 333–344.

Guttorp, P., and Sampson, P.D. (1994). Methods for estimating heterogeneous spatial covariance functions with environmental applications. In: *Handbook of Statistics*, vol. 12, G.P. Patil and C.R. Rao, Eds., Elsevier Science, New York, pp. 661–689.

Haas, T.C. (1990a). Kriging and automated variogram modeling within a moving window. *Atmospheric Environment*, **24A**, 1759–1769.

Haas, T.C. (1990b). Lognormal and moving window methods of estimating acid deposition. *Journal of the American Statistical Association*, **85**, 950–963.

Haas, T.C. (1995). Local prediction of a spatio-temporal process with an application to wet sulfate deposition. *Journal of the American Statistical Association*, **90**, 1189–1199.

Haslett, J. and Raftery, A.E. (1989). Space-time modelling with long-memory dependence: Assessing Ireland's wind resource (with discussion). *Journal of the Royal Statistical Society, Ser. C*, **38**, 1–50.

Higdon, D. (1998). A process-convolution approach to modeling temperatures in the North Atlantic Ocean, *Journal of Environmental and Ecological Statistics*, **5**, 173–190.

Higdon, D.M., Swall, J., and Kern, J. (1999). Non-stationary spatial modeling. In *Bayesian Statistics 6*, J.M. Bernardo, J.O. Berger, A.P. David, and A.F.M. Smith, Eds., Oxford University Press, Oxford, U.K., pp. 761–768.

Holland, D., Saltzman, N., Cox, L., and Nychka, D. (1998). Spatial prediction of dulfur dioxide in the eastern United States. In: *GeoENV II: Geostatistics for Environmental Applications*. J. Gomez-Hernandez, A. Soares, and R. Froidevaux, Eds. Kluwer, Dordrecht, the Netherlands, pp. 65–76.

Hughes-Oliver, J.M. and Gonzalez-Farías, G. (1999). Parametric covariance models for shock-induced stochastic processes. *Journal of Statistical Planning and Inference*, **77**, 51–72.

Hughes-Oliver, J.M., Gonzalez-Farías, G., Lu, J.C., and Chen, D. (1998). Parametric nonstationary spatial correlation models. *Statistics and Probability Letters*, **40**, 267–278.

Hughes-Oliver, J.M., Heo, T.Y., and Ghosh, S.K. (2009). An auto regressive point source model for spatial processes. *Envirometrics*, **20**, 575–594.

Iovleff, S. and Perrin, O. (2004). Estimating a non-stationary spatial structure using simulated annealing. *Journal of Computational and Graphical Statistics*, **13**, 90–105.

Kim, H.M., Mallick, B.K., and Holmes, C.C. (2005). Analyzing nonstationary spatial data using piecewise Gaussian processes. *Journal of the American Statistical Association*, **100**, 653–658.

Le, N.D. and Zidek, J.V. (2006). *Statistical Analysis of Environmental Space-Time Processes*. Springer, New York.

Loader, C., and Switzer, P. (1992). Spatial covariance estimation for monitoring data. In: *Statistics in Environmental and Earth Sciences*, A. Walden and P. Guttorp, Eds., Edward Arnold, London, pp. 52–70.

Mardia, K.V., and Goodall, C.R. (1992). Spatial-temporal analysis of multivariate environmental monitoring data. In: *Multivariate Environmental Statistics 6*. N.K. Bose, G.P. Patil, and C.R. Rao, Eds., North Holland, New York, pp. 347–385.

Mardia, K.V., Kent, J.T., and Walder, A.N. (1991). Statistical shape models in image analysis. In: *Computing Science and Statistics: Proceedings of the 23rd Symposium on the Interface*. E.M. Keramidas, ed., Interface Foundation of America, Fairfax, VA., pp. 550–557.

Matsuo, T., Nychka, D., and Paul, D. (2008). Nonstationary covariance modeling for incomplete data: Monte Carlo EM approach. Forthcoming. See http://www.image.ucar.edu/~nycka/man.html

Meiring, W., Monestiez, P., Sampson, P.D., and Guttorp, P. (1997). Developments in the modelling of nonstationary spatial covariance structure from space-time monitoring data. In: *Geostatistics Wallongong '96*, E.Y. Baafi and N. Schofield, eds., Kluwer, Dordrecht, the Netherland, pp. 162–173.

Nott, D.J. and Dunsmuir, W.T.M. (2002). Estimation of nonstationary spatial covariance structure. *Biometrika*, **89**, 819–829.

Nychka, D. and Saltzman, N. (1998). Design of air quality networks. In: *Case Studies in Environmental Statistics*. D. Nychka, W. Piegorsch, and L. Cox, eds. Springer-Verlag, New York, pp. 51–76.

Nychka, D., Wikle, C., and Royle, J.A. (2002). Multiresolution models for nonstationary spatial covariance functions. *Statistical Modelling*, **2**, 315–331.

Obled, Ch. and Creutin, J.D. (1986). Some developments in the use of empirical orthogonal functions for mapping meteorological fields. *Journal of Applied Meteorology*, **25**, 1189–1204.

Oehlert, G.W. (1993). Regional trends in sulfate wet deposition. *Journal of the American Statistical Association*, **88**, 390–399.

Paciorek, C.J. and Schervish, M.J. (2006). Spatial modelling using a new class of nonstationary covariance functions. *Environmetrics*, **17**, 483–506.

Perrin, O. and Meiring, W. (1999). Identifiability for non-stationary spatial structure. *Journal of Applied Probability*, **36**, 1244–1250.

Perrin, O. and Monestiez, P. (1998). Modeling of non-stationary spatial covariance structure by parametric radial basis deformations. In: *GeoENV II: Geostatistics for Environmental Applications*. J. Gomez-Hernandez, A. Soares, and R. Froidevaux, eds. Kluwer, Dordrecht, the Netherland, pp. 175–186.

Perrin, O. and Senoussi, R. (2000). Reducing non-stationary random fields to stationarity and isotropy using a space deformation. *Statistics and Probability Letters*, **48**, 23–32.

Pintore, A. and Holmes, C. (2004). Spatially adaptive non-stationary covariance functions via spatially adaptive spectra. Forthcoming. (Submitted manuscript available from http://www.stats.ox.ac.uk/~cholmes/).

Preisendorfer, R.W. (1988). *Principal Component Analysis in Meteorology and Oceanography*. Elsevier, Amsterdam.

Riishojgaard, L.P. (1998). A direct way of specifying flow-dependent background error correlations for meteorological analysis systems. *Tellus Series A-Dynamic Meteorology and Oceanography*, **50**, 42–57.

Sampson, P.D. (2001). Spatial covariance. In El-Shaarawi, A.H., Pierorcsh W.W., Eds. *Encyclopedia of Environmetrics*. New York: John Wiley & Sons, 2002, vol. 4, pp. 2059–2067.

Sampson, P.D., and Guttorp, P. (1992). Nonparametric estimation of nonstationary spatial covariance structure. *Journal of the American Statistical Association*, **87**, 108–119.

Schmidt, A. and O'Hagan, A. (2003). Bayesian inference for nonstationary spatial covariance structure via spatial deformations. *Journal of the Royal Statistical Society*, Series B, **65**, 743–758.

Smith, R.H. (1996). Estimating nonstationary spatial correlations. Preprint (http://www.stat.unc.edu/faculty/rs/papers/RLS_Papers.html).

Stein, M.L. (2001). Local stationarity and simulation of self-affine intrinsic random functions. *IEEE Transactions on Information Theory*, **47**, 1385–1390.

Stephenson, J., Holmes, C., Gallagher, K. and Pintore, A. (2005). A statistical technique for modelling non-stationary spatial processes. In: *Geostatistics Banff 2004*, O. Leuangthong and C.V. Deutsch, Eds., pp. 125–134.

Wikle, C.K., and Cressie, N. (1999). A dimension-reduced approach to space-time Kalman filtering. *Biometrika*, **86**, 815–829.

Xiong, Y., Chen, W., Apley, D., and Ding, X.R. (2007). A non-stationary covariance-based kriging method for metamodelling in engineering design. *International Journal for Numerical Methods in Engineering*, **71**, 733–756.

10

Monitoring Network Design

James V. Zidek and Dale L. Zimmerman

CONTENTS

10.1 Monitoring Environmental Processes

Important environmental processes have been monitored for a variety of purposes for a very long time. Concerns about climate change have led to the measurement of sea levels and the extent to which polar ice caps have receded. Concern for human health and welfare and the need to regulate airborne pollutants by the U.S. Environmental Protection Agency has led to the development of urban airshed monitoring networks; cities out of compliance with air quality standards suffer serious financial penalties. The degradation of landscapes, lakes, and monuments led to the establishment of networks for monitoring acidic precipitation as well as to surveys of water quality. Exploratory drilling to find oil reserves on the northern slopes of Alaska generated concern for the health of benthic organisms that feed the fish that feed human populations. The result was the National Oceanic and Atmospheric Agency's (NOAA) decision to monitor the concentrations of trace metals in the seabed before and after the startup of drilling (Schumacher and Zidek, 1993). Predicting the height of tsunamis following earthquakes in the Indian Ocean has led NOAA to install monitoring buoys ("tsunameters") that can help assess the type of earthquake that has occurred. Hazardous waste sites also must be monitored. Mercury, a cumulative poison, must be monitored and that poses a challenge for designing a monitoring network because mercury can be transported in a variety of ways. Concerns about flooding, along with the need for adequate supplies of water for irrigation, has resulted in the monitoring of precipitation as well as snow-melt; not surprisingly, hydrologists were among the earliest to develop rational approaches to design. Ensuring water quality leads to programs that collect water samples at specific locations along seafronts popular with swimmers, with red flags appearing on bad days.

These examples illustrate the importance of monitoring networks, the oldest one being perhaps that constructed in ancient times along the Nile River to measure its height for the purpose of forecasting the extent of the annual flood. It consisted of a series of instruments called "nilometers," each essentially a staircase in a pit next to the river to measure the height of the river at a specific location. In contrast, modern technology has produced networks of cheap-to-deploy sensors that automatically upload their measurements to central electronic data recorders. Air pollution concentrations can be recorded in this way as well as soil moisture content and snow water equivalent.

Networks have societally important purposes. Moreover, the data they provide are important to modelers, such as forecasters of future climate. However, in practice, they are seldom designed to maximize the "bang for the buck" from their product. Instead, their construction will often be influenced by administrative, political, and other pragmatic considerations. Moreover, they and their purpose may evolve and change over time (Zidek, Sun, and Le, 2000).

However, establishing these sites and maintaining them can be expensive. For tsunameters in NOAA's Pacific Ocean National Tsunami Hazard Mitigation Program, for example, Gonzalez, Bernard, Meinig, Eble et al. (2005) estimate a cost of $250,000 to install a new system and a cost of $30,000 per year to maintain it. Designers, therefore, must find a defensible basis for a design recommendation even if pragmatic considerations ultimately lead to modifications of the "ideal" design for the intended purpose.

In this chapter, we explore a variety of approaches designers have developed, along with their rationales. The choice of an approach depends on such things as context, the objective, "discipline bias," and designer background.

Before we begin, a note on vocabulary: monitored sites are sometimes called "gauged" sites, especially in hydrology. The devices placed at such sites may be called "monitors" or "gauges." Monitoring sites are sometimes called "stations." Optimizing a design requires a "design objective."

10.2 Design Objectives

As the examples in Section 10.1 show, networks can be used for a variety of purposes, such as

1. Determining the environmental impact of an event, such as a policy-induced intervention or the closure of an emissions source
2. Assessing trends over space or over time
3. Determining the association between an environmental hazard and adverse health outcomes
4. Detecting noncompliance with regulatory limits on emissions
5. Issuing warnings of impending disaster
6. Monitoring a process or medium, such as drinking water, to ensure quality or safety
7. Monitoring an easy-to-measure surrogate for a process or substance of real concern
8. Monitoring the extremes of a process

However, a network's purpose may change over time, as in an example of Zidek et al. (2000) where a network now monitoring air pollution was formed by combining three networks originally set up to monitor acidic precipitation. Thus, the sites were originally placed in rural rather than urban areas where air pollution is of greatest concern. Hence, additional sites had to be added.

Moreover, the network's objectives may conflict. For example, noncompliance detection suggests siting the monitors at the places where violations are seen as most likely to occur. But, an environmental epidemiologist would want to divide the sites equally between areas of high risk and areas of low risk to maximize the power of their health effects analyses. Even when the objective seems well-defined, such as monitoring to detect extreme values of a process, it may lead to a number of objectives on examination for implementation (Chang, Fu, Le, and Zidek, 2007).

Often many different variables of varying importance are to be concurrently measured at each monitoring site. The challenges now compound, hence, different importance weights may need to be attached. To minimize cost, the designer could elect to measure different variables at different sites. Further savings may accrue from making the measurements less frequently, forcing the designer to consider the intermeasurement times. In combination, these many choices lead to a bewildering set of objective functions to optimize simultaneously. That has led to the idea of designs based on multi-attribute theory, ones that optimize an objective function that embraces all the purposes (Zhu and Stein, 2006; Sampson, Guttorp, and Holland, 2001; Müller and Stehlik, 2008).

However, that approach will not be satisfactory for long-term monitoring programs when the network's future uses cannot be foreseen, as in the example of Zidek et al. (2000). Moreover in some situations the "client" may not be able to even specify the network's purposes precisely (Ainslie, Reuten, Steyn, Le et al., 2009). Yet, as noted above, the high cost of network construction and maintenance will require the designer to select a defensible justification for the design she or he eventually proposes. This chapter presents a catalog of approaches that may provide such a justification.

10.3 Design Paradigms

In practice, the domains in which the monitors are to be sited are "discretized," meaning the possible choices lie in a set $\mathcal{D}$ of finite size N. Practical considerations may make this set quite small. For example, the expensive equipment involved will have to be put in a secure, easily accessible location, one that is away from contaminating sources, such as heavy traffic flows. The sites may be located on a geographical grid that follows the contours of a catchment area, for example.

Most approaches to design fall into one of the following categories (Müller, 2005; Le and Zidek, 2006; Dobbie, Henderson, and Stevens, 2007):

1. Geometry-based: The approach goes back a long way (Dalenius, Hajek, Zubrzycki, 1960). It involves heuristic arguments and includes such things as regular lattices, triangular networks, or space-filling designs (Cox, Cox, and Ensor, 1997; Royle and Nychka, 1998; Nychka and Saltzman, 1998). The heuristics may reflect prior knowledge about the process. These designs can be especially useful when the design's purpose is exploratory (Müller, 2005). In their survey of spatial sampling, Cox et al. (1997) provide support for their use when certain aggregate criteria are used, for example when the objective function is the average of kriging variances. In fact, Olea (1984) finds in a geological setting, when universal kriging is used

to produce a spatial predictor, that a regular hexagonal pattern optimally reduces the average standard error among a slew of geometrical patterns for laying out a regular spatial design. Cox et al. (1997) conjecture that geometric designs may be good enough for many problems and treat this matter as a "research issue." However, since the approaches in this category tend to be somewhat specialized, our need for brevity precludes a detailed discussion.

2. Probability-based (see Section 10.4): This approach to design has been used for well over half a century by a variety of organizations, such as opinion-polling companies and government statistical agencies. It has the obvious appeal that sample selection is based on the seemingly "objective" technique of sampling at random from a list of the population elements (the *sampling frame*). Thus, in principle (though not in practice) the designers need not have any knowledge of the population or the distribution of its characteristics. Moreover, they may see competing methods as biased because they rely on prior knowledge of the population, usually expressed through models. Those models involve assumptions about the nature of the process under investigation, none of which can be exactly correct, thus skewing selection and biasing inference about the process being monitored. Nevertheless, for reasons given below, this approach has not been popular in constructing spatial (network) designs.
3. Model-based (see Section 10.5): The majority of designs for environmental monitoring networks rely on the model-based approach. For although the models do indeed skew the selection process, they do so in accord with prior knowledge and can make the design maximally efficient in extracting relevant information for inferences about the process. In contrast, the probability-based approach may be seen as gambling on the outcome of the randomization procedure and, hence, risking the possibility of getting designs that ignore aspects of the process that are important for inference.

Since each paradigm has merits and deficiencies, the eventual choice will depend on such things as context and the designer's scientific background. However, once selected the paradigms do provide a rational basis for selecting the monitoring sites. In the rest of this chapter, we will see how they can be implemented along with their strengths and weaknesses.

10.4 Probability-Based Designs

This paradigm has been widely used for such things as public opinion polling and the collection of official statistics in national surveys. Hence, a large literature exists for it, largely outside the domain of network design. Thus, we restrict our review to a few of these designs in order to bring out some of the issues that arise, referring the reader interested in details to a more comprehensive recent review (Dobbie et al., 2007).

10.4.1 Simple Random Sampling (SRS)

In the simplest of these designs, sites are sampled at random from a list of sites called the "sampling frame" with equal probability and without replacement. Responses at each site would then be measured and could even be vector valued, to include a sequence of values collected over time. They would be (approximately) independent in this paradigm where the responses at each site, both in the sample and out, are regarded as fixed and the probabilities derive entirely from the randomization process. This is an important point. It means that responses from two sites quite close to one another would be (approximately) unrelated, not something that would seem reasonable to a model-based designer.

Randomization in this paradigm yields a basis for an inferential theory including testing, confidence intervals, and so on. Moreover, these products of inference are reasonably easy to derive and can be quite similar to those from the other theories of spatial design.

10.4.2 Stratified Random Sampling

However, in practice, SRS designs are almost never used and often stratified random sampling is used instead. For one thing, budgetary constraints often limit the number of sites that can be included in the network and with SRS these could, by chance, end up in the very same geographic neighborhood, an outcome that would generally be considered undesirable. Moreover, practical considerations can rule out SRS designs. For example, maintenance and measurement can entail regular visits to the site and long travel times. Thus, spreading the selected sites out over subregions can help divide the sampling workload equitably. Sites monitoring hazardous substances may have to be placed in all of a number of administrative jurisdictions, such as counties, in response to societal concerns and resulting political pressures. Legislation may also force such a division of sites.

Finally, statistical issues may lead to a subdivision of sites into separate strata. For example, gains in statistical efficiency can be achieved when a region consists of a collection of homogeneous subregions (called strata). Then only a small number of sites need to be selected from each stratum. Although this is a form of model-based sampling in disguise, the appeal of stratified sampling designs has led to their use in important monitoring programs, e.g., a survey of U.S. lakes (Eilers, Kanciruk, McCord, Overton et al., 1987) and in EMAP (http://www.epa.gov/emap).

However, even though stratification forces the sites to be spread out geographically, it does not rule out adjacent pairs of sites being close together across stratum boundaries. Moreover, like all designs that rely on a model of some kind, stratified ones may produce no statistical benefits if that model fails to describe the population well. Practical considerations also can rule out their use, leading to other more complex designs.

10.4.3 Variable Probability Designs

Complex designs with multiple stages can sample sites with varying probabilities. Consider for example, a hypothetical survey of rivers (including streams). Stage 1 begins with the construction of a list (a sampling frame), perhaps with the help of aerial photos, of catchment areas called the "primary sampling units" or PSUs. With the help of that sampling frame, a sample of PSUs is selected using SRS, a process that guarantees equally likely selection of all items on the list, large and small. Then at Stage 2, for each PSU selected during Stage 1 a list (subsampling frame) is constructed of rivers in that PSU. Items on these lists are called "secondary sampling units" or SSUs. From these subsampling frames, a random sample of SSUs is selected with an SRS design.

Naive estimates of population characteristics, using such a design could well be biased. To illustrate, suppose an equal number of elements, say 10, are sampled from each of the selected PSUs selected in Stage 1. Further imagine that one of those PSUs, call it A, contained 1,000,000 rivers, while another, B, had just 1,000. Then every single member of A's sample would represent 100,000 of A's streams, while each member of B's would represent just 100 streams. The result: Responses measured on each of B's elements would (without adjustment) grossly over-contribute to estimates of overall population characteristics compared to A's. In short, naive estimates computed from the combined sample would be biased in favor of the characteristics of the over-represented small PSUs. How can this flaw be fixed?

The ingenious answer to that question is embraced in an inferential procedure called the Horvitz–Thompson (HT) estimator, named after its inventors. In fact, the procedure contends with the potential estimator bias in a very large class of practical designs.

To describe an HT estimator, suppose a fixed number, say n, of sites are to be sampled. Any design can be specified in terms of its sample selection probabilities, $P(S)$, for all $S = \{s_1, \ldots, s_n\} \subseteq \mathcal{D}$. Bias can now be assessed in terms of the chances that any given population element s, such as a river in our hypothetical example, is included in our random sample S, is $\pi_s = P(s \in S)$. Now denote the response of interest at location s by $Y(s)$, a quantity regarded as nonrandom in this context, even though it is, in fact, unknown. Suppose the quantity of inferential interest to be the population mean $\bar{Y} = \sum_{s\in\mathcal{D}} Y(s)/N$. Then a "design-unbiased" estimator (meaning one that is unbiased over all possible samples S that might be selected under the specified design) is given by the HT estimator:

$$
\begin{aligned}
\hat{\bar{Y}} &= \sum_{i=1}^{n} Y(s_i)/(N\pi_{s_i}) \\
&= \sum_{s\in S} Y(s)/(N\pi_s) \\
&= \sum_{s\in\mathcal{D}} I_S(s)Y(s)/(N\pi_s)
\end{aligned}
\tag{10.1}
$$

where $I_S(s)$ is 1 or 0 according as $s \in S$ or not. It follows that $\pi_s = E\{I_S(s)\}$ and, therefore, that $\hat{\bar{Y}}$ is unbiased.

One can estimate other quantities, such as strata means, when spatial trends are of interest in a similar way to compensate for the selection bias. As a more complicated example, when both $Y(s)$ and $X(s)$ are measured at each of the sample sites, one could compensate for selection bias in estimating the population level regression line's slope using

$$
\frac{\sum_{i=1}^{n}[Y(s_i) - \hat{\bar{Y}}][X(s_i) - \hat{\bar{X}}]/(\pi_{s_i})}{\sum_{i=1}^{n}[X(s_i) - \hat{\bar{X}}]^2/(\pi_{s_i})}.
$$

In short, the HT estimator addresses the problem of selection bias for a very large class designs.

We now turn to the competing paradigm for which a very rich collection of approaches to design have been developed.

10.5 Model-Based Designs

Broadly speaking, model-based designs optimize some form of inference about the process or its model parameters. We now describe a number of these approaches that get at one or another of these objectives.

10.5.1 Estimation of Covariance Parameters

Designs may need to provide estimates about the random field's model parameters, like those associated with its spatial covariance structure or variogram, for example, which play a central role in the analysis of geostatistical data. There a valid variogram model is selected and the parameters of that model are estimated before kriging (spatial prediction) is performed. These inference procedures are generally based upon examination of the empirical variogram, which consists of average squared differences of data taken at sites lagged the same distance apart in the same direction. The ability of the analyst to estimate variogram parameters efficiently is affected significantly by the design, particularly by the spatial configuration of sites where measurements are taken.

This leads to design criteria that emphasize the accurate estimation of the variogram. Müller and Zimmerman (1999) consider, for this purpose, modifications of design criteria that are popular in the context of (nonlinear) regression models, such as the determinant of the covariance matrix of the weighted or generalized least squares estimators of variogram parameters. Two important differences in the present context are that the addition of a single site to the design produces as many new lags as there are existing sites and, hence, also produces that many new squared differences from which the variogram is estimated. Second, those squared differences are generally correlated, which precludes the use of many standard design methods that rest upon the assumption of uncorrelated errors. Nevertheless, several approaches to design construction that account for these features can be devised. Müller and Zimmerman (1999) show that the resulting designs are much different from random designs on the one hand and regular designs on the other, as they tend to include several groups of tightly clustered sites. The designs depend on the unknown parameter vector θ, however, so an initial estimate or a sequential design and sampling approach is needed in practice. Müller and Zimmerman (1999) also compare the efficiency of their designs to those obtained by simple random sampling and to regular and space-filling designs, among others, and find considerable improvements.

Zhu and Stein (2005) and Zimmerman (2006) consider designs optimal for maximum likelihood estimation of the variogram under an assumption of a Gaussian random field. Since the inverse of the Fisher information matrix approximates the covariance of the variogram's maximum likelihood estimators, they use the determinant of that inverse as their criterion function. This function, like the criterion function of Müller and Zimmerman (1999), depends not only on the set of selected design points S that is to be optimized, but also on the unknown variogram parameter vector θ. Zhu and Stein (2005) offer various proposals to address this difficulty:

- Locally optimal design: Plug a preliminary estimate of θ into the criterion function.
- Minimax design: A variation of assuming nature makes the worst possible choice of θ and the designer then chooses the best possible S under the circumstances.
- Bayesian design: Put a distribution on θ.

Zhu and Stein (2005) propose a simulated annealing algorithm for optimization of their design criterion. They assess these proposals with simulation studies based on use of the Matérn spatial covariance model and make the following conclusions:

1. Although the inverse information approximation to the covariance matrix of maximum likelihood estimators is accurate only if samples are moderately large, the approximation yields a very similar ordering of designs, even for relatively small samples, as when the covariance matrix itself is used; hence, the approximation can serve generally as a useful design criterion.
2. The locally optimal approach (which is applicable only when a preliminary estimate of θ is available) yields designs that provide for much more precise estimation of covariance parameters than a random or regular design does.

3. The more widely applicable Bayesian and minimax designs are superior to a regular design, especially when the so-called "range parameter" (which measures the rate at which the spatial correlation between sites declines with distance) is known or a preliminary estimate of it is available.

Zimmerman (2006) makes similar proposals and obtains similar results. In addition, he focuses attention on the finding that for purposes of good estimation of covariance parameters, a design should have a much greater number of small lags than will occur in either a regular or random arrangement of sites. The design should have many large lags as well. Such a distribution of lags is achieved by a design consisting of regularly spaced clusters.

10.5.2 Estimation of Mean Parameters: The Regression Model Approach

The regression modeling approach to network design focuses on optimizing the estimators of coefficients of a regression model. Development of that approach has taken place outside the context of network design (Smith, 1918; Elfving, 1952; Kiefer, 1959) and an elegant mathematical theory for this problem has emerged (Silvey, 1980; Fedorov and Hackl, 1997; Müller, 2007) along with numerical optimization algorithms.

The approach as originally formulated concerns continuous sampling domains, $\mathcal{X}$, and optimal designs, ξ, with finite supports $x_1, \ldots, x_m \in \mathcal{X}$ ($\sum_{i=1}^m \xi(x_i) = 1$). In all, $n\xi(x_i)$ (suitably rounded) responses would then be measured at x_i for all $i = 1, \ldots, m$ to obtain $y_1, \ldots, y_n$. Key elements include a regression model, $y(x) = \eta(x, \beta) + \varepsilon(x)$ relating y to the selected (and fixed) xs; the assumption of independence of the εs from one sample point x to another. Optimality means maximally efficient estimation of β, that is, designs ξ that optimize $\Phi(M(\xi))$, $M(\xi)$ denoting the Fisher information matrix, and Φ, a positive-valued function depending on the criterion adopted.

As an example, in simple linear regression, $y(x) = \alpha + \beta x + \varepsilon(x)$, $x \in [a, b]$, and $M(\xi) = \sigma^2[\mathbf{X}'\mathbf{X}]^{-1}$. There the optimal design that minimizes the variance of the least squares estimator of β, has the intuitively appealing form, $x_1 = a$, $x_2 = b$ while $\xi(x_1) = \xi(x_2) = 1/2$.

However, the approach does not apply immediately to network design. For one thing, possible site locations are usually quite restricted. Moreover, once sited, the monitors must measure the process of interest for an indefinite period. Finally, to measure n responses of a random field at a single time point would require n monitors, so that $\xi \equiv 1/n$ would be completely determined once its support was specified, rendering the theory irrelevant.

Nevertheless, an attempt has been made to adapt the approach for network design (Gibrik, Kortanek, and Sweigart, 1976; Fedorov and Müller, 1988; 1989) on the grounds (Fedorov and Müller, 1989) that, unlike other approaches, which possess only algorithms for finding suboptimal designs, truly optimal designs could be found with this one. That attempt (Fedorov and Müller, 1987) assumes regression models for times $t = 1, \ldots, T$ that capture both temporal and spatial covariance: $y_t(x_i) = \eta(x_i, \beta_t) + \varepsilon_t(x_i)$ where the εs are independent and the β_ts are both random as well as autocorrelated. Moreover, $\eta(x_i, \beta_t) = g^T(x_i)\beta_t$ for a known vector-valued g.

The celebrated Karhunen–Loève (K–L) expansion makes their model more general than it may appear at first glance (Federov and Müller, 2008). K–L tells us that η has an infinite series expansion in terms of the orthogonal eigenvectors of its spatial covariance function. These eigenvectors become the gs in the expansion of η once the series has been truncated (although in practice they may involve covariance parameters whose estimates need to be plugged in.)

However, using an eigenfunction expansion of the spatial covariance to validate the regression model presents technical difficulties when the proposed network is large (Fedorov,

1996). Moreoever, Spöck and Pilz (2009) point to the difficulty of finding "numerically reliable" algorithms for solving K–L's eigenvalue problems as a serious practical limitation to the approach.

Spöck and Pilz go on to propose a different way of bringing the regression approach into spatial design. Their method is based on the polar spectral representation of isotropic random fields due to Yaglom (1987). That representation equates the spatial covariance (assumed known) as an integral of a Bessel function of the first with respect to a spectral distribution function. That equation, in turn, yields a representation of the ε process for the regression model postulated above in terms of sines and cosines of arbitrary high precision depending how many terms are kept. That adds a second parametric regression term to the regression model above and, like the K–L approach, puts the spatial design problem into a form susceptible to analysis by methods in the broader domain of Bayesian design (Pilz, 1991). This approach also faces practical challenges since commonly environmental processes are not isotropic and, moreover, their covariance matrices are not known. The authors circumvent the latter by finding designs that are minimax against the unknown covariance, where spatial prediction provides the objective function on which the minimax calculation can be based.

Overall, the attempt to move the regression modeling theory into spatial design faces a number of challenges in applications:

1. As noted above, the feasible design region will usually be a discrete set, not a continuum and use of the so-called "continuous approximation" that has been proposed to address that difficulty leads to further difficulties. For one thing, the result will not usually be a feasible solution to the original problem. For another, the solution may be hard to interpret (Fedorov and Müller, 1988, 1989). Although (Müller, 2001) has helped clarify the nature of that approximation, the value of substituting it for the hard-to-solve exact discrete design problem remains somewhat unclear.
2. The design objective inherited from the classical approach to regression-based design, the one based on inference about the regression parameters (the βs) will not always seem appropriate especially when they are mere artifacts of the orthogonal expansion described above.
3. Even when the regression model is genuine (as opposed to one from an eigenfunction expansion) and the objective function is meaningful, the range of spatial covariance kernels will be restricted unless the εs are allowed to be spatially correlated. That need is met in the extensions of the model (Fedorov, 1996; Müller, 2001). However, the resulting design objective function does not have much in common with the original besides notation (Fedorov, 1996, p. 524).
4. The assumed independence of the εs also proves a limitation in this context, although a heuristic way around this difficulty offers some promise (Müller, 2005).
5. The complexity of random environmental space–time processes renders their random response fields only crudely related to spatial site coordinates. Moreover, its shape can vary dramatically over time and season. In other words, finding a meaningful, known vector-valued function g would generally be difficult or impossible.

To summarize, although the regression modeling approach to network design comes with an evolved theory and a substantial toolbox of algorithms, using that approach in network design will prove challenging in practice.

10.5.3 Spatial Prediction

Instead of estimating model parameters as in the two approaches discussed above, the prediction of the random field at unmonitored sites based on measurements at the monitored sites may be taken as the design objective. In some cases the spatial covariance has been taken as known (McBratney, Webster, and Burgess, 1981; Yfantis, Flatman, and Behar, 1987; Benhenni and Cambanis, 1992; Su and Cambanis, 1993; Ritter, 1996). In others it was not and in one such case, a Bayesian approach was used to estimate the unknown parameters (Currin, Mitchell, Morris, and Ylvisaker, 1991).

That is the spirit underlying the approach taken in geostatistical theory that has some natural links with the regression modeling approach described above. That theory has traditionally been concerned with spatial random fields, not space–time fields until very recently (Myers, 2002) and has a large literature devoted to it (see, for example, Wackernagel, 2003). So, we will not describe this approach in detail here.

Two methods are commonly employed, cokriging and universal kriging. The first concerns the prediction of an unmeasured coordinate of the response vector, say $y_1(x_0)$, using an optimal linear predictor based on the observed response vectors at all the sampling sites. The coefficients of that optimal predictor are found by requiring it to be unbiased and to minimize the mean square prediction error. They will depend on the covariances between responses and the covariances between the prediction and the responses, covariances that are unrealistically assumed to be known and later estimated from the data usually without adequately accounting for the additional uncertainty thereby introduced. In contrast to the first method, the second relies on a regression model precisely of the form given in the previous subsection, i.e., $y(x) = g^T(x)\beta + \varepsilon(x)$ where the εs are assumed to have a covariance structure of known form. However, unlike the regression modeling approach above, the goal is prediction of the random response (possibly a vector) at a point where it has not been measured. Moreover, g (which can be a matrix in the multivariate case) can represent an observable covariate process. Optimization again relies on selecting coefficients by minimizing mean squared prediction error subject to the requirement of unbiasedness. Designs are commonly found, iteratively, one future site at a time, by choosing the site x_0 where the mean squared prediction error of the optimum predictor proves to be greatest. The designs tend to be very regular in nature, strongly resembling space-filling designs.

10.5.4 Prediction and Process Model Inference

Why not combine the goals of predicting the random field at unmeasured sites and the estimation of the field's model parameters in a single design objective criterion? Zhu and Stein (2006) and Zimmerman (2006) make that seemingly natural merger.

They focus on the case of an (isotropic) Gaussian field so if θ, the vector of covariance model parameters, were known the best linear predictor $\hat{Y}(s; S, \theta)$ of the unmeasured response at location s, $Y(s)$, could be explicitly computed as a function of the responses at points in S, $Y = \{Y(s),\ s \in S\}$, and θ. So could its mean squared prediction error (MSPE) $M(s; S, \theta)$. That quantity could then be maximized over s to get the worst case and a criterion $M(S, \theta)$ to maximize in finding an optimum S. (Alternatively, it could be averaged.) This criterion coincides with that described in the previous subsection.

Since θ is unknown, it must be estimated by, say, the restricted maximum likelihood (REML) or maximum likelihood (ML) estimator, $\hat{\theta}$. The optimal predictor could then be replaced by $\hat{Y}(s; S, \hat{\theta})$ in the manner conventional in geostatistics. But then $M(s; S, \theta)$ would not correctly reflect the added uncertainty in the predictor. Moreover, the designer may wish to optimize the performance of the plug-in predictor as well as the performance of $\hat{\theta}$, where each depends on the unknown θ. If only the former were of concern the empirical kriging

(EK)-optimality criterion of Zimmerman (2006) could be used. This criterion is given by

$$EK(\theta) = \max_{s \in S}\{M(s; S, \theta) + \text{tr}[A(\theta)B(\theta)]\},$$

where $A(\theta)$ is the covariance matrix of the vector of first-order partial derivatives of $\hat{Y}(s; S, \theta)$ with respect to θ, and $B(\theta)$ is the inverse of the Fisher information matrix associated with $\hat{\theta}$. If one is also interested in accurately estimating the MSPE of $\hat{Y}(s; S, \hat{\theta})$, the EA (for "estimated adjusted") criterion of Zhu and Stein (2006) is more appropriate. The EA criterion is an integrated (over the study region) weighted linear combination of $EK(\theta)$ and the variance of the plug-in kriging variance estimator.

Zimmerman (2006) makes the important observation that the two objectives of optimal covariance parameter estimation and optimal prediction with known covariance parameters are actually antithetical; they lead to very different designs in the cases he considers (although in some special cases, they may agree). It seems that, in general, compromise is necessary. Indeed, Zimmerman's examples and simulations indicate that the EK-optimal design resembles a regular or space-filling design with respect to overall spatial coverage, but that it has a few small clusters and is in this sense "intermediate" to the antithetical extremes. Furthermore, the EK-optimal design most resembles a design optimal for covariance parameter estimation when the spatial dependence is weak, whereas it most resembles a design optimal for prediction with known parameters when the spatial dependence is strong. The upshot of this for the designer is that placing a few sites very close together, while of no benefit for prediction with known covariance parameters, may substantially improve prediction with unknown covariance parameters.

Overall, the EA and EK approaches seem to work quite well and are well worth using when their objectives seem appropriate, albeit with the caveat that their complexity may make them difficult to explain to nonexperts and, hence, to "sell" to them.

10.5.5 Entropy-Based Design

Previous subsections covered design approaches that can be viewed as optimally reducing uncertainty about model parameters or about predictions of unmeasured responses. To these we now add another, optimally reducing the uncertainty about responses by measuring them. The question is: Which are the ones to measure? Surprisingly, these three can be combined in a single framework. Not only that, achieving the third objective, simultaneously achieves a combination of the other two. The use of entropy to represent these uncertainties is what makes this possible and ties these objectives together.

As noted earlier, specifying exact design objectives can be difficult or impossible while the high cost of monitoring demands a defensible design strategy. Design objectives have one thing in common, the reduction of uncertainty about some aspect of the random process of interest. Bayesian theory equates uncertainty with a probability distribution while entropy says the uncertainty represented by such a distribution may be quantified as entropy. Thus, selecting a design to minimize uncertainty translates into the maximization of entropy reduction.

That observation leads to the entropy-based theory of network design and provides objective functions for it (Caselton and Husain, 1980; Caselton and Zidek, 1984; Shewry and Wynn, 1987; Le and Zidek, 1994; Sebastiani and Wynn, 2000; Bueso, Angulo, and Alonso, 1998; Bueso, Angulo, Curz-Sanjuliàn, and García-Aróstegui, 1999b; Angulo et al., 2000; Angulo and Bueso, 2001; Fuentes, Chaudhuri, and Holland 2007). The idea of using entropy in experimental design goes back even farther (Lindley, 1956).

The approach can be used to reduce existing networks in size (Caselton, Kan, and Zidek, 1992; Wu and Zidek, 1992; Bueso et al., 1998) or extend them (Guttorp, Le, Sampson, and Zidek, 1993). The responses can be vector-valued as when each site is equipped with gauges to measure several responses (Brown, Le, and Zidek, 1994a). Costs can be included with the possibility that gauges can be added or removed from individual sites before hitherto unmonitored sites are gauged (Zidek et al., 2000). Data can be missing in systematic ways as, for example, when some monitoring sites are not equipped to measure certain responses (Le, Sun, and Zidek, 1997) or when monitoring sites commence operation at different times (giving the data a monotone or staircase pattern) or both (Le, Sun, and Zidek, 2001; Kibria, Sun, Zidek, and Le, 2002). Software for implementing an entopy-based design approach can be found at http://enviro.stat.ubc.ca, while a tutorial on its use is given by Le and Zidek (2006) who describe one implementation of the theory in detail.

Although the general theory concerns processes with a fixed number $k = 1, 2, \ldots,$ of responses at each site, we assume $k = 1$ for simplicity. Moreover, we concentrate on the problem of extending the network; the route to reduction will then be clear. Suppose g of the sites are currently gauged (monitored) and u are not. The spatial field thus lies over $u + g$ discrete sites.

Relabel the site locations as $\{s_1, \ldots, s_u, s_{u+1}, \ldots, s_{u+g}\}$ and let

$$\begin{aligned} \mathbf{Y}_t^{(1)} &= (Y(t, s_1), \ldots, Y(t, s_u))' \\ \mathbf{Y}_t^{(2)} &= (Y(t, s_{u+1}), \ldots, Y(t, s_{u+g}))' \\ \mathbf{Y}_t &= (\,\mathbf{Y}_t^{(1)\prime},\ \mathbf{Y}_t^{(2)\prime})'. \end{aligned}$$

Assuming no missing values, the dataset D is comprised of the measured values of $\mathbf{Y}_t^{(2)}$, $t = 1, \ldots, T$.

Although in some applications (e.g., environmental epidemiology) predicted values of the unobserved responses (past exposure) $\{\,\mathbf{Y}_t^{(1)},\ t = 1, \ldots, T\}$ may be needed, here we suppose interest focuses on the $u \times 1$ vector, $\mathbf{Y}_{T+1}^{(1)} = (Y(T+1, s_1), \ldots, Y(T+1, s_u))'$, of unmeasured future values at the currently "ungauged" sites at time $T+1$.

Our objective of extending the network can be interpreted as that of optimal partitioning of $\mathbf{Y}_{T+1}^{(1)}$, which for simplicity and with little risk of confusion, we now denote by $Y^{(1)}$. After reordering its coordinates, the proposed design would lead to the partition $Y^{(1)} = (Y^{(rem)\prime}, Y^{(add)\prime})$, $Y^{(rem)\prime}$ being a u_1-dimensional vector representing the future ungauged site responses and $Y^{(add)\prime}$ is a u_2-dimensional vector representing the new future gauged sites to be added to those already being monitored. If that proposed design were adopted, then at time $T+1$, the set of gauged sites would yield measured values of the coordinates in the vector $(Y^{(add)\prime}, Y^{(2)\prime}) = (Y_{T+1}^{(add)\prime}, Y_{T+1}^{(2)\prime}) \equiv G$ of dimension u_2+g. But, which of these designs is optimal?

Suppose $\mathbf{Y}_t$ has the joint probability density function, f_t, for all t. Then the total uncertainty about $\mathbf{Y}_t$ may be expressed by the entropy of its distribution, i.e., $H_t(\,\mathbf{Y}_t) = E[-\log f_t(\,\mathbf{Y}_t)/h(\,\mathbf{Y}_t)]$, where $h(\cdot)$ is a so-called reference density (Jaynes, 1963). It need not be integrable, but its inclusion makes the entropy invariant under one-to-one transformations of the scale of $\mathbf{Y}_t$. Note that the distributions involved in H_t may be conditional on certain covariate vectors, $\{\,\mathbf{x}_t\}$, that are regarded as fixed.

Usually in a hierarchical Bayesian model, $\mathbf{Y}_{T+1}$'s probability density function, $f_{(T+1)}(\cdot) = f_{(T+1)}(\cdot \mid \theta)$ will depend on a vector of unspecified model parameters θ in the first stage of modeling. Examples in previous subsections have included parameters in the spatial covariance model. Therefore, using that density to compute $H_{T+1}(\,\mathbf{Y}_{T+1})$ would make it an unknown and unable to play the role in a design objective function. To turn it into a usable objective function we could use instead $\mathbf{Y}_{T+1}$'s marginal distribution obtained by averaging $f_{(T+1)}(\cdot \mid \theta)$ with respect to θ's prior distribution. However, as we have seen in previous subsections, inferences about θ may well be a second design objective (Caselton et al., 1992). Thus, we turn from $H_{T+1}(\mathbf{Y}_{T+1})$ to $H_{T+1}(\mathbf{Y}_{T+1}, \theta) = H_{T+1}(Y, \theta)$ in our simplified notation for $\mathbf{Y}_{T+1}$.

Conditional on D, the total a priori uncertainty may now be decomposed as

$$H(Y, \theta) = H(Y \mid \theta) + H(\theta).$$

Assuming for simplicity that we take the reference density to be identically 1 (in appropriate units of measurement), we have

$$H(Y \mid \theta) = E[-\log(f(Y \mid \theta, D)) \mid D]$$

and

$$H(\theta) = E[-\log(f(\theta \mid D)) \mid D].$$

However, for purposes of optimizing design, we need a different decomposition that reflects the partitioning of future observations into ungauged and gauged sites, $Y' = (U, G)$, where $U \equiv Y^{(rem)'}$ and G is defined above. Now represent $H(Y, \theta) = TOT$ as

$$TOT = PRED + MODEL + MEAS$$

where with our unit reference density:

$$\begin{aligned} PRED &= E[-\log(f(U \mid G, \theta, D)) \mid D]; \\ MODEL &= E[-\log(f(\theta \mid G, D)/) \mid D]; \\ MEAS &= E[-\log(f(G \mid D)) \mid D]. \end{aligned}$$

Coming back to an earlier observation above, measuring G will eliminate all uncertainty about it, driving $MEAS$ to 0. (This would not be strictly true in practice because of measurement error, which could be incorporated in this framework at the expense of greater complexity. However, we ignore that issue here for expository simplicity.) Thus, it is optimal to choose $Y^{(add)'}$ to maximize $MEAS$ and thereby gain the most from measuring G at time $T+1$. However, since TOT is fixed, that optimum will simultaneously minimize $PRED + MODEL$ representing the combined objective of prediction and inference about model parameters. Incidentally, had we started with $H(Y)$ instead of $H(Y, \theta)$, and made a decomposition analogous to that given above, we would have arrived at the same optimization criterion: the maximization of $MEAS$.

In the above, we have presented a very general theory of design. But, how can it be implemented in practice? One solution, presented in detail in Le and Zidek (2006), assumes that the responses can, through transformation, be turned into a Gaussian random field approximately, when conditioned on $\theta = (\beta, \Sigma)$, β representing covariate model coefficients and Σ the spatial covariance matrix. In turn, θ has a so-called generalized inverted Wishart distribution (Brown, Le, and Zidek, 1994b), which has a hypercovariance matrix Ψ and a vector of degrees of freedom as hyperparameters. These assumptions lead to a multivariate-t posterior predictive distribution for Y, among other things. However, the hyperparameters, such as those in Ψ in the model, are estimated from the data, giving this approach an empirical Bayesian character. Finally $MEAS$ can be explicitly evaluated in terms of those estimated hyperparameters and turns out to have a surprisingly simple form.

In fact, for any proposed design, the objective function (to be minimized) becomes (Le and Zidek, 2006), after multiplication by a minus sign among other things, the seemingly natural:

$$det(\hat{\Psi}_{U|G}), \tag{10.2}$$

the determinant of the estimated residual hypercovariance for U given G (at time $T+1$, since G is not yet known at time T), where det denotes the determinant. The seeming simplicity of this criterion function is deceptive; in fact, it combines the posterior variances of the proposed unmeasured sites and the posterior correlations between sites in a complicated way and the optimum design, G^{opt}, often selects surprising sites for inclusion (Ainslie et al., 2009).

However, finding the optimum design is a very challenging computational problem. In fact, the exact optimal design in Equation 10.2 cannot generally be found in reasonable time, it being an NP-hard problem. This makes suboptimal designs an attractive alternative (Ko, Lee, and Queyranne, 1995). Among the alternatives are the "exchange algorithms," in particular the (DETMAX) procedure of Mitchell (1974a; 1974b) cited by Ko et al. (1995). Guttorp et al. (1993) propose a "greedy algorithm," which at each step, adds (or subtracts if the network is being reduced in size) the station that maximally improves the design objective criterion. Ko, Lee, and Queyranne, (1995) introduce a greedy plus exchange algorithm. Finally, Wu and Zidek (1992) cluster prospective sites into suitably small subgroups before applying an exact or inexact algorithm so as to obtain suboptimal designs that are good at least within clusters.

Exact algorithms for problems moderate in size are available (Guttorp et al., 1993; Ko et al., 1995). These results have been extended to include measurement error (Bueso et al., 1998) where now the true responses at both gauged and ungauged sites need to be predicted. Linear constraints to limit costs have also been incorporated (Lee, 1998), although alternatives are available in that context (Zidek et al., 2000).

While the entropy approach offers a unified approach to network design especially when unique design objectives are hard to specify, like all approaches it has shortcomings. We list some of these below:

1. When a unique objective can be specified, the entropy optimal design will not be optimal. Other approaches like those covered above would be preferred.
2. Except in the case of Gaussian fields with conjugate priors, computing the entropy poses a challenging problem that is the subject of current research, but is not yet solved. However, progress has recently been made there as well (Fuentes et al., 2007).
3. Although the field can often be transformed to make it more nearly Gaussian, that transformation may also lead to new model parameters that are difficult to interpret, making the specification of realistic priors difficult. (The entropy itself, however, would be invariant under such transformations.)
4. Computing the exact entropy optimum design in problems of realistic size is challenging. That could make simpler designs, such as those based on geometry and other methods described above, more appealing especially when the goals for the network are short term.

A general problem with approaches to design is their failure to take cost into consideration. That will always be a limiting factor in practice. However, some progress has been made for the entropy theory designs (Zidek et al., 2000). Moreover, Fuentes et al. (2007) provide a very general theory for entropy based spatial design that allows for constrained optimization that is able to incorporate the cost of monitoring in addition to other things.

10.6 Concluding Remarks

This chapter has comprehensively reviewed principled statistical approaches to the design of monitoring networks. However, the challenges involved transcend the mere application of these approaches as described here. Some of these challenges are mentioned in our introduction and more detailed reviews consider others (Le and Zidek, 2006). For example, although statistical scientists would well recognize the importance of measurement and data quality, they may not accept or even see their vital role in communicating that to those who will build and maintain the network. And while the normative approaches do provide defensible proposals for design or redesign, these cannot be used before thoroughly reviewing things like the objectives and available data. The latter will suggest important considerations not captured in more formal design approaches. For example, the case study presented in Ainslie et al. (2009) benefited greatly from learning this network's evolutionary history and analyzing the network's spatial correlation structure. They helped to understand the results of the entropy-based analysis and strengthened the eventual recommendations based partly on it.

Our review has not covered quite a number of special topics, such as mobile monitors (required, for example, in the event of the failure of a nuclear power generator and the spread downwind of its radioactive cloud) and microsensor monitoring networks. The latter, reflecting changing technology, involve a multitude of small monitors that may cost as little as a few cents that can transmit to each other as well as to their base station, which uploads its data to a global monitoring site where users can access the data. Although these networks pose new design challenges, the principles set out in this chapter can still be brought to bear. Furthermore, the reader will have recognized that the "future" in our formulation of the entropy-based design is defined as time $T + 1$. Thus the optimum design there, in principle, will depend on T, although it should remain stable for a period of time. However, this observation points to the important general point: that at least all long-term monitoring networks, however designed, must be revisited periodically and redesigned if necessary, something that is seldom done in practice.

References

Ainslie, B., Reuten, C., Steyn, D., Le, N.D., and Zidek, J.V. (2009). Application of an entropy-based optimization technique to the redesign of an existing air pollution monitoring network. *J Environ Man*, 90, 2715–2729.

Angulo, J.M., and Bueso, M.C. (2001). Random perturbation methods applied to multivariate spatial sampling design. *Environmetrics*, 12, 631–646.

Angulo, J.M., Bueso, M.C., and Alonso, F.J. (2000). A study on sampling design for optimal prediction of space-time stochastic processes. *Stochast Environ Res Risk Assess*, 14, 412–427.

Benhenni, K., and Cambanis, S. (1992). Sampling designs for estimating integrals of stochastic processes. *Ann Statist*, 20, 161–194.

Brown, P.J., Le, N.D., and Zidek, J.V. (1994a). Multivariate spatial interpolation and exposure to air pollutants. *Can Jour Statist*, 22, 489–510.

Brown, P.J., Le, N.D., and Zidek, J.V. (1994b). Inference for a covariance matrix. In *Aspects of Uncertainty: A Tribute to D.V. Lindley*, Eds. A.F.M. Smith and P.R. Freeman. New York: John Wiley & Sons.

Bueso, M.C., Angulo, J.M., and Alonso, F.J. (1998). A state-space model approach to optimum spatial sampling design based on entropy. *Environ Ecol Statist*, 5, 29–44.

Bueso, M.C., Angulo, J.M., Qian, G., and Alonso, F.J. (1999a). Spatial sampling design based on stochastic complexity. *J Mult Anal*, 71, 94–110.

Bueso, M.C., Angulo, J.M., Curz-Sanjuliàn, J., and García-Aróstegui, J.L. (1999b). Optimal spatial sampling design in a multivariate framework. *Math Geol*, 31, 507–525.

Caselton, W.F., and Husain, T. (1980). Hydrologic networks: information transmission. *Water Resources Planning and Management Division, ASCE*, 106(WR2), 503–520.

Caselton, W.F., and Zidek, J.V. (1984). Optimal monitoring network designs. *Statist & Prob Lett*, 2, 223–227.

Caselton, W.F., Kan, L., and Zidek, J.V. (1992). Quality data networks that minimize entropy. In *Statistics in the Environmental and Earth Sciences*, Eds. P. Guttorp and A. Walden. London: Griffin.

Chang, H., Fu, A.Q., Le, N.D., and Zidek, J.V. (2007). Designing environmental monitoring networks to measure extremes. *Ecol Environ Statist*, 14, 301–321.

Cox, D.D., Cox, L.H., and Ensor, K.B. (1997). Spatial sampling and the environment: some issues and directions. *Environ Ecol Statis*, 4, 219–233.

Currin, C., Mitchell, T., Morris, M., and Ylvisaker, D. (1991). Bayesian prediction of deterministic functions, with applications to the design and analysis of computer experiments. *J Amer Statist Assoc*, 86, 953–963.

Dalenius, T., Hajek, J., and Zubrzycki, S. (1960). On plane sampling and related geometrical problems. *Proceedings of the Fourth Berkeley Symposium on Mathematical Statistics and Probability, Vol I*, Berkeley: University of California Press, 125–150.

Dobbie, M.J., Henderson, B.L., and Stevens Jr., D.L. (2007). Sparse sampling: spatial design for aquatic monitoring. *Statis Surv*, 2, 113–153.

Eilers, J.M., Kanciruk, P., McCord, R.A., Overton, W.S., Hook, L., Blick, D.J., Brakke, D.F., Lellar, P.E., DeHans, M.S., Silverstein, M.E., and Landers, D.H. (1987). Characteristics of lakes in the western United States. *Data Compendium for Selected Physical and Chemical Variables, Vol 2,*. Washington, D.C.: Environmental Protection Agency, EP/600/3-86-054b.

Elfving, G. (1952). Optimum allocation in linear regression theory. *Ann Math Statist*, 23, 255–262.

Fedorov, V.V. (1996). Design for spatial experiments: model fitting and prediction. In *Handbook of Statistics*. Eds. S. Ghosh and C.R. Rao, 13, 515–553.

Fedorov, V.V., and Hackl, P. (1997). *Model-oriented design of experiments*, vol. 125 of *Lecture Notes in Statistics*. New York: Springer-Verlag.

Fedorov, V.V., and Müller, W.G. (1988). Two approaches in optimization of observing networks. In *Optimal Design and Analysis of Experiments*, Eds. Y. Dodge, V.V. Fedorov, and H.P. Wynn. New York: North Holland.

Fedorov, V.V., and Müller, W. (1989). Comparison of two approaches in the optimal design of an observation network. *Statistics*, 3, 339–351.

Fedorov, V.V., and Müller, W. (2008). Optimum Design for Correlated Fields via Covariance Kernel Expansions. *mODa 8–Advances in Model-Oriented Design and Analysis Proceedings of the 8th International Workshop in Model-Oriented Design and Analysis*, Almagro, Spain, June 4–8, 2007. Physica-Verlag HD, 57–66.

Fuentes, M., Chaudhuri, A., and Holland, D.M. (2007). Bayesian entropy for spatial sampling design of environmental data. *Environ. Ecol. Stat.* 14, 323–340.

Gaudard, M., Karson, M., Linder, E., and Sinha, D. (1999). Bayesian spatial prediction (with discussion). *J Ecolog Environ Statist*.

Gonzalez, F.I., Bernard, E.N., Meinig, C., Eble, M.C., Mofjeld, H.O., and Stalin, S. (2005). The NTHMP Tsunameter Network. *Nat Hazards*, 35, 25–39.

Gribik, P., Kortanek, K., and Sweigart, I. (1976). Designing a regional air pollution monitoring network: An appraisal of a regression experimental design approach. In *Proc Conf Environ Model Simul*, 86–91.

Guttorp, P., Le, N.D., Sampson, P.D., and Zidek, J.V. (1993). Using entropy in the redesign of an environmental monitoring network. In *Multivariate Environmental Statistics*. Eds. G.P. Patil, C.R. Rao, and N.P. Ross. New York: North Holland/Elsevier Science, 175–202.

Jaynes, E.T. (1963). Information theory and statistical mechanics. In *Statistical Physics*. Ed. K.W. Ford, New York: Benjamin, 102–218.

Kibria, G.B.M., Sun L., Zidek V., and Le, N.D. (2002). Bayesian spatial prediction of random space-time fields with application to mapping $PM_{2.5}$ exposure. *J Amer Statist Assoc*, 457, 101–112.

Kiefer, J. (1959). Optimum experimental design. *JRSS, Ser B*, 21, 272–319.

Ko, C.W., Lee, J., and Queyranne, M. (1995). An exact algorithm for maximum entropy sampling. *Oper Res*, 43, 684–691.

Le, N.D., and Zidek, J.V. (1994). Network designs for monitoring multivariate random spatial fields. In *Recent Advances in Statistics and Probobility*, Eds. J.P. Vilaplana and M.L. Puri, Utrecht, the Netherlands: VSP, 191–206.

Le, N.D., and Zidek, J.V. (2006). *Statistical Analysis of Environmental Space-Time Processes*. New York: Springer.

Le, N.D., Sun, W., and Zidek, J.V. (1997). Bayesian multivariate spatial interpolation with data missing-by-design. *JRSS, Ser B* 59, 501–510.

Le, N.D., Sun, L., and Zidek, J.V. (2001). Spatial prediction and temporal backcasting for environmental fields having monotone data patterns. *Can J Statist*, 29, 516–529.

Lee, J. (1998). Constrained maximum entropy sampling. *Oper Res*, 46, 655–664.

Lindley, D.V. (1956). On the measure of the information provided by an experiment. *Ann. Math. Statist.*, 27, 968–1005.

McBratney, A.B., Webster, R., and Burgess, T.M. (1981). The design of optimal sampling schemes for local estimation and mapping of reginalized variables. I—Theory and method. *Comp Geosci*, 7, 331–334.

Mitchell, T.J. (1974a). An algorithm for the construction of "D-Optimal" experimental designs. *Technometrics*, 16, 203–210.

Mitchell, T.J. (1974b). Computer construction of "D-Optimal" first-order designs. *Technometrics*, 16, 211–221.

Müller, W.G. (2001). *Collecting Spatial Data: Optimum Design of Experiments for Random Fields*, second ed. Heidelberg: Physica-Verlag.

Müller, W.G. (2005). A comparison of spatial design methods for correlated observations. *Environmetrics*, 16, 495–506.

Müller, W.G. (2007). *Collecting Spatial Data: Optimum Design of Experiments for Random Fields, 3rd ed.* Heidelberg: Physica-Verlag.

Müller, W.G., and Stehlik (2008). Compound optimal spatial designs. IFAS Research Paper Series 2008-37, Department of Applied Statistics, Johannes Kepler University, Linz.

Müller, W.G., and Zimmerman, D.L. (1999). Optimal design for variogram estimation. *Environmetrics*, 10, 23–37.

Myers, D.E. (2002) Space-time correlation models and contaminant plumes. *Environmetrics*, 13, 535–553.

Nychka, D., and Saltzman, N. (1998). Design of air-quality monitoring designs. In *Case Studies in Environmental Statistics*, Eds. D. Nychka, W. Piegorsch, and L. Cox. New York: Springer, 51–76.

Olea, R.A. (1984). Sampling design optimization for special functions. *Mathematical Geology*, 16, 369–392.

Pilz, J. (1991). *Bayesian Estimation and Experimental Design in Linear Regression Models*. New York: John Wiley & Sons.

Ritter, K. (1996). Asymptotic optimality of regular sequence designs. *Ann Statist*, 24, 2081–2096.

Royle, J.A., and Nychka, D. (1998). An algorithm for the construction of spatial coverage designs with implementation in Splus. *Comp and Geosci*, 24, 479–488.

Sampson, P.D., Guttorp, P. and Holland, D.M. (2001). Air quality monitoring network disign using Pareto optimality methods for multiple objective criteria. http://www.epa.gov

Schumacher, P., and Zidek, J.V. (1993). Using prior information in designing intervention detection experiments. *Ann Statist*, 21, 447–463.

Sebastiani, P., and Wynn, H.P. (2000). Maximum entropy sampling and optimal Bayesian experimental design. *JRSS. Soc. B*, 62, 145–157.

Shewry, M., and Wynn, H. (1987). Maximum entropy sampling. *J Applied Statist*, 14, 165–207.

Silvey, S.D. (1980). *Optimal Design*. London: Chapman & Hall.

Smith, K. (1918). On the standard deviations of adjusted and interpolated values of an observed polynomial function and its constants and guidance they give towards a proper choice of the distribution of observations. *Biometrika*, 12, 1–85.

Spöck, G., and Pilz, J. (2008). Spatial sampling design and covariance—robust minimax predication based on convex design ideas. *Stoch Environ Res Risk Assess*, DOI 10.1007

Su, Y., and Cambanis, S. (1993). Sampling designs for estimation of a random process. *Stoch Proc Appl*, 46, 47–89.

Ucinski, D. (2004). *Optimal Measurement Methods for Distributed Parameter System Identification*. CRC Press: Boca Raton, FL.

Wackernagel, H. (2003) *Multivariate Geostatistics: An Introduction with Applications*, 2nd ed. Berlin: Springer.

Wu, S., and Zidek, J.V. (1992). An entropy-based review of selected NADP/NTN network sites for 1983–1986. *Atmosph Environ*, 26A, 2089–2103.

Yaglom, A.M. (1987). *Correlation Theory of Stationary and Related Random Functions*. New York: Springer.

Yfantis, E.A., Flatman, G.T., and Behar, J.V. (1987). Efficiency of kriging estimation for square, triangular and hexagonal grids. *Math Geol*, 19, 183–205.

Zhu, Z., and Stein, M.L. (2005). Spatial sampling design for parameter estimation of the covariance function. *J Statist Plan Infer*, 134, 583–603.

Zhu, Z., and Stein, M.L. (2006). Spatial sampling design for prediction with estimated parameters. *J Agri, Biol Environ Statist*, 11, 24–44.

Zimmerman, D.L. (2006). Optimal network design for spatial prediction, covariance parameter estimation, and empirical prediction. *Environmetrics*, 17, 635–652.

Zidek, J.V., Sun, W., and Le, N.D. (2000). Designing and integrating composite networks for monitoring multivariate Gaussian pollution fields. *Appl Statist*, 49, 63–79.

Zidek, J.V., Sun, L., and Le, N.D. (2003). *Designing Networks for Monitoring Multivariate Environmental Fields Using Data with Monotone Pattern*. TR 2003–5, Statistical and Applied Mathematical Sciences Institute, North Carolina.

11

Non-Gaussian and Nonparametric Models for Continuous Spatial Data

Mark F.J. Steel and Montserrat Fuentes

CONTENTS

Statistical modeling of continuous spatial data is often based on Gaussian processes. This typically facilitates prediction, but normality is not necessarily an adequate modeling assumption for the data at hand. This has led some authors to propose data transformations before using a Gaussian model: in particular, De Oliveira, Kedem, and Short (1997) propose to use the Box–Cox family of power transformations. An approach based on generalized linear models for spatial data is presented in Diggle, Tawn, and Moyeed (1998). In this chapter, we present some flexible ways of modeling that allow the data to inform us on an appropriate distributional assumption. There are two broad classes of approaches we consider: first, we present a purely parametric modeling framework, which is wider than the Gaussian family, with the latter being a limiting case. This is achieved by scale mixing a Gaussian process with another process, and is particularly aimed at accommodating heavy tails. In fact, this approach allows us to identify spatial heteroscedasticity, and leads to relatively simple inference and prediction procedures. A second class of models is based on Bayesian nonparametric procedures. Most of the approaches discussed fall within the family of stick-breaking priors, which we will discuss briefly. These models are very flexible, in that they do not assume a single parametric family, but allow for highly non-Gaussian behavior. A perhaps even more important property of the models discussed in this chapter is that they accommodate nonstationary behavior.

We will adopt a Bayesian framework throughout this chapter. In order to focus on the non-Gaussian properties in space, we shall only consider spatial processes. Extensions to spatial-temporal settings are, in principle, straightforward. Throughout, we denote a

k-variate normal distribution on a random vector $\mathbf{y}$ with mean $\boldsymbol{\mu}$ and variance–covariance matrix $\boldsymbol{\Sigma}$, as $\mathbf{y} \sim \mathrm{N}_k(\boldsymbol{\mu}, \boldsymbol{\Sigma})$ with density function $f_N^k(\mathbf{y}|\boldsymbol{\mu}, \boldsymbol{\Sigma})$.

11.1 Non-Gaussian Parametric Modeling

There are a number of parametric approaches leading to non-Gaussian models for continuous data that will not be discussed in detail in this section. In particular, we shall not deal with models that use a transformation of the data (as in De Oliveira et al., 1997) to induce Gaussian behavior. Another approach that is not discussed here is the use of Gaussian processes as a component within a nonlinear model for the observations, such as a generalized linear model (see Diggle et al., 1998 and further discussed in Chapter 4) or a frailty model for survival data, such as used in Banerjee, Wall, and Carlin (2003) and Li and Ryan (2002). In addition, we shall omit discussion of some application-specific ways of modeling non-Gaussian data, such as the approach of Brown, Diggle, and Henderson (2003) to model the opacity of flocculated paper.

Let $Y(\mathbf{s})$ be a random process defined for locations $\mathbf{s}$ in some spatial region $\mathcal{D} \subset \Re^d$. We assume the model

$$Y(\mathbf{s}) = \mathbf{x}(\mathbf{s})^T \boldsymbol{\beta} + \eta(\mathbf{s}) + \epsilon(\mathbf{s}), \tag{11.1}$$

where the mean surface is assumed to be a linear function of $\mathbf{x}(\mathbf{s})^T = (x_1(\mathbf{s}), \ldots, x_k(\mathbf{s}))$, a vector of k variables, which typically include known functions of the spatial coordinates, with unknown coefficient vector $\boldsymbol{\beta} \in \Re^k$. Further, $\eta(\mathbf{s})$ is a second-order stationary error process with zero mean and variance σ^2 and with an isotropic correlation function (depending only on the distance between points) $\mathrm{corr}\,[\eta(\mathbf{s}_i), \eta(\mathbf{s}_j)] = C_\theta(\|\mathbf{s}_i - \mathbf{s}_j\|)$, where $C_\theta(d)$ is a valid correlation function of distance d, parameterized by a vector $\boldsymbol{\theta}$. Finally, $\epsilon(\mathbf{s})$ denotes an uncorrelated Gaussian process with mean zero and variance τ^2, modeling the so-called "nugget effect" (or "pure error," allowing for measurement error and small-scale variation). The ratio $\omega^2 = \tau^2/\sigma^2$ indicates the relative importance of the nugget effect.

We assume that we have observed a single realization from this random process at n different locations $\mathbf{s}_1, \ldots, \mathbf{s}_n$ and we denote the vector observation by $\mathbf{y} = (y_1, \ldots, y_n)^T$, where we use the notation $y_i = Y(\mathbf{s}_i)$. As mentioned above, the most commonly made distributional assumption is that $\eta(\mathbf{s})$ is a Gaussian process, which implies that $\mathbf{y}$ follows an n-variate Normal distribution with $E[\mathbf{y}] = X^T\boldsymbol{\beta}$, where $X = (\mathbf{x}(\mathbf{s}_1), \ldots, \mathbf{x}(\mathbf{s}_n))$, and $\mathrm{var}[\mathbf{y}] = \sigma^2\mathbf{C}_\theta + \tau^2\mathbf{I}_n$, where $\mathbf{C}_\theta$ is the $n \times n$ correlation matrix with $C_\theta(\|\mathbf{s}_i - \mathbf{s}_j\|)$ as its (i, j)th element. Note that even though we only have one observation per location, we are still able to criticize the normality assumption; in particular, the n elements of $B(\mathbf{y} - X^T\boldsymbol{\beta})$ where $B^{-1}(B^{-1})^T = \sigma^2\mathbf{C}_\theta + \tau^2\mathbf{I}_n$ are assumed to be independent draws from a standard Normal, given all model parameters.

11.1.1 The Gaussian-Log-Gaussian Mixture Model

In Palacias and Steel (2006), an alternative stochastic specification based on scale mixing the Gaussian process $\eta(\mathbf{s})$ is proposed. In particular, a mixing variable $\lambda_i \in \Re_+$ is assigned to each observation $i = 1, \ldots, n$, and the sampling model for the ith location, $i = 1, \ldots, n$, is now changed to

$$y_i = \mathbf{x}(\mathbf{s}_i)^T\boldsymbol{\beta} + \frac{\eta_i}{\sqrt{\lambda_i}} + \epsilon_i, \tag{11.2}$$

where we have used the notation $\eta_i = \eta(\mathbf{s}_i)$ and $\epsilon_i = \epsilon(\mathbf{s}_i)$, and $\epsilon_i \sim \mathrm{N}_1(0, \tau^2)$, iid and independent of $\boldsymbol{\eta} = (\eta_1, \ldots, \eta_n)' \sim \mathrm{N}_n(\mathbf{0}, \sigma^2\mathbf{C}_\theta)$. The mixing variables λ_i are independent of ϵ_i and $\boldsymbol{\eta}$. In order to verify that the sampling model described above is consistent with a well-defined stochastic process, we can check that the Kolmogorov consistency conditions are satisfied. In Palacios and Steel (2006), it is shown that Equation (11.2) does support a stochastic process provided that the distribution of the mixing variables satisfies a very weak symmetry condition under permutation.

In this spatial model, scale mixing introduces a potential problem with the continuity of the resulting random field Y. Let us, therefore, consider a stationary process $\lambda(\mathbf{s})$ for the mixing variables, so that $\lambda_i = \lambda(\mathbf{s}_i)$. The representation in Equation (11.2) makes clear that we are now replacing the Gaussian stochastic process $\eta(\mathbf{s})$ by a ratio of independent stochastic processes $\eta(\mathbf{s})/\sqrt{\lambda(\mathbf{s})}$. Mean square continuity of the spatial process $\eta(\mathbf{s})/\sqrt{\lambda(\mathbf{s})}$ is defined by $\mathrm{E}[\{\eta(\mathbf{s}_i)/\sqrt{\lambda(\mathbf{s}_i)} - \eta(\mathbf{s}_j)/\sqrt{\lambda(\mathbf{s}_j)}\}^2]$ tending to zero as $\mathbf{s}_i \to \mathbf{s}_j$. Assuming that $\mathrm{E}[\lambda_i^{-1}]$ exists, we obtain

$$\mathrm{E}\left[\left\{\frac{\eta_i}{\sqrt{\lambda_i}} - \frac{\eta_j}{\sqrt{\lambda_j}}\right\}^2\right] = 2\sigma^2\left\{\mathrm{E}[\lambda_i^{-1}] - C_\theta(\|\mathbf{s}_i - \mathbf{s}_j\|)\mathrm{E}[\lambda_i^{-1/2}\lambda_j^{-1/2}]\right\},$$

which in the limit as $\|\mathbf{s}_i - \mathbf{s}_j\| \to 0$ tends to $2\sigma^2\{\mathrm{E}[\lambda_i^{-1}] - \lim_{\|\mathbf{s}_i-\mathbf{s}_j\|\to 0}\mathrm{E}[\lambda_i^{-1/2}\lambda_j^{-1/2}]\}$. If λ_i and λ_j are independent, then $\lim_{\|\mathbf{s}_i-\mathbf{s}_j\|\to 0}\mathrm{E}[\lambda_i^{-1/2}\lambda_j^{-1/2}] = \{\mathrm{E}[\lambda_i^{-1/2}]\}^2 \le \mathrm{E}[\lambda^{-1}]$ from Jensen's inequality and, thus, $\eta(\mathbf{s})/\sqrt{\lambda(\mathbf{s})}$ is not mean square continuous. This also can be seen immediately by considering the logarithm of the process $\log\{\eta(\mathbf{s})/\sqrt{\lambda(\mathbf{s})}\} = \log\{\eta(\mathbf{s})\} - (1/2)\log\{\lambda(\mathbf{s})\}$. This discontinuity essentially arises from the fact that two separate locations, no matter how close, are assigned independent mixing variables. Thus, in order to induce mean square continuity of the process (in the version without the nugget effect), we need to correlate the mixing variables in $\boldsymbol{\lambda}$, so that locations that are close will have very similar values of λ_i. In particular, if $\lambda^{-1/2}(\mathbf{s})$ is itself mean square continuous, then $\eta(\mathbf{s})/\sqrt{\lambda(\mathbf{s})}$ is a mean square continuous process.

Therefore, Palacios and Steel (2006) consider the following mixing distribution:

$$\ln(\boldsymbol{\lambda}) = (\ln(\lambda_1), \ldots, \ln(\lambda_n))^T \sim \mathrm{N}_n\left(-\frac{\nu}{2}\mathbf{1}, \nu\mathbf{C}_\theta\right), \tag{11.3}$$

where $\mathbf{1}$ is a vector of ones, and we correlate the elements of $\ln(\boldsymbol{\lambda})$ through the same correlation matrix as $\boldsymbol{\eta}$. Equivalently, we assume a Gaussian process for $\ln(\lambda(\mathbf{s}))$ with constant mean surface at $-\nu/2$ and covariance function $\nu\mathbf{C}_\theta(\|\mathbf{s}_i - \mathbf{s}_j\|)$. One scalar parameter $\nu \in \Re_+$ is introduced in Equation (11.3), which implies a lognormal distribution for λ_i with $\mathrm{E}[\lambda_i] = 1$ and $\mathrm{var}[\lambda_i] = \exp(\nu) - 1$. Thus, the marginal distribution of λ_i is tight around unity for very small ν (of the order $\nu = 0.01$) and as ν increases, the distribution becomes more spread out and more right skewed, while the mode shifts toward zero. For example, for $\nu = 3$, the variance is 19.1 and there is a lot of mass close to zero. It is exactly values of λ_i close to zero that will lead to an inflation of the scale in Equation (11.2) and will allow us to accommodate heavy tails. On the other hand, as $\nu \to 0$, we retrieve the Gaussian model as a limiting case. Figure 11.1 illustrates this behavior.

In Palacios and Steel (2006) the mixture model defined by Equations (11.2) and (11.3) is called the Gaussian-log-Gaussian (GLG) model. This approach is similar to that of Damian, Sampson, and Guttorp (2001, 2003), where an additional space deformation, as in Sampson and Guttorp (1992), is used to introduce nonstationarity. Note that the latter complication requires repeated observations for reliable inference.

11.1.2 Properties and Interpretation

The correlation structure induced by the GLG model is given by

$$\mathrm{corr}\,[y_i, y_j] = C_\theta(\|\mathbf{s}_i - \mathbf{s}_j\|)\frac{\exp\left(\nu\left\{1 + \frac{1}{4}[C_\theta(\|\mathbf{s}_i - \mathbf{s}_j\|) - 1]\right\}\right)}{\exp(\nu) + \omega^2}. \tag{11.4}$$

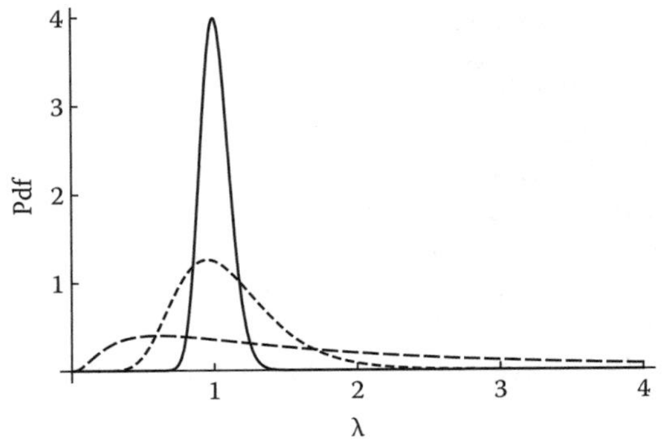

FIGURE 11.1
Marginal probability density function of mixing variables λ_i for various values of ν. Solid line $\nu = 0.01$, short dashes: $\nu = 0.1$, long dashes: $\nu = 1$.

Thus, in the case without nugget effect ($\omega^2 = 0$), we see that if the distance between $\mathbf{s}_i$ and $\mathbf{s}_j$ tends to zero, the correlation between y_i and y_j tends to one, so that the mixing does not induce a discontinuity at zero. It can also be shown (see Palacios and Steel, 2006) that the smoothness of the process is not affected by the mixing, in the sense that without the nugget effect the process $Y(\mathbf{s})$ has exactly the same smoothness properties as $\eta(\mathbf{s})$.

The tail behavior of the finite-dimensional distributions induced by the GLG process is determined by the extra parameter ν. In particular, Palacios and Steel (2006) derive that the kurtosis of the marginal distributions is given by $\text{kurt}[y_i] = 3\exp(\nu)$, again indicating that large ν corresponds to heavy tails, and Gaussian tails are the limiting case as $\nu \to 0$.

Our chosen specification for mixing the spatially dependent process as in Equation (11.2) requires a smooth $\lambda(\mathbf{s})$ process, which means that observations with particularly small values of λ_i will tend to cluster together. Thus, what we are identifying through small values of λ_i are regions of the space where the observations tend to be relatively far away from the estimated mean surface. Therefore, we can interpret the presence of relatively small values of λ_i in terms of spatial heteroscedasticity, rather than the usual concept of outlying observations. However, for convenience we will continue to call observations with small λ_i "outliers."

It may be useful to have an indication of which areas of the space require an inflated variance. Indicating regions of the space where the Gaussian model fails to fit the data well might suggest extensions to the underlying trend surface (such as missing covariates) that could make a Gaussian model a better option. The distribution of λ_i is informative about the outlying nature of observation i. Thus, Palacios and Steel (2006) propose to compute the ratio between the posterior and the prior density functions for λ_i evaluated at $\lambda_i = 1$, i.e.,

$$R_i = \frac{p(\lambda_i|\mathbf{y})}{p(\lambda_i)}|_{\lambda_i=1}\,. \tag{11.5}$$

In fact, this ratio R_i is the so-called Savage–Dickey density ratio, which would be the Bayes factor in favor of the model with $\lambda_i = 1$ (and all other elements of $\boldsymbol{\lambda}$ free) versus the model with free λ_i (i.e., the full mixture model proposed here) if $C_\theta(\|\mathbf{s}_i - \mathbf{s}_j\|) = 0$ for all $j \neq i$. In this case, the Savage–Dickey density ratio is not the exact Bayes factor, but has to be adjusted as in Verdinelli and Wasserman (1995). The precise adjustment in this case is explained in Palacios and Steel (2006). Bayes factors convey the relative support of the data for one model versus another and immediately translate into posterior probabilities of rival models since the posterior odds (the ratio of two posterior model probabilities) equals the Bayes factor times the prior odds (the ratio of the prior model probabilities).

11.1.3 Prediction

An important reason for geostatistical modeling is prediction at unsampled sites. We wish to fully incorporate all uncertainty in the problem, including the covariance function.

Let $\mathbf{y} = (\mathbf{y}_o^T, \mathbf{y}_p^T)^T$ where $\mathbf{y}_o$ correspond to the $n - f$ observed locations and $\mathbf{y}_p$ is a vector of values to predict at f given sites. We are interested in the posterior predictive distribution of $\mathbf{y}_p$, i.e.,

$$p(\mathbf{y}_p|\mathbf{y}_o) = \int p(\mathbf{y}_p|\mathbf{y}_o, \boldsymbol{\lambda}, \boldsymbol{\zeta}) p(\boldsymbol{\lambda}_p|\boldsymbol{\lambda}_o, \boldsymbol{\zeta}, \mathbf{y}_o) p(\boldsymbol{\lambda}_o, \boldsymbol{\zeta}|\mathbf{y}_o) d\boldsymbol{\lambda} d\boldsymbol{\zeta}, \tag{11.6}$$

where we have partitioned $\boldsymbol{\lambda} = (\boldsymbol{\lambda}_o^T, \boldsymbol{\lambda}_p^T)^T$ conformably with $\mathbf{y}$ and $\boldsymbol{\zeta} = (\boldsymbol{\beta}, \sigma, \tau, \boldsymbol{\theta}, \nu)$. The integral in Equation (11.6) will be approximated by Monte Carlo simulation, where the draws for $(\boldsymbol{\lambda}_o, \boldsymbol{\zeta})$ are obtained directly from the Markov chain Monte Carlo (MCMC) inference algorithm (which is described in some detail in Palacios and Steel, 2006) and, because $p(\boldsymbol{\lambda}_p|\boldsymbol{\lambda}_o, \boldsymbol{\zeta}, \mathbf{y}_o) = p(\boldsymbol{\lambda}_p|\boldsymbol{\lambda}_o, \nu)$ we can evaluate Equation (11.6) by using drawings for $\boldsymbol{\lambda}_p$ from

$$p(\ln \boldsymbol{\lambda}_p|\boldsymbol{\lambda}_o, \nu) = f_N^f\left(\ln \boldsymbol{\lambda}_p \mid \frac{\nu}{2}[\mathbf{C}_{po}\mathbf{C}_{oo}^{-1}\mathbf{1}_n - \mathbf{1}_f] + \mathbf{C}_{po}\mathbf{C}_{oo}^{-1}\ln \boldsymbol{\lambda}_o, \nu[\mathbf{C}_{pp} - \mathbf{C}_{po}\mathbf{C}_{oo}^{-1}\mathbf{C}_{op}]\right), \tag{11.7}$$

where we have partitioned

$$\mathbf{C}_{\boldsymbol{\theta}} = \begin{pmatrix} \mathbf{C}_{oo} & \mathbf{C}_{op} \\ \mathbf{C}_{po} & \mathbf{C}_{pp} \end{pmatrix}$$

conformably with $\mathbf{y}$. Thus, for each posterior drawing of $(\boldsymbol{\lambda}_o, \boldsymbol{\zeta})$, we will generate a drawing from Equation (11.7) and evaluate

$$p(\mathbf{y}_p|\mathbf{y}_o, \boldsymbol{\lambda}, \boldsymbol{\zeta}) = f_N^f\left(\mathbf{y}_p|(\mathbf{X}_p - \mathbf{A}\mathbf{X}_o)\boldsymbol{\beta} + \mathbf{A}\mathbf{y}_o, \sigma^2\left(\boldsymbol{\Lambda}_p^{-\frac{1}{2}}\mathbf{C}_{pp}\boldsymbol{\Lambda}_p^{-\frac{1}{2}} + \omega^2\mathbf{I}_f - \mathbf{A}\boldsymbol{\Lambda}_o^{-\frac{1}{2}}\mathbf{C}_{op}\boldsymbol{\Lambda}_p^{-\frac{1}{2}}\right)\right), \tag{11.8}$$

where $\mathbf{I}_f$ is the f-dimensional identity matrix, $\mathbf{A} = \boldsymbol{\Lambda}_p^{-\frac{1}{2}}\mathbf{C}_{po}\boldsymbol{\Lambda}_o^{-\frac{1}{2}}\left[\boldsymbol{\Lambda}_o^{-\frac{1}{2}}\mathbf{C}_{oo}\boldsymbol{\Lambda}_o^{-\frac{1}{2}} + \omega^2\mathbf{I}_n\right]^{-1}$ and $\mathbf{X}$ and $\boldsymbol{\Lambda} = \text{Diag}(\lambda_1, \ldots, \lambda_n)$ are partitioned conformably to $\mathbf{y}$. Averaging the densities in Equation (11.8) will give us the required posterior predictive density function.

11.1.4 Correlation Function and Prior Distribution

For the correlation function $C_{\boldsymbol{\theta}}(d)$, where d is the Euclidean distance, we use the flexible Matérn class:

$$C_{\boldsymbol{\theta}}(d) = \frac{1}{2^{\theta_2 - 1}\Gamma(\theta_2)}\left(\frac{d}{\theta_1}\right)^{\theta_2}\mathcal{K}_{\theta_2}\left(\frac{d}{\theta_1}\right), \tag{11.9}$$

where $\boldsymbol{\theta} = (\theta_1, \theta_2)^T$ with $\theta_1 > 0$ the range parameter and $\theta_2 > 0$ the smoothness parameter and where $\mathcal{K}_{\theta_2}(\cdot)$ is the modified Bessel function of the third kind of order θ_2. As a consequence, $\eta(\mathbf{s})$ and thus $Y(\mathbf{s})$ are q times mean square differentiable if and only if $\theta_2 > q$.

In order to complete the Bayesian model, we now need to specify a prior distribution for the parameters $(\boldsymbol{\beta}, \sigma^{-2}, \omega^2, \nu, \boldsymbol{\theta})$. The prior distribution used in Palacios and Steel (2006) is a carefully elicited proper prior. They argue against the use of reference priors as used in Berger, De Oliveira, and Sansó (2001) for a simpler Gaussian model with fixed smoothness parameter θ_2 and without the nugget effect. In addition, such a reference prior would be extremely hard to derive for the more general GLG model discussed here. The prior used has a product structure with a normal prior for $\boldsymbol{\beta}$, a gamma prior for σ^{-2}, and a generalized inverse Gaussian (GIG) prior for both ω^2 and ν. The prior on $\boldsymbol{\theta}$ either imposes prior independence between θ_1 and θ_2 or between the alternative range parameter

TABLE 11.1

Temperature Data: Posterior Means (Standard Deviation) of the Trend Parameters

Model	β_1	β_2	β_3	β_4	β_5	β_6	β_7
Gaussian	3.19(0.22)	−0.20(0.23)	0.19(0.31)	−0.20(0.23)	0.37(0.45)	−0.24(0.28)	−0.40(0.18)
GLG	3.23(0.06)	−0.08(0.11)	0.12(0.13)	−0.19(0.09)	0.09(0.24)	−0.17(0.14)	−0.42(0.07)

$\rho = 2\theta_1\sqrt{\theta_2}$ (see Stein, 1999, p. 51) and θ_2. In both cases, the prior on the Matérn parameters consists of the product of two judiciously chosen exponential distributions.

An extensive sensitivity analysis in Palacios and Steel (2006) suggests that a dataset of small (but typical) size is not that informative on certain parameters. In particular, the parameters $(\nu, \boldsymbol{\theta}, \omega^2)$ are not that easily determined by the data and thus require very careful prior elicitation. In general, spatial models do suffer from weak identification issues and, thus, prior specification is critical. Chapter 4 (Section 4.4) discusses the fact that some parameters are not consistently estimated by classical maximum likelihood methods under infill asymptotics.

11.1.5 An Application to Spanish Temperature Data

We analyze the maximum temperatures recorded in an unusually hot week in May 2001 in 63 locations within the Spanish Basque country. So, $\mathcal{D} \subset \Re^2$ and, for the trend function $\mathbf{x}(\mathbf{s})$, we use a quadratic form in the coordinates (with linear terms and the cross-product). As this region is quite mountainous (with the altitude of the monitoring stations in between 16 and 1,188 meters (52 and 3,897 feet)), altitude is added as an extra explanatory variable (corresponding to regression coefficient β_7). Table 11.1 presents some posterior results for $\boldsymbol{\beta}$, using both the Gaussian and the GLG model. The Gaussian model tends to higher absolute values for β_2 and β_5 and the inference on $\boldsymbol{\beta}$ is generally a lot less concentrated for this model. In both models, higher altitude tends to reduce the mean temperature, as expected. Posterior inference on the other parameters in the models is presented in Table 11.2. Clearly, the Gaussian model assigns a larger importance to the nugget effect (see the difference in $\tau/[\sigma \exp(\nu/2)]$, which is the ratio of standard deviations between the process inducing the nugget effect and the spatial process), while making the surface a lot smoother than the GLG model. In order to accommodate the outlying observations (discussed later), the Gaussian model needs to dramatically increase the values of both σ and τ. Since most of the posterior mass for ν is well away from zero, it is not surprising that the evidence in favor of the GLG model is very strong indeed. In particular, the Bayes factor in favor of the GLG model is $3.4 \cdot 10^{20}$, a lot of which is attributable to three very extreme observations: observations 20, 36, and 40, which are all close together. Table 11.3 presents the Bayes factors in favor of

TABLE 11.2

Temperature Data: Posterior Means (Standard Deviation) for Some Nontrend Parameters

	Gaussian	GLG
σ	0.32 (0.11)	0.09 (0.03)
ω^2	1.22 (0.79)	1.27 (1.12)
τ	0.31 (0.06)	0.08 (0.02)
θ_1	5.71 (10.33)	4.02 (12.70)
θ_2	1.87 (2.03)	0.61 (0.98)
ρ	8.64 (8.20)	2.35 (2.97)
ν	0 (0)	2.51 (0.76)
$\sigma^2 \exp(\nu)$	0.11 (0.09)	0.12 (0.15)
$\tau/[\sigma \exp(\nu/2)]$	1.05 (0.35)	0.30 (0.12)

TABLE 11.3

Temperature Data: Bayes Factors in Favor of $\lambda_i = 1$ for Selected Observations; the Entries 0.000 Indicate Values Less than 0.0005

obs.#	$E[\lambda_i\|z]$	S.Dev.$[\lambda_i\|z]$	(11.5)	corr	BF for $\lambda_i = 1$
20	0.020	0.024	0.000	0.74	0.000
36	0.015	0.020	0.000	0.79	0.000
40	0.016	0.020	0.000	0.62	0.000
41	0.059	0.085	0.006	0.57	0.004

$\lambda_i = 1$ for the four observations with smallest mean λ_i. The column labeled "corr" is the multiplicative correction factor to the Savage–Dickey density ratio mentioned in Section 11.1.2. Clearly, all observations listed in Table 11.3 are outliers, indicating two regions with inflated variance.

Figure 11.2 displays the predictive densities, computed as in Section 11.1.3 for five unobserved locations, ranging in altitude from 53 to 556 meters (174 to 1,824 feet). The GLG model leads to heavier extreme tails than the Gaussian model as a consequence of the scale mixing. Nevertheless, in the (relevant) central mass of the distribution, the GLG predictives clearly are more concentrated than the Gaussian ones, illustrating that the added uncertainty due to the scale mixing is more than offset by changes in the inference on other aspects of the model. In particular, the nugget effect is much less important for the non-Gaussian model. From Equation (11.8) it is clear that the predictive standard deviation is bounded from below by τ (in order to interpret the numbers in Table 11.2 in terms of observables measured in degrees centigrade, we need to multiply τ by a factor 10, due to scaling of the data). Clearly, a lot of the predictive uncertainty in the Gaussian case is due to the nugget effect.

11.2 Bayesian Nonparametric Approaches

In this section we will use nonparametric models for the spatial components, which can accommodate much more flexible forms and can also easily deal with skewness, multimodality etc. In addition, even though the prior predictive distributions induced by these models are stationary, the posterior predictives can accommodate very nonstationary behavior. As the Bayesian nonparametric methods presented are all based on the broad class of stick-breaking priors, we will first briefly explain this class of priors. See Müller and Quintana (2004) for an excellent overview of nonparametric Bayesian inference procedures, while Dunson (forthcoming) provides an insightful and very up-to-date discussion of nonparametric Bayesian methods, specifically aimed at applications in biostatistics. One approach that we will not discuss here is that of transforming the space corresponding to a Gaussian parametric model, as introduced in Sampson and Guttorp (1992) and developed in Schmidt and O'Hagan (2003) in a Bayesian framework.

11.2.1 Stick-Breaking Priors

Bayesian nonparametric methods avoid dependence on parametric assumptions by working with probability models on function spaces; in other words, by using (in principle) infinitely many parameters. A useful and broad class of such random probability measures is the class of stick-breaking priors. This class was discussed in some detail by Ishwaran and James (2001) and is at the basis of many recent studies.

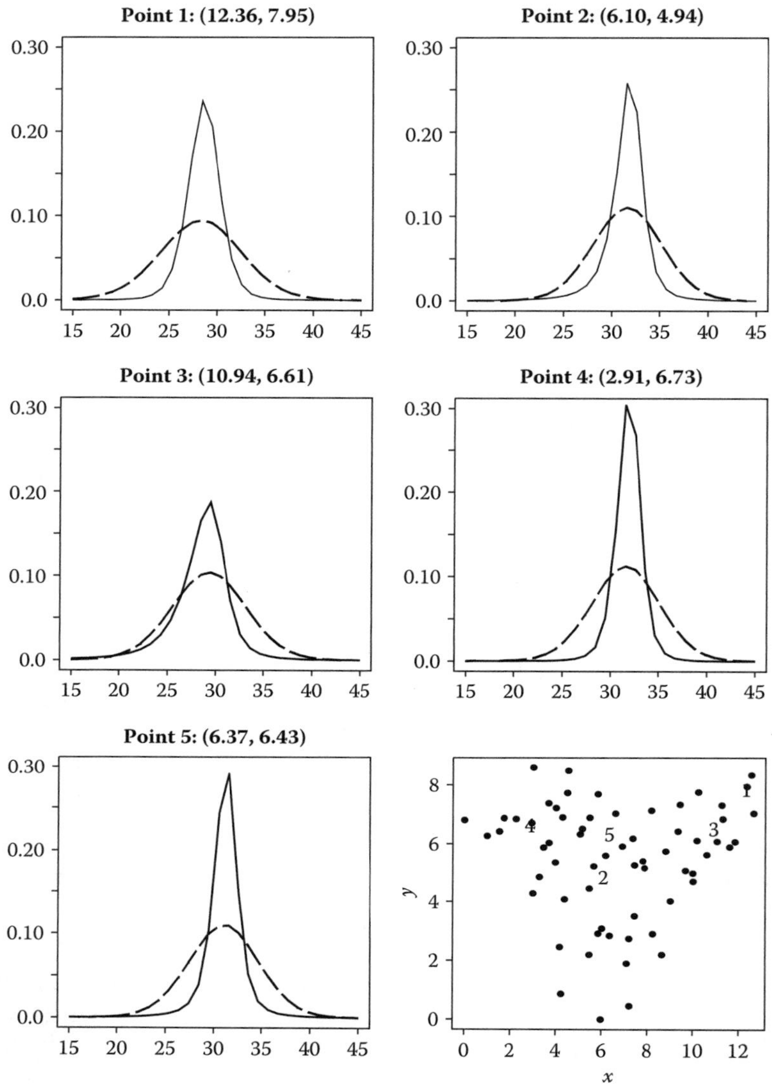

FIGURE 11.2
Temperature data: Predictive densities at five unobserved locations. The observables are measured in degrees centigrade, and the elevations at the predicted sites range from 53 (point 2) to 556 (point 3) meters. Dashed line: Gaussian, solid line: GLG. The lower right panel indicates the locations of the observed sites by dots and the five unobserved sites by their respective numbers.

A random probability distribution, F, has a stick-breaking prior if

$$F \stackrel{d}{=} \sum_{i=1}^{N} p_i \delta_{\boldsymbol{\theta}_i}, \tag{11.10}$$

where δ_z denotes a Dirac measure at z, $p_i = V_i \prod_{j<i}(1 - V_j)$ where $V_1, \ldots, V_{N-1}$ are independent with $V_i \sim \text{Beta}(a_i, b_i)$ and $\boldsymbol{\theta}_1, \ldots, \boldsymbol{\theta}_N$ are independent draws from a centering (or base) distribution H.

The definition in Equation (11.10) allows for either finite or infinite N (with the latter corresponding to the conventional definition of nonparametrics). For $N = \infty$, several interesting and well-known processes fall into this class:

1. The Dirichlet process prior (see Ferguson, 1973) characterized by MH, where M is a positive scalar (often called the mass parameter) arises when V_i follows a Beta(1, M) for all i. This representation was first given by Sethuraman (1994).
2. The Pitman–Yor (or two-parameter Poisson–Dirichlet) process occurs if V_i follows a Beta($1 - a, b + ai$) with $0 \leq a < 1$ and $b > -a$. As special cases, we can identify the Dirichlet process for $a = 0$ and the stable law when $b = 0$.

Stick-breaking priors, such as the Dirichlet process, almost surely lead to discrete probability distributions. This is often not desirable for directly modeling observables that are considered realizations of some continuous process. To avoid this problem, the mixture of Dirichlet process model (introduced in Antoniak, 1974) is now the most commonly used specification in practice. Such models assume a continuous model for the observables, given some unknown parameters, and then use a stick-breaking prior as in Equation (11.10) to model these parameters nonparametrically.

An important aspect of these models is that they tend to cluster the observations by assigning several observations to the same parameter values (or atoms of the nonparametric distribution).

Conducting inference with such models relies on MCMC computational methods. One approach corresponds to marginalizing out F and using a Pólya urn representation to conduct a Gibbs sampling scheme. See MacEachern (1998) for a detailed description of such methods. Another approach (see Ishwaran and James, 2001) directly uses the stick-breaking representation in Equation (11.10) and either truncates the sum or avoids truncation through slice sampling or the retrospective sampler proposed in Papaspiliopoulos and Roberts (2008). An accessible and more detailed discussion of computational issues can be found in, e.g., Dunson (forthcoming).

In order to make this wide class of nonparametric priors useful for our spatial context, we need to somehow index it by space. More generally, we can attempt to introduce dependencies on time or other covariates (leading to nonparametric regression models). Most of the (rather recent) literature in this area follows the ideas in MacEachern (1999), who considered allowing the masses, $\mathbf{V} = (V_1, V_2, \ldots)$, or the locations, $\boldsymbol{\theta} = (\boldsymbol{\theta}_1, \boldsymbol{\theta}_2, \ldots)$, of the atoms to follow a stochastic process defined over the domain. This leads to so-called dependent Dirichlet processes (DDPs) and a lot of this work concentrates on the "single-p" DDP model where only the locations, $\boldsymbol{\theta}$, follow stochastic processes. An application to spatial modeling is developed in Gelfand, Kottas, and MacEachern (2005) by allowing the locations $\boldsymbol{\theta}$ to be drawn from a random field (a Gaussian process). A generalization of this idea is briefly explained in Section 11.2.2.

11.2.2 Generalized Spatial Dirichlet Process

The idea in Gelfand et al. (2005) is to introduce a spatial dependence through the locations, by indexing $\boldsymbol{\theta}$ with the location $\mathbf{s}$ and making $\boldsymbol{\theta}(\mathbf{s})$ a realization of a random field, with H being a stationary Gaussian process. Continuity properties of these realizations will then follow from the choice of covariance function. In the simple model $Y(\mathbf{s}) = \eta(\mathbf{s}) + \epsilon(\mathbf{s})$ where $\eta(\mathbf{s})$ has this spatial Dirichlet prior and $\epsilon(\mathbf{s}) \sim \mathrm{N}(0, \tau^2)$ is a nugget effect, the joint density of the observables $\mathbf{y} = [Y(\mathbf{s}_1), \ldots, Y(\mathbf{s}_n)]^T$ is almost surely a location mixture of Normals with density function of the form $\sum_{i=1}^{N} p_i f_N^n(\mathbf{y}|\boldsymbol{\eta}_i, \tau^2\mathbf{I}_n)$, using Equation (11.10). This allows for a large amount of flexibility and is used in Gelfand et al. (2005) to analyze a dataset of precipitation data consisting of 75 replications at 39 observed sites in the south of France.

However, the joint distribution over any set of locations uses the same set of weights $\{p_i\}$, so the choice between the random surfaces in the location mixture is not dependent on location. In Duan, Guindani, and Gelfand (2007) and Gelfand, Guindani, and Petrone (2007), this framework is extended to allow for the surface selection to vary with the location, while still preserving the property that the marginal distribution at each location is generated from the usual Dirichlet process. This extension, called the generalized spatial Dirichlet process model, assumes that the random probability measure on the space of distribution functions for $\eta(\mathbf{s}_1), \ldots, \eta(\mathbf{s}_n)$ is

$$F^{(n)} \stackrel{d}{=} \sum_{i_1=1}^{\infty} \cdots \sum_{i_n=1}^{\infty} p_{i_1,\ldots,i_n} \delta_{\theta_{i_1}} \ldots \delta_{\theta_{i_n}}, \tag{11.11}$$

where $i_j = i(s_j)$, $j = 1, \ldots, n$, the locations θ_{i_j} are drawn from the centering random field H and the weights $p_{i_1,\ldots,i_n}$ are distributed independently from the locations on the infinite unit simplex. These weights allow for the site-specific selection of surfaces and are constrained to be consistent (to define a proper random process) and continuous (in the sense that they assign similar weights for sites that are close together). This will induce a smooth (mean square continuous) random probability measure.

From Equation (11.11) it is clear that we can think of this generalization in terms of a multivariate stick-breaking representation. In Duan et al. (2007), a particular specification for $p_{i_1,\ldots,i_n}$ is proposed, based on thresholding of auxiliary Gaussian random fields. This leads to processes that are non-Gaussian and nonstationary, with nonhomogeneous variance.

As the locations associated with the sampled sites effectively constitute one single observation from the random field, we need replications in order to conduct inference using this approach. However, replications over time do not need to be independent and a dynamic model can be used.

11.2.3 Hybrid Dirichlet Mixture Models

The idea of the previous section is further developed in Petrone, Guindani, and Gelfand (2009) in the context of functional data analysis. Here, we have observations of curves (e.g., in space or in time) and often it is important to represent the n observed curves by a smaller set of canonical curves. The curves $\mathbf{y}_i = [Y_i(\mathbf{x}_1), \ldots, Y_i(\mathbf{x}_m)]^T$, $i = 1, \ldots, n$ are assumed to be observed at a common set of coordinates $\mathbf{x}_1, \ldots, \mathbf{x}_m$.

Petrone et al. (2009) start from a more general class than stick-breaking priors, namely the species sampling prior, which can be represented as Equation (11.10) with a general specification on the weights. This leads them to a different way of selecting the multivariate weights in Equation (11.11). In Petrone et al. (2009), these weights are interpreted as the distribution of a random vector of labels, assigning curves to locations. They use mixture modeling where the observations are normally distributed with a nonparametric distribution on the locations, i.e.,

$$\begin{aligned} \mathbf{y}_i \mid \boldsymbol{\theta}_i &\stackrel{ind}{\sim} \mathrm{N}_m(\boldsymbol{\theta}_i, \sigma^2 \mathbf{I}_m), \\ \boldsymbol{\theta}_i \mid F_{\mathbf{x}_1,\ldots,\mathbf{x}_m} &\stackrel{iid}{\sim} F_{\mathbf{x}_1,\ldots,\mathbf{x}_m}, \end{aligned} \tag{11.12}$$

where

$$F_{\mathbf{x}_1,\ldots,\mathbf{x}_m} \stackrel{d}{=} \sum_{j_1=1}^{k} \cdots \sum_{j_m=1}^{k} p(j_1, \ldots, j_m)\, \delta_{\theta_{j_1,1},\ldots,\theta_{j_m,m}}, \tag{11.13}$$

where $p(j_1, \ldots, j_m)$ represents the proportion of (hybrid) species $(\theta_{j_1,1}, \ldots, \theta_{j_m,m})$ in the population, $p(j_1, \ldots, j_m) \geq 0$, $\sum_{j_1=1}^{k} \cdots \sum_{j_m=1}^{k} p(j_1, \ldots, j_m) = 1$, and $\boldsymbol{\theta}_j = (\theta_{j,1}, \ldots, \theta_{j,m}) \stackrel{iid}{\sim} H$

(again typically chosen to be an m-dimensional distribution of a Gaussian process), independently of the $p(j_1, \ldots, j_m)$s.

This generates a location mixture of normals with local random effects. Hybrid Dirichlet process mixtures are obtained as limits of the finite mixture framework above for $k \to \infty$. Functional dependence in the (hidden) label process is modeled through an auxiliary Gaussian copula, which contributes to the simplicity and the flexibility of the approach.

An application to magnetic resonance imaging (MRI) brain images in Petrone et al. (2009) illustrates the modeling of the species recombination (hybridization) through the labeling prior and the improvement over simple mixtures of Dirichlet processes.

11.2.4 Order-Based Dependent Dirichlet Process

An alternative approach to extending the framework in Equation (11.10) is followed by Griffin and Steel (2006), who define the ranking of the elements in the vectors $\mathbf{V}$ and $\boldsymbol{\theta}$ through an ordering $\boldsymbol{\pi}(\mathbf{s})$, which changes with the spatial index (or other covariates). Since weights associated with atoms that appear earlier in the stick-breaking representation tend to be larger (i.e., $\mathrm{E}[p_i(\mathbf{s})] < \mathrm{E}[p_{i-1}(\mathbf{s})]$), this induces similarities between distributions corresponding to similar orderings. The similarity between $\boldsymbol{\pi}(\mathbf{s}_1)$ and $\boldsymbol{\pi}(\mathbf{s}_2)$ will control the correlation between $F_{\mathbf{s}_1}$ and $F_{\mathbf{s}_2}$, the random distributions at these spatial locations. The induced class of models is called order-based dependent Dirichlet processes (πDDPs).

This specification also preserves the usual Dirichlet process for the marginal distribution at each location, but, in contrast with the single-p approaches, leads to local updating, where the influence of observations decreases as they are farther away.

The main challenge is to define stochastic processes $\boldsymbol{\pi}(\mathbf{s})$, and Griffin and Steel (2006) use a point process Φ and a sequence of sets $U(\mathbf{s})$, which define the region in which points are relevant for determining the ordering at $\mathbf{s}$. The ordering, $\boldsymbol{\pi}(\mathbf{s})$, then satisfies the condition

$$\|\mathbf{s} - \mathbf{z}_{\pi_1(\mathbf{s})}\| < \|\mathbf{s} - \mathbf{z}_{\pi_2(\mathbf{s})}\| < \|\mathbf{s} - \mathbf{z}_{\pi_3(\mathbf{s})}\| < \ldots,$$

where $\|\cdot\|$ is a distance measure and $\mathbf{z}_{\pi_i(\mathbf{s})} \in \Phi \cap U(\mathbf{s})$. We assume there are no ties, which is almost surely the case for, e.g., Poisson point processes. Associating each atom $(V_i, \boldsymbol{\theta}_i)$ with the element of the point process $\mathbf{z}_i$ defines a marked point process from which we can define the distribution $F_{\mathbf{s}}$ for any $\mathbf{s} \in \mathcal{D}$. Using a stationary Poisson process for Φ, the autocorrelation function between random probability measures at different locations can be expressed in the form of deterministic integrals, as explained in Griffin and Steel (2006). More specific constructions can even lead to analytical expressions for the autocorrelation structure.

In particular, Griffin and Steel (2006) define a practical proposal for spatial models through the so-called permutation construction. This is obtained through defining $\mathcal{D} \subset \Re^d$ and $U(\mathbf{s}) = \mathcal{D}$ for all values of $\mathbf{s}$. In one dimension ($d = 1$), we can derive an analytic form for the autocorrelation function. Let Φ be Poisson with intensity λ, $\mathcal{D} \subset \Re$ and $U(s) = \mathcal{D}$ for all s. Then we obtain

$$\mathrm{corr}(F_{s_1}, F_{s_2}) = \left(1 + \frac{2\lambda h}{M+2}\right) \exp\left\{\frac{-2\lambda h}{M+1}\right\},$$

where $h = |s_1 - s_2|$ is the distance between s_1 and s_2, and M is the mass parameter of the marginal Dirichlet process.

Note the unusual form of the correlation structure above. It is the weighted sum of a Matérn correlation function with smoothness parameter 3/2 (with weight $(M+1)/(M+2)$) and an exponential correlation function (with weight $1/(M+2)$), which is a less smooth member of the Matérn class, with smoothness parameter 1/2. So, for $M \to 0$, the correlation function will tend to the arithmetic average of both and for large M the correlation structure

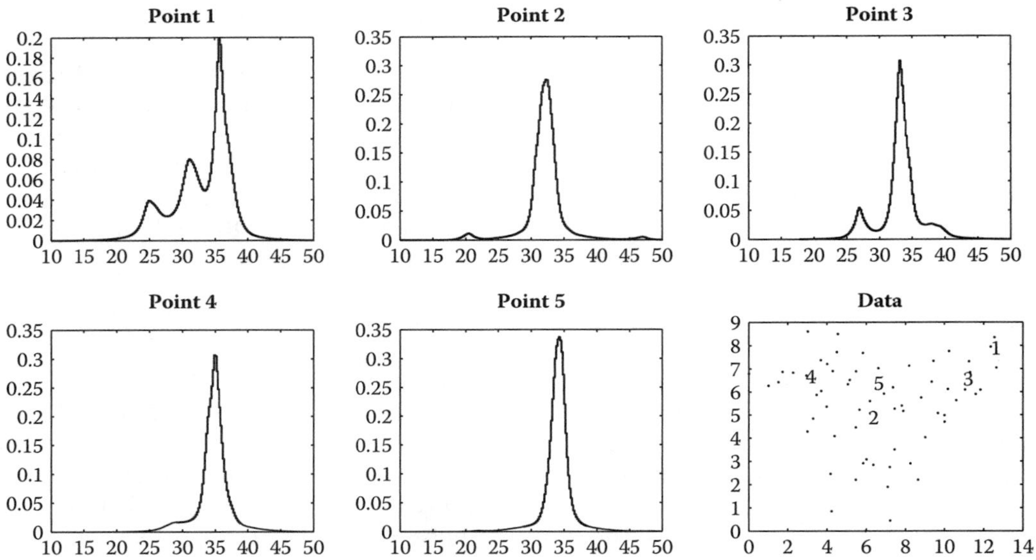

FIGURE 11.3
Temperature data: The posterior predictive distribution at five unobserved locations. The latter are indicated by numbers in the lower right-hand panel, where the observed locations are denoted by dots.

will behave like a Matérn with smoothness parameter 3/2. In higher dimensions, for $d \geq 2$, the autocorrelation function can be expressed as a two-dimensional integral, as detailed in Ishwaran and James (2001).

Griffin and Steel (2006) suggest prior distributions for (M, λ) and use the order-based dependent Dirichlet process with the permutation construction to analyze the Spanish temperature data as described in Section 11.1.5. In particular, the model used is (11.1) with the sum of $\eta(\mathbf{s})$ and the intercept (β_1) modeled through a πDDP process, using as centering distribution H a $\mathrm{N}(\bar{y}, \tau^2/\kappa)$, where $\bar{y}$ is the observation mean and an inverted gamma prior is adopted for κ. In Figure 11.3 we display the posterior predictive distributions at the same unsampled locations as in Figure 11.2. The lower right panel indicates the location of these unobserved locations (with numbers), as well as the observed ones (with dots). While some of the predictives are similar to those obtained with the parametric GLG model, there is now clearly a much larger variety of predictive shapes, with multimodality and skewness illustrating the large flexibility of such nonparametric approaches. Of course, the πDDP part of the model does not only allow for departures from Gaussianity, but also serves to introduce the spatial correlation.

Finally, note that we do not require replication of observations at each site for inference with πDDP models. In principle, extensions to spatial-temporal models can be formulated easily by treating time as a third dimension in defining the ordering, but it is not obvious that this would be the most promising approach. Implementation of these ideas to nonparametric modelling in time is the subject of current research.

11.2.5 Spatial Kernel Stick-Breaking Prior

An extension of the stick-breaking prior of Sethuraman (1994) to the multivariate spatial setting is proposed in Reich and Fuentes (2007). The stick-breaking prior can be extended to the univariate spatial setting by incorporating spatial information into either the model for the locations $\boldsymbol{\theta}_i$ or the model for the weights p_i. As explained in Section 11.2.2, Gelfand et al. (2005) model the locations as vectors drawn from a spatial distribution. This approach is generalized by Duan et al. (2007) to allow both the weights and locations to vary spatially. However, we have seen that these models require replication of the spatial process. As discussed in the previous section, Griffin and Steel (2006) propose a spatial Dirichlet model that does not require replication. The latter model permutes the V_i based on spatial location, allowing the occurrence of $\boldsymbol{\theta}_i$ to be more or less likely in different regions of the spatial domain. The nonparametric multivariate spatial model introduced by Reich and Fuentes (2007) has multivariate normal priors for the locations $\boldsymbol{\theta}_i$. We call this prior process a spatial kernel stick-breaking (SSB) prior. Similar to Griffin and Steel (2006), the weights p_i vary spatially. However, rather than random permutation of V_i, Reich and Fuentes (2007) introduce a series of kernel functions to allow the masses to change with space. This results in a flexible spatial model, as different kernel functions lead to different relationships between the distributions at nearby locations. This model is similar to that of Dunson and Park (2008), who use kernels to smooth the weights in the nonspatial setting. This model is also computationally convenient because it avoids reversible jump MCMC steps and the inversion of large matrices.

In this section, first, we introduce the SSB prior in the univariate setting, and then we extend it to the multivariate case. Let $Y(\mathbf{s})$, the observable value at site $\mathbf{s} = (s_1, s_2)$, be modeled as

$$Y(\mathbf{s}) = \eta(\mathbf{s}) + \mathrm{x}(\mathbf{s})^T\boldsymbol{\beta} + \epsilon(\mathbf{s}), \tag{11.14}$$

where $\eta(\mathbf{s})$ is a spatial random effect, $\mathrm{x}(\mathbf{s})$ is a vector of covariates for site $\mathbf{s}$, $\boldsymbol{\beta}$ are the regression parameters, and $\epsilon(\mathbf{s}) \stackrel{iid}{\sim} N(0, \tau^2)$.

The spatial effects are assigned a random prior distribution, i.e., $\eta(\mathbf{s}) \sim F_{\mathbf{s}}(\eta)$. This SSB modeling framework introduces models marginally, i.e., $F_{\mathbf{s}}(\eta)$ and $F_{\mathbf{s}'}(\eta)$, rather than jointly, i.e., $F_{\mathbf{s},\mathbf{s}'}(\eta)$, as in the referenced work of Gelfand and colleagues (2005). The distributions $F_{\mathbf{s}}(\eta)$ are smoothed spatially. Extending (11.10) to depend on $\mathbf{s}$, the prior for $F_{\mathbf{s}}(\eta)$ is the potentially infinite mixture

$$F_{\mathbf{s}}(\eta) \stackrel{d}{=} \sum_{i=1}^{N} p_i(\mathbf{s})\delta_{\boldsymbol{\theta}_i}, \tag{11.15}$$

where $p_i(\mathbf{s}) = V_i(\mathbf{s})\prod_{j=1}^{i-1}(1 - V_j(\mathbf{s}))$, and $V_i(\mathbf{s}) = w_i(\mathbf{s})V_i$. The distributions $F_{\mathbf{s}}(\eta)$ are related through their dependence on the V_i and $\boldsymbol{\theta}_i$, which are given the priors $V_i \sim \text{Beta}(a, b)$ and $\boldsymbol{\theta}_i \sim H$, each independent across i. However, the distributions vary spatially according to the functions $w_i(\mathbf{s})$, which are restricted to the interval $[0, 1]$. $w_i(\mathbf{s})$ is modeled using a kernel function, but alternatively $\log(w_i(\mathbf{s})/(1 - w_i(\mathbf{s})))$ could be modeled as a spatial Gaussian

TABLE 11.4

Examples of Kernel Functions and the Induced Functions $\gamma(\mathbf{s}, \mathbf{s}')$, Where $\mathbf{s} = (s_1, s_2)$, $h_1 = |s_1 - s_1'| + |s_2 - s_2'|$, $h_2 = \sqrt{(s_1 - s_1')^2 + (s_2 - s_2')^2}$, $I(\cdot)$ is the Indicator Function, and $x^+ = \max(x, 0)$

Name	$w_i(\mathbf{s})$	Model for κ_{1i} and κ_{2i}	$\gamma(\mathbf{s}, \mathbf{s}')$
Uniform	$\prod_{j=1}^{2} I\left(\|s_j - \psi_{ji}\| < \frac{\kappa_{ji}}{2}\right)$	$\kappa_{1i}, \kappa_{2i} \equiv \lambda$	$\prod_{j=1}^{2}\left(1 - \frac{\|s_j - s_j'\|}{\lambda}\right)^+$
Uniform	$\prod_{j=1}^{2} I\left(\|s_j - \psi_{ji}\| \frac{<\kappa_{ji}}{2}\right)$	$\kappa_{1i}, \kappa_{2i} \sim \text{Exp}(\lambda)$	$\exp(-h_1/\lambda)$
Exponential	$\prod_{j=1}^{2} \exp\left(-\frac{\|s_j - \psi_{ji}\|}{\kappa_{ji}}\right)$	$\kappa_{1i}, \kappa_{2i} \equiv \lambda$	$0.25\left[\prod_{j=1}^{2}\left(1 + \frac{\|s_j - s_j'\|}{\lambda}\right)\right]\exp\left(-\frac{h_1}{\lambda}\right)$
Squared exp.	$\prod_{j=1}^{2} \exp\left(-\frac{(s_j - \psi_{ji})^2}{\kappa_{ji}^2}\right)$	$\kappa_{1i}, \kappa_{2i} \equiv \lambda^2/2$	$0.5\exp\left(-\frac{h_2^2}{\lambda^2}\right)$
Squared exp.	$\prod_{j=1}^{2} \exp\left(-\frac{(s_j - \psi_{ji})^2}{\kappa_{ji}^2}\right)$	$\kappa_{1i}, \kappa_{2i} \sim \text{InvGa}\left(\frac{3}{2}, \frac{\lambda^2}{2}\right)$	$0.5/\left(1 + (\frac{h_2}{\lambda})^2\right)$

process. Other transformations could be considered, but we use kernels for simplicity. There are many possible kernel functions and Table 11.4 gives three examples. In each case, the function $w_i(\mathbf{s})$ is centered at knot $\boldsymbol{\psi}_i = (\psi_{1i}, \psi_{2i})$ and the spread is controlled by the bandwidth parameter $\boldsymbol{\kappa}_i = (\kappa_{1i}, \kappa_{2i})$. Both the knots and the bandwidths are modeled as unknown parameters. The knots $\boldsymbol{\psi}_i$ are given independent uniform priors over the spatial domain. The bandwidths can be modeled as equal for each kernel function or varying independently following distributions given in the third column of Table 11.4.

To ensure that the stick-breaking prior is proper, we must choose priors for κ_i and V_i so that $\sum_{i=i}^{N} p_i(\mathbf{s}) = 1$ almost surely for all $\mathbf{s}$. Reich and Fuentes (2007) show that the SSB prior with infinite N is proper if $E(V_i) = a/(a+b)$ and $E\left[w_i(\mathbf{s})\right]$ (where the expectation is taken over $(\boldsymbol{\psi}_i, \boldsymbol{\kappa}_i)$) are both positive. For finite N, we can ensure that $\sum_{i=i}^{N} p_i(\mathbf{s}) = 1$ for all $\mathbf{s}$ by setting $V_N(\mathbf{s}) \equiv 1$ for all $\mathbf{s}$. This is equivalent to truncating the infinite mixture by attributing all of the mass from the terms with $i \geq N$ to $p_N(\mathbf{s})$.

In practice, allowing N to be infinite is often unnecessary and computationally infeasible. Choosing the number of components in a mixture model is notoriously problematic. Fortunately, in this setting the truncation error can easily be assessed by inspecting the distribution of $p_N(\mathbf{s})$, the mass of the final component of the mixture. The number of components N can be chosen by generating samples from the prior distribution of $p_N(\mathbf{s})$. We increase N until $p_N(\mathbf{s})$ is satisfactorily small for each site $\mathbf{s}$. Also, the truncation error is monitored by inspecting the posterior distribution of $p_N(\mathbf{s})$, which is readily available from the MCMC samples.

Assuming a finite mixture, the spatial stick-breaking model can be written as a finite mixture model where $g(\mathbf{s}) \in \{1, \ldots, N\}$ indicates the particular location allocated to site $\mathbf{s}$, i.e.,

$$\begin{aligned}
Y(\mathbf{s}) &= \theta_{g(\mathbf{s})} + \mathrm{x}(\mathbf{s})^T\boldsymbol{\beta} + \epsilon(\mathbf{s}), \text{ where } \epsilon(\mathbf{s}) \overset{iid}{\sim} N(0, \tau^2) \qquad (11.16)\\
\theta_j &\overset{iid}{\sim} N(0, \sigma^2),\ j = 1, \ldots, N\\
g(\mathbf{s}) &\sim \text{Categorical}(p_1(\mathbf{s}), \ldots, p_N(\mathbf{s}))\\
p_j(\mathbf{s}) &= w_j(\mathbf{s})V_j \prod_{k<j}[1 - w_k(\mathbf{s})V_k], \text{ where } V_j \overset{iid}{\sim} \text{Beta}(a, b),
\end{aligned}$$

where $\eta(\mathbf{s}) = \theta_{g(\mathbf{s})}$. The regression parameters $\boldsymbol{\beta}$ are given vague normal priors. This model can also easily be fitted if we have an infinite mixture model with $N = \infty$, using retrospective sampling ideas from Papaspiliopoulos and Roberts (2008).

Understanding the spatial correlation function is crucial for analyzing spatial data. Although the SSB prior foregoes the Gaussian assumption for the spatial random effects, we can still compute and investigate the covariance function. Conditional on the probabilities $p_j(s)$ (but not the locations $\boldsymbol{\theta}_j$), the covariance between two observations is

$$\text{Cov}(Y(\mathbf{s}), Y(\mathbf{s}')) = \sigma^2 P(\eta(\mathbf{s}) = \eta(\mathbf{s}')) = \sigma^2 \sum_{j=1}^{N} p_j(\mathbf{s}) p_j(\mathbf{s}'). \qquad (11.17)$$

For a one-dimensional spatial domain, integrating over (V_i, ψ_i, κ_i) and letting $N \to \infty$ gives

$$\text{var}\,(Y(s)) = \sigma^2 + \tau^2 \qquad (11.18)$$

$$\text{Cov}(Y(s), Y(s')) = \sigma^2\gamma(s, s')\left[2\frac{a+b+1}{a+1} - \gamma(s, s')\right]^{-1}, \qquad (11.19)$$

where

$$\gamma(s, s') = \frac{\int\int w_i(s) w_i(s') p(\psi_i, \kappa_i) d\psi_i d\kappa_i}{\int\int w_i(s) p(\psi_i, \kappa_i) d\psi_i d\kappa_i} \in [0, 1]. \qquad (11.20)$$

Since (V_i, ψ_i, κ_i) have independent priors that are uniform over the spatial domain, integrating over these parameters gives a stationary prior covariance. However, the conditional covariance can be nonstationary. More importantly, the posterior predictive distribution can accommodate nonstationarity. Therefore, we conjecture the SSB model is more robust to nonstationarity than traditional stationary kriging methods.

If $b/(a+1)$ is large, i.e., the V_i are generally small and there are many terms in the mixture with significant mass, the correlation between $Y(\mathbf{s})$ and $Y(\mathbf{s}')$ is approximately proportional to $\gamma(\mathbf{s}, \mathbf{s}')$. Table 11.4 from Reich and Fuentes (2007) gives the function $\gamma(\mathbf{s}, \mathbf{s}')$ for several examples of kernel functions and shows that different kernels can produce very different correlation functions.

To introduce a multivariate extension of the SBB prior, let $\boldsymbol{\eta}(\mathbf{s}) = (\eta_1(\mathbf{s}), \ldots, \eta_p(\mathbf{s}))^T$ be a p-dimensional spatial process. The prior for $\boldsymbol{\eta}(\mathbf{s})$ is

$$\boldsymbol{\eta}(\mathbf{s}) \sim F_{\mathbf{s}}(\boldsymbol{\eta}), \text{ where } F_{\mathbf{s}}(\boldsymbol{\eta}) \stackrel{d}{=} \sum_{i=1}^{N} p_i(\mathbf{s})\delta_{\boldsymbol{\theta}_i}. \tag{11.21}$$

The weights $p_i(\mathbf{s})$ are shared across components, the p-dimensional locations $\boldsymbol{\theta}_i \stackrel{iid}{\sim} N(0, \Sigma)$, and Σ is a $p \times p$ covariance matrix that controls the association between the p spatial processes. The covariance Σ could have an inverse Wishart prior.

11.2.6 A Case Study: Hurricane Ivan

In Reich and Fuentes (2007), the multivariate SSB prior is used to model the complex spatial patterns of hurricane wind vectors, with data from Hurricane Ivan as it passes through the Gulf of Mexico at 12 p.m. on September 15, 2004. Three sources of information are used in the analysis and are plotted in Figure 11.4. The first source is gridded satellite data (Figure 11.4a) available from NASA's SeaWinds database (http://podaac.jpl.nasa.gov/products/product109.html). These data are available twice daily on a 0.25×0.25 degree, global grid. Due to the satellite data's potential bias, measurement error, and coarse temporal resolution, the wind fields analysis is supplemented with data from the National Data Buoy Center of the National Oceanic and Atmospheric Administration (NOAA). Buoy data are collected every 10 minutes at a relatively small number of marine locations (Figure 11.4b). These measurements are adjusted to a common height of 10 meters above sea level using the algorithm of Large and Pond (1981).

In addition to satellite and buoy data, the deterministic Holland model (Holland, 1980) is incorporated in the analysis. NOAA currently uses this model alone to produce wind fields for their numerical ocean models. The Holland model predicts that the wind velocity at location $\mathbf{s}$ is

$$H(\mathbf{s}) = \left(\frac{B}{\rho} \left(\frac{Rmax}{r} \right)^B (P_n - P_c) \exp\left[-\left(\frac{Rmax}{r} \right)^B \right] \right)^{1/2}, \tag{11.22}$$

where r is the radius from the storm center to site $\mathbf{s}$, P_n is the ambient pressure, P_c is the hurricane central pressure, ρ is the air density, $Rmax$ is the radius of maximum sustained winds, and B controls the shape of the pressure profile.

We decompose the wind vectors into their orthogonal west/east (u) and north/south (v) vectors. The Holland model for the u and v components is

$$H_u(\mathbf{s}) = H(\mathbf{s}) \sin(\phi) \text{ and } H_v(\mathbf{s}) = H(\mathbf{s}) \cos(\phi), \tag{11.23}$$

where ϕ is the inflow angle at site $\mathbf{s}$, across circular isobars toward the storm center, rotated to adjust for the storm's direction. We fix the parameters $P_n = 1010$ mb, $P_c = 939$ mb,

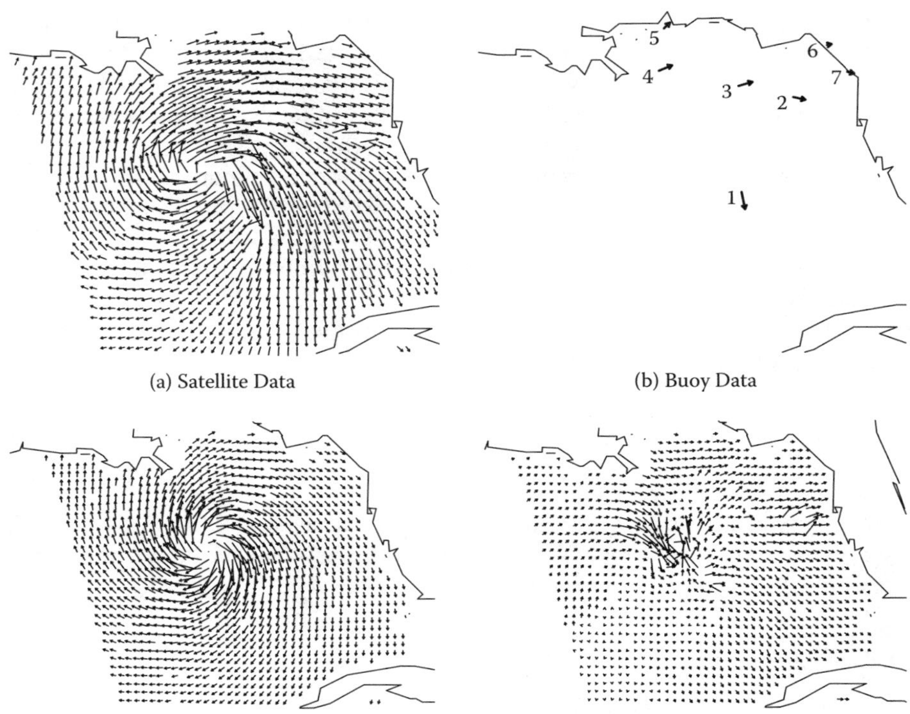

(a) Satellite Data (b) Buoy Data

(c) Holland Model Output (d) Satellite — Holland Model Output

FIGURE 11.4
Plot of various types of wind field data/output for Hurricane Ivan on September 15, 2004.

$\rho = 1.2$ kg m^{-3}, and $Rmax = 49$, and $B = 1.9$ using the meteorological data from the national hurricane center (http://www.nhc.noaa.gov) and recommendations of Hsu and Yan (1998). The output from this model for Hurricane Ivan is plotted in Figure 11.4c. By construction, Holland model output is symmetric with respect to the storm's center, which does not agree with the satellite observations in Figure 11.4a.

Let $u(\mathbf{s})$ and $v(\mathbf{s})$ be the underlying wind speed in the west/east and north/south directions, respectively, for spatial location $\mathbf{s}$. We distinguish the different sources of wind data: $u_T(\mathbf{s})$ and $v_T(\mathbf{s})$ are satellite measurements and $u_B(\mathbf{s})$ and $v_B(\mathbf{s})$ are buoy measurements. The model used by Reich and Fuentes (2007) for these data is

$$u_T(\mathbf{s}) = a_u + u(\mathbf{s}) + e_{uT}(\mathbf{s}) \quad v_T(\mathbf{s}) = a_v + v(\mathbf{s}) + e_{vT}(\mathbf{s}) \tag{11.24}$$
$$u_B(\mathbf{s}) = u(\mathbf{s}) + e_{uB}(\mathbf{s}) \quad v_B(\mathbf{s}) = v(\mathbf{s}) + e_{vB}(\mathbf{s}),$$

where $\{e_{uT}, e_{vT}, e_{uB}, e_{vB}\}$ are independent (from each other and from the underlying winds), zero mean, Gaussian errors, each with its own variance, and $\{a_u, a_v\}$ account for additive bias in the satellite and aircraft data. Of course, the buoy data may also have bias, but it is impossible to identify bias from both sources, so all the bias is attributed to the satellite measurements.

The underlying orthogonal wind components $u(\mathbf{s})$ and $v(\mathbf{s})$ are modeled as a mixture of a deterministic wind model and a semiparametric multivariate spatial process

$$u(\mathbf{s}) = H_u(\mathbf{s}) + R_u(\mathbf{s}) \tag{11.25}$$
$$v(\mathbf{s}) = H_v(\mathbf{s}) + R_v(\mathbf{s}), \tag{11.26}$$

where $H_u(\mathbf{s})$ and $H_v(\mathbf{s})$ are the orthogonal components of the deterministic Holland model in Equation (11.23) and $\mathbf{R}(\mathbf{s}) = (R_u(\mathbf{s}), R_v(\mathbf{s}))'$ follows a multivariate extension of the SSB prior as in Equation (11.21). The prior for $\mathbf{R}(\mathbf{s})$ is

$$\mathbf{R}(\mathbf{s}) \sim F_{\mathbf{s}}(\eta), \text{ where } F_{\mathbf{s}}(\eta) \stackrel{d}{=} \sum_{i=1}^{N} p_i(\mathbf{s})\delta_{\boldsymbol{\theta}_i}. \tag{11.27}$$

The masses $p_i(\mathbf{s})$ are shared across components, the two-dimensional locations $\boldsymbol{\theta}_i \stackrel{iid}{\sim} N(0, \Sigma)$, and Σ is a 2×2 covariance matrix that controls the association between the two wind components. The inverse covariance Σ^{-1} has a Wishart prior with 3.1 degrees of freedom and inverse scale matrix $0.1\,\mathbf{I}_2$. After transforming the spatial grid to be contained in the unit square, the spatial knots ψ_{1i} and ψ_{2i} have independent Beta(1.5,1.5) priors to encourage knots to lie near the center of the hurricane where the wind is most volatile.

In Reich and Fuentes (2007), this multivariate SSB model is fitted to 182 satellite observations and 7 buoy observations for Hurricane Ivan. To illustrate the effect of relaxing the normality assumption, Reich and Fuentes (2007) also fits a fully Gaussian model that replaces the stick-breaking prior for $\mathbf{R}(\mathbf{s})$ in Equation (11.27) with a zero-mean Gaussian prior

$$\text{var}(\mathbf{R}(\mathbf{s})) = \Sigma \quad \text{and} \quad \text{Cov}(\mathbf{R}(\mathbf{s}), \mathbf{R}(\mathbf{s}')) = \Sigma \times \exp(-||\mathbf{s} - \mathbf{s}'||/\lambda), \tag{11.28}$$

where Σ controls the dependency between the wind components at a given location and λ is a spatial range parameter. The covariance parameters Σ and λ have the same priors as the covariance parameters in the SSB model.

Because the primary objective is to predict wind vectors at unmeasured locations to use as inputs for numerical ocean models, statistical models are compared in terms of expected mean squared prediction error (EMSPE) (Laud and Ibrahim, 1995) see e.g., Gelfand and Ghosh (1998). The EMSPE is smaller for the semiparametric EMSPE; see e.g., model with uniform kernels (EMSPE = 3.46) than for the semiparametric model using squared exponential kernels (EMSPE = 4.19) and the fully Gaussian model (EMSPE = 5.17).

Figure 11.5 summarizes the posterior from the SSB prior with uniform kernel functions. The fitted values in Figure 11.5a and Figure 11.5b vary rapidly near the center of the storm and are fairly smooth in the periphery. After accounting for the Holland model, the correlation between the residual u and v components $R_u(\mathbf{s})$ and $R_v(\mathbf{s})$ ($\Sigma_{12}/\sqrt{\Sigma_{11}\Sigma_{22}}$, where

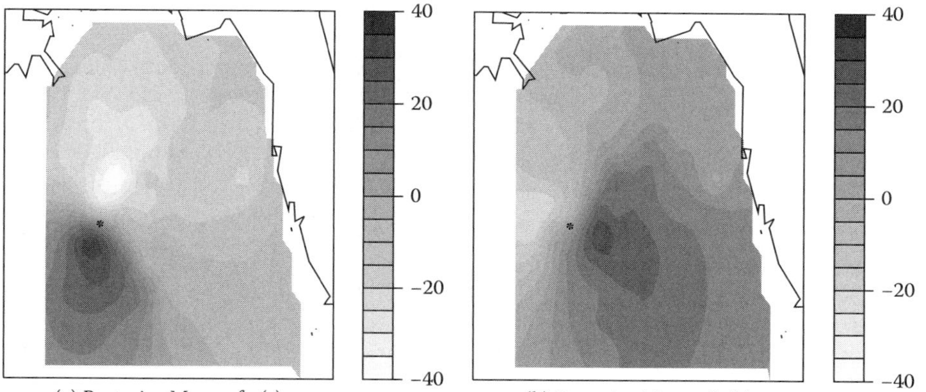

(a) Posterior Mean of u(s) (b) Posterior Mean of v(s)

FIGURE 11.5
Summary of the posterior of the spatial stick-breaking model with uniform kernels. Panels (a) and (b) give the posterior mean surface for the u and v components.

Σ_{kl} is the (k, l) element of Σ) is generally negative, confirming the need for a multivariate analysis.

To show that the multivariate SSB model with uniform kernel functions fits the data well, 10% of the observations are randomly set aside (across u and v components and buoy and satellite data) throughout the model fitting and the 95% predictive intervals for the missing observations are obtained. The prediction intervals contain 94.7% (18/19) of the deleted u components and 95.2% (20/21) of the deleted v components. These statistics suggest that the model is well calibrated.

References

Antoniak, C.E. (1974). Mixtures of Dirichlet processes with applications to non-parametric problems, *Annals of Statistics*, **2**, 1152–1174.

Banerjee, S., Wall, M. and Carlin, B.P. (2003). Frailty modeling for spatially correlated survival data with application to infant mortality in Minnesota, *Biostatistics*, **4**, 123–142.

Berger, J.O., De Oliveira, V. and Sansó, B. (2001). Objective Bayesian analysis of spatially correlated data, *Journal of the American Statistical Association*, **96**, 1361–1374.

Brown, P.E., Diggle, P.J. and Henderson, R. (2003). A non-Gaussian spatial process model for opacity of flocculated paper, *Scandinavian Journal of Statistics*, **30**, 355–368.

Damian, D., Sampson, P.D. and Guttorp, P. (2001). Bayesian estimation of semi-parametric non-stationary spatial covariance structures, *Environmetrics*, **12**, 161–178.

Damian, D., Sampson P.D. and Guttorp, P. (2003). Variance modeling for nonstationary processes with temporal replications, *Journal of Geophysical Research (Atmosphere)*, **108**, 8778.

De Oliveira, V., Kedem, B. and Short, D.A. (1997). Bayesian prediction of transformed Gaussian random fields, *Journal of the American Statistical Association*, **92**, 1422–1433.

Diggle, P.J., Tawn, J.A. and Moyeed, R.A. (1998). Model-based geostatistics (with discussion), *Applied Statistics*, **47**, 299–326.

Dunson, D.B. (Forthcoming). Nonparametric Bayes applications to biostatistics, in *Bayesian Nonparametrics*, Cambridge: Cambridge Univ. Press.

Dunson, D.B. and Park, J.H. (2008). Kernel stick-breaking processes, *Biometrika*, **95**, 307–323.

Duan, J.A., Guindani, M. and Gelfand, A.E. (2007). Generalized spatial Dirichlet process models, *Biometrika*, **94**, 809–825.

Ferguson, T.S. (1973). A Bayesian analysis of some nonparametric problems, *Annals of Statistics*, **1**, 209–230.

Gelfand, A.E. and Ghosh, S.K. (1998). Model choice: A minimum posterior predictive loss approach. *Biometrika*, **77**, 1–11.

Gelfand, A.E., Guindani, M. and Petrone, S. (2007). Bayesian nonparametric modeling for spatial data analysis using Dirichlet processes, in *Bayesian Statistics*, **8**, J.M. Bernardo, M.J. Bayarri, J.O. Berger, A.P. Dawid, D. Heckerman, A.F.M. Smith and M. West, eds., Oxford: Oxford Univ. Press.

Gelfand, A.E., Kottas, A. and MacEachern, S.N. (2005). Bayesian nonparametric spatial modelling with Dirichlet processes mixing, *Journal of the American Statistical Association*, **100**, 1021–1035.

Griffin, J.E. and Steel, M.F.J. (2006). Order-based dependent Dirichlet processes, *Journal of the American Statistical Association*, **101**, 179–194.

Holland, G.J. (1980). An analytic model of the wind and pressure profiles in hurricanes. *Monthly Weather Review*, **108**, 1212–1218.

Hsu, S.A. and Yan, Z. (1998). A note on the radius of maximum wind for hurricanes. *Journal of Coastal Research*, **14**, 667–668.

Ishwaran, H. and James, L. (2001). Gibbs-sampling methods for stick-breaking priors, *Journal of the American Statistical Association*, **96**, 161–173.

Large, W.G. and Pond, S. (1981). Open ocean momentum flux measurements in moderate to strong winds. *Journal of Physical Oceanography*, **11**, 324–336.

Laud, P. and Ibrahim, J. (1995). Predictive model selection. *Journal of the Royal Statistical Society, Series B*, **57**, 247–262.

Li, Y. and Ryan, L. (2002). Modeling spatial survival data using semiparametric frailty models, *Biometrics*, **58**, 287–297.

MacEachern, S.N. (1998). Computational methods for mixture of Dirichlet process models, in *Practical Nonparametric and Semiparametric Bayesian Statistics*, D. Dey, P. Müller and D. Sinha, eds. New York: Springer-Verlag, pp. 23–44.

MacEachern, S.N. (1999). Dependent nonparametric processes, in *ASA Proceedings of the Section on Bayesian Statistical Science*, Alexandria, VA: American Statistical Association.

Müller, P. and Quintana, F.A. (2004). Nonparametric Bayesian data analysis, *Statistical Science*, **19**, 95–110.

Palacios, M.B. and Steel, M.F.J. (2006). Non-Gaussian Bayesian geostatistical modelling, *Journal of the American Statistical Association*, **101**, 604–618.

Papaspiliopoulos, O. and Roberts, G. (2008). Retrospective MCMC for Dirichlet process hierarchical models, *Biometrika*, **95**, 169–186.

Petrone, S., Guindani, M. and Gelfand, A.E. (2009). Hybrid Dirichlet mixture models for functional data, *Journal of the Royal Statistical Society B*, **71**, 755–782.

Reich, B. and Fuentes, M. (2007). A multivariate semiparametric Bayesian spatial modeling framework for hurricane surface wind fields, *Annals of Applied Statistics*, **1**, 249–264.

Sampson, P.D. and Guttorp, P. (1992). Nonparametric estimation of nonstationary spatial covariance structure, *Journal of the American Statistical Association*, **87**, 108–119.

Schmidt, A.M. and O'Hagan, A. (2003). Bayesian inference for nonstationary spatial covariance structure via spatial deformations, *Journal of the Royal Statistical Society*, B, **65**, 745–758.

Sethuraman, J. (1994). A constructive definition of Dirichlet priors, *Statistica Sinica*, **4**, 639–650.

Stein, M.L. (1999). *Interpolation of Spatial Data. Some Theory of Kriging*. Springer-Verlag, New York.

Verdinelli, I. and Wasserman, L. (1995). Computing Bayes factors by using a generalization of the Savage–Dickey density ratio, *Journal of the American Statistical Association*, **90**, 614–618.

Part III

Discrete Spatial Variation

The objective of Part III is to present a thorough discussion of the literature up to the present on the analysis of what can be called discrete spatial variation. By this we mean analysis that only envisions a finite collection of spatial random variables and only seeks to model this finite collection. Examples include lattice data, pixel (and voxel) data, and areal unit data (where we allow for irregular areal units both in size and shape). Applications include image analysis, agricultural field trials, disease mapping, environmental processes, spatial econometrics, and approximation for finite-dimensional distributions associated with large datasets arising through a spatial stochastic process specification.

Inference for discrete spatial variation is strikingly different from that for continuous spatial variation. The goals are explanation and smoothing rather than interpolation and prediction. The spatial modeling works with the notion of neighbors rather than with the notion of a covariance function. Inverse covariance matrices are specified rather than covariance matrices themselves. And, computation with large datasets, at least in the Gaussian case, is much faster than with the continuous case.

The most widely used tool for building models for discrete spatial variation is the Markov random field. In Chapter 12, working with graphical models, Rue and Held develop the general theory for such fields in both the Gaussian and non-Gaussian cases. They also focus on modern computing strategies for fitting models employing such fields. In Chapter 13, Held and Rue take on the development of conditionally autoregressive (CAR) and intrinsically autoregressive (IAR) models, illuminating their use within hierarchical modeling with discussion of applications. In Chapter 14, Waller and Carlin focus on the most prominent application of CAR modeling: the context of disease mapping. Here, there is a rich literature, evidencing the importance of spatial modeling in this setting. Static and dynamic models are discussed and a novel example is presented. Finally, in Chapter 14, Pace and Lesage take us to the world of spatial econometrics, developing simultaneous autoregressive (SAR) models as spatial versions of customary autoregressive time series models. Here, likelihood analysis is computationally more convenient than Bayesian methods and sparse matrix methods enable rapid analysis even with datasets involving as many as 10^5 units.

12

Discrete Spatial Variation

Hävard Rue and Leonhard Held

CONTENTS

12.1 Markov Random Fields

Statistical modeling of a finite collection of spatial random variables is often done through a *Markov random field* (MRF). An MRF is specified through the set of conditional distributions of one component given all the others. This enables one to focus on a single random variable at a time and leads to simple computational procedures for simulating MRFs, in particular for Bayesian inference via Markov chain Monte Carlo (MCMC). The main purpose of this chapter is to give a thorough introduction to the Gaussian case, so-called *Gaussian* MRFs (GMRFs), with a focus toward general properties and efficient computations. Examples and applications appear in Chapters 13 and 14. At the end, we will discuss the general case where the joint distribution is not Gaussian, and, in particular, the famous Hammersley–Clifford theorem. A modern and general reference to GMRFs is the monograph by Rue and Held (2005), while for MRFs in general, one can consult Guyon (1995) and Lauritzen (1996) for the methodology background and Li (2001) for spatial applications in image analysis. The seminal papers by J. Besag (1974, 1975) are still worth reading.

12.1.1 Notation

We denote by $\boldsymbol{x} = (x_1, \ldots, x_n)^T$ the n-dimensional vector, x_i the ith element, $\boldsymbol{x}_A = \{x_i : i \in A\}$, $\boldsymbol{x}_{-A} = \{x_i : i \notin A\}$, and $\boldsymbol{x}_{i:j} = (x_i, \ldots, x_j)^T$ for $j \geq i$. We use generically $\pi(\cdot)$ and $\pi(\cdot|\cdot)$, as the (probability) density for its arguments, like $\pi(\boldsymbol{x})$ and $\pi(x_i|\boldsymbol{x}_{-i})$. The Gaussian distribution is denoted by $\mathcal{N}(\boldsymbol{\mu}, \boldsymbol{\Sigma})$ and its density value at $\boldsymbol{x}$ is $\mathcal{N}(\boldsymbol{x}; \boldsymbol{\mu}, \boldsymbol{\Sigma})$; here $\boldsymbol{\mu}$ is the expected value and $\boldsymbol{\Sigma}$ the covariance matrix. We use the abbreviation SPD to indicate a symmetric and positive definite matrix. The bandwidth of a matrix $\boldsymbol{A}$ is $\max|i-j|$ over all i, j with $A_{i,j} \neq 0$.

12.1.2 Gaussian Markov Random Fields

12.1.2.1 Definition

We will first discuss GMRFs in general and then return to the conditional specification later on. A GMRF is simply a Gaussian distributed random vector $\boldsymbol{x}$, which obeys some conditional independence properties. That is, for some $i \neq j$, then

$$x_i \perp x_j \mid \boldsymbol{x}_{-\{i,j\}}, \tag{12.1}$$

meaning that conditioned on $\boldsymbol{x}_{-\{i,j\}}$, x_i and x_j are independent. This conditional independence is represented using an (undirected) labeled graph $\mathcal{G} = (\mathcal{V}, \mathcal{E})$, where $\mathcal{V} = \{1, \ldots, n\}$ is the set of vertices, and $\mathcal{E} = \{\{i, j\} : i, j \in \mathcal{V}\}$ is the set of edges in the graph. For all $i, j \in \mathcal{V}$, the edge $\{i, j\}$ is not included in $\mathcal{E}$ if (12.1) holds, and included otherwise. Figure 12.1 displays such a graph, where $n = 4$ and $\mathcal{E} = \{\{1,2\}, \{2,3\}, \{3,4\}, \{4,1\}\}$. From this graph we deduce that $x_2 \perp x_4|\boldsymbol{x}_{\{1,3\}}$ and $x_1 \perp x_3|\boldsymbol{x}_{\{2,3\}}$. A central goal is now to specify a GMRF $\boldsymbol{x}$

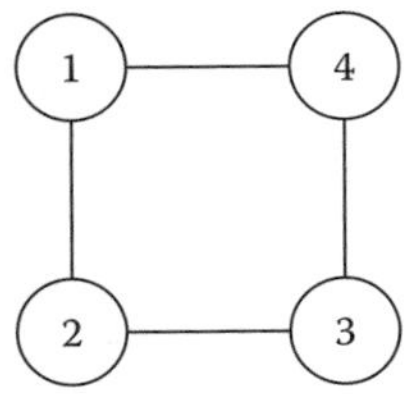

FIGURE 12.1
A conditional independence graph.

with conditional independence properties in agreement with some given graph $\mathcal{G}$. Using the precision matrix $\boldsymbol{Q} = \boldsymbol{\Sigma}^{-1}$ of $\boldsymbol{x}$, this turns out to be particularly simple.

Theorem 12.1
Let $\boldsymbol{x}$ be Gaussian distributed with a symmetric and positive definite (SPD) precision matrix $\boldsymbol{Q}$, then for $i \neq j$

$$x_i \perp x_j \mid \boldsymbol{x}_{-\{i,j\}} \iff Q_{i,j} = 0.$$

So any SPD precision matrix $\boldsymbol{Q}$ with $Q_{2,4} = Q_{4,2} = Q_{1,3} = Q_{3,1} = 0$ has conditional independence properties as displayed in Figure 12.1. We then say that $\boldsymbol{x}$ is a GMRF with respect to $\mathcal{G}$. A formal definition follows.

Definition 12.1 (GMRF)
A random vector $\boldsymbol{x} = (x_1, \ldots, x_n)^T \in \mathbb{R}^n$ is called a GMRF wrt (with regard to) the labeled graph $\mathcal{G} = (\mathcal{V}, \mathcal{E})$ with mean $\boldsymbol{\mu}$ and SPD precision matrix $\boldsymbol{Q}$, iff its density has the form

$$\pi(\boldsymbol{x}) = (2\pi)^{-n/2}|\boldsymbol{Q}|^{1/2} \exp\left(-\frac{1}{2}(\boldsymbol{x}-\boldsymbol{\mu})^T \boldsymbol{Q}(\boldsymbol{x}-\boldsymbol{\mu})\right) \tag{12.2}$$

and

$$Q_{i,j} \neq 0 \iff \{i, j\} \in \mathcal{E} \quad \text{for all} \quad i \neq j.$$

The case where $\boldsymbol{Q}$ is singular still provides a GMRF with an explicit form for its joint density, but the joint density is improper. Such specifications cannot be used as data models, but can be used as priors as long as they yield proper posteriors. Examples include intrinsic autoregression, which is discussed in Chapters 13 and 14. Here is a simple example of a (proper) GMRF.

Example 12.1
Let $\{x_t\}$ be a stationary autoregressive process of order one, i.e., $x_t|x_{t-1} = \phi x_{t-1} + \epsilon_t$, for $t = 2, \ldots, n$, where $|\phi| < 1$ and ϵ_t are independent normally distributed zero mean innovations with unit variance. Further assume that x_1 is normal with mean zero and variance $1/(1-\phi^2)$, which is simply the stationary distribution of this process. Then $\boldsymbol{x}$ is a GMRF wrt to $\mathcal{G}$ where $\mathcal{E} = \{\{1, 2\}, \{2, 3\}, \ldots, \{n-1, n\}\}$. The precision matrix has nonzero elements $Q_{i,j} = -\phi$ for $|i - j| = 1$, $Q_{1,1} = Q_{n,n} = 1$ and $Q_{i,i} = 1 + \phi^2$ for $i = 2, \ldots, n-1$.

This example nicely illustrates why GMRFs are so useful. First note that the lag-k autocorrelation is $\phi^{|k|}$, so the covariance matrix of $\boldsymbol{x}$ is dense, i.e., is nonzero everywhere. This is in contrast to the sparse precision matrix; only $n + 2(n-1) = 3n - 2$ of the n^2 terms in $\boldsymbol{Q}$ are nonzero. The sparse precision matrix makes fast $\mathcal{O}(n)$ algorithms for the simulation of autoregressive processes possible.

12.1.3 Basic Properties

12.1.3.1 Conditional Properties

Although a GMRF can be seen as a general multivariate Gaussian random variable, some properties simplify and some characteristics are easier to compute. For example, conditional distributions are easier to compute due to the sparse precision matrix. To see this, we split $\mathcal{V}$ into the nonempty sets A and $B = -A$. Partition $\boldsymbol{x}$, $\boldsymbol{\mu}$ and $\boldsymbol{Q}$ accordingly, i.e.,

$$\boldsymbol{x} = \begin{pmatrix} \boldsymbol{x}_A \\ \boldsymbol{x}_B \end{pmatrix}, \quad \boldsymbol{\mu} = \begin{pmatrix} \boldsymbol{\mu}_A \\ \boldsymbol{\mu}_B \end{pmatrix} \quad \text{and} \quad \boldsymbol{Q} = \begin{pmatrix} \boldsymbol{Q}_{AA} & \boldsymbol{Q}_{AB} \\ \boldsymbol{Q}_{BA} & \boldsymbol{Q}_{BB} \end{pmatrix}.$$

We also need the notion of a subgraph $\mathcal{G}^A$, which is the graph restricted to A: the graph we obtain after removing all nodes not belonging to A and all edges where at least one node does not belong to A. Then the following theorem holds.

Theorem 12.2

Let $\boldsymbol{x}$ be a GMRF wrt $\mathcal{G}$ with mean $\boldsymbol{\mu}$ and SPD precision matrix $\boldsymbol{Q}$. Let $A \subset \mathcal{V}$ and $B = \mathcal{V} \setminus A$ where $A, B \neq \emptyset$. The conditional distribution of $\boldsymbol{x}_A|\boldsymbol{x}_B$ is then a GMRF wrt the subgraph $\mathcal{G}^A$ with mean $\boldsymbol{\mu}_{A|B}$ and SPD precision matrix $\boldsymbol{Q}_{A|B}$, where

$$\boldsymbol{\mu}_{A|B} = \boldsymbol{\mu}_A - \boldsymbol{Q}_{AA}^{-1}\boldsymbol{Q}_{AB}(\boldsymbol{x}_B - \boldsymbol{\mu}_B) \quad \textit{and} \quad \boldsymbol{Q}_{A|B} = \boldsymbol{Q}_{AA}.$$

Remark 12.1

Note that this result is just an alternative view of conditional distributions for Gaussians, which is commonly expressed using the partitioned covariance matrix

$$\boldsymbol{\Sigma} = \begin{pmatrix} \boldsymbol{\Sigma}_{AA} & \boldsymbol{\Sigma}_{AB} \\ \boldsymbol{\Sigma}_{BA} & \boldsymbol{\Sigma}_{BB} \end{pmatrix}.$$

We have that $\mathrm{Cov}(\boldsymbol{x}_A|\boldsymbol{x}_B) = \boldsymbol{\Sigma}_{AA} - \boldsymbol{\Sigma}_{AB}\boldsymbol{\Sigma}_{BB}^{-1}\boldsymbol{\Sigma}_{BA}$, which is identical to $\boldsymbol{Q}_{AA}^{-1}$ and similarly for the conditional mean.

The first "striking" feature is that the conditional precision matrix $\boldsymbol{Q}_{A|B}$ is a submatrix of $\boldsymbol{Q}$ and, therefore, explicitly available, just remove the rows and columns in $\boldsymbol{Q}$ that belong to B. Sparseness of $\boldsymbol{Q}$ will be inherited to $\boldsymbol{Q}_{A|B}$. The expression for the conditional mean $\boldsymbol{\mu}_{A|B}$ involves the inverse $\boldsymbol{Q}_{AA}^{-1}$, but only in a way such that we can write $\boldsymbol{\mu}_{A|B} = \boldsymbol{\mu}_A - \boldsymbol{b}$, where $\boldsymbol{b}$ is the solution of a sparse linear system $\boldsymbol{Q}_{AA}\boldsymbol{b} = \boldsymbol{Q}_{AB}(\boldsymbol{x}_B - \boldsymbol{\mu}_B)$. Note that the term $\boldsymbol{Q}_{AB}$ is nonzero only for those vertices in A that have an edge to a vertex in B, so usually only a few terms will enter in this matrix–vector product. In the special case $A = \{i\}$, the expressions simplify to

$$\mu_{i|-i} = \mu_i - \sum_{j:j\sim i} \frac{Q_{i,j}}{Q_{i,i}}(x_j - \mu_j) \quad \text{and} \quad Q_{i|-i} = Q_{i,i}. \tag{12.3}$$

Here we used the notation $j : j \sim i$ to indicate a sum over all vertices j that are neighbors to vertex i, i.e., $\{i, j\} \in \mathcal{E}$. So $Q_{i,i}$ is the conditional precision of x_i and the conditional expectation of x_i is a weighted mean of neighboring x_js with weights $-Q_{i,j}/Q_{i,i}$.

Example 12.2

We continue with Example 12.1. From (12.3) we obtain the conditional mean and precision of $x_i|\boldsymbol{x}_{-i}$,

$$\mu_{i|-i} = \frac{\phi}{1+\phi^2}(x_{i-1} + x_{i+1}) \quad \text{and} \quad Q_{i|-i} = 1 + \phi^2, \qquad 1 < i < n.$$

12.1.3.2 Markov Properties

The graph $\mathcal{G}$ of a GMRF is defined through looking at which x_i and x_j are conditionally independent, the so-called *pairwise* Markov property. However, more general Markov properties can be derived from $\mathcal{G}$.

A *path* from vertex i_1 to vertex i_m is a sequence of distinct nodes in $\mathcal{V}$, $i_1, i_2, \ldots, i_m$, for which $(i_j, i_{j+1}) \in \mathcal{E}$ for $j = 1, \ldots, m-1$. A subset $C \subset \mathcal{V}$ *separates* two nodes $i \notin C$ and

$j \notin C$, if every path from i to j contains at least one node from C. Two disjoint sets $A \subset \mathcal{V} \setminus C$ and $B \subset \mathcal{V} \setminus C$ are separated by C, if all $i \in A$ and $j \in B$ are separated by C. In other words, we cannot walk on the graph starting somewhere in A ending somewhere in B without passing through C. The global Markov property, is that

$$\boldsymbol{x}_A \perp \boldsymbol{x}_B \mid \boldsymbol{x}_C \tag{12.4}$$

for all mutually disjoint sets A, B and C where C separates A and B, and A and B are nonempty.

Theorem 12.3
Let $\boldsymbol{x}$ be a GMRF wrt $\mathcal{G}$, then $\boldsymbol{x}$ obeys the global Markov property.

Note that $A \cup B \cup C$ can be a subset of $\mathcal{V}$; hence, this result gives information about conditional independence properties for the marginal $\boldsymbol{x}_{A \cup B \cup C}$ as well.

Example 12.3
We continue Example 12.1. Using conditional independence, we know that $x_1 \perp x_n | \boldsymbol{x}_{-\{1,n\}}$, but from the global Markov property, we also know that $x_1 \perp x_n | x_j$, for all $1 < j < n$.

12.1.4 Conditional Specification

Following the seminal work of J. Besag (1974, 1975), it is common to specify a GMRF implicitly through the so-called full conditionals $\{\pi(x_i | \boldsymbol{x}_{-i})\}$. If required, we can derive from the full conditionals the mean and the precision matrix of the corresponding joint distribution. However, the full conditionals cannot be specified completely arbitrarily, as we must ensure that they correspond to a proper joint density. We will return to this issue in Section 12.1.8.

A conditional specification defines the full conditionals $\{\pi(x_i | \boldsymbol{x}_{-i})\}$ as normal with moments

$$\mathrm{E}(x_i | \boldsymbol{x}_{-i}) = \mu_i + \sum_{j \neq i} \beta_{i,j}(x_j - \mu_j) \quad \text{and} \quad \mathrm{Precision}(x_i | \boldsymbol{x}_{-i}) = \kappa_i > 0. \tag{12.5}$$

The rationale for such an approach, is that it is easier to specify the full conditionals than the joint distribution. Comparing (12.5) with (12.2), we can choose $\boldsymbol{\mu}$ as the mean, $Q_{i,i} = \kappa_i$, $\beta_{i,j} = -Q_{i,j}/Q_{i,i}$ to obtain the same full conditionals; see below for a formal proof. However, since $\boldsymbol{Q}$ is symmetric, we must require that

$$\kappa_i \beta_{i,j} = \kappa_j \beta_{j,i} \tag{12.6}$$

for all $i \neq j$. In particular, if $\beta_{i,j}$ is nonzero, then $\beta_{j,i}$ cannot be zero. The edges in the graph $\mathcal{G}$ are defined as $\{\{i, j\} : \beta_{i,j} \neq 0\}$. In addition to the symmetry constraint (12.6), there is a joint requirement that $\boldsymbol{Q}$ is SPD. Unfortunately, this is a *joint* property, which is hard to validate locally. One convenient approach that avoids this problem is to choose a diagonally dominant parametrization that ensures $\boldsymbol{Q}$ to be SPD: $Q_{i,i} > \sum_j |Q_{i,j}|$ for all i. This implies that

$$\sum_j |\beta_{i,j}| < 1, \quad \forall i.$$

However, this assumption can be restrictive (Rue and Held, 2005, Sec. 2.7 and Sec. 5.1).

Although we were able to identify a Gaussian with the same full conditionals, we need to know that the joint density is unique. This question is answered by Brook's lemma (Brook, 1964).

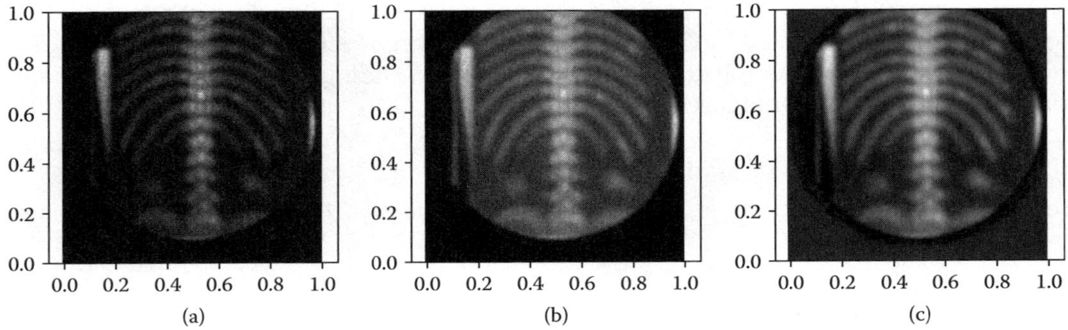

FIGURE 12.2
Panel (a) shows the raw x-ray image, (b) shows the square-root transformed image, and (c) shows the restored image (the posterior mean).

Lemma 12.1 (Brook's lemma)

Let $\pi(\boldsymbol{x})$ be the density for $\boldsymbol{x} \in \mathbb{R}^n$ and define $\Omega = \{\boldsymbol{x} \in \mathbb{R}^n : \pi(\boldsymbol{x}) > 0\}$. Let $\boldsymbol{x}, \boldsymbol{x}' \in \Omega$, then

$$\frac{\pi(\boldsymbol{x})}{\pi(\boldsymbol{x}')} = \prod_{i=1}^{n} \frac{\pi(x_i | x_1, \ldots, x_{i-1}, x'_{i+1}, \ldots, x'_n)}{\pi(x'_i | x_1, \ldots, x_{i-1}, x'_{i+1}, \ldots, x'_n)} \tag{12.7}$$

$$= \prod_{i=1}^{n} \frac{\pi(x_i | x'_1, \ldots, x'_{i-1}, x_{i+1}, \ldots, x_n)}{\pi(x'_i | x'_1, \ldots, x'_{i-1}, x_{i+1}, \ldots, x_n)}. \tag{12.8}$$

The uniqueness argument follows by keeping $\boldsymbol{x}'$ fixed, and then $\pi(\boldsymbol{x})$ is proportional to the full conditionals at the right-hand side of Equation (12.7). The constant of proportionality is found by using that $\pi(\boldsymbol{x})$ integrates to one. The consequence is that we can derive the joint density from the full conditionals, which we will illustrate in the following. For simplicity, fix $\boldsymbol{\mu} = \boldsymbol{0}$ and $\boldsymbol{x}' = \boldsymbol{0}$. Using the full conditionals in Equation (12.5), then Equation (12.7) simplifies to

$$\log \frac{\pi(\boldsymbol{x})}{\pi(\boldsymbol{0})} = -\frac{1}{2} \sum_{i=1}^{n} \kappa_i x_i^2 - \sum_{i=2}^{n} \sum_{j=1}^{i-1} \kappa_i \beta_{ij} x_i x_j. \tag{12.9}$$

and (12.8) simplifies to

$$\log \frac{\pi(\boldsymbol{x})}{\pi(\boldsymbol{0})} = -\frac{1}{2} \sum_{i=1}^{n} \kappa_i x_i^2 - \sum_{i=1}^{n-1} \sum_{j=i+1}^{n} \kappa_i \beta_{ij} x_i x_j. \tag{12.10}$$

Since (12.9) and (12.6) must be identical, then $\kappa_i \beta_{ij} = \kappa_j \beta_{ji}$ for $i \neq j$. The density of $\boldsymbol{x}$ can then be expressed as

$$\log \pi(\boldsymbol{x}) = \text{const} - \frac{1}{2} \sum_{i=1}^{n} \kappa_i x_i^2 - \frac{1}{2} \sum_{i \neq j} \kappa_i \beta_{ij} x_i x_j;$$

hence, $\boldsymbol{x}$ is zero mean GMRF provided $\boldsymbol{Q}$ is SPD.

We will now illustrate practical use of the conditional specification in a simple example.

Example 12.4

The image in Figure 12.2(a) is a 256 × 256 gamma camera image of a phantom designed to reflect structure expected from cancerous bones. Each pixel in the circular part of the image,

$\mathcal{I}$, represent the gamma radiation count, where a black pixel represents (essentially) zero counts and a white pixel the maximum count. The image is quite noisy and the task in this example is to (try to) remove the noise. The noise process is quite accurately described by a Poisson distribution, so that for each pixel i, the recorded count y_i relates to the true signal η_i, as $y_i \sim \text{Poisson}(\eta_i)$. For simplicity, we will use the approximation that

$$\sqrt{y_i} \mid \eta_i \sim \mathcal{N}\left(\sqrt{\eta_i}, \frac{1}{4}\right), \quad i \in \mathcal{I}$$

and the square-root transformed image is displayed in Figure 12.2(b). Taking a Bayesian approach, we need to specify a prior distribution for the (square-root-transformed) image $x = (x_1, \ldots, x_n)^T$, where $x_i = \sqrt{\eta_i}$. (We need η_i to be (somewhat) larger than zero for this approximation to be adequate.) Although this is a daunting problem in general, for such noise-removal tasks it is usually sufficient to specify the prior to be informative for how the true image behaves locally. Since the image itself is locally smooth, we might specify the prior through the full conditionals (12.5). Using the full conditionals we only need to answer questions like: *What if we do not know the true signal in pixel i, but all others; what is then our belief in x_i?* One choice is to set $\beta_{i,j}$ to zero unless j is one of the four nearest neighbors of i; $N_4(i)$, say. As we have no particular preference for direction, we might take for each i,

$$\beta_{i,j} = \frac{\delta}{4}, \quad j \in N_4(i)$$

where δ is considered as fixed. Further, we take κ_i to be common (and unknown) for all i, and restrict δ to $|\delta| < 1$ so that the (prior) precision matrix is diagonally dominant. (We ignore here some corrections at the boundary where a boundary pixels may have less than four neighbors.) We take further $\boldsymbol{\mu} = \mathbf{0}$ and a (conjugate) $\Gamma(a, b)$ prior for κ (with density $\propto \kappa^{a-1} \exp(-b\kappa)$), and then the posterior for (x, κ) reads

$$\begin{aligned} \pi(x, \kappa \mid y) &\propto \pi(x \mid \kappa)\, \pi(\kappa) \prod_{i \in \mathcal{I}} \pi(y_i \mid x_i) \\ &\propto \kappa^{a-1} \exp(-b\kappa)\, |Q_{prior}(\kappa)|^{1/2} \exp\left(-\frac{1}{2} x^T Q_{post}(\kappa) x + b^T x\right). \end{aligned} \tag{12.11}$$

Here, $b_i = 4\sqrt{y_i}$ for $i \in \mathcal{I}$ and zero otherwise, $Q_{post}(\kappa) = Q_{prior}(\kappa) + D$ where D is a diagonal matrix where $D_{i,i} = 4$ if $i \in \mathcal{I}$ and zero otherwise, and

$$Q_{prior}(\kappa)_{i,j} = \kappa \begin{cases} 1, & i = j \\ \delta/4, & j \in N_4(i) \\ 0, & \text{otherwise.} \end{cases}$$

Conditioned on κ and the observations, then x is a GMRF with precision matrix Q_{post} and where the mean $\boldsymbol{\mu}_{post}$ is given by the solution of

$$Q_{post} \boldsymbol{\mu}_{post} = b. \tag{12.12}$$

12.1.5 MCMC Algorithms for GMRFs

One of the attractive properties of GMRFs is that it integrates so nicely into the MCMC approach for doing Bayesian inference (see Gilks, Richardson, and Spiegelhalter, 1996; Robert and Casella, 1999 for a general background on MCMC). The nice correspondence between the conditional specification using the full conditionals and the MCMC algorithms for doing inference is one of the main reasons why GMRFs are so widely used. It turns out that GMRFs have a nice connection with numerical methods for sparse matrices, which leads to exact algorithms for GMRFs; we will return to this issue in Section 12.1.7. To set the scene, let $\pi(\boldsymbol{\theta})$ be the posterior of interest, where the task is to compute posterior marginals,

$$\pi(\theta_1), \ldots, \pi(\theta_n)$$

or from these compute summaries like $\mathrm{E}(\theta_1)$ and $\mathrm{var}(\theta_1)$, and so on. Relating back to Example 12.4, then $\boldsymbol{\theta} = (\boldsymbol{x}, \kappa)$ and the task is to compute $\mathrm{E}(x_i|\boldsymbol{y})$ for all $i \in \mathcal{I}$, and use that as an estimate of $\sqrt{\eta_i}$. We can quantify the uncertainty in our estimate from $\mathrm{var}(x_i|\boldsymbol{y})$ or from quantiles in the posterior marginal $\pi(x_i|\boldsymbol{y})$ itself.

12.1.5.1 Basic Ideas behind MCMC

The basic ideas of MCMC are simple. We will briefly present the two basic ideas, *Markov chain* and *Monte Carlo*, which together makes *Markov chain Monte Carlo*. The Monte Carlo approach for (also Bayesian) inference is mostly about Monte Carlo integration, which substitutes integrals with empirical means; for some suitable function $f(\cdot)$, we have

$$\mathrm{E}(f(\boldsymbol{\theta})) = \int f(\boldsymbol{\theta})\pi(\boldsymbol{\theta})\, d\boldsymbol{\theta} \approx \frac{1}{N}\sum_{i=1}^{N} f(\boldsymbol{\theta}^{(i)}) \tag{12.13}$$

where

$$\boldsymbol{\theta}^{(1)}, \boldsymbol{\theta}^{(2)}, \ldots, \boldsymbol{\theta}^{(N)}$$

are N samples from $\pi(\boldsymbol{\theta})$. By interpreting the integral as an expected value, we can approximate it as the empirical mean over N samples from $\pi(\boldsymbol{\theta})$. If the samples are independent, then we also control the error, as

$$\frac{1}{\sqrt{N}}\sum_{i=1}^{N} f(\boldsymbol{\theta}^{(i)}) \;\approx\; \mathcal{N}(\mathrm{E}(f(\boldsymbol{\theta})), \mathrm{var}(f(\boldsymbol{\theta})))$$

under quite general assumptions. Our estimate can be made as precise as we like, choosing N large enough. Note that the error behaves like $\mathcal{O}(1/\sqrt{N})$, so one extra digit in accuracy requires $100 \times N$ samples.

Generating samples from a distribution $\pi(\boldsymbol{\theta})$ can be difficult even for low dimensions, but less so in the one-dimensional case. The main obstacle apart from the dimensionality is the often missing normalizing constant in $\pi(\boldsymbol{\theta})$; we often only know an unnormalized version of the density and to normalize it is as hard as computing $\mathrm{E}(f(\boldsymbol{\theta}))$.

The second main idea is to circumvent the problem of generation samples from $\pi(\boldsymbol{\theta})$ directly, but do generate samples from $\pi(\boldsymbol{\theta})$ *implicitly*. This is done by constructing a Markov chain with $\pi(\boldsymbol{\theta})$ as the equilibrium distribution, and we can simulate this Markov chain to obtain a sample. It turns out that this can be done without knowing the normalizing constant of $\pi(\boldsymbol{\theta})$. In principle, we can then generate a sample from $\pi(\boldsymbol{\theta})$ by simulating the Markov chain for a long time, until it has converged and then the state of the Markov chain is one sample.

We will now present the *Gibbs sampler*, which is one of the two most popular MCMC algorithms. We will postpone our discussion of the second one, the Metropolis–Hastings algorithm, until Section 12.1.9. The Gibbs sampler was introduced into the mainstream IEEE literature by Geman and Geman (1984) and into statistics by Gelfand and Smith (1990). Later on, it became clear that the general ideas (and their implications) were already around Hastings (1970); see (Robert and Casella, 1999, Chap. 7) for a historical account.

12.1.5.2 The Gibbs Sampler

The most intuitive MCMC algorithm is called the *Gibbs sampler*. The Gibbs sampler simulates a Markov chain by repeatedly sampling from the full conditionals. The state vector in the Markov chain, $\boldsymbol{\theta}$, is in the following Gibbs sampler algorithm overwritten at each instance.

> Initialize the state vector $\boldsymbol{\theta} = \boldsymbol{\theta}_0$.
> While TRUE; do
>   For $i = 1, \ldots, n$,
>   Sample $\theta_i \sim \pi(\theta_i \mid \boldsymbol{\theta}_{-i})$
>   Output new state $\boldsymbol{\theta}$

This algorithm defines a Markov chain with equilibrium distribution $\pi(\boldsymbol{\theta})$; intuitively, if $\boldsymbol{\theta}$ is a sample from $\pi(\boldsymbol{\theta})$ and we update its ith component θ_i by a sample from $\pi(\theta_i|\boldsymbol{\theta}_{-i})$, then $\boldsymbol{\theta}$ is still distributed according to $\pi(\boldsymbol{\theta})$. Further, the algorithm will output a new state, $\boldsymbol{\theta}$ at time $t = 1, 2, \ldots$, say. Denote these samples $\boldsymbol{\theta}^{(1)}, \boldsymbol{\theta}^{(2)}, \ldots$.

Under quite general conditions, the distribution for $\boldsymbol{\theta}^{(k)}$ will converge, as $k \to \infty$ to the equilibrium distribution $\pi(\boldsymbol{\theta})$. For some large k, k_0 say, we can consider $\boldsymbol{\theta}^{(k_0)}$ to be a sample from $\pi(\boldsymbol{\theta})$. However, if $\boldsymbol{\theta}^{(k_0)}$ has the correct distribution, then so will the next state $\boldsymbol{\theta}^{(k_0+1)}$; however, they will be dependent. The Monte Carlo estimate in (12.13) will be modified into

$$\mathrm{E}(f(\boldsymbol{\theta})) = \frac{1}{N - k_0 + 1} \sum_{i=k_0}^{N} f(\boldsymbol{\theta}^{(i)}). \tag{12.14}$$

Note that $\{\boldsymbol{\theta}^{(i)}\}$ for $i = k_0, \ldots$ are now, in general, *dependent*. The variance estimate of (12.14) must take this dependence into account. We have discharged the first $k_0 - 1$ states from the Markov chain, which is named the *burn-in*. To (try to) determine the value for k_0, the main idea is to look at the trace of θ_i, say; plot $\{\theta_i^{(k)}\}$ against $k = 1, 2, \ldots$, and try to determine a k_0 for which the fluctuations around the mean value seem reasonably stable (in a distributional sense), (see Robert, 1998; Robert and Casella, 1999, for details). Choosing k_0 too large does not bias the estimate, but choosing k_0 too small will.

Example 12.5

We will now illustrate the use of a Gibbs sampler to generate a sample from a GMRF. If we specify the GMRF through the full conditionals (12.5), we immediately have the Gibbs sampler. The algorithm goes as follows:

> Initialize $\boldsymbol{x} = \boldsymbol{\mu}$ (or some other value)
> While TRUE; do
>   For $i = 1, \ldots, n$
>   Compute $\mu_{i|-i} = \mathrm{E}(x_i|\boldsymbol{x}_{-i}) = \mu_i + \sum_{j \neq i} \beta_{i,j}(x_j - \mu_j)$
>   Sample $z \sim \mathcal{N}(0, 1)$
>   Set $x_i = \mu_{i|-i} + z/\sqrt{\kappa_i}$
>   Output $\boldsymbol{x}$

The attractiveness of the Gibbs sampler for GMRFs is both the *simplicity* and the *speed*. It is simple, as we do not need to work out the joint distribution for $\boldsymbol{x}$ in order to generate samples from $\pi(\boldsymbol{x})$. As long as the full conditionals define a valid joint distribution, then the Gibbs sampler will converge to the correct distribution. It is customary that the number of neighbors to each site i does not depend on n. In this case the computational cost of running the Gibbs sampler for one iteration (updating all the n elements of $\boldsymbol{x}$) is $\mathcal{O}(n)$ operations, and this is optimal in an order sense.

Example 12.6

Let us return to Example 12.4, which is somewhat more involved than Example 12.5 for two reasons. First, we need to deal with the unknown precision κ. Second, for fixed κ, the GMRF is known only implicitly as the mean is given as the solution of (12.12). It turns out that we *do not* need to solve (12.12) in order to construct a Gibbs sampler.

In this example, $\boldsymbol{\theta} = (\kappa, \boldsymbol{x})$, and we need to compute from the posterior (12.11) all full conditionals. Using $\boldsymbol{Q}_{post}(\kappa) = \kappa \boldsymbol{Q}_{prior}(1) + \boldsymbol{D}$, the full conditional for κ is

$$\kappa \mid \boldsymbol{x}, \boldsymbol{y} \sim \Gamma\left(n/2 + a,\ b + \frac{1}{2}\boldsymbol{x}^T \boldsymbol{Q}_{prior}(1)\boldsymbol{x}\right).$$

The full conditionals for x_i are derived using

$$\begin{aligned}
\pi(x_i \mid \boldsymbol{x}_{-i}, \kappa, \boldsymbol{y}) &\propto \pi(\boldsymbol{x}, \kappa \mid \boldsymbol{y}) \\
&\propto \exp\left(-\frac{1}{2}x_i^2 Q_{post,i,i}(\kappa) - x_i \sum_{j \in N_4(i)} Q_{prior,i,j}(\kappa) x_j + b_i x_i\right) \\
&= \exp\left(-\frac{1}{2}c_i x_i^2 + d_i x_i\right),
\end{aligned}$$

where c_i depends on κ, and d_i depends on both κ and $\{x_j\}$ for $j \in N_4(i)$. The full conditional for x_i is then $\mathcal{N}(d_i/c_i, 1/c_i)$. The Gibbs sampler algorithm then becomes:

Initialize $\boldsymbol{x} = \boldsymbol{0}$ and $\kappa = 1$ (or some other values)
While TRUE; do
    Sample $\kappa \sim \Gamma\left(n/2 + a,\ b + \frac{1}{2}\boldsymbol{x}^T \boldsymbol{Q}_{prior}(1)\boldsymbol{x}\right)$
    For $i = 1, \ldots, n$
        Compute c_i and d_i
        Sample $x_i \sim \mathcal{N}(d_i/c_i, 1/c_i)$
    Output new state $(\kappa, \boldsymbol{x})$

The posterior mean for $\boldsymbol{x}$ is displayed in Figure 12.2(c), using $a = 1, b = 0.01$ and $\beta = 1/4$. It is rather amazing that we can generate samples from the posterior for $(\kappa, \boldsymbol{x})$ using this simple algorithm. Also note that conditioning on the observations $\boldsymbol{y}$ does not alter the neighborhood for x_i, it is still $N_4(i)$. The consequence is that the Gibbs sampler has the same computational cost as without data, which is $\mathcal{O}(n)$ for each iteration. This includes the cost for updating κ, which is dominated by the cost of computing $\boldsymbol{x}^T \boldsymbol{Q}_{prior}(1)\boldsymbol{x}$ (which also is $\mathcal{O}(n)$).

12.1.6 Multivariate GMRFs

To fix ideas, we will consider a generalization of Example 12.4 where the observations are now sequences of images. The sequence can either be a movie where each frame in the sequence is indexed by time, or the height where recorded a three-dimensional object as a set of two-dimensional images. Other examples include a temporal version of spatial models of disease counts in each administrative region of a country. Figure 12.3 shows five consecutive frames of three-dimensional cells taken by confocal microscopy. The first frame has a lot of noise, but the signal gets stronger farther up in the image stack. We consider the same problem as for Example 12.4; we want to estimate the true signal in the presence of the noise. The five frames represent the same three-dimensional object, but at different heights.

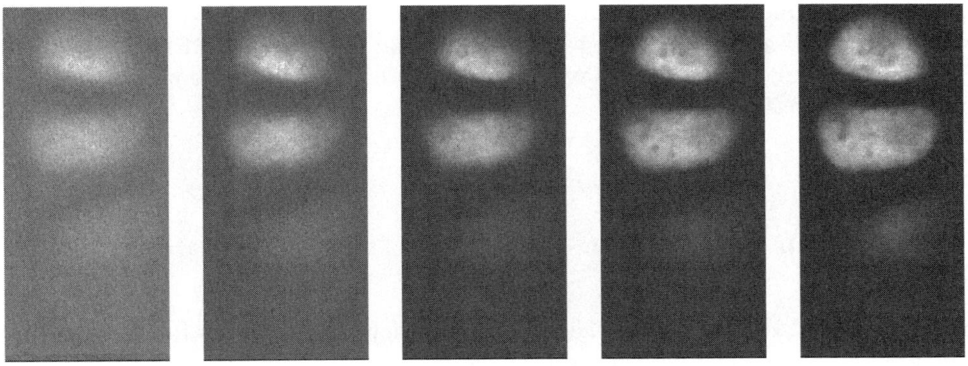

FIGURE 12.3
Panels (a) to (e) show five consecutive frames of a three-dimensional confocal microscopy image.

We can use this information when we specify the full conditionals. It is then both easier and more natural to specify a multivariate version of the full conditionals (12.5), which we now will describe. Let $\boldsymbol{x}_i$ represent all the $p = 5$ observations at pixel i

$$\boldsymbol{x}_i = (x_{i,1}, x_{i,2}, x_{i,3}, x_{i,4}, x_{i,5})^T.$$

Here, $x_{i,2}$ is the pixel at location i in frame 2 and so on. The conditional specification (12.5) extends naturally to

$$\mathrm{E}(\boldsymbol{x}_i \mid \boldsymbol{x}_{-i}) = \boldsymbol{\mu}_i - \sum_{j:j\sim i} \boldsymbol{\beta}_{i,j}(\boldsymbol{x}_j - \boldsymbol{\mu}_j), \quad \text{and} \quad \mathrm{Precision}(\boldsymbol{x}_i \mid \boldsymbol{x}_{-i}) = \boldsymbol{\kappa}_i > 0, \tag{12.15}$$

for some $p \times p$ matrices $\{\boldsymbol{\beta}_{i,j}\}$ and $\{\boldsymbol{\kappa}_i\}$ (see also Mardia, 1988). In this formulation, we can now specify that our knowledge of $x_{i,3}$ might benefit of knowing $x_{i,2}$ and $x_{i,4}$. These pixels are in the same $\boldsymbol{x}_i$ vector, although they represent the ith pixel at the previous and next frame. Additionally, we can have dependency from neighbors within the same frame, such as $\{x_{j,3},\ j \in N_4(i)\}$. In short, we can specify how $\boldsymbol{x}_i$ depends on $\{\boldsymbol{x}_j\}$, $j \neq i$, and thinking about neighbors that are (small p-) vectors.

The conditional specification in this example motivates the introduction of a multivariate GMRF, which we denote as MGMRF$_p$. Its definition is a direct extension of (12.1). Let $\boldsymbol{x} = (\boldsymbol{x}_1^T, \ldots, \boldsymbol{x}_n^T)^T$ be Gaussian distributed, where each $\boldsymbol{x}_i$ is a p-vector. Similarly, let $\boldsymbol{\mu} = (\boldsymbol{\mu}_1^T, \ldots, \boldsymbol{\mu}^T)^T$ denote the mean and $\widetilde{\boldsymbol{Q}} = (\widetilde{\boldsymbol{Q}}_{i,j})$ the precision matrix with $p \times p$ elements $\widetilde{\boldsymbol{Q}}_{i,j}$.

Definition 12.2 (MGMRF$_p$)
A random vector $\boldsymbol{x} = (\boldsymbol{x}_1^T, \ldots, \boldsymbol{x}_n^T)^T$ where $\dim(\boldsymbol{x}_i) = p$, is called a MGMRF$_p$ wrt $\mathcal{G} = (\mathcal{V} = \{1, \ldots, n\}, \mathcal{E})$ with mean $\boldsymbol{\mu}$ and a SPD precision matrix $\widetilde{\boldsymbol{Q}}$, iff its density has the form

$$\begin{aligned}\pi(\boldsymbol{x}) &= \left(\frac{1}{2\pi}\right)^{np/2} |\widetilde{\boldsymbol{Q}}|^{1/2} \exp\left(-\frac{1}{2}(\boldsymbol{x}-\boldsymbol{\mu})^T \widetilde{\boldsymbol{Q}}(\boldsymbol{x}-\boldsymbol{\mu})\right)\\ &= \left(\frac{1}{2\pi}\right)^{np/2} |\widetilde{\boldsymbol{Q}}|^{1/2} \exp\left(-\frac{1}{2}\sum_{ij}(\boldsymbol{x}_i-\boldsymbol{\mu}_i)^T \widetilde{\boldsymbol{Q}}_{i,j}(\boldsymbol{x}_j-\boldsymbol{\mu}_j)\right)\end{aligned}$$

and

$$\widetilde{\boldsymbol{Q}}_{i,j} \neq \boldsymbol{0} \Longleftrightarrow \{i, j\} \in \mathcal{E} \quad \text{for all} \quad i \neq j.$$

It is important to note that a size n MGMRF$_p$ is just another GMRF of dimension np; so all our previous results and forthcoming sparse matrix algorithms for GMRFs also apply for a MGMRF$_p$. However, some results have easier interpretation using the block formulation, such as

$$\boldsymbol{x}_i \perp \boldsymbol{x}_j \mid \boldsymbol{x}_{-\{i,j\}} \iff \widetilde{\boldsymbol{Q}}_{i,j} = \boldsymbol{0}$$

and

$$\mathrm{E}(\boldsymbol{x}_i \mid \boldsymbol{x}_{-i}) = \boldsymbol{\mu}_i - \widetilde{\boldsymbol{Q}}_{i,i}^{-1} \sum_{j:j\sim i} \widetilde{\boldsymbol{Q}}_{i,j}(\boldsymbol{x}_j - \boldsymbol{\mu}_j) \quad \text{and} \quad \mathrm{Precision}(\boldsymbol{x}_i \mid \boldsymbol{x}_{-i}) = \widetilde{\boldsymbol{Q}}_{i,i}. \tag{12.16}$$

From Equation (12.16), we can obtain the consistency requirements for the conditional specification Equation (12.15) by choosing

$$\widetilde{\boldsymbol{Q}}_{i,j} = \begin{cases} \boldsymbol{\kappa}_i \boldsymbol{\beta}_{i,j} & i \neq j \\ \boldsymbol{\kappa}_i & i = j. \end{cases}$$

Since $\widetilde{\boldsymbol{Q}}_{i,j} = \widetilde{\boldsymbol{Q}}_{j,i}^T$, then we have the requirement that $\boldsymbol{\kappa}_i \boldsymbol{\beta}_{i,j} = \boldsymbol{\beta}_{j,i}^T \boldsymbol{\kappa}_j$ for $i \neq j$, additionally to $\boldsymbol{\kappa}_i > 0$ for all i. Finally, there is also the "global" requirement that $\widetilde{\boldsymbol{Q}}$ must be SPD, which is equivalent to $(\boldsymbol{I} + (\boldsymbol{\beta}_{i,j}))$ being SPD.

12.1.7 Exact Algorithms for GMRFs

Similarly to the nice connection between GMRFs and the Gibbs sampler (and also more general MCMC algorithms), GMRFs also have a nice connection with very efficient numerical algorithms for sparse matrices. This connection allows for exact algorithms for GMRFs. We will now discuss this connection, starting with various exact algorithms to efficiently sample from GMRFs. This includes solving tasks like unconditional and conditional sampling, sampling under linear hard and soft constraints, evaluating the log-density of a (possibly constrained) GMRF at a particular value, and computing marginal variances for (possibly constrained) GMRFs. Although all these tasks are formally "just matrix algebra," we need to ensure that we take advantage of the sparse precision matrix $\boldsymbol{Q}$ in all steps so computations can make use of the efficient numerical algorithms for sparse matrices developed in the computational sciences literature. Further, we can derive all the algorithms for sparse matrices by considering conditional independence properties of GMRFs. The core of all algorithms is the Cholesky factorization $\boldsymbol{Q} = \boldsymbol{L}\boldsymbol{L}^T$ of the precision matrix $\boldsymbol{Q}$, where $\boldsymbol{L}$ is a lower-triangular matrix. We postpone details on how to compute $\boldsymbol{L}$ for the moment and simply assume that this factorization is available. We assume throughout that $\boldsymbol{x}$ is a GMRF wrt $\mathcal{G}$ with mean $\boldsymbol{\mu}$ and an SPD precision matrix $\boldsymbol{Q}$. Before we start to discuss the details, we will examine for why this is important.

12.1.7.1 Why Are Exact Algorithms Important?

Exact efficient algorithms are generally preferable when they exist, even though they apparently require algorithms that are more involved than simple iterative ones. Computational feasibility *is* important even for statistical modeling, as a statistical model is not of much use if we cannot do inference efficiently enough to satisfy the end-user.

Sampling from a GMRF can be done exactly using the Cholesky factorization of the precision matrix, which algorithmically is much more involved than the simple Gibbs sampler in Example 12.5. However, an exact algorithm is exact and automatic, whereas the Gibbs sampler algorithm is approximate and requires, in most cases, human intervention to verify (if at all possible) that the output is indeed a sample from the correct distribution. In the spatial case, it turns out that we can (typically) sample a GMRF exactly at the cost of $\mathcal{O}(n^{3/2})$ operations, which correspond to $\mathcal{O}(\sqrt{n})$ iterations with the single-site Gibbs sampler. The exact algorithm can further produce independent samples at the cost of $\mathcal{O}(n \log n)$ each.

Let us return again to Example 12.4. We will now slightly modify this example to demonstrate that Gibbs sampling algorithms are not always that simple. Earlier, the parameter δ was considered to be fixed. What if we want to do inference for δ as well? In this case, we will get another update in the Gibbs sampler algorithm for which δ has to be sampled from

$$\pi(\delta \mid \boldsymbol{x}, \kappa, \boldsymbol{y}) \propto \pi(\delta)\ |\boldsymbol{Q}_{prior}(\kappa, \delta)|^{1/2} \exp\left(-\frac{1}{2}\boldsymbol{x}^T \boldsymbol{Q}_{prior}(\kappa, \delta)\boldsymbol{x}\right), \tag{12.17}$$

where $\pi(\delta)$ is the prior distribution. The main point is that the prior precision matrix depends on both κ and δ, and we need to compute the determinant of $\boldsymbol{Q}_{prior}(\kappa, \delta)$. This quantity is not analytically tractable (except for special cases). In general, we have to *compute* the determinant of $\boldsymbol{Q}_{prior}(\kappa, \delta)$, and if we do this within the Gibbs sampler algorithm, we have to do this many times. Similar comments can be made when the task is to infer unknown parameters of the GMRF.

Exact algorithms can also be used to improve the (single-site) Gibbs sampler algorithm. The idea is to update a subset of $\boldsymbol{\theta}$, $\boldsymbol{\theta}_a$ say, from $\pi(\boldsymbol{\theta}_a|\boldsymbol{\theta}_{-a})$, instead of one component at a time. Here, $\boldsymbol{\theta}_a$ can be, for example, the GMRF part of the model. Returning again to Example 12.4, we can improve the Gibbs sampler algorithm to an algorithm that updates $(\kappa, \boldsymbol{x})$ in two blocks

$$\kappa \sim \pi(\kappa|\boldsymbol{x}, \boldsymbol{y}) \quad \text{and} \quad \boldsymbol{x} \sim \pi(\boldsymbol{x}|\kappa, \boldsymbol{y}),$$

which is possible because $\pi(\boldsymbol{x}|\kappa, \boldsymbol{y})$ is a GMRF.

In many cases, involving GMRFs, it is possible to avoid using MCMC as an exact (up to numerical integration) or approximate solution exists; see Rue, Martino, and Chopin (2009) for details, applications and extensions. The benefit of such an approach is the absence of Monte Carlo error and improved computational speed. Returning again to Example 12.4, then

$$\pi(\kappa \mid \boldsymbol{y}) \propto \frac{\pi(\boldsymbol{x}, \kappa|\boldsymbol{y})}{\pi(\boldsymbol{x}|\kappa, \boldsymbol{y})}$$

for any $\boldsymbol{x}$. To compute the right-hand side (for a given κ), we need to solve (12.12) using exact algorithms. After computing the posterior density of κ, we can use this result to compute the posterior density of any x_i, using

$$\pi(x_i \mid \boldsymbol{y}) = \int \pi(\kappa \mid \boldsymbol{y})\ \pi(x_i \mid \kappa, \boldsymbol{y})\ d\kappa,$$

which we approximate with a finite sum. In this example, $x_i|\kappa, \boldsymbol{y}$ is Gaussian, so $\pi(x_i|\boldsymbol{y})$ can be approximated with a finite mixture of Gaussians. The additional task here is to compute the marginal variance for x_i. Extending this example to unknown δ, we get

$$\pi(\kappa, \delta \mid \boldsymbol{y}) \propto \frac{\pi(\boldsymbol{x}, \boldsymbol{\kappa}, \delta|\boldsymbol{y})}{\pi(\boldsymbol{x}|\kappa, \delta, \boldsymbol{y})} \tag{12.18}$$

and

$$\pi(x_i \mid \boldsymbol{y}) = \int\int \pi(\kappa, \delta \mid \boldsymbol{y})\ \pi(x_i \mid \kappa, \delta, \boldsymbol{y})\ d\kappa\ d\delta,$$

where we also need to compute the determinant of $\boldsymbol{Q}_{\text{prior}}(\kappa, \delta)$. Again, $\pi(x_i|\boldsymbol{y})$ can be approximated by a finite mixture of Gaussians, whereas $\pi(\kappa|\boldsymbol{y})$ and $\pi(\delta, \boldsymbol{y})$ must be computed from (12.18) using numerical integration.

12.1.7.2 Some Basic Linear Algebra

Let $\boldsymbol{A}$ be SPD, then there exists a unique (Cholesky) factorization $\boldsymbol{A} = \boldsymbol{L}\boldsymbol{L}^T$, where $\boldsymbol{L}$ is lower triangular and called the Cholesky triangle. This factorization is the starting point for solving $\boldsymbol{A}\boldsymbol{y} = \boldsymbol{b}$ by first solving $\boldsymbol{L}\boldsymbol{v} = \boldsymbol{b}$ and then $\boldsymbol{L}^T\boldsymbol{y} = \boldsymbol{v}$. The first linear system $\boldsymbol{L}\boldsymbol{v} = \boldsymbol{b}$ is solved directly using forward substitution

$$v_i = \frac{1}{L_{i,i}}\left(b_i - \sum_{j=1}^{i-1} L_{i,j} v_j\right), \quad i = 1, \ldots, n,$$

whereas $\boldsymbol{L}^T\boldsymbol{y} = \boldsymbol{v}$ is solved using backward substitution

$$y_i = \frac{1}{L_{i,i}}\left(v_i - \sum_{j=i+1}^{n} L_{j,i} y_j\right), \quad i = n, \ldots, 1.$$

Computing $\boldsymbol{A}^{-1}\boldsymbol{Y}$, where $\boldsymbol{Y}$ is a $n \times k$ matrix, is done by computing $\boldsymbol{A}^{-1}\boldsymbol{Y}_j$ for each of the k columns $\boldsymbol{Y}_j$ using the algorithm above. Note that $\boldsymbol{A}$ needs to be factorized only once. Note that in the case $\boldsymbol{Y} = \boldsymbol{I}$ (and $k = n$), the inverse of $\boldsymbol{A}$ is computed, which explains the command solve(A) in R (R Development Core Team, 2007) to invert a matrix.

12.1.7.3 Sampling from a GMRF

Sampling from a GMRF can be done using the following steps: sample $\boldsymbol{z} \sim \mathcal{N}(\boldsymbol{0}, \boldsymbol{I})$, i.e., n standard normal variables, solve $\boldsymbol{L}^T\boldsymbol{v} = \boldsymbol{z}$, and compute $\boldsymbol{x} = \boldsymbol{\mu} + \boldsymbol{v}$. The sample $\boldsymbol{x}$ has the correct distribution as $\mathrm{E}(\boldsymbol{v}) = \boldsymbol{0}$ and $\mathrm{Cov}(\boldsymbol{v}) = \boldsymbol{L}^{-T}\boldsymbol{I}\boldsymbol{L}^{-1} = (\boldsymbol{L}\boldsymbol{L}^T)^{-1} = \boldsymbol{Q}^{-1}$. The log-density of $\boldsymbol{x}$,

$$\log \pi(\boldsymbol{x}) = -\frac{n}{2}\log 2\pi + \frac{1}{2}\log|\boldsymbol{Q}| - \frac{1}{2}\underbrace{(\boldsymbol{x}-\boldsymbol{\mu})^T\boldsymbol{Q}(\boldsymbol{x}-\boldsymbol{\mu})}_{q}$$

is evaluated as follows. If $\boldsymbol{x}$ is sampled using the algorithm above, then $q = \boldsymbol{z}^T\boldsymbol{z}$, otherwise, we compute $\boldsymbol{w} = \boldsymbol{x} - \boldsymbol{\mu}$, $\boldsymbol{u} = \boldsymbol{Q}\boldsymbol{w}$ and then $q = \boldsymbol{w}^T\boldsymbol{u}$. Note that $\boldsymbol{u} = \boldsymbol{Q}\boldsymbol{w}$ is a sparse-matrix vector product, which can be computed efficiently:

$$u_i = Q_{i,i}w_i + \sum_{j:j\sim i} Q_{i,j}w_j,$$

where the diagonal term is added explicitly since i is not a neighbor of i. The determinant of $\boldsymbol{Q}$ can be found from the Cholesky factorization: $|\boldsymbol{Q}| = |\boldsymbol{L}\boldsymbol{L}^T| = |\boldsymbol{L}|^2$. Since $\boldsymbol{L}$ is lower triangular, we obtain

$$\frac{1}{2}\log|\boldsymbol{Q}| = \sum_i \log L_{i,i}.$$

Conditional sampling of $\boldsymbol{x}_A$ conditioned on $\boldsymbol{x}_B$, as described in Theorem 12.2, is similar: factorize $\boldsymbol{Q}_{AA}$ and *solve*

$$\boldsymbol{Q}_{AA}(\boldsymbol{\mu}_{A|B} - \boldsymbol{\mu}_A) = -\boldsymbol{Q}_{AB}(\boldsymbol{x}_B - \boldsymbol{\mu}_B)$$

for $\boldsymbol{\mu}_{A|B}$, using forward and backward substitution. The (sparse) matrix vector product on the right-hand side is computed using only the nonzero terms in $\boldsymbol{Q}_{AB}$. The remaining steps are the same.

12.1.7.4 Sampling from a GMRF Conditioned on Linear Constraints

In practical applications, we often want to sample from a GMRF under a linear constraint,

$$\boldsymbol{Ax} = \boldsymbol{e},$$

for a $k \times n$ matrix $\boldsymbol{A}$ of rank k. The common case is that $k \ll n$. A brute-force approach is to directly compute the conditional (Gaussian) density $\pi(\boldsymbol{x}|\boldsymbol{Ax} = \boldsymbol{e})$, but this will reveal that the corresponding precision matrix is (usually) not sparse anymore. For example, if $x_i \perp x_j | \boldsymbol{x}_{-\{i,j\}}$ without the constraint, then a sum-to-zero constraint $\sum x_i = 0$ makes x_i and x_j negatively correlated. In order not to lose computational efficiency, we must approach this problem in a more subtle way by correcting an unconstrained sample $\boldsymbol{x}$ to obtain a constrained sample $\boldsymbol{x}^c$:

$$\boldsymbol{x}^c = \boldsymbol{x} - \boldsymbol{Q}^{-1}\boldsymbol{A}^T(\boldsymbol{AQ}^{-1}\boldsymbol{A}^T)^{-1}(\boldsymbol{Ax} - \boldsymbol{e}). \tag{12.19}$$

A direct calculation shows that $\boldsymbol{x}^c$ has the correct mean and covariance. A closer look at (12.19) makes it clear that all the matrix terms are easy to compute: $\boldsymbol{Q}^{-1}\boldsymbol{A}^T$ just solves k linear systems of type $\boldsymbol{Qv}_j = (\boldsymbol{A}^T)_j$, for $j = 1, \ldots, k$, whereas $\boldsymbol{AQ}^{-1}\boldsymbol{A}^T$ is a $k \times k$ matrix and its Cholesky factorization is fast to compute since k is small in typical applications. Note that the extra cost of having k constraints is $\mathcal{O}(nk^2)$, hence, negligible when k is small.

Evaluating the constrained log density perhaps needs more attention, since $\boldsymbol{x}|\boldsymbol{Ax}$ is singular with rank $n - k$. However, the following identity can be used:

$$\pi(\boldsymbol{x}|\boldsymbol{Ax}) = \frac{\pi(\boldsymbol{Ax}|\boldsymbol{x})\,\pi(\boldsymbol{x})}{\pi(\boldsymbol{Ax})}. \tag{12.20}$$

Note now that all terms on the right-hand side are easy to compute: $\pi(\boldsymbol{x})$ is a GMRF with mean $\boldsymbol{\mu}$ and precision matrix $\boldsymbol{Q}$, $\pi(\boldsymbol{Ax})$ is (k-dimensional) Gaussian with mean $\boldsymbol{A\mu}$ and covariance $\boldsymbol{AQ}^{-1}\boldsymbol{A}^T$, while $\pi(\boldsymbol{Ax}|\boldsymbol{x})$ is either 0 (when the configuration $\boldsymbol{x}$ is inconsistent with the value of $\boldsymbol{Ax}$) or equal to $|\boldsymbol{AA}^T|^{-1/2}$.

Example 12.7

Let $\boldsymbol{x}$ be n independent, zero-mean, normal random variables with variance $\{\sigma_i^2\}$. A sum-to-zero constrained sample $\boldsymbol{x}^*$ can be generated from a sample of the unconstrained $\boldsymbol{x}$ using

$$x_i^* = x_i - c\sigma_i^2, \quad \text{where} \quad c = \sum x_j / \sum \sigma_j^2, \quad i = 1, \ldots, n.$$

The above construction can be generalized to condition on so-called *soft constraints*, which we condition on the k observations $\boldsymbol{y} = (y_1, \ldots, y_k)^T$, where

$$\boldsymbol{y} \mid \boldsymbol{x} \sim \mathcal{N}(\boldsymbol{Ax}, \boldsymbol{\Upsilon}).$$

Here, $\boldsymbol{A}$ is a $k \times n$ matrix with rank k and $\boldsymbol{\Upsilon} > 0$. The conditional density for $\boldsymbol{x}|\boldsymbol{y}$ has precision matrix $\boldsymbol{Q} + \boldsymbol{A}^T\boldsymbol{\Upsilon}^{-1}\boldsymbol{A}$, which is often a dense matrix. We can use the same approach as in Equation (12.19), which now generalizes to

$$\boldsymbol{x}^c = \boldsymbol{x} - \boldsymbol{Q}^{-1}\boldsymbol{A}^T(\boldsymbol{AQ}^{-1}\boldsymbol{A}^T + \boldsymbol{\Upsilon})^{-1}(\boldsymbol{Ax} - \boldsymbol{\epsilon}), \tag{12.21}$$

where $\boldsymbol{\epsilon} \sim \mathcal{N}(\boldsymbol{y}, \boldsymbol{\Upsilon})$. Similarly, if $\boldsymbol{x}$ is a sample from $\pi(\boldsymbol{x})$ and then $\boldsymbol{x}^c$ computed from Equation (12.21) is distributed as $\pi(\boldsymbol{x}|\boldsymbol{y})$. To evaluate the conditional density, we use the same approach as in Equation (12.20), which now reads

$$\pi(\boldsymbol{x}|\boldsymbol{y}) = \frac{\pi(\boldsymbol{y}|\boldsymbol{x})\,\pi(\boldsymbol{x})}{\pi(\boldsymbol{y})}.$$

Here, $\boldsymbol{\pi}(\boldsymbol{x})$ is the density for a GMRF, $\boldsymbol{\pi}(\boldsymbol{y}|\boldsymbol{x})$ is the density for a k-dimensional Gaussian with mean $\boldsymbol{Ax}$ and covariance matrix $\boldsymbol{\Upsilon}$, whereas $\pi(\boldsymbol{y})$ is the density for a k-dimensional Gaussian with mean $\boldsymbol{A\mu}$ and covariance matrix $\boldsymbol{AQ}^{-1}\boldsymbol{A}^T + \boldsymbol{\Upsilon}$.

12.1.7.5 The Cholesky Factorization of Q

All the exact simulation algorithms are based on the Cholesky triangle $\boldsymbol{L}$, which is found by factorizing $\boldsymbol{Q}$ into $\boldsymbol{L}\boldsymbol{L}^T$. We will now discuss this factorization in more detail, show why sparse matrices allow for faster factorization, and how reordering the indices can speed up the computations.

12.1.7.6 Interpretation of the Cholesky Triangle

The Cholesky factorization of $\boldsymbol{Q}$ is explicitly available; it is just a matter of doing the computations in the correct order. By definition, we have

$$Q_{i,j} = \sum_{k=1}^{j} L_{i,k} L_{j,k}, \quad i \geq j,$$

where we have used that $\boldsymbol{L}$ is lower triangular meaning that $L_{i,k} = 0$ for all $k > i$. To fix ideas, assume $n = 2$, so that $Q_{1,1} = L_{1,1}^2$, $Q_{2,1} = L_{2,1}L_{1,1}$ and $Q_{2,2} = L_{2,1}L_{2,1} + L_{2,2}^2$. Then we see immediately that we can compute $L_{1,1}$, $L_{2,1}$ and $L_{2,2}$ in this particular order. This generalizes for $n > 2$; we can compute $L_{i,1}$ for $i = 1, \ldots, n$ (in this order), then $L_{i,2}$ for $i = 2, \ldots, n$, and so on. Due to this simple explicit structure, the Cholesky factorization can be computed quickly and efficient implementations are available in standard linear algebra libraries (Anderson, Bai, Bischof, Demmel et al., 1995). However, the complexity of the computations is of order $\mathcal{O}(n^3)$.

The natural way to speed up the Cholesky factorization is to make use of a particular structure in the precision matrix or the Cholesky triangle. For GMRFs, the issue is that sparsity in $\boldsymbol{Q}$ implies (a related) sparsity in $\boldsymbol{L}$. The implication is that if we *know* that $L_{j,i} = 0$, then we do not need to compute it. And if the main bulk of $\boldsymbol{L}$ is zero, then we can achieve great computational savings. In order to understand these issues, we need to understand what $\boldsymbol{L}$ really means in terms of statistical interpretation. The simulation algorithm for a zero mean GMRFs, which solves $\boldsymbol{L}^T\boldsymbol{x} = \boldsymbol{z}$, where $\boldsymbol{z} \sim \mathcal{N}(\boldsymbol{0}, \boldsymbol{I})$ gives the following result immediately.

Theorem 12.4

Let $\boldsymbol{x}$ be a GMRF wrt the labeled graph $\mathcal{G}$, mean $\boldsymbol{\mu}$, and an SPD precision matrix $\boldsymbol{Q}$. Let $\mathbf{L}$ be the Cholesky triangle of $\mathbf{Q}$. Then for $i \in \mathcal{V}$,

$$E\left(x_i \mid \boldsymbol{x}_{(i+1):n}\right) = \mu_i - \frac{1}{L_{i,i}} \sum_{j=i+1}^{n} L_{j,i}(x_j - \mu_j) \qquad \textit{and}$$

$$\textit{Precision}\left(x_i \mid \boldsymbol{x}_{(i+1):n}\right) = L_{i,i}^2.$$

Hence, the elements of $\boldsymbol{L}$ have an interpretation as the contribution to the conditional mean and precision for x_i, given all those x_js where $j > i$. This is in contrast to the elements of $\boldsymbol{Q}$, which have a similar interpretation, but where we condition on all other x_js. A simple consequence of this interpretation is that $Q_{i,i} \geq L_{i,i}^2$ for all i.

If we merge Equation (12.2) with Theorem 12.4, we obtain the following result.

Theorem 12.5

Let $\boldsymbol{x}$ be a GMRF wrt to the labeled graph $\mathcal{G}$, with mean $\boldsymbol{\mu}$ and SPD precision matrix $\boldsymbol{Q}$. Let $\mathbf{L}$ be the Cholesky triangle of $\mathbf{Q}$ and define for $1 \leq i < j \leq n$ the future *of i except j as*

$$F(i, j) = \{i+1, \ldots, j-1, j+1, \ldots, n\}.$$

Then

$$x_i \perp x_j \mid \boldsymbol{x}_{F(i,j)} \iff L_{j,i} = 0.$$

So, if we consider the marginal distribution of $\boldsymbol{x}_{i:n}$, then $L_{j,i} = 0$ is equivalent to x_i and x_j being conditionally independent. This is a useful result, as it indicates that if we can determine zeros in $\boldsymbol{L}$ using conditional independence in the sequence of marginals $\{\boldsymbol{x}_{i:n}\}$, the marginals $\{\boldsymbol{x}_{i:n}\}$ are easier to compute through the Cholesky triangle. However, we can use a weaker criterion, which implies that $x_i \perp x_j \mid \boldsymbol{x}_{F(i,j)}$, the global Markov property in Theorem 12.3.

COROLLARY 12.1
If $F(i, j)$ separates $i < j$ in $\mathcal{G}$, then $L_{j,i} = 0$.

This is the main result. If we can verify that $i < j$ are separated by $F(i, j)$, an operation that depends only on the graph and not the numerical values in $\boldsymbol{Q}$, then we know that $L_{j,i} = 0$ no matter what the numerical values in $\boldsymbol{Q}$ are. If $i < j$ are not separated by $F(i, j)$, then $L_{j,i}$ *can* be zero, but is in general nonzero. Since two neighbors $i \sim j$ are not separated by any set, then $L_{j,i}$ is in general nonzero for neighbors.

Example 12.8
Consider a GMRF with graph as in Figure 12.1. We then know that $L_{1,1}$, $L_{2,2}$, $L_{3,3}$, $L_{4,4}$ is nonzero, and $L_{2,1}$, $L_{3,2}$, $L_{4,3}$, and $L_{4,1}$ is in general nonzero. The two elements remaining are $L_{3,1}$ and $L_{4,2}$, which we check using Corollary 12.1; nodes 1 and 3 are separated by $F(1, 3) = \{2, 4\}$, so $L_{3,1}$ must be zero, whereas $F(4, 2) = \{3\}$ does not separate 2 and 4 due to node 1, hence $L_{4,2}$ is in general nonzero.

12.1.7.7 Cholesky Factorization of Band Matrices

Although only one element of the Cholesky triangle in Example 12.8 was necessarily zero, we obtain a larger amount of zeros for autoregressive models. Let $\boldsymbol{x}$ be a pth order autoregressive process, AR(p),

$$x_t \mid x_{t-1}, \ldots, x_1 \sim \mathcal{N}(\phi_1 x_{t-1} + \cdots + \phi_p x_{t-p}, \sigma^2), \quad t = 1, \ldots, n.$$

where we set $x_0 = x_{-1} = \cdots = x_{-p+1} = 0$ for simplicity. The precision matrix for an AR(p) process will then be a band matrix with bandwidth $b_w = p$. Using Corollary 12.1, it follows immediately that $L_{j,i} = 0$, $j > i$, for all $j - i > p$, hence $\boldsymbol{L}$ is a band matrix with the same bandwidth.

Theorem 12.6
If $\boldsymbol{Q}$ is a SPD band matrix with bandwidth b_w, then its Cholesky triangle $\mathbf{L}$ is a lower triangular band matrix with the same bandwidth.

In this example, only $\mathcal{O}(n(b_w + 1))$ of the $\mathcal{O}(n^2)$ terms in $\boldsymbol{L}$ are nonzero, which is a significant reduction. A direct consequence is that the algorithm for computing the Cholesky factorization can be simplified; two of the three loops only need to go within the bandwidth, so the complexity is reduced from $\mathcal{O}(n^3)$ to $\mathcal{O}(nb_w^2)$. For fixed b_w, this gives a computational cost, which is linear in n.

12.1.7.8 Reordering Techniques: Band Matrices

The great computational savings we obtained for band matrices naturally raise the question of who then we can use this approach also for "nonband," but sparse, matrices. A rationale for such an approach is that the indices in the graph are arbitrary, hence, we can permute the indices to obtain a small bandwidth, do the computations and perform the inverse

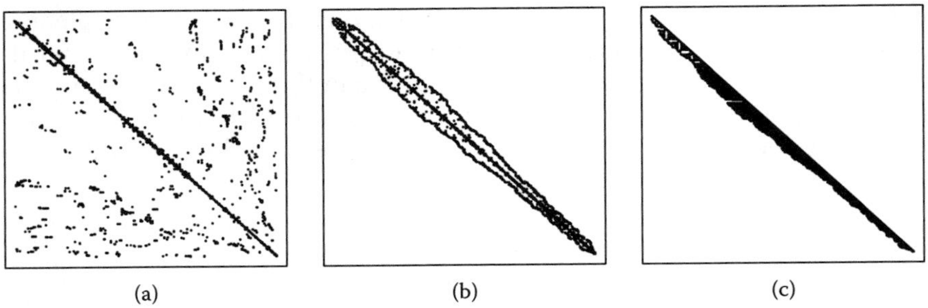

FIGURE 12.4
Panel (a) displays the nonzeros of the 380×380 precision matrix, panel (b) displays the reordered precision matrix with bandwidth 38, and panel (c) displays the Cholesky triangle of the band matrix (b).

permutation on the answer. Formally, let $\boldsymbol{P}$ by one of the $n!$, $n \times n$ permutation matrices; each row and column of $\boldsymbol{P}$ has one and only one nonzero entry, which is 1. The transpose of a permutation matrix is the inverse permutation, $\boldsymbol{P}^T\boldsymbol{P} = \boldsymbol{I}$. For example, $\boldsymbol{Q}\boldsymbol{\mu} = \boldsymbol{b}$ can be solved as follows; multiply both sides with $\boldsymbol{P}$

$$\underbrace{(\boldsymbol{P}\boldsymbol{Q}\boldsymbol{P}^T)}_{\widetilde{\boldsymbol{Q}}}\underbrace{\boldsymbol{P}\boldsymbol{\mu}}_{\widetilde{\boldsymbol{\mu}}} = \underbrace{\boldsymbol{P}\boldsymbol{b}}_{\widetilde{\boldsymbol{b}}}$$

solve $\widetilde{\boldsymbol{Q}}\widetilde{\boldsymbol{\mu}} = \widetilde{\boldsymbol{b}}$, and then apply the inverse permutation to obtain the solution $\boldsymbol{\mu} = \boldsymbol{P}^T\widetilde{\boldsymbol{\mu}}$.

The next issue is how to permute the sparse matrix in order to obtain a (possible) small bandwidth. This issue is somewhat more technical, but there is a huge literature in computer science (Duff, Erisman, and Reid, 1989; George and Liu, 1981), and good working algorithms. So any algorithm that runs quick and gives reasonable results is fine. Figure 12.4 displays an example, where in panel (a) we display the precision matrix found from a spatial application (Rue and Held, 2005, Sec. 4.2.2), in (b) the reordered precision matrix using the Gibbs–Poole–Stockmeyer reordering algorithm (Lewis, 1982), and in (c), the Cholesky triangle of the reordered precision matrix. The bandwidth after reordering is 38.

12.1.7.9 Reordering Techniques: General Sparse Matrices

Although the band matrix approach gives very efficient algorithms for certain graphs, we often encounter situations where a more general approach is required. One such example is where the graph has some "global nodes"; nodes which are neighbors to (near) all other nodes. In statistical applications, such situations occur quite frequently, as shown in the following example.

Example 12.9

Let $\mu \sim \mathcal{N}(0, 1)$ and $\{z_t\}$ be a AR(1) process of length T with mean μ; then $\boldsymbol{x} = (\boldsymbol{z}^T, \mu)^T$ is a GMRF wrt $\mathcal{G}$ where node μ is neighbor of all other nodes. The bandwidth is $n - 1$ where $n = T + 1$, for all $n!$ reorderings.

The band matrix approach is not successful in this example, but we can derive efficient factorizations by making use of a general (and complex) factorization scheme. The general scheme computes only the nonzero terms in $\boldsymbol{L}$, which requires a substantial increase of complexity. The issue then is to reorder to minimize the number of terms in $\boldsymbol{L}$ not known to be zero. Define $M(\mathcal{G})$ as the number of nonzero terms in $\boldsymbol{L}$ found using Corollary 12.1.

Then the efficiency of any reordering is usually compared using the number of *fill-ins*

$$\text{fill-ins}(\mathcal{G}) = M(\mathcal{G}) - (|\mathcal{V}| + |\mathcal{E}|/2)\,.$$

Since $L_{i,i} > 0$ for all i, and $L_{j,i}$ is in general nonzero for $i \sim j$ and $j > i$, then fill-ins $(\mathcal{G}) \geq 0$.

Autoregressive processes of order p are optimal in the sense that the precision matrix is a band matrix with bandwidth p (and dense within the band), and with identity ordering, the number of fill-ins is zero (see Theorem 12.6). For other GMRFs, the number of fill-ins is (in most cases) nonzero and different reordering schemes can be compared to find a reasonable reordering. Note that there is no need to find *the optimal* reordering, but any reasonable one will suffice.

Let us reconsider Example 12.9 where we compare two reorderings where the global node μ is ordered first and last, respectively; $\boldsymbol{x} = (\boldsymbol{z}^T, \mu)^T$ and $\boldsymbol{x}' = (\mu, \boldsymbol{z}^T)^T$. The precision matrices are

$$\boldsymbol{Q} = \begin{pmatrix} \times & \times & & & & & \times \\ \times & \times & \times & & & & \times \\ & \times & \times & \times & & & \times \\ & & \times & \times & \times & & \times \\ & & & \times & \times & \times & \times \\ & & & & \times & \times & \times \\ \times & \times & \times & \times & \times & \times & \times \end{pmatrix}, \quad \text{and} \quad \boldsymbol{Q}' = \begin{pmatrix} \times & \times & \times & \times & \times & \times & \\ \times & \times & \times & & & & \\ \times & \times & \times & \times & & & \\ \times & & \times & \times & \times & & \\ \times & & & \times & \times & \times & \\ \times & & & & \times & \times & \times \\ \times & & & & & \times & \times \end{pmatrix},$$

respectively. Here, $\times$ indicates a nonzero value. Using Corollary 12.1, we obtain the (general) nonzero structure for the Cholesky triangles

$$\boldsymbol{L} = \begin{pmatrix} \times & & & & & & \\ \times & \times & & & & & \\ & \times & \times & & & & \\ & & \times & \times & & & \\ & & & \times & \times & & \\ & & & & \times & \times & \\ \times & \times & \times & \times & \times & \times & \times \end{pmatrix} \quad \text{and} \quad \boldsymbol{L}' = \begin{pmatrix} \times & & & & & & \\ \times & \times & & & & & \\ \times & \times & \times & & & & \\ \times & \surd & \times & \times & & & \\ \times & \surd & \surd & \times & \times & & \\ \times & \surd & \surd & \surd & \times & \times & \\ \times & \surd & \surd & \surd & \surd & \times & \times \end{pmatrix}$$

where a $\surd$ indicates the fill-ins. Placing the global node μ last does not give any fill-ins, whereas placing it first gives a maximum number of fill-ins. This insight can be used to derive the reordering scheme called *nested dissection*, which goes as follows.

- Select a (small) set of nodes whose removal divides the graph into two disconnected subgraphs of almost equal size
- Order the nodes chosen *after* ordering all the nodes in both subgraphs
- Apply this procedure recursively to the nodes in each subgraph

Formally, this can be described as follows:

Lemma 12.2

Let $\boldsymbol{x}$ be a GMRF wrt to $\mathcal{G}$ and SPD precision matrix $\boldsymbol{Q}$, and partition $\boldsymbol{x}$ as $(\boldsymbol{x}_A^T, \boldsymbol{x}_B^T, \boldsymbol{x}_C^T)^T$. Partition the Cholesky triangle of $\boldsymbol{Q}$ as

$$\boldsymbol{L} = \begin{pmatrix} \boldsymbol{L}_{AA} & & \\ \boldsymbol{L}_{BA} & \boldsymbol{L}_{BB} & \\ \boldsymbol{L}_{CA} & \boldsymbol{L}_{CB} & \boldsymbol{L}_{CC} \end{pmatrix}.$$

If C separates A and B in $\mathcal{G}$, then $\boldsymbol{L}_{BA} = \boldsymbol{0}$.

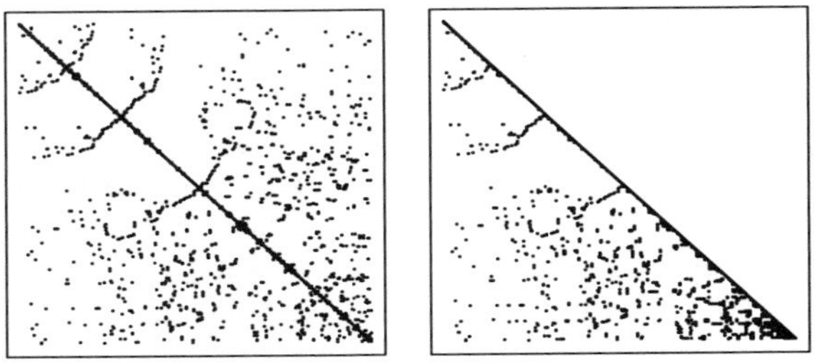

(a) (b)

FIGURE 12.5
Panel (a) displays the "optimal" reordering with 2,182 number of fill-ins, of the 380 × 380 precision matrix in Figure 12.4a. Panel (b) displays the corresponding Cholesky triangle.

The recursive approach, proceeds by partitioning A and B similarly, and so on. It turns out that the nested dissection reordering gives optimal reorderings (in the order sense) for GMRFs found from discretizing the lattice or the cube: Consider a regular square lattice with n sites where each vertex is neighbor to the nearest four vertices. The nested dissection reordering will use C as the middle column, and A and B as the left and right part. Then this process is repeated recursively. It turns out that the cost of factorization of the reordered precision matrix will be $\mathcal{O}(n^{3/2})$, which is $\sqrt{n}$ times faster than using the band approach. The number of fill-ins will (only) be $\mathcal{O}(n \log n)$. It can be shown that factorization of the precision matrix applying any reordering is larger or equal to $\mathcal{O}(n^{3/2})$; hence optimal in the order sense. For a 3D box-lattice with n vertices, the computational cost is $\mathcal{O}(n^2)$ and the number of fill-ins is $\mathcal{O}(n^{4/3})$.

In between the band reordering for long and thin graphs and the nested dissection reordering for lattice-like graphs, there are several other reordering schemes than can provide, by a case-to-case basis, better reordering. Which one to try depends on which implementation one has available; however, any reasonable choice for the reordering will suffice. Note that the number of fill-ins for a specific graph can be computed (by sparse matrix libraries) without having to do the actual factorization, as it only depends on the graph; hence, if several factorizations should be performed on the same graph, it can be of benefit to compare (a few) reorderings and choose the one with the fewest number of fill-ins.

We close this discussion by revisiting the 380 × 380 precision matrix displayed in Figure 12.4a. The band reordering gives 11,112 fill-ins, the nested dissection gives 2,460, while the "optimal" one produced 2,182 fill-ins. The optimal reordered precision matrix and the corresponding Cholesky triangle are displayed in Figure 12.5.

12.1.7.10 Exact Calculations of Marginal Variances

We will now turn to a more "statistical" issue; how to compute all (or nearly all) marginal variances of a GMRF. This is a different task than computing only one variance, say of x_i, which can simply be done by solving $\boldsymbol{Q}\boldsymbol{v} = \boldsymbol{1}_i$, where $\boldsymbol{1}_i$ is one at position i and zero otherwise, and then $\text{var}(x_i) = v_i$. Although the computation of variances is central in statistics, it is not necessarily so within computer science: we are not familiar with any sparse matrix library that provides such a feature although the general solution is known (Erisman and Tinney, 1975; Takahashi, Fagan, and Chen, 1973). We will derive the recursions from a statistical point of view, starting with the case that $\boldsymbol{Q}$ is SPD with no additional linear constraints. The case with constraints will be discussed afterwards.

12.1.7.11 General Recursions

The starting point is again that the solution of $\boldsymbol{L}^T\boldsymbol{x} = \boldsymbol{z}$ provides a sample $\boldsymbol{x}$ with precision matrix $\boldsymbol{Q}$, which implies that

$$x_i = z_i/L_{i,i} - \frac{1}{L_{i,i}} \sum_{k=i+1}^{n} L_{k,i} x_k, \qquad i = n, \ldots, 1.$$

Multiply both sides with x_j for $j \geq i$. Then the expected value reads

$$\Sigma_{i,j} = \delta_{i,j}/L_{i,i}^2 - \frac{1}{L_{i,i}} \sum_{k=i+1}^{n} L_{k,i} \Sigma_{k,j}, \quad j \geq i, \ i = n, \ldots, 1, \tag{12.22}$$

where $\delta_{i,j}$ is one if $i = j$ and zero otherwise. The sum in (12.22) only needs to be over all nonzero $L_{j,i}$s, or at least, all those ks so that i and $k > i$ are *not* separated by $F(i,k)$; see Corollary 12.1. To simplify notation, define this index set as

$$\mathcal{I}(i) = \{k > i : i \text{ and } k \text{ are not separated by } F(i,k)\}$$

and their "union,"

$$\mathcal{I} = \{\{i,k\} : k > i, i \text{ and } k \text{ are not separated by } F(i,k)\}$$

for $k, i = 1, \ldots, n$. Note that the elements of $\mathcal{I}$ are sets; hence if $\{i, j\} \in \mathcal{I}$, then so does $\{j, i\}$. $\mathcal{I}$ represent all those indices in $\boldsymbol{L}$ that are not known upfront to be zero; hence, must be computed doing the Cholesky factorization. With this notation Equation (12.22) reads

$$\Sigma_{i,j} = \delta_{i,j}/L_{i,i}^2 - \frac{1}{L_{i,i}} \sum_{k \in \mathcal{I}(i)} L_{k,i} \Sigma_{k,j}, \quad j \geq i, \ i = n, \ldots, 1. \tag{12.23}$$

Looking more closely into these equations, it turns out that we compute all the $\Sigma_{i,j}$s explicitly if we apply Equation (12.23) in the correct order:

for $i = n, \ldots, 1$
    for $j = n, \ldots, i$
        Compute $\Sigma_{i,j}$ from Equation (12.23) (recalling that $\Sigma_{k,j} = \Sigma_{j,k}$).

Although this direct procedure computes all the marginal variances $\Sigma_{n,n}, \ldots, \Sigma_{1,1}$, it is natural to ask if it is necessary to compute all the $\Sigma_{i,j}$s in order to obtain the marginal variances. Let $\mathcal{J}$ be a set of pairs of indices $\{i, j\}$, and *assume* we can compute $\Sigma_{i,j}$ from Equation (12.23) only for all $\{i, j\} \in \mathcal{J}$, and still obtain all the marginal variances. Then the set $\mathcal{J}$ must satisfy two requirements:

Requirement 1 $\mathcal{J}$ must contain $\{1, 1\}, \ldots, \{n, n\}$

Requirement 2 While computing $\Sigma_{i,j}$ from Equation (12.23), we need to have already computed all those $\Sigma_{k,j}$s that we need, i.e.,

$$\{i, j\} \in \mathcal{J} \quad \text{and} \quad k \in \mathcal{I}(i) \quad \Longrightarrow \quad \{k, j\} \in \mathcal{J}. \tag{12.24}$$

The rather surprising result is that $\mathcal{J} = \mathcal{I}$ satisfy these requirements; a result that depends only on $\mathcal{G}$ and not the numerical values in $\boldsymbol{Q}$. This result implies that we can compute all the marginal variances as follows:

for $i = n, \ldots, 1$
  for decreasing j in $\mathcal{I}(i)$
    Compute $\Sigma_{i,j}$ from Equation (12.23),

where the j-loop visits all entries in $\mathcal{I}(i)$ in decreasing order. The proof is direct. Requirement 1 is trivially true, and Requirement 2 is verified by verifying Equation (12.24): $\{i, j\} \in \mathcal{J}$, $j \geq i$, says that there must a path from i to j where all the vertices are less than i, while $k \in \mathcal{I}(i)$ says there must be a path from i to k where all the vertices are less than i. If $k \leq j$, then there must be a path from k to i to j where all indices are less than k since $k > i$, otherwise, if $k > j$, then there must be a path from k to i to j where all indices are less than j since $j \geq i$.

The computational cost for solving these recursions is smaller than factorizing the precision matrix. Consider for a 2D square lattice reordered using nested dissection reordering, then $|\mathcal{I}| = \mathcal{O}(n \log n)$, and each $\Sigma_{i,j}$ will involve about $\mathcal{O}(\log n)$ terms, giving the total cost of $\mathcal{O}(n(\log n)^2)$. Similarly, the cost will be $\mathcal{O}(n^{5/3})$ for a 3D box-lattice with nested dissection reordering.

12.1.7.12 Recursions for Band Matrices

The simplification is perhaps most transparent when $\boldsymbol{Q}$ is a band matrix with bandwidth b_w, where we have previously shown that $\mathcal{I}(i) = \{i+1, \ldots, \min(n, i+b_w)\}$; see Theorem 12.6. In this case, Requirement 2 reads (for an interior vertex and $j \geq i$),

$$0 \leq j - i \leq b_w \quad \text{and} \quad 0 < k - i \leq b_w \quad \Longrightarrow \quad -b_w \leq k - j \leq b_w,$$

which is trivially true. The algorithm then becomes

for $i = n, \ldots, 1$
  for $j = \min(i + b_w, n), \ldots, i$
    Compute $\Sigma_{i,j}$ from Equation (12.23).

Note that this algorithm is formally equivalent to Kalman recursions for smoothing. The computational cost for autoregressive models is $\mathcal{O}(n)$.

12.1.7.13 Correcting for Linear Constraints

With additional linear constraints, the constrained precision matrix will be less sparse, so we need an approach to correct marginal variances for additional linear constraints. This is similar to Equation (12.19). Let $\widetilde{\boldsymbol{\Sigma}}$ be the covariance matrix with the additional k linear constraints $\boldsymbol{A}\boldsymbol{x} = \boldsymbol{e}$, and $\boldsymbol{\Sigma}$ the covariance matrix without constraints. The two covariance matrices then relate as follows:

$$\widetilde{\boldsymbol{\Sigma}} = \boldsymbol{\Sigma} - \boldsymbol{Q}^{-1}\boldsymbol{A}^T \left(\boldsymbol{A}\boldsymbol{Q}^{-1}\boldsymbol{A}^T\right)^{-1} \boldsymbol{A}\boldsymbol{Q}^{-1}. \tag{12.25}$$

Let $\boldsymbol{W}$ be the $n \times k$ matrix solving $\boldsymbol{Q}\boldsymbol{W} = \boldsymbol{A}^T$, $\boldsymbol{V}$ the $k \times k$ Cholesky triangle of $\boldsymbol{A}\boldsymbol{W}$, and $\boldsymbol{Y}$ the $k \times n$ matrix solving $\boldsymbol{V}\boldsymbol{Y} = \boldsymbol{W}^T$, then the i, jth element of Equation (12.25) can be written as

$$\widetilde{\Sigma}_{i,j} = \Sigma_{i,j} - \sum_{t=1}^{k} Y_{t,i} Y_{t,j}. \tag{12.26}$$

All terms of $\boldsymbol{\Sigma}$ that we compute solving Equation (12.23) can now be corrected using Equation (12.26). The computational cost of this correction is dominated by computing $\boldsymbol{Y}$, which costs $\mathcal{O}(nk^2)$. Again, with not too many constraints, this correction will not require any additional computational burden.

12.1.7.14 Some Practical Issues

Although computations with GMRFs can be done using numerical methods for sparse matrices, it is not without some practical hassle. First, quite a few libraries have free binary versions but closed source code, which (in this case) can create complications. Since only a few libraries have routines for solving $\boldsymbol{L}^T\boldsymbol{x} = \boldsymbol{z}$ and none (to our knowledge) have support for computing marginal variances, we need access to the internal storage format in order to implement such facilities. However, the internal storage format is not always documented unless using an open-source code. At the current time of writing, we recommend to use one of the open-source libraries: TAUCS (Toledo, Chen, and Rotkin, 2002) or CHOLMOD (Chen, Davis, Hager, and Rajamanickam, 2006), which is written in `C`; see Gould, Scott, and Hu (2007) for a comparison and overview. Due to the open-source, it is relatively easy to provide routines for solving $\boldsymbol{L}^T\boldsymbol{x} = \boldsymbol{z}$. However, care must be taken if the computed $\boldsymbol{L}$ is used in the recursions (12.23); for example, the TAUCS library removes terms $L_{j,i}$s that turn out to be numerically zero, but we need all those that are not known to be zero. The easy way out is to disable this feature in the library, alternatively, using the fix in Rue and Martino (2007), Sec. 2.3. The `GMRFLib`-library (Rue and Held, 2005, App. B) does provide implementation of all the algorithms discussed in this chapter based on the TAUCS library.

12.1.8 Markov Random Fields

We will now leave the Gaussian case and discuss Markov random fields (MRFs) more generally. We will first study the case where each x_i is one of K different "colors" or states; i.e., $x_i \in \mathcal{S}_i = \{0, 1, \ldots, K-1\}$, and $\boldsymbol{x} \in \mathcal{S} = \mathcal{S}_1 \times \mathcal{S}_2 \times \cdots \times \mathcal{S}_n$. The case $K = 2$ is particularly important and corresponds to a binary MRF. The main result, Theorem 12.7, can then be generalized to nonfinite $\mathcal{S}$. The presentation in this section is inspired by some unpublished lecture notes by J. Besag; see Besag 1974; Geman and Geman, 1984; Guyon, 1995; Lauritzen, 1996; Li, 2001; Winkler, 1995 for more theory and applications, and Hurn, Husby, and Rue (2003) for a tutorial towards image analysis.

12.1.8.1 Background

Recall the notion of the full conditionals for a joint distribution $\pi(\boldsymbol{x})$ that is the n conditional distributions $\pi(x_i|\boldsymbol{x}_{-i})$. From the full conditionals, we can define the notion of a neighbor.

Definition 12.3 (Neighbor)
Site $j \neq i$ is called a neighbor of site i if x_j contributes to the full conditional for x_i.

Denote by ∂i, set all neighbors to site i, then

$$\pi(x_i \mid \boldsymbol{x}_{-i}) = \pi(x_i \mid \boldsymbol{x}_{\partial i})$$

for all i. In a spatial context, it is easy to visualize this by, for example, considering ∂i as those sites that are (spatially) close to site i in some sense.

The problems start when we want to specify the joint density of $\boldsymbol{x}$ through the n full conditionals and set of neighbors ∂i, using a bottom-up approach. If we know the joint distribution, then the full conditionals are found as

$$\pi(x_i \mid \boldsymbol{x}_{-i}) \propto \pi(\boldsymbol{x}).$$

But how do we derive the joint distribution if we only know the full conditionals? Let $\boldsymbol{x}^*$ denote any reference state where $\pi(\boldsymbol{x}^*) > 0$, then Brook's lemma (Lemma 12.1) gives for $n = 2$ (for simplicity only)

$$\pi(x_1, x_2) = \pi(x_1^*, x_2^*)\, \frac{\pi(x_1|x_2)}{\pi(x_1^*|x_2)}\, \frac{\pi(x_2|x_1^*)}{\pi(x_2^*|x_1^*)} \tag{12.27}$$

provided there are no zeros in the denominator, meaning that (in general)

$$\pi(x) > 0, \ \forall x \in \mathcal{S}.$$

We will refer to this condition as the *positivity condition*, and it turns out to be essential in the following. If we evaluate Equation (12.27) for all $x \in \mathcal{S}$ keeping x^* fixed to $\mathbf{0}$, say, then we know $h(x)$ where $\pi(x) \propto h(x)$. The missing constant is found from $\sum_{x \in \mathcal{S}} \pi(x) = 1$. The conclusion is that the full conditionals determine the joint distribution.

There is a small problem with this argument, which is the implicit assumption that the full conditionals are (or can be) derived from a joint distribution. If, however, we specify *candidates* for full conditionals, then we must ensure that we obtain the same joint distribution no matter the choice of ordering leading to Equation (12.27), for example,

$$\pi(x_1, x_2) = \pi(x_1^*, x_2^*) \, \frac{\pi(x_2 | x_1)}{\pi(x_1 | x_2^*)} \, \frac{\pi(x_2^* | x_1)}{\pi(x_1^* | x_2^*)}.$$

Of course, these two specifications must agree. For $n > 2$, there are quite a few such orderings, and they all have to agree. This implies that we cannot choose the full conditionals arbitrarily, but they have to satisfy some constraints to ensure that they define a valid joint distribution. Before we state the main result that defines what form the joint distribution must take under the neighbor specifications of the full conditionals, we need some new definitions.

12.1.8.2 The Hammersley–Clifford Theorem

Let $\mathcal{G} = (\mathcal{V}, \mathcal{E})$ denote the graph as defined through our specification of the neighbors to each site; $\mathcal{V} = \{1, \ldots, n\}$ and draw a directed edge from j to i if $j \in \partial i$ and $i \notin \partial j$. If i and j are mutually neighbors, draw an undirected edge. (In fact, it will turn out that if i is a neighbor of j, then also j must be a neighbor of i, although this is not known at the current stage.)

Definition 12.4 (Markov random field)
If the full conditionals of $\pi(x)$, $x \in \mathcal{S}$, honor a given graph $\mathcal{G}$, the distribution is called a Markov random field with respect to $\mathcal{G}$.

For the main result, we also need the notion of a *clique*.

Definition 12.5 (Clique)
Any single site or any set of sites, all distinct pairs of which are mutual neighbors, is called a clique.

Example 12.10
The cliques in the graph in Figure 12.6 are $\{1\}, \{2\}, \{3\}, \{4\}, \{5\}, \{1, 2\}, \{1, 3\}, \{2, 3\}, \{1, 2, 3\}, \{3, 4\}$.

The main result is the Hammersley–Clifford theorem (see Clifford, 1990, for a historical account), which states what form the joint distribution must take to honor a given graph $\mathcal{G}$.

Theorem 12.7 (Hammersley–Clifford)
Let $\pi(x) > 0$, $x \in \mathcal{S}$ denote a Markov random field with respect to a graph $\mathcal{G}$ with cliques $\mathcal{C}$, then

$$\pi(x) \propto \prod_{C \in \mathcal{C}} \Psi_C(x_C), \tag{12.28}$$

where the functions Ψ_C can be chosen arbitrarily, subject to $0 < \Psi_C(x_C) < \infty$.

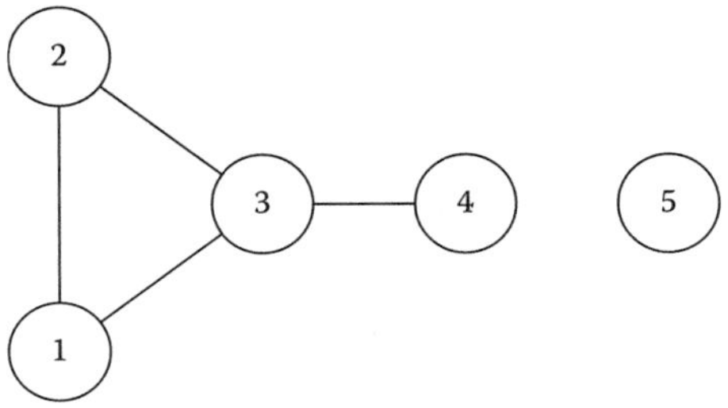

FIGURE 12.6
An example of a graph.

(This result has an interesting history; see Besag (1974) (including the discussion) and the book chapter by Clifford, 1990.) One *important* consequence of this result, is that for a given graph, the full conditionals should either be specified implicitly through the Ψ functions or verified that the chosen full conditionals can be derived from Equation (12.28) for some Ψ functions.

Example 12.11

The general form of the distribution, which honors the graph in Figure 12.6, is

$$\pi(x_1, x_2, x_3, x_4, x_5) \propto \Psi_{1,2,3}(x_1, x_2, x_3)\ \Psi_{3,4}(x_3, x_4)\ \Psi_5(x_5).$$

We will now state some corollaries to the Hammersley–Clifford theorem:

COROLLARY 12.2
The graph $\mathcal{G}$ must be undirected, so if $i \in \partial j$, then necessarily $j \in \partial i$.

COROLLARY 12.3
The MRF (12.28) also satisfies the global Markov property (12.4).

COROLLARY 12.4
Define $Q(\boldsymbol{x}) = \log(\pi(\boldsymbol{x})/\pi(\boldsymbol{0}))$, then there exists a unique representation,

$$Q(\boldsymbol{x}) = \sum_i x_i G_i(x_i) + \sum_{i<j} x_i x_j G_{i,j}(x_i, x_j) + \sum_{i<j<k} x_i x_j x_k G_{i,j,k}(x_i, x_j, x_k) + \cdots$$
$$+ x_1 x_2 \cdots x_n G_{1,2,\ldots,n}(x_1, x_2, \ldots, x_n), \qquad (12.29)$$

where $G_{i,j,\ldots,s}(x_i, x_j, \ldots, x_s) \equiv 0$ unless the sites $i, j, \ldots, s$ form a clique, and where the G functions are arbitrary but finite.

COROLLARY 12.5
The Hammersley–Clifford theorem extends to multivariate MRFs where each $\boldsymbol{x}_i$ is p-dimensional, and to nonfinite $\mathcal{S}$ subject to integrability of $\pi(\boldsymbol{x})$.

The joint distribution (12.28) can also be written as

$$\pi(\boldsymbol{x}) = \frac{1}{Z}\exp\left(-\sum_{C\in\mathcal{C}} V_C(\boldsymbol{x}_C)\right),$$

where $V_C(\cdot) = -\log \Psi_C(\cdot)$ and Z is the normalizing constant. This form is called the Gibbs distribution in statistical physics, and the $V_C(\cdot)$-functions are called the potential functions; see, for example, Geman and Geman (1984).

12.1.8.3 Binary MRFs

We will now discuss more in detail the case where $K = 2$, where we have a binary MRF. Let ∂i be the nearest four sites to site i on a regular square lattice. Then from Equation (12.29), we have that

$$Q(\boldsymbol{x}) = \sum_i \alpha_i x_i + \sum_{i<j} \alpha_{i,j} x_i x_j$$

for some α_is and $\alpha_{i,j}$s. This is the general form the distribution can take for the given graph. For $\beta_{i,j} = \beta_{j,i} = \alpha_{i,j}, i < j$ we can write the distribution of $\boldsymbol{x}$ as

$$\pi(\boldsymbol{x}) = \frac{1}{Z} \exp\left(\sum_i \alpha_i x_i + \sum_{i \sim j} \beta_{ij} x_i x_j \right). \tag{12.30}$$

The parameters $\{\alpha_i\}$ control the level whereas the parameters $\{\beta_{i,j}\}$ control the interaction. The interpretation is perhaps more transparent when we look at the full conditionals

$$\pi(x_i \mid \boldsymbol{x}_{-i}) = \frac{\exp\left(x_i(\alpha_i + \sum_{j \in \partial i} \beta_{i,j} x_j)\right)}{1 + \exp\left(\alpha_i + \sum_{j \in \partial i} \beta_{i,j} x_j\right)}. \tag{12.31}$$

The full conditional for x_i is like a logistic regression with its neighbors, and is referred to as the auto logistic model (Besag, 1974).

A interesting special case of Equation (12.31) is

$$\pi(\boldsymbol{x} \mid \beta) = \frac{1}{Z_\beta} \exp\left(\beta \sum_{i \sim j} 1[x_i = x_j] \right), \tag{12.32}$$

where $1[\cdot]$ is the indicator function and β is a common interaction parameter. This model is the famous Ising model for ferromagnestism, dating back to Ernst Ising in 1925. For later use, we write $|\beta$ and note that the normalizing constant also depends on β. More intuition from Equation (12.32) is obtained by rewriting it as

$$\pi(\boldsymbol{x} \mid \beta) = \frac{1}{Z_\beta} \exp\left(\beta \times \text{ number of equal neighbors}\right),$$

hence, realizations will favor neighbors to be equal, but is invariant for which state, 0 or 1, the neighbors should be in. This can also be seen studying the full conditionals,

$$\pi(x_i = 1 \mid \beta, \boldsymbol{x}_{\partial i}) = \frac{\exp(\beta n_1)}{\exp(\beta n_0) + \exp(\beta n_1)},$$

where n_0 is the number of neighbors to x_i that is zero, and similar with n_1. It is clear that this model favors x_i to be equal to the dominant state among its neighbors. In two or more dimensions, it can shown to exhibit phase transition; there is a critical value β^* (equal to $\log(1 + \sqrt{2}) = 0.88\ldots$ in dimension two), for which $\beta > \beta^*$. Then the distribution is severely bi-modal even as the size of the lattice tends to infinity. Further, x_i and x_j, even when arbitrarily far apart, will be positively correlated.

12.1.9 MCMC for MRFs

12.1.9.1 MCMC for Fixed β

We will now discuss how we can generate a sample from $\pi(\boldsymbol{x})$ for fixed β, focusing on the binary case and the Ising model (12.30). The Gibbs sampler in Section 12.1.5.2 is valid also for discrete MRFs.

```
Initialize x = 0 (or some other configuration)
While TRUE; do
    For i = 1, ..., n
        Sample new value for x_i from π(x_i | x_∂i)
    Output new state x
```

Again, to update x_i, we only need to take into account the neighbors $\boldsymbol{x}_{\partial i}$, which in this case are the four nearest ones.

Intuitively, in the case where β is high and one color dominates, then the Gibbs sampler will tend to propose the same color as it already has. Note that the Ising model is symmetric, so the configurations $\boldsymbol{x}$ and $\mathbf{1}-\boldsymbol{x}$ have the same probability. It is then intuitive that we could improve the Markov chain's ability to move around in the state-space, like going from a configuration with essentially all black to all white, if we propose to always *change* the color at site i (Frigessi, Hwang, Sheu, and di Stefano, 1993). If we do so, then we need to correct this proposal to maintain the equilibrium distribution. The resulting MCMC algorithm is called the *Metropolis–Hastings* algorithm.

```
Initialize x = 0 (or some other configuration)
While TRUE; do
    For i = 1, ..., n
        Propose a new value x'_i = 1 − x_i.
        Compute R_i = π(x'_i | x_∂i) / π(x_i | x_∂i)
        With probability p_i = min{1, R_i} set x_i = x'_i.
    Output new state x
```

The probability that we should accept the proposed new state, p_i, also only depends on the neighbors $\boldsymbol{x}_{\partial i}$. Note that the new proposed state is always accepted if $R_i \geq 1$.

Figure 12.7 displays samples from a 64×64 toroidal grid with $\beta = 0.6, 0.88\ldots$ and 1.0 using the Metropolis–Hastings algorithm. Below β^* the samples are more scattered, whereas above β^* they are more concentrated on one color.

12.1.9.2 MCMC for Random β

We experienced complications in our MCMC algorithm in Example 12.4 when the interaction parameter δ was random due to a normalizing constant depending on δ; see Equation (12.17). In the discrete MRF case, inference for β becomes troublesome by the same reason. In fact, it is even worse, as we cannot compute the normalizing constant exact except for small lattices.

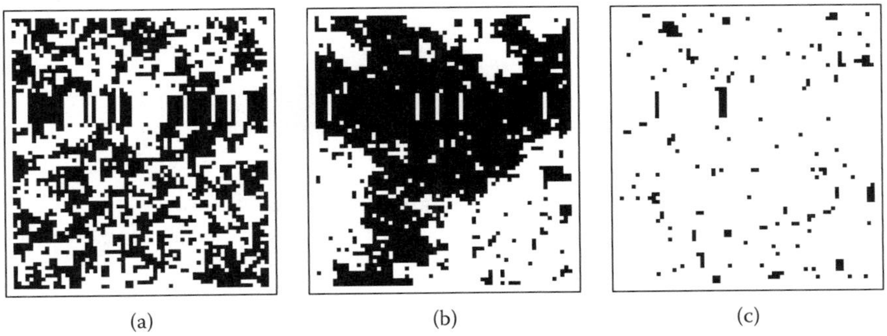

FIGURE 12.7
Samples from the Ising model on a toroidal 64×64 lattice, with (a) $\beta = 0.6$, (b) $\beta = 0.88\ldots$, and (c) $\beta = 1.0$.

Recall that $Z(\beta)$ is defined by

$$Z(\beta) = \sum_{x} \exp(\beta S(x)), \tag{12.33}$$

where $S(x)$ is the sufficient statistic $\sum_{i\sim j} 1[x_i = x_j]$. One possibility is to see whether the derivative of $Z(\beta)$ with respect to β is easier to estimate than $Z(\beta)$ itself.

$$\begin{aligned}\frac{d Z(\beta)}{d\beta} &= \sum_{x} S(x) \exp(\beta S(x)) \\ &= Z(\beta) \sum_{x} (S(x)) \exp(\beta S(x)) \,/\, Z(\beta) \\ &= Z(\beta)\, \mathrm{E}_{x|\beta} S(x).\end{aligned}$$

By solving this differential equation, we obtain

$$\log\left(Z(\beta')/Z(\beta)\right) = \int_{\beta}^{\beta'} \mathrm{E}_{x|\tilde{\beta}} S(x)\, d\tilde{\beta}. \tag{12.34}$$

As we see, this trick has reduced the difficulty of the problem to one we can tackle using the following procedure (see Hurn, Husby, and Rue, 2003, for an application).

1. Estimate $\mathrm{E}_{x|\beta} S(x)$ for a range of various β values using posterior mean estimates based on the output from a sequence of MCMC algorithms. These values will depend on the image size and so will need to be recalculated for each new problem.
2. Construct a smoothing spline $f(\beta)$ to smooth the estimated values of $\mathrm{E}_{x|\beta} S(x)$.
3. Use numerical or analytical integration of $f(\beta)$ to compute an estimate of Equation (12.34),

$$\log(\widehat{Z(\beta')/Z(\beta)}) = \int_{\beta}^{\beta'} f(\tilde{\beta})\, d\tilde{\beta}$$

for each pair (β, β') required.

The idea behind Equation (12.34) is often called thermodynamic integration in the physics literature, see Gelman and Meng (1998) for a good introduction from a statistical perspective. A similar idea also occurs in Geyer and Thompson (1992). The pseudo-likelihood method (Besag, J.E., 1974) is an early, yet still popular approach, toward inference for discrete valued MRFs (Frank and Strauss, 1986; Strauss and Ikeda, 1990), see Robins, Snijders, Wang, Handcock et al. (2007) for an update overview.

References

E. Anderson, Z. Bai, C. Bischof, J. Demmel, J.J. Dongarra, J. Du Croz, A. Greenbaum, S. Hammarling, A. McKenney, S. Ostrouchov, and D.C. Sorensen. *LAPACK Users' Guide*, 2nd ed. Philadelphia: Society for Industrial and Applied Mathematics, 1995.

J. Besag. Spatial interaction and the statistical analysis of lattice systems (with discussion). *Journal of the Royal Statistical Society, Series B*, 36(2):192–236, 1974.

J. Besag. Statistical analysis of non-lattice data. *The Statistician*, 24(3):179–195, 1975.

D. Brook. On the distinction between the conditional probability and the joint probability approaches in the specification of nearest-neighbour systems. *Biometrika*, 51(3 and 4):481–483, 1964.

Y. Chen, T.A. Davis, W.W. Hager, and S. Rajamanickam. Algorithm 8xx: CHOLMOD, supernodal sparse Cholesky factorization and update/downdate. Technical report TR-2006-005, CISE Department, University of Florida, Gainesville, 2006.

P. Clifford. Markov random fields in statistics. In G. Grimmett and D.J. Welsh, eds., *Disorder in Physical Systems*. Clarendon Press, Oxford, 1990.

I.S. Duff, A.M. Erisman, and J.K. Reid. *Direct Methods for Sparse Matrices*, 2nd ed. Monographs on Numerical Analysis. The Clarendon Press Oxford University Press, New York, 1989. Oxford Science Publications.

A.M. Erisman and W.F. Tinney. On computing certain elements of the inverse of a sparse matrix. *Communications of the ACM*, 18(3):177–179, 1975.

O. Frank and D. Strauss. Markov graphs. *Journal of the American Statistical Association*, 81:832–842, 1986.

A. Frigessi, C.R. Hwang, S.J. Sheu, and P. di Stefano. Convergence rate of the Gibbs sampler, the metropolis algorithm and other single-site updating dynamics. *Journal of the Royal Statistical Society, Series B*, 55(1):205–219, 1993.

A.E. Gelfand and A.F.M. Smith. Sampling-based approaches to calculating marginal densities. *Journal of the American Statistical Association*, 85:398–509, 1990.

A. Gelman and X.-L. Meng. Simulating normalizing contants: From importance sampling to bridge sampling to path sampling. *Statistical Science*, 13:163–185, 1998.

S. Geman and D. Geman. Stochastic relaxation, Gibbs distributions, and the Bayesian restoration of images. *IEEE Transactions on Pattern Analysis and Machine Intelligence*, 6:721–741, 1984.

A. George and J.W.H. Liu. *Computer solution of large sparse positive definite systems*. Prentice-Hall Inc., Englewood Cliffs, N.J., 1981. Prentice-Hall Series in Computational Mathematics.

C.J. Geyer and E.A. Thompson. Constrained Monte Carlo maximum likelihood for dependent data (with discussion). *Journal of the Royal Statistical Society, Series B*, 54:657–699, 1992.

W.R. Gilks, S. Richardson, and D.J. Spiegelhalter. *Markov Chain Monte Carlo in Practice*. Chapman & Hall, London, 1996.

N.I.M. Gould, J.A. Scott, and Y. Hu. A numerical evaluation of sparse direct solvers for the solution of large sparse symmetric linear systems of equations. *ACM Transactions on Mathematical Software*, 33(2):Article No. 10, 2007.

X. Guyon. *Random Fields on a Network*. Series in Probability and Its Applications. Springer-Verlag, New York, 1995.

W.K. Hastings. Monte Carlo simulation methods using Markov chains and their applications. *Biometrika*, 57:97–109, 1970.

M.A. Hurn, O.K. Husby, and H. Rue. A tutorial on image analysis. In J. Møller, Ed., *Spatial Statistics and Computational Methods*, Lecture Notes in Statistics; 173, pages 87–141. Springer-Verlag, Berlin, 2003.

S.L. Lauritzen. *Graphical Models*, vol. 17 of Oxford Statistical Science Series. The Clarendon Press Oxford University Press, New York, 1996. Oxford Science Publications.

J.G. Lewis. Algorithm 582: The Gibbs–Poole–Stockmeyer and Gibbs–King algorithms for reordering sparse matrices. *ACM Transactions on Mathematical Software*, 8(2):190–194, June 1982.

S.Z. Li. *Markov Random Field Modeling in Image Analysis*. Springer-Verlag, New York, 2nd ed., 2001.

K.V. Mardia. Multidimensional multivariate Gaussian Markov random fields with application to image processing. *Journal of Multivariate Analysis*, 24(2):265–284, 1988.

R Development Core Team. *R: A Language and Environment for Statistical Computing*. R Foundation for Statistical Computing, Vienna, Austria, 2007. ISBN 3-900051-07-0.

C.P. Robert, ed. *Discretization and MCMC Convergence Assessment*. Lecture Notes in Statistics, No. 135. Springer-Verlag, New York, 1998.

C.P. Robert and G. Casella. *Monte Carlo Statistical Methods*. Springer-Verlag, New York, 1999.

G.L. Robins, T.A.B. Snijders, P. Wang, M. Handcock, and P. Pattison. Recent developments in exponential random graph (p*) models for social networks. *Social Newworks*, 29:192–215, 2007.

H. Rue and L. Held. *Gaussian Markov Random Fields: Theory and Applications*, vol. 104 of *Monographs on Statistics and Applied Probability*. Chapman & Hall, London, 2005.

H. Rue and S. Martino. Approximate Bayesian inference for hierarchical Gaussian Markov random fields models. *Journal of Statistical Planning and Inference*, 137(10):3177–3192, 2007. Special Issue: Bayesian Inference for Stochastic Processes.

H. Rue, S. Martino, and N. Chopin. Approximate Bayesian inference for latent Gaussian models using integrated nested Laplace approximations (with discussion). *Journal of the Royal Statistical Society, Series B*, 71(2):318–392, 2009.

D. Strauss and M. Ikeda. Pseudolikelihood estimation for social networks. *Journal of the American Statistical Association*, 5:204–212, 1990.

K. Takahashi, J. Fagan, and M.S. Chen. Formation of a sparse bus impedance matrix and its application to short circuit study. In *8th PICA Conference proceedings*, pages 63–69. IEEE Power Engineering Society, 1973. Papers presented at the 1973 Power Industry Computer Application Conference in Minneapolis, MN.

S. Toledo, D. Chen, and V. Rotkin. TAUCS. A library of sparse linear solvers. Version 2.0. Manual, School of Computer Science, Tel-Aviv University, 2002. http://www.tau.ac.il/~stoledo/taucs/.

G. Winkler. *Image Analysis, Random Fields and Dynamic Monte Carlo Methods*. Springer-Verlag, Berlin, 1995.

13

Conditional and Intrinsic Autoregressions

Leonhard Held and Hävard Rue

CONTENTS

13.1 Introduction

The purpose of this chapter is to give an overview of conditional and intrinsic autoregressions. These models date back at least to Besag (1974), and have been heavily used since to model discrete spatial variation.

Traditionally, conditional autoregressions have been used to directly model spatial dependence in data that have been observed on a predefined graph or lattice structure. Inference is then typically based on likelihood or pseudo-likelihood techniques (Besag, 1974; Künsch, 1987). More recently, conditional autoregressions are applied in a modular fashion in (typically Bayesian) complex hierarchical models. Inference in this class is nearly always carried out using Markov chain Monte Carlo (MCMC), although some alternatives do exist (Breslow and Clayton, 1993; Rue, Martino, and Chopin, 2009).

In this chapter, we will describe the most commonly used conditional and intrinsic autoregressions. The focus will be on spatial models, but we will also discuss the relationship to autoregressive time series models. Indeed, autoregressive time series models are a special case of conditional autoregressions and exploring this relationship is helpful in order to develop intuition and understanding for the general class.

This chapter will not describe in detail how to build hierarchical models based on conditional autoregressive prior distributions and how to analyze them using MCMC. For a thorough discussion, see Banerjee, Carlin, and Gelfand, 2004; Higdon, 2007; Rue and Held, 2005 as well as Chapter 14.

To begin, consider a random vector $\mathbf{X} = (X_1, \ldots, X_n)$ where each component is univariate. It is convenient to imagine that each component is located at a fixed site $i \in \{1, \ldots, n\}$. These sites may refer to a particular time point or a particular point in two- or higher-dimensional space, or particular areas in a geographical region, for example.

We now wish to specify a joint distribution with density $p(\mathbf{x})$ for $\mathbf{X}$. A decomposition of the form

$$p(\mathbf{x}) = p(x_1) \cdot p(x_2|x_1) \cdot p(x_3|x_1, x_2) \cdot \ldots \cdot p(x_n|x_1, x_2, \ldots, x_{n-1}) \tag{13.1}$$

is, of course, always possible. In a temporal context, this factorization is extremely useful, and—under an additional Markov assumption—further simplifies to

$$p(\mathbf{x}) = p(x_1) \cdot p(x_2|x_1) \cdot p(x_3|x_2) \cdot \ldots \cdot p(x_n|x_{n-1}).$$

Indeed, this factorization forms the basis of so-called first-order autoregressive models and can be conveniently generalized to higher orders. However, in a spatial context, where the indices $1, \ldots, n$ are arbitrary and could, in principle, easily be permuted, Equation (13.1) is not really helpful, as it is very difficult to envision most of the terms entering the above product.

It is much more natural to specify the full conditional distribution $p(x_i|\mathbf{x}_{-i})$, the conditional distribution of X_i at a particular site i, given the values $X_j = x_j$ at all other sites $j \neq i$. In a spatial context, the Markov assumption refers to the property that the conditional distribution $p(x_i|\mathbf{x}_{-i})$ depends only on a few components of $\mathbf{x}_{-i}$, called the neighbors of site i. However, it is not obvious at all under which conditions the set of full conditionals $p(x_i|\mathbf{x}_{-i})$, $i = 1, \ldots, n$, defines a valid joint distribution. Conditions under which such a joint distribution exists are discussed in Besag (1974) using the Brook expansion (Brook, 1964), see Chapter 12 for details.

By far the most heavily studied model is the Gaussian conditional autoregression, where $p(x_i|\mathbf{x}_{-i})$ is univariate normal and $p(\mathbf{x})$ is multivariate normal. Gaussian conditional autoregressions with a Markov property are also known as Gaussian Markov random fields (Künsch, 1979; Rue and Held, 2005). Various Gaussian conditional autoregressions will be discussed in Section 13.2. However, there are also nonnormal conditional autoregressions, for example, the so-called autologistic model for binary variables X_i, as discussed in Section 13.3. In Section 13.4, we turn to intrinsic Gaussian conditional autoregressions, a limiting (improper) form of Gaussian conditional autoregressions of practical relevance in hierarchical models. Finally, Section 13.5 gives a brief sketch of multivariate Gaussian conditional autoregressions.

13.2 Gaussian Conditional Autoregressions

Suppose that, for $i = 1, \ldots, n$, $X_i|\mathbf{x}_{-i}$ is normal with conditional mean and variance

$$\mathrm{E}(X_i|\mathbf{x}_{-i}) = \mu_i + \sum_{j \neq i} \beta_{ij}(x_j - \mu_j), \tag{13.2}$$

$$\mathrm{var}(X_i|\mathbf{x}_{-i}) = \kappa_i^{-1}. \tag{13.3}$$

Here, μ_i will typically take a regression form, say, $w_i^T \alpha$ for covariates w_i associated with site i. Without loss of generality we assume that $\mu_1 = \cdots = \mu_n = 0$ in the following. Under the additional assumption that

$$\kappa_i \beta_{ij} = \kappa_j \beta_{ji}$$

for all $i \neq j$, these conditional distributions correspond to a multivariate joint Gaussian distribution with mean $\mathbf{0}$ and precision matrix $\boldsymbol{Q}$ with elements $Q_{ii} = \kappa_i$ and $Q_{ij} = -\kappa_i \beta_{ij}$, $i \neq j$, provided that $\boldsymbol{Q}$ is symmetric and positive definite.

Such a system of conditional distributions is known as an autonormal system (Besag, 1974). Usually it is assumed that the precision matrix $\boldsymbol{Q}$ is regular; however, Gaussian conditional autoregressions with singular $\boldsymbol{Q}$ are also of interest and known as *intrinsic autoregressions*, as discussed in Section 13.4.

In many applications the coefficients β_{ij} will be nonzero for only a few so-called "neighbors" of X_i. Let ∂i denote the set of "neighbors" for each site i. We can then write Equation (13.2) (using $\mu_1 = \cdots = \mu_n = 0$) as

$$\mathrm{E}(X_i | \boldsymbol{x}_{-i}) = \sum_{j \in \partial i} \beta_{ij} x_j$$

to emphasize that the conditional mean of X_i only depends on the neighbors ∂i. The random vector $\boldsymbol{X} = (X_1, \ldots, X_n)^T$ will then follow a Gaussian Markov random field, as discussed in Chapter 13.

13.2.1 Example

Suppose that the X_is follow a zero-mean Gaussian conditional autoregression with

$$\mathrm{E}(X_i | \boldsymbol{x}_{-i}) = \phi \begin{cases} \frac{1}{2}(x_2 + x_n) & \text{for } i = 1 \\ \frac{1}{2}(x_{i-1} + x_{i+1}) & \text{for } 1 < i < n \\ \frac{1}{2}(x_1 + x_{n-1}) & \text{for } i = n \end{cases} \tag{13.4}$$

where $\phi \in [0, 1)$ and $\mathrm{var}(X_i | \boldsymbol{x}_{-i}) = \kappa^{-1}$, say. At first sight, this looks like a first-order autoregressive time series model, but by linking the first "time point" x_1 with the last "time point" x_n, the model is defined on a circle. The model is called a circular first-order autoregressive model and is useful for analyzing circular data.

The precision matrix of $\boldsymbol{X} = (X_1, \ldots, X_n)^T$ is

$$\boldsymbol{Q} = \frac{\kappa}{2} \begin{pmatrix} 2 & -\phi & & & & & -\phi \\ -\phi & 2 & -\phi & & & & \\ & -\phi & 2 & -\phi & & & \\ & & \ddots & \ddots & \ddots & & \\ & & & -\phi & 2 & -\phi & \\ & & & & -\phi & 2 & -\phi \\ -\phi & & & & & -\phi & 2 \end{pmatrix} \tag{13.5}$$

with all other elements equal to zero. Thus, the precision matrix $\boldsymbol{Q}$ is a *circulant* matrix with base $\boldsymbol{d} = \kappa \cdot (1, -\phi/2, 0, \ldots, 0, -\phi/2)^T$ (the first row of $\boldsymbol{Q}$) (see Rue and Held, 2005, Sec. 2.6.1) for an introduction to circular matrices). The covariance matrix $\boldsymbol{\Sigma} = \boldsymbol{Q}^{-1}$ of $\boldsymbol{x}$ is again circular. Its base $\boldsymbol{e}$, which equals the autocovariance function of $\boldsymbol{X}$, can be calculated using the discrete Fourier transform $\mathrm{DFT}(\boldsymbol{d})$ of $\boldsymbol{d}$,

$$\boldsymbol{e} = \frac{1}{n} \mathrm{IDFT}(\mathrm{DFT}(\boldsymbol{d})^{-1}),$$

here IDFT denotes the inverse discrete Fourier transform and the power function is to be understood elementwise. See Rue and Held (2005) for a derivation.

The following R-code illustrates, how e is computed for the circulant precision matrix (13.5) with $n = 10$, $\phi = 0.9$, and $\kappa = 1$. Note that the (inverse) discrete Fourier transform is computed with the function `fft()` and that the imaginary parts of the function values are equal to zero.

```
> # function make.d computes the base d
> make.d <- function(n, phi){
+   d <- rep(0.0, n)
+   d[1] <- 1
+   d[2] <- -phi/2
+   d[n] <- -phi/2
+   return(d)
+ }
> # function e computes the base e, i.e. the autocovariance function
> # if corr=T you obtain the autocorrelation function
> e <- function(n, phi, corr=F){
+   d <- make.d(n, phi)
+   e <- Re(fft(1/Re(fft(d)), inverse=TRUE))/n
+   if(corr==F)
+     return(e)
+   else return(e/e[1])
+ }
> n <- 10
> phi <- 0.9
> result <- e(n, phi)
> print(result)

 [1] 2.3375035 1.4861150 0.9649742 0.6582722 0.4978530 0.4480677 0.4978530
 [8] 0.6582722 0.9649742 1.4861150
```

From the autocovariances e we can easily read off the autocorrelations of X. The left panel in Figure 13.1 displays the autocorrelation function for $n = 100$ and $\phi = 0.9, 0.99, 0.999, 0.9999$. Of course, the autocorrelation function must be symmetric, the correlation between

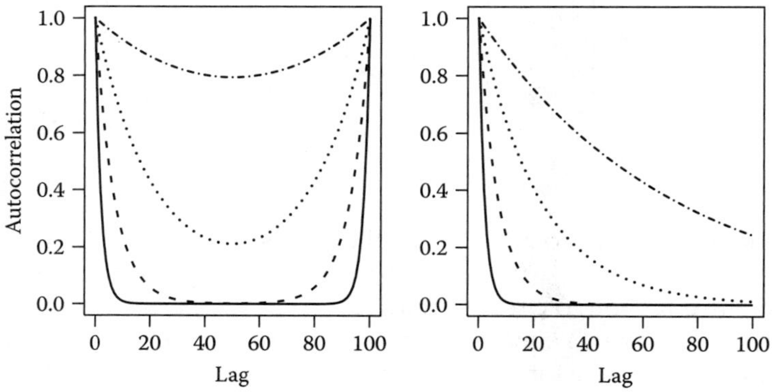

FIGURE 13.1
Autocorrelation function of the circular (left) and ordinary (right) first-order autoregressive model (13.4) and (13.6), respectively, for $n = 100$ and $\phi = 0.9$ (solid line), $\phi = 0.99$ (dashed line), $\phi = 0.999$ (dotted line), and $\phi = 0.9999$ (dot-dashed line). The corresponding coefficients of the ordinary first-order autoregressive model are $\alpha = 0.63$, $\alpha = 0.87$, $\alpha = 0.96$, and $\alpha = 0.99$; compare Equation (13.7).

x_1 and x_3 must be the same as the correlation between x_1 and x_{99}, for example. For the two smaller values of ϕ, the autocorrelation is essentially zero for lags around $n/2 = 50$. For the larger values of ϕ very close to unity, there is substantial autocorrelation between any two components of $\boldsymbol{x}$.

It is interesting to compare the autocorrelations obtained with those from the ordinary first-order autoregressive process defined through the directed definition

$$X_i | x_{i-1} \sim \mathrm{N}(\alpha x_{i-1}, \kappa^{-1}), \tag{13.6}$$

where $|\alpha| < 1$ to ensure stationarity. This model has identical neighborhood structure as the circular first-order autoregressive model, except for the missing link between X_1 and X_n. The autocorrelation function is $\rho_k = \alpha^k$ for lag k.

It is easy to show that this directed definition induces the full conditional distribution

$$X_i | \boldsymbol{x}_{-i} \sim \begin{cases} \mathrm{N}\left(\alpha x_2, \kappa^{-1}\right) & i = 1 \\ \mathrm{N}\left(\frac{\alpha}{1+\alpha^2}(x_{i-1} + x_{i+1}), \left(\kappa(1+\alpha^2)\right)^{-1}\right) & i = 2, \ldots, n-1 \\ \mathrm{N}\left(\alpha x_{n-1}, \kappa^{-1}\right) & i = n. \end{cases}$$

If we want to compare the circular autoregressive model (13.4) with the ordinary autoregressive model (13.6), we need to equate the autoregressive coefficients of the full conditional distributions. From $\phi/2 = \alpha/(1+\alpha^2)$ it follows that for a given autoregressive coefficient ϕ of the circular autoregressive model, the corresponding coefficient $\alpha = \alpha(\phi)$ of the ordinary first-order autoregressive process is

$$\alpha(\phi) = \frac{1 - \sqrt{1 - \phi^2}}{\phi}. \tag{13.7}$$

For example, $\phi = 0.99$ corresponds to $\alpha \approx 0.87$, $\phi = 0.999$ corresponds to $\alpha \approx 0.96$. This illustrates that coefficients from undirected Gaussian conditional autoregressions have a quite different meaning compared to coefficients from directed Gaussian autoregressions.

Figure 13.1 compares the autocorrelation function of the circular autoregressive model with coefficient ϕ with the corresponding autocorrelation function of the ordinary autoregressive model with coefficient $\alpha(\phi)$. A close correspondence of autocorrelations up to lag 50 can be seen for $\phi = 0.9$ and $\phi = 0.99$. The autocorrelations up to lag $n/2$ of the circular model differ from the corresponding ones from the ordinary model not more than $4.5e - 11$ and 0.00072, respectively. For $\phi = 0.999$ and $\phi = 0.9999$, the decay of the autocorrelations with increasing lag is not as pronounced as the geometric decay of the ordinary autoregressive model. This is due to the increasing impact of the link between x_n and x_1 in the circular model.

13.2.2 Gaussian Conditional Autoregressions on Regular Arrays

Suppose now that a conditional autoregressive model is defined on a lattice with $n = n_1 n_2$ nodes and let (i, j) denote the node in the ith row and jth column. In the interior of the lattice, we can now define the nearest four sites of (i, j) as its neighbors, i.e., the nodes

$$(i-1, j), (i+1, j), (i, j-1), (i, j+1).$$

A proper conditional Gaussian model with this neighborhood structure, often called *first-order autoregression*, is based on the conditional mean

$$\mathsf{E}(X_{ij} | \boldsymbol{x}_{-ij}) = \alpha(x_{i-1,j} + x_{i+1,j}) + \beta(x_{i,j-1} + x_{i,j+1}) \tag{13.8}$$

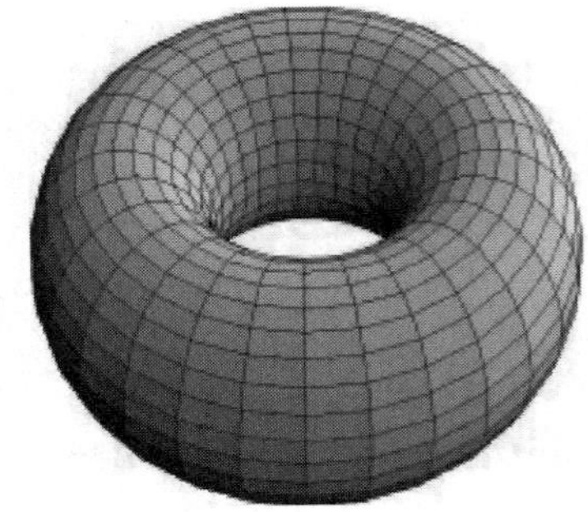

FIGURE 13.2
Illustration of a torus obtained on a two-dimensional lattice with $n_1 = n_2 = 29$ and toroidal boundary conditions.

with $|\alpha| + |\beta| < 0.5$ and $\text{var}(X_{ij}|\boldsymbol{x}_{-ij}) = \kappa^{-1}$, say. In most practical applications, both α and β will be positive. Assuming that the lattice is wrapped on a *torus*, so that every pixel has four neighbors, this process is stationary. A *torus* is a regular lattice with toroidal boundary conditions, which can be obtained in two steps. First, the lattice is wrapped to a "sausage." In a second step, the two ends of the sausage are joined such that the sausage becomes a ring. This two-stage process ensures that every pixel of the lattice has four neighbors. For example, pixel (1, 1) will have the four neighbors (1, 2), (2, 1), $(1, n_2)$ and $(n_1, 1)$. For further illustration of toroidal boundary conditions, see Figure 13.2 and the R-code in the following example. Note that an alternative way to study conditional autoregressions is on an *infinite* regular array, in which case the process will be stationary and the *spectral density* is useful. (For details, see Besag and Kooperberg, 1995; Künsch, 1987.)

13.2.3 Example

Suppose we set $\alpha = \beta = 0.2496$ in model (13.8), defined on a torus of size $n_1 = n_2 = 29$. The following R-code illustrates the computation of the autocovariance matrix of $\mathbf{X}$ by simply inverting the precision matrix of $\mathbf{X}$ using the function `solve()`. An alternative way would be to exploit the fact that the precision matrix of $\mathbf{X}$ is *block-circulant*. The two-dimensional Fourier transform can then be used to calculate the base of the autocovariance matrix (see Rue and Held, 2005, Section 2.6.2 for details).

```
> # make.prec computes the precision matrix of a toroidal first-order
> # autoregression on a two-dimensional lattice of size n1 x n2
> # with coefficient coeff
> make.prec <- function(n1, n2, coeff){
+   prec <- diag(n1*n2)
+   for(i in 1:(n1*n2)){
+     j <- ((i-1)%%n1)+1 # column index
+     k <- (n1*(n2-1))    # if i>k we are in the last row
+
+     if(j!=1) (prec[i,i-1] <- -coeff)     # left neighbor
+     else (prec[i,i+(n1-1)] <- -coeff)    # left toroidal neighbor
+
+     if(j!=n1)  (prec[i,i+1] <- -coeff) # right neighbor
+     else (prec[i,i-(n1-1)] <- -coeff)    # right toroidal neighbor
+
+     if(i>n1) (prec[i,i-n1] <- -coeff)    # top neighbor
+     else (prec[i,(j+k)] <- -coeff)       # top toroidal neighbor
```

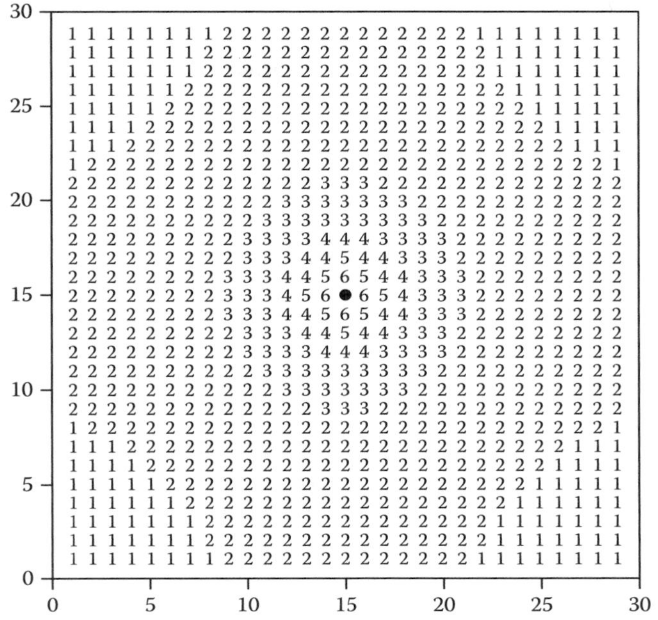

FIGURE 13.3
Plot of the correlation of a pixel x_{ij} with the pixel $x_{15,15}$ in model (13.8), defined on a torus of size $n_1 = n_2 = 29$ with coefficients $\alpha = \beta = 0.2496$. Shown is 10 times the autocorrelation, truncated to an integer.

```
+
+       if(i<=k) (prec[i,i+n1] <- -coeff)   # bottom neighbor
+       else (prec[i,j] <- -coeff)          # bottom toroidal neighbor
+     }
+     return(prec)
+ }
> prec <- make.prec(n1=29, n2=29, coeff=0.2496)
> # inversion gives the covariance matrix
> cova <- solve(prec)
```

From the autocovariance matrix, we can easily calculate autocorrelations between any pair of sites. Figure 13.3 displays the correlation of pixel x_{ij}, $1 \le i, j \le 29$, with pixel $x_{15,15}$ in the center of the plot. Although the coefficients α and β are close to the border of the parameter space, the correlation between adjacent pixels is only 0.669. The smallest correlation observed, for example, between $x_{1,1}$ and $x_{15,15}$ is 0.186.

13.3 Non-Gaussian Conditional Autoregressions

For binary or count data, direct usage of Gaussian conditional autoregressions is often not possible. Instead, conditional autoregressive models in the form of a logistic or log-linear Poisson model have been proposed. Here, we discuss the autologistic and the auto-Poisson model, which basically adopt the form (13.2) for the conditional mean of $X_i|\boldsymbol{x}_{-i}$ using a link function, as known from generalized linear modeling (McCullagh and Nelder, 1990). However, consistency requirements imply that for binary data only the logistic link, and for Poisson counts only the log link can be used (see Besag, 1972, 1974 for details). Only the autologistic model has gained some popularity in applications, the auto-Poisson has undesirable properties, which make it not suitable for most applications in spatial statistics.

13.3.1 Autologistic Models

Assume $X_i, i = 1, \ldots, n$, are binary random variables with conditional success probability $\pi_i(\boldsymbol{x}_{-i}) = \mathsf{E}(X_i|\boldsymbol{x}_{-i})$. The autologistic model specifies the (logit-transformed) conditional mean

$$\text{logit}\, \pi_i(\boldsymbol{x}_{-i}) = \mu_i + \sum_{j \in \partial i} \beta_{ij} x_j,$$

where $\beta_{ij} = \beta_{ji}$, for consistency reasons. The normalizing constant of the joint distribution, which depends on the β_{ij}s, is very difficult to compute, thus a traditional likelihood approach to estimate the coefficients is typically infeasible. Instead, a pseudo-likelihood approach has been proposed by Besag (1975), in which the product of the conditional binomial probabilities is maximized. The model can be generalized to a binomial setting with additional "sample sizes" N_i, say. Also, the model can be extended to include covariates (see Huffer and Wu, 1998, for example).

13.3.2 Auto-Poisson Models

Suppose $X_i, i = 1, \ldots, n$, are Poisson random variables with conditional mean $\lambda_i(\boldsymbol{x}_{-i}) = \mathsf{E}(X_i|\boldsymbol{x}_{-i})$. Similar to the autologistic model, the auto-Poisson model specifies the (log-transformed) conditional mean

$$\log \lambda_i(\boldsymbol{x}_{-i}) = \mu_i + \sum_{j \in \partial i} \beta_{ij} x_j.$$

It turns out that a necessary (and sufficient) condition for the existence of a joint distribution with the specified conditional distributions is that $\beta_{ij} \leq 0$ for all $i \neq j$. However, a negative coefficient β_{ij} implies negative interaction between i and j because the conditional mean of X_i decreases with an increase in x_j. This is quite opposite to the intent of most spatial modeling; however, there are applications in purely inhibitory Markov point processes (see Besag, 1976).

13.4 Intrinsic Autoregressions

Intrinsic Gaussian autoregressions arise if the precision matrix $\boldsymbol{Q}$ of the Gaussian conditional autoregression (13.2) and (13.3) is only positive semidefinite with $\text{rank}(\boldsymbol{Q}) < n$. For example, if $\beta_{ij} = w_{ij}/w_{i+}$ and $\kappa_i = \kappa w_{i+}$ where $\kappa > 0$ is a precision parameter, $w_{ij} \geq 0$ are predefined weights and $w_{i+} = \sum_{j \neq i} w_{ij}$, $\boldsymbol{Q}$ will be rank deficient. Such weights are quite common in spatial models for areal data. For example, adjacency-based weights are $w_{ij} = 1$ if regions i and j are adjacent (usually denoted by $i \sim j$) and zero otherwise. Other choices are weights based on the inverse distance between area centroids or the length of the common boundary, for example.

For adjacency-based weights, the conditional mean and variance simplify to

$$\mathsf{E}(X_i|\boldsymbol{x}_{-i}) = \sum_{j \in \partial i} x_j / m_i$$

$$\text{var}(X_i|\boldsymbol{x}_{-i}) = (\kappa \cdot m_i)^{-1},$$

here m_i denotes the number of neighbors of region i, i.e., the cardinality of the set ∂i.

The resulting joint distribution is improper, its density can be written (up to a proportionality constant) as

$$p(\boldsymbol{x}|\kappa) \propto \exp\left(-\frac{\kappa}{2}\sum_{i\sim j}(x_i - x_j)^2\right), \tag{13.9}$$

where the sum goes over all pairs of adjacent regions $i \sim j$. This is a special case of a *pairwise difference prior*, as described in Besag, Green, Higdon, and Mengersen (1995). With $\boldsymbol{x} = (x_1, \ldots, x_n)^T$, the density (13.9) can be written in the form

$$p(\boldsymbol{x}|\kappa) \propto \exp\left(-\frac{\kappa}{2}\boldsymbol{x}^T\boldsymbol{R}\boldsymbol{x}\right), \tag{13.10}$$

where the *structure matrix* $\boldsymbol{R}$ has elements

$$R_{ij} = \begin{cases} m_i & \text{if } i = j, \\ -1 & \text{if } i \sim j \\ 0 & \text{otherwise.} \end{cases}$$

We immediately see that the precision matrix $\boldsymbol{Q} = \kappa\boldsymbol{R}$ cannot be of full rank because all rows and columns of $\boldsymbol{R}$ sum up to zero.

In the special case where the index $i = 1, \ldots, n$ represents time and each time-point has the two (respectively one) nearest time-points as its neighbors, Equation (13.9) simplifies to

$$p(\boldsymbol{x}|\kappa) \propto \exp\left(-\frac{\kappa}{2}\sum_{i=2}^{n}(x_i - x_{i-1})^2\right).$$

This is a so-called *first-order random walk* model, as it corresponds to the directed formulation

$$X_i|x_{i-1} \sim \mathrm{N}(x_{i-1}, \kappa^{-1}),$$

with improper uniform prior on x_1. Obviously this is a limiting case of model (13.6) with $\alpha = 1$. The structure matrix of this model has a particularly simple form,

$$\boldsymbol{R} = \begin{pmatrix} 1 & -1 & & & & & \\ -1 & 2 & -1 & & & & \\ & -1 & 2 & -1 & & & \\ & & \ddots & \ddots & \ddots & & \\ & & & -1 & 2 & -1 & \\ & & & & -1 & 2 & -1 \\ & & & & & -1 & 1 \end{pmatrix}, \tag{13.11}$$

and forms the basis of some spatial models on regular arrays, as we will see later.

Intrinsic autoregressions are more difficult to study than ordinary (proper) conditional autoregressions. The rank deficiency of the precision matrix does not allow the computation of autocorrelation functions, for example. Similarly, it is not possible to sample from an intrinsic autoregression without imposing additional constraints, so they cannot be models for data. On infinite regular arrays, intrinsic autoregressions can be studied using the *generalized spectral density* (see Besag and Kooperberg, 1995; Künsch, 1987 for details).

13.4.1 Normalizing Intrinsic Autoregressions

An interesting question that arises is the appropriate "normalizing constant" of intrinsic Gaussian autoregressions. The constant will depend on unknown parameters in the precision matrix $\boldsymbol{Q}$ and is necessary if those need to be estimated from the data. Of course, intrinsic Gaussian autoregressions are improper, so there is no constant to normalize the density

$$p(\boldsymbol{x}|\kappa) \propto \exp\left(-\frac{1}{2}\boldsymbol{x}^T \boldsymbol{Q}\boldsymbol{x}\right) \tag{13.12}$$

if $\boldsymbol{Q}$ is not positive definite. The term "normalizing constant" has to be understood in a more general sense as the normalizing constant of an equivalent lower-dimensional *proper* Gaussian distribution.

It is now commonly accepted (Hodges, Carlin, and Fan, 2003; Knorr-Held, 2003; Rue and Held, 2005) that for the general model Equation (13.12) with $n \times n$ precision matrix $\boldsymbol{Q}$ of rank $n-k$, the correct "normalizing constant" is

$$(2\pi)^{-(n-k)/2}(|\boldsymbol{Q}|^*)^{1/2},$$

where $|\boldsymbol{Q}|^*$ denotes the *generalized determinant* of $\boldsymbol{Q}$, the product of the $n-k$ nonzero eigenvalues of $\boldsymbol{Q}$.

In the special case $\boldsymbol{Q} = \kappa\boldsymbol{R}$ of model (13.10) with known structure matrix $\boldsymbol{R}$, the "normalizing constant" simplifies to

$$\left(\frac{\kappa}{2\pi}\right)^{\frac{n-k}{2}} \tag{13.13}$$

due to the rank deficiency of $\boldsymbol{R}$ with rank $n-k$. If the neighborhood structure is nonseparable, i.e., every pixel is connected to every other by some chain of neighbors, then $k = 1$.

13.4.2 Example

Suppose data $y_i, i = 1, \ldots, n$, are observed and we assume that

$$y_i|x_i, \sigma^2 \sim \text{N}(x_i, \sigma^2) \tag{13.14}$$

are conditionally independent with known variance σ^2. Assume further that, conditional on κ, the unknown mean surface $\boldsymbol{x} = (x_1, \ldots, x_n)^T$ follows a pairwise difference prior (Equation 13.9) with a nonseparable neighborhood structure. The goal is to infer $\boldsymbol{x}$ from $\boldsymbol{y}$ in order to denoise the observed "image" $\boldsymbol{y}$ and to obtain a smoother version. A fully Bayesian analysis would place a hyperprior on κ, usually a conjugate gamma prior $\kappa \sim \text{G}(\alpha, \beta)$, i.e.,

$$f(\kappa) \propto \kappa^{\alpha-1}\exp(-\beta\kappa).$$

To implement a two-stage Gibbs sampler (see, for example, Gelfand and Smith, 1990), one would sample from $\boldsymbol{x}|\kappa, \boldsymbol{y}$ and from $\kappa|\boldsymbol{x}, \boldsymbol{y} = \kappa|\boldsymbol{x}$. Note that $\boldsymbol{R}$ is of rank $n-1$ since the graph is assumed to be nonseparable, so based on (13.9) and (13.13), it follows that

$$\kappa|\boldsymbol{x} \sim \text{G}\left(\alpha + \frac{n-1}{2}, \beta + \frac{1}{2}\sum_{i\sim j}(x_i - x_j)^2\right).$$

The other full conditional distribution is

$$\boldsymbol{x}|\kappa, \boldsymbol{y} \sim \text{N}(\boldsymbol{A}\boldsymbol{a}, \boldsymbol{A}),$$

where $\boldsymbol{A} = (\kappa\boldsymbol{R} + \sigma^2\boldsymbol{I})^{-1}$ and $\boldsymbol{a} = \sigma^2\boldsymbol{y}$.

Note that there is no need to include an intercept in (13.14), as the intrinsic autoregression $\boldsymbol{x}$ has an undefined overall level. An equivalent formulation is to include an additional intercept with a flat prior and to use an additional sum-to-zero constraint on $\boldsymbol{x}$. Note also that omission of the data error, i.e., setting $\sigma^2 = 0$, is not useful, as x_i will then equal y_i and no smoothing will be done.

13.4.3 Intrinsic Autoregressions on Regular Arrays

We now return to conditional autoregressions defined on regular arrays. When fitting model (13.8) to data, the estimated coefficients are often close to singularity (i.e., $\alpha + \beta$ will be close to 0.5) in order to obtain nonnegligible spatial autocorrelations. A limiting case of model (13.8) is obtained if $\alpha + \beta = 0.5$. For example, if $\alpha = \beta = 0.25$, the conditional mean of x_{ij} is

$$\mathsf{E}(x_{ij}|\boldsymbol{x}_{-ij}) = \frac{1}{4}(x_{i-1,j} + x_{i+1,j} + x_{i,j-1} + x_{i,j+1}).$$

This is an intrinsic autoregression and a special case of the pairwise difference prior (Equation 13.9) with conditional variance equal to $1/(4\kappa)$.

However, on regular arrays it is possible to define an anisotropic intrinsic model, which is able to weight horizontal and vertical neighbors differently. The conditional mean in this extended model is still given by Equation (13.8), but the coefficients $\alpha > 0$ and $\beta > 0$ are now allowed to vary subject to $\alpha + \beta = 0.5$. The conditional variance is still equal to $1/(4\kappa)$. This specification defines a valid intrinsic autoregression. In applications, α (or β) can be treated as an unknown parameter, so the degree of anisotropy can be estimated from the data.

To estimate α it is necessary to compute the generalized determinant of the associated precision matrix $\boldsymbol{Q}$, which can be written as a sum of two *Kronecker products*:

$$\boldsymbol{Q} = \alpha \boldsymbol{R}_{n_1} \otimes \boldsymbol{I}_{n_2} + \beta \boldsymbol{I}_{n_1} \otimes \boldsymbol{R}_{n_2}.$$

Here $\boldsymbol{R}_n$ is the structure matrix (13.11) of an n-dimensional random-walk model and $\boldsymbol{I}_n$ is the $n \times n$ identity matrix. An explicit form for the generalized determinant can be found in Rue and Held, 2005, p. 107.

13.4.4 Higher-Order Intrinsic Autoregressions

All intrinsic autoregressions up to now are of order one, in the sense that the precision matrix $\boldsymbol{Q}$ has a rank deficiency of 1. This is due to an undefined overall level of the distribution of $\boldsymbol{x}$. An equivalent representation is obtained if $\boldsymbol{x}$ is replaced by $\mu + \boldsymbol{x}$, where $\boldsymbol{x}$ has a density as described above, but under an additional sum-to-zero constraint, and the scalar μ has an improper locally uniform prior. In more complex hierarchical models with more than one intrinsic autoregression, such sum-to-zero constraints are necessary to ensure a proper posterior. Computational routines for sampling from GMRFs under linear constraints are particularly useful in this context for MCMC simulation (see Chapter 12 for details).

Intrinsic autoregressions of higher order may also be considered. On regular lattices, such autoregressions can be defined using the closest eight or twelve nearest neighbors, for example. However, appropriate weights have to be chosen with care. It is useful to start with an (improper) joint Gaussian distribution based on squared increments, similar to the squared difference prior (Equation 13.9), and to derive the full conditional from the joint distribution. For example, one might consider the increments

$$\begin{matrix} \circ & \bullet \\ \bullet & \circ \end{matrix} \;-\; \begin{matrix} \bullet & \circ \\ \circ & \bullet \end{matrix}, \qquad (13.15)$$

where the $\bullet$s enter the difference, but not the $\circ$s, which only serve to fix the spatial location. Summing over all pixels with well-defined increments, Equation (13.15) thus leads to the joint improper distribution

$$p(\boldsymbol{x}|\kappa) \propto \exp\left(-\frac{\kappa}{2} \sum_{i=1}^{n_1-1} \sum_{j=1}^{n_2-1} (x_{i+1,j+1} - x_{i+1,j} - x_{i,j+1} + x_{i,j})^2 \right). \qquad (13.16)$$

This is a special case of model (13.10) with structure matrix $\boldsymbol{R}$ defined as the *Kronecker product* of two structure matrices $\boldsymbol{R}_1$ and $\boldsymbol{R}_2$ of random-walk type (13.11) with dimension n_1 and n_2, respectively: $\boldsymbol{R} = \kappa \cdot (\boldsymbol{R}_1 \otimes \boldsymbol{R}_2)$. The rank of $\boldsymbol{R}$ is $(n_1 - 1)(n_2 - 1)$, so $\boldsymbol{R}$ has a deficiency in rank of order $n_1 + n_2 - 1$.

The conditional mean of x_{ij} in the interior of the lattice $(2 \le i \le n_1 - 1, 2 \le j \le n_2 - 1)$ now depends on its eight nearest sites and is

$$\begin{aligned} \mathsf{E}(x_{ij} | \boldsymbol{x}_{-ij}) = & \frac{1}{2}(x_{i-1,j} + x_{i+1,j} + x_{i,j-1} + x_{i,j+1}) \\ & - \frac{1}{4}(x_{i-1,j-1} + x_{i-1,j+1} + x_{i+1,j-1} + x_{i+1,j+1}), \end{aligned} \tag{13.17}$$

while the conditional precision is 4κ. In a more compact notation, the conditional mean is

$$\mathsf{E}(x_{ij} | \boldsymbol{x}_{-ij}) = \frac{1}{2} \begin{smallmatrix} \circ & \bullet & \circ \\ \bullet & \circ & \bullet \\ \circ & \bullet & \circ \end{smallmatrix} - \frac{1}{4} \begin{smallmatrix} \bullet & \circ & \bullet \\ \circ & \circ & \circ \\ \bullet & \circ & \bullet \end{smallmatrix}.$$

Anisotropic versions of this intrinsic autoregression with eight neighbors are discussed in Künsch (1987).

For illustration, we now describe how to derive the conditional mean (13.17) from (13.16). Clearly, $p(x_{ij} | \boldsymbol{x}_{-ij}, \kappa) \propto p(\boldsymbol{x} | \kappa)$, so in the interior of the lattice four terms in the double sum in Equation (13.16) depend on x_{ij}, hence,

$$\begin{aligned} p(x_{ij} | \boldsymbol{x}_{-ij}, \kappa) \propto \exp\Big(-\frac{\kappa}{2}\big(& (x_{i+1,j+1} - x_{i+1,j} - x_{i,j+1} + x_{i,j})^2 \\ & + (x_{i+1,j} - x_{i+1,j-1} - x_{i,j} + x_{i,j-1})^2 \\ & + (x_{i,j+1} - x_{i,j} - x_{i-1,j+1} + x_{i-1,j})^2 \\ & + (x_{i,j} - x_{i,j-1} - x_{i-1,j} + x_{i-1,j-1})^2\big)\Big), \end{aligned}$$

which can be rearranged to

$$\begin{aligned} p(x_{ij} | \boldsymbol{x}_{-ij}, \kappa) \propto \exp\Big(-\frac{\kappa}{2}\big(& (x_{i,j} - (x_{i+1,j} + x_{i,j+1} - x_{i+1,j+1}))^2 \\ & + (x_{i,j} - (x_{i+1,j} + x_{i,j-1} - x_{i+1,j-1}))^2 \\ & + (x_{i,j} - (x_{i-1,j} + x_{i,j+1} - x_{i-1,j+1}))^2 \\ & + (x_{i,j} - (x_{i,j-1} + x_{i-1,j} - x_{i-1,j-1}))^2\big)\Big). \end{aligned}$$

A useful identity for combining quadratic forms* eventually gives

$$\begin{aligned} p(x_{ij} | \boldsymbol{x}_{-ij}, \kappa) \propto \exp\Bigg(-\frac{4\kappa}{2}\bigg(x_{i,j} - \bigg(& \frac{1}{2}(x_{i-1,j} + x_{i+1,j} + x_{i,j-1} + x_{i,j+1}) \\ & - \frac{1}{4}(x_{i-1,j-1} + x_{i-1,j+1} + x_{i+1,j-1} + x_{i+1,j+1})\bigg)\bigg)^2\Bigg), \end{aligned} \tag{13.18}$$

from which the conditional mean (13.17) and the conditional 4κ precision can be read off.

---

* $A(x-a)^2 + B(x-b)^2 = C(x-c)^2 + \frac{AB}{C}(a-b)^2$ where $C = A + B$ and $c = (Aa + Bb)/C$.

It is easy to see that the distribution (13.16) is invariant to the addition of arbitrary constants to any rows or columns. This feature makes this distribution unsuitable as a prior for a smoothly varying surface, a defect that can be remedied by expanding the system of neighbors. Indeed, consider now the joint distribution

$$p(\mathbf{x}|\kappa) \propto \exp\left(-\frac{\kappa}{2}\sum_{i=2}^{n_1-1}\sum_{j=2}^{n_2-1}(x_{i-1,j} + x_{i+1,j} + x_{i,j-1} + x_{i,j+1} - 4x_{i,j})^2\right), \tag{13.19}$$

which is based on the increments

$$\begin{array}{ccc}\circ & \bullet & \circ\\ \bullet & \circ & \bullet\\ \circ & \bullet & \circ\end{array} \; - 4 \; \begin{array}{ccc}\circ & \circ & \circ\\ \circ & \bullet & \circ\\ \circ & \circ & \circ\end{array}.$$

The conditional mean

$$\begin{aligned}\mathsf{E}(x_{ij}|\mathbf{x}_{-ij}) = {} & \frac{8}{20}(x_{i-1,j} + x_{i+1,j} + x_{i,j-1} + x_{i,j+1})\\ & - \frac{1}{10}(x_{i-1,j-1} + x_{i-1,j+1} + x_{i+1,j-1} + x_{i+1,j+1})\\ & - \frac{1}{20}(x_{i-2,j} + x_{i+2,j} + x_{i,j-2} + x_{i,j+2})\end{aligned}$$

can be derived for pixels in the interior of the lattice ($3 \le i \le n_1 - 2, 3 \le j \le n_2 - 2$). In our compact notation, the conditional mean is, hence,

$$\mathsf{E}(x_{ij} \mid \mathbf{x}_{-ij}) = \frac{1}{20}\left(8 \begin{array}{ccccc}\circ&\circ&\circ&\circ&\circ\\ \circ&\circ&\bullet&\circ&\circ\\ \circ&\bullet&\circ&\bullet&\circ\\ \circ&\circ&\bullet&\circ&\circ\\ \circ&\circ&\circ&\circ&\circ\end{array} - 2 \begin{array}{ccccc}\circ&\circ&\circ&\circ&\circ\\ \circ&\bullet&\circ&\bullet&\circ\\ \circ&\circ&\circ&\circ&\circ\\ \circ&\bullet&\circ&\bullet&\circ\\ \circ&\circ&\circ&\circ&\circ\end{array} - 1 \begin{array}{ccccc}\circ&\circ&\bullet&\circ&\circ\\ \circ&\circ&\circ&\circ&\circ\\ \bullet&\circ&\circ&\circ&\bullet\\ \circ&\circ&\circ&\circ&\circ\\ \circ&\circ&\bullet&\circ&\circ\end{array}\right).$$

The conditional variance is $1/(20\kappa)$, while appropriate modifications for both mean and variance are necessary on the boundary of the lattice (see Rue and Held, 2005, for a detailed discussion). Anisotropic versions have also been considered (Künsch, 1987).

This conditional autoregression is based on the 12 nearest neighbors of each pixel. The distribution (13.19) is invariant to the linear transformation

$$x_{ij} \to x_{ij} + p_{ij},$$

where

$$p_{ij} = \gamma_0 + \gamma_1 i + \gamma_2 j$$

for arbitrary coefficients γ_0, γ_1, and γ_2. This is a useful property, as the prior is often used in applications for smoothing deviations from a two-dimensional linear trend p_{ij}.

This model has some drawbacks, however. First, the four corners—$x_{1,1}, x_{1,n_2}, x_{n_1,1}, x_{n_1,n_2}$—do not appear in Equation (13.19). Second, viewed as a difference approximation to a differential operator, model (13.19) induces a so-called anisotropic discretization error, i.e., the approximation error is larger along the diagonals than in the horizontal or vertical direction (for details on this issue, see page 117 in Rue and Held, 2005).

A more elaborate model is given by

$$\begin{aligned}p(\mathbf{x}|\kappa) \propto \exp\Bigg(-\frac{\kappa}{2}\sum_{i=2}^{n_1-1}\sum_{j=2}^{n_2-1}\bigg(&\frac{2}{3}(x_{i-1,j} + x_{i+1,j} + x_{i,j-1} + x_{i,j+1})\\ &+ \frac{1}{6}(x_{i-1,j-1} + x_{i-1,j+1} + x_{i+1,j-1} + x_{i+1,j+1}) - \frac{10}{3}x_{i,j}\bigg)^2\Bigg),\end{aligned} \tag{13.20}$$

based on the increments

$$\frac{2}{3}\begin{array}{ccc}\circ&\bullet&\circ\\\bullet&\circ&\bullet\\\circ&\bullet&\circ\end{array}+\frac{1}{6}\begin{array}{ccc}\bullet&\circ&\bullet\\\circ&\circ&\circ\\\bullet&\circ&\bullet\end{array}-\frac{10}{3}\begin{array}{ccc}\circ&\circ&\circ\\\circ&\bullet&\circ\\\circ&\circ&\circ\end{array}.$$

Note that the four corners—$x_{1,1}$, x_{1,n_2}, $x_{n_1,1}$, x_{n_1,n_2}—now enter the joint distribution. The full conditional of x_{ij} depends on 24 neighbors, its conditional expectation is

$$\mathsf{E}(x_{ij}\mid \boldsymbol{x}_{-ij})=\frac{1}{468}\left(144\begin{array}{ccccc}\circ&\circ&\circ&\circ&\circ\\\circ&\circ&\bullet&\circ&\circ\\\circ&\bullet&\circ&\bullet&\circ\\\circ&\circ&\bullet&\circ&\circ\\\circ&\circ&\circ&\circ&\circ\end{array}-18\begin{array}{ccccc}\circ&\circ&\bullet&\circ&\circ\\\circ&\circ&\circ&\circ&\circ\\\bullet&\circ&\circ&\circ&\bullet\\\circ&\circ&\circ&\circ&\circ\\\circ&\circ&\bullet&\circ&\circ\end{array}\right.$$
$$\left.+8\begin{array}{ccccc}\circ&\circ&\circ&\circ&\circ\\\circ&\bullet&\circ&\bullet&\circ\\\circ&\circ&\circ&\circ&\circ\\\circ&\bullet&\circ&\bullet&\circ\\\circ&\circ&\circ&\circ&\circ\end{array}-8\begin{array}{ccccc}\circ&\bullet&\circ&\bullet&\circ\\\bullet&\circ&\circ&\circ&\bullet\\\circ&\circ&\circ&\circ&\circ\\\bullet&\circ&\circ&\circ&\bullet\\\circ&\bullet&\circ&\bullet&\circ\end{array}-1\begin{array}{ccccc}\bullet&\circ&\circ&\circ&\bullet\\\circ&\circ&\circ&\circ&\circ\\\circ&\circ&\circ&\circ&\circ\\\circ&\circ&\circ&\circ&\circ\\\bullet&\circ&\circ&\circ&\bullet\end{array}\right)$$

and the conditional variance is $1/(13\kappa)$ (see Rue and Held, 2005, for further details).

13.5 Multivariate Gaussian Conditional Autoregressions

Multivariate Gaussian conditional autoregressions are a straightforward generalization of Equation (13.2) and Equation (13.3). Suppose $\boldsymbol{X}_i$, $i = 1, \ldots, n$ is a p-dimensional random vector and let the conditional distribution of $\boldsymbol{X}_i$ given $\boldsymbol{x}_{-i}$ be multivariate Gaussian with conditional mean and covariance matrix

$$\mathsf{E}(\boldsymbol{X}_i|\boldsymbol{x}_{-i}) = \boldsymbol{\mu}_i + \sum_{j\neq i} \boldsymbol{B}_{ij}(\boldsymbol{x}_j - \boldsymbol{\mu}_j) \tag{13.21}$$

$$\mathrm{Cov}(\boldsymbol{X}_i|\boldsymbol{x}_{-i}) = \boldsymbol{\Phi}_i^{-1}. \tag{13.22}$$

The matrices $\boldsymbol{B}_{ij}$ and $\boldsymbol{\Phi}_i > 0$ are all of dimension $p \times p$. Without loss of generality, we assume in the following that $\boldsymbol{\mu}_1 = \cdots = \boldsymbol{\mu}_n = \boldsymbol{0}$. As in the univariate case, the joint distribution of $\boldsymbol{X} = (\boldsymbol{X}_1, \ldots, \boldsymbol{X}_n)$ is multivariate normal with mean $\boldsymbol{0}$ and precision matrix $\boldsymbol{Q} = \boldsymbol{D}(\boldsymbol{I} - \boldsymbol{B})$, provided that $\boldsymbol{Q}$ is regular and symmetric (Mardia, 1988). Here, $\boldsymbol{D}$ is block-diagonal with entries $\boldsymbol{\Phi}_i, i = 1, \ldots, n$, $\boldsymbol{I}$ is the identity matrix and $\boldsymbol{B}$ is $np \times np$ with block-elements $\boldsymbol{B}_{ij}$ for $i \neq j$ and block-diagonal entries equal to zero. More details on this model can be found in Banerjee et al. (2004, Sec. 7.4.2).

In practice, we often encounter the situation that we have multivariate observations in each pixel with a fixed neighborhood structure between the pixels. A straightforward generalization of the adjacency-based intrinsic pairwise-difference prior (Equation 13.9) is

$$p(\boldsymbol{x}|\boldsymbol{\Phi}) \propto \exp\left(-\frac{1}{2}\sum_{i\sim j}(\boldsymbol{x}_i - \boldsymbol{x}_j)^T\,\boldsymbol{\Phi}(\boldsymbol{x}_i - \boldsymbol{x}_j)\right) \tag{13.23}$$

with conditional mean and covariance matrix equal to

$$\mathsf{E}(\boldsymbol{X}_i|\boldsymbol{x}_{-i}) = \sum_{j\sim i} \boldsymbol{x}_j/m_i$$

$$\mathrm{Cov}(\boldsymbol{X}_i|\boldsymbol{x}_{-i}) = (m_i \cdot \boldsymbol{\Phi})^{-1}.$$

Multivariate conditional autoregressive models are discussed in more detail in Gelfand and Vounatsov (2003) (see also Sec. 7.4 in Banergee et al., 2004).

Acknowledgments

Unpublished lecture notes by Julian Besag on Markov random fields have helped in preparing this chapter. Comments by the editor and a reviewer on a previous version are gratefully acknowledged, as well as final proofreading by Birgit Schrödle.

References

S. Banerjee, B.P. Carlin, and A.E. Gelfand. *Hierarchical Modeling and Analysis for Spatial Data*, vol. 101 of *Monographs on Statistics and Applied Probability*. Chapman & Hall, London, 2004.

J. Besag. Nearest-neighbor systems and the auto-logistic model for binary data. *Journal of the Royal Statistical Society, Series B*, 34(1):75–83, 1972.

J. Besag. Spatial interaction and the statistical analysis of lattice systems (with discussion). *Journal of the Royal Statistical Society, Series B*, 36(2):192–225, 1974.

J. Besag. Statistical analysis of non-lattice data. *The Statistician*, 24(3):179–195, 1975.

J. Besag. Some methods of statistical analysis for spatial data. *Bulletin of International Statistical Institute*, 47:77–92, 1976.

J. Besag and C. Kooperberg. On conditional and intrinsic autoregressions. *Biometrika*, 82(4):733–746, 1995.

J. Besag, P.J. Green, D. Higdon, and K. Mengersen. Bayesian computation and stochastic systems (with discussion). *Statistical Science*, 10(1):3–66, 1995.

N.E. Breslow and D.G. Clayton. Approximate inference in generalized linear mixed models. *Journal of the American Statistical Association*, 88(1):9–25, 1993.

D. Brook. On the distinction between the conditional probability and the joint probability approaches in the specification of nearest-neighbour systems. *Biometrika*, 51(3 and 4):481–483, 1964.

A. Gelfand and P. Vounatsou. Proper multivariate conditional autoregressive models for spatial data analysis. *Biostatistics*, 4:11–25, 2003.

A.E. Gelfand and A.F.M. Smith. Sampling-based approaches to calculating marginal densities. *Journal of the American Statistical Association*, 85:398–509, 1990.

D. Higdon. A primer on space-time modeling from a Bayesian perspective. In B. Finkenstädt, L. Held, and V. Isham, Eds., *Statistical Methods for Spatio-Temporal Systems*, p. 217–279. Chapman & Hall, London, 2007.

J.S. Hodges, B.P. Carlin, and Q. Fan. On the precision of the conditionally autoregressive prior in spatial models. *Biometrics*, 59:317–322, 2003.

F.W. Huffer and H. Wu. Markov chain Monte Carlo for autologistic regression models with application to the distribution of plant species. *Biometrics*, 54(1):509–524, 1998.

L. Knorr-Held. Some remarks on Gaussian Markov random field models for disease mapping. In P.J. Green, N.L. Hjort, and S. Richardson, Eds., *Highly Structured Stochastic Systems*, Oxford Statistical Science Series, no 27, pages 260–264. Oxford University Press, Oxford, U.K. 2003.

H.R. Künsch. Gaussian Markov random fields. *Journal of the Faculty of Science, The University of Tokyo*, 26(1):53–73, 1979.

H.R. Künsch. Intrinsic autoregressions and related models on the two-dimensional lattice. *Biometrika*, 74(3):517–524, 1987.

K.V. Mardia. Multidimensional multivariate Gaussian Markov random fields with application to image processing. *Journal of Multivariate Analysis*, 24(2):265–284, 1988.

P. McCullagh and J.A. Nelder. *Generalized Linear Models*. CRC Press, Boca Raton, FL, 2nd ed., 1990.

H. Rue and L. Held. *Gaussian Markov Random Fields: Theory and Applications*, vol. 104 of *Monographs on Statistics and Applied Probability*. Chapman & Hall, London, 2005.

H. Rue, S. Martino, and N. Chopin. Approximate Bayesian inference for latent Gaussian models using integrated nested Laplace approximations (with discussion). *Journal of the Royal Statistical Society, Series B*, 71:319–392, 2009.

14

Disease Mapping

Lance Waller and Brad Carlin

CONTENTS

14.1 Background

The mapping of disease incidence and prevalence has long been a part of public health, epidemiology, and the study of disease in human populations (Koch, 2005). In this chapter, we focus on the challenge of obtaining reliable statistical estimates of local disease risk based on counts of observed cases within small administrative districts or regions coupled with potentially relevant background information (e.g., the number of individuals at risk and, possibly, covariate information, such as the regional age distribution, measures of socioeconomic status, or ambient levels of pollution). Our goals are twofold: we want statistically precise (i.e., low variance) local estimates of disease risk for each region, and we also want the regions to be "small" in order to maintain geographic resolution (i.e., we want the map to show local detail as well as broad trends). The fundamental problem in meeting both goals is that they are directly at odds with one another; the areas are not only "small" in geographic area (relative to the area of the full spatial domain of interest) resulting in a detailed map, but also "small" in terms of local sample size, resulting in deteriorated local statistical precision.

Classical design-based solutions to this problem are often infeasible since the local sample sizes within each region required for desired levels of statistical precision are often unavailable or unattainable. For example, large national or state health surveys in the United States, such as the National Health Interview Survey, the National Health and Nutrition Examination Survey, or the Behavioral Risk Factor Surveillance System provide design-based estimates of aggregate or average values at the national or possibly the state level. But, even as large as they are, such surveys often do not include sufficient sample sizes at smaller geographic levels to allow accurate, local, design-based estimation everywhere (Schlaible, 1996).

In contrast, model-based approaches offer a mechanism to "borrow strength" across small areas to improve local estimates, resulting in the smoothing of extreme rates based on small local sample sizes. Such approaches often are expressed as mixed effects models and trace back to the work of Fay and Herriot (1979), who proposed the use of random intercepts to pool information and provide subgroup-level estimated rates. Their model forms the basis of a considerable literature in *small area estimation* (Ghosh and Rao, 1994; Ghosh, Natarajan, Stroud, and Carlin, 1998; Rao, 2003), which sees wide application in the analysis of statistical surveys, including the aforementioned health surveys (Raghunathan, Xie, Schenker, Parsons et al. 2007).

While addressing the fundamental problem of analyzing data from subsets with small sample sizes, most traditional approaches to small area estimation are nonspatial; the methods essentially borrow information equally across *all* small areas without regard to their relative spatial locations and smoothing estimates toward a global mean. In the statistical literature, "disease mapping" refers to a collection of methods extending small area estimation to directly utilize the spatial setting and assumed positive spatial correlation between observations, essentially borrowing more information from neighboring areas than from areas far away and smoothing local rates toward local, neighboring values. The term "disease mapping" itself derives from Clayton and Kaldor (1987), who defined empirical Bayesian methods building from Poisson regression with random intercepts defined with spatial correlation. This hierarchical approach provides a convenient conceptual framework wherein one induces (positive) spatial correlation across the estimated local disease rates via a *conditionally autoregressive* (CAR) (Besag, 1974, and Chapter 13, this volume) random effects distribution assigned to the area-specific intercepts. The models were extended to a fully Bayesian setting by Besag, York, and Mollié (1991) and are readily implemented via Markov chain Monte Carlo (MCMC) algorithms (Chapter 13, this volume). The framework is inherently hierarchical and almost custom-made for MCMC, allowing straightforward extensions to allow for model-based estimation of covariate effects (in spatially correlated outcomes), prediction of missing data (e.g., if a county neglects to report the number of new cases for a particular month when reports are available for neighboring counties), and spatial-temporal covariance structures.

In both the nonspatial and spatial settings, the amount of smoothing is determined by the data and the formulation of the model. This smoothing permits easy visualization of the underlying geographic pattern of disease. We remark, however, that such smoothing may *not* be appropriate if the goal is instead to identify boundaries or regions of rapid change in the response surface, since smoothing is antithetic to this purpose. For more on this area, called *boundary analysis* or *wombling*, see Banerjee and Gelfand (2006), Ma, Carlin, and Banerjee (2009), and Banerjee (Chapter 31, this volume).

In the sections below, we describe in detail the basic model structure of the CAR models typically used in disease mapping, their implementation via MCMC, and various extensions to handle more complex data structures (e.g., spatial-temporal data, multiple diseases, etc.). We also illustrate the methods using real-data examples, and comment on related issues in software availability and usage.

14.2 Hierarchical Models for Disease Mapping

In this section, we outline the essential elements and structure of the CAR-based family of hierarchical disease mapping models. Additional detailed development and further illustrations of the models appear in several texts and book chapters, including Mollié (1996), Best, Waller, Thomas, Conlon et al. (1999), Lawson (2001), Wakefield, Best, and Waller (2000), Banerjee, Carlin, and Gelfand (2004, Sec. 5.4), Waller and Gotway (2004, Sec. 9.5), Waller (2005), Carlin and Louis (2009, Sec. 7.7.2).

14.2.1 The Generalized Linear Model

To begin, suppose we observe counts of disease cases Y_i for a set of regions $i = 1, \ldots, I$ partitioning our study domain $\mathcal{D}$. We model the counts as either Poisson or binomial random variables in generalized linear models, using a log or logit link function, respectively. In some cases, we may also have observed values of region-specific covariates $\boldsymbol{x}_i$ with associated parameters $\boldsymbol{\beta}$. Other data often include either the local number of individuals at risk n_i or a local number of cases "expected" under some null model of disease transmission (e.g., constant risk for all individuals), denoted E_i. We assume the n_i (alternatively, the E_i) values are fixed and known.

We typically justify the use of a Poisson model as an approximation to a binomial model when the disease is rare (i.e., the binomial probability is small). We focus on Poisson models here, based on the relative rarity of the diseases in our examples, and refer readers to Wakefield (2001, 2003, 2004, Chapter 30, this volume) for a full discussion of the binomial approach, as well as related concerns about the *ecological fallacy*, i.e., the tendency of correlations obtained from fitting at an aggregate (say, regional) level to overstate those that would be obtained if the data allowed fitting of models based on individual levels of risk.

Our Poisson model in its most basic, fixed effects-only form is

$$Y_i|\zeta_i \overset{ind}{\sim} \text{Poisson}(E_i \exp(\boldsymbol{x}_i'\boldsymbol{\beta})), \text{ for } i = 1, \ldots, I.$$

Here we define the expected number of events in the absence of covariate effects as E_i. This expected number is often expressed as the number of cases defined by an epidemiologic "null model" of incidence, i.e., the product of n_i, the number of individuals at risk in region i, and r, a constant "baseline" risk per individual. This individual-level risk is often estimated from the aggregate population data via $\hat{r} = \sum_{i=1}^n Y_i / \sum_{i=1}^n n_i$, the global observed disease rate. The resulting Poisson generalized linear model (GLM) models the natural logarithm of the mean count as

$$\log[E(Y_i)] = \log(E_i) + \boldsymbol{x}_i'\boldsymbol{\beta},$$

with an offset E_i and multiplicative impacts on the model-based expected observation counts for each covariate, resulting in a region-specific relative risk of ζ_i, $\exp(\boldsymbol{x}_i'\boldsymbol{\beta})$.

Some discussion of the expected counts E_i, $i = 1, \ldots, I$, is in order. The estimated baseline risk defined above, known as *internal standardization*, is a bit of a "cheat" since we continue to think of the E_i as known, even though they now depend on our estimate of r. But, since the impact of this choice fades within increasing numbers of regions I, and noting that our definition of r serves only to set the relatively uninteresting grand intercept β_0, this seems a minor concern. In addition, one may wish to further standardize the risks and expectations to account for spatial variation in the distribution of known risk factors (such as age), rather than adjust for such risk factors in the region-specific covariates. Waller and Gotway (2004, Chap. 2) provide an overview of the mechanisms of and arguments for and against standardization in spatial epidemiology.

14.2.2 Exchangeable Random Effects

In order to borrow information across regions, we next define the random effects version of the model, but for the moment we describe the model without covariates for simplicity. In other words, consider an intercept-only GLM with offset E_i, but allow a random intercept v_i associated with each region, i.e.,

$$Y_i|v_i \stackrel{ind}{\sim} \text{Poisson}(E_i \exp[\beta_0 + v_i]),$$
$$\text{where } v_i \stackrel{ind}{\sim} N(0, \sigma_v^2), \text{ for } i = 1, \ldots, I.$$

The hierarchical structure allows us to build the overall (marginal) distribution of the Y_i in two stages. At the first stage, observations Y_i are conditionally independent given the values of the random effects, v_i. The second stage (the distribution of the random effects) allows a mechanism for inducing extra-Poisson variability in the marginal distribution of the Y_is. Other options exist for introducing different types of excess variability or overdispersion into generalized linear models of counts (e.g., McCullagh and Nelder, 1989; Gelfand and Dalal, 1990). Here, we focus on the exchangeable random intercept approach due to its similarity to the approach proposed for spatial random effects in the sections below.

From a Bayesian perspective, the first stage of the model defines the likelihood and the second stage a set of exchangeable prior distributions for the random effects, which are estimable provided σ_v^2 is known or is assigned a proper hyperprior. To complete the model specification, we assign a vague (perhaps even improper uniform, or "flat") prior to the "fixed" effect β_0, which is well identified by the likelihood.

The hierarchical structure allows a wide variety of options for shaping the random effects and resulting marginal correlations among the Y_is. This feature of maintaining a conditionally independent framework for observations given the random effects and defining a second-stage distribution for the random effects represents one of the primary advantages of hierarchical models, and has led to their widespread use in statistical analyses with complex correlation patterns (e.g., spatial, temporal, longitudinal, repeated measures, and so on), particularly for non-Gaussian data, such as our small area counts.

The addition of the random effects addresses the small area estimation problem by inducing a connection among the local relative risks (the ζ_is) through the random effects distribution, and transforming the estimation of I local relative risks to the estimation of only two parameters: the overall mean effect β_0, and the random effects variance σ_v^2. The approach provides a local estimate defined by a weighted average of the observed data in location i and the global overall mean. Clayton and Kaldor (1987), Marshall (1991), and Waller and Gotway (2004, Sec. 4.4.3) provide details of an empirical Bayes approach using data-based estimates of β_0 and σ_v^2. In a fully Bayesian approach, we assign a hyperprior distribution to σ_v^2 (e.g., a conjugate inverse gamma distribution) and summarize the full posterior distribution for statistical inference.

Extending the model to include region-specific, fixed-effect covariates simply involves replacing β_0 above by $\boldsymbol{x}_i'\boldsymbol{\beta}$ (including the fixed intercept β_0 within $\boldsymbol{\beta}$) and assigning vague priors to the elements of $\boldsymbol{\beta}$. As with β_0 above, the fixed-effect parameters are well identified by the likelihood and provide baseline (presmoothing) estimates of each local relative risk ζ_i.

14.2.3 Spatial Random Effects

To this point, the model induces some correlation, but does not specifically induce *spatial* correlation among the observations. All local estimates are compromises between the local data and a *global* weighted average based on all of the data, with weights based on the relative variances observed in the local and global estimates. Clayton and Kaldor (1987) introduced the idea of replacing the set of exchangeable priors at the second stage with a spatially structured prior distribution, leading to empirical Bayes estimates wherein local estimates are a weighted average of the regional data value and an average of observations in nearby or neighboring regions. This approach borrows strength *locally*, rather than globally. Besag et al. (1991) extended the approach to a fully Bayesian formulation, clarified some technical points regarding the spatial prior distribution, and proposed the use of MCMC algorithms for fitting such models.

In this vein, suppose we modify the model to

$$Y_i | u_i \overset{ind}{\sim} \text{Poisson}(E_i \exp[\beta_0 + u_i]),$$
$$\text{where } \mathbf{u} \sim MVN(\mathbf{0}, \Sigma_u).$$

Here, Σ_u denotes a spatial covariance matrix and we distinguish between the exchangeable random effects v_i above and a vector $\mathbf{u} = (u_1, \ldots, u_I)$ of spatially correlated random effects. Fixed effect covariates may be added in the same manner as before. In practice, the spatial covariance matrix typically consists of parametric functions defining covariance as a function of the relative locations of any pair of observations (e.g., geostatistical covariance functions and variograms). Cressie (1993, Secs. 2.3–2.6) and Waller and Gotway (2004, Sec. 8.2) provide introductions to such covariance functions, and Diggle, Tawn, and Moyeed (1998), Banerjee et al. (2004, Sec. 2.1), and Diggle and Ribeiro (2007) illustrate their use within hierarchical models, such as that above.

The model based on a multivariate Gaussian random effects distribution represents a relatively minor conceptual change from the small area estimation literature, and ties the field to parametric covariance models from geostatistics (Matheron, 1963; Cressie, 1993, Chap. 3; Waller and Gotway, Chap. 8). However, the goals of disease mapping (statistically stable local estimation) and geostatistics (statistical prediction at locations with no observations) differ and such models currently represent a relatively small fraction of the disease mapping literature. An alternative formulation built from Clayton and Kaldor's (1987) CAR formulation sees much broader application in the spatial analysis of regional disease rates, largely thanks to the computational advantages it offers over the multivariate Gaussian model. But, since the spatial structure induced by the CAR model is less immediately apparent, we now consider it in some detail.

Specifically, the CAR formulation replaces the multivariate Gaussian second stage above with a collection of conditional Gaussian priors for each u_i wherein the prior mean is a weighted average of the other u_j, $j \neq i$,

$$u_i | u_{j \neq i} \sim N\left(\frac{\sum_{j \neq i} c_{ij} u_j}{\sum_{j \neq i} c_{ij}}, \frac{1}{\tau_{CAR} \sum_{j \neq i} c_{ij}} \right), \quad i = 1, \ldots, I. \tag{14.1}$$

Here, the c_{ij}s are user-defined spatial dependence parameters defining which regions j are "neighbors" to region i, or more generally weights defining the influence of region u_j on the prior mean of u_i. The parameter τ_{CAR} denotes a hyperparameter related to the conditional variance of u_i given the values of the other elements of $\mathbf{u}$. By convention, one sets $c_{ii} = 0$ for all i, so no region is its own neighbor. Many applications consider *adjacency-based* weights, where $c_{ij} = 1$ if region j is adjacent to region i, and $c_{ij} = 0$ otherwise. Other weighting options also are available (e.g., Best, Waller, Thomas, Conlon et al., 1999), but are much less widely applied. Weights are typically assumed to be fixed, but see Lu, Reilly, Banerjee, and Carlin (2007b) for a spatial boundary analysis application where the weights are estimated from the data.

To define the connection between the autoregressive spatial dependence parameters $\{c_{ij}\}$ and the joint spatial covariance matrix Σ_u, Besag and Kooperberg (1995) note that, if $\mathbf{u}$ follows a multivariate Gaussian distribution with covariance Σ_u, then the density, $f(\mathbf{u})$, takes the form

$$f(\mathbf{u}) \propto \exp\left(-\frac{1}{2}\mathbf{u}'\Sigma_u^{-1}\mathbf{u}\right). \tag{14.2}$$

Standard multivariate Gaussian theory defines the associated conditional distributions as

$$u_i|u_{j\neq i} \sim N\left(\sum_{j\neq i}\left(\frac{-\Sigma_{u,ij}^{-1}}{\Sigma_{u,ii}^{-1}}\right)u_j, \frac{1}{\left(\Sigma_{u,ii}^{-1}\right)}\right), \tag{14.3}$$

where $\Sigma_{u,ij}^{-1}$ denotes the (i, j)th element of the precision matrix Σ_u^{-1}. Note the conditional mean for u_i is a weighted sum of u_j, $j \neq i$, and the conditional variance is inversely proportional to the diagonal of the inverse of Σ_u, just as it is in the CAR specification above.

Reversing direction and going from a set of conditional Gaussian distributions to the associated joint distribution is more involved, requiring constraints on the $\{c_{ij}\}$ to ensure, first, a Gaussian joint distribution and, second, a symmetric and valid covariance matrix Σ_u (c.f. Besag, 1974; Besag and Kooperberg, 1995; Arnold, Castillo, and Sarabia 1999). Results in Besag (1974) indicate the set of CAR priors defined in Equation (14.1) uniquely defines a corresponding multivariate normal joint distribution with mean zero, $\Sigma_{u,ii}^{-1} = \Sigma_j c_{ij}$, and $\Sigma_{u,ij}^{-1} = -c_{ij}$. However, for symmetric c_{ij}s, the sum of any row of the matrix Σ_u^{-1} is zero, indicating Σ_u^{-1} is singular, and the corresponding covariance matrix Σ_u is not well defined. This holds for any symmetric set of spatial dependence parameters c_{ij} (including the adjacency-based c_{ij}s appearing in many applications). Remarkably, the singular covariance does not preclude application of the model with such weight matrices, since pairwise *contrasts* $u_i - u_j$ are well identified even though the individual u_is are not (Besag, Green, Higdon, and Mengersen, 1995). These distributions are improper priors since they define contrasts between pairs of values $u_i - u_j$, $j \neq i$, but they do not identify an overall mean value for the elements of $\mathbf{u}$ (because such distributions define the value of each u_i relative to the values of the others). In this case, any likelihood function based on data allowing estimation of an overall mean also allows the class of improper pairwise difference priors to generate proper posterior distributions. In practice, one often assures this by the ad hoc addition of the constraint

$$\sum_{i=1}^{I} u_i = 0. \tag{14.4}$$

While the addition of the constraint slightly complicates formal implementation of Equation (14.1), Gelfand and Sahu (1999) note that the constraint can be imposed "on the fly" within an MCMC algorithm simply by replacing u_i by $u_i - \bar{u}$ for all i following each MCMC iteration. These authors also provide additional theoretical justification, and note that the constraint maintains attractive, full conditional distributions for most CAR models in the literature while avoiding awkward reduction to $(I-1)$-dimensional space. In contrast, Rue and Held (2005, Sec. 2.3.3) avoid the constraint altogether through block updates of the entire set of random effects. See also Richardson, Thomson, Best, and Elliott (2004), Knorr-Held and Rue (2002), and Chapter 13 (this volume) for important algorithmic advances related to this model.

As a computational aside, note that both the conditional mean and the conditional variance in Equation (14.3) depend on elements of the *inverse* of the covariance matrix Σ_u. As a result, MCMC algorithms applied to the joint specification based on straightforward updates from full conditional distributions will involve some sort of matrix inversion at each update of the covariance parameters. This reveals a computational advantage of the CAR prior formulation; it effectively limits modeling to the elements of Σ_u^{-1}, avoiding inversion and we focus attention on CAR models in the remainder of this chapter. We note, however, that computational convenience carries considerable conceptual cost (parameterizing Σ_u^{-1} rather than Σ_u). Recent algorithmic developments seek to ease this computational/conceptual trade-off by using structured covariance matrices to reduce the computational burden of directly modeling the joint distribution, an issue discussed in more detail in Chapter 13.

In addition to its (indirectly) defining the covariance structure of our model, the choice of c_{ij}s also has direct impact on the posterior variances. In the usual case where τ_{CAR} is unknown, it is conventionally assigned a conjugate gamma hyperprior distribution (see, e.g., Carlin and Louis, 2009, p. 424), since this leads to a closed form for the τ_{CAR} full conditional distribution needed by the MCMC algorithm. However, even here there is some controversy, since the impropriety of the standard CAR means that, despite our use of the proportionality sign in Equation (14.2), the joint distribution of the u_is really has no normalizing constant. Knorr-Held (2002) advocated $k = (n-1)/2$ for Gaussian Markov random fields (a set containing the CAR-specified model above) based on the rank of the resulting precision matrix. Hodges, Carlin, and Fan (2003) argue that the most sensible joint density to use in this case is

$$f(\mathbf{u}) \propto \tau_{CAR}^{(I-k)/2} \exp\left(-\frac{\tau_{CAR}}{2}\mathbf{u}'Q\mathbf{u}\right), \tag{14.5}$$

where Q is $I \times I$ with nondiagonal entries $q_{ij} = -1$ if $i \sim j$ and 0 otherwise, and diagonal entries q_{ii} equal to the number of region i's neighbors, and k is the number of disconnected "islands" in the spatial structure. Thus, in the usual case where every county is connected to every other by some chain of neighbors, the exponent on τ_{CAR} is $(I-1)/2$, as advocated earlier by Knorr-Held (2002), and not $I/2$, as originally suggested by Besag et al. (1991). In the case of multiple islands in the spatial map, this exponent drops further (reflecting the greater rank deficiency in Q), and the sum-to-zero constraint (Equation 14.4) must be applied to each island separately. See Lu, Hodges, and Carlin (2007a) for extensions of these ideas that enable "counting" degrees of freedom in spatial models that are distinct from but related to those given by Spiegelhalter, Best, Carlin, and van der Linde (2002).

14.2.4 Convolution Priors

Further extending disease mapping models, Besag et al. (1991) point out that we could include both global and local borrowing of information within the same model via a *convolution prior* including both exchangeable and CAR random effects for each region, as follows:

$$\begin{aligned} Y_i|u_i &\overset{ind}{\sim} \text{Poisson}(E_i \exp[\beta_0 + u_i + v_i]), \\ \text{where } u_i|u_{j\neq i} &\sim N\left(\frac{\sum_{j\neq i} c_{ij}u_j}{\sum_{j\neq i} c_{ij}}, \frac{1}{\tau_{CAR}\sum_{j\neq i} c_{ij}}\right) \\ \text{and } v_i &\overset{ind}{\sim} N(0, \sigma_v^2), \text{ for } i = 1, \ldots, I. \end{aligned}$$

To complete the model, we assign hyperpriors to the hyperparameters τ_{CAR} and $\tau_v \equiv \sigma_v^2$. Again, fixed-effect covariates may be added if desired. As mentioned above, typical applications define conjugate gamma hyperpriors, and Ghosh, Natarajan, Waller, and Kim (1999) and Sun, Tsutakawa, and Speckman (1999) define conditions on these distributions necessary to ensure proper posterior distributions. When we include both $\mathbf{u}$ and $\mathbf{v}$ in the model, some care is required to avoid assigning "unfair" excess prior weight to either global or local smoothing because τ_{CAR} is related to the conditional variance of $u_i|u_{j\neq i}$, but τ_v is related to the *marginal* variance of each v_i. This issue is explored by Bernardinelli, Clayton, and Montomoli (1995a), who, based on their empirical example, suggest taking the prior marginal standard deviation of v_i to be roughly equal to the conditional standard deviation of $u_i|u_{j\neq i}$ divided by 0.7. We stress that this is only a "rule of thumb" and merits closer scrutiny (Eberly and Carlin, 2000; Banerjee et al., 2004, p. 164). In any given application, a simple yet "fair" approach might be to first run the MCMC algorithm *without* including the data (e.g., by commenting the Poisson likelihood terms out of the `WinBUGS` code), and then choose hyperpriors that produce no-data "posteriors" for τ_{CAR} and τ_v that are roughly equal. Another approach based on marginal variances induced by the CAR prior appears in Rue and Held (2005, pp. 103–105). In any case, note that we cannot take *both* of these hyperpriors to be noninformative because then only the sum of the random effects $(u_i + v_i)$, and not their individual values, will be identified.

14.2.5 Alternative Formulations

It is important to note that, while arguably the most popular, the formulation proposed by Besag et al. (1991) is not the only mechanism for including both spatial and nonspatial variance components within a single hierarchical disease mapping model. For example, Leroux, Lei, and Breslow (1999) and MacNab and Dean (2000) define spatially structured and unstructured variation built on additive components of the precision matrix of a single random intercept rather than via the sum of two additive random intercepts. To contrast the approaches briefly, the Besag et al. (1991) formulation above defines a random intercept consisting of the sum of two parameters (u_i and v_i) for each region, resulting in a variance–covariance matrix for the multivariate normal sum $\mathbf{u} + \mathbf{v}$ defined by

$$\Sigma_{u+v} = \sigma_u^2 Q^{-1} + \sigma_v^2 \boldsymbol{I},$$

where $\boldsymbol{I}$ denotes the I-dimensional identity matrix and Q contains the number of neighbors for each region along the diagonal, $q_{ij} = -1$ if $i \sim j$, and 0 otherwise, as in Equation (14.5) above. The Leroux et al. (1999) formulation defines a single, random intercept w_i for each region where

$$\begin{aligned} Y_i | w_i &\overset{ind}{\sim} \text{Poisson}(E_i \exp[\beta_0 + w_i]), \\ \boldsymbol{w} &\sim MVN(\mathbf{0}, \Sigma_w), \text{ and} \\ \Sigma_w &= \sigma^2 D^{-1}. \end{aligned}$$

Here, σ^2 defines an overall dispersion parameter, and

$$D = \lambda Q + (1 - \lambda)\boldsymbol{I},$$

with $0 \leq \lambda \leq 1$ denoting a spatial dependence parameter where $\lambda = 0$ defines a nonspatial model and the level of spatial dependence increases with λ. By defining the spatial and nonspatial components for the inverse (or generalized inverse) of Σ_w, Leroux et al. (1999) allow ready definition of the conditional mean and variance for the random effects in terms of parameters λ and σ^2, i.e.,

$$\begin{aligned} E(w_i | w_{j \neq i}) &= \frac{\lambda}{1 - \lambda + \lambda q_{ii}} \text{ and} \\ \text{var}(w_i | w_{j \neq i}) &= \frac{\sigma^2}{1 - \lambda + \lambda q_{ii}}, \end{aligned}$$

recalling q_{ii} denotes the number of neighbors for region i, $i = 1, \ldots, I$. Leroux et al.'s (1999) formulation also allows parameter estimation via penalized quasi-likelihood in the manner of Breslow and Clayton (1993).

Another set of alternative approaches seeks to identify potential discontinuities in the risk surface by loosening the rather strong amount of spatial smoothing often induced by the Besag et al. (1991) general CAR formulation above. Besag et al. propose replacing the squared pairwise differences, $(u_i - u_j)^2$, inherent in the Gaussian CAR formulation, with the L1-norm-based pairwise differences, $|u_i - u_j|$, resulting in shrinkage toward neighborhood median rather than mean values; hence, yielding relatively weaker amounts of smoothing between neighboring values as illustrated in simulation studies by Best et al. (1999).

Green and Richardson (2002) take a different approach based on a hidden Markov field, thereby deferring spatial correlation to an additional, latent layer in the hierarchical model. Their formulation draws from a spatial type of cluster analysis where each region belongs to one of several classes and class assignments are allowed to be spatially correlated (Knorr-Held and Rasser, 2000; Denison and Holmes, 2001). The number of classes and the regional class assignments are unobserved, requiring careful MCMC implementation within the variable-dimensional parameter space. The approach allows for discontinuities between latent class assignments and for spatially varying amounts of spatial correlation (i.e., stronger correlation for some classes, weaker for others). The approach is not a direct extension of the CAR models above, but rather utilizes a Potts model from statistical image analysis and statistical physics to model spatial association in the regional labels. This increased flexibility (at an increased computational cost) provides inference for both group membership as well as very general types of spatial correlation. However, the advantages of the added flexibility increase with the number of regions and the complexity of subgrouping within the data. As a result, applications of the Green and Richardson (2002) approach appear more often in the analysis of high-dimensional biomedical imaging and genetic expression data than in disease mapping.

Simulation studies in Leroux et al. (1999), MacNab and Dean (2000), and Green and Richardson (2002) identify the types of situations where these alternative model formulations gain advantage over the CAR formulation. For example, the Leroux et al. (1999) model shows gains in performance over the CAR model as spatial correlation decreases to zero, and the Green and Richardson (2002) approach improves several measures of model fit from those observed in CAR formulations, particularly in cases with strong underlying discontinuities in risk. However, the near custom-fit between the Besag et al. (1991) formulation and fairly standard MCMC implementation continues to fuel its popularity in general disease mapping applications. As such, we focus on this model and its extensions below.

14.2.6 Additional Considerations

The `WinBUGS` software package (`www.mrc-bsu.cam.ac.uk/bugs/welcome.shtml`) permits ready fitting of the disease mapping models defined above with exchangeable, CAR, or convolution prior distributions on the random intercepts. `WinBUGS` also allows mapping of the fitted spatial residuals $E(u_i|y)$ or the fitted region-specific relative risks, $E(e^{\beta_0+u_i+v_i}|y)$, local estimates of the relative risk of being in region i compared to what was expected.

While hierarchical models with CAR and/or convolution priors see broad application for parameter estimation and associated small area estimation for regional data, they certainly are not the only models for such data, nor necessarily optimal in any particular way. In addition, CAR-based hierarchical models are defined only for the given set of regions and do not aggregate or disaggregate sensibly into CAR models on larger or smaller regions, respectively. Furthermore, regions on the edges of the study domain often have fewer neighbors and, hence, less information to draw from for borrowing strength locally than interior regions resulting in "edge effects" of reduced performance. The dependence on the given set of regions also implies that adjacency-based neighborhoods can correspond to very different ranges of spatial similarity around geographically large regions than around geographically small regions. For these reasons, the CAR prior cannot be viewed as a simple "discretized" version of some latent, smooth random process; see, e.g., Banerjee et al. (2004, Sec. 5.6). However, Besag and Mondal (2005) provide some connections between CAR-based models and latent de Wijs processes on smaller scales that may allow rescaling of distance-based correlation structures across zonal systems.

Finally, it is important to keep in mind that the CAR structure is applied to the random effects at the second stage of the hierarchy, not directly to the observed data themselves. Generally speaking, this ensures a proper posterior distribution for the random effects for a broad variety of likelihood structures. In applications where one assumes a Gaussian first stage, CAR random effects are especially attractive with closed form, full conditional distributions. In a generalized linear model setting (as in most disease mapping applications), the hierarchical structure allows us to maintain use of Gaussian CAR random variables within the link function, rather than attempting to work with Poisson (or binomial) CAR distributions for the counts themselves. For Poisson outcomes, the Gaussian-CAR-within-the-link-function structure avoids extreme and unfortunate restrictions (e.g., negative spatial correlation and normalizing constants defined by awkward functions of model parameters) imposed by CAR-based "autoPoisson" models (Besag, 1974). The hierarchical modeling approaches based on the CAR and convolution priors described above allow us to incorporate spatial correlation into generalized linear models of local disease rates as well as conveniently defer such correlation to the second level of the model. That is, the formulation avoids analytical complications inherent in modeling spatial correlation within non-Gaussian distributions with interrelated mean and variance structures.

14.3 Example: Sasquatch Reports in Oregon and Washington

To illustrate the disease mapping models above, we consider an admittedly unconventional (and a bit whimsical) dataset, namely the number of reported encounters with the legendary North American creature Sasquatch (Bigfoot) for each county in the U.S. states of Washington and Oregon. These data were obtained in May 2008 from the Web site of the Bigfoot Field Research Organization, `www.bfro.net`. For those unfamiliar with the story, Sasquatch is said to be a large, bipedal hominoid primarily purported to reside in remote areas in the Pacific Northwest. While reported encounters do not reflect a "disease" per se and we do not necessarily expect all individuals residing in a given county to experience the same "risk" of reporting an encounter, cryptozoologists and Sasquatch enthusiasts alike may be interested in identifying areas with higher than expected local per-person *rates* of reported encounters. For our purposes, the data serve as a general example of the type we have described; namely, regional counts of a (thankfully) rare event standardized by the local population size and with associated regional covariates. The models above allow us to explore region-specific relative risks of reporting in order to explore any underlying geographic patterns and identify where reports are higher or lower than expected if every individual were equally likely to file a report. While the null model of equal per-person risk of reporting is unlikely to be true, it nevertheless forms a point of reference for our region-to-region comparisons. As we shall see, the data also offer an opportunity to explore in some detail the behavior of the methods in the presence of a single, large, outlying observation. While the data provide a template for illustrating the models, readers in search of more traditional applications of CAR-based disease mapping models may find detailed examples in Mollié (1996), Best et al. (1999), Wakefield et al. (2000), Banerjee et al. (2004), and Waller and Gotway (2004, Chap. 9).

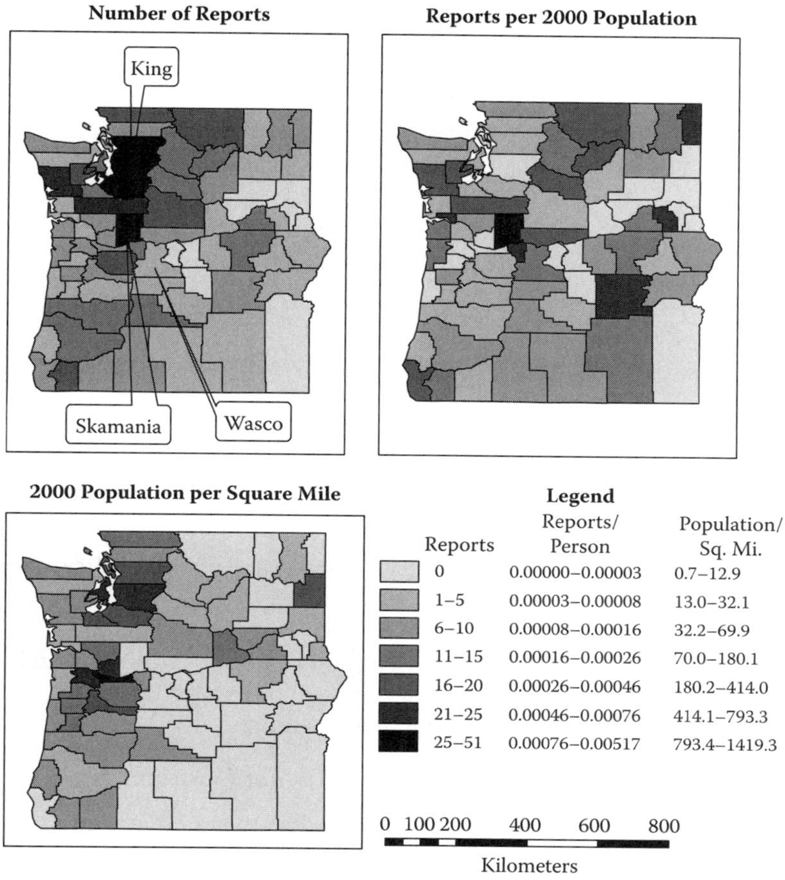

FIGURE 14.1
Maps of the number of reported encounters with Sasquatch (Bigfoot) by county (upper left), the "rate" of reports by county (upper right), and the population density per county reported as the number of residents per square mile (2000 U.S. Census).

Figure 14.1, created by linking our data to maps of U.S. county boundaries using `ArcGIS` (ESRI, Redlands, CA), displays the data. The map in the upper left shows the number of reports per county ranging from zero (light gray) to a high of 51 in Skamania County in Washington, on the border with Oregon. The map in the upper right displays the local "rate" of reporting defined as the number of reports divided by the county population as reported by the 2000 U.S. Census, displayed in intervals defined by Jenks' "natural breaks" method (Jenks, 1977; MacEachren, 1995, Chap. 4). The population adjustment is somewhat contrived as some reports in the dataset can date back to the 1970s and a few back to the 1950s, but the 2000 population counts offer a crude form of standardization. The adjustment from counts to rates is most dramatic in the counties surrounding Puget Sound, revealing that the numbers of reports in this area are quite small when computed on a roughly per-person basis. The shift is particularly striking in King County, home to much of the Seattle metropolitan area. In contrast, Skamania County is extreme for both counts and rates and clearly of interest as our analysis continues. Finally, the lower left map shows the geographic distribution of a potential covariate, the population density based on the number of residents per square mile (again, from the 2000 Census and classified by Jenks'

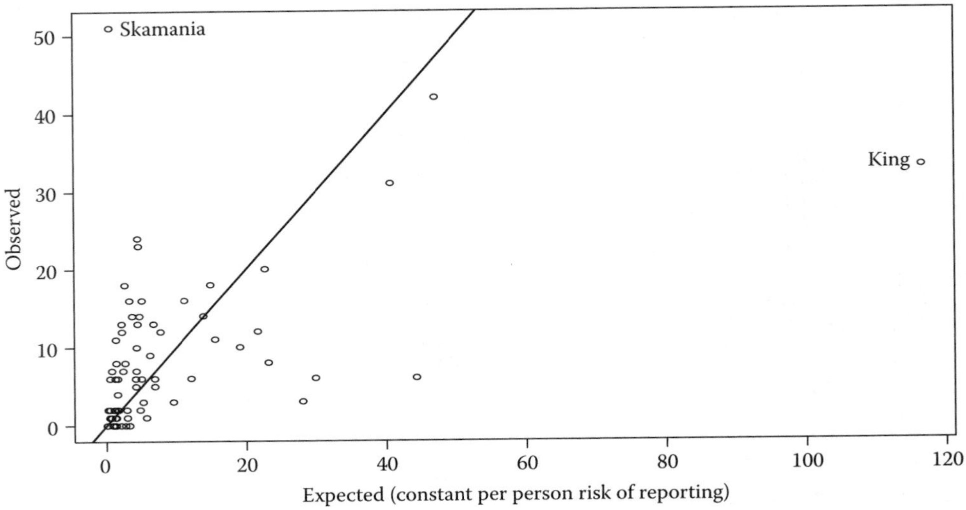

FIGURE 14.2
Observed number of Sasquatch reports per county compared to the number expected under a constant per-person probability of filing a report.

natural breaks method). We note that rural Skamania County, while high in both number and rate of reporting, is in the lowest category of (log) population density.

Our model begins with the simple Poisson regression,

$$Y_i|\boldsymbol{\beta} \overset{ind}{\sim} \text{Poisson}(E_i \exp(\beta_0 + \beta_1 x_i)), \text{ for } i = 1, \ldots, I,$$

with x_i denoting the natural logarithm of population density and E_i the internally standardized expected count $n_i(\sum_i Y_i / \sum_i n_i)$, i.e., the number of reports expected if each resident is equally likely to file a report. Figure 14.2 shows a scatterplot of the observed and expected counts, revealing (not surprisingly) a great deal of heterogeneity about the line of equality, with Skamania and King Counties again standing out for reporting considerably more and less than expected, respectively.

Figure 14.3 motivates our choice of covariate by illustrating how the county-specific rate of reporting decreases with increasing population density, with Skamania county remaining a obvious outlier. The extreme variation displayed suggests potential local instability in rates and suggests the use of random effects to adjust for the excess heterogeneity present in the data. However, we note that in most disease-mapping applications unstable high local rates are often due to very low expected numbers of cases (e.g., $E_i << 1$) and a single observed case, while here the high rate in Skamania is apparently due to an extremely high number of local reports (51).

We fit four models to the data, first a simple fixed effect model, then models with random intercepts following exchangeable (nonspatial), CAR (spatial), and convolution (both) priors. We used the program `maps2WinBUGS` (`sourceforge.net/projects/maps2winbugs`) to transfer the map data from `ArcGIS` format to `WinBUGS` format, then used `GeoBUGS` (the spatial analysis and mapping tool within `WinBUGS`) define our adjacency matrix. We note that the adjacency matrix defines $c_{ij} = 1$ for any regions i and j sharing a common boundary, but also includes some counties falling very close to one another; for example, Wasco County is included among the neighbors of Skamania County. Each model was fit using MCMC within `WinBUGS` using 100,000 iterations. To reduce correlation between parameters β_0 and β_1, we centered our covariate by subtracting the overall mean (log) population density from each value. Our MCMC samples provide posterior inference for

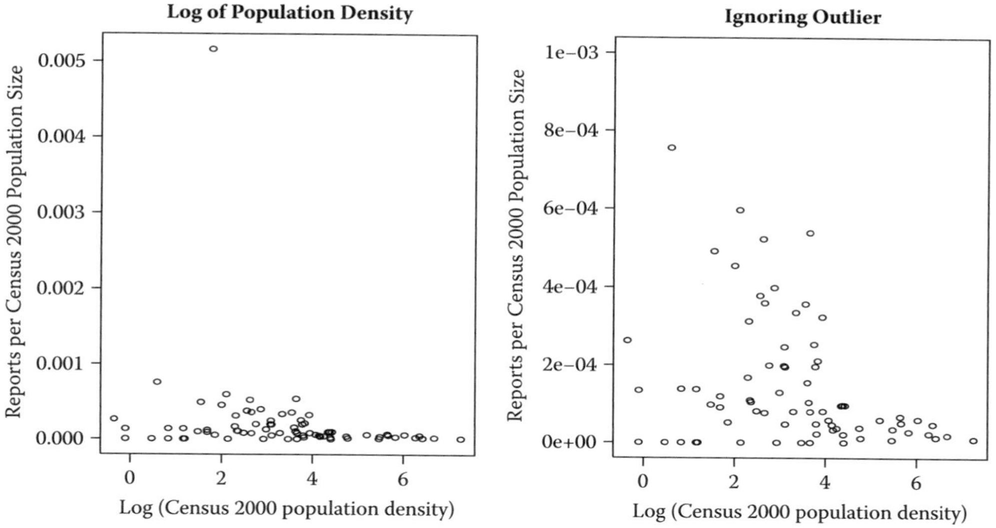

FIGURE 14.3
Scatterplots of the county-specific reporting rates versus the natural logarithm of population density with (left) and without (right) Skamania County.

model parameters and for county-specific relative risks (e.g., $RR_i = \exp(\beta_0 + \beta_1 x_i + u_i + v_i)$ for the convolution model).

In all four models, the estimated effect of population density (β_1) was negative and significantly different from 0; in the convolution model, we obtained a 95% equal-tail credible interval of (–0.68, –0.35). Thus, Bigfoot sightings are significantly more likely to arise in more thinly populated counties. One might speculate that this is due to Bigfoot's preference for habitats with fewer humans per unit area, or simply a tendency of Bigfoot afficionados to live and work in such regions. Effective model size (p_D) and deviance in formation criterion (DIC) scores (Spiegelhalter, Best, Carlin, and van der Linde 2002) do not differ appreciably across the three random effects models. This is confirmed by Figure 14.4, which shows the local relative risk (RR_i) for each county based on the maps. Counties are shaded by the same intervals to ease comparisons between maps. We note that the model with no random effects (top left) does a very poor job of predicting the local high rate in Skamania County, and that relative risks are exaggerated in the low population density counties along Oregon's southern border. The maps of relative risks based on the three random effects models are similar in general, with some subtle differences. All three are able to capture the excess variability observed in the data, especially the extreme value in Skamania County. We note that the interval containing the largest estimated relative risks covers a very large range of values, with the darkest counties in the fixed effect model representing local relative risks less than 20, but the other three maps all assigning Skamania County a relative risk near 70. The convolution prior appears to offer something of a compromise between the nonspatial exchangeable model and the spatial CAR model, particularly along the eastern border of our study area. This is sensible because with both types of random effects in our model, we would expect the fitted values to exhibit both spatial and nonspatial heterogeneity.

This compromise is seen more clearly in Figure 14.5, which shows the posterior median and 95% credible sets for the log relative risks associated with Skamania County and its neighbors for each of the four models. As noted above, for Skamania County, the neighborhood includes seven adjacent counties and one nearby county, namely Wasco County to the southeast (labeled in Figure 14.1). Figure 14.5 reveals that the model with no random effects

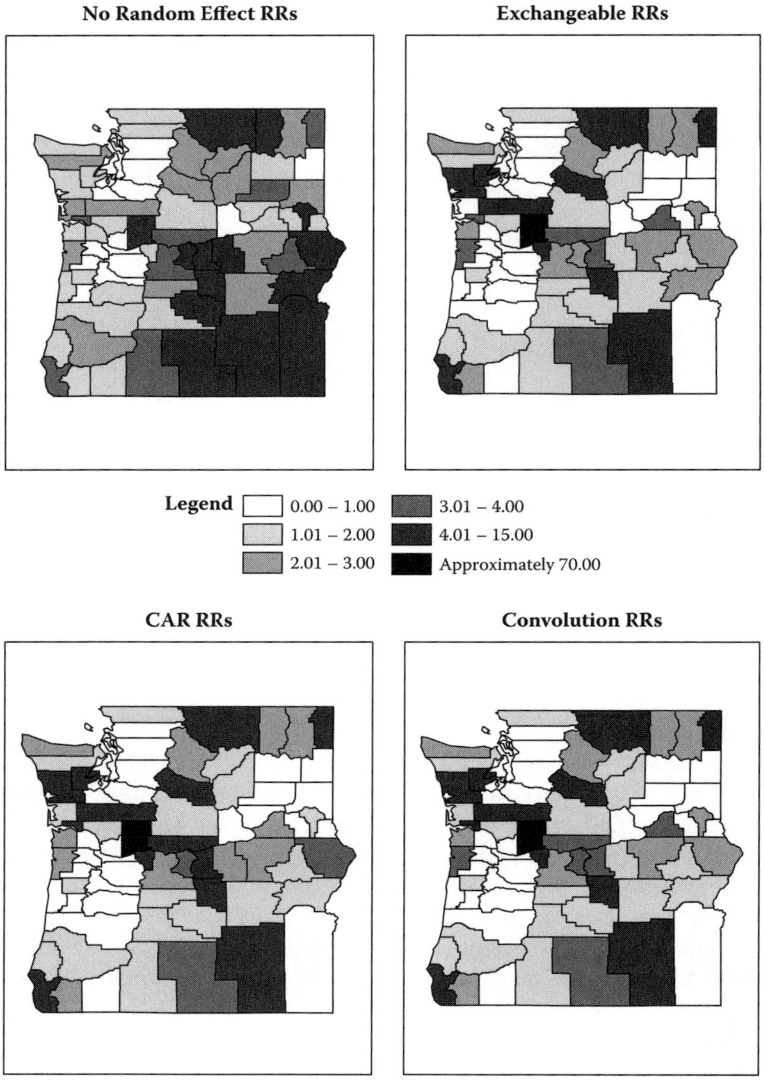

FIGURE 14.4
Maps of county-specific relative risk of reporting sitings. Each map has counties shaded by the intervals reported in the legend.

generates deceptively tight credible sets (ignoring the substantial extra-Poisson variation in the dataset) and clearly misses the increased risk of reporting observed in Skamania County (indicated by a filled circle), well above that expected based on the offset (population size) and the covariate (population density). The three random effect models are quite similar for our data, and all three capture the increased risk of reporting in Skamania County.

A closer look at the posterior distribution for Wasco County (indicated by a filled square) in Figure 14.5 highlights the subtle differences between models in our data. Note that Wasco County has a wide credible set, suggesting a locally imprecise estimate in need of input from other regions. In the CAR model, the posterior distribution of the relative risk of reporting in Wasco County is pulled (slightly) upward toward that of its neighbor Skamania when compared to the posterior distribution for Wasco County in the exchangeable model.

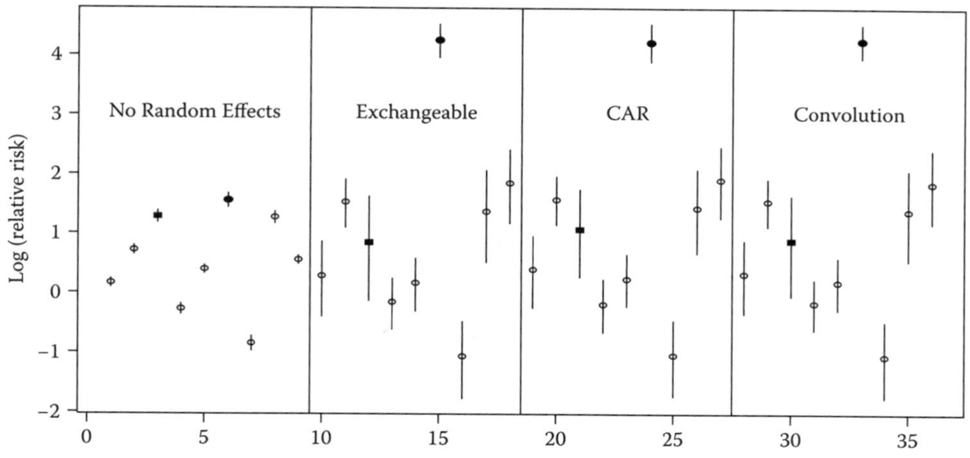

FIGURE 14.5
Posterior median and 95% credible sets for local relative risks associated with Skamania County (filled circles) and its neighboring counties. Wasco County is of interest and is labeled by a filled square.

As suggested by the maps, the convolution model represents a compromise between the posterior distributions of relative risks from the spatial CAR model and the nonspatial exchangeable model. We remark that the `WinBUGS` code for the full model (given in the Appendix) uses a relatively vague *Gamma*(0.1, 0.1) hyperprior for τ_{CAR}; a hyperprior centered more tightly on smaller values (or even a fixed, larger value of τ_{CAR}) would further smooth Wasco toward Skamania.

Our example illustrates several features of disease mapping models. First, there is often a concern that disease mapping models may oversmooth extreme rates, particularly observations that are very unlike their neighbors. Our data provide an extreme example of this with a single outlier, well identified by the model that does not overly influence its neighboring estimates. As noted in Figure 14.5, there is some impact on the most variable neighboring estimates (e.g., that from Wasco County), but this is very slight and unlikely to strongly influence conclusions. In addition, the example reveals that all three random effects models fit the data approximately equally well, suggesting a clear need for borrowing strength from other regions, but does not suggest a clear preference for global versus local assistance. Further analyses might check the robustness of the Bayesian model choice decision (via DIC or some other method) to deleting Skamania from the dataset and recomputing the posterior.

14.4 Extending the Basic Model

14.4.1 Zero-Inflated Poisson Models

Many spatial datasets feature a large number of zeros (areas with no reported disease cases) that may stretch the credibility of our Poisson likelihood. As such, a *zero-inflated Poisson* (ZIP) model may offer a sensible alternative. Lambert (1992) implemented such a model in a manufacturing-related regression context using the expectation-maximization (EM) algorithm; Agarwal, Gelfand, and Citron-Pousty (2002) offer a fully Bayes–MCMC version for spatial count data.

In our context, suppose we model the disease count in region i as a mixture of a Poisson(λ_i) distribution and a point mass at 0. That is, when there are no cases observed in region i, we assume that such zero counts arise as Poisson variates with probability $(1-\omega_i)$, and as "structural zeros" with probability ω_i. More formally, we can write the regression model as

$$Pr(Y_i = y|\omega_i, \lambda_i) = \begin{cases} \omega_i + (1-\omega_i)\exp(-\lambda_i) & \text{for } y = 0, \\ (1-\omega_i)\exp(-\lambda_i)\lambda_i^y/y! & \text{for } y > 0, \end{cases}$$

$$\text{where} \quad \log(\lambda_i) = \log(E_i) + \boldsymbol{x}_i'\boldsymbol{\beta} + u_i \quad \text{and} \quad \text{logit}(\omega_i) = \boldsymbol{z}_i'\boldsymbol{\gamma} + w_i,$$

and $\boldsymbol{x}_i$ and $\boldsymbol{z}_i$ are covariate vectors, which may or may not coincide. More parsimonious models are often used to preserve parameter identifiability, or to allow the λ_i and ω_i to be related in some way. Agarwal et al. (2002) set $\boldsymbol{x}_i = \boldsymbol{z}_i$ and eliminate the w_i random effects; a follow-up paper by Agarwal (2006) retains both the u_i and w_i, but assigns them independent proper CAR priors. Another possibility is to replace $\boldsymbol{z}_i'\boldsymbol{\gamma} + w_i$ with $\nu(\boldsymbol{x}_i'\boldsymbol{\beta} + u_i)$ in the expression for $logit(\omega_i)$ (Lambert, 1992); note that $\nu < 0$ will often be necessary to reverse the directionality of $\boldsymbol{x}_i$'s relationship when switching responses from λ_i to ω_i.

To complete the Bayesian model, we assign vague normal priors to $\boldsymbol{\beta}$ and $\boldsymbol{\gamma}$, and CAR or exchangeable priors to the u_i and w_i. Disease mapping can now proceed as usual, with the advantage of now being able to use the posterior means of the ω_i as the probability that region i is a structural zero in the spatial domain.

We remark that this model may be sensible for the data in the previous section, since 10 of the 75 counties had no reported Sasquatch sightings. In the interest of brevity, however, we leave this investigation (in `WinBUGS` or some other language) to the interested reader.

14.4.2 Spatiotemporal Models

Many spatially referenced disease count datasets are collected over time, necessitating an extension of our Section 14.2 models to the spatial-temporal case. This is straightforward if time and space are both discretely indexed, say, with space indexed by county and time indexed by year. In fact, the data may have still more discrete indexes, as when disease counts are additionally broken out by race, gender, or other sociodemographic categories.

To explicate the spatial-temporal extension as concretely as possible, we develop spatial-temporal extensions in the context of a particular dataset originally analyzed by Devine (1992, Chap. 4). Here, Y_{ijkt} is the number of lung cancer deaths in county i during year t for gender j and race k in the U.S. state of Ohio, and n_{ijkt} is the corresponding exposed population count. These data were originally taken from a public use data tape (Centers for Disease Control, 1988), and are now available online at `www.biostat.umn.edu/~brad/data2.html`. The subset of lung cancer data we consider here are recorded for $J = 2$ genders (male and female) and $K = 2$ races (white and nonwhite) for each of the $I = 88$ Ohio counties over an observation period of $T = 21$ years, namely 1968 to 1988 inclusive, yielding a total of 7,392 observations.

We begin our modeling by extending our Section 14.2 Poisson likelihood to

$$Y_{ijkt} \overset{ind}{\sim} \text{Poisson}(E_{ijkt}\exp(\mu_{ijkt})).$$

We obtain internally standardized expected death counts as $E_{ijkt} = n_{ijkt}\hat{r}$, where $\hat{r} = \bar{y} = \sum_{ijkt} y_{ijkt} / \sum_{ijkt} n_{ijkt}$, the average statewide death rate over the entire observation period. The temporal component is of interest to explore changes in rates over a relatively long period of time. Demographic issues are of interest because of possible variation in residential exposures for various population subgroups. In addition, the demographic profile of the counties most likely evolved over the time period of interest.

Devine (1992) and Devine, Louis, and Halloran (1994) applied Gaussian spatial models employing a distance matrix to the average lung cancer rates for white males over the

21-year period. Waller, Zhu, Gotway, Gorman et al. (1997) explored a full spatial-temporal CAR-based model, adopting the mean structure

$$\mu_{ijkt} = s_j\alpha + r_k\beta + s_j r_k\xi + u_i^{(t)} + v_i^{(t)}, \tag{14.6}$$

where s_j and r_k are the gender and race scores

$$s_j = \begin{cases} 0 & \text{if male} \\ 1 & \text{if female} \end{cases} \quad \text{and} \quad r_k = \begin{cases} 0 & \text{if white} \\ 1 & \text{if nonwhite.} \end{cases}$$

Letting $\mathbf{u}^{(t)} = (u_1^{(t)}, \ldots, u_I^{(t)})'$, $\mathbf{v}^{(t)} = (v_1^{(t)}, \ldots, v_I^{(t)})'$, and denoting the I-dimensional identity matrix by $\boldsymbol{I}$, we adopt the prior structure

$$\mathbf{u}^{(t)} \mid \lambda_t \overset{ind}{\sim} CAR(\lambda_t) \quad \text{and} \quad \mathbf{v}^{(t)} \mid \tau_t \overset{ind}{\sim} N\left(\mathbf{0}, \frac{1}{\tau_t}\boldsymbol{I}\right), \quad t = 1, \ldots, T, \tag{14.7}$$

so that heterogeneity and clustering may vary over time. Note that the socio-demographic covariates (gender and race) do not interact with time or location.

To complete the model specification, we require prior distributions for α, β, ξ, the τ_t and the λ_t. Since α, β, and ξ will be identified by the likelihood, we may employ a flat prior on these three parameters. Next, for the priors on the τ_t and λ_t we employed conjugate, conditionally iid *Gamma*(a, b) and *Gamma*(c, d) priors, respectively. As mentioned earlier, some precision is required to facilitate implementation of an MCMC algorithm in this setting. On the other hand, too much precision risks likelihood-prior disagreement. To help settle this matter, we fit a spatial-only (reduced) version of model (14.6) to the data from the middle year in our set (1978, $t = 11$), using vague priors for λ and τ having both mean and standard deviation equal to 100 ($a = c = 1$, $b = d = 100$). The resulting posterior 0.025, 0.50, and 0.975 quantiles for λ and τ were (4.0, 7.4, 13.9) and (46.8, 107.4, 313.8), respectively. As such, in fitting our full spatial-temporal model (14.6), we retain $a = 1$, $b = 100$ for the prior on τ, but reset $c = 1$, $d = 7$ (i.e., prior mean and standard deviation equal to 7). While these priors are still quite vague, the fact that we have used a small portion of our data to help determine them does give our approach a slight empirical Bayes flavor. Still, our specification is consistent with the aforementioned advice of Bernardinelli et al. (1995a). Specifically, recasting their advice in terms of prior precisions and the adjacency structure or our CAR prior for the $u_i^{(t)}$, we have $\lambda \approx \tau/(2\bar{m})$, where $\bar{m}$ is the average number of counties adjacent to a randomly selected county (about five to six for Ohio).

Model fitting is readily accomplished in `WinBUGS` using an assortment of univariate Gibbs and Metropolis steps. Convergence was diagnosed by graphical monitoring of the chains for a representative subset of the parameters, along with sample autocorrelations and Gelman and Rubin (1992) diagnostics. The 95% posterior credible sets (−1.10, −1.06), (0.00, 0.05), and (−0.27, −0.17) were obtained for α, β, and ξ, respectively. The corresponding point estimates are translated into the fitted relative risks for the four subgroups in Table 14.1. It is interesting that the fitted sex–race interaction ξ reverses the slight advantage white men hold over nonwhite men, making nonwhite females the healthiest subgroup, with a relative risk nearly four times smaller than either of the male groups. Many Ohio counties have very small nonwhite populations, so this result could be partly the result of our failure to model covariate-region interactions. Replacing the raw death counts Y_{ijkt} by *age-standardized* counts also serves to eliminate the nonwhite females' apparent advantage, since nonwhites die from lung cancer at slightly younger ages in our dataset; see Xia and Carlin (1998).

Turning to the spatial-temporal parameters, histograms of the sampled values (not shown) showed $v_i^{(t)}$ distributions centered near 0 in most cases, but $u_i^{(t)}$ distributions typically removed from 0. This suggests some degree of clustering in the data, but no significant

TABLE 14.1

Fitted Relative Risks for the Four Sociodemographic Subgroups in the Ohio Lung Cancer Data

Demographic Subgroup	Contribution to ε_{jk}	Fitted Log-Relative Risk	Fitted Relative Risk
White males	0	0	1
White females	α	–1.08	0.34
Nonwhite males	β	0.02	1.02
Nonwhite females	$\alpha + \beta + \xi$	–1.28	0.28

additional heterogeneity beyond that explained by the CAR prior. Use of the DIC statistic (Spiegelhalter et al., 2002) or some other Bayesian model choice statistic confirms that the nonspatial $v_i^{(t)}$ terms may be sensibly deleted from the model.

Since under our model the expected number of deaths for a given subgroup in county i during year t is $E_{ijkt} \exp(\mu_{ijkt})$, we have that the (internally standardized) expected death rate per thousand is $1000\bar{y} \exp(\mu_{ijkt})$. The first row of Figure 14.6 maps point estimates of these fitted rates for nonwhite females during the first (1968), middle (1978), and last (1988) years in our dataset. These estimates are obtained by plugging in the estimated posterior medians for the μ_{ijkt} parameters calculated from the output of the Gibbs sampler. The rates are greyscale-coded from lowest (white) to highest (black) into seven intervals: less than 0.08, 0.08 to 0.13, 0.13 to 0.18, 0.18 to 0.23, 0.23 to 0.28, 0.28 to 0.33, and greater than 0.33. The second row of the figure shows estimates of the variability in these rates (as measured by the interquartile range) for the same subgroup during these three years. These rates are also grayscale-coded into seven intervals: less than 0.01, 0.01 to 0.02, 0.02 to 0.03, 0.03 to 0.04, 0.04 to 0.05, 0.05 to 0.06, and greater than 0.06.

Figure 14.6 reveals several interesting trends. Lung cancer death rates are increasing over time, as indicated by the gradual darkening of the counties in the figure's first row.

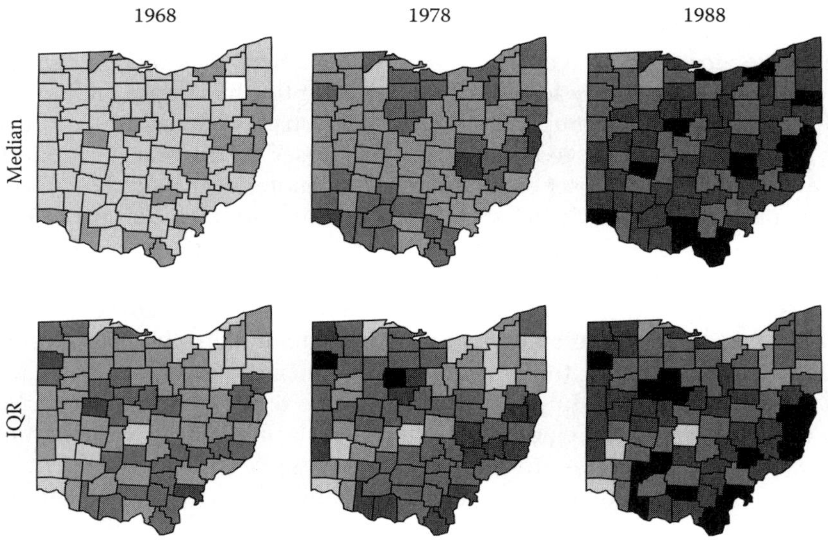

FIGURE 14.6
Posterior median and interquartile range (IQR) by county and year, nonwhite female lung cancer death rate per 1000 population (see text for grayscale key).

But their variability is also increasing somewhat, as we would expect given our Poisson likelihood. This variability is smallest for high-population counties, such as those containing the cities of Cleveland (northern border, third from the right), Toledo (northern border, third from the left), and Cincinnati (southwestern corner). Lung cancer rates are high in these industrialized areas, but there is also a pattern of generally increasing rates as we move from west to east across the state for a given year. One possible explanation for this is a lower level of smoking among persons living in the predominantly agricultural west, as compared to those in the more mining- and manufacturing-oriented east. Finally, we see increasing evidence of clustering among the high-rate counties, but with the higher rates increasing and the lower rates remaining low (i.e., increasing heterogeneity statewide). The higher rates tend to emerge in the poorer, more mountainous eastern counties, suggesting we might try adding a socioeconomic status fixed effect to the model.

Interested readers can find other variants of CAR-based spatial-temporal disease mapping models in the literature including those proposed by Bernardinelli, Clayton, Pascutto, Montomoli et al. (1995b), Knorr-Held and Besag (1998), and Knorr-Held (2000). Further extensions include spatial age–period–cohort models (Lagazio, Biggeri, and Dreassi 2003; Schmid and Held, 2004; Congdon, 2006), which incorporate temporal effects through time-varying risk and through birth-cohort–specific risks. In addition, MacNab and Dean (2001) extend the alternate formulation for convolution priors proposed by Leroux et al. (1999) (and described in Section 14.2.4) to the spatial-temporal setting through the addition of smoothing splines to model temporal and spatial-temporal trends in mortality rates.

14.4.3 Multivariate CAR (MCAR) Models

The methods illustrated so far apply to the modeling of regional counts of a single disease. However, it will often be the case that we have counts of *multiple* diseases over the same regional grid. This type of analysis has been examined in different ways (Held, Natario, Fenton, Rue et al. 2005; Knorr-Held and Best, 2001), but can be considered within a multivariate extension of the CAR models above. To adapt our notation to this case, suppose we let Y_{ij} be the observed number of cases of disease j in region i, $i = 1, \ldots, I$, $j = 1, \ldots, p$, and let E_{ij} be the expected number of cases for the same disease in this same region. As in Section 14.2, the Y_{ij} are thought of as random variables, while the E_{ij} are thought of as fixed and known. For the first level of the hierarchical model, conditional on the random effects u_{ij}, we assume the Y_{ij} are independent of each other such that

$$Y_{ij} \overset{ind}{\sim} \text{Poisson}(E_{ij} \exp(\mathbf{x}_{ij}'\boldsymbol{\beta}_j + u_{ij})), \;\; i = 1, \ldots, I, \;\; j = 1, \ldots, p, \tag{14.8}$$

where the $\mathbf{x}_{ij}$ are explanatory, region-level spatial covariates for disease j having (possibly region-specific) parameter coefficients $\boldsymbol{\beta}_j$.

Carlin and Banerjee (2003) and Gelfand and Vounatsou (2003) generalized the univariate CAR (14.2) to a joint model for the random effects u_{ij} under a *separability* assumption, which permits modeling of correlation among the p diseases while maintaining spatial dependence. Separability assumes that the association structure separates into a nonspatial and spatial component. More precisely, the joint distribution of $\mathbf{u}$ is assumed to be

$$\mathbf{u} \sim N_{np}(\mathbf{0}, \;\; [\Lambda \otimes (D - \alpha W)]^{-1}), \tag{14.9}$$

where $\mathbf{u} = (\mathbf{u}_1', \ldots, \mathbf{u}_p')'$, $\mathbf{u}_j = (u_{1j}, \ldots, u_{Ij})'$, Λ is a $p \times p$ positive definite matrix that is interpreted as the nonspatial precision (inverse dispersion) matrix between diseases, and $\otimes$ denotes the Kronecker product. Also, $\alpha \in [0, 1]$ is a *spatial autocorrelation* parameter that ensures the propriety of the joint distribution; $\alpha = 1$ returns us to the improper CAR case, while $\alpha = 0$ delivers an independence model. We denote the distribution in (14.9) by

$MCAR(\alpha, \Lambda)$. The improper $MCAR(1, \Lambda)$ model is sometimes referred to as a multivariate *intrinsic* autoregression, or *MIAR* model.

The $MCAR(\alpha, \Lambda)$ can be further generalized by allowing different smoothing parameters for each disease, i.e.,

$$\mathbf{u} \sim N_{np}(\mathbf{0}, [Diag(R_1, \ldots, R_p)(\Lambda \otimes I_{n\times n})Diag(R_1, \ldots, R_p)']^{-1}), \quad (14.10)$$

where $R_j R_j' = D - \alpha_j W$, $j = 1, \ldots, p$. The distribution in Equation (14.10) is sometimes denoted by $MCAR(\alpha_1, \ldots, \alpha_p, \Lambda)$. Note that the off-diagonal block matrices (the R_is) in the precision matrix in Equation (14.10) are completely determined by the diagonal blocks. Thus, the spatial precision matrices for each disease induce the cross-covariance structure in (14.10).

Jin, Carlin, and Banerjee (2005) developed a more flexible generalized multivariate CAR (GMCAR) model for the random effects $\mathbf{u}$. For example, in the bivariate case ($p = 2$), they specify the conditional distribution $\mathbf{u}_1|\mathbf{u}_2$ as $N((\eta_0 I + \eta_1 W)\mathbf{u}_2, [\tau_1(D - \alpha_1 W)]^{-1})$, and the marginal distribution of $\mathbf{u}_2$ as $N(\mathbf{0}, [\tau_2(D - \alpha_2 W)]^{-1})$, both of which are univariate CAR. This formulation yields the models of Kim, Sun, and Tsutakawa (2001) as a special case and recognizes explicit smoothing parameters (η_0 and η_1) for the cross-covariances, unlike the MCAR models in Equation (14.10) where the cross-covariances are not smoothed explicitly. However, it also requires the user to specify the order in which the variables (for us, diseases) are modeled, since different conditioning orders will result in different marginal distributions for $\mathbf{u}_1$ and $\mathbf{u}_2$ and, hence, different joint distributions for $\mathbf{u}$. This may be natural when one disease is a precursor to another, but in general may be an awkward limitation of the GMCAR. To overcome this, Jin, Banerjee, and Carlin (2007) developed an order-free MCAR that uses a linear model of coregionalization (Wackernagel, 2003; Gelfand et al., 2004) to develop richer spatial association models using linear transformations of much simpler spatial distributions. While computationally and conceptually more challenging, Jin et al. (2007) do illustrate the strengths of this approach over previous methods via simulation, and also offer a real-data application involving annual lung, larynx, and esophageal cancer death rates in Minnesota counties between 1990 and 2000. For more on MCAR models and underlying theory, the reader is referred to the textbooks by Banerjee et al. (2004, Sec. 7.4) and Rue and Held (2005).

Regarding computer package implementations of MCAR, `WinBUGS` offers an implementation of the MIAR case in a function called `mv.car`. While this is the only MCAR model that is built into the software itself, other more general MCAR models can be added fairly easily. For example, `WinBUGS` code to implement the GMCAR is available online at `www.biostat.umn.edu/~brad/software/GMCAR.txt`.

14.4.4 Recent Developments

The field of disease mapping and areal data modeling more generally continues to generate research interest, building on the basic models outlined and illustrated in the sections above. As elsewhere in statistics, many of these new developments have been motivated by special features of particular spatially referenced datasets. For instance, Reich, Hodges, and Carlin (2007) develop a *2NRCAR* model that can accommodate two different classes of neighbor relations, as would be needed if spatial similarity among regions that neighbor in an east–west direction is known to be different from that between north–south neighbors. The authors actually illustrate in a periodontal data setting, where many observations are taken on each tooth, and the first neighbor relation corresponds to measurements that neighbor as we go around the jaw, while the second corresponds to neighbors across each tooth (i.e., from the cheek to the tongue side).

Another important area of recent application is spatially varying coefficient models. Here the basic idea is to place a CAR (or other areal) prior on a collection of regional *regression coefficients* in a model, rather than simply on regional intercepts, thereby allowing the associations between outcomes and covariates to vary by location. So, for example, assuming a univariate covariate x_{ij} in (14.8), a CAR model would go directly onto the collection of region-specific coefficients $\boldsymbol{\beta}_j = (\beta_{1j}, \ldots, \beta_{Ij})'$ for each disease j. The spatial residuals u_{ij} might revert to an exchangeable formulation, or be deleted entirely. Such models require some care in implementation due to an increased potential for multicollinearity among the varying coefficients. However, the hierarchical approach provides a sounder, model-based inferential basis for statistical inference than more algorithmic competitors (Waller, Carlin, Xia, and Gelfand 2007; Wheeler and Calder, 2007).

14.5 Summary

The disease mapping models described and illustrated above provide a rich framework for the definition and application of hierarchical spatial models for areal data that simultaneously address our twin (but competing) goals of accurate small-area estimation and fine-scale geographic resolution. As noted, the models retain some rough edges in terms of scalability and generalization of underlying continuous phenomenon. Nevertheless, the CAR-based structure within a hierarchical generalized linear model offers a robust, flexible, and enormously popular class of models for the exploration and analysis of small area rates. Basic and even relatively advanced variations of such models are readily fit using commonly available GIS (e.g., `ArcGIS`) and Bayesian (e.g., `WinBUGS`) software tools. The coming years will no doubt bring further expansion of the hierarchial spatial modeling and software toolkits.

Appendix

`WinBUGS` code for fitting the convolution model to the Sasquatch report data.

```
model
{
  for (i in 1 : N) {
      O[i]  ~ dpois(mu[i])
      log(mu[i]) <- log(E[i]) + alpha0 + alpha1 * (X[i]-mean(X[])) +
                    c[i] + h[i]
      SMR[i] <- exp(alpha0 + alpha1 * (X[i]-mean(X[])) + c[i] + h[i])
      h[i] ~ dnorm(0,tau.h)
  }
  c[1:N] ~ car.normal(adj[], weights[], num[], tau.h)
  for(k in 1:sumNumNeigh) {
      weights[k] <- 1
  }

  alpha0  ~ d at()
  alpha1 ~ dnorm(0.0, 1.0E-5)
  tau.c  ~ dgamma(0.1, 0.1)
  tau.h ~ dgamma(0.01, 0.01)
  sigma.h <- sqrt(1/tau.h)
```

```
  sigma.c <- sqrt(1/tau.c)
  sd.c <- sd(c[])
  sd.h <- sd(h[])
  psi <- sd.c / (sd.c + sd.h)  # proportion excess variation that is spatial
}

DATA:

list(N = 75,
O = c(12, 13, 6, 1, 3, 5, 6, 8, 16, 0, 14, 9, 6, 10, 2,
      11, 16, 24, 1, 2, 7, 2, 6, 1, 1, 8, 18, 7, 23, 4,
      20, 6, 0, 18, 0, 14, 1, 2, 6, 3, 51, 8, 6, 0, 0,
      11, 13, 31, 2, 13, 33, 12, 42, 3, 1, 0, 0, 2, 14,
      0, 1, 16, 6, 10, 5, 6, 0, 2, 7, 1, 0, 2, 1, 12, 2),
E=c(21.66807,6.736003,4.18522,2.9842,5.243467,
    4.21199,12.16176,1.418131,5.080634,0.9444588,
    13.91193,6.236098,1.627794,4.329133,1.741247,
    15.56334,3.314697,4.508202,1.407866,0.2565611,
    2.3905,2.922542,29.87904,5.702312,0.4870904,
    2.688122,2.654441,0.7871273,4.602534,1.596194,
    22.70344,5.011663,2.733342,14.93348,3.310806,
    4.73323,0.7376802,1.645775,44.31355,9.55898,
    0.6623355,23.16282,1.285556,0.1284818,0.1297566,
    1.369422,4.469423,40.65957,4.800993,2.187411,
    116.5417,2.238335,47.01965,28.04050,0.6832684,
    1.102193,0.1608203,0.2726633,3.702155,1.378815,
    0.4848092,11.19194,6.909102,19.11018,6.91514,
    0.5323777,2.121124,0.5105056,4.278814,0.4979593,
    0.1037918,1.275358,1.286965,7.740242,1.123193),
X=c(4.247066,2.985682,4.428433,3.811097,4.745801,
    3.663562,4.169761,2.557227,3.830813,4.403054,
    5.640843,4.37827,3.091042,3.600048,2.660260,
    6.352978,3.927896,3.549617,3.109061,2.624669,
    3.777348,4.146304,6.419832,4.772378,1.163151,
    2.76001,2.00148,2.104134,3.339322,2.292535,
    5.193512,3.288402,2.928524,3.943522,3.663562,
    3.08191,1.686399,2.484907,7.257919,4.393214,
    1.774952,6.265301,2.312535,0.4700036,0.8329091,
    3.642836,3.104587,5.661223,5.82482,2.867899,
    6.676201,2.660260,6.025866,5.458308,1.481605,
    2.140066,1.193922,1.547563,3.749504,3.468856,
    0.8329091,4.345103,4.07244,5.47395,3.797734,
    0.5877867,1.163151,-0.3566749,2.341806,-0.1053605,
    -0.1053605,2.360854,1.856298,3.632309,1.686399),
num = c(6, 6, 6, 5, 5, 3, 3, 3, 4, 0,
        4, 5, 6, 1, 5, 6, 5, 5, 4, 5,
        5, 7, 8, 6, 3, 4, 7, 2, 8, 9,
        7, 9, 8, 8, 6, 7, 6, 4, 6, 8,
        8, 5, 8, 6, 4, 5, 8, 5, 4, 5,
        6, 6, 7, 4, 8, 5, 4, 6, 6, 3,
        7, 3, 5, 8, 6, 9, 3, 5, 5, 4,
        7, 7, 6, 7, 4),
adj = c(74, 69, 65, 5, 4, 2,
        69, 9, 8, 7, 6, 1,
        65, 64, 24, 13, 5, 4,
        24, 13, 5, 3, 1,
        65, 64, 4, 3, 1,
        9, 8, 2,
        69, 9, 2,
        9, 6, 2,
        8, 7, 6, 2,
```

```
53, 29, 18, 17,
42, 41, 29, 22, 20,
24, 23, 22, 21, 4, 3,
15,
49, 18, 17, 16, 14,
53, 51, 49, 48, 17, 15,
53, 18, 16, 15, 11,
29, 19, 17, 15, 11,
29, 21, 20, 18,
29, 22, 21, 19, 12,
23, 22, 20, 19, 13,
42, 39, 23, 21, 20, 13, 12,
64, 42, 39, 31, 24, 22, 21, 13,
64, 31, 23, 13, 4, 3,
55, 27, 26,
55, 54, 28, 25,
63, 62, 55, 50, 47, 32, 25,
54, 26,
53, 41, 34, 20, 19, 18, 12, 11,
72, 71, 64, 46, 45, 44, 43, 41, 31,
72, 64, 46, 39, 30, 24, 23,
56, 55, 52, 50, 47, 40, 35, 34, 27,
60, 59, 58, 57, 56, 55, 54, 35,
53, 52, 51, 43, 41, 40, 32, 29,
59, 58, 56, 40, 33, 32,
66, 61, 59, 58, 40, 38, 37,
71, 66, 44, 43, 40, 36,
75, 66, 61, 36,
46, 42, 41, 31, 23, 22,
59, 56, 43, 37, 36, 35, 34, 32,
46, 43, 42, 39, 34, 30, 29, 12,
41, 39, 23, 22, 12,
46, 45, 44, 41, 40, 37, 34, 30,
71, 66, 45, 43, 37, 30,
71, 44, 43, 30,
43, 41, 39, 31, 30,
63, 62, 52, 51, 50, 48, 32, 27,
63, 51, 49, 47, 16,
63, 48, 16, 15,
55, 52, 47, 32, 27,
53, 52, 48, 47, 34, 16,
53, 51, 50, 47, 34, 32,
52, 51, 34, 29, 17, 16, 11,
55, 33, 28, 26,
56, 54, 50, 33, 32, 27, 26, 25,
55, 40, 35, 33, 32,
61, 60, 58, 33,
61, 59, 57, 36, 35, 33,
61, 58, 40, 36, 35, 33,
61, 57, 33,
75, 60, 59, 58, 57, 38, 36,
63, 47, 27,
62, 49, 48, 47, 27,
72, 65, 31, 30, 24, 23, 5, 3,
74, 72, 64, 5, 3, 1,
75, 73, 71, 68, 67, 44, 38, 37, 36,
75, 68, 66,
74, 73, 70, 67, 66,
74, 70, 7, 2, 1,
74, 73, 69, 68,
73, 72, 66, 45, 44, 37, 30,
```

```
        74, 73, 71, 65, 64, 31, 30,
        74, 72, 71, 70, 68, 66,
        73, 72, 70, 69, 68, 65, 1,
        67, 66, 61, 38),
sumNumNeigh = 414)


INITIAL VALUES:

list(tau.c = 1, tau.h=1, alpha0 = 0, alpha1 = 0,
c=c(0,0,0,0,0,0,0,0,0,0,
       0,0,0,0,0,0,0,0,0,0,
       0,0,0,0,0,0,0,0,0,0,
       0,0,0,0,0,0,0,0,0,0,
       0,0,0,0,0,0,0,0,0,0,
       0,0,0,0,0,0,0,0,0,0,
       0,0,0,0,0,0,0,0,0,0,
       0,0,0,0,0),
h=c(0,0,0,0,0,0,0,0,0,0,
       0,0,0,0,0,0,0,0,0,0,
       0,0,0,0,0,0,0,0,0,0,
       0,0,0,0,0,0,0,0,0,0,
       0,0,0,0,0,0,0,0,0,0,
       0,0,0,0,0,0,0,0,0,0,
       0,0,0,0,0,0,0,0,0,0,
       0,0,0,0,0))
```

References

Agarwal, D.K. (2006). Two-fold spatial zero-inflated models for analysing isopod settlement patterns. In *Bayesian Statistics and its Applications*, S.K. Upadhyay, U. Singh and D.K. Dey, Eds., New Delhi: Anamaya Publishers.

Agarwal, D.K., Gelfand, A.E., and Citron-Pousty, S. (2002). Zero-inflated models with application to spatial count data. *Environmental and Ecological Statistics*, **9**, 341–355.

Arnold, B.C., Castillo, E., and Sarabia, J.M. (1999). *Conditional Specification of Statistical Models*. New York: Springer.

Banerjee, S. and Gelfand, A.E. (2006). Bayesian wombling: Curvilinear gradient assessment under spatial process models. *Journal of American Statistical Association*, **101**, 1487–1501.

Banerjee, S., Carlin, B.P. and Gelfand, A.E. (2004). *Hierarchical Modeling and Analysis for Spatial Data*. Boca Raton, FL: Chapman & Hall/CRC Press.

Bernardinelli, L., Clayton, D., and Montomoli, C. (1995a). Bayesian estimates of disease maps: How important are priors? *Statistics in Medicine*, **14**, 2411–2431.

Bernardinelli, L., Clayton, D., Pascutto, C., Montomoli, C., Ghislandi, M., and Songini, M. (1995b). Bayesian analysis of space-time variation in disease risk. *Statistics in Medicine*, **14**, 2433–2443.

Besag, J. (1974). Spatial interaction and the statistical analysis of lattice systems (with discussion). *Journal of the Royal Statistical Society, Ser. B*, **36**, 192–236.

Besag, J. and Kooperberg, C. (1995). On conditional and intrinsic autoregressions. *Biometrika*, **82**, 733–746.

Besag, J. and Mondal, D. (2005). First order intrinsic autoregressions and the de Wijs process. *Biometrika*, **92**, 909–920.

Besag, J., York, J.C., and Mollié, A. (1991). Bayesian image restoration, with two applications in spatial statistics (with discussion). *Annals of the Institute of Statistical Mathematics*, **43**, 1–59.

Besag, J., Green, P., Higdon, D., and Mengersen, K. (1995). Bayesian computation and stochastic systems (with discussion). *Statistical Science*, **10**, 3–66.

Best, N.G., Waller, L.A., Thomas, A., Conlon, E.M., and Arnold, R.A. (1999). Bayesian models for spatially correlated diseases and exposure data. In *Bayesian Statistics 6*, eds. J.M. Bernardo et al. Oxford: Oxford University Press, pp. 131–156.

Breslow, N.E. and Clayton, D.G. (1993). Approximate inference in generalized linear mixed models. *Journal of the American Statistical Association*, **88**, 9–25.

Carlin, B.P. and Banerjee, S. (2003). Hierarchical multivariate CAR models for spatio-temporally correlated survival data (with discussion). In *Bayesian Statistics 7*, eds. J.M. Bernardo, M.J. Bayarri, J.O. Berger, A.P. Dawid, D. Heckerman, A.F.M. Smith, and M. West. Oxford: Oxford University Press, pp. 45–63.

Carlin, B.P. and Louis, T.A. (2009). *Bayesian Methods for Data Analysis*, 3rd ed. Boca Raton, FL: Chapman & Hall/CRC Press.

Centers for Disease Control and Prevention, National Center for Health Statistics (1988). *Public Use Data Tape Documentation Compressed Mortality File*, 1968–1988. Hyattsville, MD: U.S. Department of Health and Human Services.

Clayton, D.G. and Kaldor, J.M. (1987). Empirical Bayes estimates of age-standardized relative risks for use in disease mapping. *Biometrics*, **43**, 671–681.

Congdon, P. (2006). A model framework for mortality and health data classified by age, area, and time. *Biometrics*, **62**, 269–278.

Cressie, N.A.C. (1993). *Statistics for Spatial Data, 2nd ed*. New York: John Wiley & Sons.

Denison, D.G.T. and Holmes, C.C. (2001). Bayesian partitioning for estimating disease risk. *Biometrics*, **57**, 143–149.

Devine, O.J. (1992). Empirical Bayes and constrained empirical Bayes methods for estimating incidence rates in spatially aligned areas. PhD diss., Department of Biostatistics, Emory University.

Devine, O.J., Louis, T.A., and Halloran, M.E. (1994). Empirical Bayes estimators for spatially correlated incidence rates. *Environmetrics*, **5**, 381–398.

Diggle, P.J. and Ribeiro, P.J. (2007). *Model-Based Geostatistics*. New York: Springer.

Diggle, P.J., Tawn, J.A., and Moyeed, R.A. (1998). Model based geostatistics (with discussion). *Applied Statistics*, **47**, 299–350.

Eberly, L.E. and Carlin, B.P. (2000). Identifiability and convergence issues for Markov chain Monte Carlo fitting of spatial models. *Statistics in Medicine*, **19**, 2279–2294.

Fay R.E. and Herriot, R.A. (1979). Estimates of income for small places: An application of James-Stein procedures to census data. *Journal of the American Statistical Association*, **74**, 269–277.

Gelfand, A.E. and Dalal, S.R. (1990). A note on overdispersed exponential families. *Biometrika*, **77**, 55–64.

Gelfand, A.E. and Sahu, S.K. (1999). Identifiability, improper priors, and Gibbs sampling for generalized linear models. *Journal of the American Statistical Association*, **94**, 247–253.

Gelfand, A.E. and Vounatsou, P. (2003). Proper multivariate conditional autoregressive models for spatial data analysis. *Biostatistics*, **4**, 11–25.

Gelfand, A.E., Schmidt, A.M., Banerjee, S., and Sirmans, C.F. (2004). Nonstationary multivariate process modelling through spatially varying coregionalization (with discussion). *Test*, **13**, 263–312.

Gelman, A. and Rubin, D.B. (1992). Inference from iterative simulation using multiple sequences (with discussion). *Statistical Science*, **7**, 457–511.

Ghosh, M. and Rao, J.K. (1994). Small area estimation: An appraisal (with discussion). *Statistical Science*, **9**, 55–93.

Ghosh, M., Natarajan, K., Stroud, T.W.F., and Carlin, B.P. (1998). Generalized linear models for small area estimation. *Journal of the American Statistical Association*, **93**, 273–282.

Ghosh, M., Natarajan, K., Waller, L.A., and Kim, D. (1999). Hierarchical GLMs for the analysis of spatial data: An application to disease mapping. *Journal of Statistical Planning and Inference*, **75**, 305–318.

Green, P.J. and Richardson, S. (2002). Hidden Markov models and disease mapping. *Journal of the American Statistical Association*, **97**, 1055–1070.

Held, L., Natario, I., Fenton, S.E., Rue, H., and Becker, N. (2005). Toward joint disease mapping. *Statistical Methods in Medical Research*, **14**, 61–82.

Hodges, J.S. (1998). Some algebra and geometry for hierarchical models, applied to diagnostics. *Journal of the Royal Statistical Society, Series B*, **60**, 497–536.

Hodges, J.S. and Sargent, D.J. (2001). Counting degrees of freedom in hierarchical and other richly parameterised models. *Biometrika*, **88**, 367–379.

Hodges, J.S., Carlin, B.P., and Fan, Q. (2003). On the precision of the conditionally autoregressive prior in spatial models. *Biometrics*, **59**, 317–322.

Jenks, G.F. (1977). *Optimal Data Classification for Choropleth Maps.* Occasional Paper No. 2. Lawrence, KS: Department of Geography, University of Kansas.

Jin, X., Carlin, B.P., and Banerjee, S. (2005). Generalized hierarchical multivariate CAR models for areal data. *Biometrics*, **61**, 950–961.

Jin, X., Banerjee, S., and Carlin, B.P. (2007). Order-free co-regionalized areal data models with application to multiple-disease mapping. *Journal of the Royal Statistics Society, Ser. B.*, **69**, 817–838.

Kim, H., Sun, D., and Tsutakawa, R.K. (2001). A bivariate Bayes method for improving the estimates of mortality rates with a twofold conditional autoregressive model. *Journal of the American Statistical Association*, **96**, 1506–1521.

Knorr-Held, L. (2000). Bayesian modelling of inseparable space-time variation in disease risk. *Statistics in Medicine*, **36**, 2555–2567.

Knorr-Held, L. (2002). Some remarks on Gaussian Markov random field models for disease mapping. In *Highly Structured Stochastic Systems*, eds., P. Green, N. Hjort, and S. Richardson, Oxford: Oxford University Press, pp. 260–264.

Knorr-Held, L. and Besag, J. (1998). Modelling risk from a disease in time and space. *Statistics in Medicine*, **17**, 2045–2060.

Knorr-Held, L. and Rasser, G. (2000). Bayesian detection of clusters and discontinuities in disease maps. *Biometrics*, **56**, 13–21.

Knorr-Held, L. and Best, N. (2001). A shared component model for detecting joint and selective clustering of two diseases. *Journal of the Royal Statistical Society, Series A*, **164**, 73–85.

Knorr-Held, L. and Rue, H. (2002). On block updating in Markov random field models for disease mapping. *Scandinavian Journal of Statistics*, **29**, 597–614.

Koch, T. (2005). *Cartographies of Disease: Maps, Mapping, and Medicine*, Redlands, CA: ESRI Press.

Lagazio, C., Biggeri, A., and Dreassi, E. (2003). Age-period-cohort models and disease mapping. *Environmetrics*, **14**, 475–490.

Lambert, D. (1992). Zero-inflated Poisson regression, with an application to random defects in manufacturing. *Technometrics*, **34**, 1–14.

Lawson, A.B. (2001). *Statistical Methods in Spatial Epidemiology*. New York: John Wiley & Sons.

Leroux, B.G., Lei X., and Breslow, N. (1999). Estimation of disease rates in small areas: A new mixed model for spatial dependence. In *Statistical Models in Epidemiology, the Environment and Clinical Trials*, eds., M.E. Halloran and D. Berry, New York: Springer-Verlag, pp. 179–192.

Lu, H., Hodges, J.S., and Carlin, B.P. (2007a). Measuring the complexity of generalized linear hierarchical models. *Canadian Journal of Statistics*, **35**, 69–87.

Lu, H., Reilly, C.S., Banerjee, S., and Carlin, B.P. (2007b). Bayesian areal wombling via adjacency modeling. *Environmental and Ecological Statistics*, **14**, 433–452.

Ma, H., Carlin, B.P., and Banerjee, S. (2009). Hierarchical and joint site-edge methods for Medicare hospice service region boundary analysis. Research report, Division of Biostatistics, University of Minnesota, Minneapolis.

MacEachren, A.M. (1995). *How Maps Work: Representation, Visualization, and Design.* New York: The Guilford Press.

MacNab, Y.C. and Dean, C.B. (2000). Parametric bootstrap and penalized quasi-likelihood inference in conditional autoregressive models. *Statistics in Medicine*, **19**, 2421–2435.

MacNab, Y.C. and Dean, C.B. (2001). Autoregressive spatial smoothing and temporal spline smoothing for mapping rates. *Biometrics*, **57**, 949–956.

Marshall, R.J. (1991). Mapping disease and mortality rates using empirical Bayes estimators. *Applied Statistics*, **40**, 283–294.

Matheron, G. (1963). Principles of geostatistics. *Economic Geology*, **58**, 1246–1266.

McCullagh, P. and Nelder, J.A. (1989). *Generalized Linear Models*, 2nd ed. Boca Raton, FL: Chapman & Hall/CRC Press.

Mollié, A. (1996). Bayesian mapping of disease. In *Markov Chain Monte Carlo in Practice*, Gilks, W., Richardson, S., and Spiegelhalter, D., Eds., Boca Raton, FL: Chapman & Hall/CRC Press.

Raghunathan, T.E., Xie, D., Schenker, N., Parsons, V.L., Davis, W.W., Dodd, K.W., and Feuer, E.J. (2007). Combining information from two surveys to estimate county-level prevalence rates of cancer risk factors and screening, *Journal of American Statistical Association*, **102**, 474–486.

Rao, J.N.K. (2003). *Small Area Estimation*, New York: John Wiley & Sons.

Reich, B.J., Hodges, J.S., and Carlin, B.P. (2007). Spatial analyses of periodontal data using conditionally autoregressive priors having two classes of neighbor relations. *Journal of the American Statistical Association*, **102**, 44–55.

Richardson, S., Thomson, A., Best, N., and Elliott, P. (2004). Interpreting posterior relative risk estimates in disease-mapping studies. *Environmental Health Perspectives*, **112**, 1016–1025.

Rue, H. and Held, L. (2005). *Gaussian Markov Random Fields: Theory and Applications*. Boca Raton, FL: Chapman & Hall/CRC Press.

Schlaible, W.L. (1996). *Indirect Estimators in U.S. Federal Programs*, New York: Springer.

Schmid, V. and Held, L. (2004). Bayesian extrapolation of space-time trends in cancer registry data. *Biometrics*, **60**, 1034–1042.

Spiegelhalter, D.J., Best, N., Carlin, B.P., and van der Linde, A. (2002). Bayesian measures of model complexity and fit (with discussion). *Journal of the Royal Statistical Society, Ser. B*, **64**, 583–639.

Sun, D., Tsutakawa, R.K., and Speckman, P.L. (1999). Posterior distribution of hierarchical models using CAR(1) distributions. *Biometrika*, **86**, 341–350.

Wackernagel, H. (2003). *Multivariate Geostatistics: An Introduction with Applications*, 3rd ed. New York: Springer-Verlag.

Wakefield, J. (2001). A critique of ecological studies. *Biostatistics*, **1**, 1–20.

Wakefield, J. (2003). Sensitivity analyses for ecological regression. *Biometrics*, **59**, 9–17.

Wakefield, J. (2004). Ecological inference for 2×2 tables (with discussion). *Journal of the Royal Statistics Society, Ser. A*, **167**, 385–445.

Wakefield, J.C., Best, N.G., and Waller, L. (2000). Bayesian approaches to disease mapping. In *Spatial Epidemiology: Methods and Applications*, Elliott, P., et al. Eds. Oxford: Oxford University Press. pp. 104–127.

Waller, L.A. (2005). Bayesian thinking in spatial statistics. In *Handbook of Statistics, Volume 25: Bayesian Thinking: Modeling and Computation*, Dey, D.K. and Rao, C.R., Eds. Amsterdam: Elsevier. pp. 589–622.

Waller, L.A. and Gotway, C.A. (2004). *Applied Spatial Statistics for Public Health Data*. New York: John Wiley & Sons.

Waller, L.A., Carlin, B.P., Xia, H., and Gelfand, A.E. (1997). Hierarchical spatio-temporal mapping of disease rates. *Journal of the American Statistical Association*, **92**, 607–617.

Waller, L.A., Zhu, L., Gotway, C.A., Gorman, D.M., and Gruenewald, P.J. (2007). Quantifying geographic variations in associations between alcohol distribution and violence: A comparison of geographically weighted regression and spatially varying coefficient models. *Stochastic Environmental Research and Risk Assessment*, **21**, 573–588.

Wheeler, D.C. and Calder, C.A. (2007). An assessment of coefficient accuracy in linear regression models with spatially varying coefficients. *Journal of Geographical Systems*, **9**, 145–166.

Xia, H. and Carlin, B.P. (1998). Spatio-temporal models with errors in covariates: Mapping Ohio lung cancer mortality. *Statistics in Medicine*, **17**, 2025–2043.

15

Spatial Econometrics

R. Kelley Pace and James LeSage

CONTENTS

15.1 Introduction

A long-running theme in economics is how individuals or organizations following their own interests result in benefits or costs to others (Brueckner, 2003, López-Bazo, Vayá, and Arts 2004, Ertur and Koch, 2007). These benefits or costs are labeled externalities and often termed spillovers in a spatial setting. A technological innovation provides an example of a positive externality or spillover while pollution provides an example of a negative externality or spillover. Although both innovation and pollution have global impacts, these often result in more geographically concentrated impacts (e.g., the computer industry in Silicon Valley, the Exxon Valdez oil spill in Alaska).

Spatial econometric models can quantify how changes in explanatory variables of the model directly impact individuals, regions, local governments, etc., as well as the associated spillover impacts. Quantifying these effects provides useful information for policy purposes. For example, decisions of local government officials focus on direct or own-region benefits associated with increased spending on infrastructure, whereas national government officials focus on the total benefits that include direct and indirect or spillover benefits to neighboring regions. Specifically, if state universities provide small, direct benefits, but large spillovers that accrue to other states, this could lead legislatures to under invest in higher education.

Also, other economic forces can lead to spatial interdependence among observations. For example, competitive forces compel individuals and organizations to react to others, and again these interactions are often local. Changes in relative prices or wages lead individuals or organizations to change behavior, and part of the price includes transportation costs, which depend on distance and spatial connectivity. In summary, economic forces lead to interdependence among entities, data from these entities is generated at a particular time and place and, therefore, it seems reasonable to examine these data using tools that can account for spatial and spatial-temporal dependence.

15.2 Common Models

Although a wide variety of models have been introduced in spatial econometrics, most have roots in a small number of basic models. For static models, the basic models are the conditional autoregression (CAR), the simultaneous autoregression (SAR), the spatial autoregressive model (SAM), and the spatial Durbin model (SDM). Many dynamic models stem from some form of spatial temporal autoregession (STAR).

15.2.1 Static Models

Given a set of data collected over space at a given time (or over a period of time), a static regression model relates the n observations on the dependent variable y to the n observations on the k explanatory variables X whose importance is governed by the k by 1 parameter vector β with error captured by the n disturbances ε, as shown by the data generating process (DGP) in Equation (15.1). In addition, let $X = [\iota_n \;\; Z]$ where ι_n is an n by 1 constant vector and Z contains p nonconstant explanatory variables. We assume that ε is multivariate normal with a zero mean and covariance matrix Ω $(N(0, \sigma^2\Omega))$.

$$y = X\beta + \varepsilon \tag{15.1}$$

$$E(y) = X\beta \tag{15.2}$$

Since the observations occur over space, in many cases it seems implausible that the disturbances are independent. For example, assessments for property taxes often come from a regression model. Even a model containing numerous property characteristics may still yield spatial clusters of underassessed houses (positive residuals) and overassessed houses (negative residuals). Individual house values may depend on unmeasured or difficult-to-collect variables, such as the landscaping in neighboring yards, the quality of views, access to employment and shopping (since all houses on the same block share the same access), patterns of crime, and nearby greenspace. If the omitted variables have a spatial nature, but are not correlated with the explanatory variables, this suggests using Equation (15.1) with some means of specifying the spatial dependence across the disturbances arising from omitted variables.

One way of specifying interdependence among observations is to use a spatial weight matrix. The spatial weight matrix W is an $n \times n$ exogenous nonnegative matrix where $W_{ij} > 0$ indicates that observation i depends upon neighboring observations j. For example, $W_{ij} > 0$ if region i is contiguous to region j. Also, $W_{ii} = 0$, so observations cannot be neighbors to themselves. Measures of proximity, such as cardinal distance (e.g., kilometers), ordinal distance (e.g., the m nearest neighbors), and contiguity ($W_{ij} > 0$, if region i shares a border with region j), have been used to specify W.

Two models for error covariance have been commonly employed. The conditional autoregression (CAR) specifies $\Omega = (I_n - \phi W)^{-1}$ while the simultaneous autoregression (SAR) specifies $\Omega = [(I_n - \rho W)'(I_n - \rho W)]^{-1}$ where ϕ and θ are scalar parameters and W is an $n \times n$

spatial weight matrix. For CAR, the spatial weight matrix W is symmetric, but SAR can employ nonsymmetric W.*

The simultaneous approach to modeling was first devised by Whittle (1954), and Brook (1964) made the distinction between conditional and simultaneous approaches, while Besag (1974) derived elegant conditional models. Ord (1975) provided a means for estimating simultaneous autoregressions. In addition Ripley (1981), Cressie (1993), Banerjee, Carlin, and Gelfand (2004) derive and discuss the similarities and differences between the simultaneous and conditional approaches. In particular, Banerjee et al. (2004, pp. 79–87) set forth a coherent approach to derive CAR and SAR in a multivariate normal regression context. The conditional approach implies a distribution of y that after a spatial transformation yields the CAR covariance specification. However, the simultaneous approach begins with iid disturbances that after a spatial transformation leads to the SAR covariance specification. Empirically, the SAR approach has been used more often than the CAR approach in spatial econometrics (Anselin, 2003). See Chapter 13 for a full development of CAR modeling.

For simplicity, we restrict our discussion to cases where the matrix W has real eigenvalues. The spatial weight matrix may be symmetric or nonsymmetric and is often scaled to have a maximum eigenvalue of 1.† This can be achieved by making W row stochastic or doubly stochastic (for symmetric W). Alternatively, scaling some valid candidate weight matrix by its principal eigenvalue produces a W with a principal eigenvalue of 1. In this case, the CAR or SAR variance-covariance matrices will be symmetric positive definite if the parameter lies in the open interval defined by the reciprocal of the minimum eigenvalue of W and 1.

Contiguity and ordinal distance can be used to specify sparse matrices W where the number of nonzero elements increases linearly with n, but the number of elements in W increases with the square of n. This has computational implications that we will discuss in Section 15.7. Moreover, the number of nonzero entries in each row of W constructed from contiguity or ordinal distance does not depend upon n, which can potentially simplify asymptotic analysis. Finally, sparse W leads to sparse precision matrices. In the case of CAR, the precision matrix $\Psi = I_n - \phi W$ while the SAR precision matrix is $\Psi = (I_n - \rho W)'(I_n - \rho W)$. Zeros in the ij element ($j \neq i$) of a precision matrix indicate conditional independence between observations i and j. This has interpretive advantages (especially for CAR).

In the restrictive context just set forth, CAR and SAR are alternative covariance specifications. When consistently estimated, both specifications should yield identical estimates of the regression parameter vector β for large samples under misspecification of the disturbances (but correct specification of X and orthogonality between X and the disturbances). For example, with a sufficiently large sample and a CAR DGP, use of a consistent SAR estimation method will yield the same estimates for β as doing the converse, that is, using a consistent CAR estimation method with sample data generated by a SAR process. However, the standard errors will not be consistently estimated under misspecification of the disturbances. In fact, ordinary least square (OLS) will also consistently estimate the regression parameters β under these conditions, but will inconsistently estimate the associated standard errors under misspecification of the disturbances. Pace and LeSage (2008) used this to devise a Hausman test for misspecification of spatial models. In the literature, there are many examples of spatial error model estimates that seem similar to the corresponding OLS estimates and other examples where OLS and the spatial error model estimates materially

* More general versions of these specifications involve an $n \times n$ diagonal matrix in specifying the covariance. For simplicity, we assume this equals I_n. This presentation of CAR and SAR minimizes the differences between the two approaches, and this is easier to do in a regression context with multinormal disturbances than in many other situations.

† Nonsymmetric matrices that are similar to symmetric matrices have real eigenvalues. See Ord (1975, p. 125) for a discussion.

differ. Such differences point to some form of misspecification, such as the incorrect model or omitted variables correlated with the explanatory variables. Since adding some form of spatial information led to the divergence between the estimates, this suggests examining a broader range of spatial specifications.

Another standard alternative to the error models is SAM in Equation (15.3) where ε is $N(0, \sigma^2 I_n)$. The associated DGP appears in (15.4) and the expectation appears in Equation (15.5) (Ord, 1975; Anselin, 1988).

$$y = \rho W y + X\beta + \varepsilon \tag{15.3}$$

$$y = (I_n - \rho W)^{-1} X\beta + (I_n - \rho W)^{-1}\varepsilon \tag{15.4}$$

$$E(y) = (I_n - \rho W)^{-1} X\beta \tag{15.5}$$

Note, the expansion of $(I_n - \rho W)^{-1}$ has an interesting interpretation in terms of the powers of W as shown in Equation (15.6).

$$(I_n - \rho W)^{-1} = I_n + \rho W + \rho^2 W^2 + \rho^3 W^3 + \cdots \tag{15.6}$$

$$E(y) = X\beta + WX(\rho\beta) + W^2 X(\rho^2\beta) + \cdots \tag{15.7}$$

Just as W specifies the neighbors to each observation, W^2 specifies the neighbors of the neighbors to each observation, and this is true for any power of W. Even though W may be sparse (containing mostly zeros), W raised to a sufficiently large power will be dense if W is fully connected (a path exists between any region i and all other regions). In practice, regions that are far apart will likely have small effects on one another and nearby regions will have larger effects on one another. However, effects do not monotonically decline with distance as feedback loops exist among observations. If observations i and j are both neighbors to each other, then observation i, j are second-order neighbors to themselves, and similar loops exist for higher-order neighboring relations. The complex nature of network relations is either a feature or a disadvantage depending upon one's perspective (Bavaud, 1998, Wall, 2004, Martellosio, 2007).

To make this clearer, examine Figure 15.1. This figure gives four plots of different order neighboring relations contained in W, W^2, W^3, and W^{10}. In other words, this figure depicts first-order neighbors, second-order neighbors, third-order neighbors, and tenth-order neighbors. In each plot the points represent nodes or observations and the line segments represent edges or dependence relations between observations. The plots show the various-order neighboring relations for 20 observations. The vertical axis represents north–south and the horizontal axis represents east–west.

The upper left plot shows the connections when using three nearest neighbors for each observation (W). The upper right plot shows the connections for the neighbors of the neighbors, the bottom left plot shows the connections for the neighbors of the neighbors of the neighbors, and finally the lower right plot shows the tenth-order neighboring relations. Naturally, as the relations become higher order, the number of possible connections among observations increase so that the plots (just as the associated matrices) become denser and the average connection becomes weaker.

Returning to Equation (15.7), $E(y)$ for SAM does not just involve the own explanatory variables X as with CAR, SAR, and OLS, but also involves neighboring values of the explanatory variables WX, W^2X, and so on as shown in (15.7). In the SAM, decisions made at a single location may affect all other locations. This provides an integrated way of modeling externalities in contrast to the error specifications where externalities must be modeled in X.*

* Nothing precludes using terms, such as $X = [\iota_n (I_n - \theta W)^{-1} Z]$, in an error model. This would lead to two spatial parameters, which could improve model performance. See Chapter 7 of LeSage and Pace (2009) for a further discussion of models with separate specification of simultaneous spillovers and simultaneous disturbances.

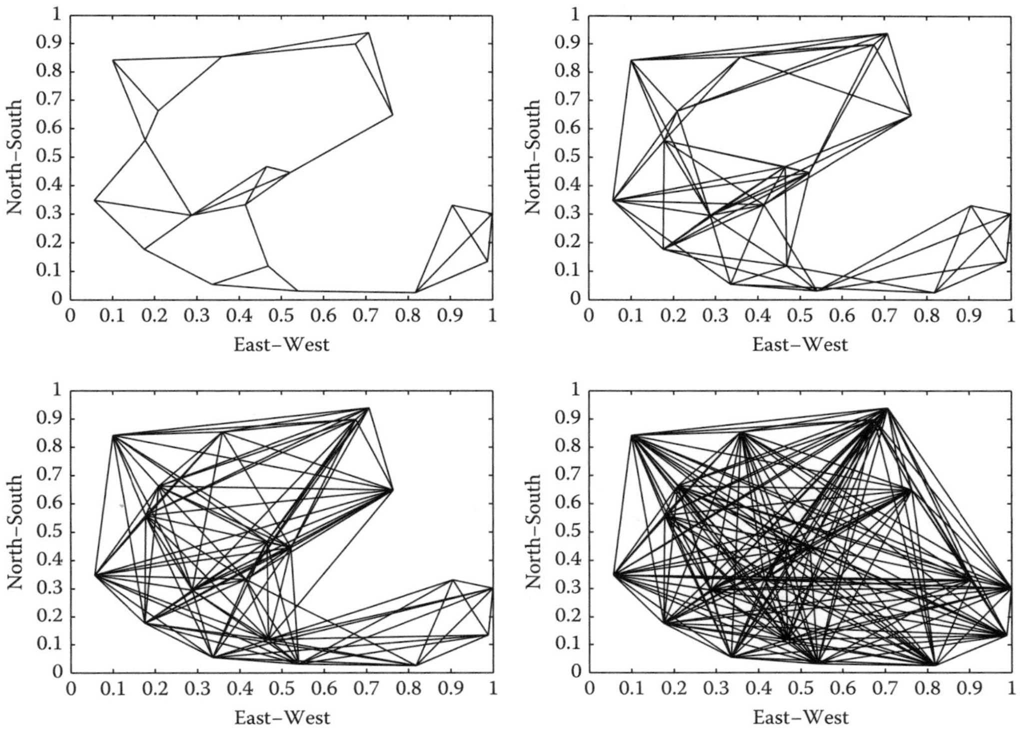

FIGURE 15.1
Graphs of various order neighbors.

The spatial Durbin model (SDM) is shown in Equation (15.8) where ε is $N(0, \sigma^2 I_n)$, and the associated DGP and expectation appear in Equation (15.9) and Equation (15.10), respectively.* The SDM provides another general alternative (Ord, 1975; Anselin, 1988) that nests the SAR error model, the SAM, and OLS. LeSage and Pace (2009) motivate the SDM through omitted variable arguments. Just as spatial error models arise in the presence of omitted variables that are independent of the explanatory variables, the SDM arises in the presence of omitted variables that follow a spatial process and are not independent of the explanatory variables. In addition, LeSage and Pace (2009) show that the SDM emerges when averaging across SAR, SAM, and OLS, so model uncertainty provides another motivation for the SDM.

$$y = \rho W y + Z\beta_1 + WZ\beta_2 + \iota_n\kappa + \varepsilon \tag{15.8}$$

$$y = (I_n - \rho W)^{-1}(Z\beta_1 + WZ\beta_2 + \iota_n\kappa) + (I_n - \rho W)^{-1}\varepsilon \tag{15.9}$$

$$E(y) = (I_n - \rho W)^{-1}(Z\beta_1 + WZ\beta_2 + \iota_n\kappa) \tag{15.10}$$

The SDM introduces an even richer means to specify externalities relative to the spatial autoregressive model. The SDM provides a convenient way of examining many of the previously introduced models since it nests SAR, SAM, and OLS. Specifically, it nests SAR when $\beta_2 = -\rho\beta_1$, SAM when $\beta_2 = 0$, and OLS when $\rho = 0$ and $\beta_2 = 0$.

* Since $W\iota_n = \iota_n$, we do not include both X and WX in the model to maintain full rank.

15.2.2 Dynamic Models

Often data have both spatial and temporal identification as well as enough variation in both space and time to call for a spatial-temporal approach. The different nature of the nondirectional, two dimensions of space versus the directional, single dimension of time increases the difficulty of the task. However, various implementations of the STAR have appeared that provide a straightforward way of approaching the problem (Pfeifer and Deutsch, 1980; Cressie, 1993; Pace, Barry, Gilley, and Sirmans 2000; Elhorst, 2001).

Both spatial and temporal models often use some form of quasi-differencing or filtering, such as $y_t - \rho W y_t$ in the spatial autoregressive model where y is now an nT by 1 vector composed of vectors of length n from each of T periods on the same n locations and t is the time period ($T \geq t > 1$). Therefore, $y_t, y_{t-1}, \ldots$ are n by 1 vectors. Similarly, the temporal autoregressive model employs quasi-differences, such as $y_t - \tau y_{t-1}$, where τ is a scalar temporal parameter. In terms of transforming y, it is not clear whether to filter y for time first and then space or space first and then time.

Before making a decision on the order of operations, consider Equation (15.11), which takes the temporal quasi-difference of the spatial quasi-difference and contrast this with Equation (15.12), which takes the spatial quasi-difference of the temporal quasi-difference.

$$[y_t - \rho W y_t] - \tau[y_{t-1} - \rho W y_{t-1}] = y_t - \rho W y_t - \tau y_{t-1} + \tau\rho W y_{t-1} \quad (15.11)$$

$$[y_t - \tau y_{t-1}] - \rho W[y_t - \tau y_{t-1}] = y_t - \rho W y_t - \tau y_{t-1} + \rho\tau W y_{t-1} \quad (15.12)$$

For this case where locations stay fixed over time and are observed at discrete intervals, the order of quasi-differencing does not matter. However, for data that are irregularly distributed over time and space, the temporal lag of a spatial lag typically differs from the spatial lag of a temporal lag, and both terms would enter the model (Pace et al., 2000).

Given the filtering of y, a similar transformation for the nonconstant explanatory variables in Z defines a spatial-temporal autoregression in Equation (15.13) that includes all lag terms associated with both the dependent variable and explanatory variables, where β_1 to β_4 are p by 1 parameter vectors, λ is a scalar parameter, and ε is $N(0, \sigma^2 I_n)$.

$$y_t = \rho W y_t + \tau y_{t-1} + \lambda W y_{t-1} + \iota_n \kappa + Z_t \beta_1 + W Z_t \beta_2 + Z_{t-1}\beta_3 + W Z_{t-1}\beta_4 + \varepsilon \quad (15.13)$$

To make the STAR model more concrete, consider influences on housing prices. Given the mantra of "location, location, location," the prices of houses in the neighborhood around each house are important ($W y_t$). In addition, recent market conditions (y_{t-1}) for that location and competing locations ($W y_{t-1}$) have a great influence on the price today (y_t). Of course, prices in the absence of characteristics are not too informative. Therefore, controlling for the characteristics of the house, its neighbors, and their recent values greatly aids in house price valuation.

Returning to Figure 15.1, moving to the past also increases the order of the neighboring relations. The previous period may only involve three nearest neighbors as in the upper left plot ($W y_{t-1}$). However, two periods ago may involve neighbors of the neighbors as shown in the upper right plot ($W^2 y_{t-2}$), and almost all the observations may have an effect when going back many periods, as depicted by the bottom right plot ($W^{10} y_{t-10}$).

One can envision these plots stacked with the highest-order neighboring relations on the bottom and the first-order neighboring relations on the top. This set of stacked plots helps visualize the diffusion of past influences since time allows higher-order connections to arise with an extended reach. However, we note that on average the magnitude of influence associated with a typical relation is reduced for higher-order connections.

15.3 Relations between Spatial and Spatiotemporal Models

In this section, we begin with a variant of the STAR model that uses only past data and the restrictive assumption that explanatory variables do not change over time to: (1) motivate how simultaneous spatial dependence arises in cross-sectional data, (2) provide a way to interpret simultaneous spatial models, and (3) to show the relation between spatial and spatial-temporal models.

We show how the expectation of the dependent variable vector from a simple spatial-temporal specification that conditions on past data can be viewed as a steady-state equilibrium that displays simultaneous spatial dependence (spillovers). Note, we only use previous data in each period for the various lags because we wish to explain cross-sectional simultaneous dependence as arising over time. Using contemporaneous Wy_t would be effectively assuming what we are trying to prove. Specifically, we wish to show that the expected value of the last n observations of the dependent variable from the associated DGP converges over time to a steady-state equilibrium, which provides a motivation for simultaneous spatial dependence among cross-sectional elements of $E(y)$ (spillovers).

We begin with the model in (15.14), where y_t is an $n \times 1$ dependent variable vector at time t ($t = 1, \ldots, T$), and X represents n observations on k explanatory variables that remain fixed over time. This is a form of STAR that relies on past data and contains no simultaneous spatial interaction where the locations of the data remain the same in each time period and observations occur with a constant frequency.

$$y_t = Gy_{t-1} + X\beta + \varepsilon_t \tag{15.14}$$

$$G = \tau I_n + \rho W \tag{15.15}$$

The scalar parameter τ governs dependence between each region i at time t and $t-1$, while the scalar parameter ρ reflects spatial dependence between region i at time t and neighboring regions $j \neq i$ at time $t-1$. The disturbance vector ε_t is distributed $N(0_n, \sigma^2 I_n)$. We assume t is sufficiently large so that $G^t \approx 0_n$. This convergence condition indirectly restricts the parameters ρ and τ (given W) to ensure stationarity. To make this more direct, assume $\tau \geq 0$, $\rho \geq 0$, and W has a principal eigenvalue of 1. In this case, stationarity requires that $\tau + \rho < 1$.

As with Elhorst (2001), we use the recursive relation $y_{t-1} = Gy_{t-2} + X\beta + \varepsilon_{t-1}$ implied by the model in Equation (15.14) to consider the state of the dynamic system after passage of t time periods, as shown in Equation (15.16).

$$\begin{aligned} y_t &= G^t y_0 + \left(I_n + G + G^2 + \cdots + G^{t-1}\right) X\beta + u \\ u &= G^{t-1}\varepsilon_1 + \cdots + G\varepsilon_{t-1} + \varepsilon_t \end{aligned} \tag{15.16}$$

The steady-state equilibrium for the dynamic process in (15.16) can be found by examining large t and taking the expectation. Using the stationarity assumption $G^t \approx 0_n$ and $(I - G)^{-1} = (I + G + G^2 + \ldots)$ in conjunction with a zero expectation for the disturbance terms in each period leads to Equation (15.17).

$$E(y_t) \approx (I_n - G)^{-1} X\beta \tag{15.17}$$

$$E(y_t) \approx (I_n(1-\tau) - \rho W)^{-1} X\beta \tag{15.18}$$

$$E(y_t) \approx \left(I_n - \frac{\rho}{1-\tau} W\right)^{-1} X \frac{\beta}{1-\tau} \tag{15.19}$$

There is a relation between the expression in Equation (15.19) and a cross-sectional spatial regression based on a set of time t cross-sectional observations shown in Equation (15.20), with the associated expectation shown in Equation (15.21) (a reparameterization of the SAM model shown in Equation (15.5)).

$$y_t = \rho^* W y_t + X\beta^* + \varepsilon_t \tag{15.20}$$

$$E(y_t) = (I - \rho^* W)^{-1} X\beta^* \tag{15.21}$$

The underlying parameters in Equation (15.19) and Equation (15.21) are related as shown in Equation (15.22) and Equation (15.23).

$$\rho^* = \frac{\rho}{1-\tau} \tag{15.22}$$

$$\beta^* = \frac{\beta}{1-\tau} \tag{15.23}$$

The term $(1-\tau)^{-1}$ from Equation (15.22) is the long-run multiplier from the time series literature. This relation between ρ^*, ρ, and τ shows that positive time dependence ($\tau > 0$) will result in a larger estimate of spatial dependence when the simultaneous cross-sectional specification is used instead of the spatial-temporal specification. Weak spatial dependence in conjunction with strong temporal dependence leads to a long-run, steady-state equilibrium where variables can show strong spatial dependence. Intuitively, high temporal dependence allows a small degree of spatial dependence per period to evolve into a final equilibrium that exhibits high levels of spatial dependence. This has implications for STAR model applications since a large temporal dependence estimate implies a small spatial parameter estimate (given stationarity). Such a small spatial parameter estimate might suggest ignoring the spatial term in the model, but this would have major implications for the steady-state equilibrium.

This same situation arises for the parameters of the explanatory variables β^* from the single cross section, which are also inflated by the long-run multiplier when $\tau > 0$ relative to β from the spatial-temporal model. This is a well-known result from time series analysis of long-run multiplier impacts arising from autoregressive models.

The relations in Equation (15.20) to Equation (15.23) facilitate interpretation of cross-sectional spatial autoregressive models. Since these models have no explicit role for time, they should be interpreted in terms of an equilibrium or steady-state outcome. This also has implications for calculating the impact of changes in the explanatory variables of these models. The model in Equation (15.20) and Equation (15.21) states that y_i and y_j for $i \neq j$ simultaneously affect each other. However, viewing changes in X_i as setting in motion a series of changes in y_i and y_j that will lead to a new steady-state equilibrium involving spillovers at some unknown future time has more intuitive appeal.

In this analysis, we do not examine the equilibrium distribution of the disturbances. However, LeSage and Pace (2009) examined this in more detail. If the disturbances are location specific and time persistent, this can lead to simultaneous dependence among the disturbances. In contrast, independent disturbances over time in conjunction with large equilibrium amounts of spatial dependence can lead to CAR disturbances.

Also, we did not analyze the full STAR model with both simultaneous and temporal dependence, which is frequently used in applied practice (Cressie, 1993, pp. 449–450). For example, variants of this model have been examined in the panel data model literature (Elhorst, 2001). We briefly turn attention to this model. As before, we assume periodic data from spatial locations that are fixed over time. We introduce the scalar parameter λ to capture the simultaneous spatial component and this leads to the STAR model in Equation (15.24) with a simplified version in Equation (15.25).

$$y_t = \lambda W y_t + \rho W y_{t-1} + \tau y_{t-1} + X\beta + \varepsilon_t \tag{15.24}$$

$$y_t = \lambda W y_t + G y_{t-1} + X\beta + \varepsilon_t \tag{15.25}$$

We rewrite the model in Equation (15.25) in terms of the matrix A as specified in Equation (15.27). Using the recursive technique that conditions on the past data (but treats A simultaneously) yields Equation (15.28) and Equation (15.29).

$$y_t = AGy_{t-1} + AX\beta + A\varepsilon_t \tag{15.26}$$
$$A = (I_n - \lambda W)^{-1} \tag{15.27}$$
$$y_t = (AG)^t y_0 + (I_n + AG + (AG)^2 + \cdots + (AG)^{t-1})AX\beta + u \tag{15.28}$$
$$u = (AG)^{t-1}\varepsilon_1 + \cdots + AG\varepsilon_{t-1} + \varepsilon_t \tag{15.29}$$

Taking expectations and rearranging Equation (15.28) gives Equation (15.30). Separating out A in Equation (15.31) shows that this term partially drops out in Equation (15.32). Finally, Equation (15.33) shows that the long-run equilibrium of the STAR model with simultaneous spatial dependence looks very similar to the STAR model that relied only on past period data.

$$E(y_t) \approx (I_n - AG)^{-1}AX\beta \tag{15.30}$$
$$E(y_t) \approx [A(A^{-1} - G)]^{-1}AX\beta \tag{15.31}$$
$$E(y_t) \approx [(A^{-1} - G)]^{-1}X\beta \tag{15.32}$$
$$E(y_t) \approx [I_n(1 - \tau) - (\lambda + \rho)W]^{-1}X\beta \tag{15.33}$$

This is a simplification afforded by constant X. For constant X with fixed locations occurring at constant frequency over time, terms such as temporal lag of the spatial lag of X or the spatial lag of the temporal lag of X both equal WX. Therefore, the resulting equilibrium still has the form $(I_n - \delta W)^{-1}X\beta$ where δ now involves a different parameterization. However, we note that in a situation with X changing each period, the STAR model that conditions solely on past data will likely differ from a STAR model that uses both past and current data.

15.4 Interpretation

Many spatial statistical applications use error models as in Equation (15.1) with expectation Equation (15.2), and we note that all error models (including OLS) have the same expectation. For sufficiently large sample sizes and the correct specification for X with no correlation between disturbances and X, applying consistent estimators to the various models should yield identical estimates for the parameters β. For small samples, estimates could vary, and use of models that differ from the true DGP could lead to inconsistent estimates of dispersion for the model parameters. Interpretation of this type of error model is straightforward because:

$$\frac{\partial y_i}{\partial x_i^{(v)}} = \beta^{(v)} \tag{15.34}$$

where i and j represent indices of the n observations and v represents an index to one of the k explanatory variables ($i, j = 1, \ldots, n$ and $v = 1, \ldots, k$).

In fact, error models may produce an overly simple interpretation for spatial econometric applications since spatial spillover effects are absent in these models by construction (unless spatial terms appear in X). Spillover effects could be introduced with a spatially complex specification for the variables X, but there are other alternatives. For example, the SDM model subsumes error models as a special case. We can express the DGP for the SDM using Equation (15.36), which allows us to examine the impact of changing elements of $x^{(v)}$

on y where $\beta_1^{(v)}$ and $\beta_2^{(v)}$ are the scalar parameters associated with the variable v and its spatial lag.

$$y = \sum_{v=1}^{k} S^{(v)} x^{(v)} + \varepsilon \tag{15.35}$$

$$S^{(v)} = (I_n - \rho W)^{-1} \left(I_n \beta_1^{(v)} + W \beta_2^{(v)}\right) \tag{15.36}$$

The partial derivatives from the SDM are shown in (15.37) and exhibit more complexity than in Equation (15.34). A change in a single observation j of variable v could possibly affect all observations i in y.

$$\frac{\partial y_i}{\partial x_j^{(v)}} = S_{ij}^{(v)} \tag{15.37}$$

To analyze the large amount of information pertaining to the partial derivatives from these models, one could summarize these. For example, the average own or direct effect can be summarized using the scalar $n^{-1} tr(S^{(v)})$, and the average of all effects (direct plus indirect) is represented by the scalar $n^{-1} \iota_n' S^{(v)} \iota_n$. The indirect effect would be the total effect less the direct effect. If spillovers or externalities exist, the indirect effects will not equal zero. LeSage and Pace (2009) discuss computationally feasible methods for summarizing the impacts for large samples.

15.5 Spatial Varying Parameters, Dependence, and Impacts

Another strand of spatial literature focuses on using varying parameters to model space. Although spatial dependence models typically estimate a parameter for each variable, spatial variable parameter models may estimate up to an $n \times n$ matrix of parameters for each variable. Some examples include Casetti's expansion method (Casetti, 1997), geographically weighted regression in Fotheringham, Brunsdon, and Charlton (2002), and the spatially varying coefficient approach in Banerjee et al. (2004).

As Cressie (1993, p. 25) points out, modeling the mean can substitute for modeling dependence and vice versa. We explore the connection between modeling dependence using spatial lags of the dependent variable versus a spatially varying parameter model. To begin this development, consider the usual linear model Equation (15.38) with the parameters written in matrix form in Equation (15.39) to Equation (15.40) where $\beta^{(v)}$ is a scalar.

$$E(y) = x^{(1)}\beta^{(1)} + x^{(2)}\beta^{(2)} + \cdots + x^{(k)}\beta^{(k)} \tag{15.38}$$

$$B^{(v)} = I_n \beta^{(v)} \quad v = 1, \ldots, k \tag{15.39}$$

$$E(y) = B^{(1)}x^{(1)} + B^{(2)}x^{(2)} + \cdots + B^{(k)}x^{(k)} \tag{15.40}$$

In the usual linear model, the impact of changing the explanatory variable is the same across observations and a change in the explanatory variable for one observation does not lead to changes in the value of the dependent variable for other observations.

As an alternative, consider a model that gives geometrically declining weights to the values of the parameters at neighboring locations, including parameters at neighbors to the neighbors, and so forth for higher-order neighboring relations as shown in Equation (15.41). Given the infinite series expansion from Equation (15.6), this yields (15.42). Note, the matrix of parameters implied by this process equals the matrix of impacts ($S^{(v)}$) discussed previously in (15.36) for the SAM (where $\beta_2^{(v)}$ equals 0 in Equation (15.36)). The expected value of the dependent variable is the sum of the impacts from all the explanatory variables, as in Equation (15.43).

$$S^{(v)} = I_n B^{(v)} + \rho W B^{(v)} + \rho^2 W^2 B^{(v)} + \cdots \quad (15.41)$$
$$S^{(v)} = (I_n - \rho W)^{-1} B^{(v)} \quad (15.42)$$
$$E(y) = S^{(1)} x^{(1)} + S^{(2)} x^{(2)} + \cdots + S^{(k)} x^{(k)} \quad (15.43)$$

To summarize, spatial dependence involving a spatial lag of the dependent variable is equivalent in the systematic part of the model to a form of a spatial varying parameter model. This provides another way of interpreting spatial models containing a spatial lag of the dependent variable.

15.6 Estimation

We briefly discuss maximum likelihood (or Bayesian) estimation of the SAM, which includes the more general SDM model when X in the following development includes both the explanatory variables as well as spatial lags of these.

Given the SAM model in Equation (15.44),

$$y = \rho W y + X\beta + \varepsilon \quad (15.44)$$

we can form the profile log likelihood in (15.45) as a function of a single model parameter ρ,

$$L(\rho) = K + \ln |I_n - \rho W| - \frac{1}{2} \ln(e(\rho)' e(\rho)) \quad (15.45)$$

where K is a constant and $e(\rho)$ are the residuals expressed as a function of ρ. Moreover,

$$e(\rho)' e(\rho) = y' M y - 2\rho y' M W y + \rho^2 y' W' M W y \quad (15.46)$$

where $M = I_n - X(X'X)^{-1}X'$. Note, $y'My$, $y'MWy$, and $y'W'MWy$ are scalars that do not depend on ρ, so these only need be computed once. Given this likelihood, one could estimate the parameters using either maximum likelihood or Bayesian techniques. We discuss ways to minimize the computational resources required to evaluate this likelihood in the next section.

15.7 Computation

A number of features of spatial econometric models coupled with special techniques can make estimation of problems involving large samples possible. For example, specifying the matrix W as a sparse matrix based on some number m of contiguous or nearest neighbors facilitates computation. We briefly discuss two alternative specifications for W. The first relies on a set of m nearest neighbors, which leads to a matrix where each observation (row) i contains m positive entries for the observations that are the m nearest neighbors to observation i. By construction, W would have nm positive entries and the proportion of positive elements equals m/n, which decreases with n. Use of a dense matrix with n^2 positive elements to model the sample of 300, 000 Census block groups would require over 670 gigabytes of storage for the $9 \cdot 10^{10}$ elements of W. In contrast, a weight matrix based on 10 nearest neighbors takes less than 23 megabytes of storage. A second sparse weight matrix specification relies on contiguity. For random points on a plane, on average each region (observation) has six contiguous regions (neighbors).

Finding m nearest neighbors or contiguous regions requires $O(n \ln(n))$ computations. Geographic information system (GIS) software can be used to identify contiguous regions, or a Delaunay triangle routine can be applied to centroid points for the regions. A convenient way of finding m nearest neighbors relies on identifying the contiguous neighbors and the neighbors to the contiguous neighbors as candidate nearest neighbors. Given a relatively short list of candidate neighbors, ordinary or partial sorting routines can be used to find the m nearest neighbors. There are also a number of specialized computational geometry routines that directly address the m nearest neighbor problem (Goodman and O'Rourke, 1997).

Given W, the main challenge in evaluating the log likelihood is to compute $\ln |I_n - \rho W|$, which requires $O(n^3)$ operations using dense matrix routines. Sparse matrix routines can be used to find the sparse LU decomposition, $LU = I_n - \rho W$, and the log-determinant calculated as the sum of the logs of the pivots (elements of the main diagonal) of U (Pace and Barry, 1997). Calculating the log-determinant in this fashion for a grid of points over the interval for ρ allows interpolation of log-determinant values for finer grids. Interpolation can be carried out any number of times to achieve the desired degree of accuracy because this operation requires almost no additional computing time. In practice, one could also rely on $O(n)$ approximations to the log-determinant term, such as those proposed by Barry and Pace (1999) and LeSage and Pace (2009) that produce quick approximations since the optimal value of ρ is not sensitive to small errors that arise from approximating the log-determinant.

15.8 Example

In this section, we illustrate interpretation of parameter estimates using the SDM model and a sample of census tract level information involving 62, 226 U.S. Census tracts over the continental United States aligned to have common boundaries in both 1990 and 2000. The model involved regressing median census tract house prices in 2000 (dependent variable) on explanatory variables from 1990, which included: the quantity of rental and owner-occupied housing units (Quantity), median household income (Income), and median years of education (Education). The use of 1990 data as explanatory variables avoids simultaneity among variables (as opposed to simultaneity among observations that arises from spatial dependence). Simultaneous relations between dependent and independent variables in economics has generated a vast literature in econometrics and would greatly complicate this model relationship. An equation where price and quantity appear as dependent and independent variables is a classic example of simultaneously determined variables in economics. Our use of temporally lagged values avoids this issue.

All variables were logged, so partial derivatives showing the direct and indirect effects of changes in the explanatory variables represent elasticities. We use a constant term as well as spatial lags of the dependent and explanatory variables for the SDM model specification. A doubly stochastic, contiguity-based, weight matrix specification was used for W. The estimated model is shown in Equation (15.47).

$$\begin{aligned} \ln(P_{2000}) = {} & \beta_1 \ln + \ln(\text{Income}_{1990})\beta_2 + \ln(\text{Education}_{1990})\beta_3 + \\ & + W\ln(\text{Income}_{1990})\beta_4 + W\ln(\text{Education}_{1990})\beta_5 \\ & + \ln(\text{Quantity})\beta_6 + W\ln(\text{Quantity})\beta_7 + \rho W \ln(P_{2000}) + \varepsilon \end{aligned} \tag{15.47}$$

TABLE 15.1

Parameter Estimates

Variables	β	t
Intercept	−0.435	−13.489
ln(Quantity 90)	−0.016	−18.573
ln(Income 90)	0.385	105.585
ln(Median Years Education 90)	0.530	46.197
W ln(Quantity 90)	0.010	7.400
W ln(Income 90)	−0.196	−39.107
W ln(Median Years Education 90)	−0.330	−20.453
ρ	0.833	399.384
Residual Variance	0.051	

As commonly occurs with large samples, estimates for the coefficients of all explanatory variables shown in Table 15.1 are significant. The tract-level year 2000 housing prices show strong spatial dependence with a value of $\rho = 0.833$ and the associated t-statistic is the largest of any variable.

As previously discussed, the parameter estimates from this model do not represent partial derivatives, and thus are difficult to interpret. To address this issue, Table 15.2 shows the average own partial derivative (β direct) along with associated t-statistics and the average of the cross-partials (β indirect) and their t-statistics.

As shown in Table 15.2, a higher quantity of housing units in 1990 leads to lower future housing prices in the direct census tract, and also has a negative influence on the future housing prices in neighboring tracts. This is the usual supply relation where a negative relation exists between price and quantity. Since housing in one area competes with housing in neighboring areas, an increase in the stock in one tract has spillovers to nearby tracts. Similarly, income and education have strong positive impacts on future housing prices in own and other tracts (positive spillovers).

From the table, we see that indirect impacts (spillovers) are larger in magnitude than the direct impacts in the case of income and education, and approximately equal for the quantity of housing variable. The indirect impacts reported in the table represent a cumulation of these impacts over all other sample tracts (observations). This means that the indirect effect falling on any single neighboring tract is of much lower magnitude than the cumulative impact shown in the table. In addition, from a steady-state equilibrium perspective, the cumulative indirect impact reflects changes that will arise in future time periods in the move to the new steady-state equilibrium. Again, the impact for any single time period is smaller than the reported scalar summary measure from the table. One can partition these impacts in a number of ways to address specific issues, as discussed in LeSage and Pace (2009).

Practitioners have often incorrectly interpreted the parameter estimates on the spatial lags of the explanatory variables presented in Table 15.1 as partial derivatives that reflect spillover effects. This incorrect interpretation would lead to an inference that the positive

TABLE 15.2

Direct and Indirect Estimates of Variables Affecting House Price Appreciation

Variables	*b* Direct	*t* Direct	*b* Indirect	*t* Indirect
ln(Quantity 90)	−0.016	−16.794	−0.014	−1.906
ln(Income 90)	0.428	115.622	0.700	43.536
ln(Median Years Education 90)	0.568	47.844	0.624	9.562

spillovers were negative, and negative spillovers were positive for the variables in this model.

In terms of computation, we used MATLAB® routines from www.spatial-statistics.com to find the contiguous neighbors using a Delaunay triangle routine and place them in an $n \times n$ sparse matrix, which took 2.7 seconds, convert the contiguity weight matrix into a doubly stochastic weight matrix (2.3 seconds), estimate the traces from the first 100 powers of W used to compute the log-determinant via the Barry and Pace (1999) approach (11 seconds), and estimate the model (1.06 seconds). This was done using a machine with an Athlon 3.2 GHz dual processor. Most of these calculations scale linearly or at an $n \ln(n)$ rate, and so with current technology estimation of large spatial systems is quite feasible.

15.9 Extensions

Spatial regression models can also be estimated using analytical or Bayesian MCMC estimation. LeSage (1997) discusses the latter approach along with extensions involving heteroscedastic variances, and MCMC methods can take advantage of the computationally efficient approaches discussed for maximum likelihood estimation. Estimating models, such as the SDM, allows us to rely on slightly extended results from standard Bayesian regression where a *normal-gamma prior* distribution is assigned for the parameters β and σ. The conditional distributions for the parameters β and σ^2 take standard forms, which can be easily sampled.

The conditional distribution for the parameter ρ requires special treatment as it does not take the form of a known distribution. A uniform prior is typically assigned for this parameter, and sampling can rely on a Metropolis–Hastings method (LeSage, 1997) or a griddy Gibbs sampler (Smith and LeSage, 2004). It is noteworthy that the log-determinant term from the log likelihood function appears in the conditional distribution for ρ, and the computationally efficient methods for calculating this term over a grid of values for ρ can be used to speed MCMC estimation. Bayesian model comparison procedures that are based on posterior model probabilities constructed from the log-marginal likelihood also benefit from these efficient methods for calculating the log-determinant. The log-marginal likelihood can be derived by analytically integrating out the parameters β and σ followed by univariate numerical integration over the parameter ρ, with details provided in LeSage and Parent (2007).

Variants of the standard SAM/SDM models that deal with cases involving missing values (?) and truncated or limited dependent variables (LeSage and Pace, 2009) can proceed by adding an additional conditional distribution to the standard MCMC sampler. For the case of missing or truncated dependent variables, we can treat these as latent values and add a sampling step that samples the missing (or truncated) values conditional on the nonmissing (or truncated) values. Replacing the missing or truncated zero values of the dependent variables with the continuous sampled latent values allows sampling the remaining model parameters based on the same distributions as in the case of continuous dependent variable values.

If we partition the vector y into vectors of nonmissing y_1 and missing (or truncated) values $y_2 = 0$, MCMC estimation requires that we sample from the conditional distribution of $(y_2|y_1)$, which takes the form of a truncated multivariate normal distribution. That is, zero-valued observations are sampled conditional on the nonzero values in y_1, and other parameters of the model. In essence, the zero-valued observations are being treated as additional parameters in the model that require estimation. This can pose computational challenges discussed in LeSage and Pace (2009). They provide details regarding computational savings that can be obtained by avoiding inversion of the $n \times n$ sparse matrix $(I_n - \rho W)$, since this would produce "fill-in" of zero values with nonzero values, increasing memory requirements as well as the time required for matrix calculations.

15.10 Conclusion

Spatial econometric models that include lags of the dependent variable provide a useful tool that can be used to quantify the magnitude of direct and indirect or spillover effects. The level of direct and indirect impacts that arise from changes in the model explanatory variables are often the focus of economic models when interest centers on individual incentives or benefits versus those of society at large. Use of spatial models that contain a spatial lag of the dependent variable appears to be the most distinctive feature that separates spatial econometrics from spatial statistical models.

Estimation and interpretation of these models pose additional challenges that are not encountered in many spatial statistical settings. An applied illustration demonstrated partitioning of impacts that arise from changes in the explanatory variables associated with the own- and cross-partial derivatives reflecting direct and indirect or spatial spillover effects.

References

Anselin, L. (1988). *Spatial Econometrics: Methods and Models*, Dordrecht: Kluwer Academic Publishers.

Anselin, L. (2003). Spatial externalities, spatial multipliers and spatial econometrics, *International Regional Science Review*, 26, 153–166.

Banerjee, S., B.P. Carlin and A.E. Gelfand (2004). *Hierarchical Modeling and Analysis for Spatial Data*, Boca Raton, FL: Chapman & Hall/CRC Press.

Barry, R. and R.K. Pace (1999). A Monte Carlo estimator of the log determinant of large sparse matrices, *Linear Algebra and Its Applications*, 289, 41–54.

Bavaud, F. (1998). Models for spatial weights: A systematic look, *Geographical Analysis*, 30, 153–171.

Besag, J. (1974). Spatial interaction and the statistical analysis of lattice systems, *Journal of the Royal Statistical Society B*, 36, 192–236.

Brook, D. (1964). On the distinction between the conditional probability and the joint probability approaches in the specification of nearest-neighbor systems, *Biometrika*, 51, 481–483.

Brueckner, J. (2003). Strategic interaction among governments: An overview of empirical studies, *International Regional Science Review*, 26, 175–188.

Casetti, E. (1997). The expansion method, mathematical modeling, and spatial econometrics, *International Regional Science Review*, 20, 9–33.

Cressie, N. (1993). *Statistics for Spatial Data*, Revised edition, New York: John Wiley & Sons.

Elhorst, J.P. (2001). Dynamic models in space and time, *Geographical Analysis*, 33, 119–140.

Ertur, C. and W. Koch (2007). Convergence, human capital and international spillovers, *Journal of Applied Econometrics*, 22:6, 1033–1062.

Fotheringham, A.S., C. Brunsdon, and M. Charlton (2002). *Geographically Weighted Regression: The Analysis of Spatially Varying Relationships*, West Sussex, U.K.: John Wiley & Sons.

Goodman, J.E. and J. O'Rourke (1997). *Handbook of Discrete and Computational Geometry*, Boca Raton: CRC Press.

LeSage, J.P. (1997). Bayesian estimation of spatial autoregressive models, *International Regional Science Review*, 20, 113–129.

LeSage, J.P. and R.K. Pace (2009). *Introduction to Spatial Econometrics*, Boca Raton, FL: Taylor & Francis.

LeSage, J.P. and O. Parent (2007). Bayesian model averaging for spatial econometric models, *Geographical Analysis*, 39:3, 241–267.

LeSage, J.P., M.M. Fischer, and T. Scherngell (2007). Knowledge spillovers across Europe: Evidence from a Poisson spatial interaction model with spatial effects, *Papers in Regional Science*, 86:3, 393–421.

López-Bazo, E., E. Vayá, and M. Arts (2004). Regional externalities and growth: evidence from European regions, *Journal of Regional Science*, 44:1, 43–73.

Martellosio, F. (2007). Some correlation properties of spatial autoregressions, MPRA paper No. 17254, Munich University Library.

Ord, J.K. (1975). Estimation methods for models of spatial interaction, *Journal of the American Statistical Association*, 70, 120–126.

Pace, R.K. and R. Barry, (1997). Quick computation of spatial autoregressive estimators, *Geographical Analysis*, 29, 232–246.

Pace, R.K. and J.P. LeSage (2008). A spatial hausman test, *Economics Letters*, 101, 282–284.

Pace, R.K., R. Barry, O.W. Gilley, and C.F. Sirmans (2000). A method for spatial-temporal forecasting with an application to real estate prices, *International Journal of Forecasting*, 16, 229–246.

Pfeifer, P.E. and S.J. Deutsch (1980). A three-stage iterative procedure for space-time modeling, *Technometrics*, 22, 35–47.

Ripley, B. (1981). *Spatial Statistics*, New York: John Wiley & Sons.

Smith, T.E. and J.P. LeSage (2004). A Bayesian probit model with spatial dependencies, in *Advances in Econometrics: Volume 18: Spatial and Spatiotemporal Econometrics*, (Oxford: Elsevier Ltd), J.P. LeSage and R.K. Pace (eds.), 127–160.

Wall, M.M. (2004). A close look at the spatial correlation structure implied by the CAR and SAR models, *Journal of Statistical Planning and Inference*, 121:2, 311–324.

Whittle P. (1954). On stationary processes in the plane, *Biometrika*, 41, 434–449.

Part IV

Spatial Point Patterns

A spatial point process is a stochastic process each of whose realizations consists of a finite or countably infinite set of points in the plane. The overall goal of this part of the handbook is to review the theory and practice of statistical methods for analyzing data that can be treated as a realization of a spatial point process.

Chapter 16, by Marie-Colette van Lieshout, sets out the underpinning mathematical theory of spatial point processes. After some motivation, several ways to characterize the distribution of a point process are described, including the important notions of Campbell and moment measures and interior and exterior conditioning by means of Palm theory.

Chapter 17, by Valerie Isham, describes various useful classes of parametric point process models. In any particular application, there will generally be an underlying physical mechanism that generates the points that are observed, and which should ideally be taken into account when formulating a model. For example, if the point events are the locations of seedling trees, a model might take into account the positions of parent trees, the clustering of the seedlings around these, and perhaps also the prevailing wind direction. Such a model is often termed "mechanistic." In contrast, "descriptive" models aim to represent the statistical properties of the data and their dependence on explanatory variables without necessarily worrying about the physical mechanisms involved.

Chapter 18, by Peter Diggle, describes statistical methods for analyzing spatial point patterns without imposing parametric assumptions about the underlying spatial point process. Topics covered include: Monte Carlo tests, nearest-neighbor and related methods, nonparametric smoothing methods, and moment-based functional summary statistics. This chapter also includes a discussion of methods of analyzing replicated patterns where, in the absence of a parametric model, inference relies on simple ideas of exchangeability between replicates.

Chapter 19, by Jesper Møller, considers inference for parametric models. The central message here is that, for many classes of model, principled methods of inference based on the likelihood function are now accessible through the use of modern Monte Carlo methods, and are often preferable to the widespread use of sensible but ad hoc methods based on functional summary statistics of the kind described in Chapter 18. Nevertheless, the chapter also discusses parametric model-fitting using estimation methods based on composite likelihoods or pseudo-likelihoods that do not require Monte Carlo methods, and acknowledges that these may continue to be of importance because they allow many dierent candidate models to be investigated quickly. Chapter 20, by Adrian Baddeley, explores the proposition that, as in other areas of applied statistics, the statistical analysis of spatial point pattern data should involve a succession of stages, including exploratory analysis, model-fitting and model selection, and diagnostic checking. This chapter describes strategies for each stage of that process, with examples of their application.

Chapter 21, also by Adrian Baddeley, extends the framework of the earlier chapters to include multitype point processes, in which points are of several qualitatively different types, or more generally marked point processes in which each point of the process carries extra information, called a *mark*, which may be a random variable, several random variables, a geometric shape, or some other information. Many of the statistical methods for spatial point patterns described in earlier chapters extend reasonably easily and naturally to marked point patterns, although marked point patterns also raise new and interesting questions concerning the appropriate way to formulate models and pursue analyses for particular applications. For example, should points precede marks, or vice versa, or must they be considered jointly?

Finally, in Chapter 22, Lance Waller describes how the models, properties, and methods discussed in earlier chapters can address specific questions encountered in the field of spatial epidemiology, i.e., the study of spatial patterns of disease morbidity and mortality within their natural setting. This particular area of application has been singled out in part because of its importance, but also because it shows how methods developed in one context can be adapted to meet the needs of new areas of application. One very good reason why spatial point process methods were not widely used in epidemiology until recently was that the models and methods available were unable to cope with the degree of complexity that typifies the spatial distribution of human settlement patterns. However, during the last two decades, these limitations have been overcome, at least in part, by importing ideas from design-based inference for case-control studies and by the development of new, often computationally intensive, methods for analyzing spatially heterogeneous point process data.

16

Spatial Point Process Theory

Marie-Colette van Lieshout

CONTENTS

16.1 Spatial Point Process Theory

16.1.1 Introduction

A *spatial point process* is a stochastic process each of whose realizations consists of a finite or countably infinite set of points in the plane. This chapter sets out the mathematical theory of such processes. After some motivation, several ways to characterize the distribution of a point process are described in Section 16.1.2. The important notions of Campbell and moment measures are introduced in Section 16.1.3, and generalized to higher orders in Section 16.1.4. Interior and exterior conditioning by means of Palm theory form the topic of Section 16.1.5. The section closes with a discussion of finite point processes and local specifications (Sections 16.1.6 to 16.1.7).

Tracing the roots of modern point process theory is not an easy task. One may refer to Poisson (1837), or to the pioneering works by Erlang (1909) and Neyman (1939) in the respective contexts of telephone networks and spatial cluster processes. Influential classic textbooks include Janossy (1948), Khinchin (1960), Matheron (1975), Matthes, Kerstan, and Mecke (1978), and Srinivasan (1974). A rich scholarly reference book is the second edition of Daley and Vere-Jones (1988) published in two volumes (2003, 2008). A more accessible, yet rigorous, course on finite point processes is van Lieshout (2000). See also Cox and Isham (1980) or Reiss (1993).

As a motivating example, consider the pattern of 270 tree locations in a 75×75 meter plot in the Kaluzhskie Zaseki Forest in central Russia depicted in the left-hand panel of Figure 16.1. The data were kindly provided by Dr. P. Grabarnik and previously analyzed in Smirnova (1994) and Grabarnik and Chiu (2002).

In a forestry context such as this, competition for nutrients and space may result in patterns in which large trees do not occur close to each other and smaller trees fill up the remaining gaps. On the other hand, seedlings tend to grow up close to mature trees, which

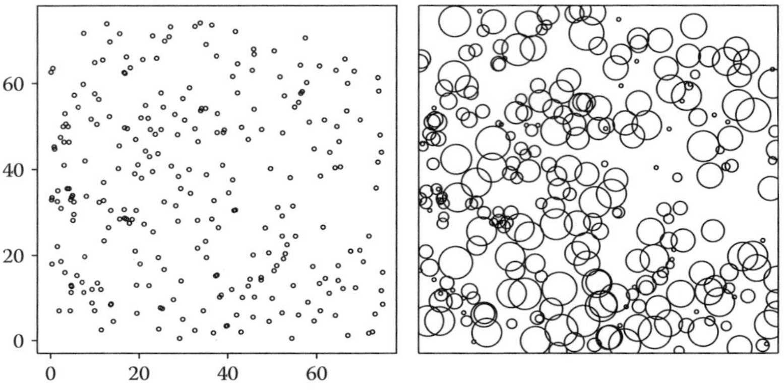

FIGURE 16.1
Trees in a broad-leaved, multispecies old-growth forest in the Kaluzhskie Zaseki Forest in central Russia (Smirnova, 1994; Grabarnik and Chiu, 2002) in a square plot of side length 75 m. (Left) Locations of the 270 trees, (right) graphical representation in which each tree is located at the center of a disk of radius equal to 0.103 times the tree diameter.

would result in a cluster of nearby trees. Such apparent clustering can also result from variations in soil fertility, as trees are more likely to flourish in regions of high fertility.

For these data, additional measurements in the form of tree diameters are available, graphically represented by the disk radii in the right-hand side panel of Figure 16.1. Such a pattern with extra information attached to each point (location) is referred to as a "marked point pattern." Clusters of small trees in the gaps left by larger trees are apparent. This clustering effect can be seen even more clearly if we plot the subpatterns of small and large trees separately (Figure 16.2). The large (old) trees seem, apart from some gaps, to have a preferred distance of about 4 meters from each other, possibly due to a planting design.

Mapped data such as those depicted in Figure 16.1 and Figure 16.2 can be described mathematically as a finite, integer-valued measure

$$\sum_{i=1}^{I} k_i \, \partial_{\mathbf{x}_i}$$

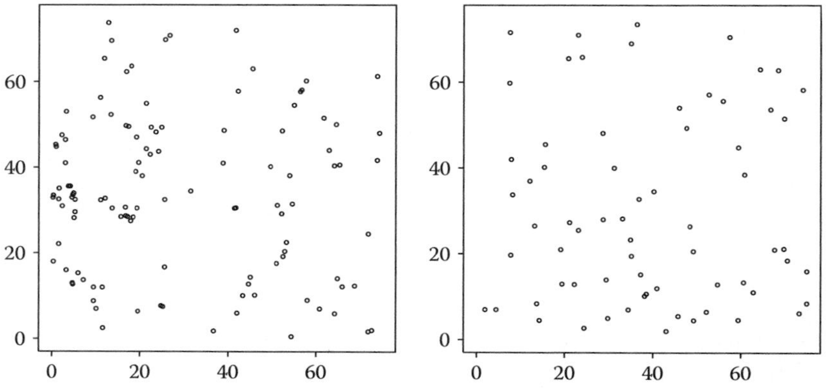

FIGURE 16.2
Locations of trees in a broad-leaved, multispecies old-growth forest (Smirnova, 1994; Grabarnik and Chiu, 2002) in a square plot of side length 75 m. (Left) Trees with diameter less than 15 cm, (right) those with diameter larger than 25 cm.

where $I \in \mathbb{N}_0$, the $\mathbf{x}_i$ are distinct (marked) points in some topological space D, and each weight k_i is a strictly positive integer, the *multiplicity* of $\mathbf{x}_i$. An equivalent description is in terms of a finite, unordered set of not necessarily distinct (marked) points

$$\mathcal{X} = \{\mathbf{x}_1, \ldots, \mathbf{x}_n\} \in D^n/\equiv, \qquad n = 0, 1, \ldots.$$

The $\mathbf{x}_i$ lie in a topological space D, and two or more $(\mathbf{x}_1, \ldots, \mathbf{x}_n) \in D^n$ that are identical up to permutation are indistinguishable. We shall refer to $\mathcal{X}$ as a *configuration*.

For example, for the unmarked pattern depicted in the left-hand panel of Figure 16.1, $D = [0, 75]^2$ equipped with the Euclidean distance topology, while marked points represented graphically in the right-hand panel of Figure 16.1 lie in the product space $[0, 75]^2 \times [0, \infty)$ generated by the Euclidean topologies on $[0, 75]^2$ and $[0, \infty)$. More precisely, if U and V are open neighborhoods of $\mathbf{x} \in [0, 75]^2$ and $m \in [0, \infty)$, respectively, then we may define an open neighborhood of $(\mathbf{x}, m)$ by $U \times V$. The class of such product sets defines a topology, the *product topology*. Equivalently, the product topology can be defined by the supremum metric. In order to obtain Figure 16.2, trees were marked by a type label in the set {large, small} equipped with the discrete metric.

If D is not bounded, the requirement that I or n be finite is relaxed to allow configurations that contain a countable number of points, each with finite multiplicity. Such a configuration $\mathcal{X} \subseteq D$ is said to be *locally finite* if it places at most a finite number of points in any bounded Borel set $A \subseteq D$, that is, if

$$N_{\mathcal{X}}(A) = \sum_{i=1}^{\infty} k_i \, \partial_{\mathbf{x}_i}(A) < \infty.$$

The family of all locally finite configurations will be denoted by N^{lf}.

Definition 16.1

Let (D, d) be a complete, separable metric space equipped with its Borel σ-algebra $\mathcal{B}$. A point process *on D is a mapping X from a probability space $(\Omega, \mathcal{A}, P)$ into N^{lf} such that for all bounded Borel sets $A \subseteq D$, the number $N(A) = N_X(A)$ of points falling in A is a (finite) random variable.*

From a practical point of view, it is convenient to work with metric spaces. Theoretically, the topological structure is needed to ensure the existence of regular versions of Palm distributions and related characteristics that will be discussed later.

Definition 16.1 may be rephrased as follows. A point process X is a random variable with values in the measurable space $(N^{\text{lf}}, \mathcal{N}^{\text{lf}})$, where $\mathcal{N}^{\text{lf}}$ is the smallest σ-algebra such that for all bounded Borel sets $A \subseteq D$, the mapping $\mathcal{X} \mapsto N_{\mathcal{X}}(A)$ is measurable. The induced probability measure $\mathcal{P}$ on $\mathcal{N}^{\text{lf}}$ is called the *distribution* of the point process.

When the $\mathbf{x}_i \in \mathcal{X}$ have a location component, i.e., when D is of the form $\mathbb{R}^d \times L$ for some mark space L, assumed to be a complete, separable metric space, and the marginal point process of locations is well-defined, X is said to be a *marked point process*. If $\mathcal{P}$ is translation invariant in the sense that it does not change if all marked points $\mathbf{x}_i = (\mathbf{a}_i, m_i) \in X$ are translated over some vector $\mathbf{y} \in \mathbb{R}^d$ into $(\mathbf{a}_i + \mathbf{y}, m_i)$, X is *stationary*. If, additionally, $\mathcal{P}$ is not affected by rotations of the location component, X is *isotropic*.

Example 16.1

Let D be a compact subset of the plane of strictly positive area $|D|$. A *binomial point process* is defined as the union $X = \{X_1, \ldots, X_n\}$ of a fixed number $n \in \mathbb{N}$ of independent, uniformly distributed points $X_1, \cdots, X_n$ in D. In other words, $\mathbf{P}(X_i \in A) = |A|/|D|$ for all Borel subsets $A \subseteq D$. Now,

$$N(A) = \sum_{i=1}^{n} 1\{X_i \in A\}$$

is a sum of random variables; hence, a random variable itself, and, as $N(A) \leq N(D) = n$, X takes values in N^{lf}.

Example 16.2

Further to the previous example, by replacing the fixed number $n \in \mathbb{N}$ by an integer-valued, random variable following a Poisson distribution with parameter $\lambda|D|$ proportional to the area $|D|$ of D, we obtain a *homogeneous Poisson process* on D. Now,

$$\mathbf{P}(N(A) = k) = \sum_{n=k}^{\infty} e^{-\lambda|D|} \frac{(\lambda|D|)^n}{n!} \binom{n}{k} \left(\frac{|A|}{|D|}\right)^k \left(1 - \frac{|A|}{|D|}\right)^{n-k} = e^{-\lambda|A|} \frac{(\lambda|A|)^k}{k!},$$

for all $k \in \mathbb{N}_0$, so that $N(A)$ is Poisson distributed with parameter $\lambda|A|$, and, by similar arguments, for disjoint Borel sets $A, B \subseteq D$, and $k, l \in \mathbb{N}_0$,

$$\mathbf{P}(N(A) = k; N(B) = l) = \mathbf{P}(N(A) = k)\,\mathbf{P}(N(B) = l).$$

Hence, the random variables $N(A)$ and $N(B)$ are independent. It should be noted that the definition of a Poisson process may be extended to the plane by tiling $\mathbb{R}^2$ and defining independent Poisson processes on each tile as before.

16.1.2 Characterization of Point Process Distributions

The distribution of a real-valued random variable may be characterized by its distribution function, characteristic function, moment generating function, or Laplace transform. Recall that the distribution $\mathcal{P}$ of a point process is that induced by the integer-valued random variables $N(A)$ counting the number of objects placed in bounded Borel sets A. Thus, it is natural to characterize $\mathcal{P}$ by properties of the distributions of the random variables $N(A)$.

Definition 16.2

The family of finite-dimensional distributions *(fidis) of a point process X on a complete, separable metric space (D, d) is the collection of the joint distributions of $(N(A_1), \ldots, N(A_m))$, where $(A_1, \ldots, A_m)$ ranges over the bounded Borel sets $A_i \subseteq D$, $i = 1, \ldots, m$, and $m \in \mathbb{N}$.*

The following uniqueness theorem holds.

Theorem 16.1

The distribution of a point process X on a complete, separable metric space (D, d) is completely specified by its finite-dimensional distributions.

In other words, if two point processes have identical fidis, they also share the same distribution. The result follows from the observation that the family of sets

$$\{\omega \in \Omega : N_{X(\omega)}(A_i) \in B_i, \quad i = 1, \ldots, m\},$$

where $A_i \subseteq D$ are bounded Borel sets and $B_i \subseteq \mathbb{R}$ are Borel, is a semiring generating $\mathcal{N}^{\text{lf}}$.

A point processes on (D, d) is called *simple* if its realizations contain no points with multiplicity $k_i \geq 2$, so that $N(\{\mathbf{x}\}) \in \{0, 1\}$ almost surely for all $\mathbf{x} \in D$. Surprisingly, as shown by Rényi (1967) for the Poisson process and by Mőnch (1971) more generally, if two simple point processes X and Y assign the same *void probabilities*, $v(A) = \mathcal{P}(N(A) = 0)$, to all bounded Borel sets A, their distributions coincide.

Theorem 16.2
The distribution of a simple point process X on a complete, separable metric space (D, d) is uniquely determined by the void probabilities of bounded Borel sets $A \subseteq D$.

In the above theorem, the collection of bounded Borel sets may be replaced by a smaller class, such as the compact sets. For details, we refer to Berg, Christensen, and Ressel (1984), Choquet (1953), Daley and Vere-Jones (1988), Matheron (1975), McMillan (1953), and Norberg (1989).

Example 16.3
The binomial point process introduced in Example 16.1 is simple and has void probabilities $v(A) = (1 - |A|/|D|)^n$, $A \in \mathcal{B}(D)$. For the homogeneous Poisson process considered in Example 16.2, $v(A) = \exp[-\lambda|A|]$.

For any point process X, one may define a simple point process X^s by ignoring the multiplicity. It follows from Theorem 16.2 that if point processes X and Y have identical void probabilities, the distributions of X^s and Y^s must be the same. Thus, X and Y exhibit the same interaction structure except for the multiplicity of their points.

Trivial cases apart, it is not possible to plot the void probabilities as a function of $A \subseteq D$. Nevertheless, a graphical representation may often be achieved. Indeed, suppose that the distribution $\mathcal{P}$ that produced a Euclidean point pattern, such as that depicted in Figure 16.1, is invariant under translations. Then $v(A) = v(T_{\mathbf{x}}A)$ for all $T_{\mathbf{x}}A = \{\mathbf{a} + \mathbf{x} : \mathbf{a} \in A\}$, $\mathbf{x} \in \mathbb{R}^d$. In particular, if $A = \bar{b}(\mathbf{x}, r) = \{\mathbf{y} \in \mathbb{R}^d : ||\mathbf{y} - \mathbf{x}|| \leq r\}$ is a closed ball, the stationarity of $\mathcal{P}$ implies that $v(\bar{b}(\mathbf{x}, r))$ is a function of r only. Based on this observation, the *empty space function* is defined as

$$F(r) = \mathcal{P}(d(\mathbf{x}, X) \leq r) = 1 - v(\bar{b}(\mathbf{x}, r)), \qquad r \geq 0, \tag{16.1}$$

that is, F is the distribution function of the distance from an arbitrary point $\mathbf{x} \in \mathbb{R}^d$ to the nearest point of X. Although $\mathcal{P}$ is not uniquely specified by the void probability it assigns to balls, plots of the estimated empty space function do provide valuable information on the interaction structure.

Example 16.4
Estimated empty space functions related to the Kaluzhskie Zaseki Forest data of Figure 16.2 are given in Figure 16.3. On biological grounds, it may be expected that the patterns formed by young and established trees are qualitatively different, and this is confirmed by Figure 16.3. Using the observed size mark as a substitute for the nonrecorded age of trees, estimated F-functions for both types of trees are shown and compared to the empty space function of a Poisson process with the same expected number of points (cf. Example 16.3). For small trees, large empty spaces occur with a higher probability than under the Poisson model. Such a behavior is typical for clustered patterns with groups of nearby points separated by gaps. The graph for larger trees is closer to that of its Poisson counterpart.

16.1.3 Campbell and Moment Measures

In the previous section, we saw that the distribution of a simple point process is described fully by its void probabilities and that a graphical representation is available in the empty space function. For point processes that may contain multiple points, the finite-dimensional distributions of Definition 16.2 can be used to specify the distribution, and their moments provide a useful complementary collection of summary statistics (see, e.g., Matthes et al., 1978).

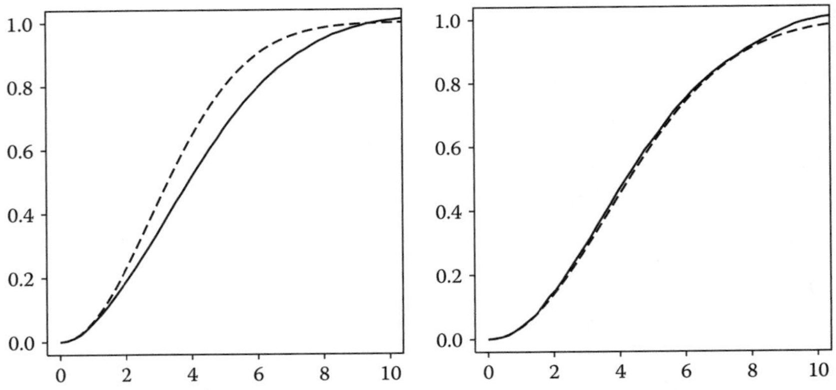

FIGURE 16.3
Estimated empty space functions of Kaluzhskie Zaseki data (solid lines). (Left) Locations of trees with diameter less than 15 cm, (right) those with diameter larger than 25 cm. The dashed lines are the empty space functions for homogeneous Poisson processes with, on average, the same number of points per unit area as the data.

Definition 16.3

Let X be a point process on a complete, separable metric space (D, d). Define a set function M by $M(A) = \mathbf{E}N(A)$ for all bounded Borel sets $A \subseteq D$.

Example 16.5

Consider the binomial point process of Example 16.1. Then,

$$M(A) = n\,\mathbf{P}(X_1 \in A) = n\,|A|/|D|$$

for any Borel set $A \subseteq D$. In particular, $M(D) = n < \infty$. For the homogeneous Poisson process of Example 16.2, $M(A) = \lambda\,|A|$.

By standard measure theoretic arguments (see, e.g., Daley and Vere-Jones (1988) or van Lieshout (2000)), the set function M can be extended to a Borel measure.

Theorem 16.3

If the function M introduced in Definition 16.3 is finite for all bounded Borel sets, it can be extended uniquely to a σ-finite measure on the Borel sets of D, the (first-order) moment measure.

Let us consider a stationary point process X on the Euclidean space $\mathbb{R}^d$, and denote by $T_{\mathbf{x}}A = \{\mathbf{a} + \mathbf{x} : \mathbf{a} \in A\}$ the translation of A over the vector $\mathbf{x} \in \mathbb{R}^d$. Then, for any Borel set A,

$$M(A) = \mathbf{E}N(A) = \mathbf{E}N(T_{\mathbf{x}}A) = M(T_{\mathbf{x}}A).$$

Consequently, provided M is finite for bounded A, a constant $0 \leq \lambda < \infty$ can be found such that $M(A) = \lambda\,|A|$. The scalar constant λ is called the *intensity* of X. An example of a stationary point process is the homogeneous Poisson process on the plane, (cf. Example 16.2).

More generally, if M is absolutely continuous with respect to Lebesgue measure, then

$$M(A) = \int_A \lambda(\mathbf{x})\,d\mathbf{x}$$

for Borel sets A, and $\lambda(\cdot)$ is referred to as an *intensity function* of the point process X.

In integral terms, Theorem 16.3 asserts that

$$\int_A dM(\mathbf{a}) = \mathbf{E}\left[\sum_{\mathbf{x}\in X} 1\{\mathbf{x} \in A\}\right]$$

for all $A \in \mathcal{B}(D)$. Linearity and monotonicity arguments imply that

$$\mathbf{E}\left[\sum_{\mathbf{x}\in X} g(\mathbf{x})\right] = \int_D g(\mathbf{x})\, dM(\mathbf{x}) \tag{16.2}$$

for any measurable function $g : D \to \mathbb{R}$ that is either nonnegative or integrable with respect to M, provided the moment measure M exists, i.e., is finite on bounded Borel sets. Identities of the form Equation (16.2) are usually referred to as *Campbell theorems* in honor of Campbell's (1909) paper. Intuitively, suppose one takes a measurement $g(\mathbf{x})$ at each $\mathbf{x}$ in some data configuration $\mathcal{X}$. Then the grand total of these measurements is equal in expectation to the spatial integral of g with respect to the moment measure of the stochastic process that generated the data.

Example 16.6

Let us revisit the forestry data discussed in Section 16.1.1, and assume it can be seen as a realization of a stationary marked point process on $D = \mathbb{R}^2 \times \mathbb{R}^+$, observed within the window $W = [0, 75]^2$. Then, for bounded Borel sets $A \subseteq \mathbb{R}^2$, $B \subseteq \mathbb{R}^+$, and $\mathbf{x} \in \mathbb{R}^2$,

$$M(T_{\mathbf{x}} A \times B) = M(A \times B).$$

If we additionally assume that the first-order moment measure of the marginal point process of locations exists, the translation invariance implies that M can be factorized as $\lambda\, \ell \times \nu_M$ where $\lambda \geq 0$ is the *intensity* of the marked point process, ℓ Lebesgue measure on $\mathbb{R}^2$, and ν_M a probability measure on $\mathbb{R}^+$, the *mark distribution*. By the Campbell theorem

$$\mathbf{E}\left[\sum_{(\mathbf{a},m)\in X} m\, 1\{\mathbf{a} \in W\}\right] = 75^2\, \lambda \int_0^\infty m\, d\nu_M(m).$$

In other words, a ratio-unbiased estimator of the mean tree diameter with respect to ν_M is the average observed diameter. The smoothed empirical mark distribution function ν_M is depicted in Figure 16.4. Note that ν_M has two clear modes.

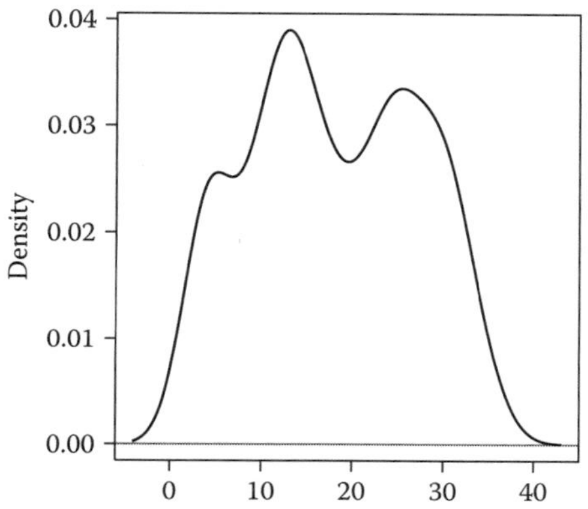

FIGURE 16.4
Estimated mark probability density function of Kaluzhskie Zaseki Forest data.

Moment measures may be refined by restricting attention to configurations with specified properties.

Definition 16.4
Let X be a point process on a complete, separable metric space (D, d). Define

$$C(A \times F) = \mathbf{E}[N(A)\,1\{X \in F\}]$$

for all bounded Borel sets $A \subseteq D$ and all $F \in \mathcal{N}^{\mathrm{lf}}$.

Moment measures may not exist, but C always defines a proper measure, the Campbell measure of Kummer and Matthes (1970).

Theorem 16.4
The function C of Definition 16.4 can be extended uniquely to a σ-finite measure on the product σ-algebra of $\mathcal{B}(D)$ and $\mathcal{N}^{\mathrm{lf}}$, the (first-order) Campbell measure.

Since $C(A \times N^{\mathrm{lf}}) = \mathbf{E}N(A) = M(A)$ for any Borel set A, it follows that C is finite if and only if M is also finite.

Rephrased in integral terms, Theorem 16.4 states that

$$\mathbf{E}\left[\sum_{\mathbf{x}\in X} g(\mathbf{x}, X)\right] = \int_D \int_{N^{\mathrm{lf}}} g(\mathbf{x}, \mathcal{X})\, dC(\mathbf{x}, \mathcal{X}) \tag{16.3}$$

for any measurable function $g : D \times N^{\mathrm{lf}} \to \mathbb{R}$, which is either nonnegative or integrable with respect to the Campbell measure.

16.1.4 Reduced and Higher-Order Campbell Measures

Second- and higher-order moment measures are defined by considering $N(A_1), N(A_2), \ldots$ jointly.

Definition 16.5
Let X be a point process on a complete, separable metric space (D, d). Define, for $n \in \mathbb{N}$, and bounded Borel subsets $A_1, \ldots, A_n$ of D,

$$M_n(A_1 \times \cdots \times A_n) = \mathbf{E}[N(A_1) \cdots N(A_n)].$$

If all $A_i \equiv A$ are equal, $M_n(A \times \cdots \times A) = M(A)^n$. Since the Borel rectangles form a semiring generating the Borel product σ-algebra, provided M_n is finite, it can be extended uniquely to a σ-finite measure on the product σ-algebra, the *nth order moment measure* of X. Its integral representation states that

$$\mathbf{E}\left[\sum_{\mathbf{x}_1,\ldots,\mathbf{x}_n \in X} g(\mathbf{x}_1, \ldots, \mathbf{x}_n)\right] = \int_D \cdots \int_D g(\mathbf{x}_1, \ldots, \mathbf{x}_n)\, dM_n(\mathbf{x}_1, \ldots, \mathbf{x}_n)$$

for any measurable function $g : D^n \to \mathbb{R}$ that is either nonnegative or integrable with respect to M_n, provided that M_n exists as a σ-finite measure.

Example 16.7
Let A and B be bounded Borel sets. The covariance of the random variables counting the number of points in A and B can be written in terms of the second-order moment

measure as

$$\mathrm{Cov}(N(A), N(B)) = M_2(A \times B) - M(A)\, M(B).$$

In particular,

$$\mathrm{var}(N(A)) = M_2(A \times A) - M(A)^2.$$

Example 16.8

For the binomial point process of Example 16.1, as $N(A_1)\, N(A_2) = \sum_{i=1}^{n} 1\{X_i \in A_1 \cap A_2\} + \sum_{i=1}^{n} \sum_{j \neq i} 1\{X_i \in A_1;\, X_j \in A_2\}$,

$$M_2(A_1 \times A_2) = n\,\frac{|A_1 \cap A_2|}{|D|} + n(n-1)\,\frac{|A_1|\,|A_2|}{|D|^2}.$$

Next, consider the homogeneous Poisson process of Example 16.2. As the total number of points is Poisson distributed with mean $\lambda\,|D|$,

$$M_2(A_1 \times A_2) = \mathbf{E}\left[\mathbf{E}\left[N(A_1) \times N(A_2)\right] \mid N(D)\right] = \lambda\,|A_1 \cap A_2| + \lambda^2\,|A_1|\,|A_2|.$$

In the above example, M_2 was computed by distinguishing between pairs of identical and pairs of distinct points. If only the latter type are taken into consideration, we obtain the nth order *factorial moment measure* μ_n defined by the integral representation

$$\mathbf{E}\left[\sum_{\mathbf{x}_1,\ldots,\mathbf{x}_n \in X}^{\neq} g(\mathbf{x}_1, \ldots, \mathbf{x}_n)\right] = \int_D \cdots \int_D g(\mathbf{x}_1, \ldots, \mathbf{x}_n)\, d\mu_n(\mathbf{x}_1, \ldots, \mathbf{x}_n) \tag{16.4}$$

for all measurable functions $g : D^n \to \mathbb{R}^+$ (see, e.g., Mecke (1976)). Standard measure-theoretic arguments imply Equation (16.4) holds true for any g that is integrable with respect to μ_n. The sum is over all n-tuples of distinct points. For $n = 1$, $\mu_1 = M_1 = M$, the first-order moment measure. We shall say that the nth order factorial moment measure exists if $\mu_n(A)$ is finite for all bounded Borel sets $A \subseteq D$.

Example 16.9

Let A_1, A_2 be Borel subsets of some compact set $D \subseteq \mathbb{R}^2$. Further to Example 16.8, the homogeneous Poisson process with intensity λ has second-order factorial moment measure

$$\mu_2(A_1 \times A_2) = \lambda^2\,|A_1|\,|A_2|.$$

For the binomial point process with n points,

$$\mu_2(A_1 \times A_2) = n(n-1)\,|A_1|\,|A_2|/|D|^2.$$

If μ_n is absolutely continuous with respect to some n-fold product measure ν^n on D^n, then Equation (16.4) can be written as

$$\mathbf{E}\left[\sum_{\mathbf{x}_1,\ldots,\mathbf{x}_n \in X}^{\neq} g(\mathbf{x}_1, \ldots, \mathbf{x}_n)\right] = \int_D \cdots \int_D g(\mathbf{x}_1, \ldots, \mathbf{x}_n)\, \rho_n(\mathbf{x}_1, \ldots, \mathbf{x}_n)\, d\nu(\mathbf{x}_1) \ldots d\nu(\mathbf{x}_n).$$

The Radon–Nikodym derivative ρ_n of μ_n is referred to as its *product density*, and $\rho_n(\mathbf{x}_1, \ldots, \mathbf{x}_n)\, d\nu(\mathbf{x}_1) \ldots d\nu(\mathbf{x}_n)$ may be interpreted as the joint probability of a point falling in each of the infinitesimal regions centered at $\mathbf{x}_1, \ldots, \mathbf{x}_n$.

Example 16.10

Further to Example 16.9, the homogeneous Poisson process with intensity λ has constant second-order product density $\rho_2(\mathbf{x}_1, \mathbf{x}_2) = \lambda^2$, while for the binomial point process with

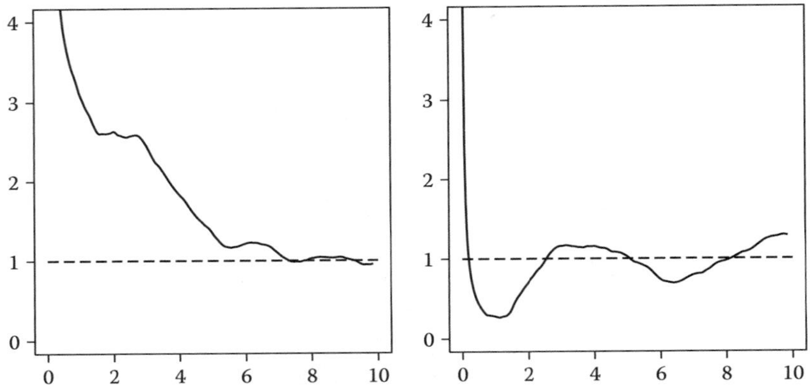

FIGURE 16.5
Estimated pair correlation functions of Kaluzhskie Zaseki Forest data (solid lines). (Left) Trees with diameter less than 15 cm, (right) those with diameter larger than 25 cm. The dashed lines are the pair correlation functions of a homogeneous Poisson process.

n points,

$$\rho_2(\mathbf{x}_1, \mathbf{x}_2) \equiv n(n-1)/|D|^2.$$

Example 16.11

We return to the setup of Example 16.4, and make the further assumptions that the (unmarked) point patterns X_s and X_l of small and large trees, respectively, are isotropic and that first- and second-order product densities exist. Then $\rho_1 \equiv \lambda$, the intensity, and $\rho_2(\mathbf{x}_1, \mathbf{x}_2) = \rho_2(||\mathbf{x}_1 - \mathbf{x}_2||)$ is a function only of the distance $r = ||\mathbf{x}_1 - \mathbf{x}_2||$ between the points. Scaled by intensity, the estimated *pair correlation functions*

$$g(r) = \rho_2(r)/\lambda^2$$

of both patterns are plotted in Figure 16.5. As $g(r)$ greater than 1 indicates a higher probability of finding a pair of r-close points than for a homogeneous Poisson process with the same intensity, the left-hand panel of Figure 16.5 indicates clustering of small (young) trees. The trees having larger diameters avoid each other up to about 2 cm. The mode around 4 cm reflects the typical interpoint distance, which also explains the cyclic behavior of the graph. Note that this interpretation differs somewhat from our earlier interpretation of the empty space function for these data (right-hand panel of Figure 16.3). Later chapters will discuss in more detail the way in which different summary descriptions of point processes can yield different insights when used as tools for data analysis.

Higher-order Campbell measures can be defined similarly (Kallenberg, 1975), for instance

$$C_2(A_1 \times A_2 \times F) = \mathbf{E}[N(A_1)\, N(A_2)\, 1\{X \in F\}]$$

for bounded Borel sets $A_i, i = 1, 2$, and $F \in \mathcal{N}^{\text{lf}}$.

To conclude this survey of Campbell-type measures, let us consider the concept of reduction. The basic idea is to single out each point in a configuration and consider the remaining pattern from this point's perspective. Therefore, reduced Campbell measures are useful tools to describe in a rigorous way the conditional distribution of a point process given that it places mass on a certain (marked) point and, dually, its behavior at a given location conditioned on the pattern elsewhere. A more detailed development of this idea will be given in the next section.

Definition 16.6
Let X be a simple point process on the complete, separable metric space (D, d). Define

$$C^!(A \times F) = \int_{N^{\text{lf}}} \sum_{\mathbf{x} \in \mathcal{X}} 1\{\mathbf{x} \in A\}\, 1\{\mathcal{X} \setminus \{\mathbf{x}\} \in F\}\, d\mathcal{P}(\mathcal{X}) = \mathbf{E}\left[\sum_{\mathbf{x} \in X \cap A} 1\{X \setminus \{\mathbf{x}\} \in F\} \right]$$

for all bounded Borel sets $A \subseteq D$ and all $F \in \mathcal{N}^{\text{lf}}$.

As the Campbell measure of Theorem 16.4, $C^!$ can be extended uniquely to a σ-finite measure on the product σ-algebra of $\mathcal{B}(D)$ and $\mathcal{N}^{\text{lf}}$, the *first-order reduced Campbell measure.* Moreover, the integral representation

$$\mathbf{E}\left[\sum_{\mathbf{x} \in X} g(\mathbf{x}, X \setminus \{\mathbf{x}\}) \right] = \int_D \int_{N^{\text{lf}}} g(\mathbf{x}, \mathcal{X})\, dC^!(\mathbf{x}, \mathcal{X}) \tag{16.5}$$

holds for any measurable function $g\colon D \times N^{\text{lf}} \to \mathbb{R}$ that is either nonnegative or $C^!$-integrable. As for higher orders, Mecke (1979) (see also Hanisch (1982)) defined the *nth-order reduced Campbell measure* of a simple point process X by

$$C_n^!(A_1 \times \cdots A_n \times F) = \mathbf{E}\left[\sum_{\mathbf{x}_1, \ldots, \mathbf{x}_n \in X}^{\neq} 1_{A_1}(\mathbf{x}_1) \cdots 1_{A_n}(\mathbf{x}_n)\, 1\{X \setminus \{\mathbf{x}_1, \ldots, \mathbf{x}_n\} \in F\} \right],$$

or in integral terms

$$\mathbf{E}\left[\sum_{\mathbf{x}_1, \ldots, \mathbf{x}_n \in X}^{\neq} g(\mathbf{x}_1, \ldots, \mathbf{x}_n, X \setminus \{\mathbf{x}_1, \ldots, \mathbf{x}_n\}) \right]$$
$$= \int_D \cdots \int_D \int_{N^{\text{lf}}} g(\mathbf{x}_1, \ldots, \mathbf{x}_n, \mathcal{X})\, dC_n^!(\mathbf{x}_1, \ldots, \mathbf{x}_n, \mathcal{X})$$

for any measurable function $g\colon D^n \times N^{\text{lf}} \to \mathbb{R}^+$, cf. (16.4).

16.1.5 Palm Theory and Conditioning

Henceforth, assume the first-order moment measure M exists and is σ-finite. Fix $F \in \mathcal{N}^{\text{lf}}$ and note that if $C(A \times N^{\text{lf}}) = M(A) = 0$, so is $C(A \times F)$. In other words, the marginal Campbell measure with second argument fixed at F is absolutely continuous with respect to M, so that for all $A \in \mathcal{B}(D)$,

$$C(A \times F) = \int_A \mathcal{P}_{\mathbf{x}}(F)\, dM(\mathbf{x})$$

for some nonnegative Borel measurable function $\mathbf{x} \mapsto \mathcal{P}_{\mathbf{x}}(F)$ on D defined uniquely up to an M-null set. As the exceptional null set may depend on F, it is not immediately clear that the $\mathcal{P}_{\mathbf{x}}(\cdot)$ are countably additive. However, the topological structure imposed on D implies that a version of the Radon–Nikodym derivatives $\mathcal{P}_{\mathbf{x}}(F)$ can be found such that

- For fixed $\mathbf{x} \in D$, $\mathcal{P}_{\mathbf{x}}(F)$ is a probability distribution on $(N^{\text{lf}}, \mathcal{N}^{\text{lf}})$.
- For fixed F, $\mathcal{P}_{\mathbf{x}}(F)$ is a Borel measurable function on D.

The probability distributions $\mathcal{P}_{\mathbf{x}}(\cdot)$ thus defined are *Palm distributions* of X at $\mathbf{x} \in D$, named in honor of Palm (1943). The Radon–Nikodym approach discussed above is due to Ryll-Nardzewski (1961).

Palm distributions are especially useful in simplifying Campbell-type formulas. Indeed, if $\mathbf{E_x}$ denotes expectation with respect to $\mathcal{P}_\mathbf{x}$, (16.3) can be rewritten as

$$\mathbf{E}\left[\sum_{\mathbf{x}\in X} g(\mathbf{x}, X)\right] = \int_D \int_{N^{\text{lf}}} g(\mathbf{x}, \mathcal{X})\, d\mathcal{P}_\mathbf{x}(\mathcal{X})\, dM(\mathbf{x}) = \int_D \mathbf{E_x}\left[g(\mathbf{x}, X)\right] dM(\mathbf{x})$$

for any measurable function $g\colon D \times N^{\text{lf}} \to \mathbb{R}$ that is either nonnegative or C-integrable.

Heuristically speaking, recall that the distribution of a simple point process is fully determined by its void probabilities. Pick some small $\epsilon > 0$ and consider

$$\mathcal{P}(N(A) = 0 \mid N(b(\mathbf{x}, \epsilon)) > 0) \approx \frac{C(b(\mathbf{x}, \epsilon) \times \{N(A) = 0\})}{\mathcal{P}(N(b(\mathbf{x}, \epsilon)) > 0)} = \frac{\int_{b(\mathbf{x},\epsilon)} \mathcal{P}_\mathbf{y}(N(A) = 0)\, dM(\mathbf{y})}{\mathcal{P}(N(b(\mathbf{x}, \epsilon)) > 0)}.$$

For small ϵ, the numerator is approximately equal to $\mathcal{P}_\mathbf{x}(N(A) = 0)\, M(b(\mathbf{x}, \epsilon))$, the denominator to $M(b(\mathbf{x}, \epsilon))$. Hence, $\mathcal{P}_\mathbf{x}(N(A) = 0)$ may be seen as the conditional void probability of A given that X places mass on $\mathbf{x}$.

Example 16.12

The lack of dependence between the points of a Poisson process (cf. Example 16.2) implies a particularly simple form for its Palm distribution. Indeed, let X be a homogeneous Poisson process with intensity $\lambda > 0$ on $\mathbb{R}^2$ and write $\mathcal{P}$ for its distribution. Then, a Palm distribution of X at $\mathbf{x} \in \mathbb{R}^2$ is given by $\mathcal{P} * \delta_\mathbf{x}$, the convolution of $\mathcal{P}$ with an atom at $\mathbf{x}$.

To see this, note that since both $\mathcal{P}_\mathbf{x}$ and $\mathcal{P} * \delta_\mathbf{x}$ are simple, by Theorem 16.2 it is sufficient to prove that their void probabilities coincide. To do so, let B be a bounded Borel set and write $v_\mathbf{x}(B)$ for the probability under $\mathcal{P} * \delta_\mathbf{x}$ that there are no points in B, $v(B)$ for $\mathcal{P}(N(B) = 0)$. For any bounded Borel set $A \subseteq \mathbb{R}^2$, by definition

$$C(A \times \{N(B) = 0\}) = \lambda \int_A \mathcal{P}_\mathbf{x}(N(B) = 0)\, d\mathbf{x}.$$

On the other hand,

$$\begin{aligned} C(A \times \{N(B) = 0\}) &= C(A \setminus B \times \{N(B) = 0\}) = \lambda\, |A \setminus B|\, v(B) = \lambda \int_{A\setminus B} v(B)\, d\mathbf{x} \\ &= \lambda \int_A v_\mathbf{x}(B)\, d\mathbf{x}, \end{aligned}$$

using the fact that for a Poisson process the random variables $N(A \setminus B)$ and $N(B)$ are independent. Since A was chosen arbitrarily, $\mathcal{P}_\mathbf{x}(N(B) = 0) = v_\mathbf{x}(B)$ for almost all $\mathbf{x}$.

Example 16.13

For a simple, stationary marked point process on $D = \mathbb{R}^2 \times \mathbb{R}^+$, as considered in Example 16.6, for all Borel subsets B of $\mathbb{R}^+$ and $\mathbf{y} \in \mathbb{R}^2$, let $F_B^\mathbf{y} = \{\mathcal{X} : (\mathbf{y}, m) \in \mathcal{X} \text{ for some } m \in B\}$ be the event of finding a point at $\mathbf{y}$ with mark in B. Because of the stationarity, a version of Palm distributions can be found that are translates of a single probability distribution, $\mathcal{P}_{(\mathbf{y},m)}(\{T_\mathbf{y}\mathcal{X} : \mathcal{X} \in F\}) = \mathcal{P}_{(0,m)}(F)$ (almost everywhere). An application of the Campbell theorem for $g((\mathbf{y}, m), \mathcal{X}) = 1\{\mathbf{y} \in A; \mathcal{X} \in F_B^\mathbf{y}\}$ yields

$$\lambda \int_A \int_0^\infty \mathcal{P}_{(\mathbf{y},m)}\left(F_B^\mathbf{y}\right) d\nu_M(m)\, d\mathbf{y} = \lambda \int_A \int_0^\infty \mathcal{P}_{(0,m)}\left(F_B^0\right) d\nu_M(m)\, d\mathbf{y} = \lambda\, \nu_M(B)\, |A|$$

for all bounded Borel sets A. Thus, the mark distribution ν_M may be interpreted as the probability distribution of the mark at an arbitrarily chosen point.

Higher-order and reduced Palm distributions can be defined in a straightforward fashion. For example, suppose that the second-order moment measure M_2 exists. Then, Jagers (1973) defined second-order Palm distributions $\mathcal{P}_{\mathbf{x}_1,\mathbf{x}_2}(F)$ that satisfy

$$C_2(A_1 \times A_2 \times F) = \int_{A_1}\int_{A_2} \mathcal{P}_{\mathbf{x}_1,\mathbf{x}_2}(F)\, dM_2(\mathbf{x}_1, \mathbf{x}_2)$$

for all Borel sets $A_i, i = 1, 2$, and $F \in \mathcal{N}^{\text{lf}}$. To obtain a reduced Palm distribution, replace M_n by μ_n and C_n by $C_n^!$ (Mecke, 1979; Hanisch, 1982). The equivalent integral representations hold as well. In particular, Equation (16.5) can be rephrased as

$$\mathbf{E}\left[\sum_{\mathbf{x}\in X} g(\mathbf{x}, X \setminus \{\mathbf{x}\})\right] = \int_D \int_{N^{\text{lf}}} g(\mathbf{x}, \mathcal{X})\, d\mathcal{P}_{\mathbf{x}}^!(\mathcal{X})\, dM(\mathbf{x}) = \int_D \mathbf{E}_{\mathbf{x}}^! \left[g(\mathbf{x}, X)\right] dM(\mathbf{x}),$$

where $\mathcal{P}_{\mathbf{x}}^!$ is a reduced Palm distribution at $\mathbf{x} \in D$.

Example 16.14

From Example 16.12, it follows immediately that a reduced Palm distribution of a homogeneous planar Poisson process with intensity λ is its distribution $\mathcal{P}$. This fundamental property is a consequence of the *Slivnyak–Mecke theorem* stating that the homogeneous Poisson process is characterized by reduced Palm distributions that are translates of a single probability distribution $\mathcal{P}_{\mathbf{x}}^!(\{T_{\mathbf{x}}\mathcal{X} : \mathcal{X} \in F\}) = \mathcal{P}_0^!(F)$ for almost all $\mathbf{x}$ and $\mathcal{P}_0^! = \mathcal{P}$ (Jagers, 1973; Kerstan and Matthes, 1964; Mecke, 1967; Slivnyak, 1962). Hence, the reduced Campbell measure

$$C^!(A \times F) = \lambda \int_A \mathcal{P}_{\mathbf{x}}^!(F)\, d\mathbf{x} = \lambda\, \mathcal{P}(F)\, |A|,$$

where $A \subseteq \mathbb{R}^2$ is a bounded Borel set and $F \in \mathcal{N}^{\text{lf}}$ is a product measure.

The analog of the empty space function (16.1) in a Palm context is the *nearest-neighbor distance distribution function* G. Let X be a simple, stationary point process with locations in $\mathbb{R}^d$. Then, for any $r \geq 0$,

$$G(r) = \mathcal{P}_{\mathbf{x}}^!(d(\mathbf{x}, X) \leq r)$$

is the probability that X places at least one point within distance r of some arbitrarily chosen $\mathbf{x} \in \mathbb{R}^d$. As translation invariance is inherited by Palm distributions, $G(r)$ is well defined and does not depend on the choice of $\mathbf{x}$.

Example 16.15

Estimated nearest-neighbor distance distribution functions for young and established trees in the Kaluzhskie Zaseki Forest data are given in Figure 16.6 and are compared to the G-function of a Poisson process with the same expected number of points (cf. Example 16.12). For small (young) trees, small distances to the nearest neighbor occur with a higher probability than under the Poisson model. Such a behavior is typical for clustered patterns with groups of nearby points separated by gaps. In contrast to the empty space distribution, the G-graph for larger trees differs from its Poisson counterpart in that up until about $r = 4$ cm, $G(r)$ is smaller than under the Poisson model, indicating repulsion at such range; beyond this range, the nearest-neighbor distance distribution is larger than what would be expected if there were no spatial interaction with the steep increase beyond $r = 4$ cm reflecting the preferred interpoint separation.

The ratio

$$J(r) = \frac{1 - G(r)}{1 - F(r)},$$

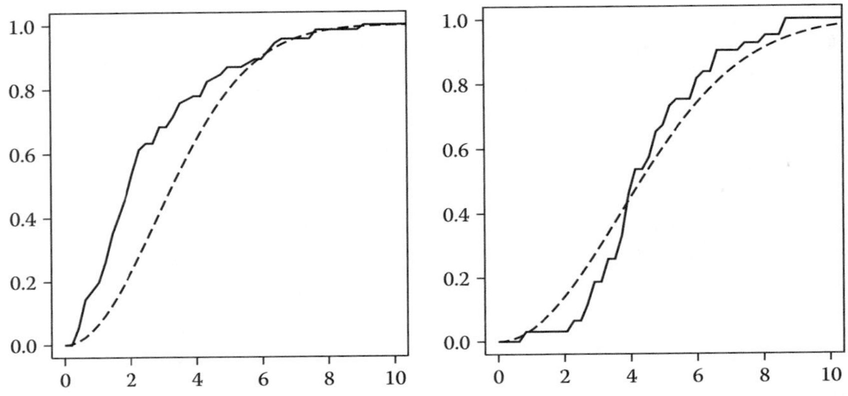

FIGURE 16.6
Estimated nearest-neighbor distance distribution functions of Kaluzhskie Zaseki Forest data (solid lines). (Left) Locations of trees with diameter less than 15 cm, (right) those with diameter larger than 25 cm. The dashed lines are the nearest-neighbor distance distribution functions for homogeneous Poisson processes with, on average, the same number of points-per-unit area as the data.

defined for all $r \geq 0$ such that $F(r) < 1$, compares the void probabilities of closed balls $\bar{b}(\mathbf{x}, r)$ under $\mathcal{P}^!_\mathbf{x}$ and $\mathcal{P}$. The advantage of considering the ratio is that no reference to the intensity of some Poisson process needs to be made. Indeed, values less than 1 indicate that the size of empty spaces tends to be larger than the distance between nearest-neighbor pairs (clustering), whereas values exceeding 1 suggest a more regular pattern (van Lieshout and Baddeley, 1996). The left and center panels of Figure 16.7 confirm the interaction structure implied by the nearest-neighbor distance distribution function. By way of illustration, the J-function of all locations regardless of the mark is shown as well. It indicates clustering at the smallest scales, has a repulsion peak around 2 to 3 cm, then decreases.

Functions, such as F, g, G, or J, provide valuable information and their empirical counterparts are useful tools for data analysis, as discussed in later chapters. Nevertheless, no such low-dimensional function fully characterizes the distribution of the process and it is wise to plot several to grasp different aspects of the underlying distribution.

As we have seen, Palm distributions can be interpreted as conditional distributions given that there is mass at some fixed point. We now turn to describing the dual notion of the

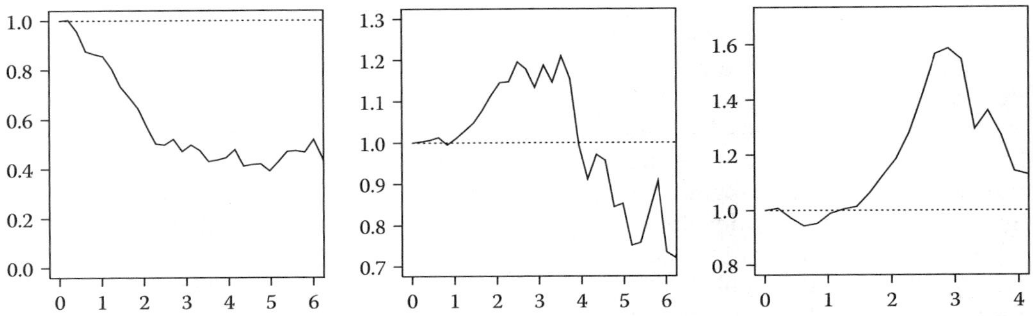

FIGURE 16.7
Estimated J-functions of Kaluzhskie Zaseki Forest data (solid lines). (Left) Locations of trees with diameter less than 15 cm, (center) those with diameter larger than 25 cm, and (right) all tree locations. The dashed lines are the J-functions of a homogeneous Poisson process.

probability mass of some fixed point, conditional on knowing the full realization of the point process elsewhere.

Let X be a simple point process on a complete, separable metric space for which the first-order moment measure $M(\cdot)$ exists as a σ-finite measure. Furthermore, assume that for any fixed bounded Borel set A, the marginal reduced Campbell measure $C^!(A \times \cdot)$ is absolutely continuous with respect to the distribution $\mathcal{P}$ of X. Then

$$C^!(A \times F) = \int_A P_{\mathbf{x}}^!(F)\, dM(\mathbf{x}) = \int_F \Lambda(A; \mathcal{X})\, d\mathcal{P}(\mathcal{X}) \tag{16.6}$$

for some $\mathcal{N}^{\mathrm{lf}}$-measurable function $\Lambda(A; \cdot)$, specified uniquely up to a $\mathcal{P}$-null set. If $\Lambda(\cdot; \mathcal{X})$ admits a Radon–Nikodym derivative $\lambda(\cdot; \mathcal{X})$ with respect to some Borel measure ν on D,

$$C^!(A \times F) = \int_F \int_A \lambda(\mathbf{x}; \mathcal{X})\, d\nu(\mathbf{x})\, d\mathcal{P}(\mathcal{X}) = \int_A \mathbf{E}\left[1_F(X)\, \lambda(\mathbf{x}; X)\right] d\nu(\mathbf{x}).$$

Replacing the indicator function $1_A(\mathbf{x})\, 1_F(\mathcal{X})$ by arbitrary nonnegative functions, we obtain the following definition.

Definition 16.7

Let X be a simple point process on a complete, separable metric space (D, d) equipped with Borel measure ν. If for any measurable function $g\colon D \times N^{\mathrm{lf}} \to \mathbb{R}^+$,

$$\mathbf{E}\left[\sum_{\mathbf{x} \in X} g(\mathbf{x}, X \setminus \{\mathbf{x}\})\right] = \int_D \mathbf{E}\left[g(\mathbf{x}, X)\, \lambda(\mathbf{x}; X)\right] d\nu(\mathbf{x})$$

for some measurable function $\lambda : D \times N^{\mathrm{lf}} \to \mathbb{R}^+$, then X is said to have Papangelou conditional intensity *λ (Papangelou, 1974).*

From a heuristic point of view, by Equation (16.6), $d\, P_{\mathbf{x}}^!(\mathcal{X})\, dM(\mathbf{x}) = \lambda(\mathbf{x}; \mathcal{X})\, d\mathcal{P}(\mathcal{X})\, d\nu(\mathbf{x})$; hence, $\lambda(\mathbf{x}; \mathcal{X})\, d\nu(\mathbf{x})$ can be interpreted as the infinitesimal probability of finding a point of X at the infinitesimal region centered at $\mathbf{x}$ conditional on the event that X and $\mathcal{X}$ agree on the complement of this region. As we shall see in the next section, conditional intensities are especially useful for finite point processes.

Example 16.16

Further to Example 16.14, note that for a homogeneous planar Poisson process with intensity λ,

$$C^!(A \times F) = \lambda\, \mathcal{P}(F)\, |A| = \int_F [\lambda\, |A|]\, d\mathcal{P}(\mathcal{X})$$

so that $\Lambda(A; \mathcal{X}) \equiv \lambda\, |A|$ is a Papangelou kernel. This kernel has density λ with respect to Lebesgue measure and does not depend on either $\mathbf{x}$ or $\mathcal{X}$, as indeed one would have anticipated because of the strong independence properties of the Poisson process.

Example 16.17

Let X be a stationary, simple point process on $\mathbb{R}^d$ with intensity $\lambda > 0$. Suppose X admits a conditional intensity and let $g\colon N^{\mathrm{lf}} \to \mathbb{R}^+$ be a measurable function. Then,

$$\lambda \int_A E_{\mathbf{x}}^!\, g(X)\, d\mathbf{x} = \int_A \mathbf{E}\left[g(X)\, \lambda(\mathbf{x}; X)\right] d\mathbf{x}$$

for all Borel sets A. In fact the *Georgii–Nguyen–Zessin formula* (Georgii, 1976; Nguyen and Zessin, 1979) states that

$$\lambda\, E_{\mathbf{x}}^!\, g(X) = \mathbf{E}\left[g(X)\, \lambda(\mathbf{x}; X)\right] \tag{16.7}$$

for almost all $\mathbf{x} \in \mathbb{R}^d$. Thus, expectations with respect to $\mathcal{P}$ may be translated into reduced Palm expectations and vice versa.

16.1.6 Finite Point Processes

Most point patterns encountered in practice are observed in a bounded region. Sometimes this region is dictated by the application; more often, the spatial process of interest extends over a space that is too large to be mapped exhaustively and data are recorded in a smaller "window" chosen for convenience. In any case, the resulting map contains a *finite* number of points.

The distribution of point processes whose realizations are almost surely finite can be described as follows (Daley and Vere-Jones, 1988; Reiss, 1993). For convenience, suppose that (D, d) is equipped with a Borel measure ν and assume $0 < \nu(D) < \infty$. Then, it suffices to specify

- A discrete probability distribution $(p_n)_{n\in\mathbb{N}_0}$ for the number of points
- A family of symmetric probability densities $\pi_n(\mathbf{x}_1, \ldots, \mathbf{x}_n)_{n\in\mathbb{N}}$ with respect to the n-fold product of ν for the points themselves (i.e., for point locations, and marks if applicable)

The symmetry requirement for the π_n is needed to make sure that the patterns generated by the π_n are permutation invariant, in other words, do not depend on the order in which their points are listed.

It should be noted that for many point processes, p_n cannot be expressed in closed form. If it can, the model specification is algorithmic, so that realizations are easily obtained. Note that if ν is diffuse, the point process is simple.

Janossy densities (Janossy, 1950) are defined in terms of $(p_n, \pi_n)_{n\in\mathbb{N}_0}$, as

$$j_n(\mathbf{x}_1, \ldots, \mathbf{x}_n) = n!\, p_n\, \pi_n(\mathbf{x}_1, \ldots, \mathbf{x}_n), \quad n \in \mathbb{N}_0. \tag{16.8}$$

In an infinitesimal sense, $j_n(\mathbf{x}_1, \ldots, \mathbf{x}_n)\, d\nu(\mathbf{x}_1) \ldots d\nu(\mathbf{x}_n)$ is the probability of finding exactly n points, one at each of infinitesimal regions centered at $\mathbf{x}_1, \ldots, \mathbf{x}_n$. For $n = 0$, Equation (16.8) is conventionally read as $j_0(\emptyset) = p_0$. As $\int_D \cdots \int_D j_n(\mathbf{x}_1, \ldots, \mathbf{x}_n)\, d\nu(\mathbf{x}_1) \ldots d\nu(\mathbf{x}_n) = n!\, p_n$, one may retrieve p_n and, hence, π_n from j_n.

Example 16.18

For the binomial point process introduced in Example 16.1, $p_m = 1\{m = n\}$ for some fixed $n \in \mathbb{N}$. As the points are uniformly and independently scattered over some compact set D, $\pi_n \equiv |D|^{-n}$ with respect to Lebesgue measure. For $m \neq n$, π_n may be defined arbitrarily.

The homogeneous Poisson process of Example 16.2 is described by $p_n = \exp[-\lambda\, |D|]$ $(\lambda\, |D|)^n / n!$ and, for each $n \in \mathbb{N}_0$, $\pi_n \equiv |D|^{-n}$ with respect to the n-fold product of Lebesgue measures. Consequently, $j_n = \lambda^n \exp[-\lambda\, |D|]$.

Although in most applications it is not realistic to assume that points are scattered randomly, Poisson processes are supremely useful as benchmarks. Indeed, one may construct a wide range of point process models by specifying their probability density (Radon–Nikodym derivative) with respect to a Poisson process with finite, diffuse intensity measure ν. However, not all finite point processes are absolutely continuous with respect to a given Poisson model; counterexamples on the plane equipped with Lebesgue measure are those that place points deterministically or in equidistant pairs.

Let X be a point process defined by its probability density f. The probability of the event $\{N(D) = n\}$ is

$$p_n = \frac{e^{-\nu(D)}}{n!} \int_D \cdots \int_D f(\mathbf{x}_1, \ldots, \mathbf{x}_n)\, d\nu(\mathbf{x}_1) \ldots d\nu(\mathbf{x}_n)$$

for each $n \in \mathbb{N}$, with $p_0 = e^{-\nu(D)} f(\emptyset)$ equal to the probability of the empty configuration $\emptyset$. If $p_n > 0$, conditionally on X containing exactly n points, their joint probability distribution is proportional to $f(\mathbf{x}_1, \ldots, \mathbf{x}_n)$. It follows that $j_n(\mathbf{x}_1, \ldots, \mathbf{x}_n) = e^{-\nu(D)} f(\mathbf{x}_1, \ldots, \mathbf{x}_n)$ is also proportional to f.

Example 16.19

Return to the setup of Example 16.2, and let ν be Lebesgue measure restricted to a compact set $D \subseteq \mathbb{R}^2$. A comparison of the Janossy densities of the homogeneous Poisson process X with intensity λ to those of a unit intensity Poisson process yields that X must have density

$$f(\mathbf{x}_1, \ldots, \mathbf{x}_n) = \exp\left[(1 - \lambda)|D|\right] \lambda^n, \qquad n \in \mathbb{N}_0,$$

with respect to the distribution of the unit intensity Poisson process (see also Example 16.18). It is not hard to verify that f indeed defines the distribution of X.

The binomial point process has density

$$f(\mathbf{x}_1, \ldots, \mathbf{x}_n) = n!\, e^{|D|}\, |D|^{-n}$$

for configurations consisting of n points, and $f \equiv 0$ on $\{N(D) \neq n\}$ with respect to the distribution of a unit intensity Poisson process on D.

For finite point processes that are defined in terms of a density with respect to some Poisson process, conditional intensities exist and are easy to compute (Ripley and Kelly, 1977).

Theorem 16.5

Let X be a finite point process on a complete, separable metric space (D, d) with probability density f with respect to the distribution of a Poisson process on D with finite, diffuse intensity measure ν and assume that $f(\mathcal{X}) > 0$ implies $f(\mathcal{X}') > 0$ for all $\mathcal{X}' \subseteq \mathcal{X}$. Then, X has Papangelou conditional intensity

$$\lambda(\mathbf{x}; \mathcal{X}) = \frac{f(\mathcal{X} \cup \{\mathbf{x}\})}{f(\mathcal{X})} \tag{16.9}$$

for $\mathbf{x} \in D \setminus \mathcal{X}$ and configurations $\mathcal{X}$ such that $f(\mathcal{X}) > 0$.

For processes that are absolutely continuous, the conditional intensity provides a fourth way of defining the distribution. Indeed, by Equation (16.9),

$$f(\{\mathbf{x}_1, \ldots, \mathbf{x}_n\}) \propto \prod_{i=1}^{n} \lambda(\mathbf{x}_i; \{\mathbf{x}_1, \ldots, \mathbf{x}_{i-1}\})$$

regardless of the order in which the points are labeled.

In summary, finite point processes are usually easier to deal with than most infinite ones. Their distribution is known once we have specified either $(p_n, \pi_n)_{n \in \mathbb{N}_0}$ or $j_n, n \in \mathbb{N}_0$. If the point process is absolutely continuous with respect to a Poisson process, alternative modeling strategies are to specify f or $\lambda(\cdot; \cdot)$.

16.1.7 Gibbs Measures by Local Specification

Infinite point processes are not easily specified by a density with respect to a (homogeneous) Poisson process. Indeed, even homogeneous Poisson processes on $\mathbb{R}^d$ with different intensities are not absolutely continuous with respect to each other (see, e.g., van Lieshout (2000)). Nevertheless, one may try to define a point process X by specification of a family of conditional densities for the finite restrictions of X to bounded Borel sets.

Again, assume that (D, d) is a complete, separable metric space equipped with a diffuse, locally finite Borel measure ν. Define a family of probability densities $f_B(\mathcal{X} \cap B \mid \mathcal{X} \cap B^c)$, B a bounded Borel set, with respect to the law of a Poisson process with intensity measure ν restricted to B. Here $\mathcal{X}$ ranges through N^{lf}. Then, for $F \in \mathcal{N}^{\mathrm{lf}}$,

$$\sum_{n=0}^{\infty} \frac{e^{-\nu(B)}}{n!} \int_B \cdots \int_B 1_F(\{\mathbf{x}_1, \ldots, \mathbf{x}_n\} \cup (\mathcal{X} \cap B^c))\, f_B(\mathbf{x}_1, \ldots, \mathbf{x}_n \mid \mathcal{X} \cap B^c)\, d\nu(\mathbf{x}_1) \cdots d\nu(\mathbf{x}_n)$$

can be interpreted as the probability that X falls in F conditional on the event $\{X \cap B^c = \mathcal{X} \cap B^c\}$.

It is not at all obvious that the family $f_B(\cdot \mid \cdot)$ defines a point process distribution $\mathcal{P}$. A necessary condition is that the *Dobrushin–Landford–Ruelle equations*

$$\mathcal{P}(F) = \int \sum_{n=0}^{\infty} \frac{e^{-\nu(B)}}{n!} \int_B \cdots \int_B 1_F(\{\mathbf{x}_1, \ldots, \mathbf{x}_n\} \cup (\mathcal{X} \cap B^c))$$
$$f_B(\mathbf{x}_1, \ldots, \mathbf{x}_n \mid \mathcal{X} \cap B^c)\, d\nu(\mathbf{x}_1) \cdots d\nu(\mathbf{x}_n)\, d\mathcal{P}(\mathcal{X}) \qquad (16.10)$$

are satisfied for all $F \in \mathcal{N}^{\mathrm{lf}}$ and all bounded Borel sets $B \subseteq \mathbb{R}^d$, in which case $\mathcal{P}$ is said to be a *Gibbs point process* with *local specification* $f_B(\cdot \mid \cdot)$.

An extensive overview of conditions for the existence of a solution to Equation (16.10) was given in Preston (1976), typically phrased in terms of bounds on the log conditional densities or intensities, and restrictions on the points in B^c that may affect f_B. For later developments, see, for example, Georgii and Häggström (1996) or Glőtzl (1980). Note that even if there is a solution, it may not be unique, a phenomenon known as "phase transition" (see Preston (1976)).

Example 16.20

Let ν be Lebesgue measure on $\mathbb{R}^2$ and consider the *area interaction process* defined by

$$f_B(\mathcal{X} \cap B \mid \emptyset) = \alpha_B(\emptyset)\, \beta^{n(\mathcal{X})} \exp\left[-\beta\, |U_r(\mathcal{X})|\right]$$

for bounded Borel sets B in the plane. Here, the parameter β is assumed to be strictly positive, $U_r(\mathcal{X}) = \cup_{\mathbf{x}_i \in \mathcal{X}} \bar{b}(\mathbf{x}_i, r)$ is the union of closed balls around each of the points in $\mathcal{X}$, and $\alpha_B(\emptyset)$ is the normalizing constant that makes $f_B(\cdot \mid \emptyset)$ a probability density.

Note that a ball of radius r intersects B only if its center is at most r away from the boundary of B or, in other words, falls in $B_{\oplus r} = \{\mathbf{b} + \mathbf{c} : \mathbf{b} \in B; \mathbf{c} \in \bar{b}(0, r)\}$. Hence, we may define, for any configuration $\mathcal{X}$ and bounded Borel set $B \subseteq \mathbb{R}^2$,

$$f_B(\mathcal{X} \cap B \mid \mathcal{X} \cap B^c) = \alpha_B(\mathcal{X} \cap B^c)\, \beta^{n(\mathcal{X})} \exp\left[-\beta\, |U_r(\mathcal{X}) \cap B_{\oplus r}|\right],$$

where again $\alpha_B(\mathcal{X} \cap B^c)$ is the normalizing constant. It can be shown that there exists a stationary solution to the Dobrushin–Landford–Ruelle equations (Baddeley and van Lieshout, 1995). Moreover, it is known that multiple solutions exist when β is large, whereas the solution is unique for β small (Ruelle, 1971; see also Chayes, Chayes, and Kotecky 1995; Dereudre, 2008; and Georgii and Häggström, 1996). It is plausible that the occurrence of a phase transition is monotone in β, but to the best of our knowledge no rigorous proof exists to date.

References

A.J. Baddeley and M.N.M. van Lieshout. Area-interaction point processes. *Annals of the Institute of Statistical Mathematics*, 46:601–619, 1995.

C. Berg, J.P.R. Christensen, and P. Ressel. *Harmonic Analysis on Semigroups*. Springer, Berlin, 1984.

N.R. Campbell. The study of discontinuous phenomena. *Proceedings of the Cambridge Philosophical Society*, 15:117–136, 1909.

J.T. Chayes, L. Chayes, and R. Kotecky. The analysis of the Widom–Rowlinson model by stochastic geometric methods. *Communications of Mathematical Physics*, 172:551–569, 1995.

G. Choquet. Theory of capacities. *Annals of the Fourier Institute*, V:131–295, 1953/1954.

D.R. Cox and V. Isham. *Point Processes*. Chapman & Hall, London, 1980.

D. Daley and D.J. Vere-Jones. *An Introduction to the Theory of Point Processes*, 1st ed., Springer, New York, 1988. *Second edition, Volume I: Elementary Theory and Methods*, 2003. *Second edition, Volume II: General Theory and Structure*, 2008.

D. Dereudre. Existence of quermass processes for nonlocally stable interaction and nonbounded convex grains. Preprint, Université de Valenciennes et du Hainaut-Cambrésis, France, 2008.

A.K. Erlang. The theory of probabilities and telephone conversations (in Danish). *Nyt Tidsskrift for Matematik*, B20:33–41, 1909.

H.-O. Georgii. Canonical and grand canonical Gibbs states for continuum systems. *Communications of Mathematical Physics*, 48:31–51, 1976.

H.-O. Georgii and O. Häggström. Phase transition in continuum Potts models. *Communications of Mathematical Physics*, 181:507–528, 1996.

E. Glőtzl. Lokale Energien und Potentiale fűr Punktprozesse (in German). *Mathematische Nachtrichten*, 96:198–206, 1980.

P. Grabarnik and S.N. Chiu. Goodness-of-fit test for complete spatial randomness against mixtures of regular and clustered spatial point processes. *Biometrika*, 89:411–421, 2002.

K.-H. Hanisch. On inversion formulae for n-fold Palm distributions of point processes in LCS-spaces. *Mathematische Nachrichten*, 106:171–179, 1982.

J. Illian, A. Penttinen, H. Stoyan, and D. Stoyan. *Statistical analysis and Modelling of Spatial Point Patterns*. Wiley, Chichester, U.K., 2008.

L. Janossy. *Cosmic Rays*. Oxford University Press, Oxford, U.K., 1948.

L. Janossy. On the absorption of a nucleon cascade. *Proceedings of the Royal Irish Academy of Sciences*, A53:181–188, 1950.

O. Kallenberg. *Random Measures*. Akademie-Verlag, Berlin, 1975.

J. Kerstan and K. Matthes. Verallgemeinerung eines Satzes von Sliwnjak. (in German). *Académie de la République Populaire Roumaine. Revue Roumaine de Mathématiques Pures et Appliquées*, 9:811–830, 1964.

A.Y. Khinchin. *Mathematical Methods in the Theory of Queueing*. Griffin, London, 1960.

G. Kummer and K. Matthes. Verallgemeinerung eines Satzes von Sliwnjak. II (in German). *Académie de la République Populaire Roumaine. Revue Roumaine de Mathématiques Pures et Appliquées*, 15:845–870, 1970.

G. Matheron. *Random Sets and Integral Geometry*. John Wiley & Sons, New York, 1975.

K. Matthes, J. Kerstan, and J. Mecke. *Infinitely Divisible Point Processes*. Wiley, Chichester, U.K., 1978.

B. McMillan. Absolutely monotone functions. *Annals of Mathematics*, 60:467–501, 1953.

G. Mőnch. Verallgemeinerung eines Satzes von A. Rényi (in German). *Studia Scientiarum Mathematicarum Hungarica*, 6:81–90, 1971.

J. Neyman. On a new class of "contagious" distributions applicable in entomology and bacteriology. *Annals of Mathematical Statistics*, 10:35–57, 1939.

X.X. Nguyen and H. Zessin. Integral and differential characterization of the Gibbs process. *Mathematische Nachrichten*, 88:105–115, 1979.

T. Norberg. Existence theorems for measures on continuous posets, with applications to random set theory. *Mathematica Scandinavica*, 64:15–51, 1989.

C. Palm. Intensitătsschwankungen im Ferngesprechverkehr (in German). *Ericsson Techniks*, 44:1–189, 1943.

F. Papangelou. The conditional intensity of general point processes and an application to line processes. *Zeitschrift für Wahrscheinlichkeitstheorie und verwandte Gebiete*, 28:207–226, 1974.

S.D. Poisson. *Recherches sur la Probabilité des Jugements en Matière Criminelle et en Matière Civile, Précédées des Règles Générales du Calcul des Probabilités* (in French). Bachelier, Paris, 1837.

C. Preston. *Random Fields*. Springer, Berlin, 1976.

R.-D. Reiss. *A Course on Point Processes*. Springer, New York, 1993.

A. Rényi. Remarks on the Poisson process. *Studia Scientiarum Mathematicarum Hungarica*, 2:119–123, 1967.

B.D. Ripley and F.P. Kelly. Markov point processes. *Journal of the London Mathematical Society*, 15:188–192, 1977.

D. Ruelle. Existence of a phase transition in a continuous classical system. *Physical Review Letters*, 27:1040–1041, 1971.

C. Ryll-Nardzewski. Remarks on processes of calls. *Proceedings of the 4th Berkeley Symposium on Mathematical Statistics*, 2:455–465, 1961.

Y.M. Slivnyak. Some properties of stationary flows of homogeneous random events (in Russian). *Teorija Verojatnostei i ee Primenenija*, 7:347–352, 1962.

O.V. Smirnova (Ed.) *East-European Broad Leaved Forests* (in Russian). Nauka, Moscow, 1994.

S.K. Srinivasan. *Stochastic Point Processes and Their Applications*. Griffin, London, 1974.

M.N.M. van Lieshout. *Markov Point Processes and Their Applications*. Imperial College Press, London, 2000.

M.N.M. van Lieshout and A.J. Baddeley. A nonparametric measure of spatial interaction in point patterns. *Statistica Neerlandica*, 50:344–361, 1996.

17

Spatial Point Process Models

Valerie Isham

CONTENTS

17.1 Model Construction

This chapter describes various classes of parametric point process models that are useful for scientific work. In any particular application, there will generally be an underlying physical mechanism that generates the point events that are observed, and which may be fully or partially understood. In building a model for these events, the modeler often seeks to represent or reflect that physical process, albeit in a highly simplified way. Thus, for example, if the point events are the locations of seedling trees, it will be natural to build a model that takes into account the positions of parent trees, the clustering of the seedlings around these, and perhaps also the prevailing wind direction, even if the exact process of seed generation and dispersal is not represented. Such a model is often termed "mechanistic," and has interpretable parameters that relate to physical phenomena. In contrast, "descriptive" models aim to represent the statistical properties of the data and their dependence on explanatory variables without necessarily worrying about the physical mechanisms involved. For example, a model that involves inhibition between nearby events can be used to model the positions of ants' nests (Harkness and Isham, 1983; see also Chapter 19, Section 19.4.2 of this volume for further discussion). The inhibition reflects competition for resources, but is not modeled directly.

It will often be simplest to construct mechanistic models of complex phenomena in terms of simpler point process models whose properties are well understood, and which are used as "building blocks." The definitions of these simple models may reflect either particular generic mechanisms for generating events (as in the hypothetical seedling example above), or particular statistical properties (for example, that there is inhibition between events, or that numbers of events in disjoint sets are independent). Thus, in this chapter, we consider some simple point process models that can be used either individually, or in combination, to represent spatial point events. Extensive further discussion and examples of the use of these models in applications can be found in many published overviews on spatial and point processes, some of which are mentioned here. Cox and Isham (1980) is a relatively informal account concentrating mostly on point processes in a single dimension, but much of the material extends easily to the multidimensional case and is at a level comparable to this chapter. Other early books include Daley and Vere-Jones (1988), which has recently been revised and expanded to two volumes (Daley and Vere-Jones, 2003, 2007, with spatial processes included in volume II) on rigorous mathematical theory; Karr (1991) and Cressie (1993), both of which focus on statistical inference (the latter for spatial data); and Stoyan, Kendall, and Mecke (1995) on modeling of spatial processes. More recent volumes have a strong emphasis on spatial processes and address mathematical theory (van Lieshout, 2000), the methodology of statistical inference (Møller and Waagepetersen, 2004, 2007), and data analysis in a range of applied fields (Baddeley, Møller, and Waagepetersen 2006; Diggle, 2003; Illian, Penttinen, Stoyan, and Stoyan 2008; Ripley, 2004), although the distinction between these three areas is far from absolute and there are substantial overlaps in coverage between the cited references.

In this chapter, we will consider only processes where events occur singly, noting that processes with multiple events are usually most easily modeled in two stages, by first modeling the process of distinct event locations and then adding a *mark* to each event to specify its multiplicity. Processes with more general marks, representing the characteristics of each event, are also of interest and are discussed in Chapter 21.

While the specific models of spatial point processes to be described below will often be specified via a density or a method of construction, for theoretical purposes point processes in spaces of more than one dimension are normally specified in terms of their joint (finite dimensional, or *fidi*) distributions of counts. Mathematically, we consider a point process, N, as a random counting measure on a space $S \subseteq \mathbb{R}^d$, where for spatial applications d is most usually 2 or 3. Thus, N takes non-negative integer values, is finite on bounded sets, and is countably additive, i.e., if $A = \cup_{i=1}^{\infty} A_i$, then $N(A) = \sum_{i=1}^{\infty} N(A_i)$. In order that the process is well-defined, the definition is restricted to Borel subsets (A) of S. The fidi distributions $\mathrm{P}(N(A_i) = n_i, i = 1, \ldots, k)$ must then be specified consistently (for all $n_i = 0, 1, \ldots; i = 1, \ldots, k; k = 1, 2, \ldots$) for arbitrary Borel sets A_i. For more details, see Section 16.2 above or, for example, Daley and Vere-Jones (2003).

Example 17.1

The homogeneous Poisson process: The most fundamental point process is the homogeneous Poisson process. For this, if A, A_i $(i = 1, \ldots, k)$ are bounded Borel subsets of S, and are disjoint, then

(i) $N(A)$ has a Poisson distribution with mean $\mu(A) = \lambda|A|$; where $|A|$ is the volume (Lebesgue measure) of A.

(ii) $N(A_1), \ldots, N(A_k)$ are independent random variables, for all k, and all $A_i (i = 1, \ldots, k)$.

If we consider the number of events in a ball centered at x and having volume ϵ, in the limit as $\epsilon \to 0$, then it follows from (i) above that the probability that there is exactly one event in the ball is $\lambda\epsilon + o(\epsilon)$ and, hence, λ is the *intensity (rate)* of the process (i.e., the limiting probability of an event per unit volume; see also Section 17.3). Since this limit does not depend on the location x, the Poisson process is said to be *homogeneous*. Similarly, it follows from (ii) above that the *conditional intensity function*, $h(x)$, (i.e., the limiting probability of an

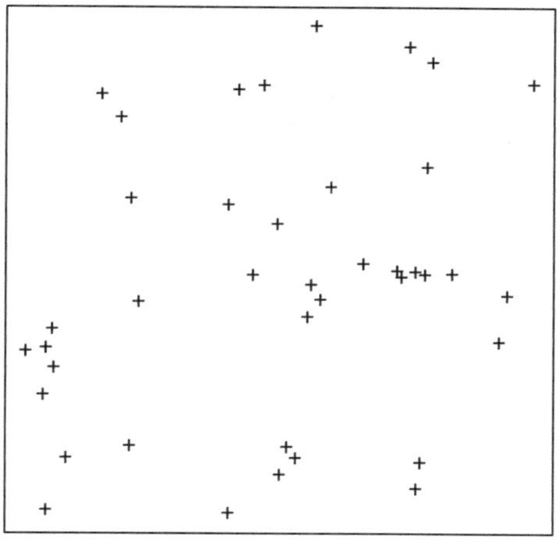

FIGURE 17.1
Realisation of a two-dimensional Poisson process.

event in a ball centered at x conditionally upon there being an event at $\mathbf{0}$; see Section 16.3) is a constant, so that $h(x) = \lambda$ for all x.

The lack of dependence between events in disjoint sets, specified in (ii) above and sometimes known as the property of *complete spatial randomness (CSR)*, is an important one for applications and provides a baseline against which to measure clustering or inhibition of events. For example, the pair correlation function $g(x, u)$ (defined as the ratio of the joint product density, $\rho^{(2)}(x, u)$ to the product of the intensities at x and u; see Section 16.4) is identically one for Poisson processes, with values greater than one indicating clustering of events and values less than one indicating inhibition. An equivalent way to view this property is that, given the number of events in A, these events are independently and uniformly distributed over the set. Because, the joint probability of the event $N(A) = n$ and the density of the locations $x_1, \ldots, x_n$ is $\lambda^n e^{-\lambda|A|}$, and thus the conditional density of the locations is $n!/|A|^n$, which is the joint density of an ordered set of n independent variables, each uniformly distributed on A. This result provides a simple mechanism for simulating realizations of a spatial Poisson process in a set S. It is only necessary first to generate an observation n, say, from a Poisson distribution with the known mean $\mu(S) = \lambda|S|$, and then to distribute the n events independently and uniformly over S. When $N(S) = n$ is fixed, the process obtained by locating events independently and at random over S is known as a *binomial* process. In this case, if A_i $(i = 1, \ldots, k)$ are disjoint subsets of S, then the joint distribution $N(A_1), \ldots, N(A_k)$ is a multinomial distribution with index n and probabilities $|A_1|/|S|, \ldots, |A_k|/|S|$ (see also Section 16.1).

It is worth noting that this conditional uniform distribution does not mean that the events will appear evenly distributed over A; see Figure 17.1 from which it can be seen that a typical realization of a homogeneous Poisson process has apparent clusters as well as large holes. Poisson processes, including the nonhomogeneous and doubly stochastic variants introduced in the following two examples, will be discussed further in Section 17.2.

Example 17.2

Nonhomogeneous Poisson process: The Poisson process defined in Example 17.1 above has a constant rate function and, if $S = \mathbb{R}^d$, is stationary. A nonhomogeneous Poisson process is defined similarly, but has a spatially varying intensity $\lambda(x)$, so that property (i) is replaced by

(i′) N(A) has a Poisson distribution with mean $\mu(A) = \int_A \lambda(x)\, dx$, where it is assumed that the function λ is such that $\mu(A)$ is finite for all bounded Borel sets $A \subseteq S$, while property (ii) specifying independence of counts in disjoint sets is retained. One possibility is that $\lambda(\boldsymbol{x})$ might depend on some observed explanatory variable, e.g., $\lambda(\boldsymbol{x}) = cz(\boldsymbol{x})$ for a known spatial function $z(\boldsymbol{x})$. For example, the rate of cases of an infection would likely depend on the density of the underlying population.

Example 17.3

Doubly stochastic Poisson (Cox) process: Now suppose that the rate function of the nonhomogeneous Poisson process is allowed to be random. In particular, let Λ be a real, nonnegative valued stochastic process on S. Typically $\{\Lambda(\boldsymbol{x})\}$ is not observed. Then a Cox process (Cox, 1955) driven by Λ is such that, given Λ, the process is a Poisson process with intensity function Λ (and is nonhomogeneous unless Λ is a single random variable—i.e., is spatially constant—in which case it is a mixed Poisson process; see Section 17.2). In this case,

$$\mu(A) = \mathrm{E}(N(A)) = \mathrm{E}_\Lambda\left(\int_A \Lambda(\boldsymbol{x})\, d\boldsymbol{x}\right)$$

Some further examples of spatial processes can be found in Sections 6.2 and 6.3 of Cox and Isham (1980). The discussions of these are mainly confined to $d = 2$ and based on rather specific constructions.

Many of the standard models for spatial point processes that will be described in the rest of this chapter, and are used in applications, are based on one of the three variants of the Poisson process described above. To obtain further models, it is often useful to start with a standard model, and to operate on it in some way to create a new one. For example, one could start with a process that is completely deterministic and has events in a regular lattice, and then introduce randomness and remove some of the regularity by a combination of randomly removing events (*thinning*), moving them (*translation*), or adding extra events (*superposition*). In the following sections, each of these three operations will be discussed in turn. First though, we consider the *probability generating functional*, which provides a convenient way of representing the finite-dimensional distributions and, thus, specifying the process.

17.1.1 The Probability Generating Functional

If Y is a nonnegative, integer-valued, random variable, the probability generating function (pgf) $G_Y(z) = \mathrm{E}(z^Y)$, defined for $|z| \leq 1$, is a useful tool that characterizes the distribution of the random variable. For a point process, N, the pgf is generalized to the *probability generating functional* (pgfl) defined by

$$G_N(\xi) = \mathrm{E}\left[\exp\left\{\int_S \ln \xi(\boldsymbol{x}) dN(\boldsymbol{x})\right\}\right] = \mathrm{E}\left[\prod \xi(\boldsymbol{x}_n)\right] \tag{17.1}$$

where the $\boldsymbol{x}_n$ are the locations of the events, and the function ξ belongs to a suitable family—we require $0 \leq \xi(\boldsymbol{x}) \leq 1$ for all $\boldsymbol{x} \in S$, and ξ to be identically 1 outside some bounded Borel set (so that there are only a finite number of nonunit terms in the product above, which, thus, converges). Use of the probability generating functional in point processes goes back at least to Bartlett (1955, for the Cox process); see also Westcott (1972).

For a homogeneous Poisson process, the pgfl is given by

$$G_N(\xi) = \exp\left[-\lambda \int_S \{1 - \xi(\boldsymbol{x})\}\, d\boldsymbol{x}\right],$$

which reduces to the familiar pgf, $\exp\{-\lambda|A|(1-z)\}$, of a Poisson variable with mean $\lambda|A|$ when $\xi(x) = z$ for $x \in A$ and $\xi(x) = 1$ otherwise, and $A \subseteq S$ is bounded. The form of the pgfl can be seen as follows. Suppose that ξ is identically 1 outside the bounded set A, and condition on the number of events in A, $N(A) = n$, and their locations $x_1, \ldots, x_n$. Then

$$
\begin{aligned}
\mathrm{E}&\left[\exp\left\{\int_S \ln \xi(x) dN(x)\right\}\right] \\
&= \sum_{n=0}^{\infty} \frac{e^{-\lambda|A|}\{\lambda|A|\}^n}{n!} \int_A dx_1 \ldots \int_A dx_n \frac{\xi(x_1)\ldots\xi(x_n)}{|A|^n} \\
&= e^{-\lambda|A|} \sum_{n=0}^{\infty} \left\{\int_A dx\lambda\xi(x)\right\}^n / n! \\
&= \exp\left[-\lambda \int_A \{1 - \xi(x)\}\, dx\right]. \qquad (17.2)
\end{aligned}
$$

Finally, the range of the integral can be extended from A to S as $\xi \equiv 1$ outside A.

The pgfl provides a complete description of a point process via its fidi distributions. For example, if $A_1, \ldots, A_k$ are disjoint, bounded subsets of S, then setting $\xi(x) = z_i$ when $x \in A_i$ $(i = 1, \ldots, k)$ in $G_N(\xi)$, gives the joint pgf of $N(A_1), \ldots, N(A_k)$. It plays a particularly important role in proving theoretical results for point processes, for example, in deriving the limiting results on thinning, translation, and superposition to be described below.

If we consider two independent random variables Y_1 and Y_2, then properties of their sum $Y = Y_1 + Y_2$ can be easily obtained from its pgf, and it follows immediately from the definition of a pgf, that the pgf of Y is simply the product of the pgfs of Y_1 and Y_2. In the same way, if N is the *superposition* $N = N_1 + N_2$ of two independent point processes N_1 and N_2, (i.e., for all Borel sets A, $N(A) = N_1(A) + N_2(A)$), then it follows immediately that $G_N(\xi) = G_{N_1}(\xi)\, G_{N_2}(\xi)$, allowing the properties of N to be directly related to those of N_1 and N_2.

17.1.2 Thinning

Suppose that some of the events of a point process N are deleted to produce a thinned process, the simplest scheme being to delete each event independently with probability $1 - p$. Denote the process of retained events by N_p. Then, with probability p an event at x contributes a factor $\xi(x)$ to the product on the right-hand side of Equation (17.1), while with probability $1-p$ the factor is 1. Thus, the thinned process has pgfl $G_{N_p}(\xi) = G_N(1-p(1-\xi))$, where G_N is the pgfl of N.

It follows immediately that a homogeneous Poisson process of rate λ thinned in this way is another homogeneous Poisson process with rate $p\lambda$. More interestingly, the process of retained events is independent of the (Poisson) process of deleted events. To see this, let $N_1 (= N_p)$ and N_2 denote the processes of retained and deleted events, define the joint probability generating functional of N_1 and N_2 by

$$
G_{N_1,N_2}(\xi_1, \xi_2) = \mathrm{E}\left[\exp\left\{\int_S [\ln \xi_1(x) dN_1(x) + \ln \xi_2(x) dN_2(x)]\right\}\right] \qquad (17.3)
$$

and use a conditioning argument similar to that used for Equation (17.2), and where $N_1(A)$ given $N(A)$ has a binomial distribution.

If N is stationary (assume here that $S = \mathbb{R}^d$) with intensity λ and conditional intensity function h (see Example 1 above and Section 16.3), then N_p is also stationary, with intensity

$\lambda_p = p\lambda$ and conditional intensity $h_p(x) = ph(x)$. For bounded Borel sets A and B, it is straightforward to show that

$$\begin{aligned} \mathrm{Cov}(N(A), N(B)) &= \int_A \int_B c(u - v)\, du\, dv \\ &= \lambda |A \cap B| + \lambda \int_A \int_B h(u - v)\, du\, dv - \lambda^2 |A|\, |B| \end{aligned} \qquad (17.4)$$

where $c(x) = \lambda\{\delta(x) + h(x) - \lambda\}$ is the *covariance density* of N and δ is the Dirac delta function; see Cox and Isham (1980, Section 2.5) for the case $d = 1$, and Daley and Vere-Jones (2003) for a more general exposition. It follows that

$$\mathrm{Cov}(N_p(A), N_p(B)) = p^2 \mathrm{Cov}(N(A), N(B)) + p(1 - p)\lambda |A \cap B|.$$

Suppose now that N is not only thinned, but also rescaled so that the intensity, λ, of the process is unchanged. Then, since we want the volume element to scale with p, the thinned-rescaled process must have an event at $p^{1/d}\, x$ if N has an event at x that is retained. It follows that the thinned-rescaled process $\tilde{N}_p$, say, has conditional intensity $\tilde{h}_p(x) = h(x/p^{1/d})$ and probability generating functional

$$G_{\tilde{N}_p}(\xi) = \exp\left\{ \int_S p \ln[1 - p\,(1 - \xi(p^{1/d}\, x))]\, dN(x) \right\}.$$

An important property of the thinned-rescaled process is that, under rather general conditions (see Daley and Vere-Jones, 2007, Section 11.3, and references therein), as $p \to 0$, $\tilde{N}_p$ tends to a Poisson process with rate λ. This result can be obtained from the pgfl via expansions of the integrand. Intuitively, when the events are deleted independently and p is close to zero, retained events will tend to be far apart in the original process so that, unless there are long-range dependencies in N, there will be little dependence left in $\tilde{N}_p$.

17.1.3 Translation

The Poisson process also can be obtained as a limit if the events of a fairly arbitrary point process on $S = \mathbb{R}^d$ are translated independently, and the displacements of distinct events are identically distributed with a nonlattice distribution, as long as the distribution of the displacements is sufficiently dispersed that the events of N are thoroughly mixed and the dependencies between them lost. The same result can be obtained by repeatedly translating each event a large number of times. Proofs, again based on the use of pgfls, and further references can be found in Daley and Vere-Jones (2007, Section 11.4).

The translated process, N_d say, can be viewed as a rather trivial kind of cluster process (see Section 17.3) in which there is just one event per cluster. The pgfl is given by $G_{N_d}(\xi) = G_N(\mathrm{E}\{\xi(\cdot + X)\})$ where X is a random displacement.

If N is stationary with intensity λ and conditional intensity h, then the translated process N_d is also stationary with rate λ, and has conditional intensity

$$h_d(x) = \int_S h(x - u) f_D(u)\, du,$$

where f_D is the density of the difference $D = X_1 - X_2$ between two independent displacements. Second-order properties follow. If N is a homogeneous Poisson process, then the translated process is another Poisson process.

17.1.4 Superposition

The third operation that gives rise in the limit to a Poisson process is that of superposition. Taken together, these limiting results on thinning, translation, and superposition explain the central role of the Poisson process in point process theory and as a simple approximating model in applications.

The limiting results on superpositions (see Daley and Vere-Jones, 2007, Section 11.2, for careful discussion and further references) are direct analogs for point processes of the central limit theorem for sums of random variables. Suppose, first, that n independent copies of a stationary point process, N, on $S = \mathbb{R}^d$ and having intensity λ, are superposed, and that the superposed process is rescaled to keep the intensity of the superposition constant as n increases (the scale must be dilated by $n^{1/d}$). Then, as $n \to \infty$, the rescaled superposition tends to a homogeneous Poisson process with intensity λ.

As with the central limit theorem for random variables, this limiting Poisson result holds not only if the component processes are independent and identically distributed, but more generally. In broad terms, the processes do not have to be identically distributed, but must be such that they are suitably sparse (in the rescaled superposition, it must be sufficiently unlikely that there will be more than one point from any component process in any bounded set) with no one process dominating the rest. The processes do not have to be independent as long as the dependencies between them are relatively weak. If, however, the component processes *are* independent, then, in the rescaled superposition outlined above, the number of events in A is, approximately, a sum of independent indicator variables and, thus, has, approximately, a Poisson distribution.

Suppose that $N_1, \ldots, N_k$ are independent, stationary processes with intensities λ_i, conditional intensity functions $h_i(x)$, and corresponding covariance densities $c_i(x) = \lambda\{\delta(x) + h(x) - \lambda\}$ for $i = 1, \ldots, k$ (see Section 17.1.2). Then their superposition $N = N_1 + \cdots + N_k$ has rate $\lambda = \sum_{i=1}^{k} \lambda_i$ and covariance density

$$c(x) = \sum_{i=1}^{k} c_i(x) = \sum_{i=1}^{k} \rho_i\{\delta(x) + h_i(x) - \lambda_i\}.$$

It follows that

$$h(x) = \lambda + \lambda^{-1} \sum_{i=1}^{k} \lambda_i\{h_i(x) - \lambda_i\},$$

from which second-order properties can be deduced. It can be seen that if $N_1, \ldots, N_k$ are Poisson processes, then their superposition is also a Poisson process.

17.2 Poisson Processes

In Example 17.1, the homogeneous Poisson process was introduced. This is the most fundamental point process, in part because it has the property of complete spatial randomness in that there are no interactions between the events. It can therefore be used as the base against which clustering and inhibition of events are measured. In addition, its "central limit"-like role as the limiting process under wide classes of thinning, translation, and superposition operations, means that it plays an important part in point process theory. The lack of dependence between events means that the properties of the Poisson process are generally well understood and simple to derive, and, therefore, it is used as a starting point in the construction of many more complex models.

For a Poisson process, the summary statistics often estimated in applications are particularly simple. For a general stationary process, satisfying appropriate regularity conditions and with intensity λ, the K function (Ripley, 1976, 1977; and see also Section 18.6) is such that $\lambda K(r)$ is the mean number of events within a distance r of an arbitrary event. Alternatively, K can be defined in terms of an integral of the pair correlation function $g(x, u) = \rho^{(2)}x, u)/\lambda^2$. Thus, suppose that N is a stationary Poisson process with $S = \mathbb{R}^d$, and intensity λ, so that $N(A)$ has a Poisson distribution with mean $\lambda|A|$. Let $b_d = \pi^{d/2}/\Gamma(1 + d/2)$ denote the volume of a ball of unit radius in $\mathbb{R}^d$, so that for $d = 1, 2$, and 3, $b_d = 2$, π and $4\pi/3$, respectively. Then, the K function is given by $K(r) = b_d r^d$, and defining the L function by $L(r) = \sqrt[d]{K(r)/b_d}$, it follows that $L(r) = r$, for all $r > 0$. As a basis for comparison, therefore, an L function that increases more slowly than linearly, at least for small values of r, indicates inhibition between neighboring events, while faster than linear increases indicate clustering.

The conditional independence property of the Poisson process means that the distribution of the distance R_1 to the nearest event is the same from either an arbitrary point or an arbitrary event and, therefore, that the empty space function F, and nearest neighbor distribution function G, coincide and are given by $F(r) = G(r) = 1 - \exp\{-\lambda b_d r^d\}$. It follows that the J function, defined by $J(r) = (1 - G(r))/(1 - F(r))$, is identically one for all values of its argument (see Sections 16.2 and 16.5 for further discussion).

More generally, let R_k be the distance to the kth nearest event ($k = 1, 2, \ldots$) from some arbitrary origin, and let $A_k = b_d R_k^d$ be the corresponding volume. It follows from the definition of the Poisson process that the volumes A_1, and $A_{k+1} - A_k$, ($k = 1, 2, \ldots$) are independent and exponentially distributed with parameter λ, and therefore, that A_k and R_k^d have gamma distributions with the same shape parameter k, and scale parameters λ and λb_d respectively, from which the distribution of R_k can be easily deduced. This result provides an alternative way of constructing realizations of a spatial Poisson process: first generate a sequence of independent exponential variates that can be used to determine the distances of the events from some arbitrary origin, and then locate these events independently and uniformly over spheres of radius R_k, $k = 1, 2, \ldots$. For example, when $d = 2$, the events are uniformly distributed on a sequence of concentric circles, where the areas of the annuli between the circles are exponentially distributed. This construction is the spatial analog of that for a one-dimensional Poisson process in terms of a sequence of independent exponentially distributed intervals; see Cox and Isham (1980, Section 6.2) and Quine and Watson (1984).

17.2.1 Nonhomogeneous Poisson Process

The nonhomogeneous Poisson process (see Example 17.2) shares many of the nice properties of the homogeneous Poisson process. In particular, there are still no interactions between events, but it has the great advantage that the intensity can vary over the underlying space S. It is straightforward to extend the derivations given in Section 17.1 to show that

- given $N(A) = n$, the n points are independently distributed over A with density $\lambda(x)/\mu(A)$ for $x \in A$;
- the pgfl is given by

$$G_N(\xi) = \exp\left[-\int_S \{1 - \xi(x)\}\, \lambda(x)\, dx\right].$$

As in the homogeneous case, the former result provides a convenient mechanism for simulating realizations of a nonhomogeneous Poisson process in a set A, by first generating the number of events (from a Poisson distribution with mean $\mu(A)$) and then distributing these events independently over A with density $\lambda(\cdot)/\mu(A)$.

The definition of Ripley's K function has been extended to nonstationary processes by Baddeley, Møller, and Waagepetersen (2000). In particular, they define the concept of second-order intensity reweighted stationarity, together with a corresponding intensity reweighted K function, effectively obtained by replacing the product density $\rho^{(2)}(\boldsymbol{x}, \boldsymbol{u})$ by the pair correlation $g(\boldsymbol{x}, \boldsymbol{u}) = \rho^{(2)}(\boldsymbol{x}, \boldsymbol{u})/(\lambda(\boldsymbol{x})\lambda(\boldsymbol{u}))$ (see also Section 18.6.2). For a nonhomogeneous Poisson process, the intensity reweighted K function, K_I say, has the same form as the K function for the homogeneous Poisson process, i.e., $K_I(r) = b_d r^d$.

17.2.2 The Cox Process

For this process, defined in Example 17.3, properties are obtained by first conditioning on the random rate process Λ. Thus, for example,

$$\mathrm{E}(N(A)|\{\Lambda(\boldsymbol{x}) = \lambda(\boldsymbol{x})\}) = \mathrm{var}(N(A)|\{\Lambda(\boldsymbol{x}) = \lambda(\boldsymbol{x})\}) = \int_A \lambda(\boldsymbol{x})\, d\boldsymbol{x},$$

so that

$$\mathrm{E}(N(A)) = \int_A \mathrm{E}_\Lambda(\Lambda(\boldsymbol{x}))\, d\boldsymbol{x}.$$

Also,

$$\mathrm{var}(N(A)) = \int_A \mathrm{E}_\Lambda(\Lambda(\boldsymbol{x}))\, d\boldsymbol{x} + \mathrm{var}_\Lambda\left(\int_A \Lambda(\boldsymbol{x})\, d\boldsymbol{x}\right)$$

from which it follows that the counts in a Cox process are always overdispersed unless $\int_A \Lambda(\boldsymbol{x})\, d\boldsymbol{x}$ is a constant for all A. The probability generating functional is given by

$$G_N(\xi) = \mathrm{E}_\Lambda\left(\exp\left[-\int_S \{1 - \xi(\boldsymbol{x})\}\, \Lambda(\boldsymbol{x})\, d\boldsymbol{x}\right]\right).$$

There are many special cases, and some examples follow.

1. $\Lambda(\boldsymbol{x})$ does not depend on $\boldsymbol{x}$. In this case, the Cox process is a *mixed* Poisson process. Each realization is then indistinguishable from a homogeneous Poisson process, but with an intensity that varies between realizations. More generally, if $\{A_i, i = 1, 2, \ldots\}$ is a fixed and known partition of S, let $\Lambda(\boldsymbol{x}) = \Lambda_i$ for $\boldsymbol{x} \in A_i$ and $i = 1, 2, \ldots$, where the Λ_i are independent random variables.
2. $\Lambda(\boldsymbol{x}) = \exp Z(\boldsymbol{x})$ where Z is a Gaussian random field on $S = \mathbb{R}^d$. In this case, N is a *log Gaussian Cox* process (Møller, Syversveen, and Waagepetersen, 1998); its properties are completely determined by the first- and second-order properties of Z (which must be such that $\mathrm{E}(N(A)) = \int_A \mathrm{E}_Z(e^{Z(\boldsymbol{x})})\, d\boldsymbol{x} < \infty$ for all $A \subseteq S$).
3. Λ is a stationary process with mean μ_Λ and autocovariance function γ_Λ. Then

$$\mathrm{E}(N(A)) = \mu_\Lambda |A| \quad \text{and} \quad \mathrm{var}(N(A)) = \mu_\Lambda |A| + \int_A\int_A \gamma_\Lambda(\boldsymbol{u} - \boldsymbol{v})\, d\boldsymbol{u}\, d\boldsymbol{v},$$

and thus the Cox process has intensity μ_Λ and, by comparison with Equation (17.4) (or directly from the definition), conditional intensity function

$$h(\boldsymbol{u}) = \left[\gamma_\Lambda(\boldsymbol{u}) + \mu_\Lambda^2\right]/\mu_\Lambda.$$

Thus we see that, when Λ is stationary, its autocovariance function can be determined from the conditional intensity of the point process. This result may be of interest in applications where scientific interest focuses on properties of the unobserved driving process Λ. Note, however, that if only one realization of the process is observed, then the Cox process cannot be distinguished from the nonhomogeneous Poisson process, so that the choice between these models in a particular application will depend on subject matter considerations (see Møller and Waagepetersen, 2004, Section 5.1 for a brief discussion).

4. Λ is a *shot noise process* driven by an unobserved Poisson process N_0, with

$$\Lambda(\boldsymbol{x}) = \gamma + \alpha \int_S e^{-\kappa||\boldsymbol{x}-\boldsymbol{u}||} d\,N_0(\boldsymbol{u}),$$

where $||\cdot||$ denotes Euclidean distance in $\mathbb{R}^d$, and α, γ, and κ are nonnegative parameters.

For a more general shot noise process, the exponentially decaying function attached to each event in the underlying Poisson process can be replaced by an alternative kernel function. For example, for the *Matérn process* (Matérn, 1960, 1986), $\gamma = 0$ and the kernel is a nonzero constant if $||\boldsymbol{x} - \boldsymbol{u}|| \leq r_0$ and is zero otherwise. For a yet more general version, the overall weights attached to each of these underlying events can be random (see Møller and Waagepetersen (2004, Section 5.4) and Møller and Torrisi (2005)).

17.3 Poisson Cluster Processes

In their most general formulation, cluster processes are constructed as a superposition of simpler processes. We start with an unobserved point process N_c of cluster centers, attached to each of which is a subsidiary process or cluster of events, where $N_{\boldsymbol{x}}$ denotes the cluster generated by a center at $\boldsymbol{x}$. The observed process, N, is the superposition of all the clustres (which can, if required, include events at the cluster centers). We assume for simplicity that all the processes are in $S = \mathbb{R}^d$. It follows, for example, that

$$\mathrm{E}(N(A)) = \int_S \mathrm{E}\{N_{\boldsymbol{x}}(A)\}\mathrm{E}\{d\,N_c(\boldsymbol{x})\}.$$

The processes resulting from the thinning and translation operations discussed in Sections 17.1.2 and 17.1.3 above can both be considered as simple cluster processes. With thinning, each cluster either consists of a single event at the cluster center (with probability p) or is empty. For translation, all clusters are of size one with independent and identically distributed displacements of the events from their cluster centers.

To make useful progress, we will assume that the clusters, relative to their cluster centers, are independent and identically distributed, and are independent of N_c. In this case, conditional upon the process of cluster centers, $\{\boldsymbol{x}_i\}$, the superposition has pgfl

$$\prod_i \mathrm{E}\left[\exp\left\{\int_S \ln \xi(\boldsymbol{x}) d\,N_{\boldsymbol{x}_i}(t)\right\}\right] = \prod_i G_s(\xi; \boldsymbol{x}_i)$$

where $G_s(\xi; \boldsymbol{x}_i)$ is the pgfl for a cluster centered at $\boldsymbol{x}_i$ and, thus, unconditionally,

$$G_N(\xi) = G_c(G_s(\xi; \cdot)).$$

In some applications (for example, relating to the locations of bacteria or larval stages of insects), the focus of interest may be the distribution of the total number of events in a particular region or quadrant, rather than the individual positions of those events. This focus led to the formulation of the so-called "contagious distributions" (see, for example, Neyman, 1939; Thomas, 1949).

17.3.1 The Compound Poisson Process

A very simple way to construct a process with multiple occurrences is to now take the cluster of events, $N_{\mathbf{x}}$, to consist of a random number $M_{\mathbf{x}}$ of events all located at $\mathbf{x}$. Given the assumption that the numbers of events at distinct locations are mutually independent and identically distributed, $G_s(\xi, ; \mathbf{x}) = G_M(\xi(\mathbf{x}))$, where G_M is the probability generating function of the event multiplicity M. When, in addition, N_c is a Poisson process of rate λ_c, then

$$G_N(\xi) = \exp\left\{-\lambda_c \int_S [1 - G_M(\xi(\mathbf{x}))] d\mathbf{x}\right\},$$

from which it follows that the number of events in an arbitrary bounded Borel set A has a compound Poisson distribution (given $N_c(A)$, $N(A)$ has the form $M_1 + M_2 + \cdots + M_{N_c(A)}$) with probability generating function

$$\mathrm{E}(z^{N(A)}) = \exp\{-\lambda_c |A| (1 - G_M(z))\}.$$

17.3.2 The Neyman–Scott Poisson Cluster Process

The compound Poisson process above is a simple example of a *Poisson cluster process*, for which the cluster centers are a homogeneous Poisson process with rate λ_c. In this case, the cluster process is stationary and has PGFL

$$G_N(\xi) = \exp\left[-\lambda_c \int_S \{1 - G_s(\xi; \mathbf{x})\}\, d\mathbf{x}\right].$$

The *Neyman–Scott process* (Neyman and Scott, 1958) is a tractable extension of the compound Poisson process in which it is assumed that the clusters are independent and identically distributed relative to their cluster centers, with a random number M of events per cluster, and that these events are independently and identically displaced from the cluster center with density f. Without loss of generality, we can assume $M \geq 1$ with probability 1. This follows from the results on thinning Poisson processes (see Section 17.1.2): A process with cluster centers occurring at rate λ_c' and $G_M(0) = \mathrm{P}(M = 0) > 0$ is equivalent to one having a Poisson process of centers with rate $\lambda_c = (1 - G_M(0))\lambda_c'$ and all clusters nonempty. For the Neyman–Scott process, by first conditioning on M, the pgfl of a cluster with center at $\mathbf{x}$ can easily be shown to be

$$G_s(\xi, \mathbf{x}) = G_M\left\{\int_S \xi(\mathbf{x} + \mathbf{u}) f(\mathbf{u})\, d\mathbf{u}\right\}.$$

The rate of the process is $\lambda_c \mathrm{E}(M)$, and its second-order properties can be determined from the corresponding conditional intensity function. In particular, the second-order product density function $\rho^{(2)}(\mathbf{x}_1, \mathbf{x}_2)$ satisfies

$$\rho^{(2)}(\mathbf{x}_1, \mathbf{x}_2) = (\lambda_c \mathrm{E}(M))^2 + \lambda_c \mathrm{E}(M(M-1)) \int_S f(\mathbf{x}_1 - \mathbf{v}) f(\mathbf{x}_2 - \mathbf{v})\, d\mathbf{v}$$

where the first term on the right-hand side comes from events in distinct clusters, while the second term comes from pairs of events in the same cluster, centered at $\boldsymbol{v}$, of which there are $M(M-1)$ ordered pairs if the cluster is of size M. Thus, the conditional intensity function is given by

$$h(\boldsymbol{u}) = \lambda_c \mathrm{E}(M) + \frac{\mathrm{E}(M(M-1))}{\mathrm{E}(M)} \int_S f(\boldsymbol{x}) f(\boldsymbol{x}+\boldsymbol{u})\, d\boldsymbol{x}.$$

Note that the second-order properties of the Neyman–Scott process involve only the second-order properties of M, and not its complete distribution.

It is immediately clear that the Neyman–Scott process is overdispersed, unless $M = 1$ with probability 1, i.e., unless the process is a Poisson process. This result is likely to hold, at least asymptotically, for a wide range of cluster processes. For, intuitively, if A is large and the chance that cluster events are far from their centers is sufficiently small that edge effects can be ignored, then $N(A)$ is the sum of all the cluster sizes for clusters with centers in A, i.e., $N(A) \simeq M_1 + \cdots + M_{N_c(A)}$, and

$$\begin{aligned} \mathrm{var}(M_1 + \cdots + M_{N_c(A)}) &= \mathrm{var}(N_c(A)\mathrm{E}(M)) + \mathrm{E}(N_c(A)\mathrm{var}(M)) \\ &= \lambda_c |A| \{(\mathrm{E}(M))^2 + \mathrm{var}(M)\} \\ &= \lambda_c |A|\, \mathrm{E}(M^2), \end{aligned}$$

while $\mathrm{E}(M_1 + \cdots + M_{N_c(A)}) = \lambda_c |A| \mathrm{E}(M)$.

It is important to note that the same point process can have more than one apparently different representation. A good example of this is a correspondence between some Poisson cluster processes and Cox processes. Suppose that the cluster associated with a cluster center at $\boldsymbol{x}_i$ is a Poisson process with rate $\lambda_s(\boldsymbol{x} - \boldsymbol{x}_i)$, where $\int_S \lambda_s(\boldsymbol{x})\, d\boldsymbol{x} < \infty$. Then, because the superposition of Poisson processes is again a Poisson process, the cluster process can also be regarded as a Cox process driven by the process $\Lambda(\boldsymbol{x}) = \int_S \lambda_s(\boldsymbol{x} - \boldsymbol{u}) d N_c(\boldsymbol{u})$, which is a shot noise process driven by N_c. Conversely, if a shot noise process has $\gamma = 0$, the contribution from each event in the unobserved driving process can be regarded as the cluster corresponding to that event, where the clusters, relative to their centers, will be independent and identically distributed Poisson processes. Thus, for example, the Matérn process can be thought of as a Poisson cluster process, in which the number of events per cluster has a Poisson distribution, and the events are independently and uniformly located within a ball of radius r_0.

17.4 Markov Point Processes

Models, such as the Cox and Poisson cluster processes discussed above, that are constructed from a Poisson process by adding additional variability will inevitably be overdispersed. Overdispersion is a characteristic property of empirical data in many applications and these models are applied widely. However, in this section we focus on some processes where interactions between neighboring events are modeled explicitly. Both attraction and inhibition of events are possible and, therefore, the models can capture underdispersion. These processes are specified in terms of their densities relative to that for a homogeneous Poisson process. We will restrict our discussion to point processes on a bounded set $S \subset \mathbb{R}^d$. For the unbounded case see, for example, the discussion in Stoyan et al. (1995, Section 5.5.3), Møller and Waagepetersen (2004, Section 6.4), or Daley and Vere-Jones (2007, Section 10.4).

For a homogeneous Poisson process of rate λ in a bounded region S ($S \subset \mathbb{R}^d$), the probability of no events in S is $\omega(\emptyset) = e^{-\lambda|S|}$, and the joint probability density that there are exactly n events ($n = 1, 2, \ldots$) at locations $x_1, \ldots, x_n$ is $\omega(x_1, \ldots, x_n) = \lambda^n e^{-\lambda|S|}$. This density is defined over all *distinct sets* of locations $\{x_1, \ldots, x_n\}$, and there is no significance in the order the events are listed. In some instances, it may be more convenient to suppose that the points have some arbitrary numbering so that the density is $\lambda^n e^{-\lambda|S|}/n!$ for and is defined over the set $S \times \cdots \times S$, for $n = 1, 2, \ldots$.

Now consider a point process with density

$$\begin{aligned} \omega(x_1, \ldots, x_n) &= g_n(x_1, \ldots, x_n)\lambda^n e^{-\lambda|S|} \text{ for } n = 1, 2, \ldots, \\ \omega(\emptyset) &= g_0\, e^{-\lambda|S|} \text{ for } n = 0, \end{aligned} \tag{17.5}$$

where the functions g_n, $n = 0, 1, \ldots$ are invariant under permutation of their arguments and are such that the distribution of $N(S)$ is normalized to 1. Then $g_n(x_1, \ldots, x_n)$ gives the likelihood of a particular configuration of points relative to a Poisson process and, up to a constant of proportionality, is the conditional density of the n events given $N(S) = n$. If we write

$$g_n(x_1, \ldots, x_n) = \exp\{-\psi_n(x_1, \ldots, x_n)\} \tag{17.6}$$

(allowing formally infinite functions ψ_n if $g_n = 0$), then ψ_n is a *potential function* for the process, which is often called a *Gibbs* process.

Special families of point processes are obtained by choosing particular forms for the functions ψ_n. In many contexts it is natural to look for models where the functions ψ_n have the same structure as n varies, and to assume that, for $n \geq 1$,

$$\begin{aligned} \psi_n(x_1, \ldots, x_n) = &\sum_i \alpha_1(x_i) + \sum_{i_1<i_2} \alpha_2(x_{i_1}, x_{i_2}) + \sum_{i_1<i_2<i_3} \alpha_3(x_{i_1}, x_{i_2}, x_{i_3}) + \cdots \\ &+ \sum_{i_1<i_2<\ldots<i_n} \alpha_n(x_{i_1}, \ldots, x_{i_n}). \end{aligned} \tag{17.7}$$

A further simplification is to assume that the model is translation invariant so that the function α_1 takes a constant value and the functions α_k depend only on the vector separations $x_{i_j} - x_{i_k}$ or, further, that they depend only on the distances $||x_{i_j} - x_{i_k}||$ between the events. In these cases, given that the density of the point process depends only on the parameter λ through the combination $\lambda e^{-\alpha_1}$, we may take $\lambda = 1$ without loss of generality.

The special case where $\alpha_k = 0$ for $k \geq 3$ is a *pairwise interaction process*. Some particular examples assume that the points interact only when they are within some critical distance, r_0, say, so that $\alpha_2(x_{i_1}, x_{i_2}) = 0$ if $||x_{i_1} - x_{i_2}|| > r_0$. The Strauss process (Strauss, 1975) is a simple special case where $\alpha_2(x_{i_1}, x_{i_2})$ takes a constant value α_2, say, for all $||x_{i_1} - x_{i_2}|| \leq r_0$ and, thus, the density of the process depends only on the number of events in a particular realization and the number of neighboring pairs. In this case, it is necessary that $\alpha_2 > 0$, i.e., there is inhibition between neighboring events, as otherwise the density cannot be properly normalized. As an extreme case, the *hard core* Gibbs process takes $\alpha_2(x_{i_1}, x_{i_2})$ to be infinite when $||x_{i_1} - x_{i_2}|| \leq r_0$, so that realizations of the process are those of a Poisson process conditioned so that all pairs of points are separated by at least r_0.

Suppose more generally that there is a neighbor structure on S, i.e., a reflexive and symmetric relation, $\sim$, specifying whether or not two elements of S are neighbors (each element is defined to be its own neighbor). For example, we could have $v \sim u$ if $||v - u|| \leq r_0$. The *boundary* $\partial\Delta$ of a set $\Delta \subset S$ is then defined to consist of those elements of $S \setminus \Delta$ that are neighbors of elements in Δ. In addition, a set $C = \{u_1, \ldots, u_k\}$ of elements of S is said to be a *clique* if $u_i \sim u_j$ for all $u_i, u_j \in C$.

Then, it is natural to consider potential functions of the form (17.7), but in which only contributions from cliques are nonzero. In this case, the conditional probability density of the realization $\{x_1, \ldots, x_\ell\}$ on a subset Δ of S given the configurations $\{u_1, \ldots, u_m\}$ on $\partial\Delta$ and $\{v_1, \ldots, v_n\}$ on $S \setminus (\Delta \cup \partial\Delta)$, is proportional to

$$\frac{\exp\{-\psi_{\ell+m+n}(x_1, \ldots, x_\ell, u_1, \ldots, u_m, v_1, \ldots, v_n)\}}{\sum_p \int \cdots \int \exp\{-\psi_{p+m+n}(x_1, \ldots, x_p, u_1, \ldots, u_m, v_1, \ldots, v_n)\}\, dx_1, \ldots, dx_p}$$

where the integral in the denominator is over distinct p-tuples of points in Δ and the summation in the denominator is over all nonnegative integer values of p. This conditional density does not depend on the realization $\{v_1, \ldots, v_n\}$ on $S \setminus (\Delta \cup \partial\Delta)$ because the events on Δ only interact with those on $\Delta \cup \partial\Delta$, and all other interactions are between events on $S \setminus \Delta$ and their contributions cancel out in the numerator and denominator of the fraction above. A point process with this property, that the conditional density of the events on a subset Δ of S given the configuration on $S \setminus \Delta$ is the same as the conditional density given only the configuration on the boundary $\partial\Delta$, is a *Markov random field,* and is often termed a *Markov point process* (see, for example, Isham (1981) and Clifford (1990) for historical accounts, and van Lieshout (2000) for a recent exposition).

The converse of this result can also be shown to hold for point processes that satisfy a positivity condition requiring that the distribution $\omega(x_1, \ldots, x_n) > 0$ is *hereditary*, that is, for all n, if $\omega(x_1, \ldots, x_n) > 0$, then all subsets of $\{x_1, \ldots, x_n\}$ also have positive density. If a point process has a density $\omega(x_1, \ldots, x_n)$ of the form given in Equation (17.5), then the point process is a Markov point process if and only if the functions g_n can be written in the form

$$g_n(x_1, \ldots, x_n) = \prod \phi(x_{i_1}, \ldots, x_{i_k})$$

where the product is over all subsets $i_1, \ldots, i_k$ of $1, \ldots, n$ for $k = 1, \ldots, n$ (Ripley and Kelly, 1977). The function ϕ is nonnegative and takes the value 1 unless $\{x_{i_1}, \ldots, x_{i_k}\}$ is a clique. In this case, the corresponding potential functions ψ_n defined in Equation (17.6) satisfy

$$\psi_n(x_1, \ldots, x_n) = -\sum \ln \phi(x_{i_1}, \ldots, x_{i_k}) \tag{17.8}$$

where the sum is over all subsets of $\{x_1, \ldots, x_n\}$ that are cliques. Thus, Equation (17.7) has the form (17.7) if $\alpha_k(x_{i_1}, \ldots, x_{i_k})$ is nonzero only when $\{x_1, \ldots, x_k\}$ form a clique. This result is the point process analog of the celebrated Hammersley–Clifford theorem for Markov random fields on discrete spaces (see Besag (1974), and Part III for discussion and references).

The conditional spatial independence property satisfied by Markov point processes is intuitively natural for many spatial point process applications and the Hammersley–Clifford representation provides a nice way to construct increasingly complex models by starting perhaps with pairwise interaction processes and gradually incorporating higher-order interactions. However, of itself, the product form for the distribution of the process does not provide an obvious means of simulating realizations. For this, it is necessary to add in a temporal aspect, and to consider a spatial point process that evolves in time by the deletion of events and the addition of new ones, and where this evolution has a temporally homogeneous Markov property such that, given the history of the process, the birth and death rates at time t depend only on that history and on t through the locations of the events at time t.

Suppose that, given the configuration of events $\{x_1, \ldots, x_n\}$ at time t, the space–time birth rate at x is $\beta(x; x_1, \ldots, x_n)$. That is, the probability that a new event is created in a ball centered at x and having volume ϵ, during $(t, t + \tau)$, is given by $\beta(x; x_1, \ldots, x_n)\epsilon\tau + o(\epsilon\tau)$ in the limit, as ϵ and τ tend to zero. Similarly, let $\delta(x; x_1, \ldots, x_n)$ denote the temporal rate at which the event at x is deleted, given that at time t the process has events at $\{x, x_1, \ldots, x_n\}$. Further, suppose that this birth and death process is stationary and reversible in time so that the detailed balance equations

$$\omega(x_1, \ldots, x_n)\beta(x; x_1, \ldots, x_n) = \omega(x, x_1, \ldots, x_n)\delta(x; x_1, \ldots, x_n), \tag{17.9}$$

are satisfied, where $\omega(x_1, \ldots, x_n)$ is the spatial equilibrium distribution of the space–time point process, and is hereditary. These detailed balance equations can be written, for $n \geq 0$, in the form

$$\gamma(x; x_1, \ldots, x_n) := \frac{\beta(x; x_1, \ldots, x_n)}{\delta(x; x_1, \ldots, x_n)} = \frac{\omega(x, x_1, \ldots, x_n)}{\omega(x_1, \ldots, x_n)} \tag{17.10}$$

where the ratio on the right-hand side is defined to be zero if the denominator (and, hence, also numerator) is zero. The ratio of densities on the right-hand side of Equation (17.10) is the *Papangelou conditional intensity* of the point process see, for example, Møller and Waagepetersen (2004). If $\omega(x_1, \ldots, x_n)$ is expressed in terms of a potential ψ_n as defined by Equation (17.5) and Equation (17.6), then these detailed balance equations can be written in the form

$$\gamma(x; x_1, \ldots, x_n) = \exp\{-\psi_{n+1}(x, x_1, \ldots, x_n) + \psi_n(x_1, \ldots, x_n)\}. \tag{17.11}$$

It follows that, for $n \geq 1$, the density $\omega(x_1, \ldots, x_n)$ can be constructed iteratively, in terms of $\omega(\emptyset)$ and the functions $\gamma(x; x_1, \ldots, x_n)$ so that, after suitable normalization, the equilibrium distribution ω is determined.

Suppose now that S is equipped with a neighbor relation $\sim$ and that, given the complete configuration of events in S, the birth and death rates at x depend only on the events in the rest of the configuration that are neighbors of x, i.e., on events that lie in ∂x. Then the same will be true of $\gamma(x; x_1, \ldots, x_n)$ and, hence, the form of $\ln \omega(x_1, \ldots, x_n)$ will be a summation over the subsets of $\{x_1, \ldots, x_n\}$ that are cliques. It follows that the equilibrium distribution ω is a Markov point process. Thus, realizations of a particular Markov point process can be simulated by using Equation (17.10) to determine the functions $\gamma(x; x_1, \ldots, x_n)$, and then simulating a spatial-temporal point process that evolves with suitable birth and death rates (see Illian et al. (2008, Section 3.6) for further discussion). These rates are not uniquely determined, but must be such that the rates at x given the remaining configuration of events depends only on events in ∂x, and have the required ratio.

References

Baddeley, A., Gregori, P., Mateu, P., Stoica, R. and Stoyan, D. (2006), *Case Studies in Spatial Point Process Modeling, Lecture Notes in Statistics 185*, Springer-Verlag, Berlin–Heidelberg–New York.

Baddeley, A.J., Møller, J. and Waagepetersen, R. (2000), Non- and semi-parametric estimation of interaction in inhomogeneous point patterns, *Statistica Neerlandica* **54**, 329–350.

Bartlett, M.S. (1955), In discussion of paper by D.R. Cox, *J. R. Statist. Soc.* **B17**, 159–160.

Besag, J. (1974), Spatial interaction and the statistical analysis of lattice systems (with discussion), *J. R. Statist. Soc.* **B36**, 192–236.

Clifford, P. (1990), Markov random fields in statistics, *in* G.R. Grimmett and D.J.A. Welsh, eds., *Disorder in Physical Systems*. A volume in honour of John M. Hammersley on the occasion of his 70th birthday, Clarendon Press, Oxford, pp. 19–32.

Cox, D. (1955), Some statistical methods connected with series of events (with discussion), *J. R. Statist. Soc.* **B17**, 129–164.

Cox, D. and Isham, V. (1980), *Point Processes*, Chapman & Hall, London.

Cressie, N.A. (1993), *Statistics for Spatial Data (revised edition)*, John Wiley & Sons, New York.

Daley, D. and Vere-Jones, D. (1988), *An Introduction to the Theory of Point Processes*, Springer, New York.

Daley, D. and Vere-Jones, D. (2003), *An Introduction to the Theory of Point Processes, Volume I: Elementary Theory and Methods*, Springer, New York.

Daley, D. and Vere-Jones, D. (2007), *An Introduction to the Theory of Point Processes, Volume II: General Theory and Structure*, Springer, New York.

Diggle, P. (2003), *Statistical Analysis of Spatial Point Patterns (2nd ed.)*, Edward Arnold, London.

Harkness, R. and Isham, V. (1983), A bivariate spatial point process of ants' nests, *Appl. Statist.* **82**, 293–303.

Illian, J., Penttinen, A., Stoyan, H. and Stoyan, D. (2008), *Statistical Analysis and Modelling of Spatial Point Patterns*, Wiley, Chichester, U.K.

Isham, V. (1981), An introduction to spatial point processes and Markov random fields, *Int. Statist. Rev.* **49**, 21–43.

Karr, A.F. (1991), *Point Processes and their Statistical Inference, 2nd ed.*, Marcel Dekker, New York.

Matérn, B. (1960), Spatial variation, *Meddelanden från Statens Skogforskningsinstitut* **49(5)**.

Matérn, B. (1986), *Spatial Variation, Lecture Notes in Statistics 36*, Springer-Verlag, Berlin.

Møller, J. and Waagepetersen, R. (2004), *Statistical Inference and Simulation for Spatial Point Processes*, Chapman & Hall, Boca Raton, FL.

Møller, J. and Torrisi, G.L. (2005), Generalised shot noise Cox processes, *Adv. Appl. Prob.* **37**, 48–74.

Møller, J. and Waagepetersen, R. (2007), Modern statistics for spatial point processes (with discussion), *Scand. J. Statist.* **34**, 643–711.

Møller, J., Syversveen, A. and Waagepetersen, R. (1998), Log Gaussian Cox processes, *Scand. J. Statist.* **25**, 451–482.

Neyman, J. (1939), On a new class of 'contagious' distributions, applicable in entomology and bacteriology, *Ann. Math. Statist.* **10**, 35–57.

Neyman, J. and Scott, E. (1958), A statistical approach to problems in cosmology (with discussion), *J. R. Statist. Soc.* **B20**, 1–43.

Quine, M.P. and Watson, D. (1984), Radial simulation of n-dimensional Poisson processes, *J. Appl. Probab.* **21**, 548–557.

Ripley, B. (1976), The second-order analysis of stationary point processes, *J. Appl. Probab.* **13**, 255–266.

Ripley, B. (1977), Modelling spatial patterns (with discussion), *J. R. Statist. Soc.* **B39**, 172–212.

Ripley, B. (2004), *Spatial Statistics*, John Wiley & Sons, New York.

Ripley, B. and Kelly, F. (1977), Markov point processes, *J. Lond. Math. Soc.* **15**, 188–192.

Stoyan, D., Kendall, W. and Mecke, J. (1995), *Stochastic Geometry and Its Applications, 2nd ed.*, Wiley, Chichester, U.K.

Strauss, D. (1975), A model for clustering, *Biometrika* **62**, 467–475.

Thomas, M. (1949), A generalisation of Poisson's binomial limit for use in ecology, *Biometrika* **36**, 18–25.

van Lieshout, M.N.M. (2000), *Markov Point Processes and Their Applications*, Imperial College Press, London.

Westcott, M. (1972), The probability generating functional, *J. Austral. Math. Soc.* **14**, 448–466.

18

Nonparametric Methods

Peter J. Diggle

CONTENTS

18.1 Introduction

Nonparametric methods play two distinct, but related, roles in the exploratory analysis of spatial point process data. First, they are used to test benchmark hypotheses about the underlying process. For example, most analyses of univariate spatial point process data begin with one or more tests of the hypothesis of *complete spatial randomness* (CSR), by which we mean that the data form a partial realization of a homogeneous Poisson process. Although this is rarely tenable as a scientific hypothesis, formally testing it serves to establish whether the data contain sufficient information to justify any more elaborate form of analysis. The second role for nonparametric methods is to estimate properties of the underlying process with a view to suggesting suitable classes of parametric model.

The two roles are linked by using test statistics derived from summaries of the data that indicate what kind of general alternative to complete spatial randomness might be operating. A first, simple goal is to describe an observed point pattern as regular or aggregated relative to the completely random benchmark. The former would be exemplified by inhibitory interactions among the points of the process, as is typically the case for the Markov point process models described in Chapter 17, Section 17.4, the latter by cluster point processes as described in Section 17.3. More subtle descriptions of an observed point pattern might include combinations of regularity and aggregation at different spatial scales or an assessment of possible nonstationarity.

Many of the statistics that are used for nonparametric testing and estimation can also be adapted to provide methods for fitting, and conducting diagnostic checks on, a proposed parametric model. These aspects are considered in Chapters 19 and 20, respectively.

The primary focus throughout this book is on *mapped* spatial data. In the point process setting, this means that the data consist of the actual locations of the points of the process within a designated study region. Historically, several examples of what we would now call nonparametric methods for spatial point patterns were first developed as methods for sampling point processes *in situ*, typically in plant ecology, or forestry where each event refers to the location of an individual plant or tree. These *sparse sampling methods* include counts of the numbers of points in small subregions, known as *quadrats*, or measurements of the distances between pairs of near-neighboring points of the process. We give a short review of sparse sampling methods in Section 18.2. Section 18.3 discusses the pervasive idea of Monte Carlo significance testing. Sections 18.4 to 18.8 consider different aspects of nonparametric methods for mapped point patterns.

Throughout this chapter, we use the term *event* to denote a point of the process, to distinguish this from an arbitrary *point* in $\mathbb{R}^2$.

18.2 Methods for Sparsely Sampled Point Patterns

All of the methods described in this section are concerned with estimating the intensity, λ, of a homogeneous spatial Poisson process, or with testing for departures from this process (henceforth, CSR), in either case using data obtained by sparse sampling. We consider two forms of sparse sampling: quadrat sampling and distance sampling.

18.2.1 Quadrat Sampling

Quadrat sampling consists of counting the number of events in each of a set of small subareas, called quadrats, laid out over the study region of interest. This sampling method has a long history in plant ecology. Originally, a *quadrat* was a 1 meter square, hence its name, but we use the term to refer to any convenient shape, a circle being the obvious alternative to a square.

The relevant properties of the homogeneous Poisson process are that the number of events in any prescribed region D follows a Poisson distribution with mean $\lambda|D|$, where $|\cdot|$ denotes area, and numbers of events in disjoint areas are mutually independent. Let Y_i denote the number of events in the ith of n quadrats, and consider two sampling designs for locating the quadrats: spatially random sampling over D, or sampling at the points of a regular grid to span D. The essence of sparse sampling is that the total area sampled should be a very small proportion of the area of D, otherwise complete enumeration would be feasible. Hence, in either case we can assume that the quadrats are not only disjoint, but spatially well separated and therefore, give rise to independent counts Y_i.

To estimate λ, the maximum likelihood estimator under CSR is the observed mean count per unit area,

$$\hat{\lambda} = \left(\sum_{i=1}^{n} Y_i\right) / (n|D|). \tag{18.1}$$

The associated standard error is $\text{SE}(\hat{\lambda}) = \surd\{\lambda/(n|D|)\}$, showing that what matters is the total area sampled. More importantly, Equation (18.1) is unbiased for the mean number of events per unit area whatever the underlying point process. Also, its standard error can be estimated robustly from the sample variance of the observed counts y_i.

To test for departure from an underlying homogeneous Poisson process, the classic test statistic is the index of dispersion, $I = s^2/\bar{y}$, where $\bar{y}$ and s^2 are the sample mean and variance of the counts. Under CSR, $(n-1)I \sim \chi^2_{n-1}$ approximately, the approximation improving with increasing $\text{E}[Y_i]$ (Fisher, Thornton, and Mackenzie, 1922). Significantly small or large values of I indicate spatial regularity or aggregation, respectively.

With regard to the study design for quadrat sampling, the only conceivable disadvantages of grid sampling are that it may induce unconscious bias in the location of the grid origin, or fall foul of a cyclic pattern in the underlying process at a frequency that matches the grid spacing. Otherwise, a grid layout is usually easier to implement in the field and opens up the possibility of more interesting secondary analyses of the resulting lattice data (for extensive discussion of stochastic models and associated statistical methods for spatially discrete data, see Part III). A possible compromise between completely random and grid-based sampling would be to locate one quadrat at random within each of a set of grid cells.

18.2.2 Distance Sampling

So-called distance sampling consists of measuring the distances from aribtrary points to near neighboring events, or between pairs of near neighboring events, for various definitions of "near neighboring." During the 1950s and 1960s, an extensive literature of such methods developed, mostly in botany, ecology, or forestry journals, in which context the methods are still sometimes used (see, for example, Pruetz and Isbel, 2000; Stratton, Goldstein, and Meinzer, 2000). When complete mapping of a spatial point pattern is feasible, sparse sampling methods are at best inefficient and may also be invalid because violation of the sparseness requirement leads to nonindependence among the different measurements.

The distribution theory associated with almost all of the distance sampling methods that have been proposed can be subsumed under the following result, coupled with the independence between partial realizations of a homogeneous Poisson process in disjoint regions. Let $Y_{k,t}$ denote the distance from either an arbitrary point or an arbitrary event to the kth nearest event within an arc of included angle t centered on the origin of measurement. Then, under CSR,

$$t\lambda Y^2_{k,t} \sim \chi^2_{2k}. \tag{18.2}$$

Suppose that our data consists of the observed values $y_1, \ldots, y_n$ of $Y_{k,t}$ measured from n origins of measurement selected either at random or, in the case of an arbitrary sampling origin, in a regular grid. Then, under CSR, the maximum likelihood estimator for λ is

$$\hat{\lambda} = (2nk)/\left(t\sum y_i^2\right). \tag{18.3}$$

Like its quadrat-based counterpart Equation (18.1) the estimator Equation (18.3) has the intuitive interpretation that it equates the theoretical and observed numbers of events per unit area searched. However, unlike Equation (18.1) the distance-based estimator Equation (18.3) does not remain unbiased when CSR does not hold. The resulting bias can be reduced by noting that estimates using arbitrary points and arbitrary events as origins of measurement tend to be biased in opposite directions, and an average of the two is typically less biased than either one alone. Note, however, that selecting an arbitrary event in practice requires enumeration of all of the points in the study region, which rather defeats the purpose of the exercise. We return to this point below.

The simplicity of Equation (18.2), coupled with the independence of measurements made from different, well-separated points or events, lends itself to the construction of a variety of test statistics whose null sampling distributions have standard forms. For example, and with the caveat already noted, if we set $k = 1$ and $t = 2\pi$ in Equation (18.2), and measure distances $x_1, \ldots, x_n$ from n arbitrary points and distances $y_1, \ldots, y_n$ from n arbitrary events, in each case to their nearest neighboring event, then the statistic $h = (\sum x_i^2)/(\sum y_i^2)$ is distributed as F on $2n$ and $2n$ degrees of freedom under CSR and, importantly, does not require knowledge of λ.

The statistic h was proposed by Hopkins (1954) and by Moore (1954). However, and as already noted, the test is infeasible in practice because the selection of n arbitrary events requires enumeration of all of the events in the study region, in which case we would do much better also to record their locations and to analyze the data as a mapped point pattern. The solution to this problem proposed by Besag and Gleaves (1973) was to use the distances x_i as defined above, i.e., the distance from an arbitrary point, O, to the nearest event, P, but to redefine y_i as the distance from P to *its* nearest neighboring event within the half plane defined by the perpendicular to OP through P and excluding O. This gives the required independence between x_i and y_i under CSR and leads to the test statistic $t = 2(\sum x_i^2)/(\sum y_i^2)$, whose null sampling distribution is again F on $2n$ and $2n$ degrees of freedom. With regard to the design question of how to choose the n points O in practice, the comments at the end of Section 18.2.1 on the relative merits of random or systematic spatial sampling apply here also.

The T-square sampling method has recently reemerged as a method for estimating the size of a refugee camp in the aftermath of a natural or manmade disaster (Brown, Jacquier, Coulombier, Balandine et al., 2001; Bostoen, Chalabi, and Grais, 2007).

For more detailed reviews of distance sampling methods, see Diggle (2003, Chap. 3) or Illian, Penttinen, Stoyan, and Stoyan (2008, Chap. 2).

18.3 Monte Carlo Tests

A characteristic feature of mapped spatial point process data is that the distribution theory for almost any interesting statistic associated with the data is analytically intractable. As a result, Monte Carlo methods of inference are very widely used, and will feature heavily in the remainder of Part IV. Here, we consider the simplest of Monte Carlo methods of inference, namely, a Monte Carlo test of a simple null hypothesis.

Monte Carlo tests were proposed by Barnard (1963). Let T be any test statistic, large values of which cast doubt on the null hypothesis, H_0. Let t_1 be the value of T calculated from a dataset. Assume, for convenience, that the null sampling distribution of T is continuous so that ties cannot occur; if this is not the case, then, as noted by Besag and Clifford (1991), tied ranks can legitimately be broken by random unrounding. Let $t_2, \ldots, t_s$ be the values of T calculated from $s - 1$ independent simulations of H_0. Then, if H_0 is true, the $t_1, t_2, \ldots, t_s$ are exchangeable and all orderings are equally likely. Hence, if R denotes the number of $t_i > t_1$, then $\mathrm{P}(R \leq r) = (r + 1)/s$, i.e., the Monte Carlo p-value is $(r + 1)/s$ or, for a test at prescribed significance level $\alpha = (k + 1)/s$, the test rejects H_0 if and only if $r \leq k$. Note that the test statistic is not T itself, but the *rank* of t_1 among the t_i. One initially surprising consequence of this is that the loss of power relative to a classical test based on T is surprisingly small; Marriott (1979) suggests as a rule of thumb that s need be no bigger than the value required so that the rejection region is $r \leq 4$, e.g., $s = 99$ for a test at the 5% level, and pro rata for more extreme prescribed significance levels. Much larger values of s would be needed for estimating accurately the percentiles of the null distribution of T (arguably a more useful exercise), for which the usual binomial sampling theory applies.

Besag and Clifford (1989, 1991) propose two extensions of the Monte Carlo test to deal with situations in which it is difficult to draw independent random samples from the null distribution of T. In the 1989 paper, they note that the key requirement for a valid test of H_0 is exchangeability, rather than independence, of the simulated values, and propose a generalization in which a Markov chain simulation is used to generate $t_2, \ldots, t_s$ to satisfy the exchangeability requirement. In the 1991 paper, they propose a sequential variant of the test in which values of two integers r and s are prescribed, and a sequence of simulations is conducted until either at least r of the simulated t_i are larger than t_1 or $s-1$ simulations have been performed. Then, in the first case the attained significance level is r/s^* where $s^* \leq s-1$ is the number of simulations performed, while in the second case the attained significance level is $(r+1)/s$, as for the nonsequential test. The sequential procedure is valuable when the procedure is computationally expensive, for example, when a Markov chain Monte Carlo (MCMC) algorithm is needed to simulate each t_i, and it is clear from the first few simulations that the data show no significant departure from H_0. Incidentally, Besag and Clifford's rule of thumb to use $r = 10$ is not wildly out of line with Marriott's $r \geq 4$, although the arguments leading to them are different.

18.4 Nearest Neighbor Methods for Mapped Point Patterns

The basic measurements described in Section 18.2.2, specifically the distances from arbitrary points or arbitrary events to their nearest neighboring events, can also be used for mapped point patterns. However, the simple distribution theory described in Section 18.2.2 no longer holds because of the inherent, and complicated, pattern of stochastic dependence among the measured distances.

Under CSR, the distribution function of the distance from either an arbitrary point or an arbitrary event to the nearest neighoring event, say X and Y respectively, is

$$F_0(u) = 1 - \exp(-\pi\lambda u^2), \tag{18.4}$$

where λ is the intensity of the process. Echoing earlier remarks, when CSR does not hold, the distribution functions of X and Y tend to deviate from Equation (18.4) in opposite directions, and together they give a useful impression of whether, and if so in what way, CSR is untenable as a model.

For an observed pattern of n points in a study region D, let $x_i : i = 1, \ldots, m$ denote the distances from each of m arbitrary points to the nearest of the n events, and $y_i : i = 1, \ldots, n$ the distances from each of the n events to its nearest neighbor. Now denote by $\hat{F}(\cdot)$ and $\hat{G}(\cdot)$ the empirical distribution functions of the x_i and y_i, respectively. Neither $\hat{F}(u)$ nor $\hat{G}(u)$ is unbiased for $F_0(u)$ under CSR because of finite-sample effects including edge effects. Various ways to correct for the latter have been proposed, but are unnecessary if we seek only to compare the observed data with realizations of a homogeneous planar Poisson process. Also, the uncorrected versions are approximately unbiased when m and n are both large; for example, Figure 18.1 illustrates the bias for data consisting of $n = 100$ events in D the unit square. In Figure 18.1 the sampling origins for computation of the x_i formed a 20×20 square lattice to span D, i.e., $m = 400$. There is no good theoretical reason to limit the value of m, but equally the information content in the data is limited by n rather than m and little is to be gained by setting $m >> n$. For remarks on this from a time when computing was a scarce resource, see, for example, Ripley (1981, p. 154), or Diggle and Matérn (1980) who give theoretical arguments to support the use of a regular grid of sampling origins.

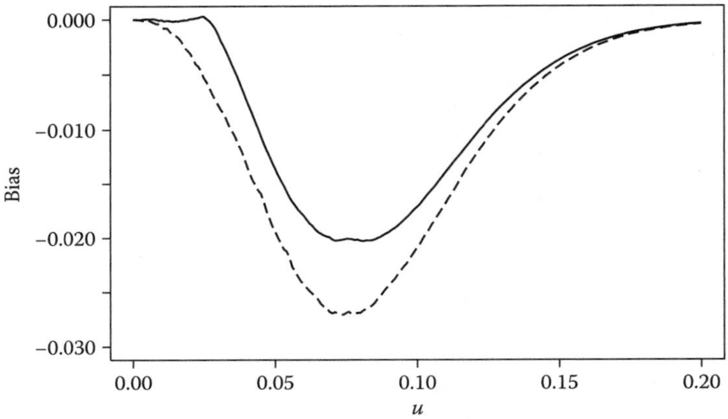

FIGURE 18.1
Bias in the empirical distribution functions of nearest neighbor distances for an underlying Poisson process conditioned to have 100 events in the unit square, estimated from 10,000 simulated realizations. The solid line denotes the bias in $\hat{F}(u)$, using a 20 × 20 square grid of sampling origins; dashed line denotes the bias in $\hat{G}(u)$.

To provide a test of CSR using either $\hat{F}(\cdot)$ or $\hat{G}(\cdot)$, or indeed any other empirical function, $\hat{H}(\cdot)$ say, a general Monte Carlo procedure is the following. Let $\hat{H}_1(\cdot)$ denote the empirical function of the data and $\hat{H}_i(\cdot) : i = 2, \ldots, s$ the empirical functions of $s - 1$ independent simulations of CSR conditonal on n, i.e., each simulation consists of n events independently and uniformly distributed on D. For any value of u, define $\bar{H}_i(u) = s^{-1} \sum_{j \neq i} \hat{H}_j(u)$. Then, the deviations $\hat{H}_i(u) - \bar{H}_i(u)$ are exchangeable under CSR and any statistic derived from these deviations, for example,

$$t = \int_0^\infty \{\hat{H}_i(u) - \bar{H}_i(u)\}^2 du, \tag{18.5}$$

provides the basis for a valid Monte Carlo test of CSR.

In practice, tests using either $\hat{F}(\cdot)$ or $\hat{G}(\cdot)$ in Equation (18.5) tend to be powerful against different kinds of alternative to CSR. This leads Diggle (2003, Chap. 2) to recommend using both routinely, and to interpret their combined significance either conservatively (quoting the attained signficance level as twice the smaller of the two individual p-values), or more informally by inspection of plots of the two empirical functions together with simulation envelopes. Figure 18.2 shows an example for a dataset giving the locations of 106 pyramidal neurons in area 24, layer 2 of the cingulate cortex; these data are one of 31 such datasets analyzed in Diggle, Lange, and Benes (1991). Visual inspection of the data suggests a pattern not dissimilar to CSR. The deviations of $\hat{F}(\cdot)$ from the mean of 99 simulations are small and, as judged by the simulation envelope, are compatible with CSR. A formal Monte Carlo test using the statistic (18.5) gave a p-value of 0.50. The deviations of $\hat{G}(\cdot)$ from the mean of the simulations are somewhat larger, both absolute and relative to the simulation envelope, and the plot suggests significant small-scale spatial regularity ($\hat{G}(u) < \bar{G}_1(u)$ for $u < 0.05$ or thereabouts). The Monte Carlo test gave $p = 0.02$. In this context, small-scale regularity is to be expected because the notional locations are reference points within cell bodies, each of which occupies a finite amount of space.

If a single summary function is required, a possible choice is the J-function proposed by van Lieshout and Baddeley (1996, see also Section 16.5) and defined as $\hat{J}(u) = \{1 - \hat{G}(u)\}/\{1 - \hat{F}(u)\}$. For CSR, the corresponding theoretical function is $J(u) = 1$ for all u and, unlike $\hat{F}(\cdot)$ and $\hat{G}(\cdot)$, $\hat{J}(\cdot)$ is insensitive to edge effects. Figure 18.3 shows $\hat{J}(\cdot)$ and its simulation envelope for the pyramdial neuron data. The left-hand panel suggests an inability to

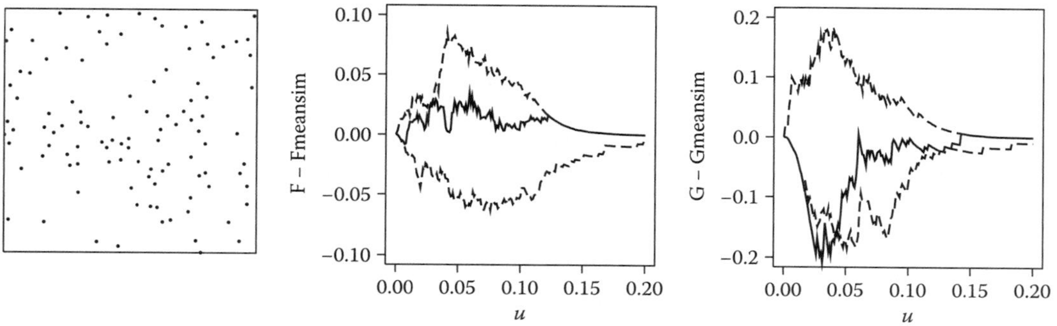

FIGURE 18.2
Example of nearest neighbor analysis. The left-hand panel shows the locations of 106 pyramidal neurons. The center and right-hand panels show the pointwise differences between empirical and theoretical (under CSR) distribution functions of point-to-event and event-to-event nearest neighbor distances, respectively, together with pointwise envelopes from 99 simulations of CSR.

discriminate between the data and CSR, but this is a by-product of the inevitable instability in the empirical function $\hat{J}(\cdot)$ as $\hat{F}(u)$ and $\hat{G}(u)$ approach 1 and $\hat{J}(u)$ becomes either zero, infinite, or indeterminate. The right-hand panel of Figure 18.3 shows that, at small values of u, the J-function tells much the same story for these data as does the G-function.

18.5 Estimating a Spatially Varying Intensity

We now turn to nonparametric estimation of the low-order moments of a spatial point process, beginning with the first-order moment, or intensity.

For a stationary process, the intensity is a constant, λ, and for an observed pattern of n points in a study region D, the natural estimator is the observed number of events per unit

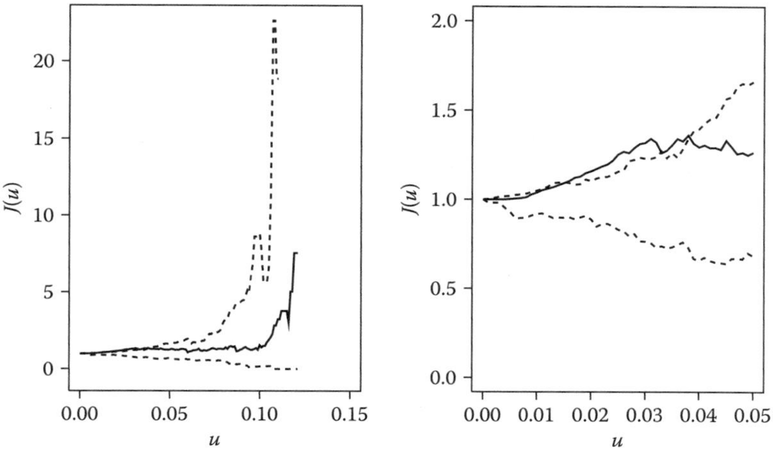

FIGURE 18.3
Empirical J-function for the pyramidal neuron data, with pointwise envelopes from 99 simulations of CSR. Right-hand panel is a magnification for distance $u \leq 0.05$.

area, $\hat{\lambda} = n/|D|$. This estimator is always unbiased. If the underlying process is a stationary Poisson process, $\hat{\lambda}$ is also the maximum likelihood estimator and has variance $\lambda/|D|$.

For nonstationary processes with spatially varying $\lambda(x)$, estimating $\lambda(x)$ from a single realization is problematic without additional assumptions. For example, and as discussed also in Chapter 17, Section 17.2, no empirical distinction can be made between a realization of a nonstationary Poisson process with deterministic intensity $\lambda(x)$ and a stationary Cox process the realization of whose stochastic intensity coincides with $\lambda(x)$.

In Section 18.8, we shall describe nonparametric methods for replicated spatial point patterns where, at least in principle, the replication gives a way of distinguishing between first-moment and second-moment effects. Here, we resolve the ambiguity by making the additional asssumption that the underlying point process is a stationary Cox process and consider initially estimators of the form

$$\tilde{\lambda}_h(x) = \sum_{i=1}^{n} I(||x - x_i|| < h)/(\pi h^2), \tag{18.6}$$

where $I(\cdot)$ is the indicator function and $||\cdot||$ denotes Euclidean distance. In other words, the estimate of $\lambda(x)$ is the observed number of events per unit area within a disk of radius h centered on x. We define the mean square error of Equation (18.6) to be $MSE(h) = \mathrm{E}[\{\tilde{\lambda}(x) - \Lambda(x)\}^2]$, where $\Lambda(\cdot)$ is the stochastic intensity of the underyling Cox process. Then,

$$MSE(h) = \lambda_2(0) + \lambda\{1 - 2\lambda K(h)\}/(\pi h^2) + (\pi h^2)^{-2} \int\int \lambda_2(||x - y||) dy dx, \tag{18.7}$$

where λ, $\lambda_2(\cdot)$ and $K(\cdot)$ are the intensity, second-order intensity, and reduced second-moment measure of the underlying Cox process, and each integration is over the disk of radius h centered at the origin (Diggle, 1985). Choosing h to minimize $MSE(h)$ then gives a practical strategy for implementing the estimator $\tilde{\lambda}(\cdot)$.

Berman and Diggle (1989) showed that by transforming to polar coordinates, the problem of evaluating the integral in (18.7) can be reduced to evaluation of the expression

$$I = K(2h)\phi(2h) - \int_0^{2h} K(s)\phi'(s) ds, \tag{18.8}$$

where $\phi(s) = 2\pi[2h^2 \cos^{-1}\{s/(2h)\} - s\sqrt{(h^2 - s^2/4)}]$ is the area of intersection of two disks, each of radius h and centers a distance s apart. The integral I therefore can be estimated by substituting the standard estimator (Equation 18.12, below) for $K(\cdot)$ into Equation (18.12) and using a one-dimensional quadrature for the integral term.

Two modifications to Equation (18.6) improve its performance. First, the indicator function can be replaced by a smoothly varying kernel function. Second, to deal with edge effects, the kernel can be scaled so that it integrates to one over the study region D. Hence,

$$\hat{\lambda}(x) = \sum_{i=1}^{n} k_h(x - x_i) \Big/ \int_D k_h(x - x_i) dx, \tag{18.9}$$

where the kernel function $k_h(x) = h^{-2}k(x/h)$, with $k(\cdot)$ a radially symmetric bivariate probability density function; for the estimator $\tilde{\lambda}(\cdot)$, $k(\cdot)$ corresponds to a uniform distribution on the disk of unit radius.

The first of the above modifications is largely cosmetic, but the second is important when the chosen value of h is relatively large. A more serious practical limitation is that the

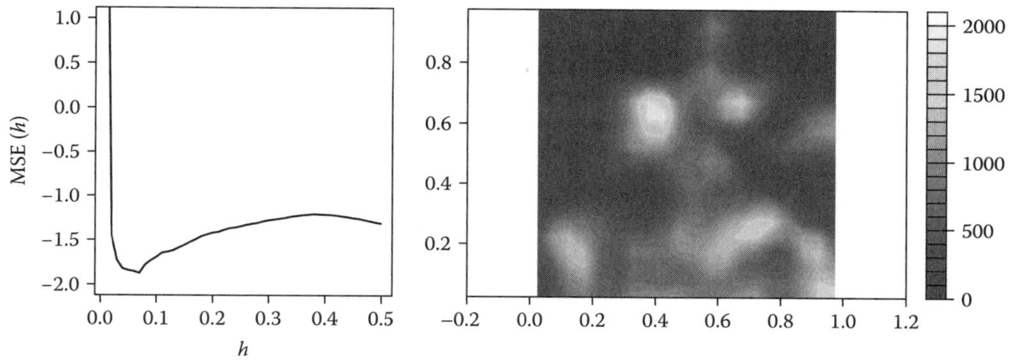

FIGURE 18.4
Kernel estimation of $\lambda(x)$ for data on the locations of 514 maple trees: Mean square error as a function of bandwidth h (left-hand panel); kernel estimate for $h = 0.07$ (right-hand panel).

working assumption of a stationary Cox process may be untenable. We therefore recommend inspecting a plot of the estimated $MSE(h)$ as a guide to the choice of h, rather than taking the minimizer of $MSE(h)$ as the automatic choice. Diggle (2003, Sec. 8.2) used a different edge correction, essentially one that scales the estimate rather than each separate kernel contribution, but results in Jones (1993) suggest that the edge correction given here as Equation (18.9) is preferable.

Figure 18.4 shows the estimated $MSE(h)$ and the edge-corrected estimate of $\lambda(x)$ for a dataset giving the locations of 514 maple trees in a 19.6-square acre plot, here rescaled to the unit square (Gerrard, 1969; Diggle, 2003, Sec. 2.6.1). The MSE-plot takes its minimum value at $h = 0.07$. The estimate $\hat{\lambda}(x)$ using $h = 0.07$ shows strong spatial heterogeneity, with a global maximum near the center around four times the average intensity, and generally elevated intensity in the lower half of the square. Recall, however, that if the underlying process is not a stationary Cox process, this method for choosing h could be misleading. For related comments, see Section 18.6.2.

Zimmerman (2008) has proposed an interesting extension to Equation (18.9) to deal with a problem that arises in some areas of application, notably spatial epidemiology as discussed in Chapter 22, whereby some of the event locations are known only to a relatively coarse spatial resolution. An example would be an epidemiological dataset in which each subject is identifed by their residential location, but for some subjects all that is known is the administrative subregion in which they reside. In this setting, suppose that within the kth subregion there are n_k events of which m_k are at known locations, then Zimmerman's modification of Equation (18.9) upweights the kernel contribution from each known location in the kth subregion by the factor n_k/m_k.

18.6 Second-Moment Methods for Mapped Point Patterns

18.6.1 Stationary Processes

As described in Chapters 16 and 17, the second-moment properties of a stationary spatial point process can be summarized either by the second-order intensity function $\lambda_2(\cdot)$ or by the reduced second-moment measure $K(\cdot)$. Under appropriate regularity conditions, the relationship between the two is that

$$K(u) = 2\pi\lambda^{-2}\int_0^u \lambda_2(v)v\,dv,$$

where λ is the intensity. From the point of view of nonparametric estimation, a more useful result is that

$$K(u) = \lambda^{-1}E(u), \tag{18.10}$$

where $E(u)$ denotes the expected number of additional events within distance u of the origin, conditional on there being an event at the origin. Expressing $\lambda K(u)$ as an expectation invites the possibility of nonparametric estimation by the method of moments, substituting the expectation in Equation (18.10) by the average observed number of additional events within distance u of each event in turn. For data $x_i : i = 1, \ldots, n$ in a region D, and using the natural estimator $\hat{\lambda} = n/|D|$ for λ, this suggests the estimator

$$\tilde{K}(u) = n^{-2}|D|\sum_{i=1}^{n}\sum_{j\neq i} I(||x_i - x_j|| \leq u). \tag{18.11}$$

The estimator (18.11) is biased by the exclusion of events that lie outside D. To counter this, Ripley (1976, 1977) proposed an edge-corrected estimator,

$$\hat{K}(u) = n^{-2}|D|\sum_{i=1}^{n}\sum_{j\neq i} w_{ij}^{-1} I(||x_i - x_j|| \leq u), \tag{18.12}$$

where w_{ij} is the proportion of the circle with center x_i and radius $||x_i - x_j||$ that is contained in D. Other edge corrections have also been proposed; see, for example, Stein (1991) or Baddeley (1999).

From a mathematical perspective, either the second-order intensity or its scale-free counterpart the pair correlation function, $g(u) = \lambda_2(u)/\lambda^2$, could be regarded as a more fundamental quantity, and can also be estimated using the array of interevent distances $u_{ij} = ||x_i - x_j||$. However, estimation is complicated by the need to smooth the empirical distribution of the u_{ij}; the analogy here is with the distinction between the straightforward estimation of a distribution function nonparametrically by its empirical counterpart and the much more complicated problem of nonparametric density estimation. For small datasets, the avoidance of the smoothing issue is a strong argument in favor of estimating $K(\cdot)$ rather than $\lambda_2(\cdot)$ or $g(\cdot)$. For large datasets, choosing a sensible, albeit nonoptimal, smoother is more straightforward and, as discussed in Chapter 19, may confer other advantages in a parametric setting.

A class of kernel estimators for $g(\cdot)$ takes the form

$$\hat{g}(u) = (2\pi u)^{-1}n^{-2}|D|\sum_{i=1}^{n}\sum_{j\neq i} w_{ij}^{-1}k_h(u - ||x_i - x_j||), \tag{18.13}$$

where now $k_h(u) = h^{-1}k(u/h)$ for some univariate probability density function $k(\cdot)$, the w_{ij} are edge-correction weights as in Equation (18.12) and h is a smoothing constant; see, for example, Møller and Waagepetersen (2004, Sec. 4.3.5). Note that $\hat{g}(0)$ is undefined and that, in practice, estimates $\hat{g}(u)$ are unstable at very small values of u.

Figure 18.5 shows the results of applying the estimators $\hat{K}(\cdot)$ and $\hat{g}(\cdot)$ to the pyramidal neuron data previously analyzed in Section 18.4. The left-hand panel of Figure 18.5 shows $\hat{K}(\cdot)$ together with its envelope from 99 simulations of CSR. As with our earlier analysis

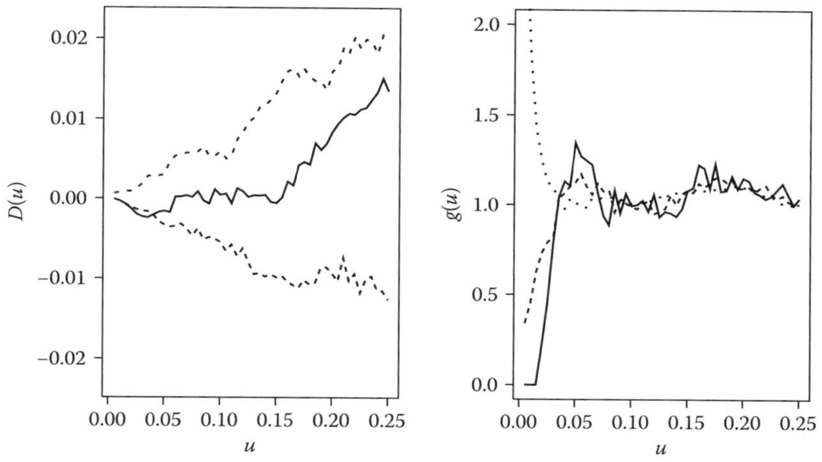

FIGURE 18.5
Estimates of the reduced second-moment measure (left-hand panel) and of the pair correlaton function (right-hand panel) for the pyramidal neuron data, in each case with pointwise envelopes from 99 simulations of CSR.

using nearest neighbor distributions, the diagram indicates small-scale spatial regularity because the estimate for the data lies below the envelope of the simulations. Notice also that subtracting πu^2 from $\hat{K}(u)$ magnifies the interesting part of the diagram and that the sampling variation in $\hat{K}(u)$ increases sharply with u. For $\hat{g}$, the right-hand panel of Figure 18.5 shows three estimates corresponding to different values of the smoothing constant, $h = 0.01, 0.025, 0.05$. Notice the radically different behaviour of the three estimates at small values of u. Oversmoothing by increasing the value of h hides the small-scale spatial regularity because the numerator in the expression (18.13) for $\hat{g}(u)$ remains positive as u approaches zero and the term u in the denominator then dominates the behavior of $\hat{g}(u)$.

18.6.2 Nonstationary Processes

Dealing with nonstationarity is difficult in the absence of independent replication. Spatial heterogeneity and spatial clustering, although phenomenologically different, can be difficult, or even impossible, to distinguish empirically. In Section 18.5, we exploited this ambiguity in describing how stationary point process theory can be used to suggest a method for estimating a spatially varying intensity. To attempt to estimate first and second moment properties from a single realization, we use the more general theoretical framework of intensity-reweighted stationarity processes, as developed in Baddeley, Møller, and Waagepetersen (2000). An *intensity-reweighted stationarity process* is one for which $\lambda(x)$ is strictly positive and the pair correlation function, $\rho(u) = \lambda_2(x, y)/\lambda(x)\lambda(y)$, depends only on $u = ||x - y||$. The corresponding intensity-reweighted K-function is

$$K_I(u) = 2\pi \int_0^u g(u) u du.$$

If $\lambda(x)$ were a known function, estimates of $K_I(u)$ and of $g(u)$ would be defined by the following natural extensions of Equation (18.12) and Equation (18.13), respectively, hence,

$$\hat{K}(u) = |D|^{-1} \sum_{i=1}^{n} \sum_{j \neq i} w_{ij}^{-1} I(||x_i - x_j|| \leq u/\{\lambda(x_i)\lambda(x_j)\} \qquad (18.14)$$

and

$$\hat{g}(u) = (2\pi u)^{-1}|D|^{-1}\sum_{i=1}^{n}\sum_{j\neq i} w_{ij}^{-1}k_h(u - ||x_i - x_j||)/\{\lambda(x_i)\lambda(x_j)\}. \tag{18.15}$$

In practice, $\lambda(x)$ must itself be estimated. A pragmatic strategy is to estimate $\lambda(x)$ by a kernel smoother, as in Section 3.5, but using a relatively large bandwidth, chosen subjectively; in effect, this amounts to treating large-scale spatial variation as a first-moment effect, smaller-scale variation as a second-moment effect and making a pragmatic decison as to what constitutes "large scale." The implied partition of the spatial variation into two components is critically dependent on the choice of the kernel smoothing bandwidth h. Baddeley et al. (2000) show that the resulting estimator $\hat{K}_I(u)$ is biased, potentially severely so, but that the bias can be reduced to some extent by leaving out the datum x_i when estimating $\lambda(x_i)$. Another strategy is to use a parametric estimate of $\lambda(\cdot)$. A third is to estimate $\lambda(x)$ from a second dataset that can be assumed to be a realization of a Poisson process with the same first-moment properties as the process of interest; one setting, perhaps the only one, in which this is a viable strategy is in an epidemiological case-control study where the control data consist of the locations of an independent random sample from the population of interest. See Chapter 22 for a description of the use of point process methods in spatial epidemiology.

18.7 Marked Point Patterns

A *marked* point process is one in which each event is accompanied by the realized values of one or more random variables in addition to its location. Here, we consider only stationary processes with a single mark attached to each event, and consider separately the cases of qualitative and quantitative marks.

18.7.1 Qualitative Marks: Multivariate Point Patterns

When the mark random variable is qualitative, an equally apt name for the process is that it is a *multivariate* process whose events are of m distinguishable *types*, $k = 1, \ldots, m$. In this setting, we refer to the set of all observed events, irrespective of type, as the *superposition*, and the corresponding underlying process as the *superposition process*. For exploratory analysis, it is useful to consider the observed pattern in relation to two benchmark hypotheses: independence and random labeling. For each, we consider how the nearest neighbor or second-moment methods described in Sections 18.4 and 18.6 can be adapted to provide diagnostic summaries.

The hypothesis of *independence* specifies that the m component patterns are realizations of independent univariate spatial point processes. Under this hypothesis, let $F_k(u)$ denote the distribution function of the distance from an arbitrary point to the nearest event of type k, and $F(u)$ the distribution of the distance from an arbitrary point to the nearest event in the superposition process. Under independence, we have the result that

$$F(u) = 1 - \prod_{k=1}^{m}\{1 - F_k(u)\}. \tag{18.16}$$

Now, denote by $G_k(u)$ the distribution function of the distance from an arbitrary type k event to the nearest other type k event, and $G(u)$ the distribution function of the distance from an arbitrary event to the nearest other event, without reference to type. If λ_k is the intensity of type k events, $\lambda = \sum_{k=1}^{m} \lambda_k$ the intensity of the superposition process and $p_k = \lambda_k/\lambda$ the proportion of type k events in the superposition process, then under independence,

$$G(u) = 1 - \sum_{j=1}^{m} p_j\{1 - G_j(u)\} \prod_{k \neq j} \{1 - F_k(u)\}. \tag{18.17}$$

Finally, consider the multivariate K-functions,

$$K_{jk}(u) = \lambda_k^{-1} E_{jk}(u), \tag{18.18}$$

where for all $j \neq k$, $E_{jk}(u)$ is the expected number of type k events within distance u of an arbitrary type j event. Then, under independence we have the strikingly simple result that

$$K_{jk}(u) = \pi u^2. \tag{18.19}$$

Any of Equation (18.16), Equation (18.17) or Equation (18.19) can be used as the basis of a statistic to measure departure from independence. Also, if the data are observed on a rectangular region D, a Monte Carlo test can be implemented by the following device, suggested in Lotwick and Silverman (1982). First, wrap the data onto a torus by identifying opposite edges of D. Then, under independence, the distribution of any statistic calculated from the data is invariant under independent random toroidal shifts of the component patterns.

We now consider our second benchmark hypothesis, random labeling. A multivariate process is a *random labeling* if the component processes are derived from the superposition process by a series of independent multinomial outcomes on the integers $k = 1, \ldots, m$ with constant cell probabilities $p_k : k = 1, \ldots, m$.

Under random labeling, there is, in general, no simple relationship that links $F(u)$ with the component-wise functions $F_k(u)$, or $G(u)$ with the $G_k(u)$. In contrast, the multivariate K-functions obey the very simple relationship that, for all $j = 1, \ldots, m$ and $k = 1, \ldots, m$

$$K_{jk}(u) = K(u), \tag{18.20}$$

where $K(\cdot)$ denotes the K-function of the superposition process.

To test for departure from random labeling, a Monte Carlo test procedure consists of calculating any suitable statistic from the data and recalculating after independent simulated random labelings of the superposition. Note also that a test of this kind remains valid even if the superposition process is nonstationary.

Independence and random labeling are equivalent if the underlying component processes are Poisson processes but, in general, can be very different in character. For example, in a multivariate Cox process with random intensities $\Lambda_k(x)$, independence corresponds to independence of the $\Lambda_k(\cdot)$, whereas random labeling corresponds to a deterministic relationship, $\Lambda_k(x) = p_k \Lambda(x)$, where $\Lambda(x)$ is the random intensity of the superposition process.

The properties of summary functions for multivariate point patterns will be explored further in Chapter 21.

18.7.1.1 Example: Displaced Amacrine Cells in the Retina of a Rabbit

These data originate from Wienawa-Narkiewicz (1983) and were made available to us by Prof. Abbie Hughes. The data were first analyzed in Diggle (1986). The presentation here closely follows Diggle (2003, Sec. 4.7). The data in Figure 18.6 consist of the locations of 294 displaced amacrine cells in a rectangular section of the retina of a rabbit, of approximate dimension 1060 by 662 µm, here rescaled to approximately 1.6 by 1.0. The cells are of two

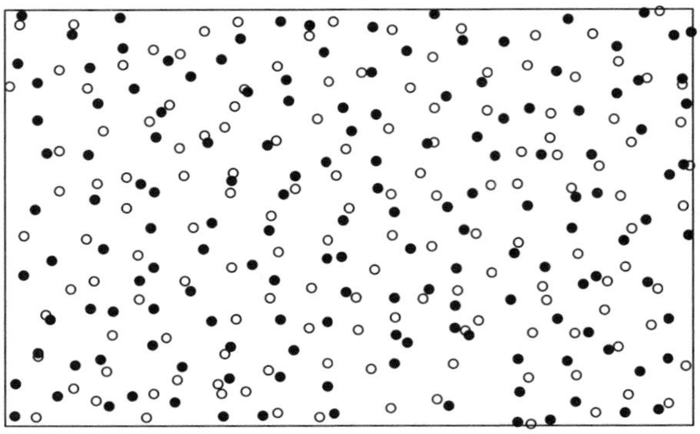

FIGURE 18.6
The displaced amacrine cell data: 152 *on* cells and 142 *off* cells are shown as closed and open circles, respectively.

types, according to whether they respond to light going *on* (152 cells) or *off* (142 cells). Two competing developmental hypotheses are the following:

H_a: The two types of cell form initially in *separate layers* that subsequently fuse in the mature retina;

H_b: Cells are initially undifferentiated in a *single layer*, with the separation into on and off cells occurring at a later developmental stage.

Independence and random labeling are natural benchmark hypotheses corresponding to H_a and H_b, respectively. As we have shown, the second-moment properties can be used to establish whether the data are compatible with either or both of these benchmarks. Figure 18.7 shows estimates of the second-moment properties of the data. Recall that under independence, $K_{12}(u) = \pi u^2$, whereas under random labeling $K_{11}(u) = K_{22}(u) = K_{12}(u) = K(u)$. Figure 18.7 suggests, first, that the second-moment properties of the two component processes are very similar. Because the two component patterns also have approximately the same intensity, they can be considered informally as if they were replicates, and the difference between $\hat{K}_{11}(u)$ and $\hat{K}_{22}(u)$ gives a rough indication of the sampling variation in either one. It then follows that $\hat{K}_{12}(u)$ differs from either of $\hat{K}_{11}(u)$ or $\hat{K}_{22}(u)$ by far more than

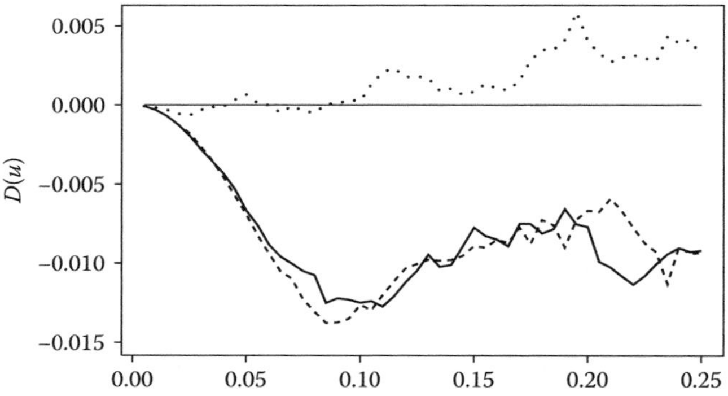

FIGURE 18.7
Second-order analysis of the displaced amacrine cell data. Estimates of $K(u) - \pi u^2$ are shown for the *on* cells (solid line), and the *off* cells (dashed line). The dotted line shows the estimate of $K_{12}(u) - \pi u^2$.

can be explained by sampling fluctuations; hence, the data are incompatible with random labeling. However, the estimate $\hat{K}_{12}(u)$ is fairly close to πu^2, at least for small u where the estimate is most precise. These observations together suggest a preliminary conclusion in favor of the separate layer hypothesis H_a. Note also that both $\hat{K}_{11}(u)$ and $\hat{K}_{22}(u)$ are zero at small values of u, suggesting a strict inhibitory effect within each component process, whereas no such effect is apparent in $\hat{K}_{12}(u)$. This adds to the evidence in favor of H_a.

Diggle (2003, Sec. 4.7 and Sec. 7.2.2 to 7.2.5) gives a more extensive discussion of this example. See also Hughes (1985) and Eglen et al. (2006) for an explanation of the biological background to this and related investigations.

18.7.2 Quantitative Marks

For a process with a quantitative mark attached to each event, the natural counterpart of the random labeling hypothesis is that the marks are determined by independent random sampling from an arbitrary probability distribution. A natural counterpart of the independence hypothesis is that the unmarked point process and the mark process are independent stochastic processes. With these definitions, and in contrast to the discussion in Section 18.7.1 above, random labeling becomes a special case of independence.

Another consideration for a quantitatively marked point process is whether the mark process exists throughout the study region, or only at the event locations. Geostatistical data, as discussed in Part I, can be considered as an example of the first kind; a process in which the events are birds' nests and the mark is the number of hatchlings successfully reared is an example of the second kind. This distinction has implications for how we should approach exploratory analysis. For processes of the first kind, random labeling is physically implausible, independence is the more natural benchmark hypothesis, and a suitable class of alternatives is that the intensity of events at a point x is related to the value, $Z(x)$ say, of the underlying spatially continuous mark process at x. For processes of the second kind, random labeling is a natural benchmark hypothesis, with departures from it suggesting some form of stochastic interaction between neighboring events.

Schlather, Riberiro, and Diggle (2004) proposed several summary statistics aimed at investigating departures from the independence hypothesis. They were motivated primarily by the fact that conventional geostatistical methods rely on this assumption, and their proposed statistics therefore focused on investigating how departure from independence could affect estimation of the second-moment properties of an underlying spatially continuous mark process.

In the author's opinion, the exploratory analysis of marked point process data remains a relatively underdeveloped topic of research.

18.8 Replicated Point Patterns

A replicated point pattern dataset is one consisting of the locations $\{x_{ij} : j = 1, \ldots, n_i; i = 1, \ldots, r\}$ in each of r study regions A_i. An important distinction is between *independent* replication, and *pseudo-replication*. By independent replication, we mean that the r point patterns are derived from a sample of r experimental units drawn as an independent random sample from a population of experimental units. By pseudo-replication, we mean any other form of replication that we choose to treat as if it were an independent replication. For example, if only a single point pattern is available on a region D, we could create pseudo-replicates by partitioning D into congruent subregions D_i. Under stationarity, the resulting point patterns are replicates in the sense that they are realizations of the same process, but they are not necessarily independent. In the analysis of the amacrines data, reported in Section 18.7.1, we exploited another form of pseudo-replication to assess the sampling variation in estimates of the second-moment properties of the data.

Independent replication is common in microanatomical studies where the study units i are animals or plants, and each D_i is objectively determined to correspond to a specific anatomical region; for example, we can easily imagine that the displaced amacrine cell data could have been but one of a number of independent replicates obtained from different rabbits.

Independent replication may also be a tenable assumption if the D_i are spatially separated subregions of a larger region D. For example, in forestry, we might choose to collect data on tree locations from several different parts of a forest. However, and in contrast to the microanatomical setting where the experimental units are well-defined, discrete entities, the assumption that the same process has generated the patterns in all of the subregions of D_i would then require separate justification.

In the remainder of this section, we assume that we have independent replication by design. This does not rule out applying the methods described to pseudo-replicated data, but to do so involves stronger assumptions, namely that data obtained as multiple samples from a single realization of an underlying point process have the same statistical properties as data obtained as a single sample from multiple realizations.

As in other branches of statistics, independent replication open up the possibility of a design-based approach to inference. In what follows, we assume that the D_i are congruent, meaning that they are identical in size and shape, and that they have the same position within the separate sampling units in relation to some objectively defined origin of measurement.

Now, suppose that we compute the value of a summary description, $\mathcal{F}$ say, from each of r observed patterns; typically, $\mathcal{F}$ will be an empirical function rather than a single number, for example, $\hat{K}(\cdot)$ as discussed in Section 18.6. The simplest possible inferential task might then be to estimate the expectation of $\mathcal{F}$ under random sampling from the population of interest. For observed summaries $\mathcal{F}_i : i = 1, \dots, r$, an unbiased estimator is the sample mean,

$$\bar{\mathcal{F}} = r^{-1} \sum_{i=1}^{r} \mathcal{F}_i, \tag{18.21}$$

with estimated standard error $SE(\bar{\mathcal{F}}) = s_{\mathcal{F}}/\sqrt{r}$, where

$$s_{\mathcal{F}}^2 = (r-1)^{-1} \sum_{i=1}^{r} \{\mathcal{F}_i - \bar{\mathcal{F}}\}^2 \tag{18.22}$$

is the sample variance of the $\mathcal{F}_i$. Note that means and variances are here to be interpreted pointwise if $\mathcal{F}$ is an empirical function.

In the above discussion, there is no requirement that the underlying process be stationary. However, the choice of the summary descriptor $\mathcal{F}$ should depend on whether stationarity is a reasonable working assumption. More importantly, the implicit estimand in Equation (18.21) is the expectation of $\mathcal{F}$ under repeated sampling, and this therefore, should be chosen to have a useful scientific interpretation in any specific application.

Diggle et al. (1991) and Baddeley, Moyeed, Howard, and Boyde (1993) both use the estimated K-function as summary descriptor, with replicates corresponding to different subjects (human brains and macaque monkey skulls, respectively). In Diggle et al., 31 subjects were divided into three treatment groups with 12, 10, and 9 replicates. In Baddeley et al., the point pattern of interest was three-dimensional, and four subjects were each sampled in 10 spatially separate cuboidal subregions. Baddeley et al. discussed different weighted averages of the individual estimates $\hat{K}(\cdot)$ and associated standard errors, and developed an analysis of variance to divide the variablity in the $\hat{K}(\cdot)$ into between-subject and within-subject components. Diggle et al., proposed a bootstrap testing procedure to compare the expectations of $\hat{K}(\cdot)$ in the three treatment groups as follows. Let $\hat{K}_{ij}(u)$ be the estimate of $K(u)$, and n_{ij} the number of events, for the jth subject in the ith treatment group. Also, let $\bar{K}_i(u)$ denote a suitable weighted average of the $\hat{K}_{ij}(u)$ within the ith treatment group and $\bar{K}(u)$ a weighted average over all subjects. Then, the residual functions,

$$r_{ij}(u) = n_{ij}^{1/2}\{\hat{K}_{ij}(u) - \bar{K}_i(u)\},$$

are approximately exchangeable. To test the hypothesis that $\mathrm{E}[\bar{K}_i(u)]$ is the same for all three treatment groups i, we therefore reconstruct bootstrapped K-functions, $K^*_{ij}(\cdot)$ as

$$K^*_{ij}(u) = \bar{K} + n_{ij}^{-1/2} r^*_{ij}(u), \tag{18.23}$$

where the $r^*_{ij}(\cdot)$ are a random permutation of the $r_{ij}(\cdot)$. For a Monte Carlo test, we then compare the value of any test statistic computed from the actual $\hat{K}_{ij}(\cdot)$ with the values computed from bootstrapped $K^*_{ij}(\cdot)$, for a large number of independent bootstrap resamples.

In the nonstationary case, independent replication provides an objective way of separating first-order and second-order properties. This requires that the D_i are congruent, or can be made so by a combination of shift, rotation, and scale transformations. For example, Webster, Diggle, Clough, Green et al. (2005) analyze the pattern of TSE (transmissible spongiform encephalopathy) lesions in mouse brain tissue sections taken from three functionally distinct regions of the brain, namely the paraterminal body, the thalamus, and the tectum of mid-brain. Each of these regions has an easily recognizable orientation and registration point on its boundary. The sizes of the regions vary between animals, but can easily be scaled to a common area. Webster et al. (2005) estimate a spatially varying first-order intensity $\lambda(x)$ as a weighted average of kernel smoothers applied to the shifted, rotated, and scaled point patterns from each animal, with the bandwidth h chosen using a cross-validated log likelihood. They then use the estimate (18.14) of the inhomogeneous K-function, pooled over animals, to estimate the second-order properties.

References

Baddeley, A.J. (1999). Spatial sampling and censoring. In O.E. Barndorff-Nielsen, W.S. Kendall, and M.N.M. Van Lieshout, Eds. *Stochastic Geometry: Likelihood and Computation*, pp. 37–78, London. Chapman & Hall.

Baddeley, A.J., Møller, J., and Waagepetersen R. (2000). Non and semi parametric estimation of interaction in inhomogeneous point patterns. *Statistica Neerlandica*, 54:329–350.

Baddeley, A.J., Moyeed R.A., Howard, C.V. and Boyde, A. (1993). Analysis of a three-dimensional point pattern with replication. *Applied Statistics*.

Barnard, G.A. (1963). Contribution to the discussion of Professor Bartlett's paper. *Journal of the Royal Statistical Society, Series B*, 25:294.

Berman, M., and Diggle, P. (1989). Estimating weighted integrals of the second-order intensity of a spatial point process. *Journal of the Royal Statistical Society, Series B*, 51:81–92.

Besag, J. and Clifford, P. (1989). Generalized Monte Carlo significance tests. *Biometrika*, 76:633–642.

Besag, J. and Clifford, P. (1991). Sequential Monte Carlo *p*-values. *Biometrika*, 78:301–304.

Besag, J.E. and Gleaves, J.T. (1973). On the detection of spatial pattern in plant communities. *Bulletin of the International Statistical Institute*, 45 (1):153–158.

Bostoen, K., Chalabi, Z., and Grais, R.F. (2007). Optimisation of the t-square sampling method to estimate population sizes. *Emerging Themes in Epidemiology*, 4:X.

Brown, V., Jacquier, G., Coulombier, D., Balandine, S., Belanger, F., and Legros, D. (2001). Rapid assessment of population size by area sampling in disaster situations. *Disasters*, 25:164–171.

Diggle, P.J. (1985). A kernal method for smoothing point process data. *Applied Statistics*, 34:138–147.

Diggle, P.J. (1986). Displaced amacrine cells in the retina of a rabbit: Analysis of a bivariate spatial point pattern. *Journal of Neuroscience Methods*, 18:115–125.

Diggle, P.J. (2003). *Statistical Analysis of Spatial Point Patterns* (2nd ed.). Edward Arnold, London.

Diggle, P.J., Lange, N., and Benes, F.M. (1991). Analysis of variance for replicated spatial point patterns in clinical neuroanatomy. *Journal of the American Statistical Association*, 86:618–625.

Diggle P.J. and Matérn, B. (1980). On sampling designs for the study of nearest neighbor distributions in R^2. *Scandinavian Journal of Statistics*, 7:80–84.

Eglen, S.J., Diggle, P.J. and Troy, J.B. (2006). Functional independence of on- and off-center beta retinal ganglion cell mosaics. *Visual Neuroscience*, 22:859–871.

Fisher, R.A., Thornton, H.G. and Mackenzie, W.A. (1922). The accuracy of the plating method of estimating the density of bacterial populations, with particular reference to the use of Thornton's agar medium with soil samples. *Annals of Applied Botany*, 9:325–359.

Gerrard, D.J. (1969). Competition quotient: A new measure of the competition affecting individual forest trees. Research bulletin no. 20, Agricultural Experiment Station, Michigan State University.

Hopkins B. (1954). A new method of determining the type of distribution of plant individuals. *Annals of Botany*, 18:213–227.

Hughes, A. (1985). New perspectives in retinal organisation. *Progress in Retinal Research*, 4:243–314.

Illian, J., Penttinen, A., Stoyan, H. and Stoyan D. (2008). *Statistical Analysis and Modelling of Spatial Point Patterns*. Wiley, Chichester, U.K.

Jones, M.C. (1993). Simple boundary corrections for kernal density estimation. *Statistics and Computing*, 3:135–146.

Lotwick, H.W. and Silverman, B.W. (1982). Methods for analysing spatial processes of several types of points. *Journal of the Royal Statistical Society, Series B*, 44:406–413.

Marriott, F.H.C. (1979). Monte Carlo tests: How many simulations? *Applied Statistics*, 28:75–77.

Møller, J. and Waagepetersen, R. (2004). *Statistical Inference and Simulation for Spatial Point Processes*. Chapman & Hall, Boca Raton, FL.

Moore, P.G. (1954). Spacing in plant populations. *Ecology*, 35:222–227.

Pruetz, J.D. and Isbel, L.A. (2000). Correlations of food distribution and patch size with agonistic interactions in female vervets (*chlorocebus aethiops*) and patas monkeys (*erythrocebus patas*) living in simple habitats. *Behavioral Ecology and Sociobiology*, 49:38–47.

Ripley, B.D. (1976). The second-order analysis of stationary point processes. *Journal of Applied Probability*, 13:255–266.

Ripley, B.D. (1977). Modelling spatial patterns (with discussion). *Journal of the Royal Statistical Society*, B 39:172–212.

Ripley, B.D. (1981). *Spatial Statistics*. John Wiley & Sons, New York.

Schlather, M.S., Riberiro, P.J. and Diggle, P.J. (2004). Detecting dependence between marks and locations of marked point processes. *Journal of the Royal Statistical Society, Series B*, 66:79–93.

Stein, M.L. (1991). A new class of estimators for the reduced second moment measure of point processes. *Biometrika*, 78:281–286.

Stratton, L.C., Goldstein, G. and Meinzer, F.C. (2000). Temporal and spatial partitioning of water resources among eight woody species in a Hawaiian dry forest. *Oecologia*, 124:309–317.

Webster, S., Diggle, P.J., Clough, H.E., Green, R.B. and French, N.P. (2005). Strain-typing transmissible spongiform encephalopathies using replicated spatial data. In Baddeley, A., Gregori P., Mateu J., Stoica R., and Stoyan, D., Eds., *Case Studies in Spatial Point Processes*, Springer, New York, pp. 197–214.

Wieniawa-Narkiewicz, E. (1983). *Light and Electron Microscopic Studies of Retinal Organisation*. PhD thesis, Australian National University, Canberra.

Zimmerman, D.L. (2008). Estimating the intensity of a spatial point process from locations coarsened by incomplete geocoding. *Biometrics*, 64:262–270.

19

Parametric Methods

Jesper Møller

CONTENTS

19.1 Introduction

This chapter considers *inference* procedures for *parametric spatial point process models*. The widespread use of sensible but ad hoc methods based on functional summary statistics has through the past two decades been complemented by *likelihood-based methods* for parametric spatial point process models. The increasing development of such likelihood-based methods, whether frequentist or Bayesian, has led to more objective and efficient statistical procedures for parametric inference. When checking a fitted parametric point process model, summary statistics and residual analysis play an important role in combination with simulation procedures, as discussed in Chapter 5.

Simulation-free estimation methods based on *composite likelihoods* or *pseudo-likelihoods* are discussed in Section 19.3. *Markov chain Monte Carlo (MCMC)* methods have had an increasing impact on the development of simulation-based likelihood inference, in which context we describe *maximum likelihood inference* in Section 19.4, and *Bayesian inference* in Section 19.5. On one hand, as the development in computer technology and computational statistics continues, computationally intensive, simulation-based methods for likelihood inference probably will play an increasing role for statistical analysis of spatial point patterns. On the other hand, since larger and larger point pattern datasets are expected to be collected in the future, and the simulation-free methods are much faster, they may continue to be of importance, at least at a preliminary stage of a parametric spatial point process analysis, where many different parametric models may quickly be investigated.

Much of this review draws on the monograph of Møller and Waagepetersen (2004) and the discussion paper of Møller and Waagepetersen (2007). Other recent textbooks related to the topic of this chapter include Baddeley, Gregori, Matev, Stoica et al. (2006), Diggle (2003), Illian, Penttinen, Stoyan, and Stoyan (2008), and van Lieshout (2000). Readers interested in background material on MCMC algorithms for spatial point processes are referred to Geyer and Møller (1994), Geyer (1999), Møller and Waagepetersen (2004), and the references therein. Note that the comments and corrections to Møller and Waagepetersen (2004) can be found at www.math.aau.dk/~ jm.

19.2 Setting and Notation

The methods in this chapter will be applied to *parametric models of Poisson, Cox, Poisson cluster, and Gibbs (or Markov) point processes*. These models were described in Chapter 17, but the reader will be reminded about the definitions and some of the basic concepts of these models. Often spatio-temporal point process models specified in terms of a conditional intensity (of another kind than the Papangelou conditional density, which is of fundamental importance in the present chapter), while other kinds of spatio-temporal point process models, which are closely related to the Cox point process models considered in this chapter, can be found in, e.g., Brix and Diggle (2001) and Brix and Møller (2001).

We mostly confine attention to planar point processes, but many concepts, methods, and results easily extend to $\mathbb{R}^d$ or a more general metric space, including multivariate and marked point process models. Chapter 27 treats statistics for multivariate and marked point process models.

We illustrate the statistical methodology with various examples, most of which concern inhomogeneous point patterns. Often the R package `spatstat` has been used (see Baddeley and Turner (2005, 2006) and www.spatstat.org). Software in R and C, developed by Rasmus Waagepetersen in connection with our paper Møller and Waagepetersen (2007), is available at www.math.aau.dk/~rw/sppcode.

We consider a planar spatial point process X, excluding the case of multiple points, meaning that X can be viewed as a random subset of $\mathbb{R}^2$. We assume also that X is locally finite, i.e., $X \cap B$ is finite whenever $B \subset \mathbb{R}^2$ is finite.

We let $W \subset \mathbb{R}^2$ denote a bounded observation window of area $|W| > 0$. In most examples given in this chapter, W is a rectangular region. Usually we assume that just a single realization $X \cap W = x$ is observed, i.e., the data

$$x = \{\mathbf{s}_1, \ldots, \mathbf{s}_n\}$$

is a spatial point pattern. Here the number of points, denoted $n(x) = n$, is finite and considered to be a realization of a nonnegative discrete random variable (if $n = 0$, then x is the empty point configuration). Sometimes, as is the case in two of our examples, two or more spatial point patterns are observed, and sometimes a hierarchical point process model may then be appropriate as illustrated in Sections 19.4.2 and 19.5.1; see also Chapter 18 where nonparametric methods for multivariate point patterns are discussed.

In order to account for edge-effects, we may assume that $X \cap W = x \cup y$ is observed so that "x conditional on y" is conditionally independent of X outside W. The details are given in Sections 19.3.4 and 19.4.1.

Finally, $\mathbb{I}[\cdot]$ is an indicator function, and $\|\cdot\|$ denotes the usual distance in $\mathbb{R}^2$.

19.3 Simulation-Free Estimation Methods

This section reviews simple and quick estimation procedures based on various *estimating equations* for parametric models of spatial point processes. The methods are *simulation-free* and the estimating equations are derived from a composite likelihood (Sections 19.3.1 to 19.3.2), or by a minimum contrast estimation procedure (Section 19.3.2), or by considering a pseudo-likelihood function (Section 19.3.4).

19.3.1 Methods Based on First-Order Moment Properties

Consider a spatial point process X with a parametric *intensity function* $\rho_\beta(\mathbf{s})$, where $\mathbf{s} \in \mathbb{R}^2$ and β is an unknown real d-dimensional parameter, which we want to estimate. We assume that $\rho_\beta(\mathbf{s})$ is expressible in closed form. This is the case for many parametric Poisson, Cox and Poisson cluster point process models, while it is intractable for Gibbs (or Markov) point processes (Chapter 17), see, e.g., Møller and Waagepetersen (2004). Below we consider a composite likelihood function (Lindsay, 1988) based on the intensity function.

Recall that we may interpret $\rho_\beta(\mathbf{s})\,\mathrm{d}\mathbf{s}$ as the probability that precisely one point falls in an infinitesimally small region containing the location $\mathbf{s}$ and of area $\mathrm{d}\mathbf{s}$. Let $C_i,\ i \in I$, be a finite partitioning of the observation window W into disjoint cells C_i of small areas $|C_i|$. Define $N_i = \mathbb{I}[X \cap C_i \neq \emptyset]$ and

$$p_i(\beta) = \mathrm{P}_\beta(N_i = 1) \approx \rho_\beta(\mathbf{u}_i)|C_i|,$$

where $\mathbf{u}_i$ denotes a representative point in C_i. Consider the product of marginal likelihoods for the Bernoulli trials N_i,

$$\prod_{i\in I} p_i(\beta)^{N_i}(1 - p_i(\beta))^{1-N_i} \approx \prod_{i\in I}(\rho_\beta(\mathbf{u}_i)|C_i|)^{N_i}(1 - \rho_\beta(\mathbf{u}_i)|C_i|)^{1-N_i}. \tag{19.1}$$

In the right-hand side of Equation (19.1), we may neglect the factors $|C_i|$ in the first part of the product, since they cancel when we form likelihood ratios. Then, as the cell sizes $|C_i|$ tend to zero, under suitable regularity conditions the limit of the product of marginal likelihoods becomes, omitted the constant factor $\exp(|W|)$,

$$L_c(\beta; x) = \exp\left(-\int_W \rho_\beta(\mathbf{s})\,\mathrm{d}\mathbf{s}\right)\prod_{i=1}^{n} \rho_\beta(\mathbf{s}_i). \tag{19.2}$$

We call $L_c(\beta; x)$ the *composite likelihood* function based on the intensity function. If X is a Poisson point process with intensity function $\rho_\beta(\mathbf{s})$, then $L_c(\beta; x)$ coincides with the likelihood function.

If there is a unique β that maximizes $L_c(\beta; x)$, we call it the *maximum composite likelihood estimate* (based on the intensity function). The corresponding estimating function $s_c(\beta; x)$ is given by the derivative of $\log L_c(\beta; x)$ with respect to β,

$$s_c(\beta; x) = \sum_{i=1}^{n} \mathrm{d}\log\rho_\beta(\mathbf{s}_i)/\mathrm{d}\beta - \int_W (\mathrm{d}\log\rho_\beta(\mathbf{s})/\mathrm{d}\beta)\rho_\beta(\mathbf{s})\,\mathrm{d}\mathbf{s}. \tag{19.3}$$

The estimating equation $s_c(\beta; x) = 0$ is unbiased (assuming in Equation (19.3) that $(\mathrm{d}/\mathrm{d}\beta)\int_W \cdots = \int_W (\mathrm{d}/\mathrm{d}\beta)\cdots$). Asymptotic properties of maximum composite likelihood estimators are investigated in Waagepetersen (2007) and Waagepetersen and Guan (2009). For a discussion of asymptotic results for maximum likelihood estimates of Poisson process models, see Rathbun and Cressie (1994) and Waagepetersen (2007).

By exploiting the fact that Equation (19.2) coincides with the likelihood for a Poisson process, the maximum composite likelihood estimate can easily be determined using `spatstat`, provided the intensity function is of the *log-linear form*

$$\log \rho_\beta(\mathbf{s}) = a(\mathbf{s}) + \beta^T z(\mathbf{s}), \tag{19.4}$$

where $a(\mathbf{s})$ and $z(\mathbf{s})$ are known real functions with $z(\mathbf{s})$ of the same dimension as β. In practice, $z(\mathbf{s})$ is often a *covariate*. See also Berman and Turner (1992). This covariate may only be partially observed on a grid of points and, hence, some interpolation technique may be needed (Rathbun, 1996; Rathbun, Shiffman, and Gwaltney, 2007; Waagepetersen, 2008). An example is considered in Section 19.3.3.

We refer to a *log-linear Poisson process* when X is a Poisson process with intensity function of the form (19.4). For many *Cox process* models, the intensity function is also of the log-linear form (19.4). Specifically, let $Y = \{Y(\mathbf{s}) : \mathbf{s} \in \mathbb{R}^2\}$ be a spatial process where each $Y(\mathbf{s})$ is a real random variable with mean one, and let X conditional on $Y(\mathbf{s})$ be a Poisson process with intensity function

$$\Lambda(\mathbf{s}) = \exp(\beta^T z(\mathbf{s}))Y(\mathbf{s}). \tag{19.5}$$

Then Equation (19.4) is satisfied. Usually Y is not observed, and the distribution of Y may depend on another parameter ψ, which may be estimated by another method, as discussed in Section 19.3.2.

19.3.2 Methods Based on Second-Order Moment Properties

Let the situation be as in the first paragraph of Section 19.3.1 and, in addition, suppose that the spatial point process X has a parametrically specified *pair correlation function* g_ψ or other second-order characteristic, such as the *(inhomogeneous) K-function* K_ψ (Baddeley, Møller, and Waagepetersen, 2000; see also Chapter 18). We assume that g_ψ or K_ψ is expressible in closed form, which is the case for many parametric Poisson, Cox and Poisson cluster point process models. We assume also that β and ψ are variation independent, that is, $(\beta, \psi) \in B \times \Psi$, where $B \subseteq \mathbb{R}^p$ and $\Psi \subseteq \mathbb{R}^q$.

Recall that $\rho^{(2)}_{\beta,\psi}(\mathbf{s}, \mathbf{t}) = \rho_\beta(\mathbf{s})\rho_\beta(\mathbf{t})g_\psi(\mathbf{s}, \mathbf{t})$ is the *second-order product density*, and we may interpret $\rho^{(2)}_{\beta,\psi}(\mathbf{s}, \mathbf{t})\, d\mathbf{s}\, d\mathbf{t}$ as the probability of observing a point in each of two infinitesimally small regions containing $\mathbf{s}$ and $\mathbf{t}$ and of areas $d\mathbf{s}$ and $d\mathbf{t}$, respectively. Using the same principle as in Section 19.3.1, but considering now pairs of cells C_i and C_j, $i \neq j$, we can derive a *composite likelihood* $L_c(\beta, \psi)$ based on the second-order product density. Plugging in an estimate $\hat{\beta}$, e.g., the maximum composite likelihood estimate based on the intensity function, we obtain a function $L_c(\hat{\beta}, \psi)$, which may be maximized to obtain an estimate of ψ. See Møller and Waagepetersen (2007).

Minimum contrast estimation is a more common estimation procedure, where the idea is to minimize a "contrast" (or "distance") between, e.g., K_ψ and its nonparametric empirical counterpart $\hat{K}(r)$, as defined in Section 18.6, thereby obtaining a minimum contrast estimate. For instance, ψ may be estimated by minimizing the contrast

$$\int_a^b \left(\hat{K}(r)^\alpha - K_\psi(r)^\alpha\right)^2 dr, \tag{19.6}$$

where $0 \leq a < b < \infty$ and $\alpha > 0$ are chosen on an ad hoc basis (see, e.g., Diggle (2003) or Møller and Waagepetersen (2004)). Theoretical properties of minimum contrast estimators are studied in Heinrich (1992).

These simulation-free estimation procedures are fast and computationally easy, but a disadvantage is that we have to specify tuning parameters, such as a, b, α in (19.6).

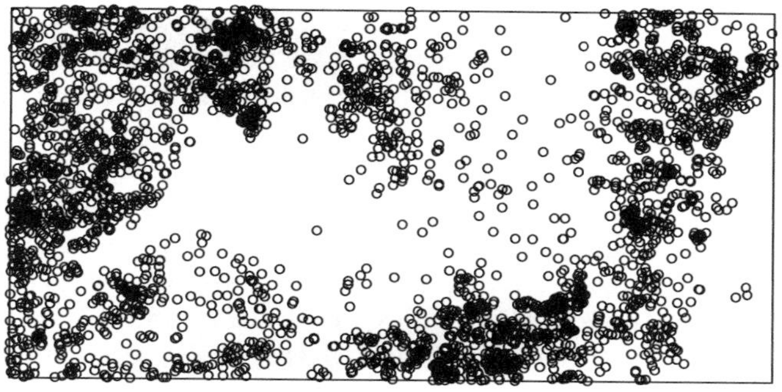

FIGURE 19.1
Locations of 3605 *Beilschmiedia pendula Lauraceae* trees observed within a 500 × 1000 m region at Barro Colorado Island (Panama Canal).

19.3.3 Example: Tropical Rain Forest Trees

Figure 19.1 provides an example of an *inhomogeneous point pattern* where the methods described in Sections 19.3.1 and 19.3.2 can be applied. The figure shows the locations of rain forest trees in a rectangular observation window W of size 500 × 1000 m. This point pattern, together with a second point pattern of another species of trees, has previously been analyzed in Waagepetersen (2007) and Møller and Waagepetersen (2007). The data are just a small part of a much larger dataset comprising hundreds of thousands of trees belonging to hundreds of species (Hubbell and Foster, 1983; Condit et al., 1996; Condit, 1998). Figure 19.2 shows two kinds of covariates z_1 (altitude) and z_2 (norm of altitude gradient), which are measured on a 100 × 200-square grid, meaning that we approximate the altitude and the norm of altitude gradient to be constant on each of 100 × 200 squares of size 5 × 5 m.

A plot of a nonparametric estimate of the inhomogeneous K-function (omitted here) confirms that the point pattern in Figure 19.1 is clustered. This clustering may be explained by the covariates in Figure 19.2, by other unobserved covariates, and by tree reproduction by seed dispersal. We, therefore, assume an inhomogeneous Cox process model as specified by (19.5) with $\beta = (\beta_0, \beta_1, \beta_2)^T$ and $z = (z_0, z_1, z_2)^T$, where $z_0 \equiv 1$ so that β_0 is interpreted as an intercept. Moreover, Y in (19.5) is modeled by a stationary shot noise process with mean one, that is,

$$Y(\mathbf{s}) = \frac{1}{\omega\sigma^2} \sum_{\mathbf{t}\in\Phi} k((\mathbf{s}-\mathbf{t})/\sigma), \tag{19.7}$$

where Φ is a stationary Poisson process with intensity $\omega > 0$, $k(\cdot)$ is a density function with respect to Lebesgue measure, and $\sigma > 0$ is a scaling parameter. We call X an *inhomogeneous*

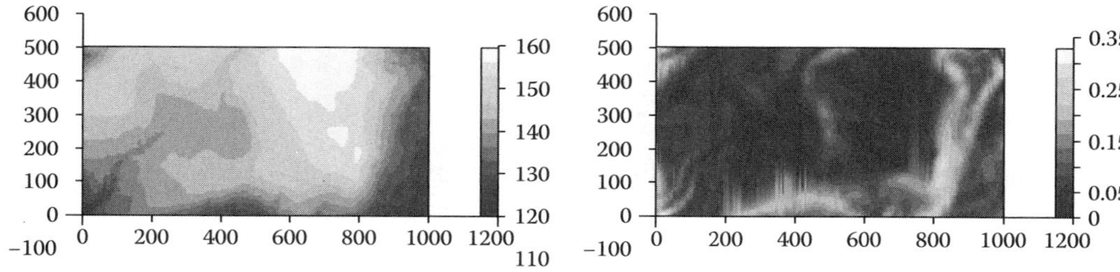

FIGURE 19.2
Rain forest trees: The covariates z_1 (altitude, left panel) and z_2 (norm of altitude gradient, right panel) are recorded on a 5 × 5 m grid (the units on the axes are in meters).

shot noise Cox process (Møller, 2003; Waagepetersen, 2007; Møller and Waagepetersen, 2007). Finally, as in a modified Thomas process (Thomas, 1949; Neyman and Scott, 1958), we assume that $k(x) = \exp(-\|x\|^2/2)/(2\pi)$ is a bivariate Gaussian kernel. For short, we then refer to X as an *inhomogeneous Thomas process*.

For β, we obtain the maximum composite likelihood estimate $(\hat{\beta}_0, \hat{\beta}_1, \hat{\beta}_2) = (-4.989, 0.021, 5.842)$ (under the Poisson model, this is the maximum likelihood estimate). Assuming asymptotic normality (Waagepetersen, 2007), 95% confidence intervals for β_1 and β_2 under the fitted inhomogeneous Thomas process are $[-0.018, 0.061]$ and $[0.885, 10.797]$, respectively. Much narrower 95% confidence intervals for β_1 and β_2 are obtained under the fitted Poisson process, namely $[0.017, 0.026]$ and $[5.340, 6.342]$, respectively. This difference arises because in the Cox process there is an extra variability due to the process Y appearing in Equation (19.5).

An unbiased estimate of the inhomogeneous K-function at distance $r > 0$ is given by

$$\sum_{i,j=1,\ldots,n:i\neq j} \frac{\mathbb{I}[\|\mathbf{s}_i - \mathbf{s}_j\| \leq r]}{\rho(\mathbf{s}_i)\rho(\mathbf{s}_j)|W \cap (W + \mathbf{s}_i - \mathbf{s}_j)|},$$

where $W + \mathbf{s}$ denotes W translated by $\mathbf{s}$, and $|W \cap (W + \mathbf{s}_i - \mathbf{s}_j)|$ is an *edge-correction factor*, which is needed since we sum over all pairs of points observed within W. In practice, we need to plug in an estimate of $\rho(\mathbf{s}_i)\rho(\mathbf{s}_j)$. We use the parametric estimate $\rho_{\hat{\beta}}(\mathbf{s}_i)\rho_{\hat{\beta}}(\mathbf{s}_j)$ with $\hat{\beta}$ the estimate obtained above. Let $\hat{K}(r)$ denote the resulting estimate of $K(r)$. Using the minimum contrast estimation procedure based on (19.6) with $a = 0, b = 100$, and $\alpha = 1/4$, we obtain $(\hat{\omega}, \hat{\sigma}) = (8 \times 10^{-5}, 20)$.

Estimation of this inhomogeneous Thomas process and an inhomogeneous *log-Gaussian Cox process*, i.e., when $\log Y$ in (19.5) is a Gaussian process (see Møller, Syversveen, and Waagepetersen, 1998, and Chapter 17), and their corresponding estimated K-functions are further considered in Møller and Waagepetersen (2007).

19.3.4 Pseudo-Likelihood

The *maximum pseudo-likelihood estimate* is a simple and computationally fast but less efficient alternative to the maximum likelihood estimate. In the special case of a parametric Poisson point process model, the two kinds of estimates coincide. Since the pseudo-likelihood function is expressed in terms of the Papangelou conditional intensity, pseudo-likelihood estimation is particularly useful for *Gibbs (or Markov) point processes*, but generally not so for Cox and Poisson cluster processes.

We recall first from Section 17.4 the definition of the Papangelou conditional intensity in the case where X restricted to W has a *parametric density* $f_\theta(x)$ with respect to the Poisson process on W with unit intensity. Let $x = \{\mathbf{s}_1, \ldots, \mathbf{s}_n\} \subset W$ denote an arbitrary finite point configuration in W, and $\mathbf{s}$ an arbitrary location in $W \setminus x$. Assume that $f_\theta(x)$ is *hereditary*, meaning that $f_\theta(x \cup \{\mathbf{s}\}) > 0$ implies that $f_\theta(x) > 0$. For $f_\theta(x) > 0$, define the *Papangelou conditional intensity* by

$$\lambda_\theta(\mathbf{s}, x) = f_\theta(x \cup \{\mathbf{s}\})/f_\theta(x). \tag{19.8}$$

We may interpret $\lambda_\theta(\mathbf{s}, x)\,\mathrm{d}\mathbf{s}$ as the conditional probability that there is a point of the process in an infinitesimally small region containing $\mathbf{s}$ and of area $\mathrm{d}\mathbf{s}$ given that the rest of the point process coincides with x. How we define $\lambda_\theta(\mathbf{s}, x)$ if $f_\theta(x) = 0$ turns out not to be that important, but for completeness let us set $\lambda_\theta(\mathbf{s}, x) = 0$ if $f_\theta(x) = 0$. In the special case of a Poisson process with intensity function $\rho_\theta(\mathbf{s})$, we simply have $\lambda_\theta(\mathbf{s}, x) = \rho_\theta(\mathbf{s})$. In the case of a Gibbs (or Markov) point process, $\lambda_\theta(\mathbf{s}, x)$ depends on x only through the neighbors to $\mathbf{s}$ (see Section 17.4), and the intractable normalizing constant of the density cancels in (19.8).

The pseudo-likelihood function was first introduced in Besag (1977). It can be derived by a limiting argument similar to that used for deriving the composite likelihood in (19.2), the only difference being that we replace $p_i(\beta)$ in Equation (19.1) by the conditional probability

$$p_i(\theta) := \mathrm{P}_\theta(N_i = 1 | X \setminus C_i = x \setminus C_i) \approx \lambda_\theta(\mathbf{u}_i, x \setminus C_i)|C_i|.$$

Under mild conditions (Besag, Milne, and Zachary, 1982; Jensen and Møller, 1991) the limit becomes the *pseudo-likelihood* function

$$L_p(\theta; x) = \exp\left(-\int_W \lambda_\theta(\mathbf{s}, x)\,\mathrm{d}\mathbf{s}\right) \prod_{i=1}^n \lambda_\theta(\mathbf{s}_i, x), \tag{19.9}$$

where we have again omitted the constant factor $\exp(|W|)$. Clearly, for a Poisson process with a parametric intensity function, the pseudo-likelihood is the same as the likelihood. The *pseudo-score* is the derivative of $\log L_p(\theta; x)$ with respect to θ, that is,

$$s(\theta; x) = \sum_{i=1}^n \mathrm{d}\log\lambda_\theta(\mathbf{s}_i, x)/\mathrm{d}\theta - \int_W (\mathrm{d}\log\lambda_\theta(\mathbf{s}, x)/\mathrm{d}\theta)\lambda_\theta(\mathbf{s}, x)\,\mathrm{d}\mathbf{s}. \tag{19.10}$$

This provides an unbiased estimating equation $s(\theta; x) = 0$ (assuming in (19.10) that $(\mathrm{d}/\mathrm{d}\theta)\int_W \cdots = \int_W (\mathrm{d}/\mathrm{d}\theta)\cdots$). When finding the maximum pseudo-likelihood estimate, it is useful to notice that Equation (19.9) can be viewed as the likelihood for a Poisson process with "intensity function" $\lambda_\theta(\cdot; x)$. The maximum pseudo-likelihood estimate can then be evaluated using `spatstat` if λ_θ is of a log-linear form similar to that in Equation (19.4), that is,

$$\log\lambda_\theta(\mathbf{s}, x) = a(\mathbf{s}, x) + \beta^T t(\mathbf{s}, x), \tag{19.11}$$

where $a(\mathbf{s}, x)$ and $t(\mathbf{s}, x)$ are known functions (Baddeley and Turner, 2000).

Suppose that X may have points outside W, and that we do not know its marginal density $f_\theta(x)$ on W. To account for *edge effects*, assume a *spatial Markov property* is satisfied. Specifically, suppose there is a region $W_{\ominus R} \subset W$ such that conditional on $X \cap (W \setminus W_{\ominus R}) = y$, we have that $X \cap W_{\ominus R}$ is independent of $X \setminus W$, and we know the conditional density $f_\theta(x|y)$ of $X \cap W_{\ominus R}$ given $X \cap (W \setminus W_{\ominus R}) = y$, where $f_\theta(\cdot|y)$ is hereditary. Here the notation $W_{\ominus R}$ refers to the common case where X is a Gibbs (or Markov) point process with a finite interaction radius R (see Chapter 17), in which case $W_{\ominus R}$ is naturally given by the W eroded by a disc of radius R, that is,

$$W_{\ominus R} = \{\mathbf{s} \in W : \|\mathbf{s} - \mathbf{t}\| \leq R \text{ for all } \mathbf{t} \in W\}. \tag{19.12}$$

For $\mathbf{s} \in W_{\ominus R}$, exploiting the spatial Markov property, the Papangelou conditional intensity is seen not to depend on points from $X \setminus W$, and is given by replacing $f_\theta(x)$ by $f_\theta(x|y)$ in the definition (19.8). We denote this Papangelou conditional intensity by $\lambda_\theta(\mathbf{s}, x \cup y)$. Note that $\lambda_\theta(\mathbf{s}, x \cup y)$ depends only on $x \cup y$ through its neighbors to $\mathbf{s}$, and all normalizing constants cancel. Consequently, we need only specify $f_\theta(\cdot|y)$ up to proportionality, and the pseudo-likelihood $L_p(\theta; x \cup y)$ is given by Equation (19.9) when $\lambda_\theta(\mathbf{s}, x)$ is replaced by $\lambda_\theta(\mathbf{s}, x \cup y)$. The pseudo-score $s(\theta; x \cup y)$ is obtained as the derivative of $\log L_p(\theta; x \cup y)$ with respect to θ, and provides an unbiased estimating equation $s(\theta; x \cup y) = 0$.

We give an application of maximum pseudo-likelihood in Section 19.4.2. Asymptotic results for maximum pseudo-likelihood estimates are established in Jensen and Møller (1991), Jensen and Künsch (1994), and Mase (1995, 1999). Alternatively, a *parametric bootstrap* can be used, see, e.g., Baddeley and Turner (2000).

19.4 Simulation-Based Maximum Likelihood Inference

For *Poisson process* models, computation of the likelihood function is usually easy (cf. Section 19.3.1). For Gibbs (or Markov) point process models, the likelihood contains an unknown normalizing constant, while for Cox process models, the likelihood is given in terms of a complicated integral. Using MCMC methods, it is now becoming quite feasible to compute accurate approximations of the likelihood function for Gibbs and Cox process models as discussed in Sections 19.4.1 and 19.4.3. However, the computations may be time-consuming and standard software is yet not available.

19.4.1 Gibbs Point Processes

Consider a parametric model for a spatial point process X, where X restricted to W has a *parametric density* $f_\theta(x)$ with respect to the Poisson process on W with unit intensity. For simplicity and specificity, assume that $f_\theta(x)$ is of *exponential family* form

$$f_\theta(x) = \exp(t(x)^T\theta)/c_\theta, \tag{19.13}$$

where $t(x)$ is a real function of the same dimension as the real parameter θ, and c_θ is a normalizing constant. In general, apart from the special case of a Poisson process, c_θ has no closed form expression. Equation (19.13) holds if the Papangeleou conditional intensity $\lambda_\theta(\mathbf{s}, x)$ is of the log-linear form (19.11). This is the case for many *Gibbs (or Markov) point processes* when the interaction radius $R < \infty$ is known. Examples include most *pairwise interaction point processes*, such as the Strauss process, and more complicated interaction point processes, such as the area-interaction point process see Chapter 17.

From Equation (19.13), we obtain the score function $u(\theta; x)$ and the observed information $j(\theta)$,

$$u(\theta; x) = t(x) - E_\theta t(X), \quad j(\theta) = \text{var}_\theta t(X),$$

where E_θ and var_θ denote expectation and variance with respect to $X \sim f_\theta$. Let θ_0 denote a fixed reference parameter value. The score function and observed information may be evaluated using the *importance sampling formula*

$$E_\theta k(X) = E_{\theta_0}\left[k(X)\exp\left(t(X)^T(\theta - \theta_0)\right)\right]/(c_\theta/c_{\theta_0}) \tag{19.14}$$

with $k(X)$ given by $t(X)$ or $t(X)t(X)^T$. For $k \equiv 1$, we obtain

$$c_\theta/c_{\theta_0} = E_{\theta_0}\left[\exp\left(t(X)^T(\theta - \theta_0)\right)\right]. \tag{19.15}$$

Approximations of the likelihood ratio $f_\theta(x)/f_{\theta_0}(x)$, score, and observed information can be obtained by Monte Carlo approximation of the expectations $E_{\theta_0}[\cdots]$ using MCMC samples from f_{θ_0}. Here, to obtain an approximate maximum likelihood estimate, Monte Carlo approximations may be combined with Newton–Raphson updates. Furthermore, if we want to test a submodel, approximate p-values based on the likelihood ratio statistic or the Wald statistic can be derived by MCMC methods (see Geyer and Møller (1994), Geyer (1999), and Møller and Waagepetersen (2004)).

The *path sampling identity* (Gelman and Meng, 1998),

$$\log(c_\theta/c_{\theta_0}) = \int_0^1 E_{\theta(s)}t(X)(\mathrm{d}\theta(s)/\mathrm{d}s)^T\mathrm{d}s, \tag{19.16}$$

provides an alternative and often numerically more stable way of computing a ratio of normalizing constants. Here $\theta(s)$ is a differentiable curve, e.g., a straight line segment, connecting $\theta_0 = \theta(0)$ and $\theta = \theta(1)$. The log ratio of normalizing constants is approximated by evaluating the outer integral in Equation (19.16) using, e.g., the trapezoidal rule and the expectation using MCMC methods (Berthelsen and Møller, 2003; Møller and Waagepetersen, 2004).

For a Gibbs point process with unknown interaction radius R, the likelihood function is usually not differentiable as a function of R. Therefore, maximum likelihood estimates of R are often found using a profile likelihood approach, where for each fixed value of R we maximize the likelihood as discussed above. Examples are given in Møller and Waagepetersen (2004).

If X may have points outside W, and we do not know its marginal density $f_\theta(x)$ on W, we may account for *edge effects* by exploiting the spatial Markov property (Section 19.3.4), using the smaller observation window $W_{\ominus R}$ given by Equation (19.12). If $f_\theta(x|y)$ denotes the conditional density of $X \cap W_{\ominus R} = x$ given $X \cap (W \setminus W_{\ominus R}) = y$, the likelihood function,

$$L(\theta; x) = E_\theta f_\theta(x | X \cap (W \setminus W_{\ominus R})),$$

may be computed using a missing data approach, see Geyer (1999) and Møller and Waagepetersen (2004). A simpler but less efficient alternative is the *border method*, considering the conditional likelihood function

$$L(\theta; x|y) = f_\theta(x|y),$$

where the score, observed information, and likelihood ratios may be computed by analogy with the case above based on Equation (19.14). (See Møller and Waagepetersen (2004) for a discussion of these and other approaches for handling edge effects.)

Asymptotic results for maximum likelihood estimates of Gibbs point process models are reviewed in Møller and Waagepetersen (2004), but these results are derived under restrictive assumptions of stationarity and weak interaction. According to standard asymptotic results, the inverse observed information provides an approximate covariance matrix of the maximum likelihood estimate, and log likelihood ratio and Wald statistics are asymptotically χ^2-distributed. If one is suspicious about the validity of the asymptotic approach, an alternative is to use a parametric bootstrap. (See Møller and Waagepetersen, 2004.)

19.4.2 Example: Ants' Nests

Figure 19.3 shows a bivariate point pattern of *ants' nests* of two types, *Messor wasmanni* and *Cataglyphis bicolor* (see Harkness and Isham, 1983). The interaction between the two types of ants' nests is of main interest for this dataset. Note the irregular polygonal shape of the observation window W given in Figure 19.3.

The *Cataglyphis* ants feed on dead *Messors* and, hence, the positions of *Messor* nests might affect the choice of sites for *Cataglyphis* nests, while the *Messor* ants are believed not to be influenced by presence or absence of *Cataglyphis* ants when choosing sites for their nests. Högmander and Särkkä (1999) therefore specified a *hierarchical model* based on first a point process model for the *Messor* nests, and, second, a point process model for the *Cataglyphis* nests conditional on the *Messor* nests. Both types of models are *pairwise interaction point process* models, with the log Papangelou conditional intensity of the form

$$\log \lambda(\mathbf{s}, x) = U(\mathbf{s}) + \sum_{i=1}^{n} V(\|\mathbf{s} - \mathbf{s}_i\|)$$

for $x = \{\mathbf{s}_1, \ldots, \mathbf{s}_n\} \subset W$ and $\mathbf{s} \notin x$, where $U(\mathbf{s})$ and $V(\|\mathbf{s} - \mathbf{s}_i\|)$ are real functions called the first respective, second-order potential. In other words, if X is such a pairwise interaction

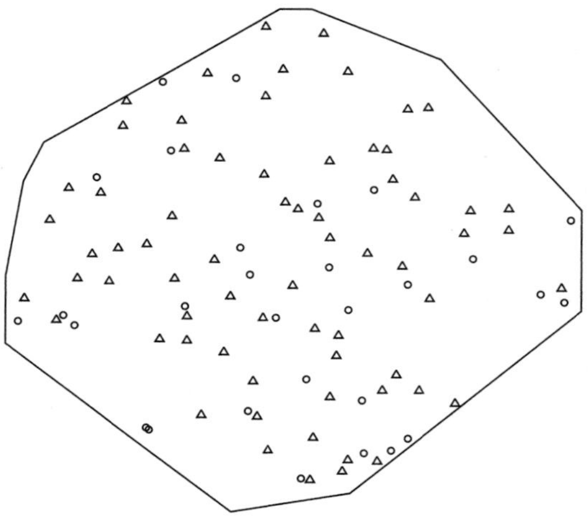

FIGURE 19.3
Locations of nests for *Messor* (triangles) and *Cataglyphis* (circles) ants. The observation window W is polygonal (solid line), and the enclosing rectangle for W (dashed line) is 414.5 × 383 ft.

point process, then X has density

$$f(x) \propto \exp\left(\sum_{i=1}^{n} U(\mathbf{s}_i) + \sum_{1\le i<j\le n} V(\|\mathbf{s}_i - \mathbf{s}_j\|)\right)$$

with respect to the Poisson process on W with intensity one. Furthermore, the pairwise interaction process models are so-called *Strauss processes with hard cores* specified as follows. For distances $t > 0$, define

$$V(t; r) = \begin{cases} -\infty & \text{if } t \le r \\ 1 & \text{if } r < t \le R \\ 0 & \text{otherwise,} \end{cases}$$

where $R \ge 0$ is the interaction range, $r \in [0, R)$ denotes a hard-core distance (or no hard core if $r = 0$), and $\exp(-\infty) = 0$. First, for the *Messor* nests, the Strauss process with hard core r_M is given by first- and second-order potentials

$$U_{M1}(\{\mathbf{s}\}) = \beta_M, \quad U_{M2}(\{\mathbf{s}_i, \mathbf{s}_j\}) = \psi_M V(\|\mathbf{s}_i - \mathbf{s}_j\|; r_M).$$

Thus, the conditional intensity for a putative *Messor* nest at a location $\mathbf{s}$ is zero if an existing *Messor* nests occur within distance r_M from $\mathbf{s}$, and otherwise the log conditional density is given by the sum of β_M and ψ_M times the number of neighboring *Messor* nests within distance R. Second, conditional on the pattern x_M of *Messor* nests, the *Cataglyphis* nests are modeled as an inhomogeneous Strauss process with one hard core r_{CM} to the *Messor* nests and another hard core r_C between the *Cataglyphis* nests, i.e., using potentials

$$U_{C1}(\{\mathbf{s}\}) = \beta_C + \psi_{CM} \sum_{i=1}^{n} V(\|\mathbf{s} - \mathbf{s}_i\|; r_{CM}), \quad U_{C2}(\{\mathbf{s}_i, \mathbf{s}_j\}) = \psi_C V(\|\mathbf{s}_i - \mathbf{s}_j\|; r_C).$$

We use the maximum likelihood estimates $r_M = 9.35$ and $r_C = 2.45$ (distances are measured in feet), which are given by the observed minimum interpoint distances in the two types of point patterns. Using positive hard cores r_M and r_C may be viewed as an ad hoc approach to obtain a model, which is well defined for all real values of the parameters β_M, β_C, ψ_M, ψ_{CM}, and ψ_C, whereby both repulsive and attractive interaction within and between the two types of ants can be modeled. However, as noted by Møller (1994) and Geyer and Thompson (1995), the Strauss hard-core process is a poor model for clustering due to the following "phase transition property": For positive values of the interaction parameter, except for a narrow range of values, the distribution will either be concentrated on point patterns with one dense cluster of points or in "Poisson-like" point patterns.

In contrast to Högmander and Särkkä (1999), we find it natural to let $r_{CM} = 0$, meaning there is no hard core between the two types of ants' nests. Further, for comparison, we fix R at the value 45 used in Högmander and Särkkä, though pseudo-likelihood computations indicate that a more appropriate interaction range would be 15. In fact, Högmander and Särkkä considered a subset of the data in Figure 19.3 within a rectangular region. They also conditioned on the observed number of points for the two species when computing maximum likelihood and maximum pseudo-likelihood estimates, whereby the parameters β_M and β_C vanish. Instead, we fit the hierarchical model to the full dataset, and we do not condition on the observed number of points.

We first correct for edge-effects by conditioning on the data in $W \setminus W_{\ominus 45}$, where $W_{\ominus 45}$ denotes the points within W with distance less than 45 to the boundary of W. Using `spatstat`, the maximum pseudo-likelihood estimate (MPLE) of (β_M, ψ_M) is $(-8.21, -0.09)$, indicating (weak) repulsion between the *Messor* ants' nests. Without edge-correction, we obtain a rather similar MPLE $(-8.22, -0.12)$. The edge-corrected MPLE of $(\beta_C, \psi_{CM}, \psi_C)$ is $(-9.51, 0.13, -0.66)$, indicating a positive association between the two species and repulsion within the *Cataglyphis* nests. If no edge-correction is used, the MPLE for $(\beta_C, \psi_{CM}, \psi_C)$ is $(-9.39, 0.04, -0.30)$. Högmander and Särkkä (1999) also found a repulsion within the *Cataglyphis* nests, but in contrast to our result a weak repulsive interaction between the two types of nests. This may be explained by the different modeling approach in Högmander and Särkkä where the smaller observation window excludes a pair of very close *Cataglyphis* nests, and where also the conditioning on the observed number of points in the two point patterns may make a difference.

No edge-correction is used for our maximum likelihood estimates (MLEs). The MLEs $\hat{\beta}_M = -8.39$ and $\hat{\psi}_M = -0.06$ again indicate a weak repulsion within the *Messor* nests, and the MLEs $\hat{\beta}_C = -9.24$, $\hat{\psi}_{CM} = 0.04$, and $\hat{\psi}_C = -0.39$ also indicate positive association between *Messor* and *Cataglyphis* nests, and repulsion within the *Cataglyphis* nests. Confidence intervals for ψ_{CM}, when the asymptotic variance estimate is based on observed information or a parametric bootstrap, are $[-0.20, 0.28]$ (observed information) and $[-0.16, 0.30]$ (parametric bootstrap).

The differences between the MLE and the MPLE (without edge-correction) seem rather minor. This is also the experience for MLEs and corresponding MPLEs in Møller and Waagepetersen (2004), though differences may appear in cases with a strong interaction.

19.4.3 Cluster and Cox Processes

This section considers maximum likelihood inference for cluster and Cox process models. This is, in general, both more complicated and computionally more demanding than for Gibbs (or Markov) point processes.

For example, consider the case of an *inhomogeneous shot noise Cox process* X as defined by (19.5) and (19.7). We can interpret this as a *Poisson cluster process* as follows. The points in the stationary Poisson process Φ in (19.7) specify the centers of the clusters. Conditional on Φ, the clusters are independent Poisson processes, where the cluster associated to $\mathbf{t} \in \Phi$

has intensity function

$$\lambda_\theta(\mathbf{s}|\mathbf{t}) = \exp(\beta^T z(\mathbf{s}))\frac{1}{\omega\sigma^2}k((\mathbf{s}-\mathbf{t})/\sigma), \quad \mathbf{s} \in \mathbb{R}^2,$$

where $\theta = (\beta, \omega, \sigma)$. Finally, X consists of the union of all cluster points.

With probability one, X and Φ are disjoint. Moreover, in applications Φ is usually unobserved. In order to deal with *edge effects*, consider a bounded region $W_{\text{ext}} \supseteq W$ such that it is very unlikely that clusters associated with centers outside W_{ext} have points falling in W (see Brix and Kendall, 2002, and Møller, 2003). We approximate then $X \cap W$ by the union of clusters with centers in $\Psi := \Phi \cap W_{\text{ext}}$. Let $f(x|\psi)$ denote the conditional density of $X \cap W$ given $\Psi = \psi$, where the density is with respect to the Poisson process on W with intensity one. For $x = \{\mathbf{s}_1, \ldots, \mathbf{s}_n\}$,

$$f_\theta(x|\psi) = \exp\left(|W| - \int_W \sum_{\mathbf{t}\in\psi} \lambda_\theta(\mathbf{s}|\mathbf{t})\,\mathrm{d}\mathbf{s}\right)\prod_{i=1}^n \lambda_\theta(\mathbf{s}_i|\mathbf{t}) \tag{19.17}$$

and the likelihood based on observing $X \cap W = x$ is

$$L(\theta; x) = E_\omega f_\theta(x|\Psi), \tag{19.18}$$

where the expectation is with respect to the Poisson process Ψ on W_{ext} with intensity ω. As this likelihood has no closed form expression, we may consider Ψ as *missing data* and use MCMC methods for finding an approximate maximum likelihood estimate (see Møller and Waagepetersen, 2004). Here one important ingredient is an MCMC simulation algorithm for the conditional distribution of Ψ given $X \cap W = x$. This conditional distribution has density

$$f_\theta(\psi|x) \propto f_\theta(x|\psi) f_\omega(\psi), \tag{19.19}$$

where

$$f_\omega(\psi) = \exp\left(|W_{\text{ext}}|(1-\omega)\right)\omega^{n(\psi)} \tag{19.20}$$

is the density of Ψ. For conditional simulation from (19.19), we use a *birth–death type Metropolis–Hastings algorithm* studied in Møller (2003).

For a *log-Gaussian Cox process* model, the simulation-based maximum likelihood approach is as above except for the following. To specify the density of the Poisson process $(X \cap W)|Y$, since $\log Y$ in (19.5) is a Gaussian process, we need only consider $Y(\mathbf{s})$ for $\mathbf{s} \in W$. Hence, in contrast to above, edge effects do not present a problem, and the conditional density of $X \cap W$ given Y is

$$f(x|Y(\mathbf{s}), \mathbf{s} \in W) = \exp\left(|W| - \int_W \exp(Y(\mathbf{s}))\,\mathrm{d}\mathbf{s} + \sum_{i=1}^n Y(\mathbf{s}_i)\right). \tag{19.21}$$

However, when evaluating the integral in (19.21) and when simulating from the conditional distribution of Y on W given $X \cap W = x$, we need to approximate Y on W by a finite-dimensional log-Gaussian random vector $Y_I = (Y(\mathbf{u}_i), i \in I)$ corresponding to a finite partition $\{C_i, i \in I\}$ of W, where $\mathbf{u}_i$ is a representative point of the cell C_i and we use the approximation $Y(\mathbf{s}) \approx Y(\mathbf{u}_i)$ if $\mathbf{s} \in C_i$. For simulation from the conditional distribution of Y_I given $X \cap W = x$, we use a *Langevin–Hastings algorithm* (also called a *Metropolis-adjusted Langevin algorithm*) (see Møller et al. (1998) and Møller and Waagepetersen (2004)).

For the shot noise Cox process model considered above, the likelihood (19.18) and its MCMC approximation are complicated functions of θ, possibly with many local modes. The same holds for a log-Gaussian Cox process model. Careful maximization procedures, therefore, are needed when finding the (approximate) maximum likelihood estimate. Further details, including examples and specific algorithms of the *MCMC missing data approach* for shot noise and log-Gaussian Cox processes, are given in Møller and Waagepetersen (2004, 2007).

19.5 Simulation-Based Bayesian Inference

A *Bayesian approach* often provides a flexible framework for incorporating prior information and analyzing spatial point process process models. Section 19.5.1 considers an application example of a Poisson process, where a Bayesian approach is obviously more suited than a maximum likelihood approach. Bayesian analysis for cluster and Cox processes is discussed in Section 19.5.2, while Section 19.5.3 considers Gibbs (or Markov) point processes. In the latter case, a Bayesian analysis is more complicated because of the unknown normalizing constant appearing in the likelihood term of the posterior density.

19.5.1 Example: Reseeding Plants

Armstrong (1991) considered the locations of 6378 plants from 67 species on a 22×22 m observation window W in the southwestern area of Western Australia. The plants have adapted to regular natural fires, where *resprouting species* survive the fire, while *reseeding species* die in the fire, but the fire triggers the shedding of seeds, which have been stored since the previous fire. See also Illian, Møller, and Waagepetersen (2009), where further background material is provided and various examples of the point patterns of resprouting and reseeding plants are shown. Figure 19.4 shows the locations of one of the reseeding plants *Leucopogon conostephioides* (called seeder 4 in Illian et al., 2009). This and five other species of reseeding plants together with the 19 most dominant (influential) species of resprouters are analyzed in Illian et al. (2009). Since it is natural to model the locations of the reseeding plants conditionally on the locations of the resprouting plants, we consider below a model for the point pattern x in Figure 19.4 conditional on the point patterns $y_1, \ldots, y_{19}$

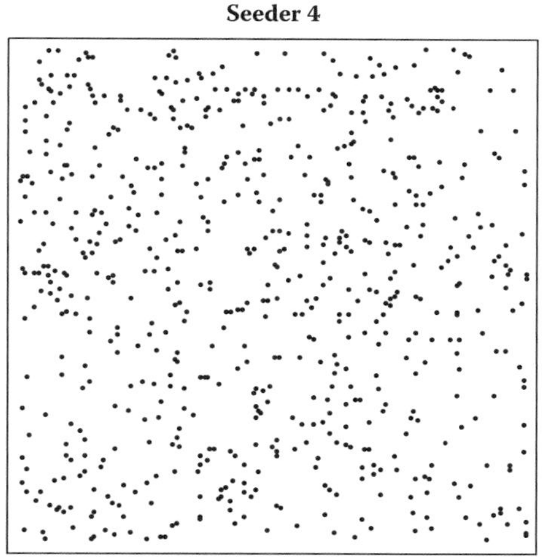

FIGURE 19.4
Locations of 657 *Leucopogon conostephioides* plants observed within a 22×22 m window.

corresponding to the 19 most dominant species of resprouters, as given in Figure 1 in Illian et al. (2009). For a discussion of possible interaction with other reseeder species, and the biological justification of the the covariates defined below, we refer again to Illian et al. (2009).

Let $\kappa_{\mathbf{t},i} \geq 0$ denote a parameter that specifies the radius of interaction of the ith resprouter at location $\mathbf{t} \in y_i$, and let κ denote the collection of all $\kappa_{\mathbf{t},i}$ for $\mathbf{t} \in y_i$ and $i = 1, \ldots, 19$. For $i = 1, \ldots, 19$, define covariates $z_i(\mathbf{s}) = z_i(\mathbf{s}; \kappa_{\mathbf{t},i}, \mathbf{t} \in y_i)$ by

$$z_i(\mathbf{s}; \kappa_{\mathbf{t},i}, \mathbf{t} \in y_i) = \sum_{\mathbf{t} \in y_i :\, \|\mathbf{s}-\mathbf{t}\| \leq \kappa_{\mathbf{t},i}} \left(1 - \left(\|\mathbf{s} - \mathbf{t}\| / \kappa_{\mathbf{t},i}\right)^2\right)^2.$$

Conditional on $y_1, \ldots, y_{19}$, we assume that $x = \{\mathbf{s}_1, \ldots, \mathbf{s}_n\}$ is a realization of a *Poisson process* with log-linear intensity function

$$\log \rho_{\theta, y_1, \ldots, y_n}(\mathbf{s}) = \beta_0 + \sum_{i=1}^{19} \beta_i z_i(\mathbf{s}; \kappa_{\mathbf{t},i}, \mathbf{t} \in y_i),$$

where $\theta = (\beta, \kappa)$ and $\beta = (\beta_0, \ldots, \beta_{19})$ is a regression parameter, where β_0 is an intercept and β_i for $i > 0$ controls the influence of the ith resprouter. The likelihood depends on κ in a complicated way, and the dimension of κ is much larger than the size of the data x. This makes it impossible to find maximum likelihood estimates.

Using a Bayesian setting, we treat $\theta = (\beta, \kappa)$ as a random variable. Based on Table 1 in Illian et al. (2009) and other considerations in that paper, we make the following prior assumptions. We let $\kappa_{\mathbf{t},i}$ follow the restriction of a Gaussian distribution $\mathrm{N}(\mu_i, \sigma_i^2)$ to $[0, \infty)$, where (μ_i, σ_i^2) is chosen so that under the unrestricted Gaussian distribution the range of the zone of influence is a central 95% interval. Furthermore, we let all the $\kappa_{\mathbf{t},i}$ and the β_i be independent, and each β_i be $\mathrm{N}(0, \sigma^2)$-distributed, where $\sigma = 8$. Combining these prior assumptions with the likelihood term, we obtain the posterior density

$$\pi(\beta, \kappa | x) \propto \exp\left(-\beta_0/(2\sigma^2) - \sum_{i=1}^{19} \left\{\beta_i^2/(2\sigma^2) + \sum_{\mathbf{t} \in y_i} (\kappa_{\mathbf{t},i} - \mu_i)^2/(2\sigma_i^2)\right\}\right)$$
$$\times \exp\left(-\int_W \rho_{\theta, y_1, \ldots, y_n}(\mathbf{s})\, \mathrm{d}\mathbf{s}\right) \prod_{i=1}^{n} \rho_{\theta, y_1, \ldots, y_n}(\mathbf{s}_i), \quad \beta_i \in \mathbb{R}, \kappa_{\mathbf{t},i} \geq 0 \qquad (19.22)$$

(suppressing in the notation $\pi(\beta, \kappa | x)$ that we have conditioned on $y_1, \ldots, y_{19}$ in the posterior distribution).

Simulations from (19.22) are obtained by a Metropolis-within-Gibbs algorithm (also called a hybrid MCMC algorithm, see, e.g., Robert and Casella, 1999), where we alternate between updating β and κ using random walk Metropolis updates (for details, see Illian et al., 2009). Thereby various posterior probabilities of interest can be estimated. For example, a large (small) value of $\mathrm{P}(\beta_i > 0 | \mathbf{x})$ indicates a positive/attractive (negative/repulsive) association to the ith resprouter, see Figure 2 in Illian et al. (2009).

The model can be checked following the idea of *posterior predictive model assessment* (Gelman, Meng, and Stern, 1996), comparing various summary statistics with their posterior predictive distributions. The posterior predictive distribution of statistics depending on X (and possibly also on (β, κ)) is obtained from simulations: We generate a posterior sample $(\beta^{(j)}, \kappa^{(j)})$, $j = 1, \ldots, m$, and for each j "new data" $x^{(j)}$ from the conditional distribution of X given $(\beta^{(j)}, \kappa^{(j)})$. For instance, the grayscale plot in Figure 19.5 is a *residual plot* based on quadrat counts. We divide the observation window into 100 equally sized quadrats and count the number of plants within each quadrat. The grayscales reflect the probabilities that counts drawn from the posterior predictive distribution are less than

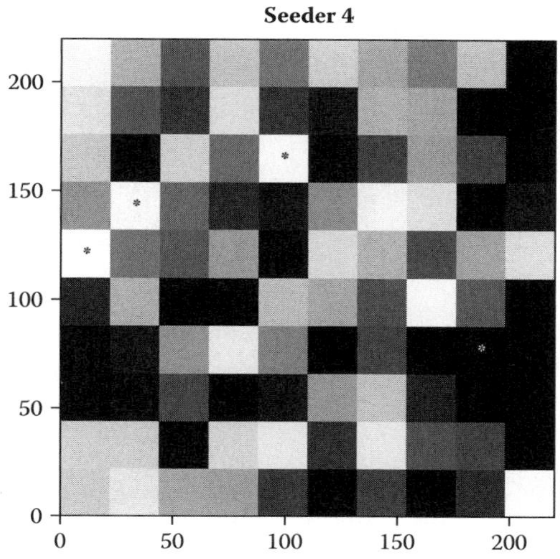

FIGURE 19.5
Residual plot based on quadrat counts. Quadrats with a "*" are where the observed counts fall below the 2.5% quantile (white "*") or above the 97.5% quantile (black "*") of the posterior predictive distribution. The grayscales reflect the probabilities that counts drawn from the posterior predictive distribution are less or equal to the observed quadrat counts (dark means small probability).

or equal to the observed quadrat counts where dark means small probability. The stars mark quadrats where the observed counts are "extreme" in the sense of being either below the 2.5% quantile or above the 97.5% quantile of the posterior predictive distribution. Figure 19.5 does not provide evidence against our model. A plot based on the L-function (Section 17.2) and the posterior predictive distribution is also given in Illian et al. (2009). This plot also shows no evidence against our model.

19.5.2 Cluster and Cox Processes

The simulation-based Bayesian approach exemplified above extends to cluster and Cox processes, where we include the "missing data" η, say, in the posterior and use a Metropolis-within-Gibbs (or other MCMC) algorithm, where we alter between updating θ and η. Examples are given below.

In the case of the *Poisson cluster process* model for X considered in Section 19.4.3, $\eta = \Psi$ is the point process of center points. Incorporating this into the posterior, we obtain the posterior density

$$\pi(\theta, \psi|x) \propto f_\theta(x|\psi) f_\omega(\psi)\pi(\theta),$$

where $f_\theta(x|\psi)$ and $f_\omega(\psi)$ are specified in Equation (19.17) and Equation (19.20), and $\pi(\theta)$ is the prior density. The Metropolis-within-Gibbs algorithm alters between updating from "full conditionals" given by

$$\pi(\theta|\psi, x) \propto f_\theta(x|\psi) f_\omega(\psi)\pi(\theta) \tag{19.23}$$

and

$$\pi(\psi|\theta, x) \propto f_\theta(x|\psi) f_\omega(\psi). \tag{19.24}$$

Yet another Metropolis-within-Gibbs algorithm may be used when updating from Equation (19.23), cf. Section 19.4.3. When updating from Equation (19.24) we use the birth–death-type Metropolis–Hastings algorithm mentioned in connection to (19.19).

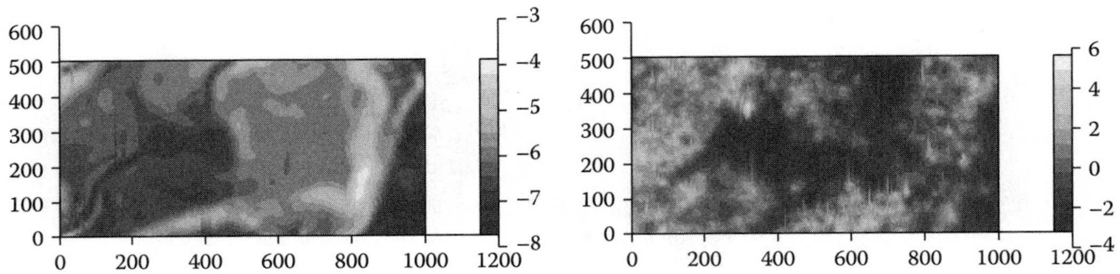

FIGURE 19.6
Posterior mean of $\beta_0 + \beta_1 z_1(\mathbf{s}) + \beta_2 z_2(\mathbf{s})$ (left panel) and $Y(\mathbf{s})$ (right panel), $\mathbf{s} \in W$, under the log-Gaussian Cox process model for the tropical rain forest trees.

The same holds for a *log-Gaussian Cox process* model for X. Then we may approximate the log-Gaussian process Y on W by the finite-dimensional log-Gaussian random variable $\eta = Y_I$ specified in Section 19.4.3, and use a Langevin–Hastings algorithm for simulating from the conditional distribution of η given (θ, x). Rue, Martino, and Chopin (2007) demonstrate that it may be possible to compute accurate Laplace approximations of marginal posterior distributions without MCMC simulations.

For instance, Møller and Waagepetersen (2007) considered a log-Gaussian Cox process model for the rain forest trees considered in Section 19.3.3. They used a 200×100 grid to index η, and imposed certain flat priors on the unknown parameters. Figure 19.6 shows the posterior means of the systematic part $\beta_0 + \beta_1 z_1(\mathbf{s}) + \beta_2 z_2(\mathbf{s})$ (left panel) and the random part $Y(\mathbf{s})$ (right panel) of the log random intensity function $\log \Lambda(\mathbf{s})$ given by (19.5). The systematic part seems to depend more on z_2 (norm of altitude gradient) than z_1 (altitude) (cf. Figure 19.2). The fluctuations of the random part may be caused by small-scale clustering due to seed dispersal and covariates concerning soil properties. The fluctuation may also be due to between-species competition.

Møller and Waagepetersen (2004, 2007), Beneš, Bodlak, Møller, and Waagepetersen (2005), and Waagepetersen and Schweder (2006) exemplified the simulation-based Bayesian approach for both Poisson cluster (or shot noise Cox) process and log-Gaussian Cox process models. Other Cox models and examples are considered in Heikkinen and Arjas (1998), Wolpert and Ickstadt (1998), Best, Ickstadt, and Wolpert (2000), and Cressie and Lawson (2000).

19.5.3 Gibbs Point Processes

For a *Gibbs (or Markov) point process*, the likelihood function depends on the unknown normalizing constant c_θ (cf. (19.13)). Hence, in a Bayesian approach to inference, the posterior distribution for θ also depends on the unknown c_θ, and in an "ordinary" Metropolis–Hastings algorithm, the Hastings ratio depends on a ratio of unknown normalizing constants. This ratio may be estimated using another method (see Section 19.4.1), but it is then unclear from which equilibrium distribution (if any) we are simulating and whether it is a good approximation of the posterior. Recently, the problem with unknown normalizing constants has been solved using an MCMC auxiliary variable method (Møller, Pettit, Berthelsen, and Reeves, 2006), which involves perfect simulations (Kendall, 1998; Kendall and Møller, 2000). The technique is applied for Bayesian inference of Markov point processes in Berthelsen and Møller (2004, 2006, 2008), where also the many technical details are discussed. Below we briefly demonstrate the potential of this technique when applied to *non/semiparametric Bayesian inference* of a *pairwise interaction point process*.

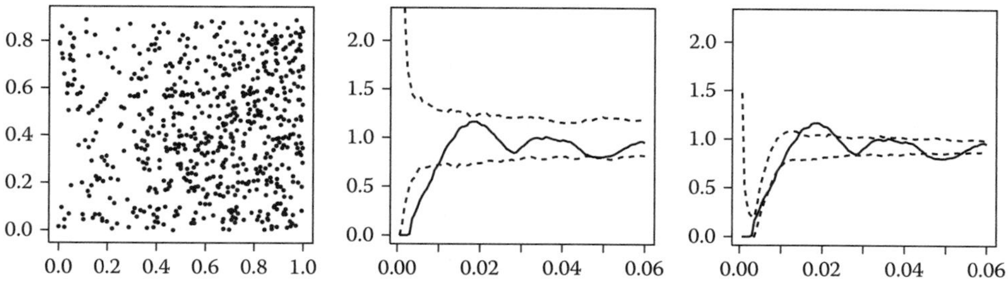

FIGURE 19.7
Left panel: Locations of 617 cells in a 2D section of the mucus membrane of the stomach of a healthy rat. Center panel: Nonparametric estimate of the pair correlation function for the cell data (full line) and 95% envelopes calculated from 200 simulations of a fitted inhomogeneous Poisson process. Right panel: Nonparametric estimate of the pair correlation function for the cell data (full line) and 95% envelopes calculated from 200 simulations of the model fitted by Nielsen (2000).

19.5.4 Example: Cell Data

The left panel of Figure 19.7 shows the location of 617 cells in a section of the mucous membrane of the stomach of a healthy rat, where (after some rescaling) $W = [0, 1] \times [0, 0.893]$ is the observation window. The left-hand side of the observation window corresponds to where the stomach cavity begins and the right-hand side to where the muscle tissue begins. The center panel of Figure 19.7 shows a nonparametric estimate $\hat{g}(r)$, $r > 0$, of the pair correlation function for the data and simulated 95% envelopes under an inhomogeneous Poisson process with a nonparametric estimate for its intensity function (Section 18.5). Under a Poisson process model, the theoretical pair correlation function is constant and equal to unity. The low values of $\hat{g}(r)$ for distances $r < 0.01$ indicate repulsion between the points. The point pattern looks inhomogeneous in the horizontal direction, and the data were originally analyzed by Nielsen (2000) using a Strauss point process model after transforming the first coordinates of the points. The right panel of Figure 19.7 shows a nonparametric estimate of the pair correlation function for the data with simulated 95% envelopes under the fitted transformed Strauss point process. The estimated pair correlation is almost within the 95% envelopes for small values of the distance r, suggesting that the transformed Strauss model captures the small-scale inhibition in the data. Overall, the estimated pair correlation function follows the trend of the 95% envelopes, but it falls outside the envelopes for some values. As the comparison with the envelopes can be considered as a multiple test problem, this is not necessarily reason to reject the model.

We consider an inhomogeneous pairwise interaction point process model for the point pattern $x = \{\mathbf{s}_1, \ldots, \mathbf{s}_n\}$ in Figure 19.7 (left panel). The density is

$$f_{\beta,\varphi}(x) = \frac{1}{c_{(\beta,\varphi)}} \prod_{i=1}^{n} \beta(\mathbf{s}_i) \prod_{1 \le i < j < \le n} \varphi(\|\mathbf{s}_i - \mathbf{s}_j\|) \tag{19.25}$$

with respect to the Poisson process on W with intensity one. Here the first-order term β is a nonnegative function that models the inhomogeneity, the second-order term φ is a nonnegative function that models the interaction, and $c_{(\beta,\varphi)}$ is a normalizing constant. A priori it is expected that the cell intensity only changes in the direction from the stomach cavity to the surrounding muscles tissue. Therefore, it is assumed that $\beta(\mathbf{s})$ depends only on $\mathbf{s} = (t, u)$ through its first coordinate t. Further, partly in order to obtain a well-defined density and partly in order to model a *repulsive* interaction between the cells, we assume that $0 \le \varphi(\|\mathbf{s}_i - \mathbf{s}_j\|) \le 1$ is a nondecreasing function of the distance $r = \|\mathbf{s}_i - \mathbf{s}_j\|$. Furthermore, we specify a flexible prior for $\beta(\mathbf{s}) = \beta(t)$ by a shot noise process and a flexible prior for

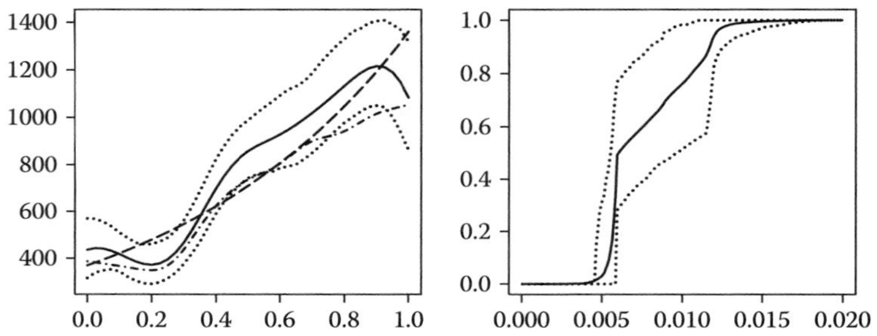

FIGURE 19.8
Posterior mean (solid line) and pointwise 95% central posterior intervals (dotted lines) for β (left panel) and φ (right panel). The left panel also shows the first-order term (dashed line) estimated by Nielsen (2000) and an estimate of the cell intensity (dot-dashed line).

$\varphi(r)$ by a piecewise linear function modeled by a marked Poisson process. For details of these priors and how the auxiliary variable method from Møller et al. (2006) is implemented to obtain simulations from the posterior distribution of (β, φ) given x, see Berthelsen and Møller (2008).

The left panel of Figure 19.8 shows the posterior mean of β, $E(\beta|x)$, together with pointwise 95% central posterior intervals. Also shown is the smooth estimate of the first-order term obtained by Nielsen (2000), where the main difference compared with $\mathrm{E}(\beta|x)$ is the abrupt change of $E(\beta|x)$ in the interval [0.2, 0.4]. For locations near the edges of W, $\mathrm{E}(\beta|x)$ is "pulled" toward its prior mean as a consequence of the smoothing prior.

The intensity $\rho_{\beta,\varphi}(\mathbf{s})$ of the point process is given by the mean of the Papangelou conditional intensity, that is,

$$\rho_{\beta,\varphi}(\mathbf{s}) = E[\lambda_{\beta,\varphi}(\mathbf{s}, Y) f_{\beta,\varphi}(Y)], \tag{19.26}$$

where the expectation is with respect to the Poisson process Y on W with intensity one (see, e.g., Møller and Waagepetersen (2004)). Define

$$\rho_{\beta,\varphi}(t) = \frac{1}{b}\int_0^b \rho_{\beta,\varphi}(t, u)\, du,$$

where $W = [0, a] \times [0, b] = [0, 1] \times [0, 0.893]$. Apart from boundary effects, since $\beta(\mathbf{s})$ only depends on the first coordinate of $\mathbf{s} = (t, u)$, we may expect that the intensity (19.26) only slightly depends on the second coordinate u, i.e., $\rho_{\beta,\varphi}(\mathbf{s}) \approx \rho_{\beta,\varphi}(t)$. We, therefore, refer to $\rho_{\beta,\varphi}(t)$ as the cell intensity, though it is more precisely the average cell intensity in W at $u \in [0, a]$. The left panel of Figure 19.8 also shows a nonparametric estimate $\hat{\rho}(t)$ of the cell intensity (the dot dashed line). The posterior mean of $\beta(t)$ is not unlike $\hat{\rho}(t)$ except that $E(\beta(t)|x)$ is higher, as would be expected due to the repulsion in the pairwise interaction point process model.

The posterior mean of φ is shown in the right panel of Figure 19.8 together with pointwise 95% central posterior intervals. The figure shows a distinct hard core on the interval from zero to the observed minimum interpoint distance $d = \min_{i\neq j} \|\mathbf{s}_i - \mathbf{s}_j\|$, which is a little less than 0.006, and an effective interaction range, which is no more than 0.015 (the posterior distribution of $\varphi(r)$ is concentrated close to one for $r > 0.015$). The corner at $r = d$ of the curve showing the posterior mean of $\varphi(r)$ arises because $\varphi(r)$ is often zero for $r < d$ (since the hard core is concentrated close to d), while $\varphi(r) > 0$ for $r > d$.

Acknowledgments

Much of this contribution is based on previous work with my collaborators, particularly, Kasper K. Berthelsen, Janine Illian, Anne R. Syversveen, and last but not least, Rasmus P. Waagepetersen. Supported by the Danish Natural Science Research Council, grant no. 272-06-0442 (point process modeling and statistical inference).

References

P. Armstrong. Species patterning in the heath vegetation of the northern sandplain. Honours thesis, University of Western Australia, 1991.

A. Baddeley and R. Turner. Practical maximum pseudolikelihood for spatial point patterns. *Australian and New Zealand Journal of Statistics*, 42:283–322, 2000.

A. Baddeley and R. Turner. Spatstat: An R package for analyzing spatial point patterns. *Journal of Statistical Software*, 12:1–42, 2005. www.jstatsoft.org, ISSN: 1548-7660.

A. Baddeley and R. Turner. Modelling spatial point patterns in R. In A. Baddeley, P. Gregori, J. Mateu, R. Stoica, and D. Stoyan, Eds., *Case Studies in Spatial Point Process Modeling*, pp. 23–74. Springer Lecture Notes in Statistics 185, Springer-Verlag, New York, 2006.

A. Baddeley, P. Gregori, J. Mateu, R. Stoica, and D. Stoyan, Eds. *Case Studies in Spatial Point Process Modeling*. Springer Lecture Notes in Statistics 185, Springer-Verlag, New York, 2006.

A. Baddeley, J. Møller, and R. Waagepetersen. Non- and semi-parametric estimation of interaction in inhomogeneous point patterns. *Statistica Neerlandica*, 54:329–350, 2000.

V. Beneš, K. Bodlak, J. Møller, and R.P. Waagepetersen. A case study on point process modelling in disease mapping. *Image Analysis and Stereology*, 24:159–168, 2005.

M. Berman and R. Turner. Approximating point process likelihoods with GLIM. *Applied Statistics*, 41:31–38, 1992.

K.K. Berthelsen and J. Møller. Likelihood and non-parametric Bayesian MCMC inference for spatial point processes based on perfect simulation and path sampling. *Scandinavian Journal of Statistics*, 30:549–564, 2003.

K.K. Berthelsen and J. Møller. An efficient MCMC method for Bayesian point process models with intractable normalising constants. In A. Baddeley, P. Gregori, J. Mateu, R. Stoica, and D. Stoyan, Eds., *Spatial Point Process Modelling and Its Applications*. Publicacions de la Universitat Jaume I, Custellóndela Plana, Spain, 2004.

K.K. Berthelsen and J. Møller. Bayesian analysis of Markov point processes. In A. Baddeley, P. Gregori, J. Mateu, R. Stoica, and D. Stoyan, eds., *Case Studies in Spatial Point Process Modeling*, pp. 85–97. Springer Lecture Notes in Statistics 185, Springer-Verlag, New York, 2006.

K.K. Berthelsen and J. Møller. Non-parametric Bayesian inference for inhomogeneous Markov point processes. *Australian and New Zealand Journal of Statistics*, 50:627–649, 2008.

J. Besag. Some methods of statistical analysis for spatial data. *Bulletin of the International Statistical Institute*, 47:77–92, 1977.

J. Besag, R.K. Milne, and S. Zachary. Point process limits of lattice processes. *Journal of Applied Probability*, 19:210–216, 1982.

N.G. Best, K. Ickstadt, and R.L. Wolpert. Spatial Poisson regression for health and exposure data measured at disparate resolutions. *Journal of the American Statistical Association*, 95:1076–1088, 2000.

A. Brix and P.J. Diggle. Spatio-temporal prediction for log-Gaussian Cox processes. *Journal of the Royal Statistical Society Series B*, 63:823–841, 2001.

A. Brix and W.S. Kendall. Simulation of cluster point processes without edge effects. *Advances in Applied Probability*, 34:267–280, 2002.

A. Brix and J. Møller. Space-time multitype log Gaussian Cox processes with a view to modelling weed data. *Scandinavian Journal of Statistics*, 28:471–488, 2001.

R. Condit. *Tropical Forest Census Plots*. Springer-Verlag and R.G. Landes Company, Berlin, and Georgetown, Texas, 1998.

R. Condit, S.P. Hubbell, and R.B. Foster. Changes in tree species abundance in a neotropical forest: Impact of climate change. *Journal of Tropical Ecology*, 12:231–256, 1996.

N. Cressie and A. Lawson. Hierarchical probability models and Bayesian analysis of minefield locations. *Advances in Applied Probability*, 32:315–330, 2000.

P.J. Diggle. *Statistical Analysis of Spatial Point Patterns*. Arnold, London, 2nd ed., 2003.

A. Gelman and X.-L. Meng. Simulating normalizing constants: From importance sampling to bridge sampling to path sampling. *Statistical Science*, 13:163–185, 1998.

A. Gelman, X.L. Meng, and H.S. Stern. Posterior predictive assessment of model fitness via realized discrepancies (with discussion). *Statistica Sinica*, 6:733–807, 1996.

C.J. Geyer. Likelihood inference for spatial point processes. In O.E. Barndorff-Nielsen, W.S. Kendall, and M.N.M. van Lieshout, Eds., *Stochastic Geometry: Likelihood and Computation*, pages 79–140, Boca Raton, FL, 1999. Chapman & Hall.

C.J. Geyer and J. Møller. Simulation procedures and likelihood inference for spatial point processes. *Scandinavian Journal of Statistics*, 21:359–373, 1994.

C.J. Geyer and E.A. Thompson. Annealing Markov chain Monte Carlo with applications to pedigree analysis. *Journal of the American Statistical Association*, 90:909–920, 1995.

R.D. Harkness and V. Isham. A bivariate spatial point pattern of ants' nests. *Applied Statistics*, 32:293–303, 1983.

J. Heikkinen and E. Arjas. Non-parametric Bayesian estimation of a spatial Poisson intensity. *Scandinavian Journal of Statistics*, 25:435–450, 1998.

L. Heinrich. Minimum contrast estimates for parameters of spatial ergodic point processes. In *Transactions of the 11th Prague Conference on Random Processes, Information Theory and Statistical Decision Functions*, pages 479–492, Prague, 1992. Academic Publishing House.

H. Högmander and A. Särkkä. Multitype spatial point patterns with hierarchical interactions. *Biometrics*, 55:1051–1058, 1999.

S.P. Hubbell and R.B. Foster. Diversity of canopy trees in a neotropical forest and implications for conservation. In S.L. Sutton, T.C. Whitmore, and A.C. Chadwick, Eds., *Tropical Rain Forest: Ecology and Management*, pages 25–41. Blackwell Scientific Publications, 1983.

J. Illian, A. Penttinen, H. Stoyan, and D. Stoyan. *Statistical Analysis and Modelling of Spatial Point Patterns*. John Wiley & Sons, Chichester, U.K. 2008.

J.B. Illian, J. Møller, and R.P. Waagepetersen. Spatial point process analysis for a plant community with high biodiversity. *Environmental and Ecological Statistics*, 16:389–405, 2009.

J.L. Jensen and H.R. Künsch. On asymptotic normality of pseudo likelihood estimates for pairwise interaction processes. *Annals of the Institute of Statistical Mathematics*, 46:475–486, 1994.

J.L. Jensen and J. Møller. Pseudolikelihood for exponential family models of spatial point processes. *Annals of Applied Probability*, 3:445–461, 1991.

W.S. Kendall. Perfect simulation for the area-interaction point process. In L. Accardi and C.C. Heyde, Eds., *Probability Towards 2000*, pages 218–234. Springer Lecture Notes in Statistics 128, Springer Verlag, New York, 1998.

W.S. Kendall and J. Møller. Perfect simulation using dominating processes on ordered spaces, with application to locally stable point processes. *Advances in Applied Probability*, 32:844–865, 2000.

B.G. Lindsay. Composite likelihood methods. *Contemporary Mathematics*, 80:221–239, 1988.

S. Mase. Consistency of the maximum pseudo-likelihood estimator of continuous state space Gibbs processes. *Annals of Applied Probability*, 5:603–612, 1995.

S. Mase. Marked Gibbs processes and asymptotic normality of maximum pseudo-likelihood estimators. *Mathematische Nachrichten*, 209:151–169, 1999.

J. Møller. Contribution to the discussion of N.L. Hjort and H. Omre (1994): Topics in spatial statistics. *Scandinavian Journal of Statistics*, 21:346–349, 1994.

J. Møller. Shot noise Cox processes. *Advances in Applied Probability*, 35:4–26, 2003.

J. Møller and R.P. Waagepetersen. *Statistical Inference and Simulation for Spatial Point Processes*. Chapman & Hall, Boca Raton, FL, 2004.

J. Møller and R.P. Waagepetersen. Modern spatial point process modelling and inference (with discussion). *Scandinavian Journal of Statistics*, 34:643–711, 2007.

J. Møller, A.N. Pettitt, K.K. Berthelsen, and R.W. Reeves. An efficient MCMC method for distributions with intractable normalising constants. *Biometrika*, 93:451–458, 2006.

J. Møller, A.R. Syversveen, and R.P. Waagepetersen. log Gaussian Cox processes. *Scandinavian Journal of Statistics*, 25:451–482, 1998.

J. Neyman and E.L. Scott. Statistical approach to problems of cosmology. *Journal of the Royal Statistical Society Series B*, 20:1–43, 1958.

L.S. Nielsen. Modelling the position of cell profiles allowing for both inhomogeneity and interaction. *Image Analysis and Stereology*, 19:183–187, 2000.

S.L. Rathbun. Estimation of Poisson intensity using partially observed concomitant variables. *Biometrics*, 52:226–242, 1996.

S.L. Rathbun and N. Cressie. Asymptotic properties of estimators for the parameters of spatial inhomogeneous Poisson processes. *Advances in Applied Probability*, 26:122–154, 1994.

S.L. Rathbun, S. Shiffman, and C.J. Gwaltney. Modelling the effects of partially observed covariates on Poisson process intensity. *Biometrika*, 94:153–165, 2007.

C.P. Robert and G. Casella. *Monte Carlo Statistical Methods*. Springer-Verlag, New York, 1999.

H. Rue, S. Martino, and N. Chopin. Approximate Bayesian inference for latent Gaussian models using integrated nested Laplace approximations. Preprint Statistics No. 1/2007, Norwegian University of Science and Technology, Trondheim, 2007.

M. Thomas. A generalization of Poisson's binomial limit for use in ecology. *Biometrika*, 36:18–25, 1949.

Van Lieshout, M.N.M. *Markov Point Processes and Their Applications*. Imperial College Press, London, 2000.

R. Waagepetersen. An estimating function approach to inference for inhomogeneous Neyman–Scott processes. *Biometrics*, 63:252–258, 2007.

R. Waagepetersen. Estimating functions for inhomogeneous spatial point processes with incomplete covariate data. *Biometrika*, 95:351–363, 2008.

R. Waagepetersen and Y. Guan. Two-step estimation for inhomogeneous spatial point processes. *Journal of the Royal Statistical Society: Series B*, 71:685–702, 2009.

R. Waagepetersen and T. Schweder. Likelihood-based inference for clustered line transect data. *Journal of Agricultural, Biological, and Environmental Statistics*, 11:264–279, 2006.

R.L. Wolpert and K. Ickstadt. Poisson/gamma random field models for spatial statistics. *Biometrika*, 85:251–267, 1998.

20

Modeling Strategies

Adrian Baddeley

CONTENTS

20.1 Fundamental Issues

Before embarking on a statistical analysis of spatial point pattern data, it is important to ask two questions. First, are point process methods appropriate to the scientific context? Second, are the standard assumptions for point process methods appropriate in the context?

20.1.1 Appropriateness of Point Process Methods

Treating a spatial point pattern as a realization of a spatial point process effectively assumes that the pattern is random (in the general sense, i.e., the locations and number of points are not fixed) and that the point pattern is the *response* or observation of interest. The statistician should consider whether this is appropriate in the scientific context. Consider the following illustrative scenarios.

Scenario 1 *A silicon wafer is inspected for defects in the crystal surface and the locations of all defects are recorded.*

This point pattern could be analyzed as a point process in two dimensions, assuming the defects are point-like at the scale of interest. Questions for study would include frequency of defects, spatial trends in intensity, and spacing between defects.

Scenario 2 *Earthquake aftershocks in Japan are detected and the epicenter latitude and longitude and the time of occurrence are recorded.*

These data could be analyzed as a point process in space–time (where space is the two-dimensional plane or the Earth's surface) or as a marked point process in two-dimensional space. If the occurrence times are ignored, it may be analyzed as a spatial point process. Spatiotemporal point processes are discussed in Chapter 25.

Scenario 3 *The locations of petty crimes that occurred in the past week are plotted on a street map of Chicago.*

This could be analyzed as a two-dimensional spatial point process. Questions for study include the frequency of crimes, spatial variation in intensity, and evidence for clusters of crimes. One issue is whether the recorded crime locations can lie anywhere in two-dimensional space, or whether they are actually restricted to locations on the streets. In the latter case, it would be more appropriate to treat the data as a point process on a network of one-dimensional lines.

Scenario 4 *A tiger shark is captured, tagged with a satellite transmitter, and released. Over the next month its location is reported daily. These points are plotted on a map.*

It is probably *not* appropriate to analyze these data as a spatial point process. A realization of a spatial point process is an unordered set of points, so the serial order in which the data were recorded is ignored by spatial point process methods.

At the very least, the date of each observation of the shark should be included in the analysis. The data could be treated as a space–time point process, except that this would be a strange process, consisting of exactly one point at each instant of time.

These data should properly be treated as a sparse sample of a continuous trajectory, and analyzed using other methods; see Chapter 26.

Scenario 5 *A herd of deer is photographed from the air at noon each day for 10 days. Each photograph is processed to produce a point pattern of individual deer locations on a map.*

Each day produces a point pattern that could be analyzed as a realization of a point process. However, the observations on successive days are dependent (e.g., having constant herd size and systematic foraging behavior). Assuming individual deer cannot be identified from day to day, this is effectively a "repeated measures" dataset where each response is a point pattern. Methods for this problem are in their infancy. A pragmatic alternative may be to treat the data as a space–time point process.

Scenario 6 *In a designed controlled experiment, silicon wafers are produced under various conditions. Each wafer is inspected for defects in the crystal surface, and the locations of all defects are recorded as a point pattern.*

This is a designed experiment in which the response is a point pattern. Methods for this problem are in their infancy. There are some methods for *replicated* spatial point patterns that apply when each experimental group contains several point patterns. See Section 18.8.

20.1.2 The Sampling Design

Data cannot be analyzed properly without knowledge of the context and the sampling design. The vast majority of statistical techniques for analyzing spatial point patterns assume what we may call the *"standard model"*:

1. The points are observed inside a region W, the "sampling window" or "study region," that is fixed and known.
2. Point locations are measured exactly.
3. No two points lie at exactly the same location.
4. The survey is exhaustive within W, i.e., there are no errors in detecting the presence of points of the random process within W.

Assumption 1 implies that the sampling region does not depend on the data. This excludes some types of experiments in which we continue recording spatial points until a stopping criterion is satisfied. For example, a natural stand of trees surrounded by open grassland, or a cluster of fossil discoveries, would typically be mapped in its entirety. This does not fit the standard model; it requires different techniques, analogous to sequential analysis.

For many statistical analyses, it is important to *know* the sampling window W. This is a fundamental methodological issue that is unique to spatial point patterns. The data do not consist solely of the locations of the observed points. As well as knowing where points were observed, we also need to know where points were **not** observed.

Difficulties arise when the sampling region W is not well defined, but we wish nevertheless to use a statistical technique that assumes the standard model and requires knowledge of W. It is a common error to take W to be the smallest rectangle containing the data points, or the convex hull of the data points. These are "underestimates" of the true region W and typically yield overestimates of the intensity λ and the K-function. Some more defensible methods exist for estimating W if it is unknown (Baillo, Cuevas, and Justel 2008; Mammen and Tsybakov, 1995; Moore, 1984; Ripley and Rosson, 1977).

An alternative to the standard model is to assume that the entire spatial point process has a finite total number of points, all of which are observed. This process cannot be spatially homogeneous, which precludes or complicates the use of many classical techniques: for example the usual interpretation of the K-function assumes homogeneity. If covariates are available, we may be able to use them to construct a reference intensity $\lambda_0(u)$ that represents the "sampling design" or the "null hypothesis." For example, in spatial epidemiology, the covariates may include (a surrogate for) the spatially varying population density, which serves as the natural reference intensity for models of disease risk.

Most current methods require Assumption 3. Duplicate points do commonly occur, through data entry errors, discretization of the spatial coordinate system, reduction of resolution due to confidentiality or secrecy requirements, and other factors. Care should be taken in interpreting the output of statistical software if any duplicate points are present. A high prevalence of duplicate points precludes the use of many current techniques for spatial point processes.

Assumption 4 is usually implicit in our analysis. However, this does not preclude experiments where there is unreliable detection of the objects of interest. Examples include studies of the abundance of wild animal species, and galaxy surveys in radioastronomy. A galaxy catalog is obtained by classifying each faint spot in an image as either a galaxy or noise. In such studies, the analysis is consciously performed on the point process of *detected* points. Certain types of inference are then possible about the underlying process of *true* points, for example, estimation of the galaxy K-function, which is invariant under independent random thinning.

The points in a spatial point pattern dataset often represent the locations of physical objects. If the physical sizes of the objects cannot be neglected at the scale of interest, we may encounter methodological problems. Point process methods may still be applied to the locations, but may lead to artifacts. If the points are the centers of regions, such as cells observed in a microscope image, then a finding that the centers are regular or inhibited could simply be an artifact caused by the nonzero size of the cells. An extreme case occurs when the points are the centers of tiles in a space-filling tessellation. The strong geometric dependence in a tessellation causes striking artifacts in point process statistics; it is more appropriate to use methods specifically developed for tessellations.

20.2 Goals of Analysis

The choice of strategy for modeling and analyzing a spatial point pattern depends on the research goals. Our attention may be focused primarily on the *intensity* of the point pattern, or primarily on the *interaction* between points, or equally on the intensity and interaction. There is a choice concerning the scope of statistical inference, that is, the "population" to which we wish to generalize.

20.2.1 Intensity

The *intensity* is the (localized) expected density of points per unit area. It is typically interpreted as the rate of occurrence, abundance, or incidence of the events recorded in the point pattern. When the prevention of these events is the primary concern (e.g., defects in crystal, petty crimes, cases of infectious disease), the intensity is usually the feature of primary interest. The main task for analysis may be to quantify the intensity, to decide whether intensity is constant or spatially varying, or to map the spatial variation in intensity. If covariates are present, then the main task may be to investigate whether the intensity depends on the covariate, for example, whether the abundance of trees depends on the acidity of soil.

The intensity is a first moment quantity (related to expectations of counts of points). Hence, it is possible to study the intensity by formulating a *model for the intensity only*, for example, a parametric or semiparametric model for the intensity as a function of the Cartesian coordinates. In such analyses, stochastic dependence between points is a nuisance feature that complicates the methodology and inference. Statistical inference is supported by the method of moments and by properties of composite likelihood (Section 19.3.1).

Alternatively, we may formulate a complete stochastic model for the observed point pattern (i.e., a spatial point process model) in which the main focus is the description of the intensity. The model should exhibit the right type of stochastic dependence, and the intensity should be a tractable function of the model parameters. If there is positive association between points, useful models include log-Gaussian Cox processes (Section 19.5.2) and inhomogeneous cluster models (Section 19.3.3). If there is negative association, Gibbs processes (Section 17.2.4; Section 19.5.3) are appropriate, although the intensity is not a simple function of the model parameters. Statistical inference is supported by properties of the likelihood (Sections 19.4.1 and 19.4.3) or pseudo-likelihood (Section 19.3.4).

20.2.2 Interaction

"Interpoint interaction" is the conventional term for stochastic dependence between points. This covers a wide range of behavior since the only point processes that do not exhibit stochastic dependence are the Poisson processes. The term "interaction" can be rather prejudicial. One *possible* cause of stochastic dependence is a direct physical interaction between the objects recorded in the point pattern. For example, if the spatial pattern of pine seedlings in a natural forest is found to exhibit negative association at short distances, this might be interpreted as reflecting biological interaction between the seedlings, perhaps due to competition for space, light or water.

The main task for analysis may be to decide whether there is stochastic dependence, to determine the type of dependence (e.g., positive or negative association), or to quantify its strength and spatial range.

Interpoint interaction is measured by second-order moment quantities, such as the K-function (Section 16.6), or by higher-order quantities, such as the distance functions G, F and J (Section 16.4). Just as we must guard against spurious correlations in numerical data by carefully adjusting for changes in the mean, a rigorous analysis of interpoint interaction requires that we take into account any spatial variation in intensity.

A popular classical approach to spatial point pattern analysis was to assume that the point pattern is *stationary*. This implies that the intensity is constant. Analysis could then concentrate on investigating interpoint interaction. It was argued (e.g., Ripley, 1976) that this approach was pragmatically justified when dealing with quite small datasets (containing only 30 to 100 points), or when the data were obtained by selecting a small subregion where the pattern appeared stationary, or when the assumption of stationarity is scientifically plausible.

Figure 20.1 shows the Swedish pines dataset of Strand (1972), presented by Ripley, (1987), as an example where the abovementioned conditions for assuming stationarity were satisfied. There is nevertheless some suggestion of inhomogeneity. Contour lines represent the fitted intensity under a parametric model in which the logarithm of the intensity is a quadratic function of the cartesian coordinates. Figure 20.2 shows the estimated K-function of the Swedish pines assuming stationarity, and the inhomogeneous K-function using the fitted log-quadratic intensity. The two K-functions convey quite a similar message, namely that there is inhibition between the saplings at distances less than one meter. The two K-functions agree because gentle spatial variation in intensity over large spatial scales is irrelevant at shorter scales.

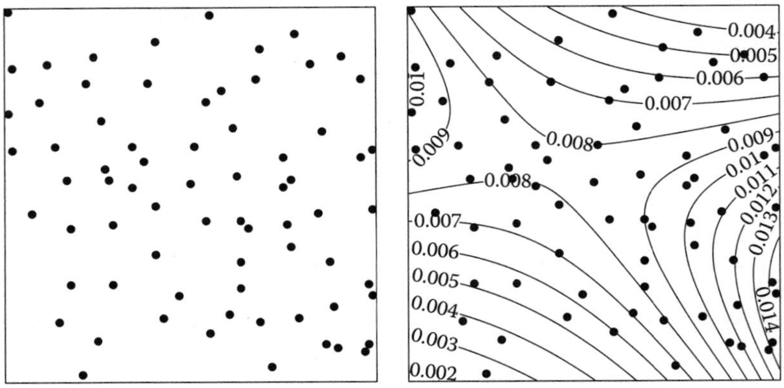

FIGURE 20.1
Swedish pines data with fitted log-quadratic intensity.

20.2.3 Intensity and Interaction

In some applications, intensity and interaction are both of interest. For example, a cluster of new cases of a disease may be explicable either by a localized increase in intensity due to etiology (such as a localized pathogen), sampling effects (a localized increase in vigilance, etc.), or by stochastic dependence between cases (due to person-to-person transmission, familial association, genetics, social dependence, etc.). The spatial arrangement of galaxies in a galaxy cluster invites complex space–time models, in which the history of the early universe is reflected in the overall intensity of galaxies, while the observed local arrangement of galaxies involves gravitational interactions in recent history.

When a point pattern exhibits both spatial inhomogeneity and interpoint interaction, several strategies are possible. An *incremental* or *marginal* modeling strategy seeks to estimate spatial trend, then "subtract" or "adjust" for spatial trend, possibly in several stages, before looking for evidence of interpoint interaction. A *joint* modeling strategy tries to fit one stochastic model that captures all relevant features of the point process, and in particular, allows the statistician to "account" for spatial inhomogeneity during the analysis of interpoint interaction.

These choices are familiar from time series analysis. Incremental modeling is analogous to seasonal adjustment of time series, while joint modeling is analogous to fitting a time

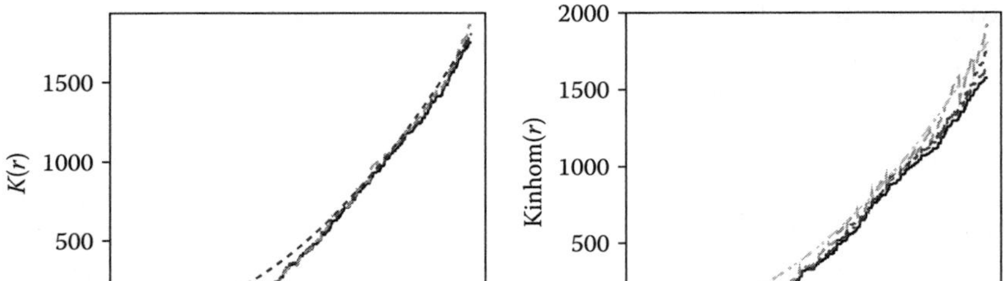

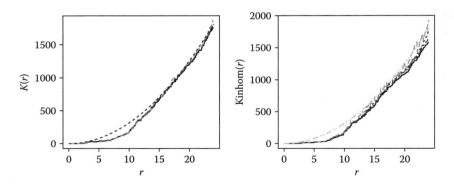

FIGURE 20.2
Left: Estimated K-function of Swedish pines assuming stationarity. Right: Estimated inhomogeneous K-function of Swedish pines using fitted log-quadratic intensity.

series model that embraces both seasonal trend and autocorrelation. Incremental modeling is less prone to the effects of model misspecification, while joint modeling is less susceptible to analogs of Simpson's paradox. Joint modeling would normally be employed in the final and more formal stages of analysis, while incremental modeling would usually be preferred in the initial and more exploratory stages.

For example, in the analysis of the Swedish pines data above, we first fitted a parametric intensity model, then computed the inhomogeneous K-function which "adjusts" for this fitted intensity. This is an incremental modeling approach. A corresponding joint modeling approach is to fit a Gibbs point process with nonstationary spatial trend. Again we assume a log-quadratic trend. Figure 20.2 suggests fitting a Strauss process model (Section 17.4) with interaction radius r between 4 and 15 units. The model selected by maximum profile pseudolikelihood has $r = 9.5$ and a fitted interaction parameter of $\gamma = 0.27$, suggesting substantial inhibition between points.

20.2.4 Confounding between Intensity and Interaction

In analyzing a point pattern, it may be impossible to distinguish between clustering and spatial inhomogeneity. Bartlett (1964) showed that a single realization of a point process model that is stationary and clustered (i.e., exhibits positive dependence between points) may be *identical* to a single realization of a point process model that has spatially inhomogeneous intensity but is not clustered. Based on a single realization, the two point process models are distributionally equivalent and, hence, unidentifiable. This represents a fundamental limitation on the scope of statistical inference from a spatial point pattern, assuming we do not have access to replicate observations. The inability to separate trend and autocorrelation, within a single dataset, is also familiar in time series.

This may be categorized as a form of *confounding*. A linear model $Y = X\beta + \epsilon$ is confounded if the columns of the design matrix X are not linearly independent, so that the parameter vector β is not identifiable. Bartlett's examples show that a point process model involving both spatial inhomogeneity and interpoint interaction may be confounded, that is, unidentifiable, given only a single realization of the spatial point process.

The potential for confounding spatial inhomogeneity and interpoint interaction is important in the interpretation of summary statistics such as the K-function. In Figure 20.3, the left panel shows a realization of a spatially inhomogeneous Poisson process, its intensity a linear function of the Cartesian coordinates. The right panel is a plot of $\widehat{L}(r) - r$ against r,

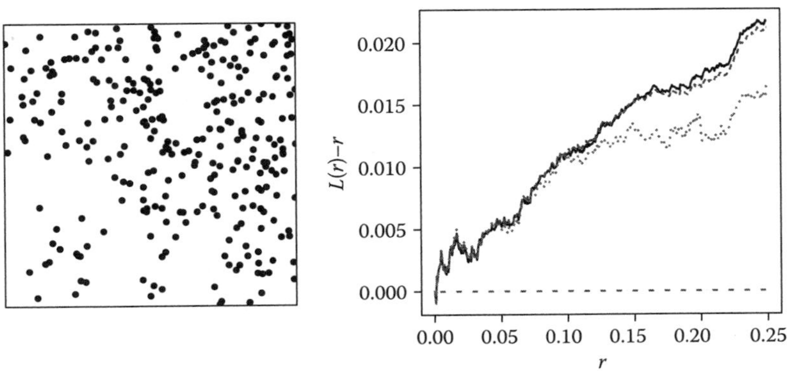

FIGURE 20.3
Illusory clustering. Left: Realization of a nonstationary Poisson process. Right: Plot of $\hat{L}(r) - r$ against r for the same point pattern, inviting the interpretation that the pattern is clustered.

where $\widehat{L}$ is the estimate of $L(r) = \sqrt{K(r)/\pi}$ assuming the point process is stationary. The right-hand plot invites the incorrect interpretation that the points are clustered.

The χ^2 test of complete spatial randomness (CSR) using quadrat counts is afflicted by a similar weakness. In the classical analysis of contingency tables, the χ^2 test of uniformity is usually justified by assuming that the cell counts N_i in the contingency table are independent with unknown mean. Then the Pearson χ^2 test statistic $X^2 = \sum_i (N_i - \overline{N})^2/\overline{N}$ (with $\overline{N} = n/m$, where there are m cells and a total count of n points) is a reasonable index of departure from uniformity, as well as a valid test statistic with null distribution that is asymptotically χ^2_{m-1}. Applying this test to the quadrat counts for a spatial point pattern effectively assumes that the quadrat counts are independent. Suppose, alternatively, that we assume the point process is stationary, and we wish to test whether it exhibits interpoint interaction. Then the mean counts $E[N_i]$ are assumed equal, and we wish to test whether they are Poisson-distributed. A common index of over- or under-dispersion is the sample variance-to-mean ratio $I = (1/(m-1)) \sum_i (N_i - \overline{N})^2/\overline{N}$. However, I is a constant multiple of X^2. Thus, large values of the statistic X^2 may be interpreted as suggesting either spatial inhomogeneity or spatial clustering, depending on our underlying assumptions.

20.2.5 Scope of Inference

There is a choice concerning the scope of statistical inference, that is, the "population" to which we wish to generalize from the data.

At the lowest level of generalization, we are interested only in the region that was actually surveyed. In applying precision agriculture to a particular farm, we might use the observed spatial point pattern of tree seedlings, which germinated in a field sown with a uniform density of seed, as a means of estimating the unobservable spatially varying fertility of the soil in the same field. Statistical inference here is a form of interpolation or prediction. The modeling approach is influenced by the prediction goals: To predict soil fertility it may be sufficient to model the point process intensity only, and ignore interpoint interaction.

At the next level, the observed point pattern is treated as a "typical" sample from a homogeneous pattern, which is the target of inference. To draw conclusions about an entire forest from observations in a small study region, we treat the forest as a spatial point process $\mathbf{X}$, effectively extending throughout the infinite two-dimensional plane. In order to draw inferences based only on a sample of $\mathbf{X}$ in a fixed bounded window W, we might assume that $\mathbf{X}$ is *stationary* and/or *isotropic*, meaning that statistical properties of the point process are unaffected by vector translations (shifts) and/or rotations, respectively. This implies that our dataset is a typical sample of the process, and supports nonparametric inference about stochastic properties of $\mathbf{X}$, such as its intensity and K-function (stationarity guaranteeing that the method of moments produces unbiased estimators). It also supports parametric inference, for example, about the interaction parameter γ of a Strauss process model for the spatial dependence between trees. If the sample domain is sufficiently large, then under additional mixing assumptions, the parameter estimates are consistent and asymptotically normal. Parametric modeling also enables finite-sample inference, although its implementation usually requires Monte Carlo methods.

At a higher level, we seek to extract general "laws" or "relationships" from the data. This involves generalizing from the observed point pattern, to a hypothetical population of point patterns, which are governed by the same "laws," but which may be very different from the observed point pattern. One important example is modeling the dependence of the point pattern on a spatial covariate (such as terrain slope). This is a form of regression. The conditional distribution of the spatial point process given the spatial covariate is modeled by a regression specification. For example, it may be assumed that the intensity $\lambda(u)$ of the point process at a location u is a function $\lambda(u) = \rho(Z(u))$ of the spatial covariate $Z(u)$. The regression function ρ is the target of inference. The scope of inference is a population of experiments where the same variables are observed and the same regression relationship is assumed to hold. A model for ρ (parametric, non- or semiparametric) is formulated and fitted. Estimation of ρ may be based solely on the method of moments. More detailed inference requires either replication of the experiment, or an assumption, such as joint stationarity of the covariates and the response, under which a large sample can be treated as containing sufficient replication.

At the highest level, we seek to capture all sources of variability that influence the spatial point pattern. Sources of variability may include "fixed effects," such as regression on an observable spatial covariate, and also "random effects," such as regression on an unobserved, random spatial covariate. For example, a Cox process (Example 17.1.3) is defined by starting with a random intensity function $\Lambda(u)$ and, conditional on the realization of Λ, letting the point process be Poisson with intensity Λ. In forestry applications, Λ could represent the unobserved, spatially inhomogeneous fertility of soil, modeled as a random process. Thus, Λ is a "random effect." Whether soil fertility should be modeled as a fixed effect or random effect depends on whether the main interest is in inferring the value of soil fertility in the study region (fixed effect) or in characterizing the variability of soil fertility in general (random effect).

20.3 Exploratory Data Analysis

Many writers on spatial statistics use the term *exploratory methods* to refer to the classical summary statistics for analyzing spatial point patterns, such as the K-function. This may be slightly misleading. Exploratory data analysis (Hoaglin, Mosteller, and Tukey 1983; Tukey, 1977; Velleman and Hoaglin, 1981) refers to a methodology that allows the statistician to investigate features of the data (trend, variability, dependence on a covariate, autocovariance) without losing sight of the peculiarities of individual observations. Tools such as the K-function involve a substantial reduction of information, analogous to the sample moments of a numerical dataset, and are more properly described as (nonparametric) "summary statistics" or "indices." These tools can be used *as part of* a full-fledged exploratory analysis of spatial point pattern data.

While there are no hard rules for exploratory methods, the most natural order of analysis is to investigate the intensity of the point pattern first, before investigating interpoint interaction. A proper analysis of interpoint interaction requires that we take into account any spatial variation in intensity.

20.3.1 Exploratory Analysis of Intensity

For a stationary point process, the intensity λ is the expected number of points per unit area. The natural estimate of λ is the observed number of points divided by the area of the study region.

For a nonstationary point process, we would often assume that the intensity is a function $\lambda(u)$ of location u, defined so that the expected number of points falling in a domain B is equal to the integral of $\lambda(u)$ over B. It is then natural to compute nonparametric estimates of the intensity function. A popular choice is the fixed-bandwidth kernel smoothed estimate discussed in Section 18.5, Equation (18.6).

The choice of smoothing bandwidth h (e.g., the standard deviation of the Gaussian kernel) involves a tradeoff between bias and variability. Bias is inherent in smoothing because we effectively replace the true intensity $\lambda(u)$ by its smoothed version $\lambda^*(u)$, suppressing

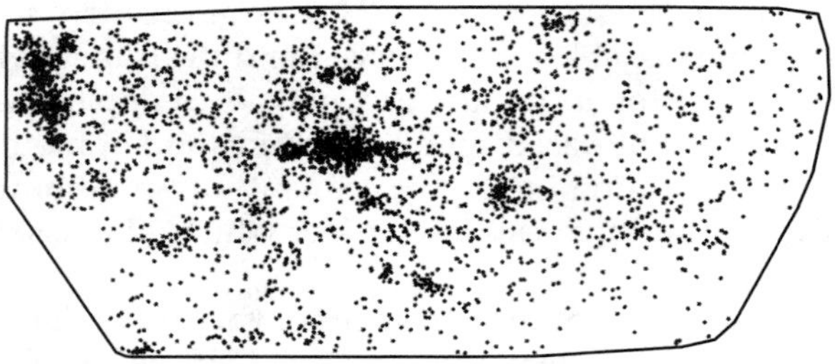

FIGURE 20.4
Shapley galaxy concentration survey. Sky position (right ascension and declination coordinates) for 4,215 galaxies.

any spatial variation at scales smaller than the bandwidth. Increasing the bandwidth h will increase the bias but decrease the variability of $\hat{\lambda}(u)$, since more data points contribute to the intensity estimate. An "optimal" bandwidth might be selected by an automatic procedure, such as minimization of the predicted mean square error (MSE) (Section 18.5), but for exploratory purposes it would usually be prudent to display several kernel estimates obtained with different bandwidths. Each choice of h focuses attention on a different spatial scale.

Kernel smoothing with a fixed kernel k and fixed bandwidth h at all spatial locations may be inappropriate, if the point pattern exhibits very different spatial scales at different locations. For example, Figure 20.4 shows the sky positions of 4,215 galaxies cataloged in a radioastronomical survey of the Shapley supercluster (Drinkwater, Parker, Proust, Slezak et al., 2004). This pattern contains two very high concentrations of points. A fixed-bandwidth kernel estimate of intensity (Figure 20.5) is inadequate, since it suppresses the small-scale variations in intensity, which are of greatest interest.

One solution is to replace the fixed kernel k by a *spatially-varying* or *adaptive* smoothing operator. Many solutions can be found in the seismology literature, because earthquake epicenters typically concentrate around a line or curve. Ogata (2003, 2004) used the Dirichlet/Voronoi tessellation determined by the data points to control the local scale of smoothing.

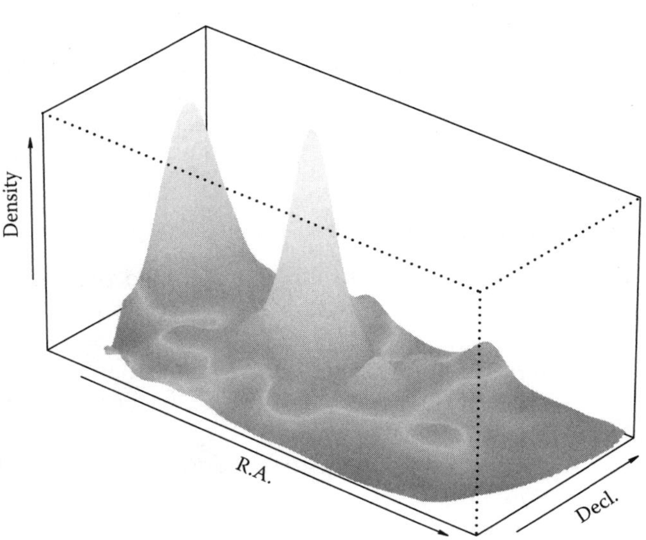

FIGURE 20.5
Fixed bandwidth kernel estimate of intensity for the Shapley galaxy pattern.

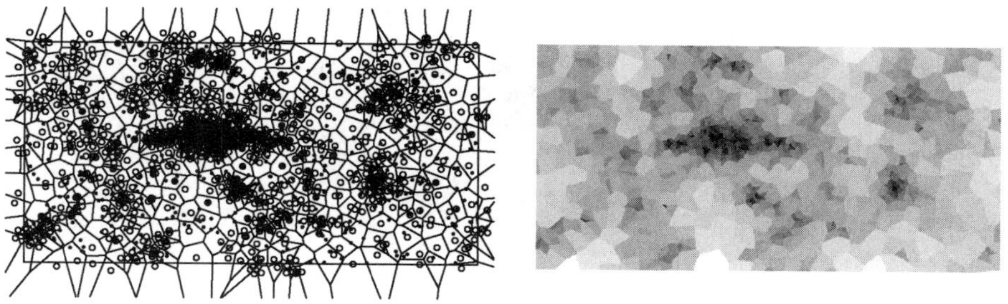

FIGURE 20.6
Adaptive estimation of intensity in the Shapley galaxy pattern. Left: Dirichlet tessellation of one half of data superimposed on other half of data. Right: Intensity estimate constant in each Dirichlet tile.

One simple version of this strategy is shown in Figure 20.6. In a large rectangular subwindow of the Shapley data, we divided the points randomly into two groups, the training data $\mathbf{x}_T$ and modeling data $\mathbf{x}_M$. This is a standard trick to avoid overfitting; in the spatial context, it is appropriate because random thinning preserves relevant properties of the pattern (such as intensity and the inhomogeneous K-function) up to a scale factor. We computed the Dirichlet tessellation of $\mathbf{x}_T$, and extracted the tile areas w_j for each $x_j \in \mathbf{x}_T$. Superimposing this tessellation on $\mathbf{x}_M$ (Figure 20.6, left panel), we formed an intensity estimate for $\mathbf{x}_M$, which is constant on each Dirichlet tile: $\kappa(u) = 1/w_{J(u)}$ where $x_{J(u)}$ is the point of $\mathbf{x}_T$ closest to u. A reasonable estimate of the intensity function of the modeling data is $\widehat{\lambda}(u) = (n(\mathbf{x}_M)/n(\mathbf{x}_T))\kappa(u)$, displayed in the right panel of Figure 20.6.

In extreme cases, there may be a "singular" concentration of intensity that cannot be described by an intensity function, for example, earthquake epicenters that are always located on a tectonic plate boundary, or galaxies that are located on a plane in space. In singular cases the point process must be described by an intensity *measure*. Quadrat counting (Section 18.2.1) has a useful role in such cases.

Closely related to kernel smoothing is the spatial *scan statistic* (Alm, 1988; Anderson and Titterington, 1997; Kulldorff, 1999; Kulldorff and Nagarwalla, 1995) designed to identify "hot spots" or localized areas of increased intensity. See Chapter 22.

20.3.2 Exploratory Analysis of Dependence on Covariates

A *spatial covariate*, in the most general sense, is any kind of spatial data that plays the role of an explanatory variable. It might be a spatial function $Z(u)$ defined at every location u, such as soil pH, terrain altitude, terrain gradient, or gravitational strength. Alternatively, it may be another form of spatial data, such as a spatial tessellation (e.g., dividing the study region into different land cover types), a spatial point pattern, a spatial pattern of lines, etc. The spatial coordinates themselves (e.g., geographical coordinates) may be treated as covariate functions $Z(u)$.

A common question is whether the *intensity* of a spatial point process depends on a spatial covariate. For example, we may wish to know whether trees in a rainforest prefer particular soil conditions (Section 19.3.3), or whether the risk of a rare disease is higher in the vicinity of a pollution source (e.g., Diggle, Morris, Elliott, and Shaddick, 1997; Elliott, Wakefield, Best, and Briggs, 1999, 2000). To assess this question, we could convert the covariate into a spatial function $Z(u)$, then investigate whether the intensity $\lambda(u)$ depends on $Z(u)$.

Figure 20.7 shows data from a geological survey of a 158 × 35 km region in Queensland, Australia. The point pattern records the locations of deposits of copper that were detected by intensive geological survey. Line segments represent the observed spatial pattern of 'lineaments,' which are geological faults or similar linear features observable in remotely

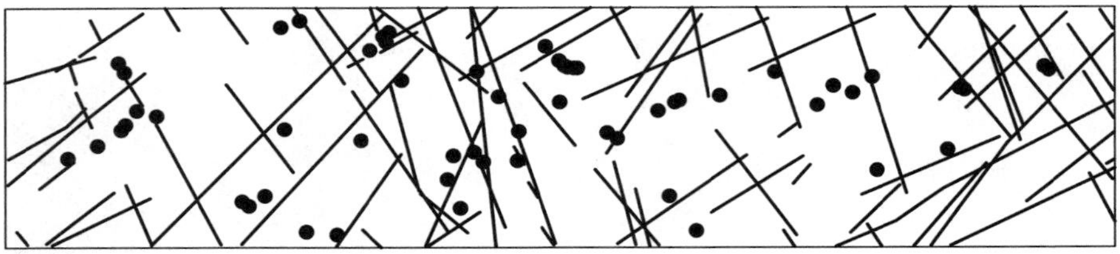

FIGURE 20.7
Queensland copper data. Pointlike deposits of copper (circles) and geological lineaments (straight lines) in a study region. (Data from Berman, M. *Applied Statistics*, 35:54–62, 1986.)

sensed imagery. It is of interest to determine whether the lineament pattern (which can be observed cheaply from satellite data) has predictive value for the point pattern of copper deposits, e.g., whether the intensity of copper deposits is relatively high in the vicinity of the lineaments (Berman, 1986).

The first step in an exploratory analysis of the copper data is to construct a spatial covariate function $Z(u)$. A natural choice is the distance transform (Borcefors, 1986; Rosenfeld and Pfalz, 1968) of the lineament pattern. That is, $Z(u)$ is the distance from the location u to the nearest lineament.

Berman (1986) proposed a diagnostic technique for assessing whether the intensity function $\lambda(u)$ of a point pattern depends on a spatial covariate function $Z(u)$. The empirical cumulative distribution function (cdf) of the covariate values at the points of the spatial point pattern $\mathbf{x}$,

$$C_{\mathbf{x}}(z) = \frac{1}{n} \sum_i 1\{Z(x_i) \leq z\}$$

and the cumulative distribution function of Z at *all* points of the study region W,

$$C_W(z) = \frac{1}{|W|} \int_W 1\{Z(u) \leq z\} \, du$$

(where $|W|$ denotes the area of W) are both plotted against r. If the true intensity does *not* depend on Z and is constant, $\lambda(u) \equiv \lambda$, then $E[n(X)C_X(z)] = \lambda|W|C_W(z)$ and $E[n(X)] = \lambda|W|$ so that $E[C_{\mathbf{x}}(z)] \approx C_W(z)$. If additionally the point process is Poisson (i.e., under CSR), then the values $Z(x_i)$ are independent and identically distributed with cdf C_W, so that $E[C_{\mathbf{x}}(z)] = C_W(z)$. Conversely, discrepancies between $C_{\mathbf{x}}$ and C_W suggest that the point process intensity does depend on $Z(u)$. The maximum discrepancy between the two curves

$$D = \sup_z |C_{\mathbf{x}}(z) - C_W(z)|$$

is the test statistic for the Kolmogorov–Smirnov test of equality of the two distributions (assuming C_W is continuous).

Figure 20.8 shows Berman's diagnostic plot for the Queensland copper data against distance from the nearest lineament. The solid line represents $C_{\mathbf{x}}$ and the dashed line is C_W. The dotted envelopes are the two-standard-error limits for $C_{\mathbf{x}}$ assuming CSR. The plot shows no evidence for dependence of intensity on the covariate.

20.3.3 Exploratory Analysis of Interpoint Interaction

The most common techniques for investigating interpoint interaction are *distance methods* (see Sections 18.2 and 18.4). Distances between points of the pattern, or from arbitrary locations to the nearest point of the pattern, are measured, summarized and displayed in

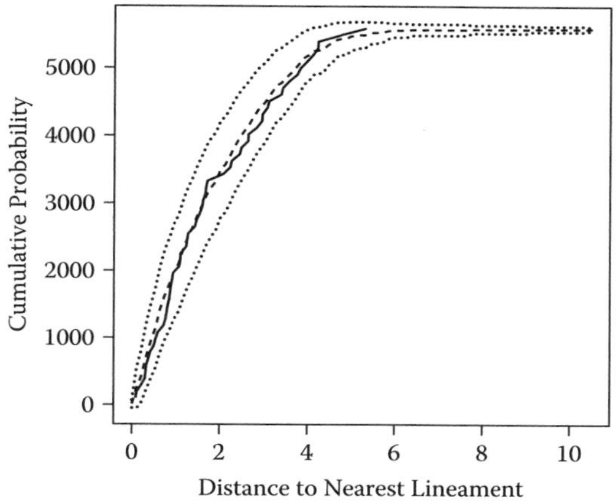

FIGURE 20.8
Berman's diagnostic plot for the Queensland copper data against distance from the nearest lineament.

some fashion. For example, the K-function is a summary of the observed distances between all pairs of points.

20.3.3.1 Analysis Assuming Stationarity

One traditional approach has been to analyze point pattern data under the assumption that the pattern is *stationary* (implying that the intensity is constant). Under this assumption, the interpretation of tools, such as the K-function, is fairly unambiguous. Among stationary processes, only the homogeneous Poisson point process (complete spatial randomness, CSR) does not exhibit interaction. If the K-function estimated from the data is found to deviate from the K-function of CSR, this is evidence of interpoint interaction (assuming stationarity). The nature of the deviation suggests the type and range of interaction. For example, the left panel of Figure 20.2 shows the K-function estimated from the Swedish pines data (Figure 20.1), together with the K-function for CSR. Since the empirical value of $K(r)$ dips below the theoretical value $K(r) = \pi r^2$ for values of r in the range from 0 to 12, we conclude that an inhibition (negative association) is present at this scale.

The K-function is a second-order moment quantity and does not fully characterize the stochastic dependence between points. There exist point processes that have the same K-function as CSR, but which are manifestly clustered (Baddeley and Silverman, 1984). It is good practice to inspect several different summary functions for a point pattern dataset in addition to K, such as the nearest neighbor distance function G and the empty space function F. Figure 20.9 shows the estimates of G and F for Swedish pines, which reinforce the conclusion obtained from the K-function.

Different plots and transformations of F, G and K may be useful for interpretive and modeling purposes. The transformation $L(r) = \sqrt{K(r)/\pi}$ proposed by Besag (1977) stabilizes the variance of the empirical K-function and transforms the Poisson case to $L(r) = r$. The *pair correlation function* $g(r) = K'(r)/(2\pi r)$ where $K'(r) = \mathrm{d}K(r)/\mathrm{d}r$ is more easily interpretable in terms of models. For the cumulative distribution functions F and G, it may be more useful to plot in P–P style (plotting the empirical $F(r)$ against the Poisson $F(r)$ for each r) and to apply Fisher's variance stabilizing transformation $\phi(F(r)) = \sin^{-1}\sqrt{F(r)}$.

The K-function has been generalized by replacing the circle of radius r by an annulus, ellipse or sector of a circle; for example, counting only pairs of points that are separated by distances between r and $r + s$. These generalized K-functions are, in turn, generalized by the *reduced second-moment measure* κ. This can be visualized as the intensity measure of

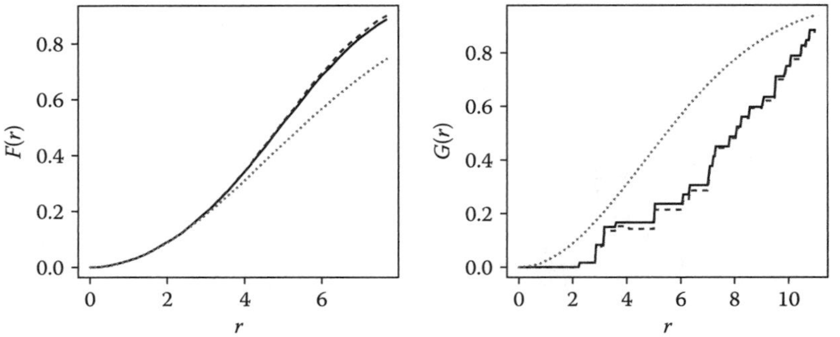

FIGURE 20.9
Estimated empty space function F (left) and nearest neighbor function G (right) for the Swedish pines data.

the process of all vectors $x_j - x_i$ between pairs of points in the pattern. For the Swedish pines data, this process is displayed in the left panel of Figure 20.10 with the origin at the center of the picture. This is a "Fry plot" (Fry, 1979; Hanna and Fry, 1979; Patterson, 1934, 1935). The right panel of Figure 20.10 shows an edge-corrected kernel estimate of the second-moment density (the density of the reduced second-moment measure κ, essentially a smoothed version of the left panel) with lighter shades representing higher density values. The marked drop in density close to the origin suggests inhibition between points at short distances. The hole has no obvious asymmetries, suggesting that an isotropic interaction model would be adequate.

20.3.3.2 Full Exploratory Analysis

When it cannot be assumed that the point pattern is stationary, the use of the K-function and other summary functions is open to critique. Various alternative tools have been suggested.

The *inhomogeneous K-function* is a simple modification of the original K-function that allows us to deal with spatial variation in the intensity (Baddeley, Møller, and Waagepetersen, 2000). It allows the intensity to have any form, but assumes that the *correlation* structure is stationary. To estimate the inhomogeneous K-function, we modify the standard estimators of $K(r)$ by weighting the contribution from each pair of points x_i, x_j by $1/(\lambda(x_i)\lambda(x_j))$ where $\lambda(u)$ is the estimated intensity function. The estimated intensity at each data point must be nonzero. The inhomogeneous K-function takes the value $K(r) = \pi r^2$ for an inhomogeneous

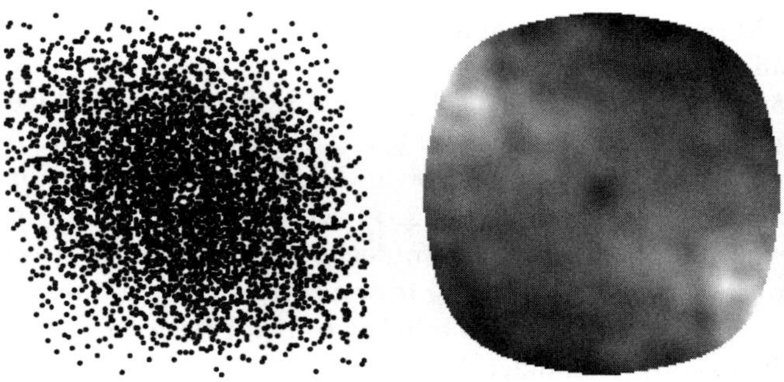

FIGURE 20.10
Fry plot (left) and kernel estimate of the density of the reduced second-moment measure (right) for the Swedish pines data.

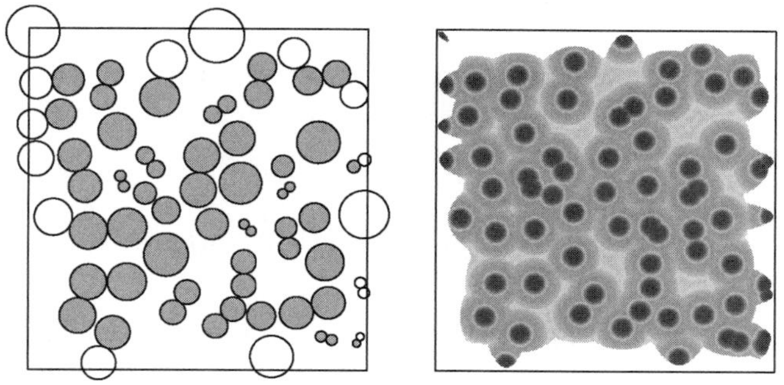

FIGURE 20.11
Exploratory analysis of distances in the Swedish pines data. Left: Stienen diagram; right: distance map.

Poisson point process, so that its interpretation is very similar to that of the original K-function.

If it is possible that there may be spatial variation in the *scale* of pattern (for example, in the spacing between points), the simple tools shown in Figure 20.11 may be useful. The left panel shows the *Stienen diagram* (Steinen, 1980; Stoyan and Stoyan, 1995) for the Swedish pines data, constructed by computing for each point x_i the distance s_i to its nearest neighbor, and drawing a circle around x_i of radius $s_i/2$. The resulting circles never overlap, by construction. Pairs of osculating circles represent pairs of points that are mutual nearest neighbors. In our diagram, the filled circles represent observations that are not biased by edge effects (s_i is less than the distance from x_i to the boundary of the study region). The right panel of Figure 20.11 shows the *distance map* or distance transform. For each location u, the value $d(u)$ is the distance from u to the nearest point of the data pattern. Distance values are displayed only when they are not biased by edge effects. The Stienen diagram and distance map are useful for detecting inhomogeneity in the scale of pattern. Inhomogeneity is visually apparent as a trend in the distance values or in the size of Stienen circles.

For more detailed exploratory analysis, we may decompose the estimator of the K-function into a sum of contributions from each data point x_i. After normalization, these functions $K_i(r)$ represent the "local" pattern around each point x_i. Exploratory analysis of the $K_i(r)$, for example principal component analysis or hierarchical cluster analysis, can identify regions or groups of points with similar patterns of local behavior. This approach of *second-order neighborhood analysis* (Getis and Franklin, 1987) or *local indicators of spatial association (LISA)* (Anselin, 1995) has been used, for example, to distinguish between explosive mines and random clutter in a minefield (Cressie and Collins, 2001b). Spatial smoothing of the individual K-functions can identify spatial trends (Getis and Franklin, 1987).

20.4 Modeling Tools

Modeling is a cyclic process in which tentative models are developed and fitted to data, compared with competing models, subjected to diagnostic checks, and refined. Introducing a new term or component into a model allows us either to capture a feature of interest in the data or to account for the effect of a covariate that is not of primary interest (Cox and Snell, 1981).

For historical reasons, this process is not yet fully developed in spatial statistics and is not widely reflected in the literature. Practical general software for model fitting emerged only in the late 1990s.

20.4.1 Parametric Models of Intensity

It may be desirable to model only the intensity of the point process. For an inhomogeneous Poisson point process with intensity function $\lambda_\theta(u)$ depending on a parameter $\theta \in \Theta$, the log likelihood is (up to a constant)

$$\log L(\theta) = \sum_{i=1}^{n} \log \lambda_\theta(x_i) - \int_W \lambda_\theta(u)\, du \tag{20.1}$$

where $\mathbf{x} = \{x_1, \ldots, x_n\}$ is the point pattern dataset observed in a study region W. If the model is a non-Poisson point process with intensity function $\lambda_\theta(u)$, this becomes the composite log likelihood.

Maximization of $L(\theta)$ typically requires numerical methods. It was pointed out by Brillinger (Brillinger, 1978, 1994; Brillinger and Preisler, 1986) and Berman and Turner (1992) that a discretized version of the Poisson point process likelihood is formally equivalent to the likelihood of a (nonspatial) Poisson generalized linear model. This makes it possible to adapt existing software for fitting generalized linear models to the task of fitting Poisson spatial point process models.

The rationale for building models is broadly similar to reliability or survival analysis. Because the canonical parameter of the Poisson distribution is the logarithm of the Poisson mean, it is natural to formulate models for $\log \lambda(u)$. Many operations on a point process have a multiplicative effect on $\lambda(u)$ and, hence, an additive effect on $\log \lambda(u)$. For example, if a Poisson process with intensity function $\kappa(u)$ is subjected to independent random thinning with retention probability $p(u)$ at location u, then the resulting process is Poisson with intensity $\lambda(u) = p(u)\kappa(u)$. A log-Gaussian Cox process is one in which $\log \lambda(u)$ is a Gaussian random process. Many models for plant growth and animal abundance lend themselves to this log-additive structure.

However, some operations on a point process have an additive effect on the intensity. The superposition of two point patterns has intensity equal to the sum of the intensities of the components (regardless of the dependence between the two components). This is a consideration in some kinds of analysis of a marked point pattern where the marks are ignored, effectively superimposing the points of different types.

The functional form of the additive terms in $\log \lambda(u)$ or $\lambda(u)$ may be arbitrary (except that $\lambda(u)$ must be nonnegative, finite and integrable). If the goal is simply to capture spatial inhomogeneity, it may be appropriate to model $\log \lambda(u)$ as a polynomial function of the Cartesian coordinates. To reduce the number of parameters one may take a smaller basis of functions, such as smoothing splines or harmonic polynomials (suggested by P. McCullagh, 2002). Figure 20.12 shows the maple trees from the Lansing Woods dataset (see Section 21.1.1) with a kernel estimate of intensity, the fitted intensity assuming a general log-cubic model (the log intensity was a general third-degree polynomial in the Cartesian coordinates), and the fitted intensity assuming the log intensity was a third-degree harmonic polynomial of x and y.

If the aim is to model the dependence of intensity on a spatial covariate Z in the form $\lambda(u) = \rho(Z(u))$, then we may choose any functional form for ρ that yields nonnegative and finite values. Figure 20.13 shows two fitted intensity models for the Queensland copper data (Figure 20.7), which assume that $\log \rho(z)$ is a polynomial in z of degree 7 (left panel) and a piecewise constant function with jumps at 2, 4 and 6 km (right panel). The apparent drop in $\rho(z)$ at values beyond $z = 6$ km is probably an artifact because fewer than 0.7% of pixels lie more than 6 km from a lineament. Refitting the models to the subset of data where $Z(u) < 6$ suggests that ρ is roughly constant.

In "singular" cases where the intensity function does not exist, such as point patterns of earthquake epicenters that are concentrated along a line, it is usually appropriate to take the

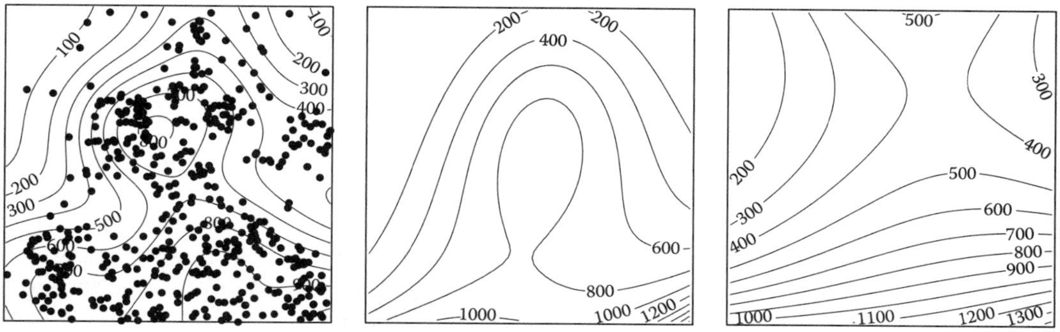

FIGURE 20.12
Intensity models for the maple trees in Lansing Woods. Left: Data with kernel estimate of intensity; middle: fitted log-cubic intensity; right: fitted log-harmonic-cubic intensity.

line of concentration as fixed and given, and to assume the process has a one-dimensional intensity along this line of concentration, which is then modeled in a similar fashion.

20.4.2 Multilevel Models for Clustering

Stochastic dependence between points can be modeled by multilevel processes derived from the Poisson process, such as Poisson cluster processes and Cox processes. Typically these models are effective for describing positive association ("attraction" or "clustering") between points.

Cox models are always "clustered" since a Cox process is conditionally Poisson given the random intensity function Λ, and, therefore, is overdispersed relative to the Poisson process with the same expected intensity $\lambda(u) = \mathbf{E}[\Lambda(u)]$. Poisson cluster models allow more freedom in the type of dependence between points because points within a cluster may be dependent in any fashion we choose. However, the most popular Poisson cluster models are Neyman–Scott models in which the points within a cluster are conditionally independent, yielding again a clustered point process.

The class of Cox models is extremely large; the random intensity function $\Lambda(u)$ may be taken to be any random spatial process. No general technique for fitting all Cox models has been established, although the E–M algorithm (treating the unobserved Λ as the missing data) would be useful in many cases.

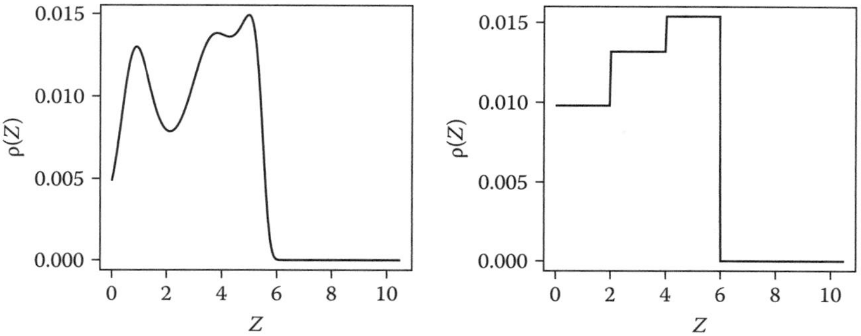

FIGURE 20.13
Fitted models for intensity of the Queensland copper point pattern as a function of distance from the nearest lineament. Left: log-polynomial of degree 7; right: piecewise constant with jumps at 2, 4 and 6 km.

In a *log-Gaussian Cox process*, (Section 19.4.3) we assume that $\log \Lambda(u)$ is a Gaussian random field $\Xi(u)$. The log likelihood can be evaluated using conditional simulation and maximized numerically. Spatial inhomogeneity in the intensity can be accommodated by introducing additive model terms in the mean function of $\Xi(u)$, which give rise to corresponding additive effects in the log intensity.

Neyman–Scott cluster models (Section 17.3) can also be used to describe clustering, and can be simulated directly. However, their likelihoods are relatively intractable, as they are effectively mixture models with a Poisson random number of components. Instead, cluster models would often be fitted using the method of minimum contrast (Section 19.3.2), using the fact that the second-moment characteristics of the process depend only on the second-moment properties of the cluster mechanism (Section 17.2.3). For example, the modified Thomas process (Stoyan, Kendall, and Mecke, 1987, p. 144), is a Poisson cluster process in which each parent has a Poisson (μ) random number of offspring, displaced from the parent location by independent random vectors with an isotropic Gaussian distribution. The modified Thomas process has K-function

$$K(r) = \pi r^2 + \frac{1}{\kappa}\left(1 - \exp\left(\frac{-r^2}{4\sigma^2}\right)\right) \tag{20.2}$$

and intensity $\lambda = \mu\kappa$, where κ is the parent intensity and σ the standard deviation of the offspring displacements. The parameters κ and σ can be estimated by fitting this parametric form to the empirical K-function using minimum contrast, and the remaining parameter μ follows.

Figure 20.14 shows a subset of the redwood seedlings data of Strauss (1975) extracted by Ripley (1976) and the result of fitting the modified Thomas process K-function to the empirical K-function of this pattern. The fitted parameters were $\hat{\kappa} = 23.5$, $\hat{\mu} = 2.6$ and $\hat{\sigma} = 0.05$.

Inhomogeneous cluster models can be used to model both spatial inhomogeneity and clustering. Consider a Poisson cluster process in which the parent points are Poisson with constant intensity κ, and for a parent point at location y_i, the associated cluster is a finite Poisson process with intensity $\lambda_i(u) = g(u - y_i)\rho(u)$ where $\kappa(u)$ is a spatially varying "fertility" or "base intensity" and g is a nonnegative integrable function. This can be interpreted as another multiplicative model in which the rate of production of offspring depends on the intensity κ of parents, the cluster dispersal $g(u - v)$ from parent v to offspring u, and the intrinsic fertility $\rho(u)$ at location u. If g is the isotropic Gaussian density and ρ is constant, the model reduces to the stationary modified Thomas process.

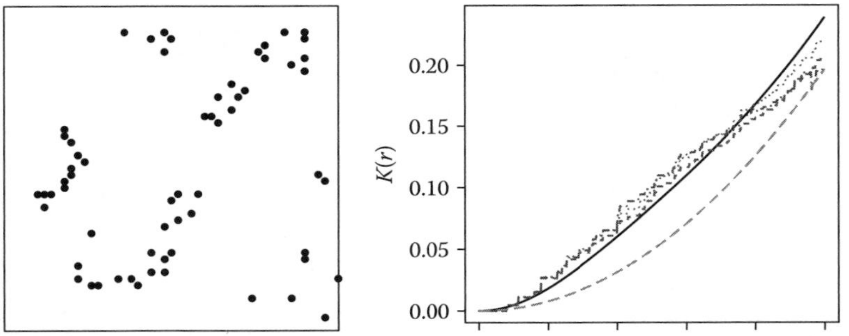

FIGURE 20.14
Redwood seedlings data rescaled to the unit square (left) and result of fitting a modified Thomas process model to the K-function (right).

The form of $\lambda_i(u)$ is judiciously chosen so that the intensity of the process of offspring is proportional to $\rho(u)$, and the pair correlation function is stationary and determined by g. Thus, the intensity and clustering parameters are "separable," making it easy to formulate and fit models that combine spatial inhomogeneity and clustering. The intensity may be modeled and fitted using any of the techniques described in the previous section. The fitted intensity is then used to compute the inhomogeneous K-function. A cluster dispersal function g is chosen and the corresponding functional form of the inhomogeneous K-function is derived. The parameters of g are then estimated by minimum contrast.

20.4.3 Finite Gibbs Models

In order to describe *negative* association ("regularity" or "inhibition") between points, the most useful models are finite Gibbs point processes (Section 17.4, Section 19.4.1) particularly Markov point processes. They also may be used to model weak positive association between points.

Recalling Section 17.4, suppose that data are observed in a study region W. A finite Gibbs process in W has a probability density $f(\mathbf{x})$ of the form

$$f(\mathbf{x}) = \exp(-\Psi(\mathbf{x})), \tag{20.3}$$

with respect to the Poisson process of unit intensity on W, where the *potential* Ψ is of the form

$$\Psi(x_1, \ldots, x_n) = \alpha_0 + \sum_i \alpha_1(x_i) + \sum_{i<j} \alpha_2(x_i, x_j) + \ldots + \alpha_n(x_1, \ldots, x_n). \tag{20.4}$$

The constant α_0 is the normalizing constant for the density, and the *interaction potentials* $\alpha_1, \alpha_2, \ldots$ are functions determining the contribution to the potential from each d-tuple of points.

The *Gibbs representation* (20.3)–(20.4) of the probability density f as a product of terms of increasing order, is a natural way to understand and model stochastic dependence between points. If the potential functions α_d are identically zero for $d \geq 2$, then the point process is Poisson with intensity $\lambda(u) = \exp(-\alpha_1(u))$. Thus, the first-order potential α_1 can be regarded as controlling *spatial trend* and the higher-order potentials α_d for $d \geq 2$ control *interpoint interaction.*

In principle, we may formulate models with any desired form of spatial trend and stochastic dependence structure simply by writing down the interaction potentials $\alpha_1, \alpha_2, \ldots$. One constraint is that the resulting probability density f must be integrable. This is guaranteed if, for example, all the higher-order potentials α_d for $d \geq 2$ are nonnegative (implying that the process exhibits inhibition or negative association between points).

The most common models are *pairwise interaction* processes (Section 17.4) where the pair potential depends only on interpoint distance, $\alpha_2(u, v) = \varphi(||u - v||)$. An example is the *Strauss process* with pairwise interaction $\varphi(t) = \theta$ if $t \leq r$ and zero otherwise, where $r > 0$ is an "interaction distance" and θ is an "interaction strength" parameter. The Strauss process density is integrable when $\theta \geq 0$, implying that it exhibits inhibition or negative association between points. The parameter θ controls the "strength" of negative association. If $\theta = 0$, the model reduces to a Poisson process, while, if $\theta = \infty$, the model is a *hard core process,* which assigns zero probability to any outcome where there is a pair of points lying closer than the critical distance r. Figure 20.15 shows simulated realizations of the Strauss process with the same first-order potential and two different values of θ.

Models with higher-order interaction can easily be constructed. For example, there is an obvious extension of pairwise interaction to an interaction between triples of points. Several

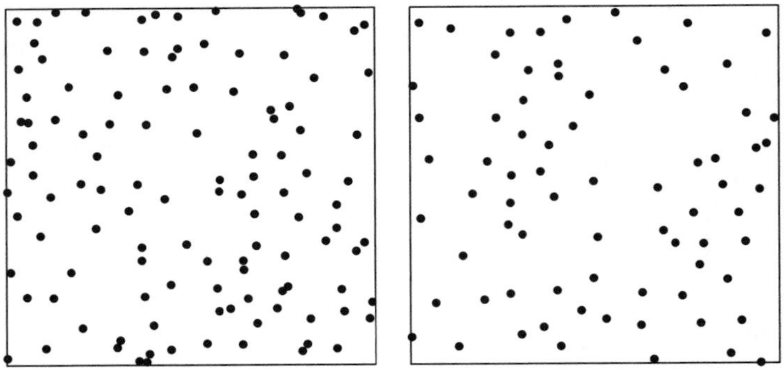

FIGURE 20.15
Simulated realizations of the Strauss process with pairwise interaction parameter $\theta = 1$ (left) and $\theta = 3$ (right). Unit square window, interaction radius $r = 0.07$, first order potential $\alpha_1 = \log 200$.

models with infinite order of interaction (that is, with nontrivial interaction potentials of every order d) are available, including the Widom–Rowlinson penetrable spheres model or "area-interaction" process (Baddeley and van Lieshout, 1995; Widom and Rowlinson, 1970) and the saturation model of Geyer (1999).

A useful tool for modeling and analysis of Gibbs processes is the *Papangelou conditional intensity* (Definition 7 of Section 16.5). The conditional intensity at a location u given the point pattern $\mathbf{x}$ is (for $u \notin \mathbf{x}$)

$$\lambda(u \mid \mathbf{x}) = \frac{f(\mathbf{x} \cup \{u\})}{f(\mathbf{x})}$$

by Theorem 5 of Section 16.6. The negative log of the conditional intensity can be interpreted as the energy required to add a new point, at location u, to the existing configuration $\mathbf{x}$. This energy can be decomposed according to the Gibbs representation into

$$-\log \lambda(u \mid \mathbf{x}) = \alpha_1(u) + T_2(u \mid \mathbf{x}) + T_3(u \mid \mathbf{x}) + \cdots, \tag{20.5}$$

where

$$T_2(u \mid \mathbf{x}) = \sum_{i=1}^{n} \alpha_2(u, x_i)$$

is the sum of the pair potentials between a new point at location u and each existing point x_i, and similarly for higher-order potentials. Thus, $\alpha_1(u)$ is the energy required to "create" a point at the location u, while $T_2(u \mid \mathbf{x})$ is the energy required to overcome pairwise interaction forces between the new point u and the existing points of $\mathbf{x}$, and so on. This can be a useful metaphor for modeling point patterns in applications such as forestry.

In Equation (20.5), the first term $\alpha_1(u)$ may be interpreted as controlling spatial trend, while the subsequent terms control interpoint interaction. Thus, we may construct point process models by choosing appropriate functional forms for α_1 to control spatial trend and $T_2, T_3, \ldots$ to control interpoint interaction. This can be regarded as a generalization of the modeling approach espoused in Section 20.4.1. The log intensity has been generalized to the log conditional intensity. Additive terms in the log conditional intensity that depend only on the location u are interpreted as multiplicative effects on spatial trend. Additive terms in the log conditional intensity that depend on the configuration $\mathbf{x}$ modify the interpoint interaction.

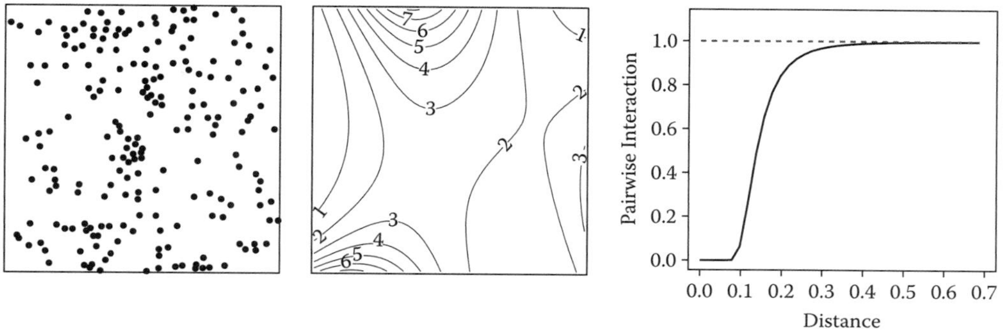

FIGURE 20.16
Japanese pines data (left), fitted log-cubic spatial trend (middle) and fitted, soft core, pairwise interaction $\exp \varphi(t)$ (right).

Gibbs models may be fitted to data by maximizing the point process *pseudolikelihood* (Section 19.3.4). Suppose the conditional intensity is $\lambda_\theta(u \mid \mathbf{x})$, where θ is a parameter. Define

$$\log \mathrm{PL}(\theta) = \sum_{i=1}^{n} \log \lambda_\theta(x_i \mid \mathbf{x}) - \int_W \lambda_\theta(u \mid \mathbf{x})\, \mathrm{d}u \tag{20.6}$$

and define the *maximum pseudolikelihood estimate* (MPLE) $\hat{\theta}$ as the value maximizing $\log \mathrm{PL}(\theta)$. Under regularity conditions, the pseudoscore (derivative of $\log \mathrm{PL}$) is an unbiased estimating function. The MPLE is typically a consistent estimator although it is inefficient compared to the maximum likelihood estimate (MLE). Numerical methods for maximizing the pseudolikelihood are very similar to those used for maximizing the Poisson likelihood (20.1): a discretized version of the log pseudolikelihood is formally equivalent to the log likelihood of a (nonspatial) Poisson generalized linear model (Baddeley and Turner, 2000).

Figure 20.16 shows the Japanese pine saplings data of Numata (1961, 1964) and the result of fitting a pairwise interaction model with log-cubic spatial trend (α_1 was a polynomial of degree 3 in the Cartesian coordinates) and soft core interaction $\varphi(t) = \theta t^{1/4}$ where θ is a parameter.

One point of confusion is the difference between the first-order potential α_1 (called the *spatial trend*) and the *intensity* of the point process. In the special case of a Poisson process, the first-order potential α_1 determines the intensity directly, through $\lambda(u) = \exp(-\alpha_1(u))$. However, for non-Poisson processes, this is not true; the intensity depends on the potentials of all orders. There is typically no simple expression for the intensity in terms of the potentials. For simplicity, assume that the higher-order potentials α_d for $d \geq 2$ are invariant under translation. Then, any spatial inhomogeneity in the point process must arise from the first-order potential α_1, so it is appropriate to call this the "spatial trend." One useful result is

$$\lambda(u) = \mathbb{E}[\lambda(u \mid X)], \tag{20.7}$$

a special case of the *Georgii–Nguyen–Zessin* formula (Example 17 of Section 16.5). The expectation on the right-hand side of Equation (20.7) is with respect to the distribution of the Gibbs process. Thus, Equation (20.7) expresses an equivalence between two properties of the same Gibbs process. For example, for a pairwise interaction process, (20.7) gives

$$\lambda(u) = \mathbb{E}[\exp(-\alpha_1(u) - T_2(u \mid \mathbf{X}))]. \tag{20.8}$$

Typically, T_2 is nonnegative, which implies $\lambda(u) \leq \exp(-\alpha_1(u))$. The intensity is a decreasing function of the pair potential. For example, Figure 20.15 showed realizations of two Strauss processes with the same first-order potential, but with different interaction parameters $\theta = 1$ and $\theta = 3$. These patterns clearly do not have equal intensity.

A quick approximation to the intensity is obtained by replacing the right-hand side of Equation (20.7) by the expectation of $\lambda(u \mid \mathbf{x})$ under a Poisson process with the same (unknown) intensity function $\lambda(u)$, and then solving for the intensity. For pairwise interaction processes with α_1 constant and $\alpha_2(u, v) = \varphi(\|u - v\|)$, this produces an analytic equation

$$\lambda = \exp(-\alpha_1(u) - G\lambda) \tag{20.9}$$

to be solved for the constant intensity λ, where

$$G = \int_{\mathbb{R}^2} [1 - \exp(-\varphi(\|u\|))] \, du.$$

For example, the Strauss process has $G = \pi r^2(1 - \exp\theta)$. With the parameters used in Figure 20.15, we obtain approximate intensities $\lambda_1 \approx 95.7$ and $\lambda_3 \approx 73.9$ for the Strauss processes with $\theta = 1$ and $\theta = 3$, respectively. These are slight overestimates; unbiased estimates obtained using perfect simulation are $\lambda_1 \approx 92.7$ (standard error 0.2) and $\lambda_3 \approx 65.7$ (standard error 0.5).

Similarly, the second moment properties of a Gibbs process are not easily accessible functions of the interaction potentials. The K-function can be estimated by simulation or using analytic approximation to a sparse Poisson process.

In principle, as we have seen, any point process that has a probability density can be expressed as a Gibbs model. In practice, Gibbs models are only convenient to apply to data when the potentials α_d are easy to compute, or at least, when the right-hand side of Equation (20.5) is easy to compute. This includes most pairwise interaction models, and some infinite order interaction models such as the Widom–Rowlinson model.

Finally, we note that there is a natural connection between Gibbs models and the classical summary statistics, such as the K-function. Suppose the probability density $f(\mathbf{x})$ depends on the parameter vector θ in the linear exponential form $f(\mathbf{x}) = \exp(\theta B(\mathbf{x}) + A(\theta))$. Then the "canonical sufficient statistic" of the model is the statistic $B(\mathbf{x})$. The maximum likelihood estimate of θ is a function of $B(\mathbf{x})$. It turns out that, ignoring edge corrections, the empirical K-function is the canonical sufficient statistic for the Strauss process (when the interaction distance r is treated as a nuisance parameter). Similarly, the empirical empty space function F is the canonical sufficient statistic for the Widom–Rowlinson penetrable spheres model, and the empirical nearest neighbor distance function G is the canonical sufficient statistic for the Geyer saturation model with saturation parameter $s = 1$.

20.5 Formal Inference

Formal statistical inference about a spatial point pattern includes parameter estimation, hypothesis tests about the value of a parameter in a point process model, formal model selection (frequentist or Bayesian), goodness-of-fit tests for a model that has been fitted to data, and tests of goodness-of-fit to the homogeneous Poisson process (complete spatial randomness). Inference typically requires the use of simulation methods, because there are very few theoretical results available about the null distribution of test statistics, except in the case of Poisson processes.

20.5.1 Generic Goodness-of-Fit Tests

Following the influential paper of Ripley (1977), the exploratory analysis of a point pattern is often accompanied by a Monte Carlo test of goodness-of-fit to the homogeneous Poisson process. The test is usually performed by plotting the estimate of the K-function (or another summary function) together with the envelopes of the K-function estimated from simulated realizations of a Poisson process of the same estimated intensity. See Section 18.3.

A general weakness of goodness-of-fit tests is that the alternative hypothesis is very broad (embracing all point processes other than the model specified in the null hypothesis), so that rejection of the null hypothesis is rather uninformative, and acceptance of the null hypothesis is unconvincing because of weak power against specific alternatives.

Another criticism of this approach is that the homogeneous Poisson process may not be appropriate (even as a null model) in the scientific context. For example, in spatial epidemiology (Chapter 22), a point pattern of cases of infectious disease should be studied relative to the spatially varying density of the susceptible population. It would be natural to test the goodness-of-fit of the *inhomogeneous* Poisson process with intensity proportional to the population density. The Monte Carlo testing technique can easily be adapted to this model.

The Japanese pines data (Figure 20.16) show clear evidence of spatial inhomogeneity. A goodness-of-fit test for CSR would not be very informative. Instead we may fit an inhomogeneous Poisson process to the data and subject this to goodness-of-fit testing. The left panel of Figure 20.17 shows a Monte Carlo test of goodness-of-fit of an inhomogeneous Poisson process with log-cubic intensity using the inhomogeneous K-function based on the fitted intensity. There is a suggestion of inhibition between points at short distances.

Another use of goodness-of-fit testing is in the final stages of modeling. After a suitable model has been developed and fitted to the data, the adequacy of the model can be confirmed by a goodness-of-fit test (assuming that the null hypothesis is accepted). The right panel of Figure 20.17 shows a Monte Carlo test of goodness-of-fit for the Gibbs model with log-cubic trend and soft core pairwise interaction, fitted to the Japanese pines data using the inhomogeneous K function.

20.5.2 Formal Inference about Model Parameters

For Poisson point process models (either homogeneous or inhomogeneous) the full technology of likelihood inference is available. In the common case of a log-linear model

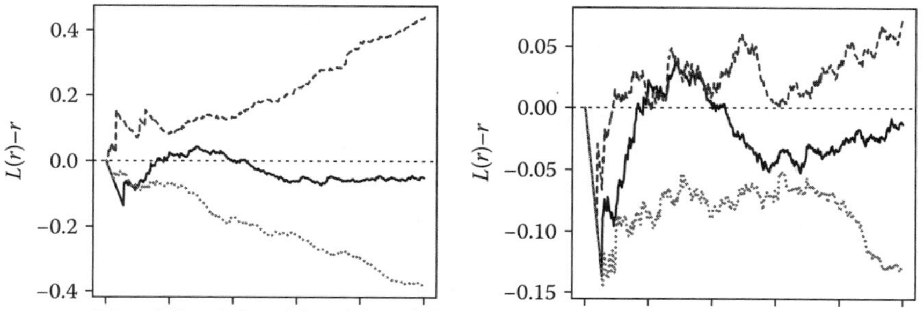

FIGURE 20.17
Monte Carlo test of goodness-of-fit for models fitted to the Japanese pines data. Left: Poisson process with log-cubic intensity. Right: Gibbs process with log-cubic spatial trend and soft core pairwise interaction. Pointwise envelopes of 19 simulations.

$\lambda_\theta(u) = \exp(\theta^{\mathrm{T}} S(u))$ where $S(u)$ is a real- or vector-valued covariate, the maximum likelihood estimator $\hat{\theta}$ is asymptotically normal with mean θ and variance–covariance matrix $I^{-1}(\theta)$, where

$$I(\theta) = \int S(u)^{\mathrm{T}} S(u) \lambda_\theta(u)\, du$$

is the Fisher information matrix. Likelihood ratio tests and confidence intervals for parameters, based on the normal distribution, are asymptotically valid. Model selection may be conducted using the standard analysis of deviance, the Akaike information criterion, or Bayesian methods.

For point process models other than the Poisson process, very little exact statistical theory is available. In Gibbs models fitted by maximum likelihood (typically using simulation to evaluate the score function), typically the MLE is still asymptotically normal, and the Fisher information can be estimated by simulation. This enables the likelihood ratio test and confidence intervals to be applied.

For large datasets or rapid modeling applications, maximum likelihood is computationally prohibitive. Instead, inference may be based on maximum pseudolikelihood. The null distribution of the log *pseudolikelihood* ratio would often be estimated by Monte Carlo simulation. However, the tails of the null distribution are poorly estimated by direct simulation. Baddeley and Turner (2000) proposed approximating the tails of the null distribution of the log *pseudolikelihood* ratio by the tails of a gamma distribution fitted to a modest number of simulations. General theory of composite likelihood suggests that the asymptotic null distribution of the log pseudolikelihood ratio is that of a weighted sum of χ^2 variables, which can then be approximated by a single gamma variable.

20.6 Model Validation

In model validation, we investigate the correctness of each component (or "assumption") of the fitted model.

Unlike "formal" model checking (such as goodness-of-fit testing), which rests on mathematical assumptions and has a well-defined mathematical interpretation, validation is an "informal" process, in which the mathematical assumptions are checked. The goal of validation is to assess whether any component of the fitted model seems to be inappropriate and, if so, to suggest the form of a more appropriate model.

A typical validation tool in classical statistics is a plot of the residuals from linear regression, plotted against the explanatory variable x. Any visual impression of a pattern, suggesting that the residuals depend on x in a systematic way, suggests that the form of the regression curve should be reconsidered. Various plots and transformations of the residuals may be useful for different purposes. The assumption that errors are normally distributed can be assessed using a normal Q–Q plot of the residuals. Model validation for spatial point processes involves analogs of these classical residual plots (Baddeley, Turner, Møller, and Hazelton, 2005). Outliers in spatial point patterns may also be studied, by adapting classical outlier techniques (Wartenberg, 1990) or by localized spatial analysis (Cressie and Collins, 2001a).

20.6.1 Intensity Residuals

A fitted model for the point process intensity $\lambda(u)$ may be checked by comparing observed and expected numbers of points using analogs of the classical residuals.

This is familiar in the context of quadrat counting. We divide the study region W into disjoint subsets or "quadrats" B_i and count the number $n_i = n(\mathbf{x} \cap B)$ of data points falling in each B_i. If $\hat{\lambda}(u)$ is the fitted intensity, then the expected value of n_i is

$$e_i = \int_{B_i} \hat{\lambda}(u)\,\mathrm{d}u$$

and the "raw residual" is

$$r_i = n_i - e_i.$$

The residuals have mean zero if the fitted intensity is exactly correct. If the fitted model is Poisson, we may also consider the "Pearson residual"

$$r_i^P = \frac{n_i - e_i}{\sqrt{e_i}}.$$

The residuals may be plotted spatially as a grayscale image and inspected for any suggestion of spatial trend.

This is a special case of the *residual intensity* R, a signed measure defined for all regions B by

$$R(B) = n(\mathbf{x} \cap B) - \int_B \hat{\lambda}(u)\,\mathrm{d}u.$$

This is a signed measure with atoms of mass 1 at the data points and a diffuse component with negative density $-\hat{\lambda}(u)$ at other locations. Again $R(B)$ has mean zero if the fitted intensity is exactly correct.

In order to visualize the residual intensity, it is useful to apply kernel smoothing. This yields a *smoothed residual intensity*

$$\begin{aligned} s(u) &= e(u) \int k(u - v)\,\mathrm{d}R(u) \\ &= e(u) \sum k(u - x_i) - e(u) \int k(u - v)\hat{\lambda}(v)\,\mathrm{d}v \\ &= \widetilde{\lambda}(u) - \lambda(u) \end{aligned} \tag{20.10}$$

equal to the difference between a standard, nonparametric, kernel-smoothed intensity estimate $\widetilde{\lambda}(u)$ and a correspondingly kernel-smoothed version of the fitted model intensity $\lambda(u)$. A contour plot or image plot of $s(u)$ may be inspected for any suggestion of spatial trend.

The left panel of Figure 20.18 shows a contour plot of the smoothed residual intensity $s(u)$ for a model with log-linear intensity (the log intensity was a linear function of the Cartesian coordinates) fitted to the Japanese pines data. The saddle shape of the surface suggests that this model is inappropriate.

Various other plots and transformations of the residual intensity may be useful. For intensity models that depend (or *should have depended*) on a spatial covariate $Z(u)$, an alternative is to plot the *cumulative residual* $C(z) = R(B(z))$ against z, where

$$B(z) = \{u \in W : \; Z(u) \leq z\} \tag{20.11}$$

is the subset of the study region where the covariate takes a value less than or equal to z. Systematic departure from the horizontal suggests that the form of dependence on Z should be modified. The right panel of Figure 20.18 plots the cumulative residual for the log-linear intensity model of Japanese pines, as a function of the x coordinate. Dotted lines show the two-standard-error limits assuming a Poisson process. The obvious pattern of

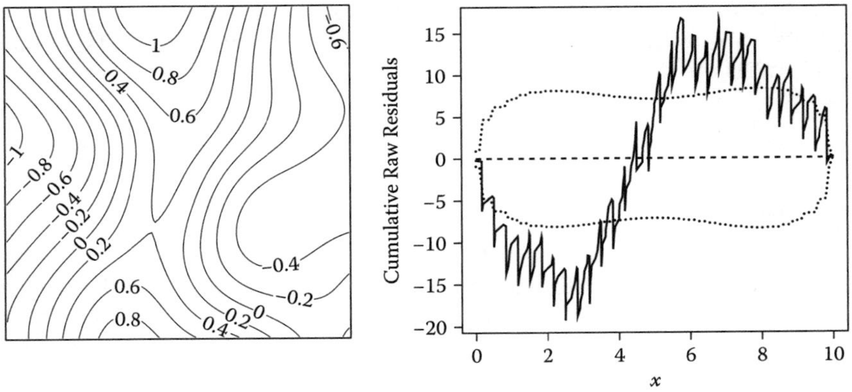

FIGURE 20.18
Residual plots for an intensity model fitted to the Japanese pines data. Left: Contour plot of smoothed residual intensity. Right: Cumulative residual as a function of the x coordinate.

deviation from zero shows that a linear log intensity model is not appropriate, and suggests a polynomial of higher order.

In the special case of homogeneous intensity, the cumulative residual plot is the difference between the two curves plotted in Berman's diagnostic plot (Section 20.3.2).

20.6.2 Validation of Poisson Models

Validation of a fitted Poisson point process model involves checking the assumption of stochastic independence between points as well as the fitted intensity. The assumption of stochastic independence also opens up more techniques for checking the fitted intensity function.

If the fitted intensity model has been judged adequate (perhaps using residual plots), then a useful diagnostic for the Poisson assumption is the inhomogeneous K-function.

An alternative technique for assessing both the intensity model and the independence assumption is to transform the process to uniformity. Suppose X is a Poisson process with intensity function $\lambda(u)$, and $Z(u)$ is a real-valued covariate function. Then the values $z_i = Z(x_i)$ of the covariate observed at the data points constitute a Poisson point process on the real line. The expected number of values z_i satisfying $z_i \leq z$ is

$$\tau(z) = \int_{B(z)} \lambda(u)\,\mathrm{d}u,$$

where again $B(u)$ is the level set region where $Z(u)$ takes values less than or equal to z. Assume τ is a continuous function. Then the values $t_i = \tau(z_i) = \tau(Z(x_i))$ constitute a Poisson process with *unit* intensity on an interval of the real line.

To exploit this in practice, we choose a covariate function $Z(u)$ judiciously, compute the function τ numerically by discretizing the study region, evaluate $z_i = Z(x_i)$, transform to $t_i = \tau(z_i)$, and apply standard goodness-of-fit tests for the uniform Poisson process to the values t_i. A good choice of Z is one that has a strong influence on the intensity, under the types of departure from the fitted model that are of interest. Thus, Z might be one of the covariates already included in the fitted model (where the form of the model is under suspicion), or a covariate that should have been included in the model, or a surrogate for a lurking variable.

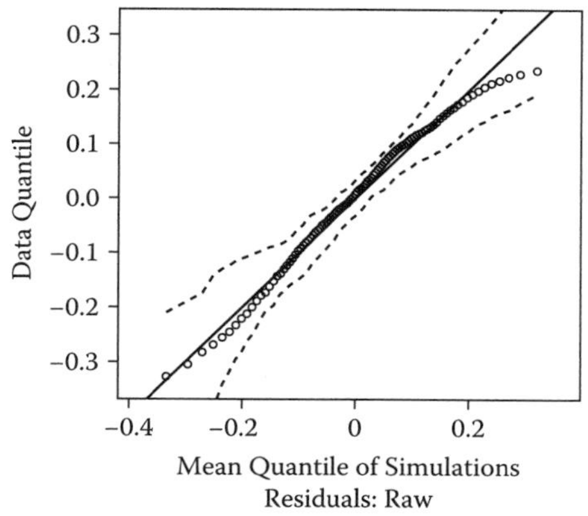

FIGURE 20.19
Q–Q plot of smoothed residuals for a Poisson model with log-cubic intensity fitted to the Japanese pines data.

This is closely related to the Kolmogorov–Smirnov test of goodness-of-fit to a probability distribution. In this context, the Kolmogorov–Smirnov test is a goodness-of-fit test for the fitted intensity function. Transformation of the point process to uniformity allows a wider range of tests to be applied, including tests of stochastic dependence.

The independence property of the Poisson process is a distributional assumption, analogous to the assumption of normally distributed errors in linear regression. To validate a fitted linear regression, we would often check the assumption of normally distributed errors using a normal Q–Q plot of the residuals. This plots the empirical quantiles of the residuals against the corresponding theoretical quantiles of the normal distribution. Baddeley et al. (2005) proposed checking the independence assumption of the Poisson process using a Q–Q plot of the smoothed residual field. Quantiles of (discretized values of) the smoothed residual field for the data are plotted against the corresponding expected quantiles estimated by simulation from the fitted Poisson model. Figure 20.19 shows the residual Q–Q plot for an inhomogeneous Poisson model fitted to the Japanese pines data, suggesting the Poisson assumption is adequate. In special cases, the residual Q–Q plot is related to summary statistics, such as the empty space function F.

20.6.3 Validation of Multilevel Models

Multilevel models, such as log-Gaussian Cox processes and inhomogeneous cluster models (Section 20.4.2), are typically fitted by the method of moments or the closely related method of minimum contrast, at least with regard to the parameters controlling interpoint interaction. The fitted model can be validated in two ways.

First, the sample moments can be compared with the fitted moments, to assess whether the choice of functional form for the model is appropriate. For example, Figure 20.14 in Section 20.4.2 showed a graphical comparison between the empirical and fitted inhomogeneous K-functions of the redwood data, suggesting that a modified Thomas model is appropriate.

Second, a summary statistic that is unrelated to the model-fitting technique can be evaluated for the data and compared with its predicted mean under the model, typically computed by simulation. This can be used to assess whether the *type* of stochastic dependence in the model is appropriate to the data. Figure 20.20 shows the nearest neighbor distance function $G(r)$ estimated from the redwood seedlings data, and the (simultaneous) envelope of 19 simulations from the fitted modified Thomas process. The sample size appears

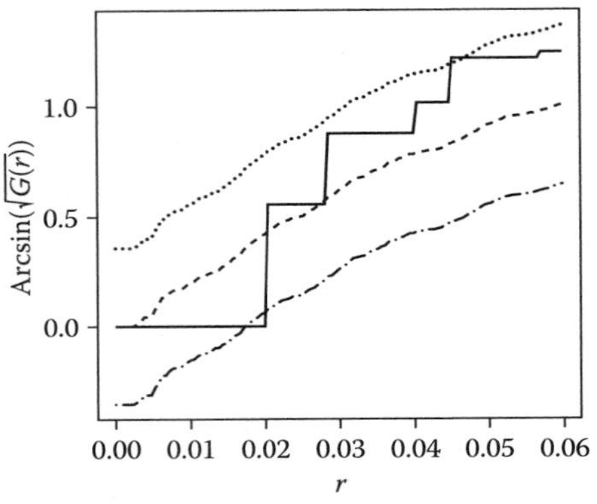

FIGURE 20.20
Monte Carlo test of goodness-of-fit for the modified Thomas process model fitted to the redwood seedlings data, based on the nearest neighbor distance function $G(r)$: variance stabilized, simultaneous envelopes, 19 simulations.

to be too small for definitive conclusions, but there is a slight suggestion that the cluster model may be inappropriate.

20.6.4 Validating Gibbs Models

Baddeley et al. (2005) proposed a generalization of residuals to the case of Gibbs models. The residual measure is defined by

$$R(B) = n(\mathbf{x} \cap B) - \int_B \hat{\lambda}(u \mid \mathbf{x})\, du \tag{20.12}$$

for any region B, where $\hat{\lambda}(u \mid \mathbf{x})$ is the fitted Papangelou conditional intensity. Thanks to the Georgii–Nguyen–Zessin formula, these generalized residuals also have mean zero if the fitted conditional intensity is exactly correct. The residuals are often easy to compute because the conditional intensity is typically very easy to compute, at least for Gibbs models in common use.

Useful diagnostic plots of the residuals include a contour plot of the smoothed residual field, which has the same form as (20.10), but which is now the difference between non-parametric and parametric estimates of the *conditional* intensity. To assess dependence on a covariate $Z(u)$ we may again plot the cumulative conditional residuals $C(z) = R(B(z))$ against the covariate value z, where $B(z)$ was defined in Equation (20.11). The stochastic dependence in the fitted model may be assessed using a Q–Q plot of the smoothed residuals.

20.7 Software

20.7.1 Packages for Spatial Statistics in R

R is a popular, open source, software environment for statistical computing and graphics (R Development Core Team, 2004). "Packages" or add-on modules for R are available from the CRAN Web site network cran.r-project.org. Packages for spatial statistics are reviewed on the Spatial task view Web page www.cran.r-project.org/web/views/Spatial.html.

All the computations for this chapter were performed using the package `spatstat`. It is designed to support a complete statistical analysis of spatial point pattern data, including graphics, exploratory analysis, model-fitting, simulation and diagnostics (Baddeley and Turner, 2005). Detailed notes on analysing spatial point patterns using `spatstat` are available on the Web (Baddeley, 2008).

The package `splancs` supports exploratory analysis of spatial and space-time point patterns (Bivand and Gephardt, 2000; Rowlingson and Diggle, 1993). The package `spatclus` extends `spatstat` by adding methods for spatial cluster detection (Demattei, Molinari, and Davrès, 2007). The package `MarkedPointProcess` provides methods for the analysis of multidimensional marks attached to points in a spatial pattern using techniques of Schlather, Ribeiro, and Diggle (2004).

20.7.2 Other Packages

Several software packages for analyzing spatial point patterns are freely available but not open-source. Important among these is (`SaTScan`™), available at www.satscan.org, which implements methods based on spatial scan statistics (Alm, 1988; Anderson and Titterington, 1997; Kulldorff, 1999; Kulldorff and Nagarwalla, 1995) that are not commonly implemented in other packages.

References

S.E. Alm. Approximation and simulation of the distributions of scan statistics for Poisson processes in higher dimensions. *Extremes*, 1(1):111–126, 1988.

N.H. Anderson and D.M. Titterington. Some methods for investigating spatial clustering, with epidemiological applications. *Journal of the Royal Statistical Society A*, 160(1):87–105, 1997.

L. Anselin. Local indicators of spatial association–LISA. *Geographical Analysis*, 27:93–115, 1995.

A. Baddeley. Analysing spatial point patterns in R. Workshop notes, CSIRO, 2008. http://www.csiro.au/resources/pf16h.html

A. Baddeley, J. Møller, and R. Waagepetersen. Non- and semiparametric estimation of interaction in inhomogeneous point patterns. *Statistica Neerlandica*, 54(3):329–350, November 2000.

A. Baddeley and R. Turner. Practical maximum pseudolikelihood for spatial point patterns (with discussion). *Australian and New Zealand Journal of Statistics*, 42(3):283–322, 2000.

A. Baddeley and R. Turner. Spatstat: An R package for analyzing spatial point patterns. *Journal of Statistical Software*, 12(6):1–42, 2005.

A. Baddeley, R. Turner, J. Møller, and M. Hazelton. Residual analysis for spatial point processes (with discussion). *Journal of the Royal Statistical Society, series B*, 67(5):617–666, 2005.

A.J. Baddeley and B.W. Silverman. A cautionary example on the use of second-order methods for analyzing point patterns. *Biometrics*, 40:1089–1094, 1984.

A.J. Baddeley and M.N.M. van Lieshout. Area-interaction point processes. *Annals of the Institute of Statistical Mathematics*, 47:601–619, 1995.

A. Baillo, A. Cuevas, and A. Justel. Set estimation and nonparametric detection. *Canadian Journal of Statistics*, 28:765–782, 2008.

M.S. Bartlett. A note on spatial pattern. *Biometrics*, 20:891–892, 1964.

M. Berman. Testing for spatial association between a point process and another stochastic process. *Applied Statistics*, 35:54–62, 1986.

M. Berman and T.R. Turner. Approximating point process likelihoods with GLIM. *Applied Statistics*, 41:31–38, 1992.

J.E. Besag. Discussion contribution to Ripley (1977). *Journal of the Royal Statistical Society, series B*, 39: 193–195, 1977.

R. Bivand and A. Gebhardt. Implementing functions for spatial statistical analysis using the R language. *Journal of Geographical Systems*, 2:307–317, 2000.

G. Borgefors. Distance transformations in digital images. *Computer Vision, Graphics and Image Processing*, 34:344–371, 1986.

D.R. Brillinger. Comparative aspects of the study of ordinary time series and of point processes. In P.R. Krishnaiah, ed., *Developments in Statistics*, pp. 33–133. Academic Press, San Diego, 1978.

D.R. Brillinger. Time series, point processes, and hybrids. *Canadian Journal of Statistics*, 22:177–206, 1994.

D.R. Brillinger and H.K. Preisler. Two examples of quantal data analysis: a) multivariate point process, b) pure death process in an experimental design. In *Proceedings, XIII International Biometric Conference, Seattle*, pp. 94–113. International Biometric Society, 1986.

D.R. Cox and E.J. Snell. *Applied Statistics: Principles and Examples*. Chapman & Hall, London, 1981.

N. Cressie and L.B. Collins. Analysis of spatial point patterns using bundles of product density LISA functions. *Journal of Agricultural, Biological and Environmental Statistics*, 6:118–135, 2001b.

N. Cressie and L.B. Collins. Patterns in spatial point locations: local indicators of spatial association in a minefield with clutter. *Naval Research Logistics*, 48:333–347, 2001a.

C. Demattei, Molinari N., and J.P. Daurès. Arbitrarily shaped multiple spatial cluster detection for case event data. *Computational Statistics and Data Analysis*, 51:3931–3945, 2007.

P. Diggle, S. Morris, P. Elliott, and G. Shaddick. Regression modeling of disease risk in relation to point sources. *Journal of the Royal Statistical Society, series A*, 160:491–505, 1997.

M.J. Drinkwater, Q.A. Parker, D. Proust, E. Slezak, and H. Quintana. The large-scale distribution of galaxies in the Shapley supercluster. *Publications of the Astronomical Society of Australia*, 21(1): 89–96, February 2004. doi:10.1071/AS03057.

P. Elliott, J. Wakefield, N. Best, and D. Briggs, eds. *Disease and Exposure Mapping*. Oxford University Press, Oxford, 1999.

P. Elliott, J. Wakefield, N. Best, and D. Briggs, eds. *Spatial Epidemiology: Methods and Applications*. Oxford University Press, Oxford, 2000.

N. Fry. Random point distributions and strain measurement in rocks. *Tectonophysics*, 60:89–105, 1979.

A. Getis and J. Franklin. Second-order neighbourhood analysis of mapped point patterns. *Ecology*, 68:473–477, 1987.

C.J. Geyer. Likelihood inference for spatial point processes. In O.E. Barndorff-Nielsen, W.S. Kendall, and M.N.M. van Lieshout, Eds., *Stochastic Geometry: Likelihood and Computation*, number 80 in Monographs on Statistics and Applied Probability, Chap. 3, pp. 79–140. Chapman & Hall, Boca Raton, FL, 1999.

S.S. Hanna and N. Fry. A comparison of methods of strain determination in rocks from southwest Dyfed (Pembrokeshire) and adjacent areas. *Journal of Structural Geology*, 1:155–162, 1979.

D. Hoaglin, F. Mosteller, and J. Tukey. *Understanding Robust and Exploratory Data Analysis*. John Wiley & Sons, New York, 1983.

M. Kulldorff. Spatial scan statistics: Models, calculations, and applications. In J. Glaz and N. Balakrishnan, eds., *Recent Advances on Scan Statistics*, pp. 303–322. Birkhauser, Boston, 1999.

M. Kulldorff and N. Nagarwalla. Spatial disease clusters: Detection and inference. *Statistics in Medicine*, 14:799–810, 1995.

E. Mammen and A.B. Tsybakov. Asymptotical minimax recovery of sets with smooth boundaries. *Annals of Statistics*, 23:502–524, 1995.

P. McCullagh. Personal communication, (2002).

M. Moore. On the estimation of a convex set. *Annals of Statistics*, 12:1090–1099, 1984.

M. Numata. Forest vegetation in the vicinity of Choshi — coastal flora and vegetation at Choshi, Chiba prefecture, IV (in Japanese). *Bulletin of the Choshi Marine Laboratory*, (3):28–48, 1961. Chiba University, Japan.

M. Numata. Forest vegetation, particularly pine stands in the vicinity of Choshi — flora and vegetation in Choshi, Chiba prefecture, VI (in Japanese). *Bulletin of the Choshi Marine Laboratory*, (6):27–37, 1964. Chiba University, Japan.

Y. Ogata. Space-time model for regional seismicity and detection of crustal stress changes. *Journal of Geophysical Research*, 109, 2004. B03308, doi:10.1029/2003JB002621.

Y. Ogata, K. Katsura, and M. Tanemura. Modeling heterogeneous space-time occurrences of earthquakes and its residual analysis. *Applied Statistics*, 52(4):499–509, 2003.

A.L. Patterson. A Fourier series method for the determination of the component of inter-atomic distances in crystals. *Physics Reviews*, 46:372–376, 1934.

A.L. Patterson. A direct method for the determination of the components of inter-atomic distances in crystals. *Zeitschrift fuer Krystallographie*, 90:517–554, 1935.

R Development Core Team. *R: A language and environment for statistical computing.* R Foundation for Statistical Computing, Vienna, Austria, 2004. http://www.R-project.org

B.D. Ripley. The second-order analysis of stationary point processes. *Journal of Applied Probability*, 13: 255–266, 1976.

B.D. Ripley. Modeling spatial patterns (with discussion). *Journal of the Royal Statistical Society, series B*, 39:172–212, 1977.

B.D. Ripley. *Spatial Statistics*. John Wiley & Sons, New York, 1981.

B.D. Ripley and J.-P. Rasson. Finding the edge of a Poisson forest. *Journal of Applied Probability*, 14: 483–491, 1977.

A. Rosenfeld and J.L. Pfalz. Distance functions on digital pictures. *Pattern Recognition*, 1:33–61, 1968.

B. Rowlingson and P. Diggle. Splancs: Spatial point pattern analysis code in S-PLUS. *Computers and Geosciences*, 19:627–655, 1993.

M. Schlather, P. Ribeiro, and P. Diggle. Detecting dependence between marks and locations of marked point processes. *Journal of the Royal Statistical Society, series B*, 66:79–83, 2004.

H. Stienen. The sectioning of randomly dispersed particles, a computer simulation. *Mikroskopie*, 37 (Suppl.):74–78, 1980.

D. Stoyan, W.S. Kendall, and J. Mecke. *Stochastic Geometry and Its Applications*. John Wiley & Sons, Chichester, U.K., 1987.

D. Stoyan and H. Stoyan. *Fractals, Random Shapes and Point Fields*. John Wiley & Sons, Chichester, U.K., 1995.

L. Strand. A model for stand growth. In *IUFRO Third Conference Advisory Group of Forest Statisticians*, pages 207–216, Paris, 1972. INRA, Institut National de la Recherche Agronomique.

D.J. Strauss. A model for clustering. *Biometrika*, 63:467–475, 1975.

J. Tukey. *Exploratory Data Analysis*. Addison-Wesley, Reading, MA, 1977.

P.F. Velleman and D.C. Hoaglin. *Applications, Basics and Computing of Exploratory Data Analysis*. Duxbury Press, Pacific Grove, CA, 1981.

D. Wartenberg. Exploratory spatial analyses: outliers, leverage points, and influence functions. In D.A. Griffith, Ed., *Spatial Statistics: Past, Present and Future*, pages 133–162. Institute of Mathematical Geography, Ann Arbor, Michigan, 1990.

B. Widom and J.S. Rowlinson. New model for the study of liquid-vapor phase transitions. *The Journal of Chemical Physics*, 52:1670–1684, 1970.

21

Multivariate and Marked Point Processes

Adrian Baddeley

CONTENTS

21.1 Motivation and Examples

A *multivariate* or *multitype* spatial point pattern is one that consists of several qualitatively different types of points. Examples include a map of the locations of trees labeled with the species classification of each tree, a spatial case/control study in spatial epidemiology where each point represents either a case or a control, and a map of locations of telephone calls to the emergency services, labeled by the nature of each emergency.

More generally, a *marked* point pattern is one in which each point of the process carries extra information called a *mark*, which may be a random variable, several random variables, a geometrical shape, or some other information. A multitype point pattern is the special case where the mark is a categorical variable.

21.1.1 Multivariate Point Patterns

The amacrine cells data were presented in Figure 18.6 of Chapter 18. The retina is a flat sheet containing several layers of cells. Amacrine cells occupy two adjacent layers, the "on" and "off" layers. In a microscope field of view, the locations of all amacrine cells were recorded, together with the type of each cell. The main question of interest is whether the "on" and "off" layers grew independently.

Such data can be approached in several ways. One way is to separate the points according to their type, yielding M distinct point patterns, where M is the number of possible types. We then have a *multivariate* observation $(X_1, X_2, \ldots, X_M)$ where X_m is the pattern of points of type m. Alternatively, the data may be treated as a single pattern of n points, in which each point location x_i is labeled by its type z_i for $i = 1, 2, \ldots, n$. This is a *multitype* point pattern. These two concepts are mathematically equivalent, but suggest slightly different statistical approaches. For example, Figure 18.6 displays the amacrine data as a "multitype" plot using different symbols to distinguish the two cell types. Alternatively we could display the two different cell types as separate point patterns in a "multivariate" plot.

Figure 21.1 shows the results of a survey of a forest plot in Lansing Woods, Michigan (Gerrard, 1969). The data give the locations of 2,251 trees and their botanical classification (into hickories, maples, red oaks, white oaks, black oaks and miscellaneous trees). This is a "multivariate" plot showing each species of tree as a separate point pattern. One question

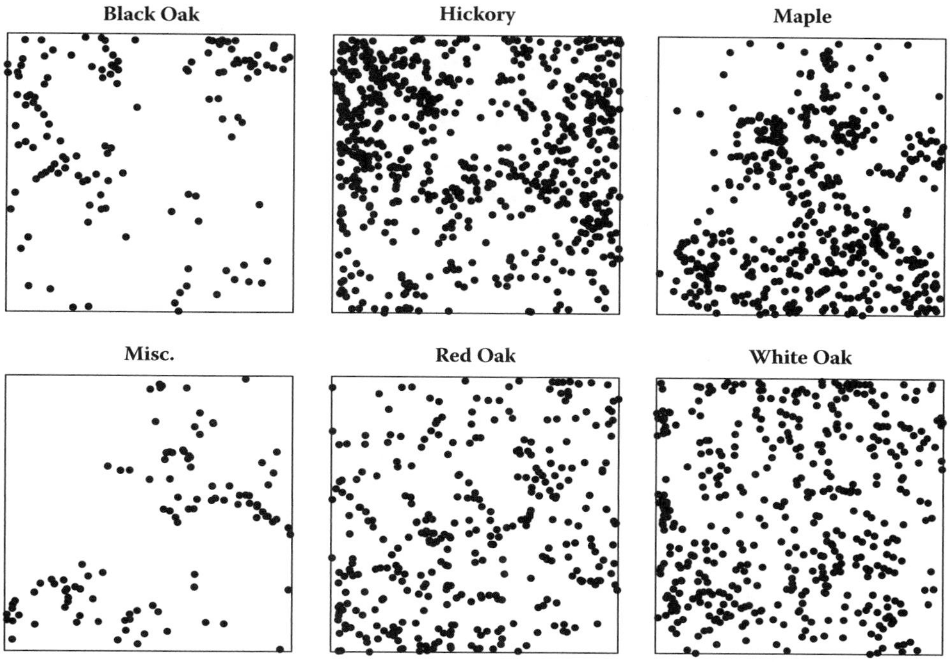

FIGURE 21.1
Lansing Woods data, separated by species.

of interest is whether the species are evenly mixed across the forest plot or are *segregated* into different subregions where one species is predominant over the others.

Figure 21.2 shows domicile addresses of new cases of cancer of the larynx (58 cases) and cancer of the lung (978 cases), recorded in the Chorley and South Ribble Health Authority (Lancashire, England) between 1974 and 1983. The location of a now-unused industrial incinerator is also shown. The data were first presented and analyzed by Diggle (1990). They have subsequently been analyzed by Diggle and Rowlingson (1994) and Baddeley, Turner, Møller, and Hazelton (2005). The aim is to assess evidence for an increase in the incidence of cancer of the larynx in the vicinity of the incinerator. The lung cancer cases serve as a surrogate for the spatially varying density of the susceptible population.

21.1.2 Marked Point Patterns

The left panel of Figure 21.3 is from a survey of 584 Longleaf pine (*Pinus palustris*) trees in a square plot region in southern state of Georgia, by Platt, Evans, and Rathbun (1988). The location of each tree was recorded together with its diameter at breast height (dbh), a convenient measure of size that is also a surrogate for age. "Adult" trees are conventionally defined as those with dbh greater than or equal to 30 cm. This is a marked point pattern with nonnegative real-valued marks. In the figure, each tree is represented by a circle with its center at the tree location and its radius proportional to the mark value. One of the many questions about this dataset is to account for spatial inhomogeneity in the ages of trees.

In a more detailed forest survey, we might record several variables for each tree: its species, its diameter, insect counts, the results of chemical assay of its leaves, and so on. These data can be regarded as a multivariate mark attached to the tree. At a still more

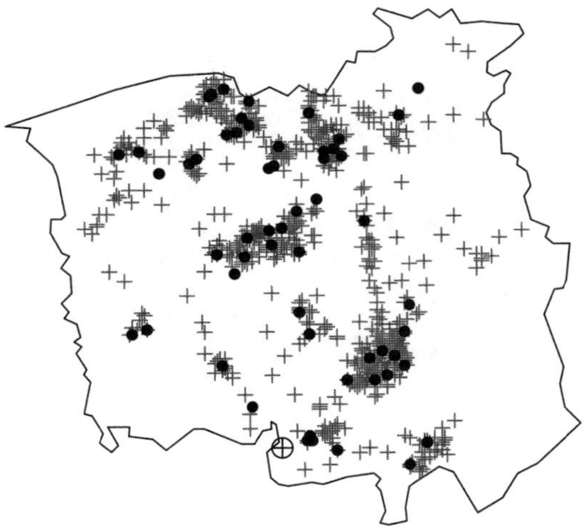

FIGURE 21.2
Chorley–Ribble data. Spatial locations of cases of cancer of the larynx (•) and cancer of the lung (+), and the location of a disused industrial incinerator (⊕).

complicated level, the mark attached to each point may be a function (such as the spectral signature of light from a galaxy, attached to the galaxy's location), a shape (such as the shape of the galaxy), and so on.

A spatial pattern of geometrical objects, such as disks or polygons of different sizes and shapes, can be treated as a marked point process where the points are the centers of the objects, and the marks are parameters determining the size and shape of the objects. The right panel of Figure 21.3 shows the spatial locations and diameters of sea anemones (beadlet anemone *Actinia equina*) in a sample plot on the north face of a boulder on the shore at Bretagne, France, collected by Kooijman (1979). Geometrical objects of arbitrary shape can be accommodated by allowing the mark to be a copy of the object translated so that its centroid is at the origin. Ripley and Sutherland (1990) used this *germ-grain model* to represent the location and shape of spiral galaxies. Stoyan (1993) used a germ-grain model to analyze a collage by the artist Hans Arp in which torn pieces of paper were arranged on a canvas "at random," according to the artist.

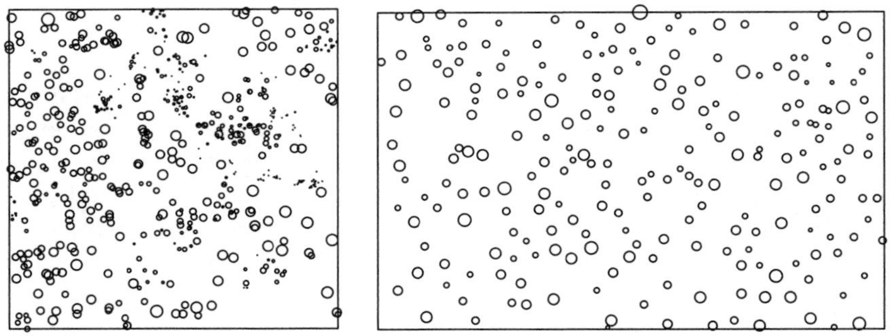

FIGURE 21.3
Marked point patterns with real-valued marks. Left: Longleaf pines data with mark values represented as radii of circles (not to scale). Right: Beadlet anemones data with diameters plotted to scale.

21.2 Methodological Issues

21.2.1 Appropriateness of Point Process Methods

Suppose we have data consisting of spatial locations $(x_1, \ldots, x_n)$ and values attached to them $(z_1, \ldots, z_n)$. Without background information, it is unclear how such data should be analyzed.

Scenario 1 *Today's maximum temperatures at 25 Australian cities are displayed on a map.* This is not a point process in any useful sense. Point process methods are only useful if the locations x_i can be regarded as random. Here the cities are fixed locations. The temperatures are observations of a spatial variable at a fixed set of locations. Geostatistical methods are more appropriate (Armstrong, 1997; Diggle et al., 1998; Journel and Huijbregts, 1978).

Scenario 2 *A mineral exploration dataset records the map coordinates where 15 core samples were drilled, and, for each core sample, the assayed concentration of iron in the sample.* Typically this would *not* be treated as a point process. The core sample locations were chosen by a geologist and are part of the experimental design. They cannot be regarded as a response, and point process models are not appropriate. The main interest is in the iron concentration at these locations. Again this should probably be analyzed using geostatistical methods. We would assume that the locations x_i were effectively arbitrary sample points, at which we measured the values $z_i = Z(x_i)$ of iron concentration. Here $Z(u)$ is the iron concentration at a location u. Our goal is to draw conclusions about the function Z.

21.2.2 Responses and Covariates

As in any statistical analysis, it is vital to decide which quantities to treat as *response* variables, and which as *explanatory* variables (e.g., Cox and Snell, 1981).

Point process methods are appropriate if the spatial locations x_i are "response" variables. *Marked* point process methods are appropriate if *both* the spatial locations x_i *and* the associated values z_i are part of the "response."

Scenario 3 *Trees in an orchard are examined and their disease status (infected/not infected) is recorded. We are interested in the spatial characteristics of the disease, such as contagion between neighboring trees.* These data probably should *not* be treated as a point process. The response is "disease status." We can think of disease status as a label applied to the trees after their locations have been determined. Since we are interested in the spatial correlation of disease status, the tree locations are effectively fixed covariate values. It would probably be best to treat these data as a discrete random field (of disease status values) observed at a finite known set of sites (the trees). However, a different approach might be required for a naturally regenerated stand of trees with endemic disease.

Scenario 4 *In an intensive geological survey of a desert region, the locations x_i of all natural deposits of a rare mineral are recorded. Deposits are effectively points at the scale of the survey. For each deposit location x_i, the estimated total yield y_i (kg) of the deposit is recorded.* This could be analyzed as a marked point pattern, in which each deposit location x_i is marked by the deposit yield y_i. We assume that the locations x_i and yields y_i, taken together, are the outcome of a random process. The pattern of points x_i carries substantial information, and our goal is to draw conclusions about the locations, the marks, and the dependence between them.

Scenario 5 *In the same geological survey, for each deposit x_i, we record whether the surrounding rock was volcanic ($v_i = 1$) or nonvolcanic ($v_i = 0$). We wish to determine whether deposits are* *more likely to occur in volcanic rock.* The deposits x_i are the "response" of interest, and should be interpreted as a point process. However, the rock type values v_i are clearly intended to serve as a *covariate* (explanatory variable) because we wish to determine whether the abundance of deposits depends on rock type v.

A major difficulty in Scenario 5 is that these data are inadequate. It is not sufficient to record the covariate values at the points of the point pattern. The covariate must also be observed at some other locations in the study region. The relative frequencies of the rock types $v = 0, 1$, observed at the deposit locations only, are not sufficient to estimate the relative frequencies of deposits in the two rock types. In schematic terms, $P(v \mid \text{deposit})$ does not determine $P(\text{deposit} \mid v)$. Bayes' formula indicates that we would need additional information about the relative frequencies $P(v)$ of the two rock types in the study area, (see Chapter 20, Section 20.3.2).

Thus, marks and covariates play different statistical roles. Marks are attributes of the individual points in the pattern and are part of the "response" in the experiment, while covariates are "explanatory" variables. A covariate must be observable at any spatial location, while a mark may be observable only at the points in the point pattern.

It may be difficult to decide whether a variable should be treated as a response or as a covariate (Cox and Snell, 1981). This issue also arises in spatial statistics. For example, the longleaf pines data (left panel of Figure 21.3) give the location and diameter at breast height (dbh) of each tree in a forest. It is surely appropriate to treat dbh as a mark; the diameter of a tree is an attribute of that tree (a surrogate for its age) not a quantity that can be observed at arbitrary spatial locations. However, Figure 21.3 and Figure 21.4 show that the age distribution is not spatially homogeneous. The forest contains some areas where most trees have relatively small dbh, and, hence, are relatively young. It is known that some forest stands were cleared decades ago, and such areas would now contain relatively young trees. A more sophisticated analysis of the longleaf pines data might use a spatially smoothed trend surface of the dbh values as a covariate — effectively a surrogate for the history of the forest.

21.2.3 Modeling Approaches

Marked point patterns raise new and interesting questions concerning the appropriate way to formulate models and pursue analyses for particular applications.

In a statistical analysis of two response variables X and Y, we have the choice of modeling the joint distribution $[X, Y]$, or one of the conditional distributions $[Y|X]$ or $[X|Y]$.

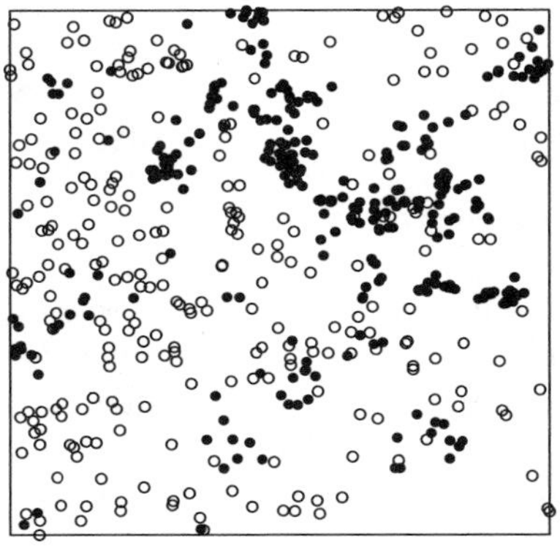

FIGURE 21.4
Longleaf pines classified into "juvenile" (dbh < 30, filled circles) and "adult" (dbh ≥ 30, open circles).

Conditioning is appropriate (and usually simpler and more efficient) if the conditioning variable does not contain information about the parameters of interest.

Similarly, in the analysis of a marked point pattern, an important choice is whether to analyze the marks and locations jointly or conditionally. Schematically, if we write X for the points and M for the marks, then we could specify a model for the marked point process $[X, M]$. Alternatively we may *condition on the locations* of the points, treating only the marks as random variables $[M|X]$.

For example, the Chorley–Ribble data (Figure 21.2) consist of domicile locations x_i of cancer cases, with marks m_i indicating the cancer type. The locations and types were analyzed jointly by Diggle (1990). A joint analysis requires estimation of the unknown, spatially varying density of the susceptible population. The same data were reanalyzed conditionally on the locations by Diggle and Rowlingson (1994). Conditioning on the locations removes the need to estimate the population density, greatly simplifying the analysis, and eliminating error due to estimation of population density.

In some cases, we may *condition on the marks*, treating the locations as a random point process $[X|M]$. This is meaningful if the mark variable is a continuous real-valued quantity, such as time, age or distance. For example, in an astronomical survey giving the sky position x_i and redshift z_i of distant galaxies, one of the issues is sampling bias. The probability of detecting a galaxy depends on its apparent brightness, hence, on its distance from Earth. Rather than estimating or guessing the detection probability as a function of redshift, it may be appropriate to condition on the redshift values.

In a multitype point process, there is a further option of *conditioning on some types of points*. For example, Chapter 19, Section 19.4.2 describes an analysis by Högmander and Särkkä (1999) of the ants' nests data in which the locations of *Cataglyphis* nests are modeled conditionally on the locations of *Messor* nests. This is appropriate for investigating whether *Cataglyphis* nests are preferentially placed near *Messor* nests. Such analysis is natural from the "multivariate" viewpoint.

One important situation is where the marks are provided by a *(random) field*. A random field is a quantity $Z(u)$ observable at any spatial location u. Our data consist of a spatial point pattern $\mathbf{X} = \{x_1, \ldots, x_n\}$ and the values $z_i = Z(x_i)$ of a random field Z observed at these random points. A typical question is to determine whether $\mathbf{X}$ and Z are independent. Techniques for this purpose are discussed in Section 21.7. If $\mathbf{X}$ and Z are independent, then we may condition on the locations and use geostatistical techniques to investigate properties of Z. However, in general, geostatistical techniques, such as the variogram, have a different interpretation when applied to marked point patterns (Wälder and Stoyan, 1996; Schlather, Rubiero, and Diggle, 2004.)

21.2.4 Kinds of Marks

The choice of statistical technique also depends on the type of mark variable, and particularly on whether the marks are continuous or discrete.

The distinction between categorical marks and continuous numerical marks is complex. At one extreme, there are multitype point patterns with only two or three types of points, such as the examples presented here, for which plotting and analysis are easily manageable. Then there are multitype point patterns involving a larger number of distinct types, where separate visualization and analysis of each type (and comparison of each pair of types) become unwieldy. Multivariate techniques, such as principal components, may be useful (Illian, Benson, Crawford, and Staines, 2006). Some types of points may occur with low frequency, so that separate analysis of each type is unreliable, and it is appropriate to pool some of the types (as occurred with the Lansing Woods data, where one category consists of "other" trees). Some multitype datasets involve an *a priori* unlimited number of types, for example, a forestry survey of a rainforest with high biodiversity. Such datasets often empirically satisfy "Zipf's Law": The frequency of the kth most frequent type is approximately proportional to $1/k$. Practical strategies include pooling the lowest-ranking types, or mapping the types to continuous numerical values.

21.3 Basic Theory

Here we give a sketch of the general theory of marked point processes in $\mathbb{R}^d$, expanding on the theory presented in Chapter 16.

21.3.1 Product Space Representation

For mathematical purposes, a mark is effectively treated as an extra spatial coordinate. A marked point at location x in $\mathbb{R}^d$, with mark m from some set $\mathcal{M}$, is treated as a point (x, m) in the space $\mathbb{R}^d \times \mathcal{M}$.

If the marks are real numbers, we take $\mathcal{M} = \mathbb{R}$ so that $\mathbb{R}^d \times \mathcal{M}$ is $\mathbb{R}^{d+1}$. A marked point pattern in $\mathbb{R}^d$ with real-valued marks is equivalent to a point pattern in $\mathbb{R}^{d+1}$ where the first d coordinates are interpreted as spatial coordinates of the point location, and the $(d+1)$th coordinate is interpreted as the mark value.

If the marks are categorical values, then $\mathcal{M}$ is a finite set containing M elements (say), and $\mathbb{R}^d \times \mathcal{M}$ is effectively a stack of M separate copies of $\mathbb{R}^d$. A marked point pattern in $\mathbb{R}^d$ with categorical marks in $\mathcal{M}$ is equivalent to M point patterns $X_1, \ldots, X_M$ in $\mathbb{R}^d$, where X_m is the pattern of points of type m.

A common technical device is to count the number of marked points that fall in the set $A \times B$, where A is a specified region of $\mathbb{R}^d$, and B is a specified subset of $\mathcal{M}$. This count is simply the number of marked points (x_i, m_i) whose location x_i falls in A *and* whose mark m_i belongs to B.

21.3.2 Marked Point Processes

A marked point process is simply defined as a point process in the product space $\mathbb{R}^d \times \mathcal{M}$, that is, a point process of pairs (x_i, m_i). The only stipulation is that, if the marks are ignored, then the point process of locations x_i must be locally finite.

Definition 21.1

Let $\mathcal{M}$ be any locally compact separable metric space. A marked point process *in $\mathbb{R}^d$ $(d \geq 1)$ with marks in $\mathcal{M}$ is defined as a point process Ψ in $\mathbb{R}^d \times \mathcal{M}$ such that, for any bounded set $A \subset \mathbb{R}^d$, the number $\Psi(A \times \mathcal{M})$ of marked points in $A \times \mathcal{M}$ is finite, with probability 1.*

Notice that $\Psi(K \times \mathcal{M})$ is just the number of marked points (x_i, m_i) whose location x_i falls in K.

Example 21.1 (Binomial marked point process)

Let $x_1, \ldots, x_n$ be independent, uniformly distributed random points in some region W in two-dimensional space. Let $m_1, \ldots, m_n$ be independent, uniformly distributed random elements of $\mathcal{M}$. Then the pairs $(x_1, m_1), \ldots, (x_n, m_n)$ can be interpreted as a marked point process in $\mathbb{R}^2$ with marks in $\mathcal{M}$.

21.3.3 Poisson Marked Point Processes

A *Poisson* marked point process is simply a Poisson process in $\mathbb{R}^d \times \mathcal{M}$ that can be interpreted as a marked point process.

Lemma 21.1

Let Ψ be a Poisson point process on $\mathbb{R}^d \times \mathcal{M}$ with intensity measure Λ. Suppose $\Lambda(A \times \mathcal{M}) < \infty$ for any bounded set $A \subset \mathbb{R}^d$. Then Ψ is a marked point process, called the Poisson marked point process *with intensity measure Λ.*

The condition on $\Lambda(A \times \mathcal{M})$ stipulates that the expected number of points falling in any bounded region of $\mathbb{R}^d$ must be finite.

Notice that this condition excludes the *homogeneous* Poisson process in $\mathbb{R}^3$ with constant intensity λ. This cannot be interpreted as a marked point process in $\mathbb{R}^2$ with marks in $\mathbb{R}$. The expected number of points in any bounded region of $\mathbb{R}^2$ is infinite. To put it another way, the projection of a uniform Poisson process in $\mathbb{R}^3$ onto $\mathbb{R}^2$ is infinitely dense, so it is not a point process.

Figure 21.5 shows simulated realizations of Poisson marked point processes. Poisson marked point processes are discussed further in Section 21.5.

21.3.4 Intensity

With a few exceptions, it is not necessary to introduce new mathematical definitions for fundamental properties of a marked point process, such as the intensity measure, moment measures, Campbell measure, Palm distribution and conditional intensity. Since a marked point process is a special case of a point process, the existing definitions (see Chapter 16) are sufficient.

Two exceptions are the definition of a *stationary* marked point process, explained in Section 21.3.5, and the choice of reference measure for intensities, explained below.

Suppose Φ is a marked point process on $\mathbb{R}^d$ with marks in $\mathcal{M}$, a general mark space. The *intensity measure* of Φ is simply the intensity measure of Φ viewed as a point process on $\mathbb{R}^d \times \mathcal{M}$. Formally, it is a measure Λ on $\mathbb{R}^d \times \mathcal{M}$ defined by $\Lambda(S) = \mathbb{E}[\Phi(S)]$ for all bounded sets $S \subset \mathbb{R}^d \times \mathcal{M}$, provided that this expectation is always finite.

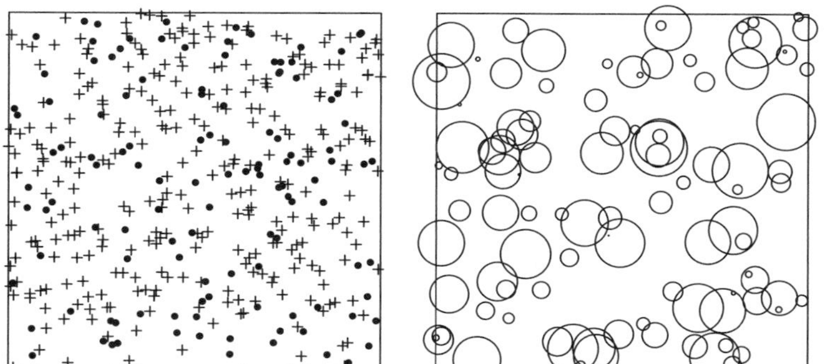

FIGURE 21.5
Simulated realizations of stationary Poisson marked point processes. Left: Multitype process with two types (plotted as symbols + and •) of intensity 70 and 30, respectively. Right: Positive real marks, uniformly distributed in [0, 1], plotted as radii of circles.

The intensity measure Λ is determined by the values

$$\Lambda(A \times B) = \mathbb{E}[\Phi(A \times B)],$$

that is, the expected number of marked points with locations x_i falling in the set $A \subset \mathbb{R}^d$ and marks m_i belonging to the set $B \subseteq \mathcal{M}$, for a sufficiently large class of sets A and B. For example, for a multitype point process, the intensity measure is equivalent to specifying the intensity measure of the process X_m of points of mark m, for each possible type m.

A marked point process may have an *intensity function*. For a point process on $\mathbb{R}^d$, the intensity function is a function $\lambda(u)$ such that

$$\mathbb{E}[\Phi(A)] = \int_A \lambda(u)\, \mathrm{d}u$$

for all bounded sets $A \subset \mathbb{R}^d$. For a marked point process on $\mathbb{R}^d$ with marks in $\mathcal{M}$, the intensity function is a function $\lambda(u, m)$ such that

$$\mathbb{E}[\Phi(A \times B)] = \int_A \int_B \lambda(x, m)\, \mathrm{d}\mu(m)\, \mathrm{d}x \tag{21.1}$$

for a sufficiently large class of sets $A \subset \mathbb{R}^d$ and $B \subseteq \mathcal{M}$. Here, μ is an additional reference measure μ on $\mathcal{M}$ that we are obliged to choose. The intensity function of a marked point process is not unambiguously defined until we specify the reference measure μ on $\mathcal{M}$. If marks are real numbers, the conventional choice of reference measure is Lebesgue measure. Then Equation (21.1) becomes

$$\mathbb{E}[\Phi(A \times B)] = \int_A \int_B \lambda(x, m)\, \mathrm{d}m\, \mathrm{d}x.$$

If marks are categorical values or integers, the reference measure is usually counting measure. Then (21.1) becomes

$$\mathbb{E}[\Phi(A \times B)] = \int_A \sum_{m \in B} \lambda(x, m)\, \mathrm{d}x.$$

Similarly, the (Papangelou) *conditional intensity* of Φ, if it exists, is defined by the equation in Definition 7 of Chapter 16. This definition depends on the choice of reference measure ν on $\mathbb{R}^d \times \mathcal{M}$, and we would normally take $\mathrm{d}\nu(u, m) = \mathrm{d}u\, \mathrm{d}\mu(m)$.

21.3.5 Stationary Marked Point Processes

A marked point process is defined to be *stationary* when its distribution is unaffected by shifting *the locations* x_i.

Definition 21.2

A marked point process in $\mathbb{R}^d$ *with marks in* $\mathcal{M}$ *is* stationary *if its distribution is invariant under translations of* $\mathbb{R}^d$, *that is, under transformations* $(x, m) \mapsto (x + v, m)$ *where* v *is any vector in* $\mathbb{R}^d$.

Under the transformation $(x, m) \mapsto (x + v, m)$, a marked point pattern is simply shifted by the vector v with the marks unchanged.

In a multitype point process with points of types $1, \ldots, M$ (say), stationarity implies that the subprocesses $X_1, \ldots, X_M$ of points of type $1, 2, \ldots, M$ are stationary point processes, but additionally it implies that they are *jointly stationary*, in the sense that $(v + X_1, v + X_2, \ldots, v + X_M)$ has the same joint distribution as $(X_1, \ldots, X_M)$ for any translation vector v.

Lemma 21.2

If Ψ is a stationary marked point process, its intensity measure Λ must satisfy

$$\Lambda(A \times B) = \lambda\,|A|\,Q(B)$$

for all bounded $A \subset \mathbb{R}^d$ and $B \subset \mathcal{M}$, where $\lambda > 0$ is a constant, $|A|$ denotes the Lebesgue volume of A, and Q is a probability measure on $\mathcal{M}$. We call λ the point intensity *and Q the* mark distribution.

This important result implies that the "marginal distribution of marks" is a well-defined concept for any stationary marked point process. It can also be used to construct estimators of the mark distribution Q.

21.3.6 Operations on Marked Point Processes

21.3.6.1 Marginal Process of Locations

If Ψ is a marked point process, discarding the marks and replacing each pair (x_i, m_i) by its location x_i yields a point process $\Xi = \pi(\Psi)$ in $\mathbb{R}^d$, known as the "projected process," the "superposition," the "marginal process of locations" or the "process of points without marks."

In Example 21.1, the projected process Ξ is the binomial process consisting of n independent uniform random points in W.

If Ψ is a stationary marked point process in $\mathbb{R}^d$ with marks in $\mathcal{M}$, then clearly $\Xi = \pi(\Psi)$ is a stationary point process in $\mathbb{R}^d$.

Note that if we were to replace each marked point (x_i, m_i) in a marked point process by its *mark* m_i, the result would usually *not* be a point process on $\mathcal{M}$.

21.3.6.2 Conditioning on Locations

It is often appropriate to analyze a marked point pattern by conditioning on the locations. Under reasonable assumptions (Last, 1990) there is a well-defined conditional distribution $P(\Psi \mid \Xi)$ of the marked points given the locations. Conditional on the locations $\Xi = \mathbf{x}$, the marks effectively constitute a random field, with values in $\mathcal{M}$, at the discrete sites in $\mathbf{x}$.

In Example 21.1, given the locations $\Xi = \mathbf{x}$, the marks are conditionally independent random variables.

21.3.6.3 Restriction

Sometimes we restrict the point process Ψ to a domain $C \subset \mathbb{R}^d \times \mathcal{M}$ (by deleting all points of Ψ that fall outside). The result is still a marked point process.

In particular, "slicing" or restricting the marks is useful. Taking $C = \mathbb{R}^d \times B$ where $B \subset \mathcal{M}$ is a subset of the mark space, the restricted point process consists of all marked points (x_i, m_i) whose marks m_i lie in the set B.

21.4 Exploratory Analysis of Intensity

The first step in exploratory analysis of a marked point pattern dataset is usually to study its intensity. The intensity *function* of a point process is the occurrence rate of events. Depending on the application, the intensity function may be interpreted as measuring abundance, fertility, productivity, accident rate, disease risk, etc. The intensity function of a marked point process (if it exists) is a function $\lambda(u, m)$ of location u and mark m. It conveys information about the abundance and density of the locations x_i, but also about the distribution of the marks m_i, and about the dependence between marks and locations. Only minimal statistical assumptions are needed to estimate some properties of the intensity using nonparametric techniques, such as kernel smoothing.

21.4.1 Intensity for Multitype Point Patterns

For a multitype point process, the interpretation of the intensity function $\lambda(u, m)$ is straightforward. For each type m, the function $\lambda_m(u) = \lambda(u, m)$ is the intensity of the process of points of type m. This can be estimated from the observed pattern of points of type m only. Figure 21.6 shows kernel estimates of intensity for each species of tree in the Lansing Woods data. The kernel bandwidth can be selected using a cross-validated likelihood method (Diggle, Zheng, and Durr, 2005).

The process of points without marks has intensity

$$\nu(u) = \sum_{m \in \mathcal{M}} \lambda_m(u) = \sum_{m \in \mathcal{M}} \lambda(u, m). \tag{21.2}$$

The conditional probability that a point at location u has mark m, given that there is a point at location u, is

$$p(m \mid u) = \frac{\lambda_m(u)}{\nu(u)} = \frac{\lambda_m(u)}{\sum_{m' \in \mathcal{M}} \lambda_{m'}(u)} \tag{21.3}$$

and is undefined where $\nu(u) = 0$.

Given estimates of the intensity functions $\lambda_m(\cdot)$ of points of each type m, plugging these into Equation (21.2) and Equation (21.3) yields estimates of the total intensity $\nu(u)$ and the conditional mark distribution $p(m \mid u)$. A plot of the estimated conditional mark probabilities $p(m \mid u)$ for each species in Lansing Woods is not shown; it looks very similar to Figure 21.6 because the total intensity $\nu(u)$ is almost constant.

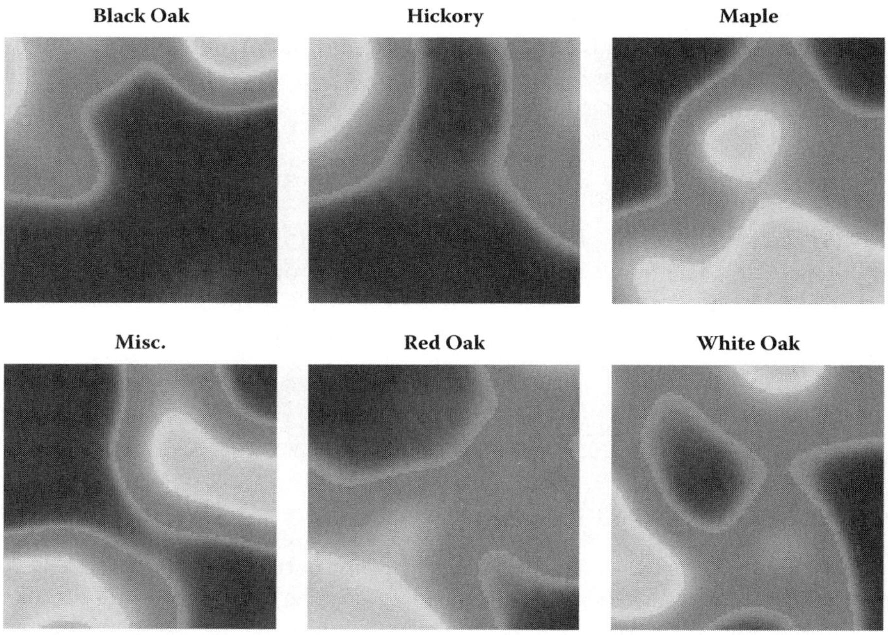

FIGURE 21.6
Kernel estimates of intensity for each species of tree in Lansing Woods.

A marked point process with intensity function $\lambda(u, m)$ is called *first order stationary* if the intensity does not depend on location u, so that $\lambda(u, m) = \beta(m)$ for some function $\beta(m)$. Thus, the intensity of the points of type m is constant, that is, the process of points of type m is first-order stationary. The total intensity of points is $\nu(u) = \sum_m \lambda(u, m) = \sum_m \beta(m) = B$, say, so that the unmarked point process is also first-order stationary. The conditional mark distribution is

$$p(m \mid u) = \frac{\lambda(u, m)}{\nu(u)} = \frac{\beta(m)}{\sum_{m' \in \mathcal{M}} \beta(m')},$$

which does not depend on u, say $p(u \mid m) = q(m)$. Thus, the marks can be considered to have marginal distribution $q(m)$. Inspection of Figure 21.6 suggests strongly that the Lansing Woods data are not first-order stationary.

More generally, we can ask if the intensity is *separable* in the sense that $\lambda(u, m) = \kappa(u)\beta(m)$, where β and κ are functions. Then the intensity functions of each type of point are proportional to $\kappa(u)$, and hence proportional to one another; the points of different types share the same "form" of spatial inhomogeneity. The total intensity of unmarked points is $\nu(u) = B\kappa(u)$ where $B = \sum_{m \in \mathcal{M}} \beta(m)$. The conditional mark distribution is $p(m \mid u) = \lambda(u, m)/\nu(u) = \beta(m)/B$, which does not depend on u. Thus, the locations of the points are spatially inhomogeneous, but the distribution of marks is spatially homogeneous.

If the separability equation $\lambda(u, m) = \beta(u)\kappa(m)$ does *not* hold, then we say that the types are *segregated*. This is clearly the appropriate description of the Lansing Woods data, since (for example) the hickories and maples are strongly segregated from each other.

It is useful, especially in epidemiological applications, to study the *relative risk* or relative intensity

$$\rho(m, m' \mid u) = \frac{\lambda(u, m)}{\lambda(u, m')} = \frac{p(m \mid u)}{p(m' \mid u)} \tag{21.4}$$

of two specified types m, m' as a function of location u. We say there is *spatial variation in relative risk* if $\rho(m, m' \mid u)$ is not constant as a function of u. The types are segregated if and only if there is spatial variation in the relative risk of at least one pair of types m, m'.

Diggle et al. (2005) studied outbreaks of bovine tuberculosis classified by the genotype of the tuberculosis bacterium. The relative risk of two genotypes of the disease gives clues about the mechanism of disease transmission (i.e., strong segregation of types would suggest that contagion is localized) and clues to appropriate management of new cases (i.e., if there is strong segregation and a new case does not belong to the locally predominant type, the infection is more likely to be the result of importation of infected animals).

The Chorley–Ribble data (Figure 21.2) are an example of a spatial *case-control study*. We have points of two types, cases and controls, in which the cases are the disease events under investigation, while the controls are a surrogate for the reference population of susceptible individuals. In this context, the ratio (21.4) of the intensity of cases to the intensity of controls is simply the *risk* $\rho(u)$ of the disease under investigation. Spatial variation in disease risk $\rho(u)$ is an important topic.

Kelsall and Diggle (1998) proposed a Monte Carlo test of spatial variation in disease risk, and Diggle et al. (2005) generalized this to a Monte Carlo test of segregation in multitype point patterns. The null hypothesis is that $p(m \mid u)$ is constant as a function of u for each m. The test statistic is

$$T = \sum_i \sum_m (\hat{p}(m \mid x_i) - \hat{p}(m))^2,$$

where $\hat{p}(m \mid u)$ is the kernel estimate of $p(m \mid u)$ based on the marked point pattern dataset, and $\hat{p}(m)$ is the relative frequency of mark m in the data ignoring locations. Randomization for the Monte Carlo test is performed by holding the locations x_i fixed while the marks are randomly permuted.

Parametric and semiparametric estimates of intensity can be obtained by model-fitting techniques, which are discussed below.

21.4.2 Intensity for Marked Point Patterns with Real-Valued Marks

For a marked point process in the plane, with real-valued marks, the intensity $\lambda(u, m)$, if it exists, is a function of three coordinates. Integrating out the mark variable yields

$$\lambda_2(u) = \int_{-\infty}^{\infty} \lambda(u, m)\, dm,$$

the intensity of the point process of locations. The ratio

$$p(m \mid u) = \frac{\lambda(u, m)}{\lambda_2(u)}$$

is the conditional probability density of the mark m at a point at location u.

Given a marked point pattern $\{(x_i, m_i)\}$ the intensity function $\lambda(u, m)$ could be estimated by kernel smoothing in the product space:

$$\hat{\lambda}(u, m) = \sum_i \kappa((x_i, m_i) - (u, m)),$$

where κ is a probability density on $\mathbb{R}^3$. The intensity of locations $\lambda_2(u)$ would then be estimated by kernel smoothing the locations x_i using the corresponding two-dimensional marginal kernel

$$\kappa_2((x, y)) = \int_{-\infty}^{\infty} \kappa((x, y, z))\, dz.$$

This ensures that $\hat{\lambda}_2$ is the marginal integral of $\hat{\lambda}$ and that the ratio $\hat{p}(m \mid u) = \hat{\lambda}(u, m)/\hat{\lambda}_2(u)$ is a probability density.

To investigate whether there is a spatial trend in the mark values, it is useful to study the expected mark of a point at location u,

$$e(u) = \int_{-\infty}^{\infty} m p(m \mid u)\, dm = \frac{\int_{-\infty}^{\infty} m\, \lambda(u, m)\, dm}{\int_{-\infty}^{\infty} \lambda(u, m)\, dm}. \tag{21.5}$$

The estimator $\hat{e}(u)$ obtained by plugging into Equation (21.5) a kernel estimator of $\lambda(u, m)$ is identical to the usual Nadaraya–Watson estimator of a smooth function. The local variance

$$v(u) = \int_{-\infty}^{\infty} (m - e(u))^2 p(m \mid u)\, dm \tag{21.6}$$

is also useful. Figure 21.7 shows kernel estimates of mean tree diameter $e(u)$ and variance of tree diameter $v(u)$ for the Longleaf pines data (left panel of Figure 21.3), strongly suggesting an area of predominantly young trees in the upper right of the field.

Simple exploratory methods, such as discretizing the marks, can also be effective, as exemplified in Figure 21.4.

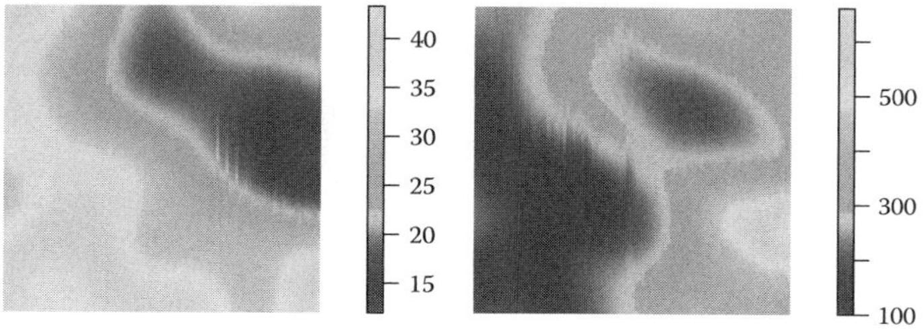

FIGURE 21.7
Kernel-weighted estimates of (left) mean tree diameter $e(u)$ and (right) variance of tree diameter $v(u)$ for the Longleaf pines.

21.5 Poisson Marked Point Processes

Poisson models are very important in the analysis of marked point pattern data. They are the most basic and most important stochastic models for marked point patterns. After exploratory analysis of the intensity, the next step in data analysis would usually be to fit a Poisson marked point process model, or to test whether a Poisson model is appropriate. It is often pragmatic to assume a Poisson process when conducting hypothesis tests, for example, testing for a covariate effect.

The *stationary* Poisson marked point process is the reference model of a completely random pattern because it is the only model in which the locations of the points are independent of each other, the marks are independent of each other, and the marks are independent of the locations (although many other models satisfy one of these statements).

A *nonstationary*, or spatially inhomogeneous, Poisson marked point process model describes a random pattern in which marked points are independent of each other, although the locations may have a spatially varying intensity, and the marks may have a spatially varying probability distribution. A Poisson marked point process model is completely determined by its intensity; fitting the model is equivalent to estimating its intensity. Thus, the exploratory techniques of Section 21.4 for estimating the intensity of a marked point process also provide nonparametric techniques for fitting Poisson models.

Parametric models for the intensity function can be fitted by maximum likelihood. This allows us to model the effect of covariates, and to perform likelihood-based inference, such as confidence intervals, hypothesis tests for a covariate effect, and goodness-of-fit tests. The numerical techniques are closely related to generalized linear models.

21.5.1 Properties of Poisson Marked Point Processes

Poisson marked point processes were introduced in Section 21.3.3. Here we discuss some of their important properties.

21.5.1.1 Random Marking

Lemma 21.3 (Random marking)
A marked point process Ψ on $\mathbb{R}^d$ with marks in $\mathcal{M}$ is Poisson if and only if

1. *The corresponding point process of locations is a Poisson process on $\mathbb{R}^d$.*
2. *Conditional on the locations x_i, the marks m_i are independent.*
3. *The conditional distribution of the mark at a location x_i depends only on x_i.*

Properties 2 and 3 above are sometimes called the *random marking property*.

Example 21.2

Consider the Poisson marked point process in $\mathbb{R}^2$ with marks in $[0, \infty)$ with intensity function $\lambda(u, m)$, $u \in \mathbb{R}^2$, $m \geq 0$ (with respect to Lebesgue measure). The marginal process of locations is the Poisson process with intensity function

$$\beta(u) = \int_0^\infty \lambda(u, m)\, \mathrm{d}m$$

on $\mathbb{R}^2$. Conditional on the locations, the marks are independent, and a mark at location u has probability density $f(m \mid u) = \lambda(u, m)/\beta(u)$.

Example 21.3

Let the marks be discrete categories $1, 2, \ldots, M$. Let Ψ be the Poisson marked point process in $\mathbb{R}^2$ with marks in $\mathcal{M} = \{1, 2, \ldots, M\}$ with intensity function $\lambda(u, m)$, $u \in \mathbb{R}^2$, $m \in \mathcal{M}$ (with respect to counting measure on $\mathcal{M}$). The marginal process of locations is the Poisson process with intensity function

$$\beta(u) = \sum_{m=1}^{M} \lambda(u, m)$$

on $\mathbb{R}^2$. Conditional on the locations, the marks are independent, and a mark m at location u has probability distribution $p(m \mid u) = \lambda(u, m)/\beta(u)$.

21.5.1.2 Slicing and Thinning

For any subset $B \subset \mathcal{M}$ of the mark space, let Ψ_B be the process consisting of marked points (x_i, m_i) with marks m_i that belong to B. We saw above that this is also a marked point process.

Lemma 21.4 (Independence of Components)

A marked point process Ψ on $\mathbb{R}^d$ with marks in $\mathcal{M}$ is Poisson if and only if, *for any subset $B \subset \mathcal{M}$,*

1. *The processes Ψ_B and $\Psi_{(B^c)}$ are independent*
2. *The point process of locations of points in Ψ_B is a Poisson point process on $\mathbb{R}^d$*

Property 1 above is sometimes called the *independence of components property*. If a Poisson marked point process is divided into two subprocesses by dividing the marks into two categories, then the corresponding subprocesses must be independent.

It also follows that, if Ψ is Poisson, then the thinned process Ψ_B is also Poisson. This is the *thinning property* of the Poisson marked point process.

In Example 21.3, the subprocesses $X_1, \ldots, X_M$ of points of each type $m = 1, \ldots, M$ are independent Poisson point processes in $\mathbb{R}^2$.

21.5.2 Stationary Poisson Marked Point Process

A *stationary* Poisson marked point process has a simple and elegant structure. By Lemma 21.2, its intensity measure must be of the form $\Lambda(A \times B) = \lambda\,|A| Q(B)$ where Q is a probability distribution on $\mathcal{M}$. The marginal process of locations is a stationary Poisson point process in $\mathbb{R}^d$ with intensity λ. Conditional on the locations, the marks are independent and identically distributed with common distribution Q.

The random marking property becomes even more elegant in the stationary case:

Lemma 21.5 (Random marking, stationary case)
A marked point process Ψ on $\mathbb{R}^d$ with marks in $\mathcal{M}$ is a stationary Poisson marked point process if and only if

1. *The corresponding point process of locations is a homogeneous Poisson process on $\mathbb{R}^d$.*
2. *Conditional on the locations x_i, the marks m_i are independent and identically distributed.*

Since the established term CSR (complete spatial randomness) is used to refer to the uniform Poisson point process, it would seem appropriate that the uniform *marked* Poisson point process be called *complete spatial randomness and independence* (CSRI).

In Example 21.2, the Poisson marked point process is stationary if and only if $\lambda(u, m)$ does not depend on u, so that $\beta(u)$ is constant and $f(m \mid u) = f(m)$ does not depend on u. In Example 21.3, the Poisson multitype point process is stationary iff $\lambda(u, m)$ does not depend on u, so that $\beta(u)$ is constant and $p(m \mid u) = p(m)$ does not depend on u.

We see that the *stationary* Poisson multitype point process has three equivalent descriptions:

1. The points constitute a uniform Poisson process with intensity λ, and the marks are iid with distribution (p_m).
2. The component point processes $Y_1, \ldots, Y_M$ consisting of points of type $1, \ldots, M$ respectively, are independent point processes, and Y_m is Poisson with intensity λp_m.
3. The (point, mark) pairs constitute a Poisson process in $R^d \times \mathcal{M}$ with intensity λp_m for points of type m.

(See Kingman, 1993.) Thus, the stationary Poisson multitype point process exhibits both the **random labeling** and **independence of components** properties. These properties are not equivalent.

21.5.3 Fitting Poisson Models

Poisson marked point process models may be fitted to point pattern data by maximum likelihood, using methods similar to those in Section 20.4.1, provided the model has an intensity function. Penalized likelihood and Bayesian methods can also be used.

Suppose we are given a marked point pattern dataset

$$\mathbf{y} = \{(x_1, m_1), \ldots, (x_n, m_n)\}, \quad x_i \in W, \quad m_i \in \mathcal{M}, \quad n \geq 0$$

of pairs (x_i, m_i) of locations x_i with marks m_i. The likelihood for a Poisson marked point process with intensity function $\lambda(u, m)$ is

$$L = \exp\left(\int_M \int_W (1 - \lambda(u, m))\, \mathrm{d}u\, \mathrm{d}\mu(m)\right) \prod_{i=1}^{n(\mathbf{y})} \lambda(x_i, m_i), \tag{21.7}$$

where μ is the reference measure on $\mathcal{M}$ that was used to define the intensity function (Equation (21.1) in Section 21.3.4). For example, for a multitype point process the conventional choice of μ is counting measure, and the likelihood is

$$L = \exp\left(\sum_{m \in \mathcal{M}} \int_W (1 - \lambda(u, m))\, \mathrm{d}u\right) \prod_{i=1}^{n(\mathbf{y})} \lambda(x_i, m_i). \tag{21.8}$$

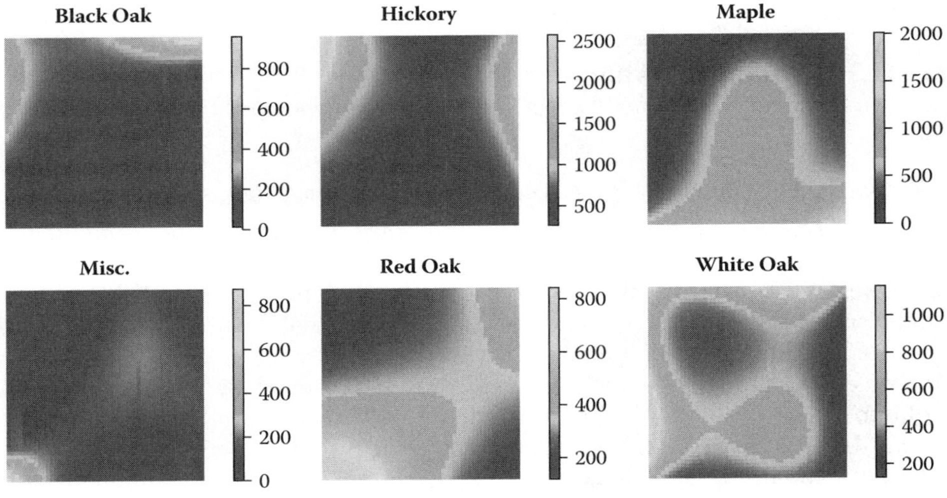

FIGURE 21.8
Maximum likelihood estimates of intensity for each species of tree in Lansing Woods assuming a separate log-cubic intensity function for each species. Compare with Figure 21.6.

Apart from the choice of reference measure μ, likelihood-based methods for marked point processes are a straightforward adaptation of likelihood methods for unmarked point processes (Section 20.4.1). The log likelihood is formally equivalent to the log likelihood of a Poisson loglinear regression, so the Berman–Turner algorithm can again be used to perform maximum likelihood estimation.

Figure 21.8 shows an application of this technique to the Lansing Woods data. Each species of tree was assumed to have an intensity of log-cubic form in the Cartesian coordinates, i.e., $\log \lambda((x, y), m)$ was assumed to be a cubic function of x, y with coefficients depending on m. This model has 60 parameters. The corresponding estimated probabilities $p(m \mid u)$ are not shown since their appearance is very similar to Figure 21.8. Confidence intervals for the coefficients and images of the estimated standard errors for $\hat{\lambda}(u, m)$ and $\hat{p}(m \mid u)$ can also be obtained from the asymptotic normal distribution of the maximum likelihood estimator.

A parametric test for segregation in multitype patterns can be performed by fitting Poisson models and testing whether certain terms in the model are nonzero. The model is separable (nonsegregated) if the log intensity is a sum $\log \lambda(u, m) = A(u) + B(m)$ of terms $A(u)$ depending only on location, and terms $B(m)$ depending only on the type of point. The presence of any terms that depend on both u and m would imply segregation.

For example, in Figure 21.8, the Lansing Woods data were modeled by a log-cubic intensity function with coefficients depending on species. This model is segregated. The null hypothesis of no segregation corresponds to a log-cubic intensity function in which only the intercept term depends on species, while other coefficients are common to all species. The likelihood ratio test yielded a test statistic of 613 on 45 df, which is extremely significant.

21.6 Exploring Interaction in Multivariate Point Patterns

Any marked point process that is not Poisson is said to exhibit stochastic dependence or interpoint interaction.

Techniques for exploratory analysis of interaction in marked point patterns depend greatly on whether the marks are categorical, continuous or otherwise. This section deals with multitype point patterns.

21.6.1 Multitype Summary Functions

The summary functions F, G, J and K (and other functions derived from K, such as L and the pair correlation function) have been extended to multitype point patterns. Like the original summary functions, these multitype summary functions rest on the assumption that the multitype point process is *stationary* (Definition 21.2 of Section 21.3.5).

21.6.1.1 A Pair of Types

Assume the multitype point process $\mathbf{X}$ is stationary. Let $\mathbf{X}_j$ denote the subpattern of points of type j, with intensity λ_j. Then, for any pair of types i and j, we define the following generalizations of the summary functions K, G and J. These are based on measuring distances from points of type i to points of type j. The *bivariate G-function* $G_{ij}(r)$ is the cumulative distribution function of the distance from a typical point of type i to the nearest point of type j. The *bivariate K-function* $K_{ij}(r)$ is $1/\lambda_j$ times the expected number of points of type j within a distance r of a typical point of type i. From the bivariate K-function, we derive the corresponding L-function

$$L_{ij}(r) = \sqrt{\frac{K_{ij}(r)}{\pi}}$$

and the bivariate analog of the pair correlation function

$$g_{ij}(r) = \frac{1}{2\pi r}\frac{\mathrm{d}}{\mathrm{d}r}K_{ij}(r).$$

The bivariate or cross-type J-function (van Lieshout and Baddeley, 1999) is

$$J_{ij}(r) = \frac{1 - G_{ij}(r)}{1 - F_j(r)},$$

where F_j is the empty space function for the process $\mathbf{X}_j$ of points of type j. Thus, $J_{ij}(r)$ is the probability that there is no point of type j within a distance r of a typical point of type i, divided by the probability that there is no point of type j within a distance r of a fixed point. The functions G_{ij}, K_{ij}, L_{ij}, g_{ij}, J_{ij} are called "bivariate," "cross-type" or "i-to-j" summary functions when $i \neq j$. Needless to say, when $i = j$ these definitions reduce to the classical summaries $G(r)$, $K(r)$, $L(r)$, $g(r)$ and $J(r)$, respectively, applied to the process of points of type i only.

The interpretation of the cross-type summary functions is different from that of the original functions F, G, K. If the component processes $\mathbf{X}_i$ and $\mathbf{X}_j$ are *independent* of each other, then we obtain $K_{ij}(r) = \pi r^2$ and consequently $L_{ij}(r) = r$ and $g_{ij}(r) = 1$, while $J_{ij}(r) = 1$. There is no requirement that X_j be Poisson (except that if this is also true, it justifies the square root transformation in $L_{ij}(r)$ as a variance-stabilizing transformation).

The cross-type G-function is anomalous here. If $\mathbf{X}_i$ and $\mathbf{X}_j$ are independent, then $G_{ij}(r) = F_j(r)$. The benchmark value $G_{ij}(r) = 1 - \exp(-\lambda_j \pi r^2)$ is true under the additional assumption that $\mathbf{X}_j$ is a uniform Poisson process (CSR).

Alternatively, suppose that the marked point process has the random labeling property (i.e., types are assigned to locations by independent identically distributed random marks; see Chapter 18, Section 21.5.1.1). Then we obtain $K_{ij}(r) = K(r)$ (and consequently $L_{ij}(r) = L(r)$ and $g_{ij}(r) = g(r)$) where $K(r)$, $L(r)$, $g(r)$ are the classical summary statistics for the

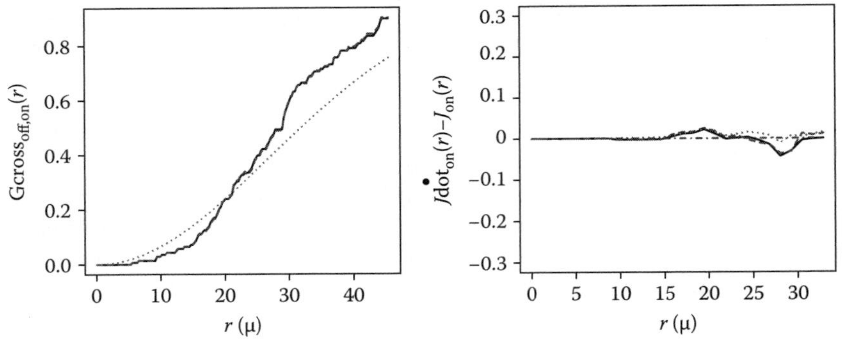

FIGURE 21.9
Assessing independence between types in the amacrine data. Left: Cross-type nearest neighbor distance distribution $G_{ij}(r)$ for the amacrine cells data (for i = "off" and j = "on"). Interpretation is inconclusive. Right: Estimated discrepancy function $J_{i\bullet}(r) - J_{ii}(r)$ for the amacrine cells data for type i = "on." Consistent with independence between the two types of cells.

point pattern regardless of marks. The functions $G_{ij}(r)$ and $J_{ij}(r)$ do not have a simple form in this case (van Lieshout and Baddeley, 1999).

The cross-type summary functions can be estimated by straightforward modifications of the techniques used for the classical summary functions. Figure 18.7 of Chapter 18, Section 18.7.1 shows the estimated bivariate K-functions $K_{ij}(r)$ for each pair of types in the amacrine cells data. This is consistent with regularity of the points of a given type, and independence between the two types of points.

The pair correlation function $g(r)$ of an (unmarked) stationary point process has the following very useful interpretation. Consider two fixed locations u and v. Let U and V be small regions containing these points, with areas $\mathrm{d}u$ and $\mathrm{d}v$, respectively. Then the probability that there will be a random point inside U and a random point inside V is

$$\mathbb{P}\left(\text{point in } U,\ \text{point in } V\right) \sim \lambda^2 g(||u - v||)\, \mathrm{d}u\, \mathrm{d}v, \tag{21.9}$$

where λ is the intensity of the process.

For a multitype point process the corresponding interpretation of the bivariate pair correlation $g_{ij}(r)$ is

$$\mathbb{P}\left(\text{point of type } i \text{ in } U,\ \text{point of type } j \text{ in } V\right) \sim \lambda_i \lambda_j g_{ij}(||u - v||)\, \mathrm{d}u\, \mathrm{d}v, \tag{21.10}$$

where λ_i, λ_j are the intensities of the points of type i and j, respectively.

21.6.1.2 One Type to Any Type

One may also generalize the classical summary functions in a different way, based on measuring distances from points of type i to points of *any* type.

The counterpart of the nearest neighbor distribution function G is $G_{i\bullet}(r)$, the distribution function of the distance from a point of type i to the nearest other point of any type. Define $K_{i\bullet}(r)$ as $1/\lambda$ times the expected number of points of any type within a distance r of a typical point of type i. Here $\lambda = \sum_j \lambda_j$ is the intensity of the entire process $\mathbf{X}$. The corresponding L-function is

$$L_{i\bullet}(r) = \sqrt{\frac{K_{i\bullet}(r)}{\pi}}$$

and the corresponding pair correlation function

$$g_{i\bullet}(r) = \frac{1}{2\pi r} \frac{\mathrm{d}}{\mathrm{d}r} K_{i\bullet}(r).$$

Finally define $J_{i\bullet}$ by

$$J_{i\bullet}(r) = \frac{1 - G_{i\bullet}(r)}{1 - F(r)},$$

where F is the empty space function of the process of points regardless of type.

Suppose the marked point process has the random labeling property. Then a typical point of type i is just a typical point of the point pattern, so $G_{i\bullet} = G(r)$, $K_{i\bullet} = K(r)$, $L_{i\bullet} = L(r)$, $g_{i\bullet} = g(r)$ and $J_{i\bullet} = J(r)$, where the functions on the right are the classical summary functions for the point process regardless of marks.

Alternatively, if the process $\mathbf{X}_i$ of points of type i is independent of $\mathbf{X}_{-i}$, the process of points of all other types, then

$$\begin{aligned}
G_{i\bullet} &= 1 - (1 - G_{ii}(r))(1 - F_{-i}(r)) \\
K_{i\bullet} &= p_i K_{ii}(r) + (1 - p_i)\pi r^2 \\
L_{i\bullet} &= (p_i K_{ii}(r)/\pi + (1 - p_i)r^2)^{1/2} \\
g_{i\bullet} &= 1 + p_i(g_{ii}(r) - 1) \\
J_{i\bullet} &= J_{ii}(r),
\end{aligned}$$

where G_{ii}, K_{ii}, g_{ii} and J_{ii} are the classical functions G, K, g, J for the points of type i only, $p_i = \lambda_i/\lambda$ is the probability of type i, and $F_{-i}(r)$ is the empty space function for the points of all types not equal to i.

21.6.2 Mark Connection Function

A simple characterization of the bivariate pair correlation g_{ij} was given in Equation (21.10) above. If we divide this by the corresponding expression for the pair correlation g in Equation (21.9), we obtain

$$\frac{\mathbb{P}\left(\text{point of type } i \text{ in } U,\ \text{point of type } j \text{ in } V\right)}{\mathbb{P}\left(\text{point in } U,\ \text{point in } V\right)} \sim p_i p_j \frac{g_{ij}(r)}{g(r)},$$

where $r = ||u - v||$ and $p_i = \lambda_i/\lambda$ is the probability of type i. The left-hand side can be interpreted as the conditional probability, given that there are points of the process at the locations u and v, that the marks attached to these points are i and j, respectively. This is sometimes called the *mark connection function*

$$p_{ij}(r) = \mathbb{P}^{u,v}\{m(u) = i,\ m(v) = j\}, \tag{21.11}$$

where $\mathbb{P}^{u,v}$ is the (second order Palm) conditional probability given that there are points of the process at the locations u and v.

Figure 21.10 shows the estimated mark connection functions $p_{ij}(r)$ for the amacrine cells data, for each pair of types i and j. The horizontal dashed lines show the values $p_i p_j$ that would be expected under random labeling. This shows that two points lying close together are more likely to be of different types than we would expect under random labeling. Although this could be termed a positive association between the cells of different types, it does not necessarily indicate dependence between the cell types; it could also be explained as an artifact of the negative association between cells of the same type.

21.6.3 Nonstationary Patterns

The exploratory summary functions defined above rest on the assumption that the point process is stationary. If this is not true, there is a risk of misinterpretation of the summary

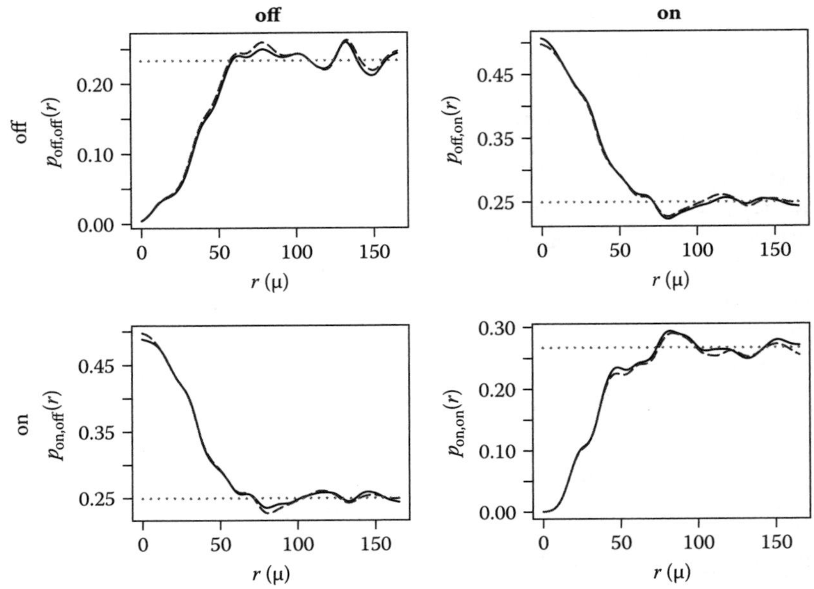

FIGURE 21.10
Array of estimated mark connection functions $p_{i,j}(r)$ of the amacrine cells data, for each pair of types i, j.

functions. This is the problem of confounding between inhomogeneity and clustering, explained in Chapter 20, Section 20.2.4.

The inhomogeneous K-function (see Baddeley, Møller, and Waagepetersen (2000) and Chapter 20, Section 20.3.2) can be generalized to inhomogeneous multitype point processes. Inhomogeneous analogs of the functions $K_{ij}(r)$ and $K_{i\bullet}(r)$ are obtained by weighting each point by the reciprocal of the appropriate intensity function, and weighting the contribution from each pair of points by the product of these weights.

21.7 Exploring Dependence of Numerical Marks

Figure 21.11 shows the locations of Norwegian spruce trees in a natural forest stand in Saxony, Germany. Each tree is marked with its diameter at breast height. The data were first analyzed by Fiksel (1984). This pattern appears to be approximately stationary. A basic question about this pattern is whether the sizes of neighboring trees are strongly dependent. Various exploratory statistics can be used.

21.7.1 Mark Correlation Function

Generalizing Equation (21.11), we may choose any "test function" $f(m_1, m_2)$ and define

$$c_f(r) = \mathbb{E}^{u,v}\left[f(m(u), m(v))\right], \tag{21.12}$$

the expected value of the test function applied to the marks at two points of the process that are separated by a distance r. We would usually normalize this by dividing by the expected

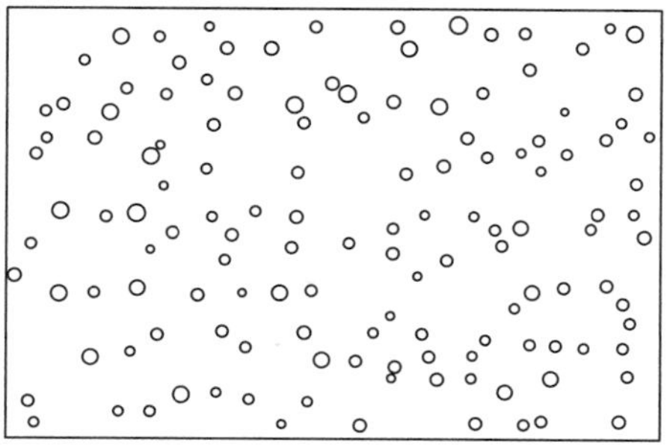

FIGURE 21.11
Spruce trees in a 56 × 38 meter sampling region. Tree diameters inflated by a factor of 4.

value under the assumption of random labeling:

$$k_f(r) = \frac{\mathbb{E}^{u,v}\left[f(m(u), m(v))\right]}{\mathbb{E}[f(M_1, M_2)]}, \tag{21.13}$$

where M_1, M_2 are independent, identically distributed random marks, which have the same distribution as the marks in the process. The function k_f is called the *mark correlation function* based on the test function f.

The test function f is any function $f(m_1, m_2)$ that can be applied to two marks $m_1, m_2 \in \mathcal{M}$, and which returns a nonnegative real value. Common choices of f are, for nonnegative real-valued marks, $f(m_1, m_2) = m_1 m_2$; for categorical marks (multitype point patterns), $f(m_1, m_2) = \mathbf{1}\{m_1 = m_2\}$; and for marks representing angles or directions, $f(m_1, m_2) = \sin(m_1 - m_2)$.

In the first case $f(m_1, m_2) = m_1 m_2$, the mark correlation is

$$k_{mm}(r) = \frac{\mathbb{E}^{u,v}\left[m(u)\, m(v)\right]}{\mathbb{E}[M_1]^2}. \tag{21.14}$$

Note that $k_f(r)$ is not a "correlation" in the usual statistical sense. It can take any nonnegative real value. The value 1 suggests "lack of correlation," under random labeling, $k_f(r) \equiv 1$. The interpretation of values larger or smaller than 1 depends on the choice of function f.

The mark correlation function $k_f(r)$ can be estimated nonparametrically. The numerator $c_f(r)$ is estimated by a kernel smoother of the form

$$\hat{c}_f(r) = \frac{\sum_{i<j} f(m_i, m_j)\kappa(||x_i - x_j|| - r) w(x_i, x_j)}{\sum_{i<j} \kappa(||x_i - x_j|| - r) w(x_i, x_j)},$$

where κ is a smoothing kernel on the real line and $w(u, v)$ is an edge correction factor. The numerator $\mathbb{E}[f(M_1, M_2)]$ is estimated by the sample average of $f(m_i, m_j)$ taken over all pairs i and j.

Figure 21.12 shows the estimated mark correlation function for the spruce trees based on $f(m_1, m_2) = m_1 m_2$. This suggests there is no dependence between the diameters of neighboring trees, except for some negative association at short distances (closer than 1 meter apart).

21.7.2 Mark Variogram

For a marked point process with real-valued marks, the *mark variogram* is

$$\gamma(r) = \frac{1}{2}\, \mathbb{E}^{u,v}\left[(m(u) - m(v))^2\right]. \tag{21.15}$$

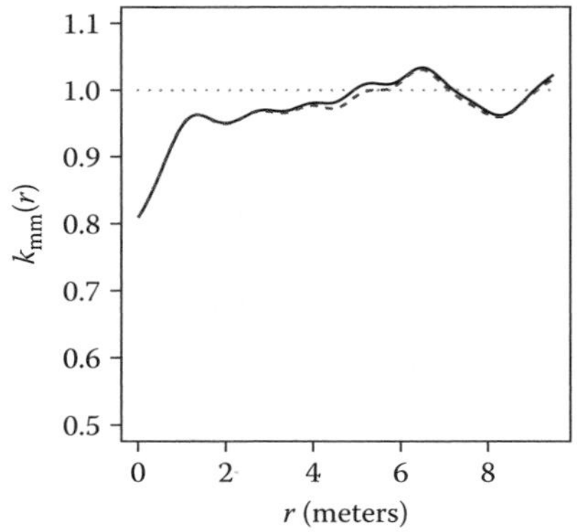

FIGURE 21.12
Estimated mark correlation function for the spruce trees.

That is, $2\gamma(r)$ is the expected squared difference between the mark values at two points separated by a distance r.

This definition is analogous to the variogram of a random field. However, the mark variogram is not a variogram in the usual sense of geostatistics (Wälder and Stoyan, 1996; Stoyan and Wälder, 2000). It may exhibit properties that are impossible or implausible for a geostatistical variogram. This occurs because the mark variogram is a conditional expectation — the expected squared difference given that there exist two points separated by a distance r — and the conditioning event is different for each value of r.

The mark variogram is of the general form (21.12) with $f(m_1, m_2) = \frac{1}{2}(m_1 - m_2)^2$. It, therefore, can be estimated using the same nonparametric smoothing methods.

21.7.3 Dependence between Marks and Locations

Another question about the spruce trees is whether the diameter of a tree depends on the spatial pattern of neighboring tree locations.

Schlather et al. (2004) defined the functions $E(r)$ and $V(r)$ to be the conditional mean and conditional variance of the mark attached to a typical random point, given that there exists another random point at a distance r away from it:

$$E(r) = \mathbb{E}^{u,v}[m(0)] \tag{21.16}$$

$$V(r) = \mathbb{E}^{u,v}[(m(0) - E(r))^2]. \tag{21.17}$$

These functions may serve as diagnostics for dependence between the points and the marks. If the points and marks are independent, then $E(r)$ and $V(r)$ should be constant (Schlather et al., 2004).

The mean mark function $E(r)$ is again of the same general form (21.12) with $f(m_1, m_2) = m_1$, and can be estimated nonparametrically using smoothing methods. Similarly $V(r)$ can be estimated by smoothing.

Figure 21.13 shows estimates of $E(r)$ and $V(r)$ for the spruces data. These graphs suggest that the diameter of a tree does not depend on the spatial pattern of surrounding trees, except possibly for a negative association at very close distances.

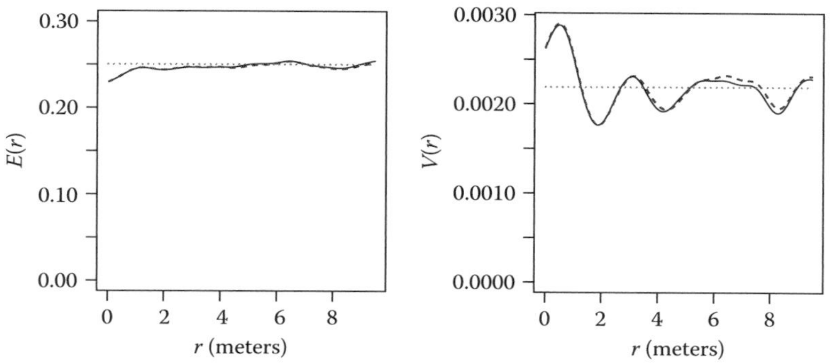

FIGURE 21.13
The functions $E(r)$ and $V(r)$ estimated for the spruce trees.

21.8 Classical Randomization Tests

When marks are present, the concept of a completely random marked point process is still well defined and unambiguous, but there are now several different types of departure from complete randomness and, consequently, several different "tests of randomness."

- *Random labeling:* Given the locations X, the marks are conditionally independent and identically distributed.
- *Independence of components:* The subprocesses $\mathbf{X}_m$ of points of each mark m are independent point processes.
- *Complete spatial randomness and independence (CSRI):* The locations $\mathbf{X}$ are a uniform Poisson point process, and the marks are independent and identically distributed.

These null hypotheses are not equivalent. CSRI implies both the random labeling property and the independence of components property. However, the properties of random labeling and independence of components are not equivalent, and they typically have different implications in any scientific application.

21.8.1 Poisson Null Hypothesis

The null hypothesis of a homogeneous Poisson marked point process can be tested by direct simulation. The left panel of Figure 21.14 shows simulation envelopes of the cross-type L function for the amacrine data. Each simulated pattern is generated by the homogeneous Poisson point process with intensities estimated from the amacrine data. The envelopes serve as the critical limits for a Monte Carlo test of the null hypothesis of CSRI.

21.8.1.1 Independence of Components

Now suppose the null hypothesis is *independence of components*; the subprocesses $\mathbf{X}_m$ of points of each mark m are independent point processes. Under the null hypothesis, we have $K_{ij}(r) = \pi r^2$, $G_{ij}(r) = F_j(r)$ and $J_{ij}(r) \equiv 1$, while the "i-to-any" functions have complicated values. Thus, we would normally use K_{ij} or J_{ij} to construct a test statistic for independence of components.

In a randomization test of the independence-of-components hypothesis, the simulated patterns X are generated from the dataset by splitting the data into subpatterns of points of one type, and randomly shifting these subpatterns independently of each other.

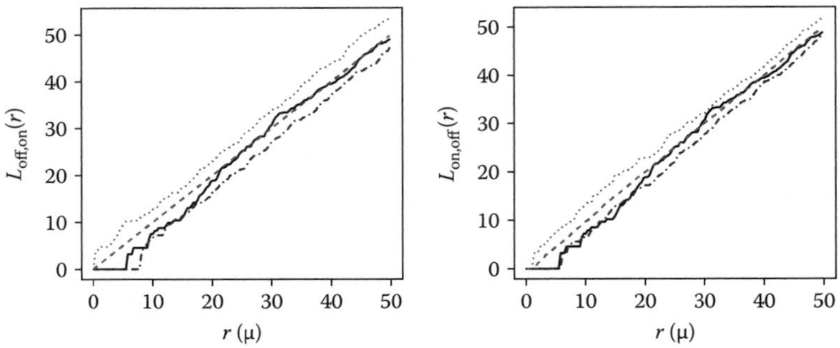

FIGURE 21.14
Monte Carlo tests for the amacrine data using the bivariate L function $L_{ij}(r)$ with i = "off" and j = "on." Left: Empirical value and envelope of $L_{ij}(r)$ from 39 simulations of a homogeneous Poisson process. Right: Empirical value and envelope of $L_{ij}(r)$ from 39 random shifts of the amacrine data.

The right panel of Figure 21.14 shows a randomization test based on envelopes of L_{ij} for the amacrine data with the simulations generated by randomly shifting the "off" cells. The outcome suggests that the independence-of-components hypothesis should be marginally rejected. Further investigation is required.

21.8.1.2 Random Labeling

In a randomization test of the random labeling null hypothesis, the simulated patterns X are generated from the dataset by holding the point locations fixed, and randomly resampling the marks, either with replacement (independent random sampling) or without replacement (randomly permuting the marks).

Under random labeling,

$$J_{i\bullet}(r) = J(r)$$
$$K_{i\bullet}(r) = K(r)$$
$$G_{i\bullet}(r) = G(r)$$

(where G, K, J are the summary functions for the point process without marks) while the other, cross-type functions have complicated values. Thus, we would normally use something like $K_{i\bullet}(r) - K(r)$ to construct a test statistic for random labeling.

Figure 21.15 shows envelopes of the function $J_{i\bullet}(r) - J(r)$ obtained from 39 random relabelings of the amacrine data. The random labeling hypothesis seems to be accepted.

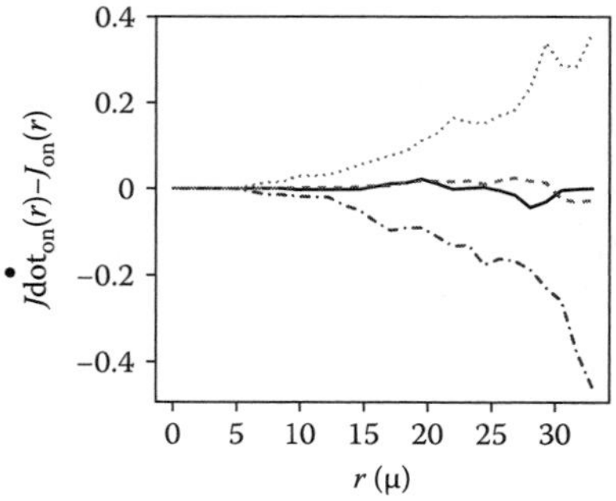

FIGURE 21.15
Envelope of $J_{i\bullet}(r) - J(r)$ from 39 random relabelings of the amacrine data (i = "on").

21.9 Non-Poisson Models

Marked point processes models that are *not* Poisson can be constructed by familiar devices (Cox processes, cluster processes, thinned processes) and also by some interesting new tricks (random field model, hierarchical models).

21.9.1 Cox and Cluster Processes

A *marked Cox process* in $\mathbb{R}^d$ with marks in $\mathcal{M}$ is simply a Cox process in $\mathbb{R}^d \times \mathcal{M}$ that satisfies the finiteness condition so that it can be interpreted as a marked point process. Similarly a *marked cluster process* is a cluster process in $\mathbb{R}^d \times \mathcal{M}$ that satisfies the finiteness condition.

In the multitype case, with marks in $\mathcal{M} = \{1, 2, \ldots, M\}$ say, a multitype Cox process is equivalent to M Cox processes in $\mathbb{R}^d$ whose random intensity functions $\Lambda_1(u), \ldots, \Lambda_M(u)$ have some arbitrary dependence structure. In a multitype cluster process, each cluster is a finite set of multitype points following some arbitrary stochastic mechanism.

Figure 21.16 shows a simulated realization of a multitype Neyman–Scott process. Each cluster contains 5 points uniformly distributed in a disk of radius $r = 0.1$, and points are independently marked as type 1 or 2 with equal probability. The marginal process of locations is a Matérn cluster process. In this particular case, the marks are conditionally independent given the locations. In other words, this is an example of a non-Poisson process that has the random marking property.

Correlated bivariate Cox models have also been studied (Diggle and Milne, 1983).

21.9.2 Mark-Dependent Thinning

Another way to generate non-Poisson marked point processes is to apply dependent thinning to a Poisson marked point process. Interesting examples occur when the thinning rule depends on both the location and the mark of each point. The left panel in Figure 21.17 shows a multitype version of Matérn's Model I obtained by generating a stationary multitype Poisson process, then deleting any point that lay closer than a critical distance r to a point of *different* type.

A slight modification to this model is a *hierarchical* version in which we first generate points of type 1 according to a Poisson process, and then generate points of type 2 according to a Poisson process conditional on the requirement that no point of type 2 lies within a distance r of a point of type 1. An example is shown in the right panel of Figure 21.17.

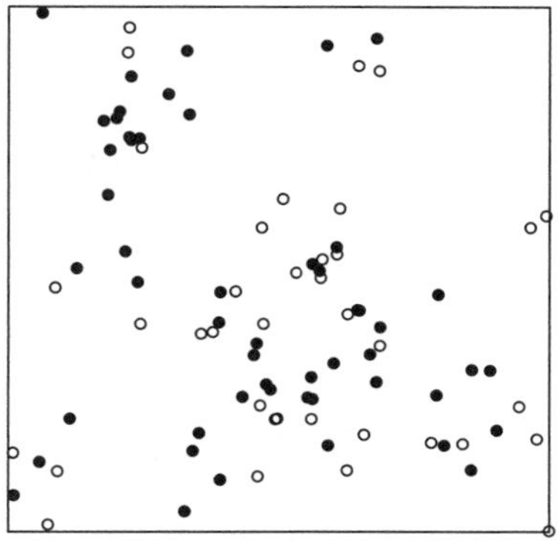

FIGURE 21.16
Simulated realization of multitype Neyman–Scott process.

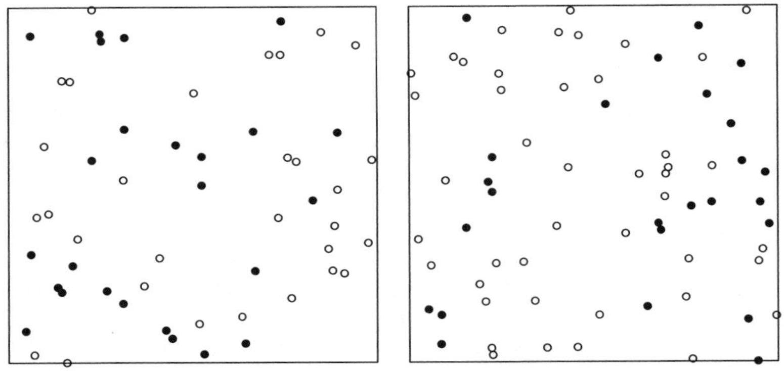

FIGURE 21.17
Multitype versions of Matérn model I. Left: Simultaneous (annihilation of each type by the other). Right: Hierarchical (annihilation of type 2 by type 1). Open circles: type 1; filled circles: type 2.

21.9.3 Random Field Marking

In a *random field model*, we start with a point process Φ in $\mathbb{R}^d$ and a random function $G(u)$ on $\mathbb{R}^d$. We assume only that X and G are independent. Then, to each point x_i in Φ we attach the mark $m_i = G(x_i)$ given by the value of the random field at that location. The result is a marked point process Ψ. The dependence structure of Ψ is complicated, even if G is a deterministic function. This model would be appropriate if the marks are values of a spatial function observed at random locations.

21.9.4 Gibbs Models

Gibbs point process models (Chapter 20, Section 20.4.3) are also available for marked point processes, and can be fitted to data using maximum pseudolikelihood. For simplicity, we restrict the discussion to *multitype* point processes.

21.9.4.1 Conditional Intensity

The conditional intensity $\lambda(u, \mathbf{X})$ of an (unmarked) point process $\mathbf{X}$ at a location u was defined in Definition 16.7 of Chapter 16. Roughly speaking $\lambda(u, \mathbf{x})\, du$ is the conditional probability of finding a point near u, given that the rest of the point process $\mathbf{X}$ coincides with $\mathbf{x}$.

For a marked point process $\mathbf{Y}$, the conditional intensity is a function $\lambda((u, m), \mathbf{Y})$ giving a value at a location u for each possible mark m. For a *finite* set of marks M, we can interpret $\lambda((u, m), \mathbf{y})\, du$ as the conditional probability finding a point *with mark* m near u, given the rest of the marked point process.

The conditional intensity is related to the probability density $f(\mathbf{y})$ by

$$\lambda((u, m), \mathbf{y}) = \frac{f(\mathbf{y} \cup \{u\})}{f(\mathbf{y})}$$

for $(u, m) \notin \mathbf{y}$.

For Poisson processes, the conditional intensity $\lambda((u, m), \mathbf{y})$ coincides with the intensity function $\lambda(u, m)$ and does not depend on the configuration $\mathbf{y}$. For example, the homogeneous Poisson multitype point process or "CSRI" has conditional intensity

$$\lambda((u, m), \mathbf{y}) = \beta_m, \tag{21.18}$$

where $\beta_m \geq 0$ are constants that can be interpreted in several equivalent ways. The subprocess consisting of points of type m *only* is Poisson with intensity β_m. The process obtained by ignoring the types, and combining all the points, is Poisson with intensity $\beta = \sum_m \beta_m$. The marks attached to the points are iid with distribution $p_m = \beta_m/\beta$.

21.9.4.2 Pairwise Interactions

A *multitype* pairwise interaction process is a Gibbs process with probability density of the form

$$f(\mathbf{y}) = \alpha \left[\prod_{i=1}^{n(\mathbf{y})} b_{m_i}(x_i)\right] \left[\prod_{i<j} c_{m_i,m_j}(x_i, x_j)\right], \tag{21.19}$$

where $b_m(u)$, $m \in \mathcal{M}$ are functions determining the "first-order trend" for points of each type, and $c_{m,m'}(u, v)$, $m, m'' \in \mathcal{M}$ are functions determining the interaction between a pair of points of given types m and m'. The interaction functions must be symmetric, $c_{m,m'}(u, v) = c_{m,m'}(v, u)$ and $c_{m,m'} \equiv c_{m',m}$. The conditional intensity is

$$\lambda((u, m); \mathbf{y}) = b_m(u) \prod_{i=1}^{n(\mathbf{y})} c_{m,m_i}(u, x_i). \tag{21.20}$$

21.9.4.3 Pairwise Interactions Not Depending on Marks

The simplest examples of multitype pairwise interaction processes are those in which the interaction term $c_{m,m'}(u, v)$ does not depend on the marks m, m'. Such processes can be constructed equivalently as follows (Baddeley and Møller, 1989):

- An *unmarked* Gibbs process is generated with first-order term $b(u) = \sum_{m \in \mathcal{M}} b_m(u)$ and pairwise interaction $c(u, v)$.
- Each point x_i of this unmarked process is labeled with a mark m_i with probability distribution $\mathbb{P}\{m_i = m\} = b_i(x_i)/b(x_i)$ independent of other points.

If additionally the intensity functions are constant, $b_m(u) \equiv \beta_m$, then such a point process has the random labeling property.

21.9.4.4 Mark-Dependent Pairwise Interactions

Various complex kinds of behavior can be created by postulating a pairwise interaction that does depend on the marks. A simple example is the *multitype hard-core process* in which $\beta_m(u) \equiv \beta$ and

$$c_{m,m'}(u, v) = \begin{cases} 1 \text{ if } \|u - v\| > r_{m,m'} \\ 0 \text{ if } \|u - v\| \le r_{m,m'}, \end{cases} \tag{21.21}$$

where $r_{m,m'} = r_{m',m} > 0$ is the hard-core distance for type m with type m'. In this process, two points of type m and m', respectively, can never come closer than the distance $r_{m,m'}$.

By setting $r_{m,m'} = 0$ for a particular pair of marks m, m', we effectively remove the interaction term between points of these types. If there are only two types, say $\mathcal{M} = \{1, 2\}$, then setting $r_{1,2} = 0$ implies that the subprocesses $\mathbf{X}_1$ and $\mathbf{X}_2$, consisting of points of types 1 and 2, respectively, are independent point processes. In other words, the process satisfies the independence-of-components property.

The *multitype Strauss process* has pairwise interaction term

$$c_{m,m'}(u, v) = \begin{cases} 1 & \text{if } \|u - v\| > r_{m,m'} \\ \gamma_{m,m'} & \text{if } \|u - v\| \le r_{m,m'}, \end{cases} \tag{21.22}$$

where $r_{m,m'} > 0$ are interaction radii as above, and $\gamma_{m,m'} \ge 0$ are interaction parameters.

In contrast to the unmarked Strauss process, which is well defined only when its interaction parameter γ is between 0 and 1, the multitype Strauss process allows some of the interaction parameters $\gamma_{m,m'}$ to exceed 1 for $m \neq m'$, provided one of the relevant types has a hard core ($\gamma_{m,m} = 0$ or $\gamma_{m',m'} = 0$).

If there are only two types, say $\mathcal{M} = \{1, 2\}$, then setting $\gamma_{1,2} = 1$ implies that the subprocesses $\mathbf{X}_1$ and $\mathbf{X}_2$, consisting of points of types 1 and 2, respectively, are independent Strauss processes.

The *multitype Strauss hard-core process* has pairwise interaction term

$$c_{m,m'}(u, v) = \begin{cases} 0 & \text{if } ||u - v|| < h_{m,m'} \\ \gamma_{m,m'} & \text{if } h_{m,m'} \leq ||u - v|| \leq r_{m,m'}, \\ 1 & \text{if } ||u - v|| > r_{m,m'} \end{cases} \tag{21.23}$$

where $r_{m,m'} > 0$ are interaction distances and $\gamma_{m,m'} \geq 0$ are interaction parameters as above, and $h_{m,m'}$ are hard core distances satisfying $h_{m,m'} = h_{m',m}$ and $0 < h_{m,m'} < r_{m,m'}$.

Care should be taken with the interpretation of the interaction parameters in multitype point process models. In a multitype Strauss or Strauss hard-core model, even if all interaction parameters γ_{ij} are less than 1, the marginal behavior of the component point processes can be spatially aggregated (Diggle, Eglen, and Troy, 2006, e.g.).

21.9.4.5 Pseudolikelihood for Multitype Gibbs Processes

Models can be fitted by maximum pseudolikelihood. For a multitype Gibbs point process with conditional intensity $\lambda((u, m); \mathbf{y})$, the log pseudolikelihood is

$$\log \mathsf{PL} = \sum_{i=1}^{n(\mathbf{y})} \log \lambda((x_i, m_i); \mathbf{y}) - \sum_{m \in \mathcal{M}} \int_W \lambda((u, m); \mathbf{y}) \, du. \tag{21.24}$$

The pseudolikelihood can be maximized using an extension of the Berman–Turner device (Baddeley and Turner, 2000).

In the multitype Strauss process (21.22), for each pair of types i and j there is an interaction radius r_{ij} and interaction parameter γ_{ij}. In simple terms, each pair of points, with marks i and j say, contributes an interaction term $\gamma_{i,j}$ if the distance between them is less than the interaction distance $r_{i,j}$. These parameters must satisfy $r_{ij} = r_{ji}$ and $\gamma_{ij} = \gamma_{ji}$. The conditional intensity is

$$\lambda((u, i), \mathbf{y}) = \beta_i \prod_j \gamma_{i,j}^{t_{i,j}(u,\mathbf{y})}, \tag{21.25}$$

where $t_{i,j}(u, \mathbf{y})$ is the number of points in $\mathbf{y}$, *with mark equal to* j, lying within a distance $r_{i,j}$ of the location u.

For illustration, a multitype Strauss process was fitted to the amacrine cells data, using the Huang–Ogatá approximate maximum likelihood method (Huang and Ogatá, 1999). All interaction radii were set to 60 microns. The fitted coefficients and their standard errors were as follows (writing 0 and 1 for the marks "off" and "on," respectively).

Parameter	$\log \beta_0$	$\log(\beta_1/\beta_0)$	$\log \gamma_{00}$	$\log \gamma_{01}$	$\log \gamma_{11}$
estimate	−6.045	0.247	−1.346	−0.100	−1.335
se	0.325	0.323	0.160	0.085	0.170

The corresponding estimates of the standard parameters are $\hat{\beta}_0 = 0.0024$, $\hat{\beta}_1 = 0.0030$, $\hat{\gamma}_{00} = 0.26$, $\hat{\gamma}_{11} = 0.26$ and $\hat{\gamma}_{01} = 0.905$. This process has strong inhibition between points of the same type, but virtually no interaction between points of different type. If γ_{01} were exactly equal to 1, the two types of points would be independent Strauss processes.

The usual t-test (of the null hypothesis that a coefficient is zero) accepts the null hypothesis $H_0 : \gamma_{01} = 1$. We conclude that the two types of points are independent.

For more detailed explanation and examples of modeling and the interpretation of model formulas for point processes, see Baddeley and Turner (2006) and Baddeley (2008).

References

M. Armstrong. *Basic Linear Geostatistics*. Springer Verlag, Berlin, 1997.

A. Baddeley. Analyzing spatial point patterns in R. Workshop notes, CSIRO, 2008. http://www.csiro.au/resources/pf16h.html

A. Baddeley and R. Turner. Practical maximum pseudolikelihood for spatial point patterns (with discussion). *Australian and New Zealand Journal of Statistics*, 42(3):283–322, 2000.

A. Baddeley and R. Turner. Modeling spatial point patterns in R. In A. Baddeley, P. Gregori, J. Mateu, R. Stoica, and D. Stoyan, Eds., *Case Studies in Spatial Point Pattern Modeling*, number 185 in Lecture Notes in Statistics, pp. 23–74. Springer-Verlag, New York, 2006.

A. Baddeley, J. Møller, and R. Waagepetersen. Non- and semiparametric estimation of interaction in inhomogeneous point patterns. *Statistica Neerlandica*, 54(3):329–350, November 2000.

A. Baddeley, R. Turner, J. Møller, and M. Hazelton. Residual analysis for spatial point processes (with discussion). *Journal of the Royal Statistical Society, series B*, 67(5):617–666, 2005.

A.J. Baddeley and J. Møller. Nearest-neighbor Markov point processes and random sets. *International Statistical Review*, 57:89–121, 1989.

D.R. Cox and E.J. Snell. *Applied Statistics: Principles and examples*. Chapman & Hall, London, 1981.

P.J. Diggle. A point process modeling approach to raised incidence of a rare phenomenon in the vicinity of a prespecified point. *Journal of the Royal Statistical Society, series A*, 153:349–362, 1990.

P.J. Diggle and R.K. Milne. Bivariate Cox processes: Some models for bivariate spatial point patterns. *Journal of the Royal Statistical Society, series B*, 45:11–21, 1983.

P.J. Diggle and B. Rowlingson. A conditional approach to point process modeling of elevated risk. *Journal of the Royal Statistical Society, series A (Statistics in Society)*, 157(3):433–440, 1994.

P.J. Diggle, S.J. Eglen, and J.B. Troy. Modeling the bivariate spatial distribution of amacrine cells. In A. Baddeley, P. Gregori, J. Mateu, R. Stoica, and D. Stoyan, Eds., *Case Studies in Spatial Point Pattern Modeling*, number 185 in Lecture Notes in Statistics, pp. 215–233. Springer, New York, 2006.

P.J. Diggle, J.A. Tawn, and R.A. Moyeed. Model-based geostatistics (with discussion). *Applied Statistics*, 47(3):299–350, 1998.

P.J. Diggle, P. Zheng, and P. Durr. Non-parametric estimation of spatial segregation in a multivariate point process. *Applied Statistics*, 54:645–658, 2005.

T. Fiksel. Estimation of parameterized pair potentials of marked and non-marked Gibbsian point processes. *Elektronische Informationsverarbeitung u. Kybernetika*, 20:270–278, 1984.

D.J. Gerrard. Competition quotient: A new measure of the competition affecting individual forest trees. Research Bulletin 20, Agricultural Experiment Station, Michigan State University, Lansing, 1969.

H. Högmander and A. Särkkä. Multitype spatial point patterns with hierarchical interactions. *Biometrics*, 55:1051–1058, 1999.

F. Huang and Y. Ogatá. Improvements of the maximum pseudo-likelihood estimators in various spatial statistical models. *Journal of Computational and Graphical Statistics*, 8(3):510–530, 1999.

J.B. Illian, E. Benson, J. Crawford, and H. Staines. Principal component analysis for spatial point processes — assessing the appropriateness of the approach in an ecological context. In A. Baddeley, P. Gregori, J. Mateu, R. Stoica, and D. Stoyan, Eds., *Case Studies in Point Process Modeling*, number 185 in Lecture Notes in Statistics, pages 135–150. Springer, New York, 2006.

A.G. Journel and C.J. Huijbregts. *Mining Geostatistics*. Academic Press, San Diego, 1978.

J.E. Kelsall and P.J. Diggle. Spatial variation in risk of disease: A nonparametric binary regression approach. *Applied Statistics*, 47:559–573, 1998.

J.F.C. Kingman. *Poisson Processes*. Oxford University Press, Oxford, 1993.

S.A.L.M. Kooijman. The description of point patterns. In R.M. Cormack and J.K. Ord, Eds., *Spatial and Temporal Analysis in Ecology*, pp. 305–332. International Co-operative Publication House, Fairland, MD, 1979.

G. Last. Some remarks on conditional distributions for point processes. *Stochastic Processes and Their Applications*, 34:121–135, 1990.

W.J. Platt, G.W. Evans, and S.L. Rathbun. The population dynamics of a long-lived conifer (*Pinus palustris*). *The American Naturalist*, 131:491–525, 1988.

B.D. Ripley and A.I. Sutherland. Finding spiral structures in images of galaxies. *Philosophical Transactions of the Royal Society of London, Series A*, 332:477–485, 1990.

M. Schlather, P. Ribeiro, and P. Diggle. Detecting dependence between marks and locations of marked point processes. *Journal of the Royal Statistical Society, series B*, 66:79–83, 2004.

D. Stoyan. A spatial statistical analysis of a work of art: Did Hans Arp make a "truly random" collage? *Statistics*, 24:71–80, 1993.

D. Stoyan and O. Wälder. On variograms in point process statistics. II: Models of markings and ecological interpretation. *Biometrical Journal*, 42:171–187, 2000.

M.N.M. van Lieshout and A.J. Baddeley. Indices of dependence between types in multivariate point patterns. *Scandinavian Journal of Statistics*, 26:511–532, 1999.

O. Wälder and D. Stoyan. On variograms in point process statistics. *Biometrical Journal*, 38:895–905, 1996.

22

Point Process Models and Methods in Spatial Epidemiology

Lance Waller

CONTENTS

22.1 Spatial Patterns in Epidemiology

In preceding chapters, authors reviewed the rich theory of spatial point process models and associated inference, including their basic probabilistic structure, first- and second-order properties, and likelihood structure. In this chapter, we explore how these models, properties, and methods can address specific questions encountered in the field of spatial epidemiology, the study of spatial patterns of disease morbidity and mortality.

22.1.1 Inferential Goals: What Are We Looking for?

The notion that the observed spatial pattern of incident (newly occurring) or prevalent (presently existing) cases could inform on underlying mechanisms driving disease is not new. Early medical geographers mapped incident cases of yellow fever near cities and docks in an effort to determine factors driving the observed patterns (Walter, 2000; Koch, 2005), but perhaps the most famous example involves the groundbreaking work of Dr. John Snow regarding cholera outbreaks in London during the middle of the nineteenth century (Snow, 1855). Dr. Snow's case maps and related calculations remain iconic examples presented in most (if not all) introductory epidemiology courses and textbooks.

As in any area of applied statistics, the definition and application of appropriate inferential techniques require a balanced understanding of the questions of interest, the data available or attainable, and probabilistic models defining or approximating the data-generating process. The central question of interest in most studies in epidemiology is the identification of factors increasing or decreasing the individual risk of disease as observed in at-risk populations or samples from the at-risk population. In spatial studies, we refine this central question to explore spatial variations in the risk of disease in order to identify locations (and more importantly, individuals) associated with higher risk of the disease.

Most epidemiologic studies are *observational* rather than experimental, that is, inference is based on associations and patterns observed within collections of individuals experiencing different sets of risk factors rather than between groups of individuals randomly assigned to different levels of the factors of primary interest, as in a randomized clinical trial (Rothman and Greenland, 1998). The observational nature of the data introduces complications in analysis and interpretation (e.g., confounding) and we explore spatial aspects of these in the sections below.

22.1.2 Available Data: Cases and Controls

In spatial studies of disease patterns, data typically present either as point locations within a given study area or as counts of cases from administrative districts partitioning the study area. Point location data often represent residence or occupation locations for cases, while district counts often arise from data collected as part of ongoing disease surveillance, disease registries, or official collections of events (Teutsch and Churchill, 1994; Brookmeyer and Stroup, 2004). We will refer to such data as *case data*. In the sections below, we focus on point-referenced data. Methods for modeling regional count data appear in Chapter 14, (this volume) and Waller and Gotway (2004, Chaps. 7 and 9).

In most epidemiologic studies, the observed pattern of case locations alone may not inform directly on the central question of the spatial pattern of individual risk since the population at risk is rarely homogeneously distributed across the study area, implying that a local concentration of the *number* of cases in a particular area does not necessarily represent a local increase in *risk*. As a result, studies in spatial epidemiology require data regarding the at-risk population. In standard terms, these data either represent "controls" (nondiseased individuals) or the entire population at risk. In our discussions below, we refer to both noncase and population-at-risk data as *control data*. As with case data, control data may present as either point or count data, but population-at-risk data most often derive from census data in the form of counts from small administrative districts.

With the central question of spatial risk and the types of available data in place, we next consider particular classes of inference for comparing the observed patterns of case and control events, with particular interest in determining whether cases cluster together in unexpected ways and, if so, where the most unusual clusters of cases occur. A very common paradigm follows a hypothesis-testing approach with spatially constant risk defining the null hypothesis and control data defining spatial heterogeneity in the population at risk. For point-level case and control event locations, Monte Carlo permutations of the observed number of cases across the entire set of case and control locations provides a computationally convenient manner to operationalize the constant risk hypothesis and generate null distributions for any summary statistic of interest. While popular, it is important to note this approach provides inference conditional on the observed case and control locations, i.e., it provides an assessment of the spatial pattern of cases over a discrete set of locations.

A less common and more computationally demanding framework for inference involves estimating or modeling the risk surface itself across spatially continuous space, allowing assessment of local peaks and valleys in order to address questions of clustering and/or clusters as well as adjustment for spatially varying risk factors. Such problems are best addressed within a spatial modeling framework, but have traditionally been hampered by likelihood functions with unwieldy normalizing constants. Recent advances in Markov chain Monte Carlo (MCMC), approaches (Møller and Waagepetersen, 2004; Diggle, Eglen, and Troy, 2006; Møller, Chapter 19; this volume) enable the required computation, and such methods see increased application in spatial epidemiology (Lawson, 2006).

The distinction between Monte Carlo-based inference of pattern across a fixed set of locations and likelihood or Bayesian model-based inference across continuous spatial support is not merely a matter of convenience and should be thoughtfully evaluated for concordance with both the assumptions of the analytic methods and the questions under study. If, as in the example below, our set of locations contain all possible locations for events, the use of a discrete support seems consistent with the spatial questions at hand. If, instead, data locations are based on a sample of potential locations (either a random sample from a larger discrete set of potential locations or a sample from a continuous spatial field), one should carefully consider whether the computational implementation effectively accounts for events that could occur at unobserved locations.

In the sections below we focus attention on statistical approaches for detecting clusters and clustering based on the data and questions outlined above.

22.1.3 Clusters and Clustering

Preceding chapters define models and indicators of clustered spatial point processes. As noted above, our underlying goals require us to seek out relative or comparative clustering. That is, we are interested in determining whether our case data appear to be more clustered *than expected under spatially constant risk*, i.e., do case data appear more clustered than our control data?

Besag and Newell (1991) provide thoughtful discussion and necessary distinctions relating to these questions. Specifically, they distinguish between methods to detect *clusters* (anomalous collections of cases) and methods to detect *clustering* (an overall tendency of cases to occur near other cases rather than to occur homogeneously among the population at risk). In addition, Besag and Newell distinguish between a search for *global* clusters/clustering occurring anywhere within the study area or *focused* clustering around predefined foci of putatively increased risk (e.g., locations or sources of suspected environmental hazards).

In the hypothesis testing paradigm, we must first define a null hypothesis of the absence of clusters and/or clustering. Such an approach builds on ideas of testing for complete spatial randomness (CSR) or other null models of spatial pattern. As noted above, in spatial epidemiology, the null model of interest is not CSR, but rather one of spatially constant risk, i.e., an individual is equally likely to experience the disease at any location. In this setting, the control data play an important role, essentially defining the local expected number of cases to which we compare our case data.

For point locations of chronic (noninfectious) disease, a heterogeneous Poisson process defines a natural probabilistic model for both case and control event locations. In this case, the null hypothesis of constant risk is equivalent to a null hypothesis equating the spatial probability density functions associated with each process. For data defined by regional counts of events rather than observed point locations, the general properties of an underlying heterogeneous Poisson process suggest a model of Poisson counts for each region with the expected number of events within a region defined by the integral of the underlying intensity surface over the region. Another option would be to model small area counts via a binomial distribution based on the local number at risk and the (null) constant risk of disease. For rare diseases, the approaches are often indistinguishable and the use of Poisson counts sees broad application in disease mapping (Chapter 14, Section 14.3 this volume) in order to maintain the link between the observed regional data and the

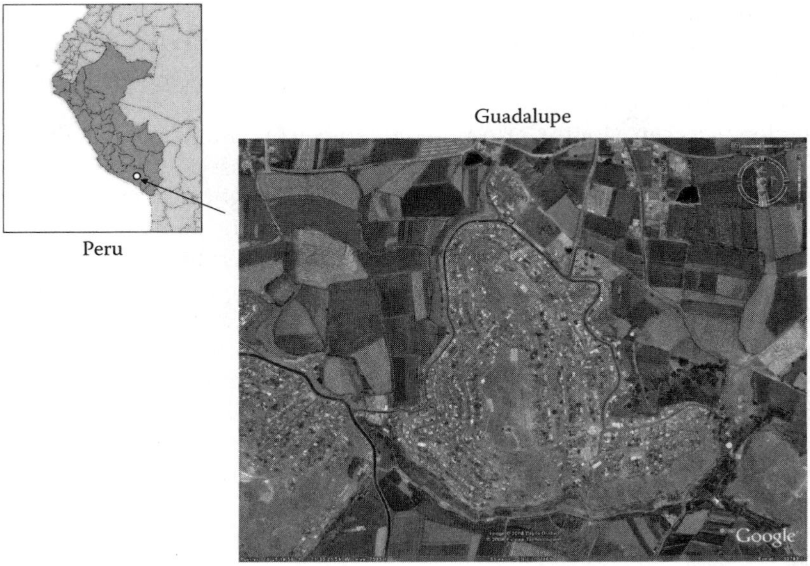

FIGURE 22.1
Google Earth image of the community of Guadalupe, Peru.

(often unobserved) underlying latent Poisson process. In many cases, the underlying point process data are unobserved and the defined set of regions represents the lowest level of geographical precision in the data with no spatial detail available within each region. Such studies are termed *ecological* in the epidemiology literature (Wakefield, 2001, 2003, 2004) and are subject to the *ecological fallacy* of inferring individual-level associations from aggregated observations (Robinson, 1950; King, 1997).

There is a large literature on various approaches for spatial epidemiology. We do not provide a comprehensive review here, but compare, contrast, and illustrate representative approaches for detecting clusters and clustering. More complete reviews of methods and applications appear in Alexander and Boyle (1996), Diggle (2000), Diggle (2003, Chap. 9), Waller and Gotway (2004, Chaps. 6–7), and Lawson (2006).

22.1.4 Dataset

To illustrate our methods, we consider a dataset originally presented in Levy, Bowman, Kawai, Waller et al. (2006). The data involve locations of households in the small community of Guadalupe, Peru. The community is perched upon a small, rocky hill between agricultural fields, as seen in Figure 22.1. Our research interest involves a study of the disease ecology of Chagas disease, a vector-borne infection caused by infection with *Trypanosoma cruzi* (*T. cruzi*) affecting 16 to 18 million people, most in Latin America (Organizacion Panamericana de la Salud, 2006; Remme, Feenstra, Lever, Médici et al., 2006). In Guadalupe, the vector for Chagas disease is *Triatoma infestans* (*T. infestans*), a member of the insect order Hemiptera ("true bugs," where one pair of wings forms a hard shell) that obtains blood meals from people and animals within the community. The adult vectors have been documented as able to fly distances of up to 450 m (or even farther on a windy day), but typically only fly under specific conditions. Locomotion by walking distances of 40 to 50 m therefore seems more probable (Vazquez-Prokopec, Ceballos, Kitron, and Gurtles 2004). The data involve global positioning system (GPS) location data for each household in the community and a label noting whether each household was infested with the vector *T. infestans* and, if so, whether any captured vectors tested positive for *T. cruzi*.

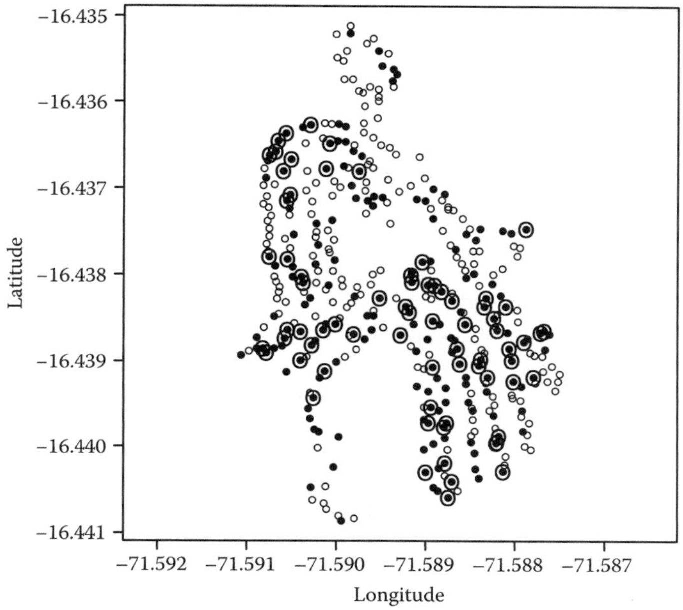

FIGURE 22.2
Locations of households in Guadalupe, Peru. Filled circles denote households infested by the Chagas disease vector *T. infestans*, and circled locations indicate households with infected vectors.

Figure 22.2 illustrates the locations of the households (all circles), the location of infested households (filled circles), and the location of households with infected vectors (circled locations). The general pattern of households roughly follows contours of the hill and illustrates the need adequately to identify controls and to define our study space in order to limit inference to the area containing houses (there are no homes in the adjacent agricultural fields).

Data collection occurred concurrently with an insecticide spraying program and two spatial questions are of primary interest:

- Are infested households randomly distributed among households in the community?
- Are households with infected vectors randomly distributed among infested houses?

These questions explore infestation patterns, and subsequent infection patterns in vectors in order to plan human surveillance, especially among children (Levy, Kawai, Bowman, Waller et al., 2007). Questions relating to clustering (Do infestation and infection tend to cluster?) and clusters (Where are unusual aggregations of infestation and/or infection of vectors?) are both of interest. In addition, the data illustrate the importance of the definition of the appropriate control group for addressing a particular question. For the first question, the locations of infested houses represent the case data and the locations of noninfested households represent the control data. For the second question, the locations of households with infected vectors represent the case data and the locations of infested households with no detected infected vectors represent the control data because we wish to test whether infection tends to cluster among households already infested with the vector. Finally, the data illustrate a setting where events clearly do relate to one another and are observed in a heterogeneous setting, generating a bit of friction between the data and our mathematical ideals of independent events in a heterogeneous setting and/or dependent events in a homogeneous setting. We will use methods from both settings in order to describe the patterns observed in the data and contrast results and conclusions.

22.2 Detecting Clusters

When testing for clusters, our goal is to identify the most unusual collections of cases, i.e., the collections of cases least consistent with the null hypothesis. Often, the identification of the particular cases (and controls) defining the suspected cluster(s) is a key point of interest for analysts. However, by definition, some collection of observations will be the most unusual in any dataset, so analysts also seek some measure of "unusualness" or statistical significance assigned to each suspected cluster.

22.2.1 Spatial Scan Statistics

We begin by considering the spatial scan statistic of Kulldorff (1997). This is one of the most widely used tests to detect clusters, primarily due to its availability in the software package `SaTScan` (Kulldorff and Information Management Services, Inc., 2006). The approach builds on the following conceptual steps:

1. Define a set of a large number of overlapping potential clusters
2. Find the most unusual of these potential clusters
3. Define a significance value associated with the most unusual clusters (i.e., determine if they are too unusual to be consistent with the null hypothesis)

An early, geographical version of the approach appears in the "geographic analysis machine" of Openshaw, Craft, Charlton, and Birch (1988). This approach used geographic information systems (GIS) to define circular potential clusters centered at each point of a fine grid covering the study area and plotted all circles containing a statistically significant excess of observed cases compared to the number of individuals at risk. In this case, the unusualness of each potential cluster was defined by a Poisson probability of observing more than the actually observed number of cases given the number expected based on a constant overall risk of disease and the number of at-risk individuals residing in the circle. The approach often resulted in maps showing suspected clusters as overlapping collections of circles (often containing many of the same cases). Statistical critiques of the approach concentrated on the large number of tests, the nonindependent nature of tests based on overlapping potential clusters, and the nonconstant variance associated with circles containing different numbers of individuals at risk. Both Turnbull, Iwano, Burnett, Howe et al. (1990) and Besag and Newell (1991) modified the general approach to address some of the statistical concerns by either observing the number of cases within spatially proximate collections of a fixed number of at-risk individuals, or observing the number of at-risk individuals associated with collections of a fixed number of observed cases, respectively. Kulldorff (1997) generalized these approaches to develop a statistical formulation of the geographical analysis machine that considered circular potential clusters of varying radii and population sizes based on the structure of scan statistics.

More specifically, the spatial scan statistic defines a set of potential clusters, each consisting of a collection of cases. The original version proposed by Kulldorff (1997) mirrors that of Openshaw et al. (1988) and the most common set of potential clusters consists of all collections of cases found within circles of radii varying from the minimum distance between two events and some maximum defined by the analysts, typically containing one-half of the study area. Recent developments extend the set of potential clusters to include elliptical sets or build collections of contiguous regions reporting unusual numbers of cases defined by spanning trees or other optimization techniques (Patil, Modarres, Myers, and Patankar, 2006; Duczmal, Cancads, and Takahashi, 2008). However, computational complexity increases as we expand to a more general set of potential clusters.

Once we define our set of potential clusters, we identify the most unusual (least consistent with the null hypothesis) cluster within the set by ranking potential clusters on some quantifiable measure of unusualness. Kulldorff (1997), building on previous work by Loader (1991) and Nagarwalla (1996), defines a likelihood ratio test statistic for each potential cluster based on an alternative hypothesis where the disease risk within the potential cluster is greater than that outside of the cluster. It is important to note that we are not defining a separate likelihood ratio *test* for each potential cluster, rather we are using the likelihood ratio *statistics* as a measure of how well the data within a given potential cluster either match or do not match the null hypothesis when compared to the hypothesis of increased risk within the potential cluster. For each potential cluster i, define $n_{i,cases,in}$ to be the number of cases inside the potential cluster and $n_{i,cases,out}$ to be the number outside, and $N_{i,all,in}$ and $N_{i,all,out}$ the numbers at risk (cases and controls) inside and outside, respectively. Under a binomial model, for potential cluster i, the likelihood ratio is proportional to

$$LR_i = \left(\frac{n_{i,cases,in}}{N_{i,all,in}}\right)^{n_{i,cases,in}} \left(\frac{n_{i,cases,out}}{N_{i,all,out}}\right)^{n_{i,cases,out}} I\left(\frac{n_{i,cases,in}}{N_{i,all,in}} > \frac{n_{i,cases,out}}{N_{i,all,out}}\right),$$

where $I(\cdot)$ denotes the indicator function (i.e., we only consider potential clusters where the empirical risk observed inside is greater than that observed outside of the potential cluster). The most likely cluster is then defined as the potential cluster maximizing LR_i over the set of all potential clusters i.

In order to assign a statistical significance level to the most likely cluster, Kulldorff (1997) avoids the multiple comparisons problem resulting from testing each potential cluster by constructing an overall permutation test of significance based on random assignment of the total number of cases among the fixed case and control locations. Such assignment is termed *random labeling* and provides an operational approach to simulating the constant risk hypothesis when conditioning on the overall set of event locations and the overall number of cases and controls. Waller and Gotway (2004, pp. 159–161) discuss some subtle differences between the null hypothesis of constant risk and that of random labeling, but note that the two are effectively equivalent after taking the conditioning into account.

Kulldorff's permutation test proceeds as follows. Let LR_s^* denote the maximum of the likelihood ratio statistics observed for each random permutation s of the cases among the full set of event locations, regardless of the location of the most likely cluster in each permutation. Using a large number of such permutations, we define the histogram of maximum likelihood ratio statistics as an approximation to the distribution of maximum likelihood ratio statistics under the random labeling hypothesis. Ranking the observed maximum likelihood ratio statistic associated with the most likely cluster provides a Monte Carlo p-value defining the statistical significance of the unusualness of the observed most likely cluster among the unusualness of the most likely cluster expected under random labeling. Again, note that the test is based on the unusualness of the most likely cluster, regardless of location, thereby avoiding the multiple comparisons problem resulting from conducting statistical tests on each potential cluster. As a result, we obtain a collection of cases representing the most unusual collection of cases, and a measure of the statistical significance of this collection, when compared to the distribution of the most unusual collection of cases arising from randomly labeled case-control data.

In some instances, a dataset contains several collections of cases that each rank above the Monte Carlo critical value for the most unusual cluster. It is common practice to report such

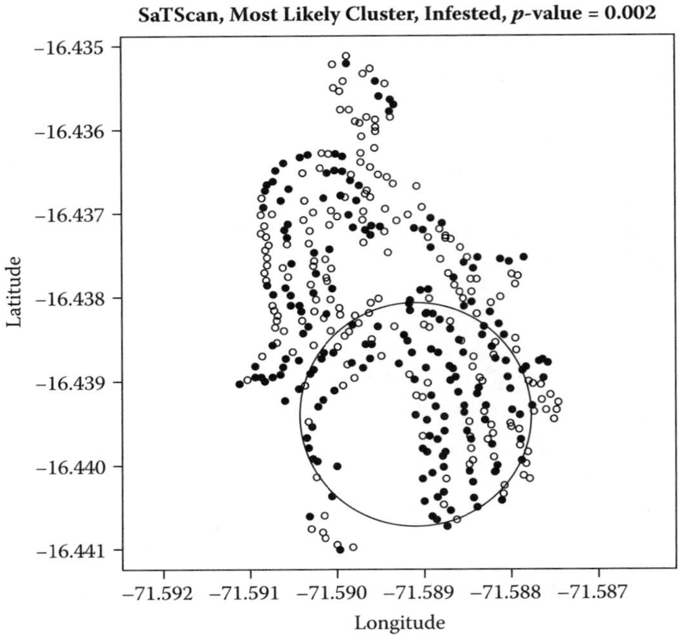

FIGURE 22.3
Most likely cluster of infested households compared to the set of noninfested households, based on a circular scan statistic.

clusters as "significant" as well, but their precise interpretation is a bit more convoluted with multiple comparisons again entering the picture.

22.2.1.1 Application to the Chagas Disease Vector Data

To illustrate the approach, we apply the spatial scan statistic to the Chagas disease vector data described above. We consider the set of potential clusters based on circular collections of households with radii ranging from the minimum between-household distance to circles containing up to one-half of the households in the study area. While conceptually simple to define in terms of distances, we note that the irregular spacing in our data implies that the radius of the largest "circular" potential cluster differs for each household, and that some "circles" will span gaps representing areas with no households, as illustrated in Figure 22.3. In concept, one could construct collections of potential clusters more closely based on particular hypothesized mechanisms of clustering (e.g., likely vector dispersal ranges or anisotropic spread adjusting for elevation); however, for simplicity of exposition, we utilize circular clusters in the example below.

Treating the set of infested households as cases (filled circles) and the set of noninfested households as controls (open circles), the most likely cluster appears in the south-central portion of the study area as indicated in Figure 22.3. The Monte Carlo p-value associated with the cluster is 0.002, based on 999 simulations under the random labeling null hypothesis. That is, this collection of households defines the highest likelihood ratio statistic of any (circular) collection of households observed in the data, and a higher statistic value than the maximum likelihood ratio statistic observed in all but two of the simulated permutations of cases among the set of all households.

Turning to the question of the pattern of infected vectors among infested households, we now compare the locations of households containing infected vectors to the locations of

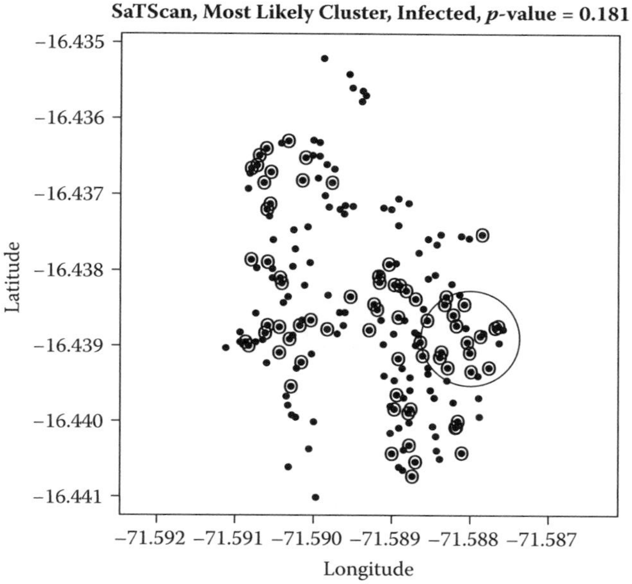

FIGURE 22.4
Most likely cluster of infected households compared to the set of infested, but not infected households, again, based on a circular scan statistic.

infested but not infected households. The most likely cluster, shown in Figure 22.4, occurs in an area adjacent to the most likely cluster of infested households. However, it is not significant at the 0.05 level, again based on 999 random permutations.

To summarize, we observe statistically significant clustering of infested households among the set of households, but not significant clustering of infected households among the set of infested households. In both cases, the spatial scan statistic identifies the most likely clusters out of our set of potential clusters. We note that in Figure 22.3 the most likely cluster contains areas with no households (near the peak of the hill). As mentioned above, there are extensions to spatial scan statistics that consider more general shapes of clusters by expanding the set of potential clusters considered, but that these come at a computational cost. We next define an alternative approach designed to identify clusters without a predefined set of potential clusters.

22.2.2 Comparing First-Order Properties

Another approach for comparing spatial point patterns of cases to that of controls involves a spatial comparison of first-order properties or intensity functions of the associated spatial point processes. Bithell (1990) and Lawson and Williams (1993) both suggest the use of estimates of case and control intensity (or density) functions to describe spatial variations in risk. Kelsall and Diggle (1995) extend the idea to focus on the local relative risk of observing a case versus observing a control at any particular location. Under the null hypothesis of spatially constant risk of disease, the spatial intensities of the case and the control should be equal up to a proportionality constant defined by the relative total number of cases and controls.

More specifically, suppose we observe n_0 control and n_1 case locations. Let $\lambda_0(x)$ and $\lambda_1(x)$ denote the intensity functions of controls and cases, respectively, and $f_0(x)$ and $f_1(x)$ the corresponding spatial density functions for an event at location x given that the event occurs within the study area of interest. We detail such an approach provided by Kelsall and Diggle (1995) who define the spatial log relative risk function based on the natural logarithm of the spatial densities of cases to that of controls, i.e.,

$$r(x) = \log\left(\frac{f_1(x)}{f_0(x)}\right),$$

so that

$$r(x) = \log\{\lambda_1(x)/\lambda_2(x)\} - \log\left\{\int_D \lambda_1(u)du \bigg/ \int_D \lambda_0(u)du\right\},$$

where integration is over our study area D. Note that the second term reflects the overall relative frequency of cases to that of controls, thus, all of the spatial information in $r(x)$ is contained in the log ratio of intensity functions. The peaks (valleys) of the $r(x)$ surface correspond to areas where cases are more (less) likely than controls.

Typically, estimation of $r(x)$ involves the logarithm of the ratio of estimated density or intensity surfaces based on the observed case and control data. Most applications to date utilize nonparametric kernel estimation, although one may extend the methods to include other covariates via generalized additive models (GAMs) (Hastie and Tibshirani, 1990). Note that the GAM approach can be implemented with the R package `mgcv`, as described in Wood (2006).

For illustration, we focus on kernel estimates here, obtaining an estimate

$$\tilde{r}(x) = \frac{n_0 \sum_{i=1}^{n_1} K[(x - x_{1,i})/b]}{n_1 \sum_{j=1}^{n_0} K[(x - x_{0,j})/b]},$$

where $K(\cdot)$ denotes a (two-dimensional) kernel function, $x_{1,i}$ the ith case location, $x_{0,j}$ the jth control location, and b the bandwidth. We use the same bandwidth for both cases and controls in order to avoid variations in the spatial log relative risk surface due solely to different levels of smoothness induced by different bandwidths.

For inference, we compare our estimated surface $\tilde{r}(x)$ to its expected value under the null hypothesis, $r(x) = 0$ for all $x \in D$. For addressing a global test of clustering anywhere within the study region, consider the statistic

$$\int_D \{\tilde{r}(u)\}^2\, du,$$

which summarizes all deviations between the estimated case and control intensities across the study area. Significance is based on comparing the observed statistic value to that obtained from random labeling permutations of the n_1 cases among the $n_0 + n_1$ locations.

For local inferences identifying areas where cases are more (or less) likely to occur than controls, we may use the data to calculate $\tilde{r}(x_{grid})$ for a fine grid of locations $x_{grid} \in D$. At each point in x_{grid}, we construct a 95% tolerance interval based on the 2.5th and 97.5th percentiles of $\tilde{r}(x)_{grid}$ obtained at that location under a large number of random labeling permutations of cases among the case and control locations. Again, we maintain a fixed bandwidth each time to avoid confounding by differing levels of smoothing.

22.2.2.1 Application to the Chagas Disease Vector Data

To illustrate the approach, Figure 22.5 shows the case and control kernel estimates for comparing infested to noninfested locations in the Guadalupe data based on a Gaussian kernel with a bandwidth of $b = 0.0005$ degrees of latitude or longitude. The proximity of Guadalupe to the equator and the small geographic size of the study area allows us to use decimal degrees as distance units without too much distortion. A quick visual comparison

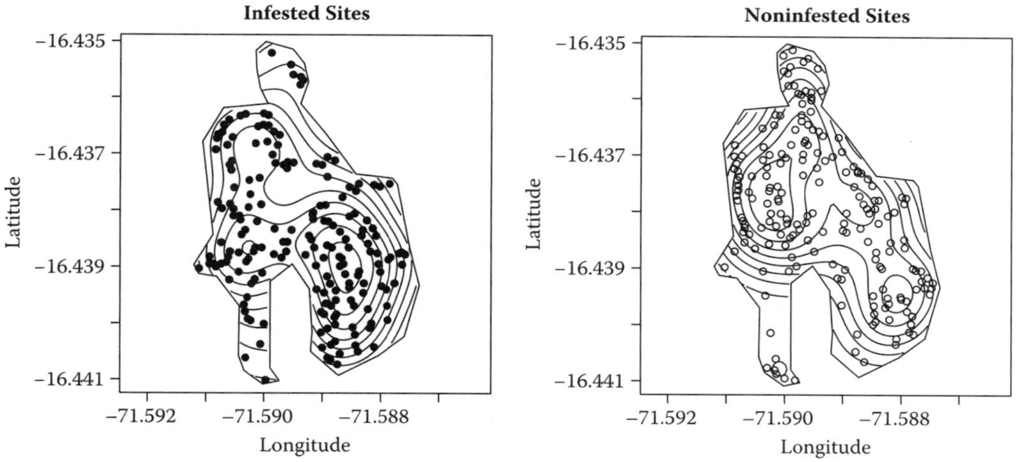

FIGURE 22.5
Kernel estimates of the intensity functions associated with households infested (left) and not infested (right) with the Chagas disease vector *T. infestans*.

indicates a mode of infested households in the southeastern portion of the study area, and a mode of noninfested households in the northeast.

Figure 22.6 shows the log relative risk surface based on the ratio of the two density estimates in Figure 22.5. The right-hand figure highlights areas falling outside the random-labeling based 95% pointwise tolerance intervals where vertical hatching indicates areas of significantly increased risk of cases and horizontal hatching indicates areas of significantly decreased risk of cases. Comparing results to the spatial scan statistic results in Figure 22.3, we find both approaches identify a cluster in the southern portion of the study area, but that the log relative risk surface is not limited to circular clusters and can identify areas of significantly increased or decreased risk with irregular boundaries, the smoothness of which depends on the user-defined bandwidth. In particular, the northeastern area of significantly decreased risk of infestation with the Chagas vector does not fit neatly into a circular boundary and remains undetected by a circle-based spatial scan statistic.

We next examine whether locations of households with *T. cruzi* positive vectors are clustered among the set of infested households. Figure 22.7 provides the kernel estimates of the spatial densities of sites with infected vectors and those with vectors, but not infection. The densities visually suggest increased spatial concentration of households with infected vectors compared to those with noninfected vectors. Figure 22.8 reveals areas of significant increased risk of infection (among vectors) in small, localized areas: one in the northwest, and two in the central-eastern concentration of infested households. We also observe an area of significantly reduced infection among households in the northeastern portion of the study area.

Taken together, the relative risk surfaces add more geographical precision to the patterns initially revealed through the spatial scan statistic results. Of particular interest is the increased risk of infestation in a large area in the south and, within that area, some pockets of increased risk of infection among these vectors. The cluster of infection in the northwest may suggest a separate focus of introduction from neighboring areas. This result invites further, more detailed comparisons and closer examination of laboratory data from the northern and southern pockets of increased risk of infection in order to determine if these areas reflect localized outbreaks (among vectors) or are simply local manifestations of the same spread of infection among vectors in the community. Also of interest is the area in

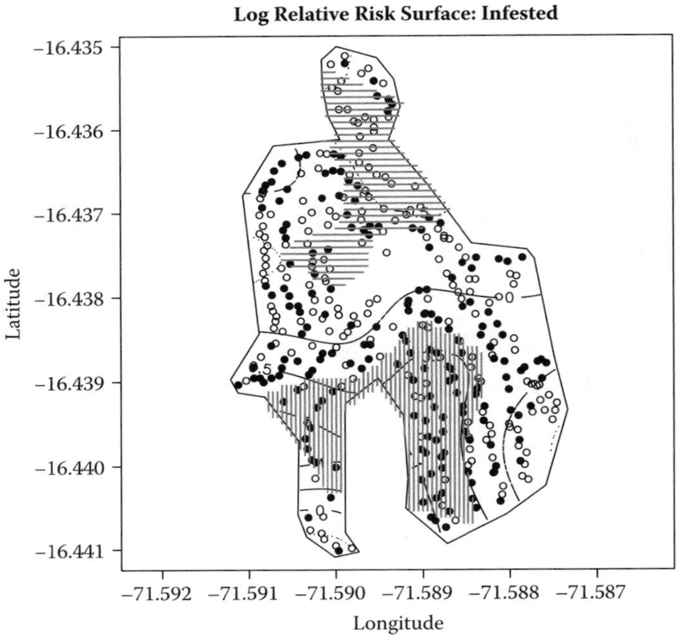

FIGURE 22.6
Log relative risk surface comparing locations of households infested with *T. infestans* to noninfested households. Solid/dashed contours indicate log relative risk values above/below the null value of zero, respectively. Vertical stripes indicate locations with log relative risk values exceeding the upper 97.5th percentile and horizontal stripes indicate locations with log relative risk values below the 2.5th percentile of log relative risk values based on random labeling.

the northeast that experiences lower than expected levels of infestation, and within those households that are infested, a lower than expected level of infection, a pattern missed by the spatial scan statistic due to its irregular geographical shape. Follow-up studies could build on the detected spatial patterns, but likely require additional data (e.g., genetic strains of the vector and/or the pathogen) in order to determine whether this area simply is the last to be infested as the vectors move into the community, or whether other factors impact the observed reduction in both infestation and infection.

22.3 Detecting Clustering

We now turn to approaches to detect *clustering*, the general tendency of cases to occur near other cases. Tests of clustering often summarize patterns across the study area and provide some sort of global measure of clustering averaged across the observed event locations. As noted above, the key point of interest in spatial epidemiology is whether the summary of clustering for cases differs from that for controls and, if so, whether this difference is consistent with differences we might expect by chance allocation of cases among the set of case and control locations (random labeling).

While a variety of approaches for summarizing clustering exists (Diggle, 2003; Waller and Gotway, 2004), one of the most popular is a comparison of second-order properties of the observed point processes as a function of distance, typically through the use of Ripley's K

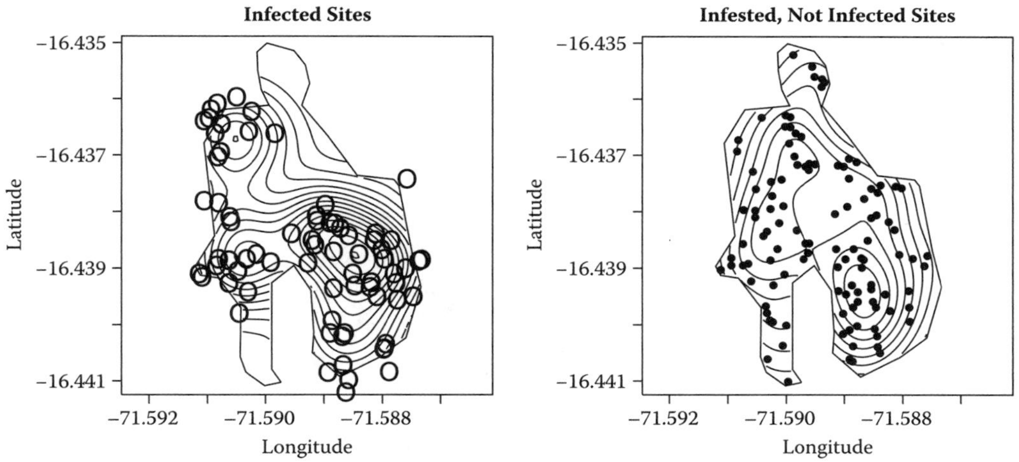

FIGURE 22.7
Kernel estimates of the intensity functions associated with infested households containing Chagas disease vectors *T. infestans* infected (left) and not infected (right) with *T. cruzi*.

function (Ripley, 1976, 1977), the scaled number of additional events expected with distance h of a randomly selected event, i.e.,

$$K(h) = \frac{E[\text{number of additional events within } h \text{ of a } \textit{randomly} \text{ chosen event}]}{\lambda}, \tag{22.1}$$

for distance $h > 0$ (Diggle, Chap. 18, this volume). Since the K-function is scaled by the assumed constant overall intensity λ, under complete spatial randomness (CSR), $K(h) = \pi h^2$, the area of a disk with radius h. Besag (1977) defines a rescaled and diagnostically convenient transformation, often termed the L function (Besag, 1977), $L(h) = (K(h)/\pi)^{1/2}$. Under CSR, $L(h) = h$, for all distances $h > 0$ so a plot of h versus $L(h) - h$ provides a convenient reference line at zero; values above 0 represent more clustering than expected under CSR and values below zero less clustering than expected under CSR. However, the interpretability of K-function as a scaled expectation of additional observed events is lost. Note that a single process can include both greater-than-expected and less-than-expected clustering at different distances. For example, one could observe regularly spaced clusters or clusters of regular patterns (Waller and Gotway, 2004, Chap. 5).

In general, spatial point pattern second-order analyses typically begin by comparing observed patterns to CSR. However, recall that our goal in spatial epidemiology is to compare the estimated K (or L) functions for the set of case locations to that of the control locations, often using random labeling permutations to define the expected variability of our estimates under a null hypothesis that the two sets of observed event locations arise from underlying stochastic processes with the same second-order structure.

Second-order analysis often uses Ripley's edge-corrected estimate of the K-function for n observed event locations, namely,

$$\widehat{K}(h) = \widehat{\lambda}^{-1} \sum_{i=1}^{n} \sum_{\substack{j=1 \\ j \neq i}}^{n} w_{ij}^{-1} \delta(d(i, j) < h), \tag{22.2}$$

where w_{ij} denotes a weight defined as the proportion of the circumference of the disk centered at event i with radius $d(i, j)$, which lies within the study area, and $\delta(d(i, j) < h) = 1$

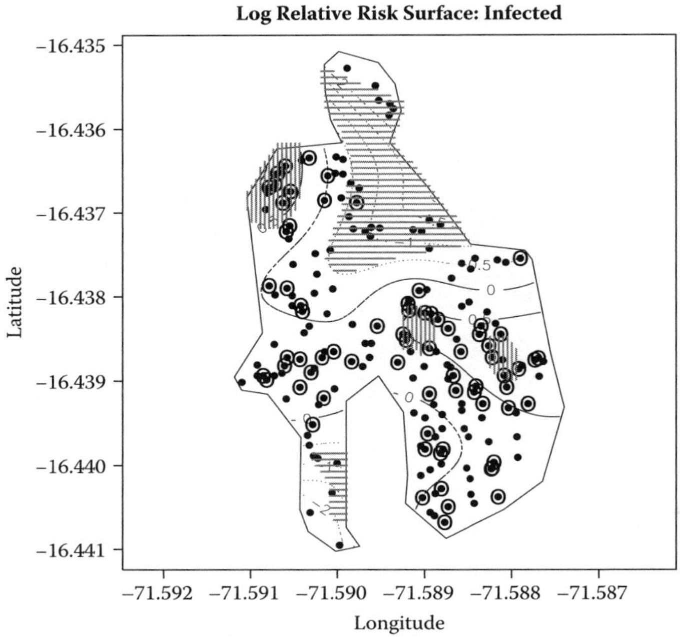

FIGURE 22.8
Log relative risk surface comparing locations of households infested with *T. cruzi*-infected *T. infestans* to infested households with no observed positive vectors. Solid/dashed contours indicate log relative risk values above/below the null value of zero, respectively. Vertical stripes indicate locations with log relative risk values exceeding the upper 97.5th percentile and horizontal stripes indicate locations with log relative risk values below the 2.5th percentile of log relative risk values based on random labeling.

when $d(i, j) < h$ and zero otherwise. The weight w_{ij} denotes the conditional probability of an event occurring at distance $d(i, j)$ from event i falling within the study area, given the location of event i and assuming a homogeneous process with fixed intensity $\lambda = \widehat{\lambda} = n/|A|$, the number of events divided by the geographic area of the study region. Note that $w_{ij} = 1$ if the distance between events i and j is less than the distance between event i and the boundary of the study area. Also, note that w_{ij} need not equal w_{ji} and that Equation (22.2) is applicable even if the study area contains "holes." The corresponding estimate of the L function becomes $\widehat{L}(h) = (\widehat{K}(h)/\pi)^{1/2}$.

In the spatial epidemiology setting, we often begin by displaying the estimated L functions for cases, controls, and all locations combined to see similarities and differences between the summarized patterns at different distances. For inference, we again rely on random labeling, conditioning on the total set of observed event locations, then randomly selecting n_1 locations to represent cases and the remaining n_0 locations to represent controls. If we estimate $\widehat{L}_{rl,1}(h)$ for each of these sets of assigned "cases," such Monte Carlo permutations of n_1 cases among the $n_0 + n_1$ locations allow us to define the expected value of $L_1(h)$ at each distance h and to construct pointwise tolerance envelopes around this expected value. In an exploratory setting, comparing $\widehat{L}_1(h)$ estimated from the observed data to the random-labeling median and tolerance envelopes illustrates distances for which the observed case pattern differs significantly from the patterns expected under random case assignment. For more formal inference, any particular distance (or range of distances) of *a priori* interest may be tested via random labeling permutations (Stoyan, Kendall, and Mecke 1995, p. 51; Waller and Gotway, 2004, pp. 139–140).

Diggle and Chetwynd (1991) propose a more direct comparison of the case and control K functions by exploring the difference of their respective estimates

$$KD(h) = \widehat{K}_1(h) - \widehat{K}_0(h),$$

itself a function of distance. By taking differences, the expected value under the null hypothesis is zero for all distances $h > 0$, obviating the need for the transformation to the L function. Inference again follows random labeling permutations of cases among all event locations and the construction of pointwise tolerance envelopes around the null value.

22.3.1 Application to the Chagas Disease Vector Data

To illustrate the approaches, we again apply the methods to the Guadalupe Chagas disease vector data. Figure 22.9 presents $\widehat{L}_1(h) - h$ based on observed case locations (infested households) as a thick line. Unlike standard introductory applications of the L-function, in our application we are not interested in comparing $L(h)$ to a null value of zero (indicating complete spatial randomness), rather we wish to compare the level of clustering observed in the infested households to range of clustering resulting from random allocation of infestation among the full set of households. To this end, the top graph in Figure 22.9 also displays the pointwise median and 95% tolerance envelopes of the estimated L-function associated with random labeling permutations of the infested households among the n_0+n_1 households in the community. After an initial dip at small distances corresponding to the average between-household spacing, the L-function for cases closely follows the upper 95% tolerance band, suggesting an observed amount of clustering on the upper end of what we might expect from random infestation of households by the disease vector, with two brief excursions outside of the upper tolerance bound for distances of approximately 0.0005 and 0.0010 decimal degrees. While suggestive, the top graph in Figure 22.9 does not provide overwhelming evidence that vectors tend to infest households closer together than we might suspect under random allocation. We note in particular that, while the L function suggests increased clustering of cases and of controls over that observed in the set of all locations, the amount of additional clustering with the set of cases appears to be fairly similar to that observed within the set of controls.

In comparison, the bottom graph in Figure 22.9 shows the difference in estimated K-functions between infested and noninfested households along with pointwise median and 95% tolerance limits based on 999 random labeling permutations. The difference between K-functions falls well within the 95% tolerance bands suggesting no evidence for significant increased clustering of infested households above and beyond that observed in the control households.

At first glance, the results shown in the two plots in Figure 22.9 seem contradictory with the L-function suggesting borderline clustering and the difference of K-functions plot suggesting very little evidence of clustering at all. This difference in inference primarily results from whether or not one uses case information alone or coupled case and control information in each random permutation of cases in order to define the null distribution of interest. In the top (L-function) plot, we compare patterns of clustering observed in the cases alone to the patterns of cases randomly sampled from the set of all households. While we draw cases from the set of all households, the measure of clustering (the L-function) does not directly include information on control locations. In the bottom (KD) plot, we summarize a contrast between clustering in cases to that in controls for the observed and each randomly labeled dataset.

More specifically, the results illustrate the importance of a clear understanding of how a particular test summarizes clustering and clear identification of the null or reference distribution used for inference (Waller and Jacquez, 1995). In the top plot, we compare the

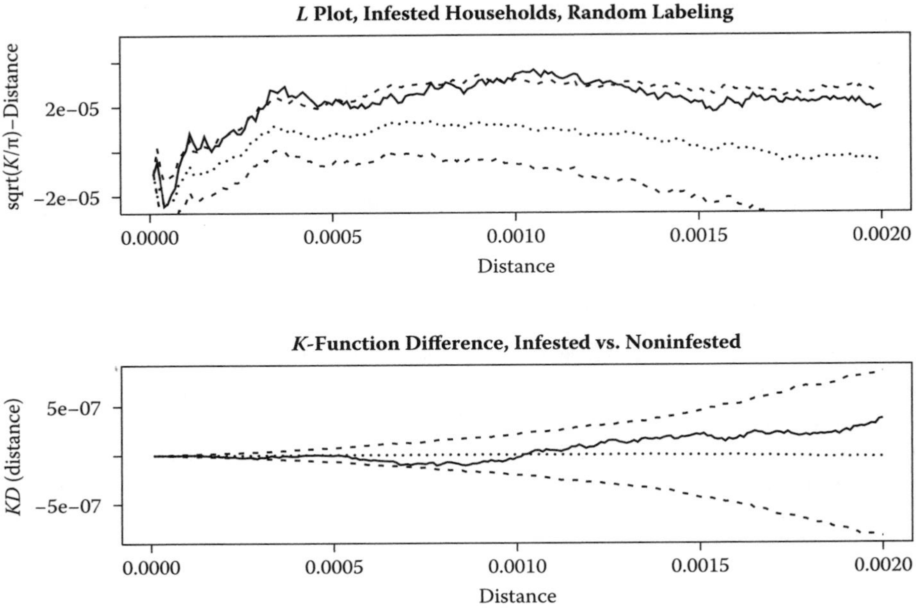

FIGURE 22.9
The top plot shows the *L*-function plot for households infested with *T. infestans*, compared to the range of *L*-function estimates for 999 random assignments of infestation status among households. The bottom plot shows the difference in *K*-functions for households infested and households not infested with *T. infestans*, compared to the range of *K*-function differences arising from 999 random assignments of infestation status among households. Distances for both plots are reported in decimal degrees of latitude and longitude.

estimated *L*-function summary of the observed case pattern to those based on randomly selected case locations from the set of all locations. It appears that the observed *L*-function is more clustered than that based on most random selections of the same number of cases from the set of all household locations. In contrast, the bottom plot compares the difference in *K*-functions between the observed set of cases to that of the observed set of controls, then compares $KD(h)$ to its randomly labeled counterparts by comparing each random selection of cases to its corresponding set of controls. That is, the difference of *K*-functions compares the observed set of case locations *relative to* its corresponding set of control locations, and uses random labeling permutations to generate inference directly comparing the pattern of cases to that of controls within each permutation. Importantly, the difference between the top and bottom plots is not due to the use of the *L* function or the *K* function (e.g., we could recreate the bottom figure based on differences in the two *L* functions), rather the difference is between using a summary of case information alone to a measure contrasting cases and controls. Finally, we also note much greater stability in the estimated median and quantiles of the $KD(h)$ than $L(h)$, and a more interpretable null value, $KD(h) = 0$ for all h, even when both cases and controls are clustered. In conclusion, we observe a set of case locations exhibiting a certain level of clustering, but not more clustering than is observed in the set of locations of controls complementing each set of cases.

Turning to the patterns of *T. cruzi* infection among vectors within the set of infested households, the estimated *L* functions in the top plot in Figure 22.10 more strongly suggest increased clustering of case locations (households with infected vectors) among all households and in case locations directly compared to control locations (households with no infected vectors). This is reflected in Figure 22.10 where the estimated *L*-function for the observed case locations and the difference between case and control *K*-functions clearly

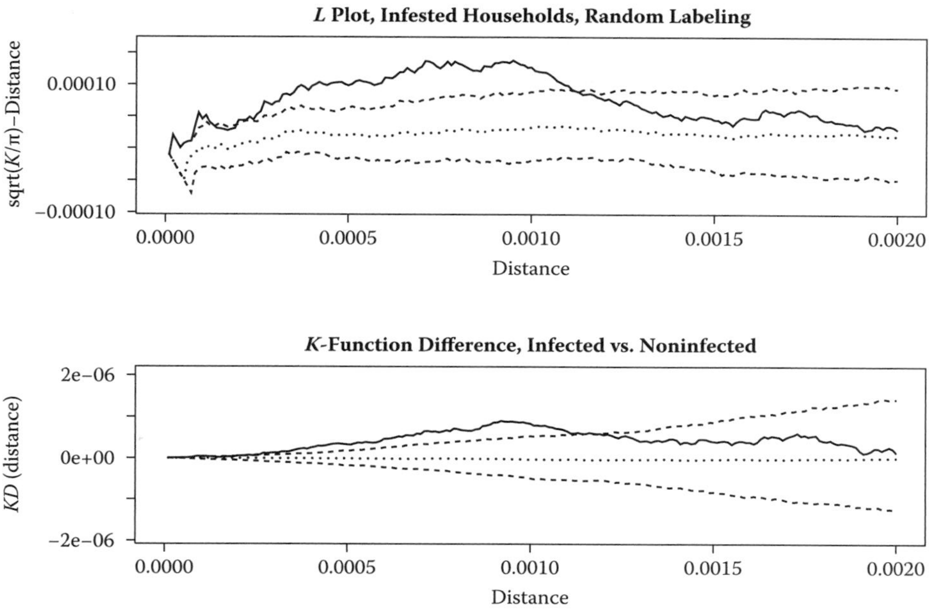

FIGURE 22.10
The top plot shows the *L*-function plot for infested households with infected vectors, compared to *L* function estimates for 999 random assignments of infection status among infested households. The bottom plot shows the difference in *K* functions for households with infected and households with noninfected vectors, compared to the range of *K* function differences arising from 999 random assignments of infestation status among households. Distances are reported in decimal degrees of latitude and longitude.

exceed their respective upper 95% tolerance bounds for the same extended range of distances. In this case, not only do case locations exhibit clustering, but the observed case locations exhibit significant clustering relative to that observed in the complementary set of controls.

22.4 Discussion

Point process methods provide valuable statistical tools for the spatial analysis of point-referenced epidemiology data. However, as noted above, the questions of primary interest diverge from many of the standard results and assumptions of point process methods in general, e.g., CSR does not represent our primary null hypothesis nor even a particularly useful comparison scenario, and most goals involve the comparison of first- or second-order properties between two separate observed processes: the pattern of cases and the pattern of controls.

The methods described above illustrate how most analytic techniques focus on one aspect of the observed processes, i.e., they search for the most likely cluster(s), or they examine summary measures of clustering. Most methods also focus on either first- or second-order properties by comparing intensity estimates assuming independent event locations, or comparing interevent interactions assuming a constant intensity. In reality, most epidemiological applications are likely to violate both assumptions. Violations of interevent independence occur due either to an inherently infectious nature of the disease or to the presence of unmeasured local environmental heterogeneity impacting disease risk, while violations of constant intensity arise due to inhomogeneous distributions of the at-risk populations. In addition, neither first- nor second-order properties uniquely summarize a spatial pattern (Baddeley and Silverman, 1984), and accurate interpretation of results requires thought and care.

Turning to inference, we find Monte Carlo random assignment of cases among the observed set of case and control locations provides a valuable approach for inference, but also carries some implicit conditioning on observed locations. Recalling the issue of discrete versus continuous spatial support raised at the beginning of the chapter, we illustrated the use of summaries originally developed for continuous spatial support (intensity functions and K functions) applied to summarize spatial patterns on a discrete set of locations. The use of permutations over a discrete sample space to provide inference need not preclude the use of summaries based on continuous support and, as with the bootstrap and other general nonparametric techniques, may allow for more accurate inference for small sample sizes than the use of continuous approximations, provided an appropriate method of sampling locations is available.

Methodological research continues to add to the analytic toolbox for spatial epidemiology. Recent developments in likelihood and Bayesian modeling for spatial point processes based on Markov chain Monte Carlo algorithms (Geyer, 1999; Møller and Waagepetersen, 2004; Diggle et al., 2006; and Møller, Chap. 19, this volume) offer opportunities for increased use of model-based inference, but also require care in the development of models directly suited for the comparison of two (or more) processes on the appropriate spatial support in order to address epidemiological questions. In addition, there is a growing literature addressing the critical need to incorporate covariate information within spatial epidemiological studies. The hypotheses of interest revolve around the respective patterns of cases and controls and this relationship is often modified by additional risk factors at the individual (e.g., age) and communitity level (e.g., neighborhood socio-economic status). One previously mentioned approach is the use of GAMs to describe associations between covariates and point process intensities via smooth nonparametric functions (Wood, 2006). Other approaches include those proposed by Diggle, Gomez-Rubio, Brown, Chetwynd et al. (2007), and Henrys and Brown (2008) for inhomogeneous point processes, parametric models for point processes (Møller, Chap. 19, this volume), and Guan, Waagepetersen, and Beale (2008) creative use of covariate information to provide asymptotic results for intensity estimation.

The applications to the spatial distribution of the Chagas disease vector *T. infestans* illustrate many of the issues involved. The results suggest a single significant cluster of infested households among circular collections of households using the spatial scan statistic, but no significant clusters among circular collections of households with vectors testing positive for *T. cruzi*. The ratio of kernel intensity estimates provides log relative risk surfaces that suggest geographically refined, locally significant increases in relative risk. These areas occur in areas roughly similar to those identified by the scan statistic when comparing infested to noninfested households, but the estimated log-relative risk surfaces also suggest a noncircular area of reduced relative risk in the northeast, which is missed by the scan statistic. The areas with suggestive increases in relative risk of infected versus noninfected vectors are small and compact and in slightly different locations than the (statistically nonsignificant) most likely circular cluster highlighted by the spatial scan statistic. The discrepancy here reveals how scan statistics may miss clusters that fall outside of the class of potential clusters under consideration. Nonparametric methods, such as the kernel-based, relative risk surfaces, allow more general shapes, but the class of potential clusters remains limited by the user-selected level of smoothness allowed in the surfaces, e.g., the bandwidth in our kernel approach. Next, our analysis of clustering suggests clustering among observed case locations compared to randomly selected cases among household locations, but no increase in clustering in the pattern of infested houses compared to the corresponding pattern of noninfested houses. Finally, we observe significant clustering of households with infected vectors versus those with no infected vectors.

The different results each partially reveal different aspects of the underlying processes. While at first glance a finding of a significant cluster without significant clustering or differing results for different methods of assessing clustering appear worrisome, further consideration reveals that the presence of a single cluster may not trigger a high summary value of clustering and that comparing the pattern of observed to randomly selected cases may differ from comparing the case and control pattern for each random selection of cases. More specifically, for the infestation data our scan statistic and log-relative risk surface results identify pockets of higher incidence of cases compared to that of controls, but our second-order analysis finds that, while both cases and controls appear clustered for a range of distances, they do not appear to be clustered with respect to interevent distance in manners different from one another. In contrast, the pattern of households with infected vectors appear to be more clustered at smaller interevent distances than those with noninfected vectors, yet the clusters resulting from this tendency are smaller and more localized.

The spatial scale of clusters and clustering helps explain some of this difference. The range of significant clustering in infection suggested by the difference of K functions in Figure 22.10 covers distances from approximately 0.0003 to 0.0012 decimal degrees, where each tick mark in Figure 22.2 represents a distance of 0.0010 decimal degrees. This range of significant clustering suggests differences in spatial clustering at a range of values reflecting aggregation at distances somewhat larger than the radius of the cluster shown in Figure 22.4.

Placed in the context of the Chagas disease vector, the results reveal increased concentrations of vector infestation broadly spread across the southern portion of Guadalupe, with small pockets of *T. cruzi* infection among vectors in these areas. The area of increased infection in the northeast suggested in Figure 22.8 merits further investigation to determine whether the cluster is stable across bandwidth values, and whether any external foci of infection exists in the region bordering this edge. In addition, the observed scale of clustering for infection is somewhat larger than the typical geographic range of individual vectors, and may suggest infection processes operating at ranges beyond that of simple vector migration.

In summary, point process summaries and measures offer insight into observed patterns in spatially referenced data in epidemiology. No single measure summarizes all aspects of the patterns of disease, but thoughtful application of a variety of measures and linkage to current knowledge of the disease of interest can reveal aspects of the process behind the patterns.

References

F.E. Alexander and P. Boyle. *Methods for Investigating Localized Clustering of Disease*. IARC Scientific Publications No. 135, International Agency for Research on Cancer, Lyon, France, 1996.

A. Baddeley and B.W. Silverman. A cautionary example for the use of second order methods for analyzing point patterns. *Biometrics*, 40:1089–1093, 1984.

J. Besag. Discussion of "Modeling Spatial Patterns" by B.D. Ripley. *Journal of the Royal Statistical Society, Series B*, 39:193–195, 1977.

J. Besag and J. Newell. The detection of clusters in rare diseases. *Journal of the Royal Statistical Society Series A*, 154:327–333, 1991.

J.F. Bithell. An application of density estimation to geographical epidemiology. *Statistics in Medicine*, 9:691–701, 1990.

R. Brookmeyer and D.F. Stroup. *Monitoring the Health of Populations: Statistical Principles and Methods for Public Health Surveillance*. Oxford University Press, Oxford, 2004.

P.J. Diggle, S.J. Eglen and J.B. Troy. Modelling the bivariate spatial distribution of amacrine cells. In P. Gregori, J. Mateau, R. Stoica, A. Baddeley, and D. Stoyan, Eds., *Case Studies in Spatial Point Process Modeling*. Springer, New York, 2006.

P.J. Diggle, V. Gomez-Rubio, P.E. Brown, A.G. Chetwynd and S. Gooding. Second-order analysis of inhomogeneous spatial point processes using case-control data. *Biometrics*, 63:550–557, 2007.

P.J. Diggle. Overview of statistical methods for disease mapping and its relationship to cluster detection. In P. Elliott, J.C. Wakefield, N.G. Best, and D.J. Briggs, Eds., *Spatial Epidemiology: Methods and Applications*. Oxford University Press, Oxford, 2000.

P.J. Diggle. *Statistical Analysis of Spatial Point Patterns, 2nd ed*. Arnold, London, 2003.

P.J. Diggle and A.G. Chetwynd. Second-order analysis of spatial clustering for inhomogeneous populations. *Biometrics*, 47:1155–1163, 1991.

L. Duczmal, A.L.F. Cancado and R.H.C. Takahashi. Delineation of irregularly shaped disease clusters through multiobjective optimization. *Journal of Computational and Graphical Statistics*, 17:243–262, 2008.

C. Geyer. Likelihood inference for spatial point processes. In O.E. Barndorff-Neilsen, W.S. Kendall, and M.N.M. van Lieshout, Eds., *Stochastic Geometry: Likelihood and Computation*, pp. 79–140. Chapman & Hall, Boca Raton, FL, 1999.

Y. Guan, R. Waagepetersen and C.M. Beale. Second-order analysis of inhomogeneous spatial point processes with proportional intensity functions. *Journal of the American Statistical Association*, 103:769–777, 2008.

T.J. Hastie and R.J. Tibshirani. *Generalized Additive Models*. Chapman & Hall, New York, 1990.

P.A. Henrys and P.E. Brown. Inference for clustered inhomogeneous spatial point processes. *Biometrics*, 65:423–430, 2008. DOI: 10.1111/j.1541-0420.2008.01070.x.

J.E. Kelsall and P.J. Diggle. Non-parametric estimation of spatial variation in relative risk. *Statistics in Medicine*, 14:2335–2342, 1995.

G. King. *A Solution to the Ecological Inference Problem*. Princeton University Press, Princeton, NJ, 1997.

T. Koch. *Cartographies of Disease: Maps, Mapping, and Medicine*. ESRI Press, Redlands, CA, 2005.

M. Kulldorff. A spatial scan statistic. *Communications in Statistics—Theory and Methods*, 26:1487–1496, 1997.

M. Kulldorff and International Management Services, Inc. *SaTScan v. 7.0: Software for the spatial and space-time scan statistics*. National Cancer Institute, Bethesda, MD, 2006.

A.B. Lawson. *Statistical Methods in Spatial Epidemiology, 2nd ed*. John Wiley & Sons, Chichester, U.K., 2006.

A.B. Lawson and F.L.R. Williams. Applications of extraction mapping in environmental epidemiology. *Statistics in Medicine*, 12:1249–1258, 1993.

M.Z. Levy, N.M. Bowman, V. Kawai, L.A. Waller, J.G. Cornejo del Carpio, E. Cordova Benzaquen, R.H. Gilman, and C. Bern. Periurban *Trypanosoma cruzi*-infected *Triatoma infestans*, Arequipa, Peru. *Emerging Infectious Diseases*, 12:1345–1352, 2006.

M.Z. Levy, V. Kawai, N.M. Bowman, L.A. Waller, L. Cabrera, V.V. Pinedo-Cancino, A.E. Seitz, F.J. Steurer, J.G. Cornejo del Carpio, E. Cordova-Benzaquen, J.H. Maguire, R.H. Gilman, and C. Bern Targeted screening strategies to detect *trypanosoma cruzi* infection in children. *PLoS Neglected Tropical Diseases*, 1(e103):doi:10.1371/journal.pntd.0000103, 2007.

C.R. Loader. Large-deviation approximations to the distribution of scan statistics. *Advances in Applied Probability*, 23:751–771, 1991.

J. Møller and R.P. Waagepetersen. *Statistical Inference and Simulation for Spatial Point Processes*. Chapman & Hall, Boca Raton, FL, 2004.

N. Nagarwalla. A scan statistic with a variable window. *Statistics in Medicine*, 15:845–850, 1996.

S. Openshaw, A.W. Craft, M. Charlton, and J.M. Birch. Investigation of leukaemia clusters by use of a geographical analysis machine. *Lancet*, 1:(8580):272–273, 1988.

Organizacion Panamericana de la Salud. Estimacion cuantitative de la enfermedad de chagas en las Americas. Technical report, Organizacion Panamericana de la Salud, Montevideo, Uruguay, 2006.

G.P. Patil, R. Modarres, W.L. Myers, and P. Patankar. Spatially constrained clustering and upper level set scan hotspot detection in surveillance geoinformatics. *Environmental and Ecological Statistics*, 13:365–377, 2006.

J.H.F. Remme, P. Feenstra, P.R. Lever, A. Médici, C.M. Morel, M. Noma, K.D. Ramaiah, F. Richards, A. Seketeli, G. Schuminis, W.H. van Brakel, and A. Vassall. Tropical diseases targeted for elimination: Chagas disease, lymphatic filariasis, onchocerciasis, and leprosy. In Breman, J.G., et al., Eds., *Disease Control Priorities in Developing Countries, 2nd ed.* pp. 433–449. The World Bank and Oxford University Press, New York, 2006.

B.D. Ripley. The second-order analysis of stationary point patterns. *Journal of Applied Probability*, 13:255–266, 1976.

B.D. Ripley. Modeling spatial patterns (with discussion). *Journal of the Royal Statistical Society, Series B*, 39:172–212, 1977.

W.S. Robinson. Ecological correlations and the behavior of individuals. *American Sociological Review,*, 15:351–357, 1950.

K.J. Rothman and S. Greenland. *Modern Epidemiology, 2nd ed.* Lippincott-Raven Publishers, Philadelphia, 1998.

J. Snow. *On the Mode of Communication of Cholera*. John Churchill, London, 1855.

D. Stoyan, W.S. Kendall, and J. Mecke. *Stochastic Geometry and Its Applications,* Second edition. John Wiley & Sons, Chichester, 1995.

S.M. Teutsch and R.E. Churchill. *Principles and Practice of Public Health Surveillance*. Oxford University Press, New York, 1994.

B.W. Turnbull, E.J. Iwano, W.S. Burnett, H.L. Howe, and L.C. Clark. Monitoring for clusters of disease: Application to leukemia incidence in upstate New York. *American Journal of Epidemiology*, 132, (suppl) S136–S143, 1990.

G.M. Vasquez-Prokopec, L.A. Ceballos, U. Kitron, and R.E. Gurtler. Active dispersal of natural populations of *triatoma infestans* (hemiptera:Reduviidae) in rural northwestern Argentina. *Journal of Medical Entomology*, 41:614–621, 2004.

J. Wakefield. A critique of ecological studies. *Biostatistics*, 1:1–20, 2001.

J. Wakefield. Sensitivity analyses for ecological regression. *Biometrics*, 59:9–17, 2003.

J. Wakefield. Ecological inference for 2×2 tables (with discussion). *Journal of the Royal Statistical Society, Series A*, 167:385–445, 2004.

L.A. Waller and G.M. Jacquez. Disease models implicit in statistical tests of disease clustering. *Epidemiology*, 6:584–590, 1995.

L.A. Waller and C.A. Gotway. *Applied Spatial Statistics for Public Health Data*. John Wiley & Sons, Hoboken, NJ, 2004.

S.D. Walter. Disease mapping: A historical perspective. In P. Elliott, J.C. Wakefield, N.G. Best, and D.J. Briggs, Eds., *Spatial Epidemiology: Methods and Applications*, pp. 223–239. Oxford University Press, Oxford, 2000.

S. Wood. *Generalized Additive Models: An Introduction with R*. Chapman & Hall/CRC Press, Boca Raton, FL, 2006.

Part V

Spatio-Temporal Processes

A probabilistic view of space–time processes, in principle, can just regard time as one more coordinate and, hence, a special case of a higher-dimensional spatial approach. Of course, this is not appropriate for dynamic spatially referenced processes, which are the topic of this section, as time has a different character than space. There has been a lot of recent work on space–time models, and a variety of ad hoc approaches have been suggested. This is perhaps one of the areas of this book where there is real scope for a unifying approach to methodology.

We begin the section with a theory Chapter 23 by Gneiting and Guttorp. This chapter contains a short description of some probabilistic models of the type of processes we have in mind, and focuses thereafter on modeling space–time covariance structures. Chapter 24, by Gamerman, focuses on dynamic models, typically state–space versions, describing temporal evolution using spatially varying hidden states. The following two chapters deal with models appropriate for some different types of data. Chapter 25 by Diggle and Gabriel extends the material in Part IV to deal with temporal evolution of spatial patterns, while Chapter 26 by Brillinger looks at trajectories in space, where potential functions and stochastic differential equations are useful tools. The final chapter in Part V Chapter 27, by Nychka and Anderson, deals with assimilating data into numerical models. Generalizing engineering approaches, such as the Kalman filter and ensemble methods using Bayesian techniques, are found productive in this context.

23

Continuous Parameter Spatio-Temporal Processes

Tilmann Gneiting and Peter Guttorp

CONTENTS

23.1 Introduction

In Chapter 2, we discussed spatial stochastic processes $\{Y(s) : s \in \mathcal{D} \subseteq \mathbb{R}^d\}$ where the domain of interest was Euclidean. Turning now to a spatio-temporal domain, we consider processes

$$\{Y(s, t) : (s, t) \in D \subseteq \mathbb{R}^d \times \mathbb{R}\}$$

that vary both as a function of the spatial location, $s \in \mathbb{R}^d$, and time, $t \in \mathbb{R}$. It is tempting to assume that the theory of such processes is not much different from that of spatial processes, and from a purely mathematical perspective this is indeed the case. Time can be considered an additional coordinate and, thus, from a probabilistic point of view, any spatio-temporal process can be considered a process on $\mathbb{R}^{d+1} = \mathbb{R}^d \times \mathbb{R}$. In particular, all technical results on spatial covariance functions (Chapter 2) and least-squares prediction or kriging (Chapters 2 and 3) in Euclidean spaces apply to space–time problems, by applying the respective result on the domain $\mathbb{R}^{d+1}$.

From a physical perspective, this view is insufficient. Time differs intrinsically from space, in that time moves only forward, while there may not be a preferred direction in space. Furthermore, spatial lags are difficult, if not impossible, to compare with temporal lags, which come in distinct units. Any realistic statistical model will make an attempt to take into account these issues. In what follows, we first describe some physically oriented approaches toward constructing space–time processes. Thereafter, we turn to covariance structures for Gaussian processes that provide geostatistical models in the spirit of Chapter 2. The section closes with a data example on wind speeds in Ireland.

While our focus is on spatio-temporal processes in $\mathbb{R}^d \times \mathbb{R}$, other spatio-temporal domains are of interest as well. Monitoring data are frequently observed at fixed temporal lags, so it may suffice to consider discrete time. In atmospheric, environmental and geophysical applications, the spatial domain of interest is frequently expansive or global, and then the curvature of the Earth needs to be taken into account. In this type of situation, it becomes critical to consider processes that are defined on the sphere, and on the sphere cross time. A simple way of constructing such a process is by defining a random function on $\mathbb{R}^3 \times \mathbb{R}$ and restricting it to the desired domain. Jones (1963), Stein (2005), and Jun and Stein (2007) use these and other approaches to construct covariance models on global spatial or spatio-temporal domains.

23.2 Physically Inspired Probability Models for Space–Time Processes

A basic, physically motivated, spatio-temporal construction derives directly from a purely spatial formulation. Consider a second-order stationary spatial stochastic process with covariance function C_S on $\mathbb{R}^d$, and suppose that its realizations move time-forward with random velocity vector $\boldsymbol{V} \in \mathbb{R}^d$. The resulting stochastic process is stationary with covariance function $C(\boldsymbol{h}, u) = \mathbb{E}\, C_S(\boldsymbol{h} - \boldsymbol{V}u)$ on the space–time domain $\mathbb{R}^d \times \mathbb{R}$, where the expectation is taken with respect to the random vector $\boldsymbol{V}$. In the case of a fixed velocity vector $\boldsymbol{v} \in \mathbb{R}^d$, we talk of the *frozen field* model, and the expectation reduces to

$$C(\boldsymbol{h}, u) = C_S(\boldsymbol{h} - \boldsymbol{v}u). \tag{23.1}$$

This general idea dates back at least to Briggs (1968) and was applied to precipitation fields by Cox and Isham (1988), as reviewed by Brillinger (1997). The frozen field model often proves useful when spatio-temporal processes are under the influence of prevailing winds or ocean currents. For a recent application see Huang and Hsu (2004).

In Chapter 2, we noted that the solution to Whittle's (1954, 1963) stochastic partial differential equation is a stationary and isotropic Gaussian spatial process with the Matérn correlation function (2.13). This approach can be extended to spatio-temporal domains. For example, Jones and Zhang (1997) generalize Whittle's equation to

$$\left[\left(\frac{\partial^2}{\partial s_1^2} + \cdots + \frac{\partial^2}{\partial s_d^2} - \frac{1}{\theta^2} \right)^{(2\nu+d)/4} - c\,\frac{\partial}{\partial t} \right] Y(\boldsymbol{s}, t) = \delta(\boldsymbol{s}, t), \tag{23.2}$$

where $\boldsymbol{s} = (s_1, \ldots, s_d)' \in \mathbb{R}^d$ is the spatial coordinate, $t \in \mathbb{R}$ is the temporal coordinate, and δ is a spatio-temporal Gaussian white noise process. The solution $\{Y(\boldsymbol{s}, t) : (\boldsymbol{s}, t) \in \mathbb{R}^d \times \mathbb{R}\}$ is a stationary Gaussian process whose spatial margins have a Matérn covariance function. The space–time covariance function on $\mathbb{R}^d \times \mathbb{R}$ is of a more complex form and allows for an integral representation not unlike Equation (23.3) below. Kelbert, Leonenko, and Ruiz-Medina (2005) and Prévôt and Röckner (2007) develop and review probabilistic theory for these and similar types of processes, and stochastic partial differential equations in general.

Brown, Kåresen, Roberts, and Tonellato (2000) construct spatio-temporal processes from physical dispersion models, which might apply to atmospheric phenomena, such as the spread of an air pollutant. The resulting space–time covariance functions on $\mathbb{R}^d \times \mathbb{R}$ allow for integral representations, with

$$C(\boldsymbol{h}, u) = \int_0^\infty e^{-\lambda(2v+|u|)}\, g(\boldsymbol{h}; u\boldsymbol{\mu}, (2v + |u|)\boldsymbol{\Sigma}_1 + \boldsymbol{\Sigma}_2)\, dv, \tag{23.3}$$

being one such example. Here, $\lambda \ge 0$, $\boldsymbol{\mu} \in \mathbb{R}^d$, and the positive definite matrices $\boldsymbol{\Sigma}_1, \boldsymbol{\Sigma}_2 \in \mathbb{R}^{d\times d}$ are parameters, and $g(\cdot\,; \boldsymbol{m}, \boldsymbol{S})$ denotes a multivariate normal density with mean vector $\boldsymbol{m}$ and covariance matrix $\boldsymbol{S}$. Again, the respective processes are generated by stochastic differential equations.

If one aims to model dynamically evolving spatial fields, an interesting notion is that of a stochastic flow (Kunita, 1990). It can be thought of as a stochastic process that describes, say, the flow of the atmosphere over a continuous domain. Technically, a homeomorphism is a bicontinuous mapping. A *stochastic flow* of homeomorphisms then is a dynamic stochastic process $\phi_{t,u}(s, \omega)$, where $t, u \in [0, T]$ represents time and $s \in \mathbb{R}^2$ represents space, with the properties that $\phi_{t,u}(\omega) = \phi_{t,u}(\cdot, \omega)$ is an onto homeomorphism from $\mathbb{R}^2$ to $\mathbb{R}^2$, $\phi_{t,t}(\omega)$ is the identity map, and

$$\phi_{t,v}(\omega) = \phi_{t,u}(\omega) \circ \phi_{u,v}(\omega)$$

for all times $t, u, v \in [0, T]$. A technical discussion requires a rather extensive set of probabilistic tools, for which we refer to Billingsley (1986) and Breiman (1968). Phelan (1996) used birth and death processes and stochastic flows to describe the atmospheric transport, formation and dissolution of rainfall cells in synoptic precipitation, and developed statistical tools for fitting these models.

While the physical background of the above stochastic process models is appealing, there have been relatively few applications to data, to our knowledge.

23.3 Gaussian Spatio-Temporal Processes

As noted in Chapter 2, Gaussian processes are characterized by their first two moments. The essential modeling issue then is to specify the mean structure and covariance structure. Similarly to spatial processes, one often employs the decomposition

$$Y(s, t) = \mu(s, t) + \eta(s, t) + \epsilon(s, t),$$

where $s \in \mathbb{R}^d$ and $t \in \mathbb{R}$. Here, $\mu(s, t)$ is a deterministic space–time trend function, the process $\eta(s, t)$ is stationary with mean zero and continuous sample paths, and $\epsilon(s, t)$ is an error field with mean zero and discontinuous realizations, which is independent of η. The error field is often referred to as a *nugget effect*, and assumed to have second-order structure

$$\text{Cov}\{\epsilon(s + h, t + u), \epsilon(s, t)\} = a\, \mathbb{I}\{(h, u) = (\mathbf{0}, 0)\} + b\, \mathbb{I}\{h = \mathbf{0}\} + c\, \mathbb{I}\{u = 0\}, \qquad (23.4)$$

where a, b and c are nonnegative constants, and $\mathbb{I}$ denotes an indicator function. The second and the third term can be referred to as a purely spatial and a purely temporal nugget, respectively.

In the simplest case, the space–time trend function, $\mu(s, t)$, decomposes as the sum of a purely spatial and a purely temporal trend component. The purely spatial component then can be modeled in the ways discussed in Chapter 14. Temporal trends are often periodic, reflecting diurnal or seasonal effects, and can be modeled with trigonometric functions or nonparametric alternatives. In addition to the spatial and temporal coordinates, the trend component might depend on environmental temporal and/or spatial covariates, such as temperature or population density, as in the air pollution study of Carroll, Chen, George, Li et al. (1997). Finally, the trend component can be modeled stochastically. Rather than pursuing this approach, we refer to Kyriakidis and Journel (1999).

A stochastic process on the space–time domain $\mathbb{R}^d \times \mathbb{R}$ is *second-order stationary* if it is a second-order stationary process on the Euclidean domain $\mathbb{R}^{d+1}$, in the sense defined in Chapter 2. Thus, a second-order stationary spatio-temporal process $\{\eta(s, t) : (s, t) \in \mathbb{R}^d \times \mathbb{R}\}$ has a constant first moment, and there exists a function C defined on $\mathbb{R}^d \times \mathbb{R}$ such that

$$\text{Cov}\{\eta(s + h, t + u), \eta(s, t)\} = C(h, u)$$

for $s, h \in \mathbb{R}^d$ and $t, u \in \mathbb{R}$, with (23.1), (23.3) and (23.4) being such examples. The function C is called the *space–time covariance function* of the process, and its margins, $C(\cdot, 0)$ and $C(\mathbf{0}, \cdot)$, are purely spatial and purely temporal covariance functions, respectively. *Strict stationarity* can be defined as in the spatial case, by translation invariance of the finite-dimensional marginal distributions. Just as in the spatial case, a Gaussian process is second-order stationary if and only if it is strictly stationary.

Evidently, the assumption of stationarity is restrictive and may not be satisfied in practice. See Chapter 9 and Sampson and Guttorp (1992) for transformations of a spatial process towards stationarity. In the spatio-temporal context, stationary covariance models often are inadequate in environmental studies of diurnal variability (Guttorp et al., 1994), but might be well suited at a coarser, say daily, temporal aggregation (Haslett and Raftery, 1989). Bruno, Guttorp, Sampson, and Cocchi (2009) introduce a method of handling temporal non-stationarity.

23.4 Bochner's Theorem and Cressie–Huang Criterion

Just as in the spatial case, the space–time covariance function $C(h, u)$ of a stationary process is *positive definite*, in the sense that all matrices of the form

$$\begin{pmatrix} C(\mathbf{0}, 0) & C(s_1 - s_2, t_1 - t_2) & \cdots & C(s_1 - s_n, t_1 - t_n) \\ C(s_2 - s_1, t_2 - t_1) & C(\mathbf{0}, 0) & \cdots & C(s_2 - s_n, t_2 - t_n) \\ \vdots & \vdots & \ddots & \vdots \\ C(s_n - s_1, t_n - t_1) & C(s_n - s_2, t_n - t_2) & \cdots & C(\mathbf{0}, 0) \end{pmatrix} \tag{23.5}$$

are valid (nonnegative definite) covariance matrices. By Bochner's theorem, every continuous space–time covariance function admits a representation of the form

$$C(h, u) = \int\int e^{i(h^T\omega + u\tau)} \, \mathrm{d}F(\omega, \tau), \tag{23.6}$$

where F is a finite, nonnegative and symmetric measure on $\mathbb{R}^d \times \mathbb{R}$, which is referred to as the *spectral measure*. If C is integrable, the spectral measure is absolutely continuous with Lebesgue density

$$f(\omega, \tau) = (2\pi)^{-(d+1)} \int\int e^{-i(h^T\omega + u\tau)} \, C(h, u) \, \mathrm{d}h \, \mathrm{d}u, \tag{23.7}$$

and f is called the *spectral density*. If the spectral density exists, the representation (23.6) in Bochner's theorem reduces to

$$C(h, u) = \int\int e^{i(h^T\omega + u\tau)} \, f(\omega, \tau) \, \mathrm{d}\omega \, \mathrm{d}\tau, \tag{23.8}$$

and C and f can be obtained from each other via the Fourier transform. As noted, these results are immediate from the respective facts for purely spatial processes in Chapters 2 and 5, by identifying the spatio-temporal domain $\mathbb{R}^d \times \mathbb{R}$ with the Euclidean space $\mathbb{R}^{d+1}$.

Cressie and Huang (1999) characterized the class of the stationary space–time covariance functions under the additional assumption of integrability. Specifically, they showed that a continuous, bounded, integrable and symmetric function on $\mathbb{R}^d \times \mathbb{R}$ is a space–time covariance function if and only if the function

$$\rho(u|\boldsymbol{\omega}) = \int e^{-i\,\boldsymbol{h}^T\boldsymbol{\omega}}\, C(\boldsymbol{h}, u)\, \mathrm{d}\boldsymbol{h} \tag{23.9}$$

is positive definite for almost all $\boldsymbol{\omega} \in \mathbb{R}^d$. In Equation (23.9) the right-hand side is considered a function of $u \in \mathbb{R}$ only, and thus Bochner's criterion for positive definiteness in the space–time domain $\mathbb{R}^d \times \mathbb{R}$ is reduced to a criterion in $\mathbb{R}$.

23.5 Properties of Space–Time Covariance Functions

The estimation of space–time covariance structures can prove highly complicated unless simplifying assumptions are employed. We focus here on second-order stationary processes with space–time covariance function $C(\boldsymbol{h}, u)$. For extensions to nonstationary settings, see Gneiting, Genton, and Guttorp (2007).

A space–time covariance function is *separable* if there exist purely spatial and purely temporal covariance functions C_{S} and C_{T} such that

$$C(\boldsymbol{h}, u) = C_{\mathrm{S}}(\boldsymbol{h}) \cdot C_{\mathrm{T}}(u)$$

for all $(\boldsymbol{h}, u) \in \mathbb{R}^d \times \mathbb{R}$. Thus, the space–time covariance function decomposes as the product of a purely spatial and a purely temporal covariance function. The covariance matrix (23.5) then admits a representation as the Kronecker product of a purely spatial and a purely temporal covariance matrix (Genton, 2007). The assumption of separability simplifies the construction of valid space–time covariance models, reduces the number of parameters, and facilitates computational procedures for large space–time datasets. However, separable models do not allow for space–time interaction (Kyriakidis and Journel, 1999; Cressie and Huang, 1999) and correspond to simplistic processes that will frequently fail to model a physical process adequately.

A space–time covariance function is *fully symmetric* (Gneiting, 2002; Stein, 2005) if

$$C(\boldsymbol{h}, u) = C(\boldsymbol{h}, -u) = C(-\boldsymbol{h}, u) = C(-\boldsymbol{h}, -u)$$

for all $(\boldsymbol{h}, u) \in \mathbb{R}^d \times \mathbb{R}$. Thus, given two spatial locations, a fully symmetric model is unable to distinguish possibly differing effects as time moves forward or backward. Atmospheric, environmental, and geophysical processes are often under the influence of prevailing air or water flows, resulting in a lack of full symmetry. For instance, if winds are predominantly westerly, then high pollutant concentrations at a westerly site today will likely result in high concentrations at an easterly site tomorrow, but not vice versa. Transport effects of this type are well known in the geophysical literature, and we refer to Gneiting et al. (2007) for an extensive list of references. Note that separable covariance functions are necessarily fully symmetric, but not vice versa.

Finally, a space–time covariance function satisfies *Taylor's hypothesis* if there exists a velocity vector $\boldsymbol{v} \in \mathbb{R}^d$ such that $C(\boldsymbol{0}, u) = C(\boldsymbol{v}u, 0)$ for all $u \in \mathbb{R}$. Originally formulated by Taylor (1938), the hypothesis postulates a particularly simple relationship between the spatial margin and the temporal margin, respectively. To give an example, the covariance function (23.1) for the frozen-field model satisfies Taylor's hypothesis. See Gneiting et al. (2007) for additional examples, and Li et al. (2009) for a recent application to precipitation fields.

These and other properties of space–time covariance functions can be tested for using methods developed by Fuentes (2005), Mitchell, Genton, and Gumpertz (2005) and Li, Genton, and Sherman (2007), among others.

23.6 Nonseparable Stationary Covariance Functions

As noted, separable covariance models have frequently been chosen for mathematical and computational convenience rather than their ability to fit physical phenomena and observational data. In this light, flexible, nonseparable covariance models have been sought, following the pioneering work of Cressie and Huang (1999). We restrict our discussion to fully symmetric functions, notwithstanding the fact that they can be combined with covariance functions of the form (23.1), to allow for general, fully symmetric or not fully symmetric formulations.

One basic approach is the *product–sum model* of De Iaco, Myers, and Posa (2001), which specifies a space–time covariance function as

$$C(\boldsymbol{h}, u) = a_0 C_S^0(\boldsymbol{h}) C_T^0(u) + a_1 C_S^1(\boldsymbol{h}) + a_2 C_T^2(u), \tag{23.10}$$

where a_0, a_1 and a_2 are nonnegative coefficients and C_S^0, C_S^1 and C_T^0, C_T^2 are purely spatial and purely temporal covariance functions. Another straightforward construction employs a space–time covariance function of the form

$$C(\boldsymbol{h}, u) = \varphi\left(\left(a_1 \|\boldsymbol{h}\|^2 + a_2 u^2\right)^{1/2}\right), \tag{23.11}$$

where a_1 and a_2 are anisotropy factors for the space and time dimensions, $\|\cdot\|$ denotes the Euclidean norm, and the function φ is a member of the class Φ_{d+1}, as introduced in Chapter 2. This corresponds to geometric anisotropy in a purely spatial framework and assumes a joint space–time metric in $\mathbb{R}^d \times \mathbb{R}$ (Christakos, Hristopulos, and Bogaert 2000). The spatial and temporal margins are constrained to be of the same functional form, and Taylor's hypothesis is satisfied.

Cressie and Huang (1999) used the test (23.9) to construct nonseparable space–time covariance functions through closed form Fourier inversion in $\mathbb{R}^d$. Gneiting (2002) gave a criterion that is based on this construction, but does not depend on Fourier inversion. Recall that a continuous function $\eta(r)$ defined for $r > 0$ is *completely monotone* if it possesses derivatives $\eta^{(n)}$ of all orders and $(-1)^n \eta^{(n)}(r) \geq 0$ for $n = 0, 1, 2, \ldots$ and $r > 0$. Suppose now that $\eta(r), r \geq 0$, is completely monotone, and $\psi(r), r \geq 0$, is a positive function with a completely monotone derivative. Then

$$C(\boldsymbol{h}, u) = \frac{1}{\psi(u^2)^{d/2}} \, \eta\left(\frac{\|\boldsymbol{h}\|^2}{\psi(u^2)}\right) \tag{23.12}$$

is a covariance function on the space–time domain $\mathbb{R}^d \times \mathbb{R}$. For instance, if a function φ belongs to the class Φ_∞, as introduced in Chapter 2, then $\eta(r) = \varphi(r^{1/2})$ is completely monotone. Examples of positive functions with a completely monotone derivative include $\psi(r) = (ar^\alpha + 1)^\beta$ and $\psi(r) = \ln(ar^\alpha + 1)$, where $\alpha \in (0, 1]$, $\beta \in (0, 1]$ and $a > 0$. The choices $\eta(r) = \sigma^2 \exp(-cr^\gamma)$, which derives from Equation (2.14) in Chapter 2, and $\psi(r) = (ar^\alpha + 1)^\beta$ yield the parametric family

$$C(\boldsymbol{h}, u) = \frac{\sigma^2}{(1 + a|u|^{2\alpha})^{\beta d/2}} \exp\left(-\frac{c\,\|\boldsymbol{h}\|^{2\gamma}}{(1 + a|u|^{2\alpha})^{\beta\gamma}}\right) \tag{23.13}$$

of space–time covariance functions. Here, a and c are nonnegative scale parameters of time and space, the smoothness parameters α and γ and the space–time interaction parameter β take values in $(0, 1]$, and σ^2 is the process variance. The spatial margin is of the powered exponential form, and the temporal margin belongs to the Cauchy class.

Quite generally, constructions of space–time covariance functions start from basic building blocks, taking advantage of the fact that the class of covariance functions is closed under products, sums, convex mixtures and limits (Ma, 2002, 2008). The product-sum model (23.10) provides one such example. In some special cases, linear combinations with negative coefficients remain valid covariance functions, which may allow for negative correlations (Ma, 2005; Gregori, Porcu, Mateu, and Sasvári, 2008). The convex mixture approach is particularly powerful; for example, it provides an alternative route to the construction in Equation (23.12), as described by Ma (2003).

23.7 Nonseparable Stationary Covariance Functions via Spectral Densities

In principle, it is easier to specify a nonseparable second-order structure in the frequency domain, since nonnegativity and integrability are the only requirements on the spectral density (23.7) of a space–time covariance function. Furthermore, a covariance function is separable if and only if the spectral density is such. Nonnegativity and integrability are often much easier to check than the notorious requirement of positive definiteness for the covariance function. However, Fourier inversion is needed to find the covariance function, and often this can only be done via numerical evaluation of the inversion formula (23.8).

Stein (2005) proposed the parametric class

$$f(\omega, \tau) = \left(c_1\left(a_1^2 + |\omega|^2\right)^{\alpha_1} + c_2\left(a_2^2 + \tau^2\right)^{\alpha_2}\right)^{-\nu} \tag{23.14}$$

of spectral densities in $\mathbb{R}^d \times \mathbb{R}$, with positive parameters satisfying $\frac{\alpha_1}{\alpha_2}(2\alpha_2\nu - 1) > d$. When $d = 2$ and $\alpha_2 = \nu = 1$, this can be shown to correspond to the solutions of the stochastic differential equation in (23.2). The covariance functions associated with the spectral density (23.14) are infinitely differentiable away from the origin, as opposed to the covariance functions in (23.12), and allow for essentially arbitrary, and potentially distinct, degrees of smoothness in space and time. A related class of nonseparable spectral densities and associated covariance functions was recently introduced by Fuentes, Chen, and Davis (2008).

23.8 Case Study: Irish Wind Data

Gneiting, Genton, and Guttorp (2007) fit a stationary space–time correlation function to the Irish wind data of Haslett and Raftery (1989). The dataset consists of time series of daily average wind speed at 11 synoptic meteorological stations in Ireland during the period 1961 to 1978. These are transformed toward stationarity via a square root transform, and the removal of a diurnal temporal trend component and a spatially varying mean. The resulting residuals are called velocity measures, for which a Gaussian assumption is plausible.

Winds in Ireland are predominantly westerly and, thus, the velocity measures propagate from west to east. Hence, we expect correlations between a westerly station today and an easterly station tomorrow to be higher than vice versa, which is clearly visible in the empirical correlations, as illustrated in Figure 23.1. In this light, Gneiting et al. (2007) fit the nonseparable and not fully symmetric correlation function

$$C(\boldsymbol{h}, u) = \frac{(1-\nu)(1-\lambda)}{1 + a\,|u|^{2\alpha}}\left(\exp\left(-\frac{c\,\|\boldsymbol{h}\|}{(1 + a\,|u|^{2\alpha})^{\beta/2}}\right) + \frac{\nu}{1-\nu}\mathbb{I}\{\boldsymbol{h} = \boldsymbol{0}\}\right) + \lambda\left(1 - \frac{1}{2v}|h_1 - vu|\right)_+$$

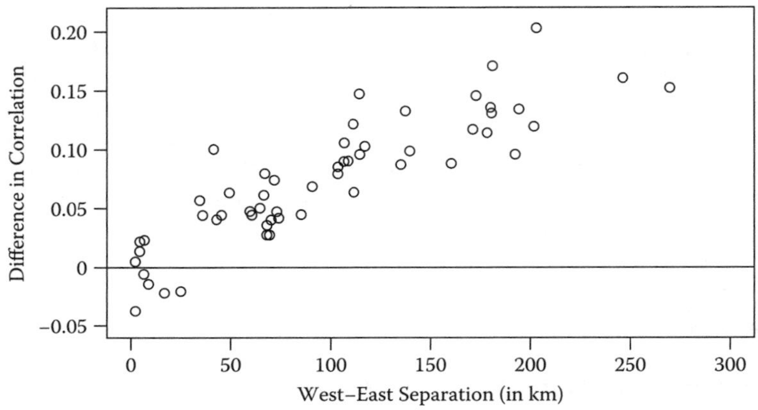

FIGURE 23.1
Difference between the empirical west-to-east and east-to-west correlations in velocity measures for the Irish wind data at a temporal lag of one day, in dependence on the longitudinal (east–west) component of the spatial lag vector.

with weighted least squares estimates $\widehat{\nu} = 0.0415$, $\widehat{\lambda} = 0.0573$, $\widehat{a} = 0.972$, $\widehat{\alpha} = 0.834$, $\widehat{c} = 0.00128$, $\widehat{\beta} = 0.681$, and $\widehat{v} = 234$, where spatio-temporal lags are measured in kilometers and days, respectively. Here, the spatial separation vector $\boldsymbol{h} = (h_1, h_2)'$ has longitudinal (east–west) component h_1 and latitudinal (north–south) component h_2, $v \in \mathbb{R}$ is a longitudinal velocity, and we write $p_+ = \max(p, 0)$. The estimates use data from 1961 to 1970 and are computed from Equation (3.9) in Chapter 3, using temporal lags of three days or less only. Given the size of the dataset, the weighted least squares estimates are unlikely to differ much from the maximum likelihood estimates, which are not computationally feasible here. Note that the fitted space–time covariance function originates from more basic building blocks, namely the nonseparable model (23.13) with powered exponential spatial margin and a temporal margin of Cauchy type, the nugget effect (23.4), and the frozen-field covariance (23.1).

Gneiting et al. (2007) reported on experiments with time-forward (out-of-sample) kriging predictors based on the fitted covariance function and simpler, separable and/or fully symmetric submodels, and find that the use of the more complex and more realistic models results in improved predictive performance.

References

Billingsley, P. (1986), *Probability and Measure*, 2nd ed., New York, John Wiley & Sons.

Breiman, L. (1968), *Probability*, Reading, MA, Addison-Wesley.

Briggs, B.H. (1968), On the analysis of moving patterns in geophysics — I. Correlation analysis, *Journal of Atmospheric and Terrestrial Geophysics*, 30, 1777–1788.

Brillinger, D.R. (1997), An application of statistics to meteorology: Estimation of motion, in Pollard, O., Jorgessen, E. and Yang, G.L. (Eds.), *Festschrift for Lucien Le Cam*, Springer, New York, pp. 93–105.

Brown, P.E., Kåresen, K.F., Roberts, G.O. and Tonellato, S. (2000), Blur-generated non-separable space–time models, *Journal of the Royal Statistical Society Ser. B*, 62, 847–860.

Bruno, F., Guttorp, P., Sampson, P.D. and Cocchi, D. (2009), A simple non-separable, non-stationary spatial-temporal model for ozone. Forthcoming. *Environmental and Ecological Statistics*. DOI 10.1007/S/0651-008-0094-8

Carroll, R.J., Chen, R., George, E.I., Li, T.H., Newton, H.J., Schmiediche, H. and Wang, N. (1997), Ozone exposure and population density in Harris county, Texas (with discussion), *Journal of the American Statistical Association*, 92, 392–415.

Christakos, G., Hristopulos, D.T. and Bogaert, P. (2000), On the physical geometry concept at the basis of space/time geostatistical hydrology, *Advances in Water Resources*, 23, 799–810.

Cox, D.R. and Isham, V. (1988), A simple spatial-temporal model of rainfall, *Proceedings of the Royal Society of London Ser. A*, 415, 317–328.

Cressie, N. and Huang, H.-C. (1999), Classes of nonseparable, spatio-temporal stationary covariance functions, *Journal of the American Statistical Association*, 94, 1330–1340, 96, 784.

De Iaco, S., Myers, D.E. and Posa, T. (2001), Space–time analysis using a general product-sum model, *Statistics & Probability Letters*, 52, 21–28.

Fuentes, M. (2005), Testing for separability of spatial-temporal covariance functions, *Journal of Statistical Planning and Inference*, 136, 447–466.

Fuentes, M., Chen, L. and Davis, J.M. (2008), A class of nonseparable and nonstationary spatial temporal covariance functions, *Environmetrics*, 19, 487–507.

Genton, M.G. (2007), Separable approximations of space–time covariance matrices, *Environmetrics*, 18, 681–695.

Gneiting, T. (2002), Nonseparable, stationary covariance functions for space–time data, *Journal of the American Statistical Association*, 97, 590–600.

Gneiting, T., Genton, M.G. and Guttorp, P. (2007), Geostatistical space–time models, stationarity, separability and full symmetry, in Finkenstädt, B., Held, L. and Isham, V. (Eds.), *Statistical Methods for Spatio-Temporal Systems*, Chapman & Hall, Boca Raton, FL, pp. 151–175.

Gregori, P., Porcu, E., Mateu, J. and Sasvári, Z. (2008), Potentially negative space time covariances obtained as sum of products of marginal ones, *Annals of the Institute of Statistical Mathematics*, 60, 865–882.

Guttorp, P., Meiring, W. and Sampson, P.D. (1994), A space–time analysis of ground-level ozone data, *Environmetrics*, 5, 241–254.

Haslett, J. and Raftery, A.E. (1989), Space–time modelling with long-memory dependence: Assessing Ireland's wind-power resource (with discussion), *Applied Statistics*, 38, 1–50.

Huang, H.-C. and Hsu, N.-J. (2004), Modeling transport effects on ground-level ozone using a non-stationary space–time model, *Environmetrics*, 15, 251–268.

Jones, R.H. (1963), Stochastic processes on a sphere, *Annals of Mathematical Statistics*, 34, 213–218.

Jones, R.H. and Zhang, Y. (1997), Models for continuous stationary space–time processes, in *Modeling Longitudinal and Spatially Correlated Data*, Lecture Notes in Statistics, 122, Gregoire, T.G., Brillinger, D.R., Diggle, P.J., Russek-Cohen, E., Warren, W.G. and Wolfinger, R.D., (Eds.), Springer, New York, pp. 289–298.

Jun, M. and Stein, M.L. (2007), An approach to producing space–time covariance functions on spheres, *Technometrics*, 49, 468–479.

Kelbert, M.Y., Leonenko, N.N. and Ruiz-Medina, M.D. (2005), Fractional random fields associated with stochastic fractional heat equations, *Advances in Applied Probability*, 37, 108–133.

Kunita, H. (1990), *Stochastic Flows and Stochastic Differential Equations*, Cambridge, U.K., Cambridge University Press.

Kyriakidis, P.C. and Journel, A.G. (1999), Geostatistical space–time models: A review, *Mathematical Geology*, 31, 651–684.

Li, B., Genton, M.G. and Sherman, M. (2007), A nonparametric assessment of properties of space–time covariance functions, *Journal of the American Statistical Association*, 102, 738–744.

Li, B., Murthi, A., Bowman, K.P., North, G.R., Genton, M.G. and Sherman, M. (2009), Statistical tests of Taylor's hypothesis: An application to precipitation fields, *Journal of Hydrometeorology*, 10, 254–265.

Ma, C. (2002), Spatio-temporal covariance functions generated by mixtures, *Mathematical Geology*, 34, 965–975.

Ma, C. (2003), Families of spatio-temporal stationary covariance models, *Journal of Statistical Planning and Inference*, 116, 489–501.

Ma, C. (2005), Linear combinations of space–time covariance functions and variograms, *IEEE Transactions on Signal Processing*, 53, 857–864.

Ma, C. (2008), Recent developments on the construction of spatio-temporal covariance models, *Stochastic Environmental Research and Risk Assessment*, 22, S39–S47.

Mitchell, M., Genton, M.G. and Gumpertz, M. (2005), Testing for separability of space–time covariances, *Environmetrics*, 16, 819–831.

Phelan, M.J. (1996), Transient tracers and birth and death on flows: Parametric estimation of rates of drift, injection, and decay, *Journal of Statistical Planning and Inference*, 51, 19–40.

Prévôt, C. and Röckner, M. (2007), *A Concise Course on Stochastic Partial Differential Equations*, New York, Springer.

Sampson, P.D. and Guttorp, P. (1992), Nonparametric estimation of nonstationary spatial covariance structure, *Journal of the American Statistical Association*, 87, 108–119.

Stein, M.L. (2005), space–time covariance functions, *Journal of the American Statistical Association*, 100, 310–321.

Taylor, G.I. (1938), The spectrum of turbulence, *Proceedings of the Royal Society of London Ser. A*, 164, 476–490.

Whittle, P. (1954), On stationary processes in the plane, *Biometrika*, 41, 434–449.

Whittle, P. (1963), Stochastic processes in several dimensions, *Bulletin of the International Statistical Institute*, 40, 974–994.

24

Dynamic Spatial Models Including Spatial Time Series

Dani Gamerman

CONTENTS

24.1 Dynamic Linear Models

This book has already presented many situations where the observation process under study contains the temporal dimension as well as the spatial dimension. This chapter is devoted to detailing a popular and fairly general framework for handling these situations. It is based on firmly established models, called dynamic or state-space models, with a nonparametric flavor. They have proved a flexible tool to handle temporal correlation (West and Harrison, 1997) in a variety of different contexts. This chapter will also provide a number of situations where these models can be applied in the context of spatial analysis.

Dynamic models are described via a p-dimensional latent process $\beta(\cdot)$ defined over time according to a temporal difference equation

$$\beta(t') = G(t', t)\beta(t) + w(t', t), \quad \text{with } w(t', t) \sim N(0, W(t', t)), \quad (t' > t), \tag{24.1}$$

where the transition matrix G conveys the deterministic part of the evolution and the system disturbance w is simply a stochastic component accounting for increased uncertainty (controlled by the disturbance variance W) over the temporal evolution. The model is completed with an initial specification for β at, say, $t = 0$. For temporally equidistant points, Equation (24.1) can be simplified to

$$\beta_t = G_t\beta_{t-1} + w_t, \quad \text{with } w_t \sim N(0, W_t). \tag{24.2}$$

This will be assumed to be the case hereafter without loss of generality; treatment of nonequidistant times involves trivial changes that only clutter the notation. Note also that this gives rise to vector autoregressive (VAR) forms of order 1 when G_t is constant over time.

Example 24.1 (First-order models)
When $G = I_p$, the identity matrix of order p, the model is the random walk $\beta_t = \beta_{t-1} + w_t$ and, therefore, model (24.2) can also be referred to as generalized random walk. This is also referred to as first-order models because they can be seen as the first order (Taylor

expansion) approximation of an arbitrary underlying smooth function β_t. Note that, unlike Gaussian processes, this model is nonstationary with var(β_t) increasing with t.

Stationary processes also may be obtained after replacing the random walk evolution matrix I_p by suitably chosen matrices P. This gives rise to VAR forms of order 1. Special cases of interest are given by $P = \rho I_p$, with $|\rho| < 1$, and $P = diag(\rho_1, \ldots, \rho_p)$, with $|\rho_i| < 1$, for $i = 1, \ldots, p$. These models are attractive for their simplicity and low dimensionality, but may be too restrictive.

Example 24.2 (Second-order models or dynamic linear trend (LT))
Assume $\beta = \begin{pmatrix} \beta_1 \\ \beta_2 \end{pmatrix}$ is a bivariate process and let $G_t = G_{LT} = \begin{pmatrix} 1 & 1 \\ 0 & 1 \end{pmatrix}$, for all t. Then clearly β_2 is undergoing a univariate random walk and β_1 is being incremented each time by β_2. Thus, β_1 plays the role of an underlying level and β_2 plays the role of its increment. Typically only β_1 is present in the data description. Both components are varying locally around their prescribed evolutions and can accommodate local changes.

The disturbance variance matrix for this model is hereafter denoted by W_{LT}. It may take any positive definite form, but there are good reasons to assume it as $W_{LT} = \begin{pmatrix} W_1 + W_2 & W_2 \\ W_2 & W_2 \end{pmatrix}$, for all t. In any case, it is recommended to assume the disturbance at the mean level to be larger than the disturbance of its increment. Thus, the first diagonal element of W_{LT} would be larger than its second diagonal element.

Example 24.3 (Seasonal models)
Assume a seasonal pattern of length p is to be described. Let $\beta = (\beta_1, \ldots, \beta_p)^T$ and $G_t = \begin{pmatrix} 0 & I_{p-1} \\ 1 & 0 \end{pmatrix}$, for all t. Clearly, G_t is a permutation matrix and the evolution over time only rearranges β components by replacing its first component by its second component in the preceding time. Thus, allowing only the first β component to be present in the data description gives a form-free pattern for the seasonality. This pattern is stochastic due to the presence of the disturbance term w_t.

Structured seasonal patterns may also be constructed. A single sine waveform is obtained by letting $\beta = \begin{pmatrix} \beta_1 \\ \beta_2 \end{pmatrix}$ and $G_t = G_S = \begin{pmatrix} c & s \\ -s & c \end{pmatrix}$, where $c = \cos(2\pi/p)$ and $s = \sin(2\pi/p)$, for all t. This evolution matrix makes β_1 take the appropriate value in the sine wave for every next time.

Once again, allowing only the first β component to be present in the data description gives a sine waveform for the seasonality. The pattern is stochastic and can accommodate variations around the sine wave due to the disturbance term w_t. This is usually associated with an additional intercept in the model for the observations since the sine wave fluctuates around 0. A combination of harmonics is obtained by allowing extra pairs of β components with different lengths and completely general forms are obtained by incorporating $[p/2]$ harmonics. See West and Harrison (1997) for details and Harvey (1989) for an alternative model for seasonal components.

The nonparametric nature of these models is easier to understand with the usual choice of a random walk. In this case, the process β is simply undergoing local changes without imposing any specific form for the temporal variation and as such is capable of (locally) tracking any smooth trajectory over time. The degree of smoothness is governed by the variances W. Models that depart from the random walk impose some structure in the mean of the process, as described in the examples above. Even for these models, the presence of the disturbance terms allows departures from this structure and accommodates data fluctuations around it.

These models can also be obtained by discretization of an underlying stochastic differential equation (Revuz and Yor, 1999), as those used above in this book for handling spatio-temporal processes in continuous time.

The typical setup for the use of dynamic models is the context of temporally correlated data $Y_t = (Y_t(s_1), \ldots, Y_t(s_n))$, for $t = 1, \ldots, T$, where it will be assumed that all temporal correlation present in the data is captured by an underlying process β. Therefore, the observations are conditionally independent (given β) leading to the likelihood for β given by $l(\beta) = \prod_{t=1}^{T} p(y_t|\beta_t)$, where T is the last observed time.

The simple, but important, case of a normal linear models gives $p(y_t|\beta_t)$ as

$$Y_t = \mu_t + v_t, \qquad \text{with } \mu_t = X_t\beta_t \quad \text{and} \quad e_t \sim N(0, V_t), \tag{24.3}$$

for $t = 1, \ldots, T$. The variance matrix V_t of the observation error e_t may be specified using any of the models previously described in this book to handle spatial correlation. The main forms are Gaussian processes, typically used in continuous space.

Also, the error e_t can be further decomposed into $e_t = \eta_t + \epsilon_t$, as before. This decomposition eases the generalization toward nonnormal observations. Assume the observational distribution is governed by parameter ξ_t. The spatially structured error term η_t is incorporated to the predictor $X_t\beta_t + \eta_t$ via $g(\mu_t) = X_t\beta_t + \eta_t$, for some differentiable function g (Diggle, Tawn, and Moyeed, 1998). An important example is the exponential family with mean μ_t. The pure noise ϵ_t retains the description of unstructured observational variation.

The $n \times m$ matrix X_t plays the role of a design matrix containing values of the explanatory variables at time t. It is typically given by known functions of location with rows $X(s_1)^T, \ldots, X(s_n)^T$, not depending on time and is thus denoted hereafter simply by X. Therefore, for any given location s, the observational predictor (mean, in the normal case) is given by

$$\mu_t(s) = X(s)^T\beta_t. \tag{24.4}$$

Thus, models are being decomposed into a deterministic part given by $X(s)^T\beta_t$ and an unexplained stochastic component e_t that may incorporate the spatial dependence. Note that, in this dynamic setting, the deterministic part of the model is only handling temporal correlation. One natural choice in the spatio-temporal setting is to let matrix X be a function of the spatial coordinates.

This approach was proposed by Stroud, Muller, and Sansó (2001). They chose to define $X(s)$ as a linear combination of basis functions of the location s. This idea is applied in related contexts by many authors. Wikle and Cressie (1999) use the same decomposition in their dimension reduction approach. They obtained it from a more general underlying process, to be described later in this section. Sansó, Schmidt, and Nobre (2008) also use this decomposition, but without the error term.*

These approaches have the common feature of considering X as a fixed function of space. This may be too restrictive to accommodate general spatial variation. Lopes, Salazar, and Gamerman (2008) allow the columns $f_1, \ldots, f_m$ of X to vary stochastically according to independent Gaussian processes. The m-dimensional, time-varying component β_t plays the role of m latent factor time series capturing the temporal variation of the data. Each of its m elements β_{tj} is associated to the observations through the space-varying vector f_j containing their loadings, for $j = 1, \ldots, m$.

* They also consider the possibility that X models time (in which case it will recover its subscript t) and β_t models space (in which case it drops its subscript t).

All models above decompose models into two groups of components: one handling space and one handling time. Even though these structures may combine into nonseparable models, richer dependence between space and time is not allowed. A description of processes that combine spatial and temporal dependence into a single structure that cannot be separated is provided below.

The key aspect in the extension is to allow state parameters β_t to vary across space. This will obviously imply a substantial increase in the parameter dimensionality and may lead to identifiability problems. The solution to keep the problem manageable and the model identifiable is to impose restrictions over the parameter space. This can be achieved through likelihood penalization from a classical perspective or through prior specifications from a Bayesian perspective.

24.2 Space Varying State Parameters

From now on, it will be assumed that the state parameter $\beta_t(\cdot)$ varies also over space. In this setting, $\beta_t(\cdot) = \{\beta_t(s) : s \in \mathcal{D}\}$. Considering n locations s_i $(i = 1, \ldots, n)$ for spatially continuous observation processes, the vector $\beta_t = (\beta_{t1}, \ldots, \beta_{tn})$ can be formed with $\beta_{ti} = \beta_t(s_i)$ denoting the state parameter at time t and location s_i.

A simple form to account for spatial and temporal variation of the state parameter is to assume that $\beta_t(\cdot)$ can be decomposed as

$$\beta_t(s) = \bar{\gamma}_t + \gamma_t(s) \tag{24.5}$$

with a trend $\bar{\gamma}_t$ common to all locations and a spatio-temporal disturbance $\gamma_t(s)$ associated with its location. Paez, Gamerman, Landim, and Salazar (2008) assumed that the common trend $\bar{\gamma}_t$ carries the temporal evolution according to (24.2). The spatio-temporal disturbance process $\gamma_t(\cdot)$ accounts for the spatial correlation through a multivariate Gaussian process, which they assume to be independent and identically distributed over time. They applied this process to the intercept and the regression coefficients of a dynamic model for pollutant measurements.

Their approach allows more generality in the description of state parameters. Despite the substantial increase in the nominal number of parameters, it achieves identifiability through the decomposition (24.5) and prior assumptions about $\bar{\gamma}_t$ and γ_t. Note that the temporal independence between the γ_ts prevents any temporal correlation between them. So, their model still separates the spatial components $\bar{\gamma}_t$ from the temporal components γ_t. Thus, their model can be useful if no temporal dependence of the spatial variation is expected.

The simplest model that does not allow for explicit separation of space and time is the spatial random walk

$$\beta_t(s) = \beta_{t-1}(s) + w_t(s), \text{ for all } s, \tag{24.6}$$

where $\beta_t(\cdot)$ is a univariate process. In model (24.6), state parameters β_t evolve in forms that are seemingly independent in space. But spatial correlation is introduced via their respective disturbance processes $w_t(\cdot)$ through their joint distribution. It does it by assuming some form of a Gaussian process. It will typically have a geostatistical model form given in Chapter 3 for spatially continuous data. The prior is completed with a Gaussian process prior for $\beta_1(\cdot)$.

Obviously, the spatial random walk can be defined for multivariate state parameters $\beta_t(\cdot)$. All it requires is an adequate multivariate representation of Gaussian processes. Some possibilities for doing it are analyzed in the areal data context by Gamerman, Moreira, and Rue (2003) and described in the spatially continuous context by Gamerman, Salazar, and Reis (2007).

The decomposition (24.5) can also be applied to (24.6). In this case, the disturbances $w_t(s)$ must also be decomposed into a purely temporal disturbance $\bar{w}_t$ and residual spatio-temporal disturbances as

$$\begin{aligned} \beta_t(s) &= \bar{\gamma}_t + \gamma_t(s), \text{ for all } s \\ \bar{\gamma}_t &= \bar{\gamma}_{t-1} + \bar{w}_t, \\ \gamma_t(s) &= \gamma_{t-1}(s) + w_t(s), \text{ for all } s. \end{aligned} \tag{24.7}$$

This decomposition was used by Gelfand, Banerjee, and Gamerman (2005). They analyzed environmental data with a normal linear regression model. Identification is ensured by setting the mean of $\gamma_t(s)$ to be 0. The first component $\bar{\gamma}_t$ accounts for purely temporal variation. The second component $\gamma_t(s)$ accounts for the remaining temporal variation that was associated with space. Both components are assumed in (24.7) to evolve according to a random walk. Huerta, Sansó, and Stroud (2004) also used it to model the effect of seasonal components in an environmental application.

More general forms can now be constructed by combining Equation (24.7) with Equation (24.2) leading to a general evolution

$$\begin{aligned} \beta_t(s) &= \bar{\gamma}_t + \gamma_t(s), \text{ for all } s \\ \bar{\gamma}_t &= \bar{G}_t \bar{\gamma}_{t-1} + \bar{w}_t \\ \gamma_t(s) &= G_t \gamma_{t-1}(s) + w_t(s), \text{ for all } s \end{aligned} \tag{24.8}$$

with transition matrices $\bar{G}_t$ and G_t and evolution disturbances $\bar{w}_t$ and $w_t(\cdot)$. When all transition matrices equal the identity matrix, the spatio-temporal random walk (24.7) is recovered.

These structures and their decomposition allow for many types of components. Thus, they give substantial flexibility to the modeler. This can be more easily appreciated with the illustrative example below.

Example 24.4

Assume that a simple linear regression model

$$Y_t(s) = \alpha_t(s) + \beta_t(s) z_t(s) + e_t(s), \text{ for } t = 1, \ldots, T \quad \text{and} \quad s = s_1, \ldots, s_n \tag{24.9}$$

is considered with a single covariate $z_t(s)$ varying in space and time. Note that both the intercept α and the (scalar) regression coefficient β are allowed to vary in space and time.

Assume also that the intercept can be completely described by a stochastic seasonal pattern that is common throughout the region of interest and can be locally described by a single wavelength of length p and has no additional spatio-temporal heterogeneity. This would imply that $\alpha_t(s) = \alpha_t$, for all (s, t). According to Example 24.3, its evolution is described with the help of additional time-varying parameter ξ_t as

$$\begin{pmatrix} \alpha_t \\ \xi_t \end{pmatrix} = G_S \begin{pmatrix} \alpha_{t-1} \\ \xi_{t-1} \end{pmatrix} + \begin{pmatrix} w_t^\alpha \\ w_t^\xi \end{pmatrix}, \text{ where } \begin{pmatrix} w_t^\alpha \\ w_t^\xi \end{pmatrix} \sim N\left[\begin{pmatrix} 0 \\ 0 \end{pmatrix}, W_\alpha \right],$$

where G_S was defined in Example 2 and W_α is a 2×2 covariance matrix.

Assume further that the regression coefficient has its common trend in the form of a random walk, but the spatial variations around this common mean are thought to undergo a location-specific, linearly local trend. The conditions stated above imply that $\beta_t(s) = \bar{\gamma}_t + \gamma_t(s)$, for all (s, t). The evolutions for $\bar{\gamma}_t$ and γ_t are given, respectively, by a univariate random walk $\bar{\gamma}_t = \bar{\gamma}_{t-1} + \bar{w}_t$, and by

$$\begin{pmatrix} \gamma_t(s) \\ \delta_t(s) \end{pmatrix} = G_{LT} \begin{pmatrix} \gamma_{t-1}(s) \\ \delta_{t-1}(s) \end{pmatrix} + w_t(s),$$

where G_{LT} was defined in Example 24.1 and additional processes had to be introduced: $\delta_t(\cdot)$ is a univariate increment (over γ_t) process and $w_t(\cdot) = (w_t^{\gamma}(\cdot), w_t^{\delta}(\cdot))$ is a bivariate process. There are many possibilities for the latter. Paez et al. (2008) assumed the same spatial correlation function for both disturbance processes. If this is assumed, then a Kronecker product representation is obtained for the covariance matrix for the components of $w_t(\cdot)$ at any given set of locations. Independence between the processes can also be assumed leading to a block diagonal representation for the covariance matrix for the components of $w_t(\cdot)$ at any given set of locations. Other forms of Gaussian processes are also possible (see Gamerman et al., 2007).

The spatial relation between the regression coefficients in Equation (24.8) enables inference about their values at unobserved locations. Consider a set of g unobserved locations $s_1^u, \ldots, s_g^u$ where the superscript u denotes unobserved. Define the collection of state parameters β_t^u at these locations as $\beta_t^u = (\beta_t^u(s_1), \ldots, \beta_t^u(s_g))$. Then clearly the evolution equations defined for β_t can be readily extended to (β_t, β_t^u). The conditional distribution for $(\beta_t^u|\beta_t)$ can be obtained from this joint specification. If all disturbances and prior distributions are normally distributed, then simple calculations show that this conditional distribution is also normal. These calculations are made conditionally on the hyperparameters. Their integration cannot typically be performed analytically. In this case, approximating methods, such as Markov chain Monte Carlo (MCMC) algorithms, must be applied.

Figure 24.1 shows an example of this interpolation for the spatio-temporal variation experienced by the regression component γ_t. This result comes from a study of the effect of precipitation on temperature (Gelfand et al., 2005) with monthly data over a single year. Relevance of spatial and temporal components in this regression setting is clear. For example, a more extreme spatial variation of the effect of temperature is observed in the months of more extreme weather.

In many situations, main interest rests in predicting unobserved data values. Data may not be observed because they are located at unobserved sites or at unobserved times. In either case, the predictive distribution for them, conditional on all observed data, must be obtained. The operation in space is referred to as interpolation or kriging. The operation in time is referred to as forecasting when interest lies in prediction into the future, hindcasting when interest lies in assessment of the performance of the model at previous times, and nowcasting when interest lies in the immediate future.

In any case, prediction is conceptually easy to perform. The structural form of the model means that all spatial and temporal correlations are contained in the state parameters. Let y^u denote the unobserved data one wants to predict and β^u denote the regression coefficient present in the observation equation for y^u. Depending on the case of interest, β^u may contain values of the state at unobserved locations and/or times. In either case, predictive distributions conditional on hyperparameters θ are obtained through

$$p(y^u|\theta, D) = \int p(y^u|\beta^u, \theta, D)\, p(\beta^u|\theta, D)\, d\beta^u, \tag{24.10}$$

where D denotes the observed data.

The first density in the integrand is the observation equation and the second is the posterior distribution of β^u. The integration in Equation (24.10) can be performed analytically in the case of normal evolution disturbances and normal observation errors. Elimination of hyperparameters is required to obtain the unconditional predictive distributions actually used for prediction. This operation is performed in very much the same way as Equation (24.10) with β^u replaced by the hyperparameters. Namely, the predictive density $p(y^u|D)$ can be obtained via

$$p(y^u|D) = \int p(y^u|\theta, D)\, p(\theta|D)\, d\theta.$$

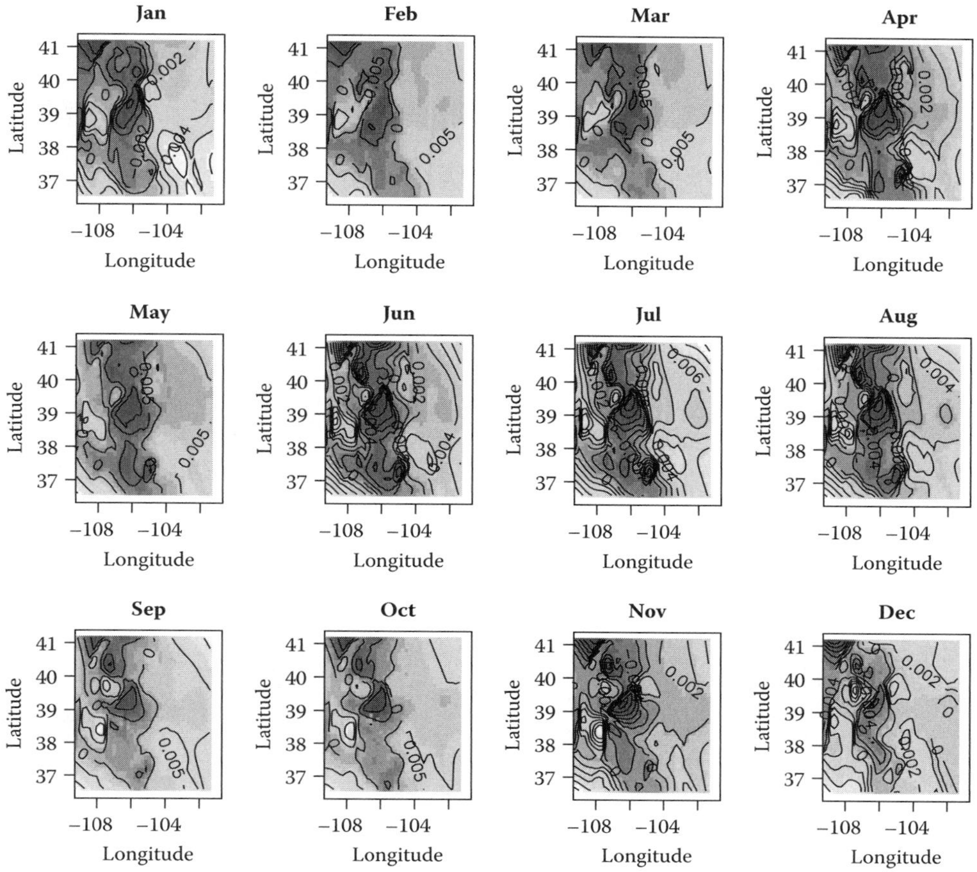

FIGURE 24.1
Posterior mean of the spatio-temporal variation γ_t of the regression coefficient of precipitation over temperature for a region of the State of Colorado. (From Gelfand, A.E., Banerjee, S. and Gamerman, D. (2005). *Environmetrics*, 16, 465–479. With permission.)

The integrand above contains the posterior density $p(\theta|D)$ of the hyperparameters and this is rarely available analytically. Thus, integration can only be performed approximately and approximating methods must be applied. In practice, MCMC/sampling methods are applied and integration with respect to β^u and θ is performed simultaneously. Figure 24.2 provides a visual summary of these prediction operations in the context of the application of Figure 24.1. Note that spatial extrapolation is more dispersed than temporal extrapolation for this application.

Models described in Equation (24.8) retain the seemingly unrelated nature because their mean structure is location-specific and correlation across space is only provided through their unstructured error terms. Correlation across space can be imposed directly through the mean structure by forms, such as

$$\beta_t(s) = \int k(u, s)\beta_{t-1}(u)du + w_t(s), \tag{24.11}$$

where $k(u, s)$ is a kernel that provides the weights with which location u influences outcomes in location s for the next time. This evolution is considered by a number of authors in a number of different contexts (see, e.g., Wikle and Cressie (1998)). When the integral can be

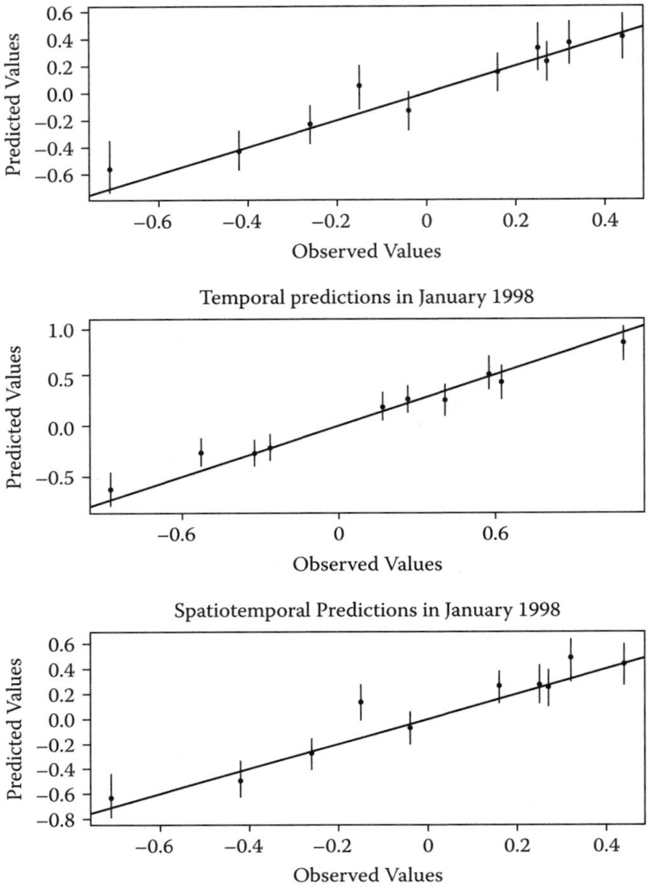

FIGURE 24.2
Predictive credible intervals for temperature values at a few unobserved locations (top panel), at a future time for a few observed locations (middle panel), and at a future time for a few unobserved locations (bottom panel) for a region of the State of Colorado. (From Gelfand, A.E., Banerjee, S. and Gamerman, D. (2005). *Environmetrics*, 16, 465–479. With permission.)

well approximated by a discrete convolution over observed locations, then Equation (24.11) falls into the general form (24.2) with the (i, j)th entry of the transition matrix given by the values of $k(s_j, s_i)$, for $i, j = 1, \ldots, n$. Evolution (24.11) has been used only for the intercept, but nothing prevents its use for more general state parameters, such as regression coefficients or seasonal components.

The presentation of this chapter is based on the Bayesian paradigm. Thus, prior distributions for the hyperparameters were also specified and inference was based on the posterior distribution. The classical paradigm may also be applied (Harvey, 1989). Its use can be illustrated in the context of prediction.

The classical approach is based on integrating out the state parameters in the operation described in Equation (24.10). This gives rise to the integrated likelihood of the hyperparameters $l(\theta) = p(y^u|\theta, D)$. Maximum likelihood estimates can be approximately obtained by numerical operations. Confidence intervals and hypotheses testing can be performed, but they require further information concerning the likelihood or about the sampling distribution of the maximum likelihood estimator. Once again, approximating methods based on asymptotic theory or on sampling techniques (such as bootstrap) must be applied to extract such information.

24.3 Applications

The class of models described above can be used in a number of contexts where space and time interact. The simplest and most obvious one is the context of a univariate component playing the role of mean of a spatio-temporal observation process, i.e., the observation process is given by $y_t(s) = \mu_t(s) + e_t(s)$, where the mean response $\mu_t(s)$ is described by a dynamic Gaussian process (24.8). One generalization of this idea is achieved by considering the regression context with spatio-temporal heterogeneity that was described above. This way, not only the intercept but also the regression coefficients may vary in space–time stochastically.

In general terms, the ideas above can be used to incorporate temporal extensions to parameters of spatial models and also into spatial extensions to parameters of dynamic models. The former accommodates temporal dependence and the latter accommodates spatial dependence.

Spatial dependence was stochastically incorporated into dynamic factor models by Lopes, Salazar, and Gamerman (2006). They used the loading matrix to achieve that. Lopes and Carvalho (2007) showed the relevance of including dynamics into the factor loadings. Thus, combination of these ideas may lead to fruitful possibilities and can be proposed with the class of models described in this chapter. Salazar (2008) implemented these ideas. Figure 24.3 shows some promising results obtained with simulations.

Another natural application area for these ideas is point process modeling. This is the observation process where events occur in a given region and their location is registered. The usual approach in this setting is to define an intensity rate, which governs the occurrence of events. Under conditional independence assumption, Poisson processes become appropriate and the intensity rate suffices to completely characterize the process. A further stochastic layer can be introduced by allowing the intensity rate to be random (Cox, 1955). A popular choice for the intensity distribution is a log Gaussian process (Møller, Syversveen, and Waagepetersen, 1998). From a Bayesian perspective, this is equivalent to a Gaussian process prior for the logarithm of the intensity rate.

Point processes also can be observed over time. In this case, the intensity rate process is a time-continuous sequence of functions over space. This is a natural framework for application of the ideas above and for the use dynamic Gaussian processes as prior distributions for the intensity rate process. Reis (2008) explores this path in a number of examples and applications.

A similar representation for the intensity rate is entertained by Brix and Diggle (2001). They considered the time-continuous differential equation specification and performed classical estimation using moment-based estimators. Calculations were performed approximately by discretizations over space and time. Details are provided in Diggle (2009).

Another area for further exploration of dynamic Gaussian processes is spatial nonstationarity. Many of the models suggested to handle nonstationarity are built over generalizations of stationary Gaussian processes. The spatial deformations of Sampson and Guttorp (1992) and the convolutions of Higdon, Swall, and Kern (1999) and Fuentes (2001) are among the most cited references in this area. Schmidt and O'Hagan (2003) and Damian, Sampson, and Guttorp (2000) have independently casted the deformation problem into the Bayesian paradigm. See also Chapter 9 for more details.

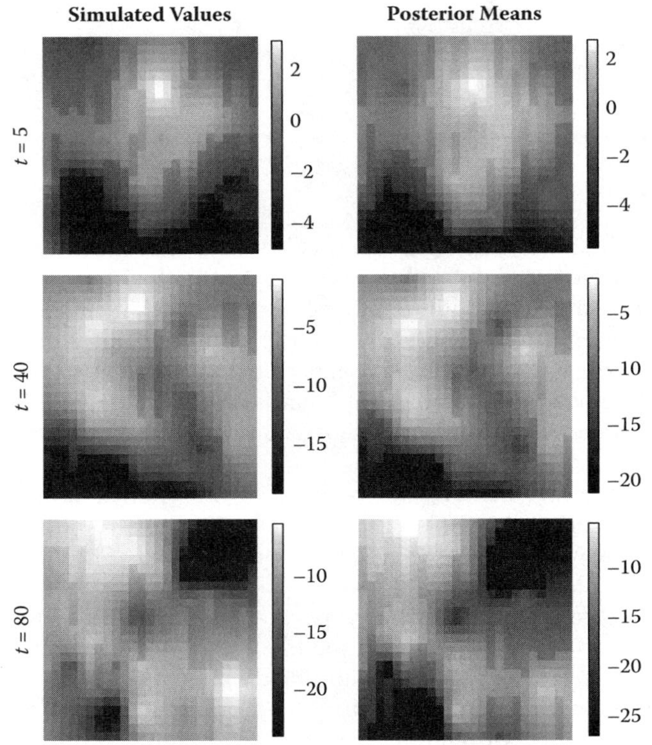

FIGURE 24.3
Results of simulation with a dynamic model with two factors and observation window of 30 locations and 80 equally spaced times. Loadings of the first factor: simulated values (left) and posterior means (right). The first, second and third rows refer to t = 5, 40 and 80, respectively. Posterior means for loadings at unobserved locations were obtained by Bayesian interpolation. (From Gamerman, D., Salazar, E., and Reis, E. (2007). *Bayesian Statistics 8*, pp. 149–174. Oxford University Press, Oxford, U.K. With permission.)

The approaches above make use of Gaussian processes for handling nonstationarity. The convolutions of Fuentes (2001) are based on kernel mixtures of Gaussian processes. The hyperparameters governing these processes may be related over space. Gaussian process prior distribution is one of the choices in this setting and this was entertained by Fuentes. The deformations approach requires the specification of a prior distribution in a Bayesian context or a penalization in a classical setting for the deformed space. In either case, a natural choice would be a Gaussian process.

There are cases when the observation process may span over a period of time. The spatial correlation structure may remain constant over the period. But the spatial nonstationarity may also vary over time. This is at the very least plausible and in many cases highly likely to occur. In this case, the time constancy is no longer valid. Alternatives must be sought to appropriately account for this variation. This is a natural setting for the consideration of dynamic Gaussian process. They can provide a starting exploration step. Due to their local behavior, they are able to describe the variation with a nonparametric flavor without imposing any specific form for the changes over time that these hyperparameter may experience. Once a specific time pattern is observed for the parameters, specific assumptions about this change can be incorporated into the model.

Consider, for example, the convolution of Gaussian processes to account for spatial nonstationarity of the data generating processes. Among the parameters defining the Gaussian processes is their sill parameters. They can be allowed to change not only over space but also over time. This would allow for many purposes: to borrow information across nearby locations and consecutive time periods and to smooth variations of this parameter over space and over time.

24.4 Further Remarks

The material of the last section is of more speculative nature. The ideas described here are just beginning to be used. They involve use of elaborate model specifications that are far from easy to be estimated from the data. This poses yet another challenge in the use of this methodology. The most common approach these days from a Bayesian perspective is MCMC (see Gamerman and Lopes, 2006). Alternatives based on noniterative approximations are also being proposed (Rue, Martino, and Chopin, 2008).

This chapter was mostly based on the so-called latent approach where spatial and temporal dependences are incorporated in the model through the use of latent structures. Other possibilities are also available. Spatial autoregression (SAR) provides a natural framework to contemplate spatial dependence directly at the observational level (Anselin, 1988). Addition of temporal components can be made separately in a different model block or jointly. Gamerman and Moreira (2004) have simply added a temporal autoregressive component to a multivariate SAR model. Kakamu and Polasek (2007) considered a SAR structure over (temporally) lagged variables thus inducing spatial and temporal dependence simultaneously. These are just a couple of the many possibilities available through this approach.

The purpose of this chapter is to draw attention to a tool that is flexible and can accommodate many patterns of spatio-temporal data variation. There are other areas that can become potential applications for this technology. It is hoped in this chapter that readers have their attention drawn to these modeling tools, find them useful, and eventually try them in their own problems at hand.

References

Anselin, L. (1988). *Spatial Econometrics: Methods and Models*, Dordrecht: Kluwer Academic Publishers.

Brix, A. and Diggle, P.J. (2001). Spatiotemporal prediction for log-Gaussian Cox processes. *Journal of the Royal Statistical Society Series B*, **63**, 823–841.

Cox, D.R. (1955). Some statistical models related with series of events. *Journal of the Royal Statistical Society Series B*, **17**, 129–164.

Damian, D., Sampson, P.D. and Guttorp, P. (2000). Bayesian estimation of semi-parametric nonstationary spatial covariance structure. *Environmetrics*, **11**, 161–176.

Diggle, P.J., Tawn, J.A. and Moyeed, R.A. (1998). Model-based geostatistics (with discussion). *Applied Statistics*, **47**, 299–350.

Fuentes, M. (2001). A high frequency kriging for nonstationary environmental processes. *Environmetrics*, **12**, 1–15.

Gamerman, D. and Lopes, H.F. (2006). *Monte Carlo Markov Chain: Stochastic Simulation for Bayesian Inference* (2nd ed.). Chapman & Hall: London.

Gamerman, D. and Moreira, A.R.B. (2004). Multivariate spatial regression models. *Journal of Multivariate Analysis*, **91**, 262–281.

Gamerman, D., Moreira, A.R.B. and Rue, H. (2003). Space-varying regression models: Specifications and simulation. *Computational Statistics and Data Analysis*, **42**, 513–533.

Gamerman, D., Salazar, E. and Reis, E. (2007). Dynamic Gaussian process priors, with applications to the analysis of space-time data (with discussion). In *Bayesian Statistics 8* (Eds. Bernardo, J.M. et al.), pp. 149–174. Oxford University Press, Oxford, U.K.

Gelfand, A.E., Banerjee, S. and Gamerman, D. (2005). Spatial process modelling for univariate and multivariate dynamic spatial data. *Environmetrics*, **16**, 465–479.

Harvey, A. (1989). *Forecasting, Structural Time Series Models and the Kalman Filter*. Cambridge, University Press, Cambridge, MA.

Higdon, D.M., Swall, J. and Kern, J. (1999). Non-stationary spatial modeling. In *Bayesian Statistics 6* (eds. Bernardo, J.M. et al.), 761–768. Oxford University Press, Oxford, U.K.

Huerta, G., Sansó, B. and Stroud, J. (2004). A spatial-temporal model for Mexico city ozone levels. *Applied Statistics*, **53**, 231–248.

Kakamu, K. and Polasek, W. (2007). Cross-sectional space-time modeling using ARNN(p, n) processes. Technical report # 203, Institute for Advanced Studies, Vienna.

Lopes, H.F. and Carvalho, C. (2007). Factor stochastic volatility with time varying loadings and Markov switching regimes, *Journal of Statistical Planning and Inference*, **137**, 3082–3091.

Lopes, H.F., Salazar, E. and Gamerman, D. (2006). Spatial dynamic factor analysis. *Bayesian Analysis*, **3**, 759–792.

Møller, J., Syversveen, A. and Waagepetersen, R. (1998). Log Gaussian Cox processes. *Scandinavian Journal of Statistics*, **25**, 451–482.

Paez, M.S., Gamerman, D., Landim, F.M.P.F. and Salazar, E. (2008). Spatially-varying dynamic coefficient models. *Journal of Statistical Planning and Inference*, **138**, 1038–1058.

Reis, E. (2008). *Dynamic Bayesian Models for Spatio-Temporal Point Processes*. Unpublished Ph.D. thesis (in Portuguese). Universidade Federal do Rio de Janeiro.

Revuz, D. and Yor, M. (1999). *Continuous Martingales and Brownian Motion* (3rd ed.). Springer, New York.

Rue, H., Martino, S. and Chopin, N. (2008). Approximate Bayesian inference for latent Gaussian models using integrated nested Laplace approximations (with discussion). *Journal of the Royal Statistical Society Series B*, **71**, 319–392.

Salazar, E. (2008). *Dynamic Factor Models with Spatially Structured Loadings*. Unpublished Ph.D. thesis (in Portuguese). Universidade Federal do Rio de Janeiro.

Sampson, P.D. and Guttorp, P. (1992). Nonparametric estimation of nonstationary spatial covariance structure, *Journal of the American Statistical Association*, **87**, 108–119.

Sansó, B., Schmidt, A.M. and Nobre, A.A. (2008). Bayesian spatio-temporal models based on discrete convolutions. *Canadian Journal of Statistics*, **36**, 239–258.

Schmidt, A.M. and O'Hagan, A. (2003). Bayesian inference for nonstationary spatial covariance structures via spatial deformations. *Journal of the Royal Statistical Society Series B*, **65**, 743–758.

Stroud, J., Muller, P. and Sansó, B. (2001). Dynamic models for spatial-temporal data. *Journal of the Royal Statistical Society, Series B*, **63**, 673–689.

West, M. and Harrison, J. (1997). *Bayesian Forecasting and Dynamic Models* (2nd ed.) Springer, New York.

Wikle, C. and Cressie, N. (1999). A dimension-reduced approach to space-time Kalman filtering. *Biometrika*, **86**, 815–829.

25

Spatio-Temporal Point Processes

Peter J. Diggle and Edith Gabriel

CONTENTS

25.1 Introduction

In their most basic form, spatio-temporal point process data consist of a time-ordered sequence of events $\{(s_i, t_i) : i = 1, \ldots, n\}$ where s denotes location, t denotes time and n is the number of events that fall within a spatial region D and a time-interval $[0, T]$. The term "point process" is usually reserved for a process that generates events in a spatial and temporal continuum, but in the current setting, we will allow either, but not both, of the spatial and temporal supports to be discrete.

Processes that are both spatially and temporally discrete are more naturally considered as binary-valued random fields, as discussed in Chapter 12. Processes that are temporally discrete with only a small number of distinct event-times can be considered initially as multivariate point processes, as discussed in Chapter 21, but with the qualification that the temporal structure of the type-label may help the interpretation of any interrelationships among the component patterns. Conversely, spatially discrete processes with only a small number of distinct event-locations can be considered as multivariate temporal point processes, but with a spatial interpretation to the component processes. In the remainder of this chapter, we consider processes that are temporally continuous and either spatially continuous or spatially discrete on a sufficiently large support to justify formulating explicitly spatio-temporal models for the data.

Our discussion will be from the perspective of methods and applications, rather than a rigorous theoretical treatment; for the latter, see, for example, Daley and Vere-Jones (2002, 2007)

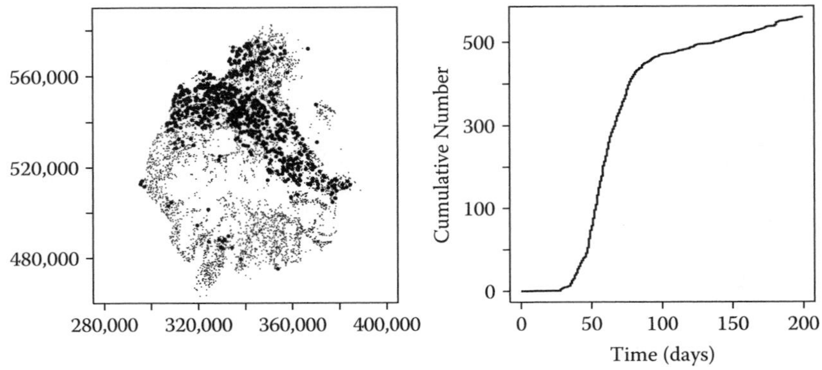

FIGURE 25.1
Data from the 2001 United Kingdom foot-and-mouth epidemic. Left: Small dots show the locations of all animal-holding farms in the county of Cumbria, larger circles show the locations of those farms that experienced foot-and-mouth during the epidemic. Right: Cumulative distribution of the times.

25.2 Exploratory Tools for Spatio-Temporal Point Processes

25.2.1 Plotting the Data

The most effective form of display for a spatio-temporal point process data is an animation, repeated viewing of which may yield insights that are not evident in static displays. As an example, we consider data from the 2001 United Kingdom foot-and-mouth epidemic.

Figure 25.1 shows two static displays of the data from the county of Cumbria. The left-hand panel shows the locations of all animal-holding farms at the start of the epidemic, with those that experienced foot-and-mouth highlighted. The right-hand panel shows the cumulative distribution of the times, in days since January 1, 2001, on which newly infected farms were reported. The spatial distribution is consistent with pronounced spatial variation in risk over the study region, while the temporal distribution shows the characteristic S-shape of an epidemic process.

An animation of the spatio-temporal pattern of incidence over the whole epidemic reveals a much richer structure than is apparent from Figure 25.1 (http://www.maths.lancs.ac.uk/~rowlings/Chicas/FMD/slider2.html). Features that become clear on repeated viewing of the animation include: a predominant pattern of spatio-temporal spread characteristic of direct transmission of infection between neighboring farms; occasional and apparently spontaneous infections occurring remotely from previously infected farms; and a progressive movement of the focus of the epidemic from its origin in the north of the county, initially to the west and later to the southeast.

As a contrasting example, the two panels of Figure 25.2 show the cumulative spatial distribution of 969 reported cases of gastrointestinal infections in the district of Preston over the years 2000 to 2002, and the cumulative distribution of the times, in days since January 1, 2000, on which cases were reported. The spatial distribution of cases largely reflects the population at risk, consistent with the endemic character of gastrointestinal infections, while the temporal distribution shows an approximately constant rate of incident cases. For these data, an animation (http://pagesperso-orange.fr/edith.gabriel/preston), adds relatively little to this description, although a detailed analysis does reveal additional structure, including seasonal effects and small-scale spatio-temporal clustering of cases, as we show in Section 25.2.2.

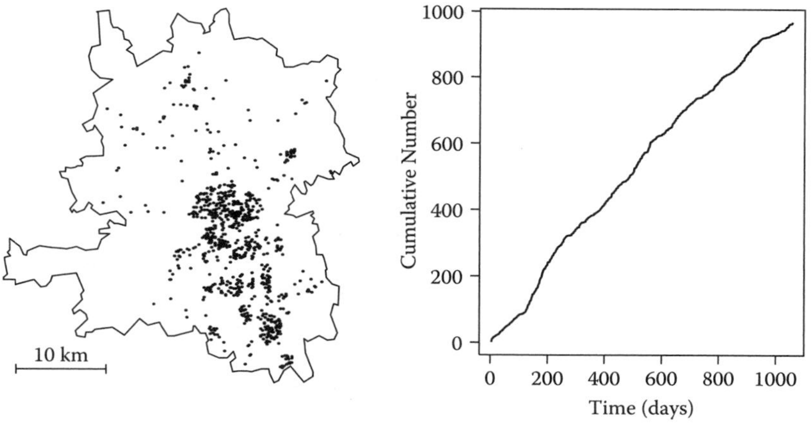

FIGURE 25.2
Gastrointestinal infections in the district of Preston, 2000 to 2002. Left: Locations of cases. Right: Cumulative distribution of the times.

25.2.2 Moment-Based Summaries

The moment-based summaries described in Chapter 18 for spatial point processes are easily extended to the spatio-temporal setting. First-moment properties are specified through the *spatio-temporal intensity* function, $\lambda(s, t)$, defined as the expected number of events per unit area per unit time, centered on location s and time t. Nonparametric estimation of $\lambda(s, t)$, for example by kernel smoothing, is straightforward in principle but difficult to visualize except, again, through animation. Note, however, that smoothing will in itself tend to introduce spatio-temporal dependence into the data, and as such can give a misleading impression of spatio-temporal structure. For data on a spatio-temporal region $A = D \times [0, T]$, estimation of the marginal spatial and temporal intensities, $\lambda_A(s) = \int_0^T \lambda(s, t)dt$ and $\mu_A(t) = \int_D \lambda(s, t)ds$, respectively, can be used to assess whether the spatio-temporal intensity is *first-order separable*, meaning that for any A, $\lambda(s, t) \propto \lambda_A(s)\mu_A(t)$. Separability, if justified, provides a useful simplification for any subsequent modeling exercise.

Second-moment summaries extend similarly. In particular, for a stationary, orderly process with $\lambda(s, t) = \lambda$, the spatio-temporal K-function is

$$K(u, v) = \lambda^{-1}\mathrm{E}[N_o(u, v)], \tag{25.1}$$

where $N_o(u, v)$ is the number of additional events whose locations lie in the disk of radius u centered on the origin and whose times lie in the interval $(0, v)$, conditional on there being an event at the spatio-temporal origin; see Diggle, Chetwynd, Haggkvist, and Morris (1995), but note that their definition includes in $N_o(u, v)$ events that follow or precede the conditioning event at time $t = 0$. With our definition, the benchmark for a homogeneous spatio-temporal Poisson process is $K(u, v) = \pi u^2 v$. More generally, for a process whose spatial and temporal components are stochastically independent, $K(u, v) = K_s(u)K_t(v)$, where $K_s(\cdot)$ and $K_t(\cdot)$ are proportional to the spatial and temporal K-functions of the component processes. Note that this interpretation of $K_s(\cdot)$ and $K_t(\cdot)$ requires us to restrict the process to a finite time-interval and a finite spatial region, respectively, otherwise the component spatial and temporal processes are undefined.

As in the spatial case considered in Section 18.6, the K-function can be defined directly in terms of a conditional expectation, as in (25.1), or as the integral of the second-order intensity function, $\lambda_2(u, v)$, such that

$$K(u, v) = 2\pi\lambda^{-2} \int_0^v \int_0^u \lambda_2(u', v')u'du'dv'$$
$$= \pi u^2 v + 2\pi\lambda^{-2} \int_0^v \int_0^u \gamma(u', v')u'du'dv', \quad (25.2)$$

where $\gamma(u, v) = \lambda_2(u, v) - \lambda^2$ is called the covariance density. A process for which $\gamma(u, v) = \gamma_s(u)\gamma_t(v)$ is called *second-order separable*. As with its first-order counterpart, separability (when justified) leads to useful simplifications in modeling. Note that second-order separability is implied by, but does not imply, independence of the spatial and temporal components.

Suppose now that we have data in the form $x_i = (s_i, t_i) : i = 1, \ldots, n$ consisting of all events x_i in a space–time region $D \times [0, T]$ and ordered so that $t_i > t_{i-1}$. Assume for the time being that the underlying process is intensity-reweighted stationary (Baddeley, Møller, and Waagpetersen, 2000) with known intensity $\lambda(s, t)$. Then, an approximately unbiased estimator for $K(u, v)$ is

$$\hat{K}(u, v) = \{n/(|D|T)\}E(u, v), \quad (25.3)$$

where

$$E(u, v) = n_v^{-1} \sum_{i=1}^{n_v} \sum_{j>i} w_{ij}^{-1} I(||s_i - s_j|| \leq u) I(t_j - t_i \leq v)/\{\lambda(s_i, t_i)\lambda(s_j, t_j)\},$$

the w_{ij} are spatial edge-correction weights as defined in Section 18.6 and n_v is the number of $t_i \leq T - v$ (Gabriel and Diggle, 2008).

25.2.3 Campylobacteriosis in Lancashire, U.K.

Campylobacter jejuni is the most commonly identified cause of bacterial gastroenteritis in the developed world. Temporal incidence of campylobacteriosis shows strong seasonal variation, rising sharply between spring and summer. Here, we analyze a dataset consisting of the locations and dates of notification of all known cases of campylobacteriosis within the Preston postcode district (Lancashire, England) between January 1, 2000 and December 31, 2002. These data can be considered as a single realization of a spatio-temporal point process displaying a highly aggregated spatial distribution. As is common in epidemiological studies, the observed point pattern is spatially and temporally inhomogeneous because the pattern of incidence of the disease reflects both the spatial distribution of the population at risk and systematic temporal variation in risk. When analyzing such spatio-temporal point patterns, a natural starting point is to investigate the nature of any stochastic interactions among the points of the process after adjusting for spatial and temporal inhomogeneity.

The three panels of Figure 25.3 show the study region, corresponding to the Preston postcode sector of the county of Lancashire, U.K.; a grayscale representation of the spatial variation in the population density, derived from the 2001 census; and the residential locations of the 619 recorded cases over the three years 2000 to 2002 in the most densely populated part of the study region.

We first estimate the marginal spatial and temporal intensities of the data. To estimate the spatial density, $m(s)$, we use a Gaussian kernel estimator with bandwidth chosen to minimize the estimated mean square error of $\hat{m}(s)$, as suggested in Berman and Diggle (1989). To estimate the temporal intensity, $\mu(t)$, we use a Poisson log-linear regression

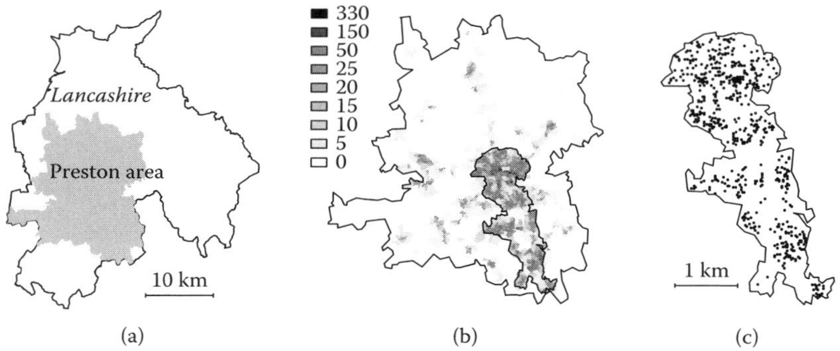

FIGURE 25.3
Campylobacteriosis data from Lancashire, U.K. (a) The study area. (b) Population density in 2001 (number of people per hectare). (c) Locations of the 619 cases of *Campylobacter jejuni* infections within the urban area.

model incorporating a time-trend, seasonal variation and day-of-the-week effects, hence,

$$\log \mu(t) = \delta_{d(t)} + \sum_{k=1}^{3} \{\alpha_k \cos(k\omega t) + \beta_t \sin(k\omega t)\} + \gamma t,$$

where $\omega = 2\pi/365$ and $d(t)$ identifies the day of the week for day $t = 1, \ldots, 1096$. The sine–cosine terms corresponding to six-month and four-month frequencies are justified by likelihood ratio criteria under the assumed Poisson model, but this would overstate their significance if, as it turns out to be the case, the data show significant spatio-temporal clustering. Figure 25.4 shows the resulting estimates of $m(s)$ and $\mu(t)$.

A comparison between Figures 25.4a and Figure 25.3b shows, unsurprisingly, that cases tend to be concentrated in areas of high population density, while Figure 25.4b shows a decreasing time-trend and a sharp peak in intensity each spring. The smaller, secondary peaks in intensity are a by-product of fitting three pairs of sine–cosine terms and their substantive interpretation is open to question; here, we are using the log-linear model only to give a reasonably parsimonious estimate of the temporal intensity as a necessary prelude to investigating residual spatio-temporal structure in the data.

To investigate spatio-temporal structure, we consider the data in relation to two benchmark hypotheses. The hypothesis of *no spatio-temporal clustering*, H_0^c, states that the data are a realization of an inhomogeneous Poisson process with intensity $\lambda(s, t) = m(s)\mu(t)$. The hypothesis of *no spatio-temporal interaction*, H_0^i, states that the data are a realization of a pair of independent spatial and temporal, reweighted second-order stationary point processes with respective intensities $m(s)$ and $\mu(t)$. Note that in formulating our hypotheses in this way, we are making a pragmatic decision to interpret separable effects as first-order, and nonseparable effects as second-order. Also, as here defined, absence of spatio-temporal clustering is a special case of absence of spatio-temporal interaction.

To test H_0^c, we compare the inhomogeneous spatio-temporal K-function of the data with tolerance envelopes constructed from simulations of a Poisson process with intensity $\hat{m}(s)\hat{\mu}(t)$. To test H_0^i, we proceed similarly, but with tolerance envelopes constructed by randomly relabeling the locations of the cases holding their notification dates fixed, thus preserving the marginal spatial and temporal structure of the data without assuming that either is necessarily a Poisson process.

Figure 25.5a shows $\hat{K}_{st}(u, v) - \pi u^2 v$ for the *C. jejuni* data. The diagonal black hatching on Figure 25.5b identifies those values of (u, v) for which the data-based estimate of $\hat{K}_{st}(u, v) - \pi u^2 v$ lies above the 95th percentile of estimates calculated from 1,000 simulations

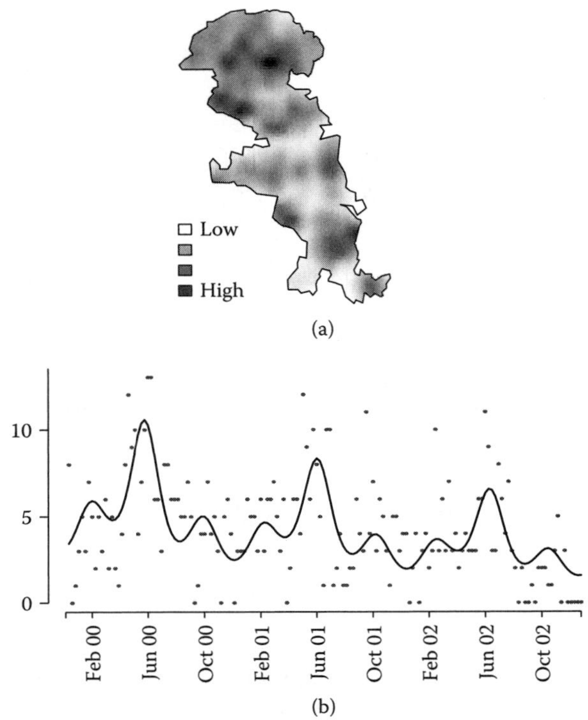

FIGURE 25.4
(a) Kernel estimate of the spatial intensity; (b) weekly numbers (dots) of notified cases compared with fitted regression curve.

of an inhomogeneous Poisson process with intensity $\hat{m}(s)\hat{\mu}(t)$. Similarly, the gray shading identifies those values of (u, v) for which $\hat{K}_{st}(u, v) - \hat{K}_s(u)\hat{K}_t(v)$ lies above the 95th percentile envelopes calculated from 1,000 random permutations of the s_i holding the t_i fixed. The results suggest spatio-temporal clustering up to a distance of 300 meters and time-lags up to 10 days, and spatio-temporal interaction at distances up to 400 meters and time-lags

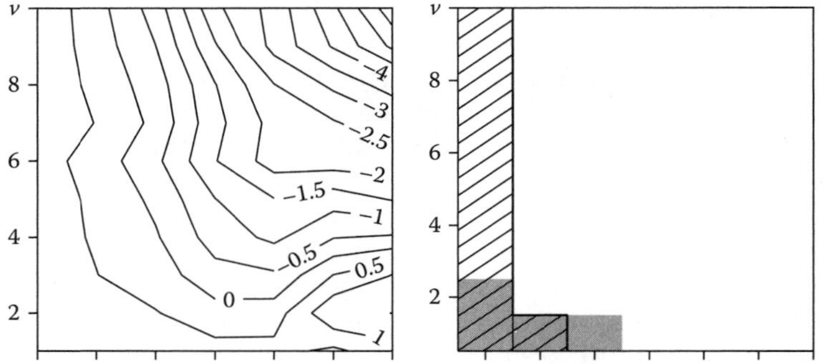

FIGURE 25.5
(a) $\hat{K}_{st}(u, v) - \pi u^2 v$ ($\times 10^6$). (b) Identification of subsets of (u, v)-space in which spatio-temporal clustering (diagonal black hatching) and/or spatio-temporal interaction (gray shading) is detected at the 5% level of significance (see text for detailed explanation).

up to 3 days. These findings are consistent with the infectious nature of the disease, which leads to multiple cases from a common source occurring relatively closely both in space and in time. The analysis also suggests the existence of stochastic structure that cannot be explained by the first-order intensity $\hat{m}(s)\hat{\mu}(t)$. Note that the relatively large negative values of $\hat{K}_{st}(u, v) - \pi u^2 v$ at large values of u and v are not significantly different from zero because the sampling variance of $\hat{K}_{st}(u, v)$ increases with u and v.

25.3 Models

A useful distinction in statistical modeling is between models whose primary purpose is to give a concise, quantitative description of the data and those whose parameters are intended to relate directly to a known or hypothesized process that generates the data. We refer to these as *empirical* and *mechanistic* models, respectively. For example, in most applications of linear regression, the model is empirical, whereas nonlinear regression models typically are more mechanistic in nature.

All of the models for spatial point processes that were discussed in Chapter 17 can be extended directly to the spatio-temporal setting. Additionally, new models can be formulated to take explicit account of time's directional nature. From a strict modeling perspective, it is sometimes argued that as spatial point patterns in nature often arise as snapshots from an underlying spatio-temporal process, purely spatial models should ideally be derived directly from underlying spatio-temporal models (for further comments, see Section 25.5).

25.3.1 Poisson and Cox Processes

A spatio-temporal Poisson process is defined by its intensity, $\lambda(s, t)$. A spatio-temporal Cox process is a spatio-temporal Poisson process whose intensity is a realization of a spatio-temporal stochastic process $\Lambda(s, t)$. A Cox process inherits the second-order properties of its stochastic intensity function. Hence, if $\Lambda(s, t)$ is stationary with mean λ and covariance function $\gamma(u, v) = \text{Cov}\{\Lambda(s, t), \Lambda(s-u, t-v)\}$, then the K-function of the corresponding Cox process is given by Equation (25.2). A convenient and relatively tractable class of Cox process models is the log-Gaussian class (Møller, Syversveen, and Waagepetersen, 1998), for which $\Lambda(s, t) = \exp\{Z(s, t)\}$, where $Z(s, t)$ is a Gaussian process. Constructions for $Z(s, t)$ are discussed in Chapter 23. As in the purely spatial case, Cox processes provide natural models when the point process in question arises as a consequence of environmental variation in intensity that cannot be described completely by available explanatory variables, rather than through direct, stochastic interactions among the points themselves.

25.3.2 Cluster Processes

We now describe two ways of modifying a Neyman–Scott cluster process to the spatio-temporal setting. Recall that in the spatial setting, the three ingredients of the Neyman–Scott process are: (1) a homogeneous Poisson process of *parent* locations; (2) a discrete distribution for the number of *offspring* per parent, realized independently for each parent; and (3) a continuous distribution for the spatial displacement of an offspring relative to its parent, realized independently for each offspring.

For our first spatio-temporal modification, parents are generated by a homogeneous spatio-temporal Poisson process with intensity ρ. Each parent gives birth to a series of offspring in a time-inhomogenous Poisson process with intensity $\alpha(t - t_0)$, where t_0 is the temporal location of the parent; hence, the number of offspring per parent follows a Poisson

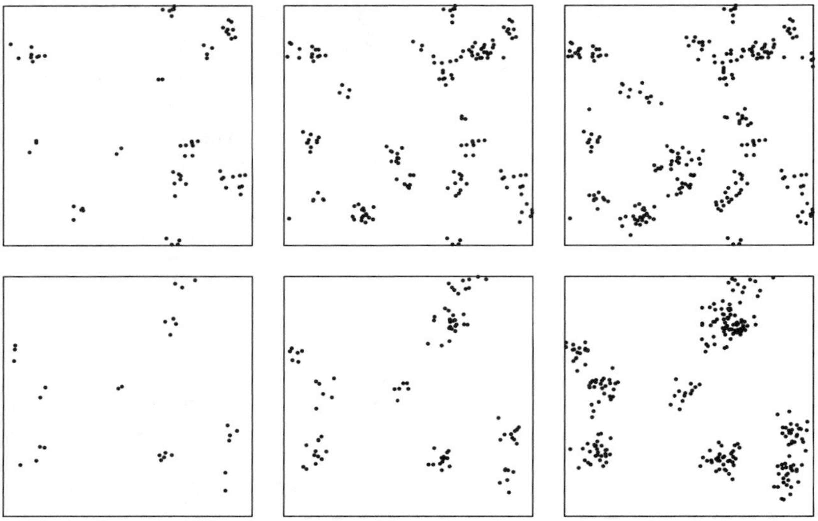

FIGURE 25.6
Top row: Cumulative spatial locations at times $t = 10, 20, 30$ for a simulated realization of the first spatio-temporal extension to the Neyman–Scott spatial cluster process. Bottom row: Cumulative spatial locations over three generations for a simulated realization of the second spatio-temporal extension to the Neyman–Scott spatial cluster process (see text for detailed explanation of the simulation models).

distribution with mean $\int_0^\infty \alpha(u)du$, which we assume to be finite. The spatial locations of the offspring of a parent at location s_0 are an independent random sample from a distribution with density $f(s - s_0)$.

A possibly more natural way to extend the Neyman–Scott process to the spatio-temporal setting is to consider a multigeneration process in which parents produce first-generation offspring as before, but now the offspring act as parents for a second generation, and so on.

These two processes have very different properties. In the first process, the pattern of events in a fixed time-window has essentially the same marginal spatial structure as a spatial Neyman–Scott process, whereas in the second the nature of the clustering becomes more diffuse with each succeeding generation.

Figure 25.6 compares realizations of the two processes, in each case with $f(s)$ describing an isotropic Gaussian density. In the first case, the mean number of parents per unit area per unit time is ρ, the rate at which each parent generates offspring is $\alpha(t) = 1$ for $0 \leq t \leq \tau$, otherwise zero, and the distribution of the displacements of offspring from their parents has standard deviation $\sigma = 0.025$ in each coordinate direction. In Figure 25.6 (top row), we set $\rho = 1$ and $\tau = 10$, and show the cumulative locations at times $t = 10, 20$ and 30, by which time the process has generated 255 events in the unit square. The three panels show clearly one feature of this process, namely, that new clusters appear continuously over time.

In the second case, ρ denotes the number of initial parents and τ denotes the mean number of offspring per event in the previous generation. In the bottom row of Figure 25.6 we set $\sigma = 0.05$ and $\rho = 10$ and show the cumulative locations of events over the first three generations, by which time the process has generated 264 events in the unit square. Note how, in contrast to the behavior illustrated in the top row of Figure 25.6, a feature of this process is that the clusters of events maintain their spatial locations, but become larger and more diffuse over successive generations.

Cluster processses of the kind described in this section provide natural models for reproducing populations in a homogeneous environment. Their empirical spatial behavior tends to be similar to, or even indistinguishable from, that of Cox processes (Bartlett, 1964).

However, as the two examples in Figure 25.6 illustrate, this ambiguity in interpretation can be resolved by their contrasting spatio-temporal properties; the top row, but not the bottom row, of these two cluster processes is also an example of a spatio-temporal Cox process.

25.3.3 Conditional Intensity Function

Let $\mathcal{H}_t$ denote the history of a spatio-temporal process up to but excluding time t, hence, in the absence of any explanatory variables, $\mathcal{H}_t = \{(x_i, t_i) : t_i < t\}$. The *conditional intensity* of the process is then defined as

$$\lambda_c(s, t|\mathcal{H}_t) = \text{limit}_{|ds|\to 0, dt\to 0} \frac{\text{P(event in } ds \times dt)}{|ds|dt}, \tag{25.4}$$

where ds denotes an infinitesimal spatial region containing s and dt an infinitesimal time interval with lower end-point t. Under the usual regularity conditions, the conditional intensity determines all properties of the underlying point process. This openes up a very rich set of models defined directly by specifying a parametric form for $\lambda_c(s, t|\mathcal{H}_t)$. The requirement for a valid specification is that $\lambda_c(s, t|\mathcal{H}_t)$ must be nonnegative valued with $\int_D \lambda(s, t|\mathcal{H}_t)ds < \infty$ for all t, all $\mathcal{H}_t$ and any $D \subset \mathbb{R}^2$ with finite area. Below, we consider two such constructions.

25.3.3.1 Pairwise Interaction Processes

Pairwise interaction processes are the spatio-temporal counterparts of the spatial pairwise interaction processes discussed in Chapter 17. An interaction function $h(s, s')$ is a nonnegative valued function of two spatial locations s and s'. A spatial-temporal pairwise interaction point process has a conditional intensity of the form

$$\lambda_c(s, t|\mathcal{H}_t) = \alpha(t) \prod_{i=1}^{n_t} h(s, s_i), \tag{25.5}$$

where n_t is the number of events in $\mathcal{H}_t$. As in the spatial case, if $h(s, s') = 1$ for all s and s', the process reduces to a Poisson process. However, in contrast to the spatial case, the only restriction on $h(\cdot)$ is that it should be nonnegative valued with $\int_D h(s)ds < \infty$ for any $D \subset \mathbb{R}^2$ with finite area. Also, the product form in Equation (25.5) is but one of many possibilities. Jensen et al. (2007) take a more empirical approach to models of this kind; they define a class of spatial-temporal Markov point processes, of which pairwise interaction processes are a special case, by their likelihood ratio with respect to a homogeneous Poisson process of unit intensity.

Figure 25.7 shows a sequence of stages in the development of a pairwise interaction process, constrained to a finite spatial region, with $\lambda(t) = 1$ and interaction function $h(s, s') = h(||s - s'||)$, where

$$h(u) = \begin{cases} 0 : u \leq \delta \\ 1 + \alpha \exp[-\{(u - \delta)/\phi\}^2] : u > \delta \end{cases} \tag{25.6}$$

with $\delta = 0.05$, $\alpha = 0.5$ and $\phi = 0.1$. Notice how the spatial character of the process changes over time. Initially, the dominant impression is of clustering induced by the attractive interactions ($h(u) > 1$) at distances greater than δ. As time progresses, the inhibitory interactions ($h(u) = 0$ for $u \leq \delta$) become more apparent as the available space fills up.

25.3.3.2 Spatial Epidemic Processes

The classic epidemic model for a closed population is the SIR model introduced by Kermack and Kendrick (1927), in which an initial population of *susceptibles* is at risk of becoming

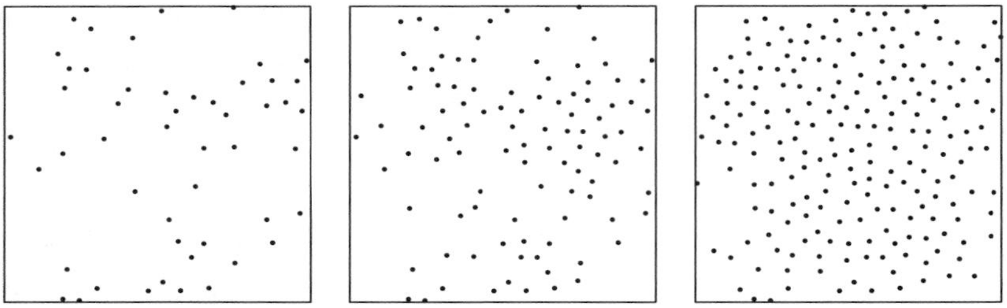

FIGURE 25.7
Cumulative spatial locations of the first 50, 100 and 200 events in a simulated realization of a spatiotemporial interaction process combining inhibitory and clustering effects (see text for detailed explanation of the simulation model).

infected by the disease in question, and later *removed* either by dying or by recovering and becoming immune to reinfection. In its simplest form, the model assumes a closed population. Extensions to open populations allow immigration and/or emigration by susceptible and/or infected individuals.

To turn the SIR model into a spatio-temporal point process, we introduce a point process (purely spatial in the closed case, spatio-temporal in the open case) for the susceptible population and a *spatial transmission kernel*, $f(s, s')$, to describe the rate at which an infectious individual at location s infects a susceptible individual at location s'. Keeling, Woolhouse, Shaw, Matthews et al. (2001) use a model of this kind to describe the 2001 United Kingdom foot-and-mouth epidemic. As is typical of real applications, they needed to extend the basic formulation of the model to take account of the particular features of the phenomenon under investigation. Firstly, the infectivity and susceptibility of individual farms depend on farm characteristics codified as explanatory variables, for example, their total stock holding. Second, the disease process includes a latent period whereby an infected farm's stock would be infectious, but asymptomatic for a variable and unknown period, thought to be of the order of several days. Finally, the model needs to take account of the main control policy during the epidemic, namely preemptive culling of the stock on farms located close to a farm known to be infected. Keeling et al. (2001) use sensible ad hoc methods of estimation to fit their model and, in particular, to identify the shape and scale of the spatial transmission kernel $f(s, s')$.

25.4 The Likelihood Function

In addition to providing the basis for model formulation, the conditional intensity function is the key to likelihood-based inference for spatio-temporal point processes. Under the usual regularity conditions, the log likelihood for data $x_i = (s_i, t_i) : i = 1, \ldots, n$ with $s_i \in D$ and $0 \leq t_1 < t_2 < \cdots < t_n \leq T$ is

$$L(\theta) = \sum_{i=1}^{n} \lambda_c(s_i, t_i | \mathcal{H}_{t_i}) - \int_0^T \int_D \lambda_c(s, t) ds dt. \tag{25.7}$$

Using Equation (25.7) as the basis for inference has the obvious attraction that the usual optimality properties of likelihood-based inference hold. However, the conditional intensity may not be tractable, and even when it is, evaluation of the integral term in Equation (25.7) is often difficult. One way around this is to use Monte Carlo likelihood methods, as described for the purely spatial case in Chapter 19. Case studies using Monte Carlo likelihood methods to fit spatio-temporal Cox process models in a forestry setting are described in Benes, Bodlak, Miller, and Waagepetersen (2005) and in Møller and Diaz-Avalos (2008).

An alternative strategy, variously proposed and advocated by Møller and Sorensen (1994), Lawson and Leimich (2000) and Diggle (2006), is to construct a partial likelihood (Cox, 1975) analogous to the method of inference widely used in the analysis of proportional hazards survival models (Cox, 1972).

The *partial likelihood* for time-ordered spatial point process data $(s_i, t_i) : i = 1, \ldots, n$ is the likelihood of the data conditional on the ordered times, t_i, hence,

$$PL(\theta) = \sum_{i=1}^{n} \log p_i, \tag{25.8}$$

where

$$p_i = \lambda_c(s_i, t_i) / \int_{\mathcal{R}_{t_i}} \lambda_c(s, t_i) ds, \tag{25.9}$$

where $\mathcal{R}_t$ is the risk set at time t, i.e., the set of locations at which future events may occur. When, as is the case for some epidemic modeling problems, the spatial support of the process is discrete, the integral in (25.9) reduces to a summation over the discrete set of potential event-locations.

Diggle, Kaimi, and Abelana (2008) compare the performance of maximum partial and full likelihood estimation using a class of inhomogeneous Poisson processes for which both methods are tractable. Their results illustrate that the loss of efficiency in using the partial likelihood by comparison with full likelihood varies according to the kind of model being fitted. Against this, the advantage of the partial likelihood method is the ease with which it can be implemented for quite complex models. In practice, the ability routinely to fit and compare realistically complex models without having to tune an iterative Markov chain Monte Carlo (MCMC) algorithm for each model is very useful.

25.5 Discussion and Further Reading

Although the structure of spatio-temporal data is superficially more complex than that of purely spatial data, the same does not necessarily apply to methods of statistical analysis. Specifically, conditioning on the past is a very powerful tool in the development of stochastic models and associated statistical methods for spatio-temporal phenomena that is not available in the purely spatial case. Another liberating feature of many applications of spatio-temporal methods is that there need be no assumption that the phenomenon under investigation is in, or even possesses, an equilibrium state. This applies, for example, to spatio-temporal epidemic phenomena, such as the foot-and-mouth epidemic illustrated in Figure 25.1. From this perspective, it is mildly ironic that some of the early work on spatial point process models placed some emphasis on deriving a spatial point process model as the equilibrium distribution of an underlying spatio-temporal process. For example, Preston (1977) derived the class of pairwise interaction spatial point processes as the equilibrium states of spatio-temporal birth-and-death processes. Ripley (1977, 1979) exploited this connection to develop a method of simulating realizations of a pairwise interaction spatial point processes that, without being named as such, was an early example of MCMC. As a second example, Kingman (1977) questioned the status of Poisson models for spatial point processes by arguing that they do not arise naturally as equilibrium distibutions of spatio-temporal processes.

Spatio-temporal point processes are widely used as models in seismology, for data consisting of the locations and times of earthquakes. See, for example, Ogata (1998), Ogata and Zhuang (2006), and Zhuang (2006).

Statistical methods for spatio-temporal point processes are experiencing a period of rapid development. This appears to have resulted from the conjunction of an explosion in the availability of rich, spatio-temporal datasets and the practical feasibility of computationally intensive methods for fitting realistically complex spatio-temporal models.

References

Baddeley, A.J., Møller, J. and Waagpetersen, R. (2000). Non- and semi-parametric estimation of interaction in inhomogeneous point patterns. *Statistica Neerlandica*, 54, 329–350.

Bartlett, M.S. (1964). The spectral analysis of two-dimensional point processes. *Biometrika*, 51, 299–311.

Becker, N.G. (1989). *Analysis of Infectious Disease Data*. London: Chapman & Hall.

Benes, V., Bodlak, K., Mller, J. and Waagepetersen R.P. (2005). A case study on point process modeling in disease mapping. *Image Analysis and Stereology*, 24, 159–168.

Berman, M. and Diggle, P. (1989). Estimating weighted integrals of the second-order intensity of a spatial point process. *Journal of the Royal Statistical Society, Series B*, 44, 406–413.

Cox, D.R. (1972). Regression models and life tables (with discussion). *Journal of the Royal Statistical Society* B, 34, 187–220.

Cox, D.R. (1975). Partial likelihood. *Biometrika*, 62, 269–275.

Daley, D.J. and Vere-Jones, D. (2002). *Introduction to the Theory of Point Processes: Elementary Theory and Methods*, Heidelberg/New York: Springer.

Daley, D.J. and Vere-Jones, D. (2007). *An Introduction to the Theory of Point Processes: General Theory and Structure*, Heidelberg/New York: Springer.

Diggle, P.J. (2006). Spatio-temporal point processes, partial likelihood, foot-and-mouth. *Statistical Methods in Medical Research*, 15, 325–336.

Diggle, P.J. (2007). Spatio-temporal point processes: Methods and applications. In *Semstat2004: Statistics of Spatio-Temporal Systems*, ed. B. Finkenstadt, L. Held, V. Isham, 1–45. Boca Raton, FL: CRC Press.

Diggle, P.J., Chetwynd, A.G., Haggkvist, R. and Morris, S. (1995). Second-order analysis of space–time clustering. *Statistical Methods in Medical Research*, 4, 124–136.

Diggle, P., Rowlingson, B. and Su, T. (2005). Point process methodology for on-line spatio-temporal disease surveillance. *Environmetrics*, 16, 423–434.

Diggle, P.J., Kaimi, I. and Abelana, R. (2008). Partial likelihood analysis of spatio-temporal point process data. Forthcoming. *Biometrics*.

Gabriel, E. and Diggle, P.J. (2008). Second-order analysis of inhomogeneous spatio-temporal point process data. Forthcoming. *Statistica Neerlandica*.

Jensen, E.B., Jónsdóttir, K., Schmiegel, J. and Barndorff-Nielsen, O. (2007). Spatio-temporal modeling—with a view to biological growth. In *Semstat2004: Statistics of Spatio-Temporal Systems*, Eds. B. Finkenstadt, L. Held, V. Isham, 47–75. Boca Raton, FL: CRC Press.

Keeling, M.J., Woolhouse, M.E.J., Shaw, D.J., Matthews, L., Chase-Topping, M., Haydon, D.T., Cornell, S.J., Kappey, J., Wilesmith, J. and Grenfell, B.T. (2001). Dynamics of the 2001 UK foot and mouth epidemic: Stochastic dispersal in a heterogeneous landscape. *Science*, 294, 813–817.

Kermack, W.O. and Kendrick, A.G. (1927). A contribution to the mathematical theory of epidemics. *Proceedings of the Royal Society*, A115, 700–721.

Kingman, J.F.C. (1977). Remarks on the spatial distribution of a reproducing population. *Journal of Applied Probability*, 14, 577–583.

Lawson, A. and Leimich, P. (2000). Approaches to the space–time modeling of infectious disease behaviour. *IMA Journal of Mathematics Applied in Medicine and Biology*, 17, 1–13.

Møller, J. and Sorensen, M. (1994). Parametric models of spatial birth-and-death processes with a view to modeling linear dune fields. *Scandinavian Journal of Statistics*, 21, 1–19.

Møller, J., Syversveen, A. and Waagepetersen, R. (1998). Log Gaussian Cox processes. *Scandinavian Journal of Statistics*, 25, 451–482.

Møller, J. and Diaz-Avalos, C. (2008). Structured spatio-temporal shot-noise Cox point process models, with a view to modeling forest fires. *Research Report R-2008-07*, Department of Mathematical Sciences, Aalborg University, Denmark.

Ogata, Y. (1998). Space–time point process model for earthquake occurrences. *Annals of the Institute of Statistical Mathematics*, 50(2), 379–402.

Ogata, Y. and Zhuang, J. (2006). Space–time ETAS models and an improved extension. *Technophysics*, 413(1-2), 13–23.

Preston, C.J. (1977). Spatial birth-and-death processes. *Bulletin of the International Statistical Institute*, 46 (2), 371–391.

Ripley, B.D. (1977). Modeling spatial patterns (with discussion). *Journal of the Royal Statistical Society* B 39, 172–212.

Ripley, B.D. (1979). AS137 Simulating spatial patterns. *Applied Statistics*, 28, 109–112.

Zhuang, J. (2006). Second-order residual analysis of spatio-temporal point processes and applications in model evaluations. *Journal of the Royal Statistical Society: Series B (Statistical Methodology)*, 68(4), 635–653.

26

Modeling Spatial Trajectories

David R. Brillinger

CONTENTS

26.1 Introduction

The study of trajectories has been basic to engineering science for many centuries. One can mention the motion of the planets, the meanderings of animals and the routes of ships. More recently there has been considerable modeling and statistical analysis of biological and ecological processes of moving particles. The models may be motivated formally by difference and differential equations and by potential functions. Initially, following Liebnitz and Newton, such models were described by deterministic differential equations, but variability around observed paths has led to the introduction of random variables and to the development of stochastic calculi. The results obtained from the fitting of such models are highly useful. They may be employed for: simple description, summary, comparison, simulation, prediction, model appraisal, bootstrapping, and also employed for the estimation of derived quantities of interest. The potential function approach, to be presented in Section 26.3.4, will be found to have the advantage that an equation of motion is set down quite directly and that explanatories, including attractors, repellers, and time-varying fields may be included conveniently.

Movement process data are being considered in novel situations: assessing Web sites, computer-assisted surveys, soccer player movements, iceberg motion, image scanning, bird navigation, health hazard exposure, ocean drifters, wildlife movement. References showing the variety and including data analyses include Brillinger, D.R. (2007a); Brillinger, Stewart, and Littnan (2006a); Eigethun, Fenske, Yost, and Paicisko (2003); Haw (2002); Ionides, Fang, Isseroff, and Oster (2004); Kendall (1974); Lumpkin and Pazos (2007); Preisler, Ager, Johnson, and Kie (2004); Preisler, Ager, and Wisdom (2006); Preisler, Brillinger, Ager, and Kie (1999); Stark, Privitera, Yang, Azzariti et al. (2001); Stewart, Antonelis, Yochem, and Baker (2006). In the chapter, consideration is given to location data $\{\mathbf{r}(t_i),\ i = 1, \ldots, n\}$ and models leading to such data. As the notation implies and practice shows, observation times, $\{t_i\}$, may be unequally spaced. The chapter also contains discussion of inclusion of explanatory variables. It starts with the presentation and discussion of two empirical examples of trajectory data. The first refers to the motion of a small particle moving about in a fluid and the second to the satellite-determined locations of a Hawaiian monk seal foraging off the island of Molokai. The following material concerns pertinent stochastic models for trajectories and some of their properties. It will be seen that stochastic differential equations (SDEs) are useful for motivating models and that corresponding inference procedures have been developed. In particular, discrete approximations to SDEs lead to likelihood functions and, hence, classic confidence and testing procedures become available.

The basic motivation for the chapter is to present a unified approach to the modeling and analysis of trajectory data.

26.2 History and Examples

26.2.1 Planetary Motion

Newton derived formal laws for the motion of the planets and further showed that Kepler's laws could be derived from these. Lagrange set down a potential function and Newton's equations of motion could be derived from it in turn. The work of Kepler, Newton and Lagrange has motivated many models in physics and engineering. For example, in a study describing the motion of a star in a stellar system, Chandrasekhar (1943) sets down equations of the form

$$\frac{d\mathbf{u}(t)}{dt} = -\beta\mathbf{u}(t) + \mathbf{A}(t) + \mathbf{K}(\mathbf{r}(t), t) \tag{26.1}$$

with $\mathbf{u}$, velocity; $\mathbf{A}$, a Brownian-like process; β, a coefficient of friction; and $\mathbf{K}$, the acceleration produced by an external force field. Chandrasekhar (1943) refers to this equation as a generalized Langevin equation. It is an example of an SDE.

Next, two examples of empirical trajectory data are presented.

26.2.2 Brownian Motion

In general science, Brownian motion refers to the movement of tiny particles suspended in a liquid. The phenomenon is named after Robert Brown, an Englishman, who in 1827 carried out detailed observations of the motion of pollen grains suspended in water (Haw, 2002). The phenomenon was modeled by Einstein. He considered the possibility that formalizing Brownian motion could support the idea that molecules existed. Langevin (1908) set down the following expression for the motion of such a particle,

$$m\frac{d^2x}{dt^2} = -6\pi\mu a\frac{dx}{dt} + X,$$

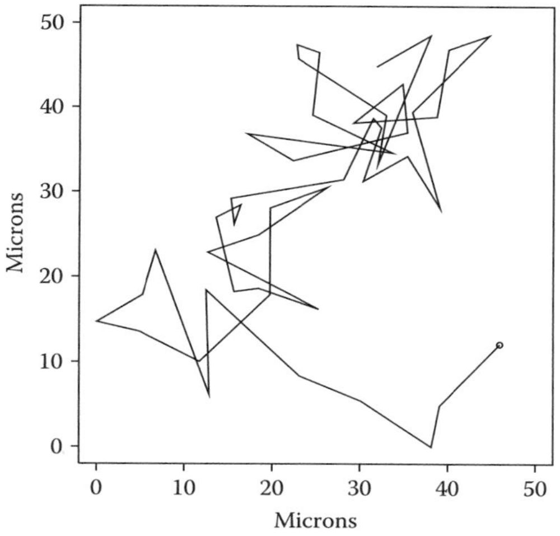

FIGURE 26.1
Perrin's measurements of the location of a mastic grain at 48 successive times. (Adapted from Guttorp, P. (1995). *Stochastic Modeling of Scientific Data*. Chapman & Hall, London.)

where m is the particle's mass, a is its radius, μ is the viscosity of the liquid, and X is the "complementary force"—a Brownian process-like term. One can view this as an example of an SDE.

A number of "Brownian" trajectories were collected by Perrin (1913). One is provided in Figure 26.1 and the results of an analysis will be presented in Section 26.6. The particles involved were tiny mastic grains with a radius of .53 microns. Supposing (x, y) refers to position in the plane, the trajectory may be written $(x(t), y(t)), t = 1, \ldots, 48$. The time interval between the measurements in this case was 30 sec.

In Figure 26.1, one sees the particle start in the lower right corner of the figure and then meander around a diagonal line running from the lower left to the upper right.

26.2.3 Monk Seal Movements

The Hawaiian monk seal is an endangered species. It numbers only about 1,400 today. They are now closely monitored, have a life span of about 30 years, weigh between 230 and 270 kilos and have lengths of 2.2 to 2.5 meters.

Figure 26.2 shows part of the path of a juvenile female monk seal swimming off the southwest coast of the island of Molokai, Hawaii. Locations of the seal as it moved and foraged were estimated from satellite detections, the animal having a radio tag glued to its dorsal fin. The tag's transmissions could be received by satellites passing overhead when the animal was on the surface. The animal's position could then be estimated.

The data cover a period of about 15 days. The seal starts on a beach on the southwest tip of Molokai and then heads to the far boundary of a reserve called Penguin Bank Reserve, forages there for a while, and then heads back to Molokai, perhaps to rest in safety. Penguin Bank Reserve is indicated by the dashed line in the figure.

An important goal of the data collection in this case was the documentation of the animals' geographic and vertical movements as proxies of foraging behavior and then to use this information to assist in the survival of the species. More detail may be found in Brillinger, Stewart, and Littnan (2006a) and Stewart, Antonelis, Yochem et al. (2006).

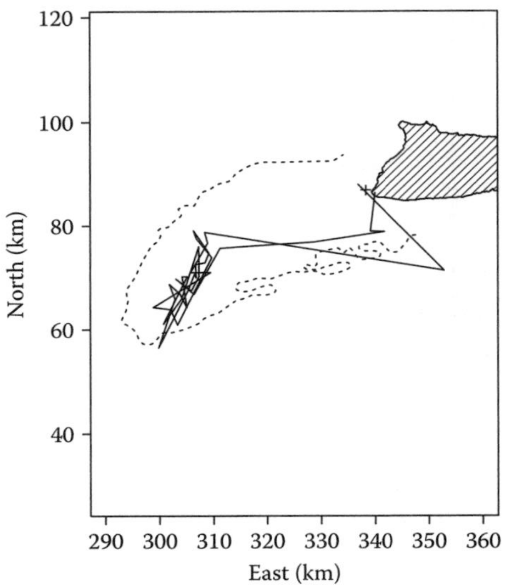

FIGURE 26.2
Estimated locations of a Hawaiian monk seal off the coast of Molokai. The dashed line is the 200 fathom line, approximately constraining an area called the Penguin Bank Reserve.

26.3 Statistical Concepts and Models

26.3.1 Displays

It is hard to improve on visual displays in studies of trajectory data. In a simple case, one shows the positions $(x(t_i), y(t_i)), i = 1, 2, \ldots,$ as a sequence of connected straight lines, as in Figure 26.1 and Figure 26.2. One can superpose other spatial information as a background. An example is Figure 26.2, which shows the outlines of Molokai, the hatched region, and Penguin Bank Reserve, the dashed line.

A related type of display results if one estimates a bivariate density function from the observed locations $(x(t_i), y(t_i)), i = 1, 2, \ldots,$ and shows the estimate in contour or image form. Such figures are used in home range estimation; however, this display loses the information on where the animal was at successive times.

A bagplot (Rousseuw, Ruts, and Tukey, 1999) is useful in processing trajectory data if estimated locations can be in serious error. It highlights the "middle 50%" of a bivariate dataset and is an extension of the univariate boxplot. An example is provided in Brillinger, Stewart, and Littnan (2006b). Before preparing the bagplot presented there, this author did not know of the existence of the Penguin Bank Reserve. Computing the bagplot of all the available locations found the reserve.

Another useful display is a plot of the estimated speed of the particle versus time. One graphs the approximate speeds,

$$\sqrt{(x(t_{i+1}) - x(t_i))^2 + (y(t_{i+1}) - y(t_i))^2}/(t_{i+1} - t_i)$$

versus the average of the times, t_i and t_{i+1}, say. It is to be remembered that this "speed" provides only the apparent speed, not the instantaneous. The particle may follow a long route getting from $\mathbf{r}(t_i)$ to $\mathbf{r}(t_{i+1})$.

Figures are presented in this chapter, but videos can assist the analyses.

26.3.2 Autoregressive Models

A bivariate time series model that is coordinate free provides a representation for processes whose realizations are spatial trajectories. One case is the simple random walk,

$$\mathbf{r}_{t+1} \;=\; \mathbf{r}_t \;+\; \epsilon_{t+1}, \qquad t = 0, 1, 2, \ldots$$

with $\mathbf{r}_0$ the starting point and $\{\epsilon_t\}$ a bivariate time series of independent and identically distributed variates.

In the same vein one can consider the bivariate order 1, autoregressive, VAR(1), given by

$$\mathbf{r}_{t+1} \;=\; \mathbf{a}\mathbf{r}_t + \epsilon_{t+1}, \qquad t = 0, 1, 2, \ldots \tag{26.2}$$

for an **a** leading to stationarity.

The second difference of the motion of an iceberg has been modeled as an autoregressive in Moore (1985).

26.3.3 Stochastic Differential Equations

The notion of a continuous time random walk may be formalized as a formal Brownian motion. This is a continuous time process with the property that disjoint increments, $d\mathbf{B}(t)$, are independent Gaussians with covariance matrix $\mathbf{I}dt$. Here $\mathbf{B}(t)$ takes values in R^2. The random walk character becomes clear if one writes

$$\mathbf{B}(t + dt) \;=\; \mathbf{B}(t) + d\mathbf{B}(t), \qquad -\infty < t < \infty.$$

The vector autoregressive of order 1 series may be seen as an approximation to a stochastic differential equation by writing

$$\mathbf{r}(t + dt) - \mathbf{r}(t) \;=\; \mu\mathbf{r}(t)dt + \sigma d\mathbf{B}(t)$$

and comparing it to Equation (26.2).

Given a Brownian process **B**, consider a trajectory **r** in R^2 that at time t has reached the position $\mathbf{r}(t)$ having started at $\mathbf{r}(0)$. Consider the "integral equation"

$$\mathbf{r}(t) \;=\; \mathbf{r(0)} + \int_0^t \mu(\mathbf{r}(s), s)ds + \int_0^t \sigma(\mathbf{r}(s), s)d\mathbf{B}(s) \tag{26.3}$$

with $\mathbf{r}$, μ, $d\mathbf{B}$ each 2 vectors and σ 2 by 2. Here, μ is called the drift and σ the diffusion coefficient. Equation (26.3) is known as Ito's integral equation.

This equation requires the formal definition of the Ito integral

$$\int_a^b \mathbf{G}(\mathbf{r}(t), t)d\mathbf{B}(t)$$

for conformal **G** and **B**. Under regularity conditions, the Ito integral can be defined as the limit in mean squared, as $\triangle \downarrow 0$, of

$$\sum_{j=1}^{N-1} \mathbf{G}(\mathbf{r}(t_j), t_j)[\mathbf{B}(t_{j+1}) - \mathbf{B}(t_j)],$$

where

$$a \;=\; t_1^{\triangle} < t_2^{\triangle} < \cdots < t_N^{\triangle} \;=\; b, \qquad \triangle = max(t_{j+1} - t_j).$$

Expressing Equation (26.3) as an "Ito integral" is a symbolic gesture, but the definition is mathematically consistent.

Equation (26.3) is often written

$$d\mathbf{r}(t) = \mu(\mathbf{r}(t), t)dt + \sigma(\mathbf{r}(t), t)d\mathbf{B}(t) \tag{26.4}$$

using differentials, but Equation (26.3) is the required formal expression. For details on Ito integrals, see Durrett (1996) or Grimmet and Stirzaker (2001).

26.3.4 Potential Function Approach

A potential function is an entity from Newtonian mechanics. It leads directly to equations of motion in the deterministic case (see Taylor, 2005). An important property is that a potential function is real-valued and thereby leads to a simpler representations for a drift function, μ, than those based on the vector-valued velocities.

To make this apparent, define a gradient system as a system of differential equations of the form

$$d\mathbf{r}(\mathrm{t})/dt = -\nabla V(\mathbf{r}(t)), \tag{26.5}$$

where $V : R^2 \rightarrow R$ is a differentiable function and $\nabla V = (\partial V/\partial x, \partial V/\partial y)^T$ denotes its gradient. ("T" here denotes transpose.) The negative sign in this system is traditional. The structure $d\mathbf{r}(t)/dt$ is called a vector field, while the function V is called a potential function.

The classic example of a potential function is the gravitational potential in R^3, $V(\mathbf{r}) = -G/|\mathbf{r} - \mathbf{r}_0|$ with G the constant of gravitation (see Chandrasekhar, 1943). This function leads to the attraction of an object at position $\mathbf{r}$ toward the position $\mathbf{r}_0$. The potential value at $\mathbf{r} = \mathbf{r}_0$ is $-\infty$ and the pull of attraction is infinite there. Other specific formulas will be indicated shortly.

In this chapter the deterministic Equation (26.5) will be replaced by a stochastic differential equation

$$d\mathbf{r}(t) = -\nabla V(\mathbf{r}(t))dt + \sigma(\mathbf{r}(t))d\mathbf{B}(t) \tag{26.6}$$

with $\mathbf{B}(t)$ a two-dimensional standard Brownian process, V a potential function, and σ a diffusion parameter. Under regularity conditions, a unique solution of such an equation exists and the solution process $\{\mathbf{r}(t)\}$ is Markov. Repeating a bit, a practical advantage of being able to write $\mu = -\nabla V$ is that V is real-valued and thereby simpler to model, to estimate, and to display.

For motion in R^2, the potential function is conveniently displayed in contour, image, or perspective form. Figure 26.3 and Figure 26.4 provide examples of image plots. If desired, the gradient may be displayed as a vector field. (Examples may be found in Brillinger, Preisler, Ager, Kie et al., 2001.)

An estimated potential function may be used for: simple description, summary, comparison, simulation, prediction, model appraisal, bootstrapping, and employed for the estimation of related quantities of interest. The potential function approach can handle attraction and repulsion from points and regions directly. While the figures of estimated potential functions usually look like what you expect a density function to be, given the tracks but there is much more to the potential surface; for example, the slopes are direction and speed of motion.

Some specific potential function forms that have proven useful are listed below. A research issue is how to choose among them and others. Subject matter knowledge can prove essential in doing this. To begin, consider the function

$$V(\mathbf{r}) = \alpha \log d + \beta d \tag{26.7}$$

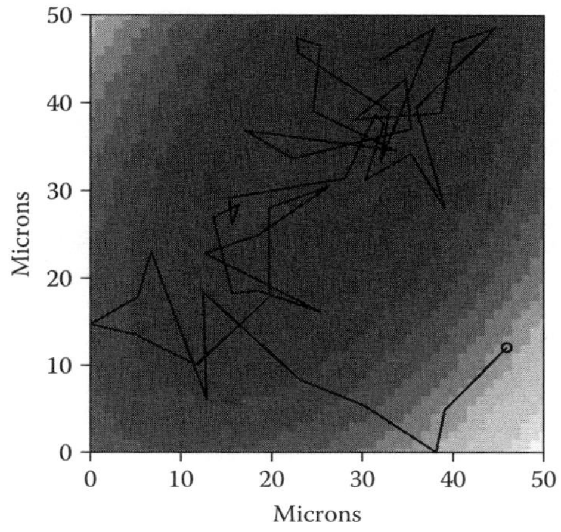

FIGURE 26.3
The estimated potential function for the Perrin data using the form (26.8) with $C = 0$. The circle represents the initial location estimate. Darker shade denotes deeper value.

with $\mathbf{r} = (x, y)^T$ the location of a particle, and $d = d(\mathbf{r})$ the distance of the particle from a specific attractor. This function is motivated by equations in Kendall (1974). The attractor may move in space in time, and then the potential function is time-dependent. Another useful functional form is

$$V(\mathbf{r}) = \gamma_1 x + \gamma_2 y + \gamma_{11} x^2 + \gamma_{12} xy + \gamma_{22} y^2 + C/d_M, \tag{26.8}$$

where $d_M = d_M(x, y)$ is the distance from location (x, y) to the nearest point of a region, M, of concern. Here, with $C > 0$, the final term keeps the trajectory out of the region. On the other hand,

$$V(\mathbf{r}) = \alpha \log d + \beta d + \gamma_1 x + \gamma_2 y + \gamma_{11} x^2 + \gamma_{12} xy + \gamma_{22} y^2, \tag{26.9}$$

where $d = d(\mathbf{r}) = d(x, y)$ is the shortest distance to a point, leads to attraction to the point as well as providing some general structure. It is useful to note for computations that the expressions (26.7) to (26.9) are linear in the parameters.

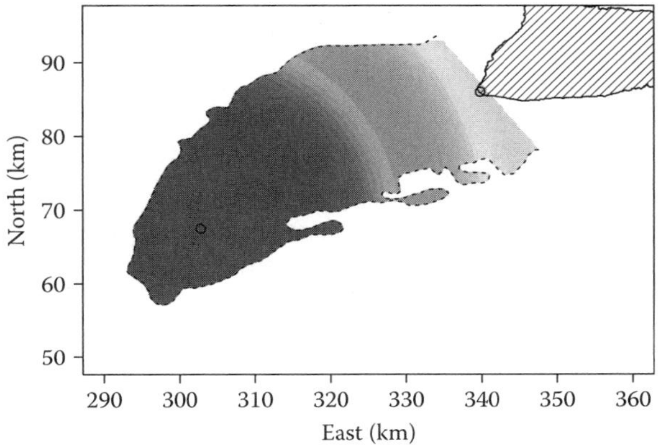

FIGURE 26.4
A potential function estimate computed to describe a Hawaiian monk seal's outbound, then inbound, foraging journeys from the southwest corner of Molokai. The circle in the southwest corner represents an assumed point of attraction.

In summary, the potential function approach advocated here is distinguished from traditional SDE-based work by the fact that μ has the special form (26.5).

26.3.5 Markov Chain Approach

Taking note of the work of Kurtz (1978) and Kushner (1974, 1977), it is possible to approximate the motion implied by an SDE, of a particle moving in R^2, by a Markov chain in discrete time and space. This can be useful for both simulations of the basic process and for intuitive understanding.

In the approach of Kushner (1974, 1977), one sets up a grid forming pixels, and then makes a Markov chain assumption. Specifically define

$$\mathbf{a}(\mathbf{r}, t) = \frac{1}{2}\sigma(\mathbf{r}, t)\sigma(\mathbf{r}, t)^T$$

and, for convenience of exposition here, suppose that $a_{ij}(\mathbf{r}, t) = 0,\ i \neq j$, i.e., the error components of the Gaussian vector are assumed statistically independent for fixed $\mathbf{r}$. Suppose further that time is discretized with $t_{k+1} - t_k = \Delta$. Write $\mathbf{r}_k = \mathbf{r}(t_k)$, and suppose that the lattice points of the grid have separation h. Let $\mathbf{e}_i$ denote the unit vector in ith coordinate direction, $i = 1, 2$. Now consider the Markov chain with transition probabilities,

$$\begin{aligned} P(\mathbf{r}_k = \mathbf{r}_0 \pm \mathbf{e}_i h | \mathbf{r}_{k-1} = \mathbf{r}_0) \\ = \frac{\Delta}{h^2}(a_{ii}(\mathbf{r}_0, t_{k-1}) + h|\mu_i(\mathbf{r}_0, t_k - 1)|^{\pm}) \\ P(\mathbf{r}_k = \mathbf{r}_0 | \mathbf{r}_{k-1} = \mathbf{r}_0) = 1 - \sum \textit{preceding}. \end{aligned}$$

Here it is supposed the probabilities are ≥ 0, which may be arranged by choice of Δ and h. In the above expressions the following notation has been employed:

$$|u|^+ = u \textit{ if } u > 0 \textit{ and } = 0 \textit{ otherwise}$$

and

$$|u|^- = -u \textit{ if } u < 0 \textit{ and } = 0 \textit{ otherwise.}$$

A discrete random walk is the simplest case of this construction.

(For results on the weak convergence of such approximations to SDEs, see Durrett (1996), Kurtz (1978), and Kushner (1974).)

With that introduction attention can turn to a different, yet related type of model. Suppose that a particle is moving along the points of a lattice in R^2 with the possibility of moving one step to the left or one to the right or one step up or one step down. View the lattice as the state space of a Markov chain in discrete time with all transition probabilities 0 except for the listed one step ones. This is the structure of the just provided approximation. The difference is that one will start by seeking a reasonable model for the transition probabilities directly, rather than coefficients for SDE.

26.4 Inference Methods

There is substantial literature devoted to the topic of inference for stochastic differential equations (references include Prakasa Rao, 1999 and Sorensen, 1997). Many interesting scientific questions can be posed and addressed involving them and their applications.

Elementary ones include: Is a motion Brownian? Is it Brownian with drift? These can be formulated in terms of the functions μ and σ of Equation (26.3) and Equation (26.4).

Consider an object at position **r**(t) in R^2 at time t. In terms of approximate velocity, Equation (26.6) leads to

$$(\mathbf{r}(t_{i+1}) - \mathbf{r}(t_i))/(t_{i+1} - t_i) = -\nabla V(\mathbf{r}(t_i)) + \sigma \mathbf{Z}_{i+1}/\sqrt{t_{i+1} - t_i} \qquad (26.10)$$

with the $\mathbf{Z}_i$ independent and identically distributed bivariate, standard normals. The reason for the $\sqrt{t_{i+1} - t_i}$ is that for real-valued Brownian $\text{var}(dB(t)) = \sigma dt$. In Equation (26.10), one now has a parametric or nonparametric regression problem for learning about V, depending on the parametrization chosen. If the t_i are equispaced, this is a parametric or nonparametric autoregression model of order 1.

If desired, the estimation may be carried out by ordinary least squares or maximum likelihood depending on the model and the distribution chosen for the $\mathbf{Z}_i$. The naive approximation (26.10) is helpful for suggesting methods. It should be effective if the time points, t_i, are close enough together. In a sense (26.10), not (26.3), has become the model of record.

To be more specific, suppose that μ has the form

$$\mu(\mathbf{r}) \;=\; g(\mathbf{r})^T \beta$$

for an L by 1 parameter β and a p by L known function g. This assumption, from Equation (26.10) leads to the linear regression model

$$\mathbf{Y}_n \;=\; \mathbf{X}_n \beta + \epsilon_n$$

having stacked the $n-1$ values $(\mathbf{r}(t_{i+1}) - \mathbf{r}(t_i))/\sqrt{t_{i+1} - t_i}$ to form the $(n-1)p$ vector $\mathbf{Y}_n$, stacked the $n-1$ matrices $\mu(\mathbf{r}(t_i), t_i)\sqrt{(t_{i+1} - t_i)}$ to form the $(n-1)p$ by L matrix $\mathbf{X}_n$ and stacked the $n-1$ values $\sigma \mathbf{Z}_{i+1}$ to form ϵ_n. One is thereby led to consider the estimate

$$\hat{\beta} = \left(\mathbf{X}_n^T \mathbf{X}_n\right)^{-1} \mathbf{X}_n^T \mathbf{Y}_n$$

assuming the indicated inverse exists. Continuing, one is led to estimate $g(\mathbf{r})^T \beta$ by $g(\mathbf{r})^T \hat{\beta}$.

Letting y_j denote the jth entry of $\mathbf{Y}_n$ and $\mathbf{x}_j^T$ denote the jth row of $\mathbf{X}_n$, one can compute

$$s_n^2 \;=\; ((n-1)p^{-1} \sum \left(y_j - \mathbf{x}_j^T \hat{\beta}\right)^T \left(y_j - \mathbf{x}_j^T \hat{\beta}\right),$$

as estimate of σ^2 and, if desired, proceed to form approximate confidence intervals for the value $g(\mathbf{r})^T \beta$ using the results of Lai and Wei (1982). In particular, the distribution of

$$(g(\mathbf{r})^T \left(\mathbf{X}_n^T \mathbf{X}_n\right)^{-1} g(\mathbf{r}))^{-1/2} g(\mathbf{r})^T (\hat{\beta} - \beta)/s_n$$

may be approximated by a standard normal for large n. (Further details may be found in Brillinger, 2007b.)

Another concern is deciding on the functional form for the drift terms μ and the diffusion coefficient σ of the motivating model (26.3). In Preisler et al. (1999) and Brillinger et al. (2001), the estimates are nonparametric.

26.5 Difficulties That Can Arise

One serious problem that can arise in work with trajectory data relates to the uncertainty of the location estimates. The commonly used Loran and satellite-based estimated locations can be in serious error. The measurement errors have the appearance of including outliers rather than coming from some smooth long-tailed distribution. In the monk seal example, the bagplot proved an effective manner to separate out outlying points. It led to the empirical discovery of the Penguin Bank Reserve in the work. Improved estimates of tracks may be obtained by employing a state space model and robust methods (see Anderson-Sprecher, 1994 and Jonsen, Flemming, and Myers, 2005).

A difficulty created by introducing the model via an SDE is that some successive pairs of time points, $t_i - t_{i-1}$, may be far apart. The concern arises because the model employed in the fitting is (26.10). One can handle this by viewing Equation (26.10) as the model of record, forgetting where it came from, and assessing assumptions, such as the normality of the errors, by traditional methods.

It has already been noted above that the speed estimate is better called the apparent speed estimate because one does not have information on the particle's movement between times t_{i-1} and t_i. Correction terms have been developed for some cases Ionides (2001).

26.6 Results for the Empirical Examples

Figure 26.3 provides the estimated potential function, $\hat{V}$, for Perrin's data assuming the functional form (26.8) with $C = 0$. The particle's trajectory has been superposed in the figure. One sees the particle being pulled toward central elliptical regions and remaining in or nearby. This nonrandom behavior could have been anticipated from the presence of viscosity in the real world (Haw, 2002). Were the process "pure" Brownian, the particle would have meandered about totally randomly and the SDE would have been

$$d\mathbf{r}(t) = \sigma d\mathbf{B}(t).$$

The Smolukowski approximation (see Chandrasekhar, 1943; Nelson, 1967) takes (26.1) into

$$d\mathbf{r}(t) = \mathbf{K}(\mathbf{r}(t), t)dt/\beta + \sigma d\mathbf{B}(t)$$

instead. The background in Figure 26.3 is evidence against the pure Brownian model for Perrin's data.

Figure 26.4 concerns the outbound foraging journeys of a Hawaiian monk seal whose outbound and inbound parts of one journey were graphed in Figure 26.2. Figure 26.4 is based on a trajectory including five journeys. The animal goes out apparently to forage and then returns to rest and be safer. The potential function employed is Equation (26.9) containing a term, $\alpha \log(d) + \beta d$, that models attraction of the animal out to the far part of Penguin Bank Reserve. More detail on this analysis may be found in Brillinger, Stewart, and Littnan (2006a, 2006b). Outbound journeys may be simulated using the fitted model and hypotheses may be addressed formally.

26.7 Other Models

Figure 26.4 shows the western coast of the island of Molokai. Coasts provide natural boundaries to the movements of the seals. In an analysis of the trajectory of a different animal, that seal is kept off Molokai in the modeling by taking $C > 0$ in the final term in (26.8), see Brillinger and Stewart, 1998.

A boundary is an example of an explanatory variable and it may be noted that there is now substantial literature on SDEs with boundaries (Brillinger, 2003). There are explanatory variables to be included. A particle may be moving in a changing field $G(\mathbf{r}(t), t)$ and one is led to write

$$d\mathbf{r} = \mu dt + \gamma \nabla G dt + \sigma d\mathbf{B}.$$

A case is provided by sea surface height (SSH) with the surface currents given by the gradient of the SSH field. It could be that $\mu = -\nabla V$ as previously noted in this chapter.

A different type of explanatory, model and analysis is provided in Brillinge, Preisler, Ager, and Wisdom (2004). The moving object is an elk and the explanatory is the changing location, $\mathbf{x}(t)$ of an all terrain vehicle (ATV). The noise of an ATV is surely a repellor when it is heard by an elk, but one wonders at what distance does the repulsion begin? The following model was employed to study that question. Let $\mathbf{r}(t)$ denote the location of an elk, and $\mathbf{x}(t)$ the location of the ATV, both at time t. Let τ be a time lag to be studied. Consider

$$d\mathbf{r}(t) = \mu(\mathbf{r}(t))dt + \nu(|\mathbf{r}(t) - \mathbf{x}(t-\tau)|)dt + \sigma d\mathbf{B}(t).$$

The times of observation differ for the elk and the ATV. They are every five min for the elk when the ATV is present and every one sec for the ATV itself. In the approach, adopted location values, $\mathbf{x}(t)$, of the ATV are estimated for the elk observation times via interpolation. One sees an apparent increase in the speed of the elk, particularly when an elk and the ATV are close to one another.

The processes described so far have been Markov. However, non-Markov processes are sometimes needed in modeling animal movement. A case is provided by the random walk with correlated increments in McCulloch and Cain (1989). One can proceed generally by making the sequence $\{\mathbf{Z}_i\}$ of Equation (26.10) an autocorrelated time series.

A more complex SDE model is described by a functional stochastic differential equation

$$d\mathbf{r}(t) = -\nabla V(\mathbf{r}(t)|H_t)dt + \sigma(\mathbf{r}(t)|H_t)d\mathbf{B}(t)$$

with $H_t = \{(t_i, \mathbf{r}(t_i)), t_i \leq t\}$ as the history up to time t. A corresponding discrete approximation is provided by

$$\mathbf{r}(t_{i+1}) - \mathbf{r}(t_i) = -\nabla V(\mathbf{r}(t_i)|H_{t_i})(t_{i+1} - t_i) + \sigma\sqrt{t_{i+1} - t_i}\mathbf{Z}_{i+1}$$

with the $\mathbf{Z}_i$ again independent standard Gaussians. With this approximation, a likelihood function may be set down directly and, thereby, inference questions addressed.

It may be that the animals are moving such great distances that the spherical shape of the Earth needs to be taken into account. One model is described in Brillinger (1997). There may be several interacting particles. In this case, one would make the SDEs of the individual particles interdependent. (References include Dyson (1963) and Spohn (1987).)

26.8 Summary

Trajectories exist in space and time. One notices them in many places and their data have become common. In this chapter, two specific approaches have been presented for analyzing such data, both involving SDE motivation. In the first approach, a potential function is assumed to exist with its negative gradient giving the SDE's drift function. The second approach involves setting up a grid and approximating the SDE by a discrete Markov chain moving from pixel to pixel. Advantages of the potential function approach are that the function itself is scalar-valued, that there are many choices for its form, and that knowledge of the physical situation can lead directly to a functional form.

Empirical examples are presented and show that the potential function method can be realized quite directly.

Acknowledgments

I thank my collaborators A. Ager, J. Kie, C. Littnan, H. Preisler, and B. Stewart. I also thank P. Diggle, P. Spector, and C. Wickham for the assistance they provided.

This research was supported by the NSF Grant DMS-0707157.

References

Anderson-Sprecher, R. (1994). Robust estimates of wildlife location using telemetry data. *Biometrics* 50, 406–416.

Brillinger, D.R. (1997). A particle migrating randomly on a sphere. *J. Theoretical Prob.* 10, 429–443.

Brillinger, D.R. (2003). Simulating constrained animal motion using stochastic differential equations. *Probability, Statistics and Their Applications*, Lecture Notes in Statistics 41, 35–48. IMS.

Brillinger, D.R. (2007a). A potential function approach to the flow of play in soccer. *J. Quant. Analysis Sports*, January.

Brillinger, D.R. (2007b). Learning a potential function from a trajectory. *IEEE Signal Processing Letters* 12, 867–870.

Brillinger, D.R., Preisler, H.K., Ager, A.A., Kie, J., and Stewart, B.S. (2001). Modelling movements of free-ranging animals. Univ. Calif. Berkeley Statistics Technical Report 610. Available as www.stat.berkeley.edu

Brillinger, D.R., Preisler, H.K., Ager, A.A., and Wisdom, M.J. (2004). Stochastic differential equations in the analysis of wildlife motion. *2004 Proceedings of the American Statistical Association, Statistics and the Environment Section*, Toronto, Canada.

Brillinger, D.R. and Stewart, B.S. (1998). Elephant seal movements: Modelling migration. *Canadian J. Statistics* 26, 431–443.

Brillinger, D.R., Stewart, B., and Littnan, C. (2006a). Three months journeying of a Hawaiian monk seal. *IMS Collections*, vol. 2, 246–264.

Brillinger, D.R., Stewart, B., and Littnan, C. (2006b). A meandering hylje. Festschrift for Tarmo Pukkila (Eds. Lipski, E.P., Isotalo, J., Niemela, J., Puntanen, S., and Styan, G.P.H.), 79–92. University of Tampere, Tampere, Finland.

Chandrasekhar, S. (1943). Stochastic problems in physics and astronomy. *Rev. Modern Physics* 15, 1–89.

Durrett, R. (1996). *Stochastic Calculus*. CRC Press, Boca Raton, FL.

Dyson, F.J. (1963). A Brownian-motion model for the eigenvalues of a random matrix. *J. Math. Phys.* 3, 1191–1198.

Eigethun, K., Fenske, R.A., Yost, M.G., and Paicisko, G.J. (2003). Time-location analysis for exposure assessment studies of children using a novel global positioning system. *Environ. Health Perspec.* 111, 115–122.

Grimmet, G. and Stirzaker, D. (2001). *Probability and Random Processes*. Oxford University, Oxford, U.K.

Guttorp, P. (1995). *Stochastic Modeling of Scientific Data*. Chapman & Hall, London.

Haw, M. (2002). Colloidal suspensions, Brownian motion, molecular reality: a short history. *J. Phys. Condens. Matter* 14, 7769–7779.

Ionides, E.L. (2001). Statistical Models of Cell Motion. PhD thesis, University of California, Berkeley.

Ionides, E.L., Fang, K.S., Isseroff, R.R. and Oster, G.F. (2004). Stochastic models for cell motion and taxis. *Math. Bio.* 48, 23–37.

Jonsen, I.D., Flemming, J.M. and Myers, R.A. (2005). Robust state-space modeling of animal movement data. *Ecology* 86, 2874–2880.

Kendall, D.G. (1974). Pole-seeking Brownian motion and bird navigation. *J. Roy. Statist. Soc. B* 36, 365–417.

Kurtz, T.G. (1978). Strong approximation theorems for density dependent Markov chains. *Stochastic Process. Applic.* 6, 223–240.

Kushner, H.J. (1974). On the weak convergence of interpolated Markov chains to a diffusion. *Ann. Probab.* 2, 40–50.

Kushner, H.J. (1977). *Probability Methods for Approximations in Stochastic Control and for Elliptic Equations.* Academic Press, New York.

Lai, T.L. and Wei, C.Z. (1982). Least squares estimation in stochastic regression models. *Ann. Statist.* 10, 1917–1930.

Langevin, P. (1908). Sur la théorie du mouvement Brownian. *C.R. Acad. Sci.* (Paris) 146, 530–533. For a translation, see Lemons, D.S. and Gythiel, A. (2007). Paul Langevin's 1908 paper. *Am. J. Phys.* 65, 1079–1081.

Lumpkin, R. and Pazos, M. (2007). Measuring surface currents with surface velocity program drifters: the instrument, its data, and some recent results. In *Lagrangian Analysis and Prediction of Coastal and Ocean Dynamics*, eds. Griffa, A., Kirwan, A.D., Mariano, A.J., Ozgokmen, T., and Rossby, H.T., Cambridge University Press, Cambridge, MA.

McCulloch, C.E. and Cain, M.L. (1989). Analyzing discrete movement data as a correlated random walk. *Ecology* 70, 383–388.

Moore, M. (1985). Modelling iceberg motion: a multiple time series approach. *Canadian J. Stat.* 13, 88–94.

Nelson, E. (1967). *Dynamical Theories of Brownian Motion*. Princeton University Press, Princeton, NJ.

Perrin, J. (1913). *Les Atomes*. Felix Alcan, Paris.

Prakasa Rao, B.L.S. (1999). *Statistical Inference for Diffusion Type Processes*. Oxford University, Oxford, U.K.

Preisler, H.K., Ager, A.A., Johnson, B.K. and Kie, J.G. (2004). Modelling animal movements using stochastic differential equations. *Environmetrics* 15, 643–647.

Preisler, H.K., Ager, A.A., and Wisdom, M.J. (2006). Statistical methods for analysing responses of wildlife to distrurbances. *J. Appl. Ecol.* 43, 164–172.

Preisler, H.K., Brillinger, D.R., Ager, A.A., and Kie, J.G. (1999). Analysis of animal movement using telemetry and GIS data. *Proc. ASA Section on Statistics and the Environment*, Alexandria, VA, April 23–25.

Rousseuw, P.J., Ruts, I. and Tukey, J.W. (1999). The bagplot: a bivariate boxplot. *Am. Statistic.* 53, 382–387.

Sorensen, M. (1997). Estimating functions for discretely observed diffusions: A review. *IMS Lecture Notes Monograph Series* 32, 305–325.

Spohn, H. (1987). Interacting Brownian particles: A study of Dyson's model. In *Hydrodynamic Behavior and Interacting Particle Systems*. (ed. G. Papanicolaou). Springer, New York.

Stark, L.W., Privitera, C.M., Yang, H., Azzariti, M., Ho, Y.F., Blackmon, T., and Chernyak, D. (2001). Representation of human vision in the brain: How does human perception recognize images? *J. Elect. Imaging* 10, 123–151.

Stewart, B., Antonelis, G.A., Yochem, P.K., and Baker, J.D. (2006). Foraging biogeography of Hawaiian monk seals in the northwestern Hawaiian Islands. *Atoll Res. Bull.* 543, 131–145.

Taylor, J.R. (2005). *Classical Mechanics*. University Science, Sausalito, CA.

27

Data Assimilation

Douglas W. Nychka and Jeffrey L. Anderson

CONTENTS

27.1 Introduction

Data assimilation refers to the statistical techniques used to combine numerical and statistical models with observations to give an improved estimate of the state of a system or process. Typically a data assimilation problem has a sequential aspect where data as it becomes available over time is used to update the state or parameters of a dynamical system. Data assimilation is usually distinguished from more traditional statistical time series applications because the system can have complicated nonlinear dynamical behavior and the state vector and the number of observations may be large. One of its primary roles is in estimating the state of a physical process when applied to geophysical models and physical measurements. Data assimilation has its roots in Bayesian inference and the restriction to linear dynamics and Gaussian distributions fits within the methods associated with the Kalman filter. Because data assimilation also involves estimating an unknown state based on possibly irregular, noisy, or indirect observations, it also has an interpretation as solving an *inverse* problem (e.g., Tarantola, 1987). One goal of this chapter is to tie these concepts back to a general Bayesian framework.

One of the most successful applications of data assimilation is in numerical weather prediction where a large and heterogeneous set of observations are combined with a sophisticated physical model for the evolution of the atmosphere to produce detailed and high resolution forecasts of weather (see, e.g., Kalnay, 2002, for an introduction). The application to weather forecasting and in general to assimilation of atmospheric and oceanographic observations has a distinctly spatial aspect as the processes of interest are typically three-dimensional fields. For this reason, it is important to include this topic in this handbook. Although there are other applications of assimilation, such as target tracking or process control, and a more general class of Bayesian filtering methods (see Wikle and Berliner, 2006; Doucet and De Freitas, 2001), such as particle filters, these topics tend not to emphasize spatial processes and so are not as relevant to this handbook. The reader is referred to more statistical treatments of state–space models in West and Harrison (1997) and Harvey, Koopman, and Shepard (2004), but again these general texts do not focus on the large spatial fields typical in geophysical data assimilation.

Spatial methods in data assimilation typically involve non-Gaussian fields and infer the spatial structure dynamically from a physical model. In this way, the dynamical model and a statistical model are connected more closely than in a standard application of spatial statistics. In addition, the sheer size of data assimilation problems requires approximate solutions that are not typical for smaller spatial datasets. In this chapter, these differences will be highlighted by reviewing the principles behind current methods. We also point out some new areas where more standard space–time statistical models might be helpful in handling model error. A large-scale example for the global atmosphere is included at the end of this chapter to illustrate some of the details of practical data assimilation for atmospheric prediction.

27.2 Bayesian Formulation of Data Assimilation

The basic ingredients of a data assimilation problem are the state vector giving a complete description of the system or process and a vector of observations made on the system. In addition, one requires conditional distributions for propagating the system forward in time and for relating the observations to the system state. This overall organization is similar in concept to a Bayesian hierarchical model (BHM) (Wikle, 2009) and is known as a state–space formulation in the context of Kalman filtering. In either interpretation, we have an observation level that relates the observed data to the (unobserved) state of the system and a supporting, process level describing the evolution of the state over time. Throughout this discussion, we also point out some parallel terminology from the geosciences. In particular, the data assimilation process is separated into an *update* or *analysis* step and a *forecast* step. This distinction is helpful because most of the spatial statistical content is in the update step. In geoscience applications, the state vector is usually a multivariate spatial field on a regular grid with observations being irregular and noisy measurements of some components, or functions of the components. Typically, the forecast step is completed by a deterministic, physically based model.

27.2.1 Bayes Theorem

To introduce the Bayesian statistical model, we will use a bracket notation for distributions where square braces denote a probability density function (pdf). Accordingly, $[Y]$ is the pdf for the random variable Y and $[Y|X]$ the conditional pdf for the random variable Y given the random variable X. Let x_t denote the state vector at time t and y_t be a vector of observations available at time t. To streamline the exposition, assume that the times are indexed by integers, $t = 1, 2, \ldots$, although handling unequally spaced times does not add any fundamental difficulty. We will assume the likelihood: $[y_t|x_t]$ and a reference or prior distribution for the state $[x_t]$. In the geosciences, $[x_t]$ is also know as the *forecast* distribution because it has been derived from forecasting the state based on data at prior times. The joint distribution of observations and the state is the product $[y_t|x_t][x_t]$ and by Bayes theorem:

$$[x_t|y_t] = \frac{[y_t|x_t][x_t]}{[y_t]}. \tag{27.1}$$

27.2.2 The Update Step

The first part of the assimilation process is to apply Bayes theorem (27.1) to obtain the posterior, $[x_t|y_t]$. In other words, this is the conditional distribution of the system state *given* the observations at time t. The prior distribution for the state is updated in light of information provided by new observations. This result is the *analysis* pdf in a geoscience context. Here the term analysis originates from the analyzed fields used to assess the current state of the atmosphere for weather forecasting. Although (27.1) is strictly a Bayesian formulation, it should be noted that if the prior, $[x_t]$, has a frequency interpretation, then the analysis will also have frequentist content. Typically, practitioners are concerned about the skill of the assimilation and so make a direct comparison between the center and spread of the analysis density and the observed state of the system. We will illustrate this point of view in the atmospheric example at the end of this chapter.

27.2.3 Forecast Step

The second part of the assimilation process is to make a forecast at a future time, e.g., $t+1$. In general, we assume that the dynamics of the state process are known and can be abstracted as $[x_{t+1}|x_t]$. This Markov property implies that future states of the system only depend on the current state and is appropriate for many physical processes. The forecast distribution is then

$$[x_{t+1}|y_t] = \int [x_{t+1}|x_t][x_t|y_t]dx_t. \tag{27.2}$$

The mean of this distribution could be used as a point forecast of the state and the pdf quantifies the uncertainty in the forecast.

27.2.4 Assimilation Cycle

At this point we have come full circle in the assimilation cycle. Confronted with new observations at time $t + 1$, say y_{t+1}, one just identifies $[x_{t+1}]$ with the forecast pdf and applies Bayes theorem. Update and forecast steps are repeated as time advances and new observations arrive. An important concept to draw from this process is that spatial information about the distribution of x_t can be generated in (27.2) from the dynamics of the process. This inheritance is explicit in considering the special case of the Kalman filter linear equations. Although one needs to prescribe a spatial prior explicitly for x_1 in the first update, often this initial information is discounted by subsequent update/forecast cycles with more observations.

One subtlety in this process is the assumption that the prior contains all information about past observations before time t. Equivalently, we are assuming that if the observations at time t are conditioned on the state at time t, then they are independent. In bracket notation,

$$[\boldsymbol{y}_t, \boldsymbol{y}_s | \boldsymbol{x}_t, \boldsymbol{x}_s] = [\boldsymbol{y}_t | \boldsymbol{x}_t][\boldsymbol{y}_s | \boldsymbol{x}_s]. \tag{27.3}$$

Given this conditional independence and updates found by Bayes theorem, the sequential assimilation process outlined above will result in a posterior that is the conditional distribution of the current state based on *all* past observations.

The reader familiar with complex Bayesian models containing hierarchies and many parameters may find this sketch incomplete because it leaves out the issues of parameters in the likelihood and hyperparameters in the priors. The parameters for ensemble inflation, however, are an example of an important set that are estimated by some methods (see Section 27.6). In data assimilation applied to geophysical problems, especially ones in forecasting, there is less emphasis on formal parameter estimation. This is due to the difficulty in applying Bayesian methods to large problems, but also due to the opportunity to tune a method to a particular problem. For example, in numerical weather prediction one is working with a particular numerical model and a predictable stream of future observations. Forecasts can be made and compared to the observations at that time and so the effects of tuning parameters can be evaluated on an ongoing basis. The forecasting skill can be quantified directly by comparison of forecasts to observations. Once the data assimilation has been tuned one would expect similar forecast skill in the future if the model and observation stream stay constant. In this context, the choice of parameters becomes more informal with forecast skill used as a guide and the experience of numerous cycles of the forecast and update steps.

Throughout this discussion, we take the perspective of numerical weather forecasting where one is interested in future predictions of the state and does not have observations at later times to update $\boldsymbol{x}_t$. However, in a retrospective analysis, one would use past, present and future observations with respect to time t for updating the state $\boldsymbol{x}_t$. This process is termed *smoothing* as opposed to forecasting, which is termed *filtering*. Given the Markov assumptions for propagating the state and conditional independence for the observations, the Bayesian computation for smoothing simplifies to a forward pass through the observation sequence followed by a backward pass through the sequence updating $[\boldsymbol{x}_t|\boldsymbol{y}_t]$ with $\boldsymbol{y}_{t+1}$. For large geophysical problems, exact smoothing is not feasible due to the difficulties of running the model backwards in time, and approximations are needed. We suggest one possible approximate smoother as an extension of the ensemble Kalman filter in Section 27.4.

27.2.5 Sequential Updating

Up to now we have assumed the full vector of observations is assimilated in a single application of Bayes theorem. An important feature of this problem is that the update can be performed sequentially on the components of $\boldsymbol{y}_t$ provided that the components are conditionally independent given $\boldsymbol{x}_t$ and that the posterior is computed exactly. To make this explicit, generically split the observation vector into two parts $\boldsymbol{y}_t = (Y^{(1)}, Y^{(2)})$ and assume the conditional independence among observations:

$$[\boldsymbol{y}_t|\boldsymbol{x}_t] = [Y^{(1)}, Y^{(2)}|\boldsymbol{x}_t] = [Y^{(1)}|\boldsymbol{x}_t][Y^{(2)}|\boldsymbol{x}_t].$$

The full posterior for the update can be rewritten as

$$[\boldsymbol{x}_t|\boldsymbol{y}_t] \propto [\boldsymbol{y}_t|\boldsymbol{x}_t][\boldsymbol{x}_t] \propto [Y^{(2)}|\boldsymbol{x}_t][Y^{(1)}|\boldsymbol{x}_t][\boldsymbol{x}_t] \propto [Y^{(2)}|\boldsymbol{x}_t]\left([Y^{(1)}|\boldsymbol{x}_t][\boldsymbol{x}_t]\right) \propto [Y^{(2)}|\boldsymbol{x}_t][\boldsymbol{x}_t|Y^{(1)}].$$

In words, this string of proportions indicates the full posterior can be obtained by first finding the posterior by updating with respect to $Y^{(1)}$ and then using this intermediate result as a subsequent prior to update with respect to $Y^{(2)}$. Since $Y^{(1)}$ and $Y^{(2)}$ are an arbitrary split of $\boldsymbol{y}_t$, by induction, one can show rigorously that the full posterior can be found by updating with each component of $\boldsymbol{y}_t$ sequentially. Each update can involve a scalar observation and the posterior computation can simplify greatly when this is done. It is important to emphasize that a sequential update is only valid under conditional independence among the components of the observation vector, but, if this holds the order of the sequential, updating does not matter. An intriguing connection with spatial statistics is that this same sequential result can be used for Bayesian spatial prediction and there is the potential to transfer the efficient parallel algorithms for this approach to find approximate solutions for large spatial problems.

27.3 The Kalman Filter and Assimilation

The Kalman filter (KF) was first developed by Kalman (1960) and Kalman and Bucy (1961) in an engineering context and as a linear filter. Although the KF can be interpreted as an optimal linear estimator, to streamline this discussion, we will add the assumption of a joint Gaussian distribution to fit into the Bayesian paradigm given above. (See Jazwinski, 1970, for more background.)

27.3.1 The KF Update Step

Assume that $[\boldsymbol{y}_t|\boldsymbol{x}_t]$ is multivariate normal with mean vector $H\boldsymbol{x}_t$ and covariance Σ_o. H is a known matrix that maps the state into the expected value of the observations and Σ_o is the observation error covariance matrix. (Both H and Σ_o can depend on t, although we will not add this additional index.) This conditional distribution can also be represented as

$$\boldsymbol{y}_t = H\boldsymbol{x}_t + \boldsymbol{e}_t, \tag{27.4}$$

where now $\boldsymbol{e}_t \sim N(0, \Sigma_o)$. As mentioned above, a subtle point assumed here is that $\boldsymbol{e}_t$ are uncorrelated over time and independent of the state. Without loss of generality Σ_o can be assumed to be diagonal by redefining $\boldsymbol{y}$ and H through a linear transformation. We will further assume that $[\boldsymbol{x}_t] \sim N(\boldsymbol{\mu}_f, \Sigma_f)$ where "f" indicates this is the forecast distribution. The update step yields a distribution that is again multivariate normal $[\boldsymbol{x}_t|\boldsymbol{y}_t] \sim N(\boldsymbol{\mu}_a, \Sigma_a)$. Here the "a" indicates the analysis distribution and the mean and covariance are

$$\boldsymbol{\mu}_a = \boldsymbol{\mu}_f + \left[\Sigma_f H^T(H\Sigma_f H^T + \Sigma_o)^{-1}\right](\boldsymbol{y}_t - H\boldsymbol{\mu}_f) \tag{27.5}$$

and

$$\Sigma_a = \Sigma_f - \Sigma_f H^T(H\Sigma_f H^T + \Sigma_o)^{-1}H\Sigma_f. \tag{27.6}$$

The matrix expression in square brackets in (27.5) is the Kalman gain and transforms a difference between the observation vector and its expectation with respect to the forecast distribution into an adjustment to the state. To derive these expressions, note that $\boldsymbol{x}_t$ and $\boldsymbol{y}_t$ are jointly distributed multivariate normal and (27.5) and (27.6) can be derived from the properties of the conditional multivariate normal.

Equation (27.5) and Equation (27.6) are the same as the equations for a spatial conditional inference. Interpreting the state vector as being values of a spatial field on a regular grid, $\boldsymbol{\mu}_a$ is the conditional mean for $\boldsymbol{x}$ given the observations and Σ_a the conditional covariance matrix. Here one interprets the "prior" $N(\boldsymbol{\mu}_f, \Sigma_f)$ as a Gaussian process model with Σ_f

being constructed from a spatial covariance function. The important distinction for data assimilation is that Σ_f is generated by the process, not from an external spatial model. This becomes clear in examining the forecast step for the linear KF.

27.3.2 The KF Forecast Step

For the forecast step for the KF, assume that the process evolves in a linear way from t to $t+1$, possibly with an additive Gaussian random component

$$x_{t+1} = Lx_t + u_t. \tag{27.7}$$

Here $u_t \sim N(0, \Sigma_m)$ independent of x_t and L is a matrix. Both L and Σ_m can depend on t. Based on all these assumptions, it is straightforward to conclude that the forecast pdf is $N(L\mu_a, L\Sigma_a L^T + \Sigma_m)$. Scrutinizing the forecast covariance matrix, one sees that $L\Sigma_a L^T$ will be based on the previous forecast covariance matrix appearing in (27.6) and will also inherit the dynamical relationship from the previous time. Thus, in the situation of assimilation for a space–time process the spatial covariance for inference is built up sequentially based on past updates with observations and propagating the posterior forward in time as a forecast distribution. It is important to realize that this spatial information is the difference or *error* between the conditional mean and the true field and is not the covariance of the process itself. For example, if the observations are both dense and accurate and Σ_m is small, the forecast covariance can be much smaller and have less structure than the covariance for x_t itself.

27.3.3 Sequential Updates

In the previous section, it was stated that the update can be done sequentially under conditional independence of the observations. It is helpful to describe how the KF update simplifies when the components of y_t are considered sequentially. Conditional independence holds among the components of y_t given our choice of Σ_o being diagonal and Gaussian distributions. Let i index the components of y_t and $\{\sigma_i^2\}$ be the diagonal elements of Σ_o. To notate the sequential aspect of the update set $\mu_a^0 = \mu_f$ and $\Sigma_a^0 = \Sigma_f$ and let μ_a^{i-1} and Σ_a^{i-1} be the prior mean and covariance used in the update with the ith component of the observation vector.

With this notation, the sequential update for the ith observation with respect to Equation (27.5) simplifies to

$$\mu_a^i = \mu_a^{i-1} + \Sigma_a^{i-1} h_i \left[\frac{y_{t,i} - h_i^T \mu_a^{i-1}}{h_i^T \Sigma_a^{i-1} h_i + \sigma_i^2} \right],$$

where h_i is the ith row of H and the expression in brackets is a scalar. The update for the covariance matrix is also simple and is a rank one correction to Σ_a^{i-1}. This form is important for the ensemble adjustment Kalman filter used in the example.

27.3.4 Problems in Implementation of the KF

There are two major difficulties in implementing the standard KF in geophysical data assimilation problems: handling large matrices and accounting for nonlinear dynamics. Typically, global or high resolution problems have large state vectors. For example, just considering an atmospheric model at medium resolution (grid cells of about 150×150 km at the equator and 26 vertical levels) the atmosphere is divided into a three-dimensional grid of $128 \times 256 \times 26$ each having at least four variables (temperature, horizontal wind components and water vapor) and a field of surface pressure. Thus, x has more than 3 million elements. In addition, the observation vector typically has on the order of 10^5 elements with operational weather forecasting systems accommodating on the order of 10^7 elements. Dimensions of this size prohibit computing or storing the elements of the covariance matrices from the update and forecast steps and so it is not possible to implement the KF exactly for large problems. One alternative is to fix Σ_f with a convenient form and this is known as three-dimensional variational data assimilation and is described below.

Besides direct problems with linear algebra, the KF poses difficulties with nonlinear models for the dynamics. For atmospheric models, evolution over time is represented as a complicated nonlinear transformation $[x_{t+1}|x_t] = \Gamma(x_t, t)$ based on the nonlinear equations of fluid dynamics and thermodynamics. Γ is implicitly defined by a computer code that implements a discretized model and usually has no closed form. Thus, even if Γ is a deterministic function, calculating a closed form for the forecast distribution, essentially $[\Gamma(x_t)|x_t]$, is not possible. This problem is compounded by the fact that x may have high dimension as well. Finally, because of the nonlinear action of Γ, one may expect that the resulting forecast distribution will not be normal and so the assumptions of multivariate normality within the update step will not hold.

In summary, although providing closed forms for the posterior and forecast distributions in a linear situation, the KF is not practical for the kinds of assimilation problems encountered in some geophysical settings. Some strategy for an approximate solution is needed. Ensemble Kalman filters approximate the posterior distribution with a discrete sample. Another approach is to avoid large covariance matrices by not updating the forecast covariance and this leads to variational methods of assimilation. We present both of these practical alternatives in the next two sections.

27.4 The Ensemble Kalman Filter

The term *ensemble* is used in the geosciences to refer to a sample either randomly drawn from a population or deliberately constructed. In data assimilation, an ensemble of system states is used as a discrete approximation to the continuous, and often high-dimensional distribution for x. The basic idea of an ensemble Kalman filter is to use a sample of states to approximate the mean vectors and covariance matrices. Each ensemble member is updated by an approximation to Bayes theorem and is propagated forward in time using Γ giving a new ensemble for approximating the forecast distribution. This idea was proposed by Evensen (1994), but has been developed by many subsequent researchers. It is a form of particle filter (Doucet and De Freitas, 2001) with the ensemble members being "particles." One departure from standard particle filtering is that the ensemble members are modified at every update step rather than just being reweighted. It should be noted at the outset that this is a rich area of research and application within the geosciences and the overview in this section cannot review many of the innovations and developments for specific problems. The details of implementing the update step with ensembles are important, especially when the ensemble size is small and one is concerned about the stability of the filter over longer periods of time. The example at the end of this chapter gives some idea of practical issues and performance.

27.4.1 Ensemble Update

The ensemble update step holds the main statistical details of the ensemble KF. By contrast, the forecast step for ensembles is both simple and explicit. To simplify notation, we will drop the time subscript because the computations are all at time t. Let $\{x_f^j\}$ for $1 \leq j \leq M$

be an M member ensemble providing a discrete representation of the forecast pdf and $\{x_a^j\}$ the corresponding ensemble for the analysis pdf (or posterior). The ensemble KF provides an algorithm using the observations and the update equations to transform the forecast ensemble into the analysis ensemble and so finesses the problem of working directly with high-dimensional and continuous pdfs. If the dimension of the state vector is N, then the storage for an ensemble is on order of $M \times N$ and can be much smaller than storing dense $N \times N$ covariance matrices. Examining the Kalman filter equations, the update Equation (27.5) and Equation (27.6) depend on the forecast mean $\boldsymbol{\mu}_f$ and the forecast covariance Σ_f. Given an ensemble, one replaces $\boldsymbol{\mu}_f$ by the *sample* mean and Σ_f by an estimate based on the *sample* covariance. Solving Equation (27.5) results in an approximate posterior mean that we will denote $\hat{\boldsymbol{\mu}}_a$ and this will be taken as the sample mean of the updated ensemble. To understand how this is different from the exact KF computations, we give some details of this solution. Let $\bar{\boldsymbol{x}}_f$ be the ensemble mean forecast vector and

$$U_f = (\boldsymbol{x}_f^1 - \bar{\boldsymbol{x}}_f, \boldsymbol{x}_f^2 - \bar{\boldsymbol{x}}_f, \ldots \boldsymbol{x}_f^M - \bar{\boldsymbol{x}}_f) \tag{27.8}$$

be a matrix of the centered ensemble members. The sample forecast covariance has the form $\hat{\Sigma}_f = \frac{1}{M-1} U_f U_f^T$. Note that this estimate has effective rank $M-1$ and, when used in the update equations, the linear algebra can exploit this reduced rank. Specifically, the full forecast covariance matrix need never be explicitly computed. Moreover, in using iterative methods to solve the linear system in Equation (27.5) the multiplication of $(H\hat{\Sigma}_f H^T + \Sigma_o)$ by an arbitrary vector can be done efficiently because of the reduced rank.

The other half of the update step involves the analysis covariance. The concept is to examine the form in Equation (27.6) and modify the ensemble members to have a sample covariance close to this expression. There are two main strategies for doing this: a Monte Carlo approach, known as perturbed observations and a deterministic approach, known as a square root filter. For perturbed observations, one generates M random vectors, $\epsilon^j \sim N(0, \Sigma_o)$ that are further constrained so that the mean across j is zero. Now, form M versions of "perturbed" observation vectors by adding these random deviates to the actual observation: $\boldsymbol{y}_t^j = \boldsymbol{y}_t + \epsilon^j$. To update each ensemble member, apply the right side of Equation (27.5) with the substitutions $\boldsymbol{x}_f^j$ for $\boldsymbol{\mu}_f$, $\hat{\Sigma}_f$ for Σ_f and $\boldsymbol{y}_t^j$ for $\boldsymbol{y}_t$ obtaining an analysis ensemble. Because the perturbed observations have zero sample mean, the ensemble mean from this method will reproduce the posterior mean $\hat{\boldsymbol{\mu}}_a$. Moreover, as M goes to infinity, the mean and covariance of the ensemble will match that of the posterior (see Furrer and Bengtsson, 2006).

Deterministic updates of the ensemble fall under the general ideas of square root Kalman filters (Tippett, Anderson, Bishop, Hamill et al., 2003). Given a matrix, A, the updated ensemble is generated through a linear transformation: $[\boldsymbol{x}_a^1| \ldots |\boldsymbol{x}_a^M] = U_f A + \hat{\boldsymbol{\mu}}_a$. Note that this ensemble will have mean vector $\hat{\boldsymbol{\mu}}_a$ and the key idea is to choose A so that the sample covariance approximates the expression in Equation (27.6). In other terms, $UAA^TU^T = \hat{\Sigma}_a$. A is only determined up to an orthogonal transformation, but the choice is important in preserving physical structure of the state vectors as realizable states of the system. Besides a postmultiplication of U there is also the option to premultiply this matrix and this is the form of the ensemble adjustment Kalman filter (Anderson, 2001) and that of Whitaker and Hamill (2002). See Livings, Dance, and Nichols (2008) for some discussion of choices of A that have an unbiasedness property. For large observation vectors the computations for an ensemble square root filter may still be extensive. One approximation is to update the components of the state vector in a moving window of observations and is termed the local ensemble transform Kalman filter (LETKF) (Szunyogh, Kostelich, and Gyarmati, 2008).

27.4.2 Ensemble Forecast Step

Given an ensemble approximating the analysis distribution, we now describe the forecast step. An elegant property of ensemble methods is that the forecast step is exact to the extent that the discrete ensemble approximates the continuous distribution of the analysis pdf. Suppose that $\{x^j_{a,t}\}$ are a random sample from $[x_t|y_t]$. Let $x^j_{f,t+1} = \Gamma(x^j_t, t)$ be the states obtained by propagating each member forward to $t+1$. By elementary probability, this forecast ensemble will be a random sample from $[x_{t+1}|y_t]$ without requiring any additional assumptions on the distribution of the posterior.

27.4.3 Practical Issues for Small Ensemble Sizes

It is well known that for small ensemble sizes the sampling variability in $\hat{\Sigma}_f$ over the course of several update/forecast cycles can induce substantial error (Mitchell and Houtekamer, 2000). A common effect is the collapse of the ensemble to the mean value. This occurs because errors in the covariance tend to produce biases that favor less spread among the ensemble members. An artificially small forecast covariance results in the Kalman gain matrix decreasing the contribution of the observations to the update. The net result is a filter that ignores information from the observations and just propagates the ensemble forward in time. This behavior is known as filter divergence.

There are two important principles to counter this behavior in practice:

- *Localization* of the ensemble covariance estimate to improve the accuracy by reducing the effects of sampling
- *Inflation* to increase the spread of the ensemble

Like other aspects of ensemble methods, these two principles are implemented in many ways and we will just review some approaches that have a statistical or spatial thread.

27.4.4 Localization of the Ensemble Covariance

The simplest form of localization is to taper the sample covariance matrix based on physical distances between an observation and state. The rationale is that beyond a certain distance scale the assimilation errors for a spatial field should not be dependent and thus the corresponding elements of the forecast error covariance matrix should be set to zero and values at intermediate distances should be attenuated. (See Houtekamer and Mitchell (2001) and Hamill, Whitaker, and Snyder (2001) for some background on localization.) Assume that each component of the state vector x_k is associated with a location u_k and $d_{k,k'}$ is the distance between u_k and $u_{k'}$. A tapered estimate is the direct (or Schur) matrix product of $\hat{\Sigma}_f$ with a correlation matrix

$$[\tilde{\Sigma}_f]_{k,k'} = [\hat{\Sigma}_f]_{k,k'}\, \phi(d_{k,k'}), \tag{27.9}$$

where ϕ is a correlation function giving a positive definite correlation matrix. Based on the properties of the direct product, $\tilde{\Sigma}_f$ will remain nonnegative definite and to improve the computational efficiency the tapering is usually done with a compactly supported kernel (Gaspari and Cohn, 1999). That is, $\phi(d)$ is identically zero for d sufficiently large, to introduce sparsity in the product covariance matrix. The result of this tapering is a covariance estimate that is biased, but has less variance. Also introducing sparsity facilitates the matrix computations.

27.4.5 Inflation of the Ensemble Spread

Inflation is the operation of adjusting the ensemble spread beyond what is prescribed by the KF update formula to give a more appropriate measure of the uncertainty in the state. This adjustment can compensate not only for sampling variation of the ensemble, but also in some cases for model error when the stochastic component, u_t from Equation (27.7) has not been explicitly included. A useful assumption is that the forecast ensemble correlation structure is correct, but that estimates of the variance of individual state vector components may be too small. After the model advance, but before the update step, the prior ensemble members are inflated so that

$$x_f^{j,\,inflated} = \sqrt{\lambda_j}(x_f^j - \mu_f^j) + \mu_f^j. \tag{27.10}$$

There are fewer strategies for this than localization and while being effective these are often global adjustments (Hamill et al., 2001; Houtekamer and Mitchell, 2001). For example, a standard approach is to multiply $\tilde{\Sigma}_f$ by a scalar that is greater than one, i.e., $\lambda_j \equiv \lambda > 1$. In the example at the end of this chapter, however, we describe a method of inflation that is based on sequentially inflating the state vector by a comparison of the forecast mean and variance to new observations (Anderson and Collins, 2007). In general, there is a need for more work on statistical models and algorithms for handling inflation. Closely related to this issue is the need for statistical models to represent model error.

27.5 Variational Methods of Assimilation

Variational methods are more established than ensemble variants of the Kalman filter or a Bayesian framework and have been successful in operational settings for making rapid and reliable weather forecasts using large numerical models and massive data streams (for example, see Rabier, Jarvinen, Klinker, Mahfouf et al., 2000). Essentially, variational methods estimate the state of the system by minimizing a cost function. As a starting point, we identify the cost function problem that is equivalent to the update step in the KF and in terms of spatial statistics this will be equivalent to kriging. However, despite this connection, variational approaches often focus on crafting a cost function without relying on a Bayesian interpretation for motivation.

27.5.1 Three-Dimensional Variational Assimilation

Under the assumptions of linearity and multivariate normality, the posterior of $[x_t|y_t]$ is also multivariate normal. Moreover, the mean of a Gaussian density is also the mode and so the posterior mode will also be the posterior mean in this case. Finally, note that maximizing $[x_t|y_t]$ is the same as minimizing minus the log of the joint distribution, $-(\log([y_t|x_t]) + \log([x_t]))$, and where terms that do not depend on x_t or y_t can be omitted. Putting these remarks together we have motivated the variational problem:

$$\min_{x} \quad (1/2)(y_t - Hx)^T \Sigma_o^{-1}(y_t - Hx) + (1/2)(x - \mu_f)^T \Sigma_f^{-1}(x - \mu_f). \tag{27.11}$$

The minimizer of this cost function is the variational estimate of the state. It is a standard exercise to show that the minimum is

$$\hat{x}_t = (H^T \Sigma_o^{-1} H + \Sigma_f^{-1})^{-1} H^T (y_t - Hx) + \mu_f. \tag{27.12}$$

Based on the Sherwood–Morrison–Woodbury formula (Golub and Van Loan, 1996), this can be shown to be the same as the KF update in (27.5). Thus, we have an alternative way of characterizing the mode of the analysis distribution. Since the kriging equations are the same as (27.5), we have also outlined how kriging can be interpreted as the solution to a variational problem.

This kind of cost function appears in many different areas and in general is characterized as a regularized solution to an inverse problem. $\boldsymbol{x}_t$ is estimated from data $\boldsymbol{y}_t$ by "inverting" H. A regularization term involving Σ_f is added to make this a well-conditioned problem and ensures a unique minimum to the cost function. Inverse problems cover a large range of applications in many different areas of science and engineering, such as tomography and remote sensing, and variational methods of assimilation are just one special case. (See Tarantola, 1987 for some background.) From the perspective of this article, we can trace the regularization to the prior distribution for the state vector. Alternatively, this term can be motivated by building in prior information about the state vector as an extra penalty in the cost function. For example, if Σ_f is a covariance function for a smooth spatial field, then as a regularization it will constrain the solution to also be smooth.

In the atmospheric sciences when $\boldsymbol{x}$ includes three-dimensional fields this is known as 3DVAR. The important difference between 3DVAR and the update from the KF is that typically Σ_f is *fixed* and termed a background error covariance. Thus, covariance information is not propagated based on the dynamics and the forecast step only involves $\Gamma(\hat{\boldsymbol{x}}_t)$. In the simplest form of 3DVAR, there are no companion measures of uncertainty. This method has the advantage that it can be easily tuned by modifying or estimating parameters in the background covariance (e.g., Dee and de Silva, 1999) and by incorporating physical constraints with additional cost terms.

An extension of 3DVAR is to add different time periods to the cost function. This is known as 4DVAR in the atmospheric sciences (Le Dimet and Talagrand, 1968; Lewis and Derber, 1985) and seeks to find a single trajectory of the model that is consistent with observations at multiple times. An example of the cost function starting at $t = 1$ and going through $t = \mathrm{T}$ is

$$\min_{\boldsymbol{x}_1} \left(\frac{1}{2}\right) \sum_{t=1}^{\mathrm{T}} \left[(\boldsymbol{y}_t - H_t\boldsymbol{x}_t)^T \Sigma_o^{-1} (\boldsymbol{y}_t - H_t\boldsymbol{x}_t)\right] + \left(\frac{1}{2}\right) (\boldsymbol{x}_1 - \boldsymbol{\mu}_b)^T \Sigma_b^{-1} (\boldsymbol{x}_1 - \boldsymbol{\mu}_b). \quad (27.13)$$

Here $\boldsymbol{\mu}_b$ and Σ_b refer to a background mean and covariance at the start of the period and provide some form of prior information on the initial state of the system. Implicit in this cost is that subsequent state vectors are found using a dynamical model, e.g., $\boldsymbol{x}_{t+1} = \Gamma(\boldsymbol{x}_t, t)$, that is deterministic. Despite the possible limitations from a high-dimensional state vector and a complex dynamical model, this solution has an appealing interpretation. Based on initial conditions, $\boldsymbol{x}_1$, the solution is a *single* trajectory through time that best fits the observations. The only regularization is done on the initial conditions and the remaining times are constrained by the dynamical model. In general form, a statistician would recognize 4DVAR as a large nonlinear ridge regression problem (where the parameters are $\boldsymbol{x}_1$). Note that, as in 3DVAR, only a point estimate is produced and measures of uncertainty need to be generated by other methods. For problems with the complexity of an atmospheric model, 4DVAR analysis is a difficult computational problem partly because finding the minimum is a nonlinear problem and also the gradient of Γ may be hard to formulate or compute. Despite these hurdles, 4DVAR systems have been implemented as the primary assimilation algorithm in large operational models for weather forecasting (e.g., Rabier et al., 2000). Current extensions of 4DVAR include adding a model error process to the variational cost function. This feature, termed a weak constraint, allows the estimated states at each time to deviate from a single trajectory based on the initial condition (see, for example, Zupanski, 1996). However, this potential benefit comes with the difficulty of specifying a space–time statistical model for the numerical model error process.

27.6 Ensemble Filter for an Atmospheric Model

This section describes a specific ensemble-based method and its application to a large numerical model. The sketch of this method, known as the ensemble adjustment Kalman filter (EAKF), includes both localization and inflation and will illustrate some of the principles common to many other practical approaches (Harlim and Hunt, 2007). The atmospheric model and observations used in this example are at the scale of an operational system used to create global and complete fields of the troposphere. The size of this problem is several orders of magnitude larger than typical applications of spatial statistics. The Data Assimilation Research Testbed (DART) (http://www.image.ucar.edu/DAReS/DART) is an open software environment that can be used to reproduce this analysis.

27.6.1 Ensemble Adjustment Kalman Filter

The EAKF is based on a sequential update algorithm where an observation vector is assimilated as a sequence of scalar problems (Anderson, 2001). It may be surprising that this is efficient. In many geophysical applications, one can take advantage of the observations errors being uncorrelated and a (great circle) distance-based tapering to induce sparsity in the elements of Σ_f. Also, sequential updating and an ensemble approximation to the posterior are amenable to parallel computation, a necessary requirement for large problems (Anderson and Collins, 2007). The EAKF is a variant of the square root filter and thus the modifications to the ensemble are deterministic. In the case of scalar updates, the adjustment to each ensemble member is done to minimize the difference between its prior and updated values. This is in contrast to an approach such as perturbed observations where independent random components are added to each ensemble member and can produce more random shuffling among the ensemble members. Other square root filters may also induce significant differences between the forecast and analysis ensemble members.

27.6.2 Covariance Localization and Inflation

The main technique for localization will be recognized by a statistician as a shrinkage and decimation of the sample correlations found from the ensemble. The elements of $\hat{\Sigma}_f$ are tapered based on distance and further attenuated based on an approximate resampling strategy where the ensemble is divided up in a small number of subsets and a shrinkage parameter is estimated by cross-validation (Anderson, 2007). In large problems with dense observations, one anticipates that the data will provide substantial information of the atmospheric state over time. Moreover, this state information can be reinforced by a physical model. From this perspective, it is more important that localization be conservative in not updating components of the state vector due to spurious correlations in $\hat{\Sigma}_f$. Over multiple assimilation cycles these errors accumulate to cause filter divergence.

Inflation of the ensemble follows by estimating a vector of inflation values $\boldsymbol{\lambda}$ that scale each component as in Equation (27.10). The spatial adaptation is implemented in a manner to be computationally feasible and parallel in the state vector components and is a good illustration of an algorithm that is effective, but does not necessarily follow directly from first statistical principles. For a new observation $\boldsymbol{y}_{t,i}$, let $\hat{\sigma}^2 = \boldsymbol{h}_i^T \hat{\Sigma}_f \boldsymbol{h}_i$ be the forecast ensemble variance and $\hat{y} = \boldsymbol{h}_i^T \hat{\boldsymbol{x}}_f$ its ensemble forecast mean. With this notation, we have the forecast error $(\boldsymbol{y}_{t,i} - \hat{y})$ with an expected variance of $\sigma^2 + \sigma_o^2$ without inflation. For the jth component of the state vector, a pseudo-likelihood is taken as

$$[\boldsymbol{y}_{t,i} | \boldsymbol{\lambda}_j] \sim N(\hat{y}, \theta^2) \tag{27.14}$$

with

$$\theta^2 = \left[1 + \gamma_j(\sqrt{\lambda_j} - 1)\right]^2 \sigma^2 + \sigma_o^2 \tag{27.15}$$

and γ_j is the correlation between x_j and $y_{t,i}$ based on the ensemble forecast covariance. A normal prior is used for λ_j and the posterior is approximated by a normal. The intuition behind the choice of Equation (27.15) is a pseudo-likelihood criterion that only links the inflation and the forecast error if the correlation between the actual state and the observation is large. Although this expression can be motivated in the case of a single-state component, it is an approximation when x has more than one element. When $\gamma = 1$ and $\boldsymbol{\lambda} \equiv \lambda$ is a constant, the variance of the forecast error is parameterized as $\lambda^2\sigma^2 + \sigma_o^2$. Thus, in this limit (27.15) will reduce to a more conventional likelihood and give an approximate Bayesian inference for a scalar inflation of the forecast variance.

27.6.3 Assimilation with the Community Atmospheric Model

The Community Atmospheric Model (CAM) 3.1 (Collins, Rasch et al., 2004) is a mature global atmospheric model that forms the atmospheric component of the Community Climate System Model, a state-of-the-art climate model. For this example, CAM is configured at a resolution of approximately 256 × 128 on a longitude/latitude grid and has 26 vertical layers. Observations consist of soundings from weather balloons, measurements made from commercial aircraft, and satellite-derived wind fields, and the initial ensemble was initialized from a climatological distribution from this season. An 80-member ensemble was used for the EAKF. Available observations were assimilated every 6 hours over the period January 1, 2007 through January 31, 2007. Overall the quality of the forecast fields are comparable to reanalysis data products produced by the U.S. National Center for Environmental Prediction and the National Center for Atmospheric Research (Kistler, Kalnay, Collins et al., 2001) and so are close to the best analyses available for the atmosphere.

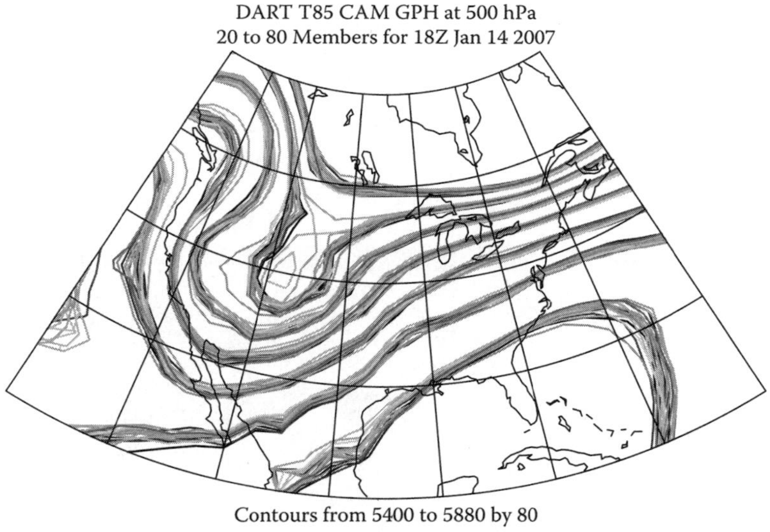

FIGURE 27.1
An illustration of the ensemble spread from an ensemble adjustment Kalman filter, CAM 3.1 and available surface and aircraft observations of the atmosphere. Plotted are the 500 hPa height contours for 20 members of an 80-member ensemble. For this day, the height levels increase from north to south. The spread in the contours at a particular height is due to variation among the ensemble members.

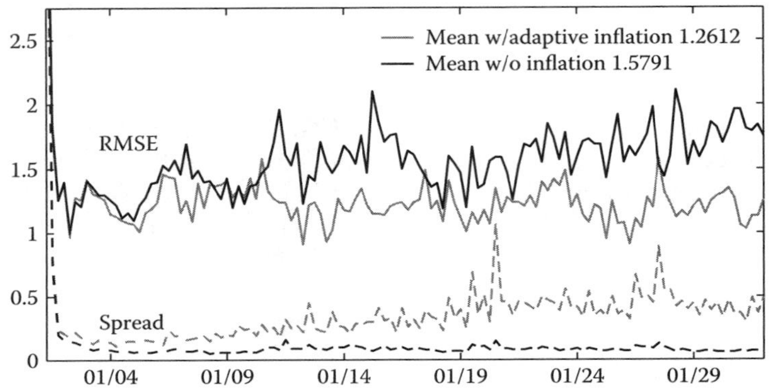

FIGURE 27.2
Root mean squared errors (RMSE) for the 500 hPa heights from data assimilation derived using CAM 3.1 over the period 1/1/2007 through 1/31/2007. RMSE for the ensemble means based on assimilation with spatial inflation of the ensemble (light solid) and without adaptive inflation (dark solid). Dashed curves are the mean spread in the ensemble members.

The following figures summarize some of the statistical results. Figure 27.1 is an example of one of the fields that describe the state of the atmosphere, the 500 hPa geopotential height field and describes the mid-level structure in the troposphere. The trough across the Rocky Mountain region of the United States and Canada indicates a large low pressure depression and was associated with a notable ice storm in the area including Texas. Based on geostrophic balance, the large-scale flow at this level is along the contours of equal height (i.e., constant pressure) and in a counterclockwise motion about the depression. The spread among the contours of the ensemble members suggests the variability in the estimated field and the variability of the ensemble members at the second lowest height contours suggest the uncertainty in the extent of the low pressure region. The tightening of contours over the eastern United States as compared to the spread over Mexico or the Pacific Ocean is related to a denser observation network in the United States. It is still an active area of research to relate the ensemble spread to frequency-based measures of confidence; however, this amount of variation is consistent with a stable assimilation method that avoids filter divergence and adjusts to different data densities. Figure 27.2 illustrates the effect of adding inflation in terms of the global root mean squared error for the height of the atmosphere at

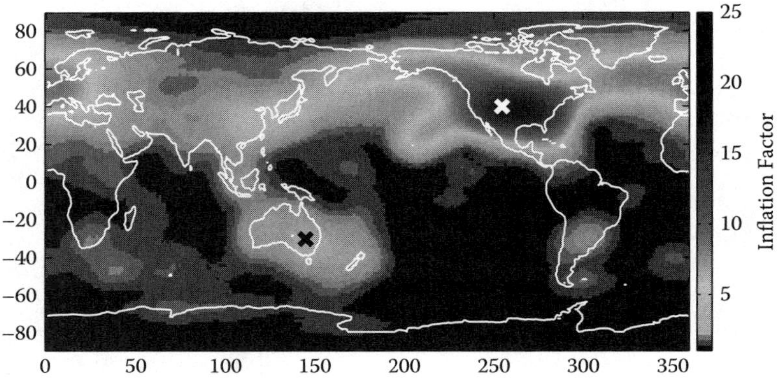

FIGURE 27.3
The field of spatial inflation values, λ, at the end of the assimilation period.

a pressure of 500 hPa. Here, adaptive inflation increases the accuracy in the ensemble mean and at the same time produces ensemble members with larger spread. It is interesting that better filter performance can be obtained with a larger range in the ensemble and suggests that the inflation may also contribute to better characterizations of the forecast uncertainty. Figure 27.3 is a snapshot of the inflation field (λ) estimated at the end of the assimilation period indicating how the inflation factor varies over space. One surprise is that a large inflation is required over North America where there is a high data density. At first it seems contradictory that a data-rich area, partly due to intensive aircraft observations, should require inflating the spread, and hence the uncertainty in the state. However, this effect can be explained by the presence of model error. Without inflation, model errors are ignored and, given a dense data region, the forecast variance will be small, especially as sampling errors accumulate through the sequential updating of the EAKF. A large amount of data in this region also allows the method to identify the discrepancy between the forecast variances and the actual forecast errors and so makes it possible to estimate an inflation field much different than unity.

It was noticed that adaptive inflation could be improved by building in some temporal correlation and the Bayesian estimation could be more formally integrated within the ensemble KF. Both of these are topics of future research. However, these results indicate that spatially varying parameters that control the assimilation process, even obtained by approximate or heuristic principles, are important and highlight an additional role for spatial statistics in data assimilation.

Acknowledgments

This work was supported by the National Science Foundation through a grant to the University Corporation for Atmospheric Research. Nychka also acknowledges the support of the Isaac Newton Institute for Mathematical Sciences.

References

Anderson, J.L. (2001). An ensemble adjustment Kalman filter for data assimilation. *Monthly Weather Review* 129, 2894–2903.

Anderson, J. and N. Collins (2007). Scalable implementations of ensemble filter algorithms for data assimilation. *Journal of Atmospheric and Oceanography Technology A*, 24 1452–1463.

Anderson, J.L. (2007). Exploring the need for localization in ensemble data assimilation using a hierarchical ensemble filter. *Physica D* 230, 99–111, doi:10.1016/j.physd.2006.02.011

Collins, W.D., P.J. Rasch et al. (2004). Description of the NCAR Community Atmosphere Model (CAM 3.0), *Technical Report NCAR/TN-464+STR*, National Center for Atmospheric Research, Boulder, CO, 210 pp.

Dee, D.P. and A.M. da Silva (1999). Maximum-likelihood estimation of forecast and observation error covariance parameters. Part I: methodology. *Monthly Weather Reveiw* 127, 1822–1834.

Doucet, A. and N. De Freitas (2001). *Sequential Monte Carlo Methods in Practice*, Springer, New York.

Evensen, G., (1994). Sequential data assimilation with a nonlinear quasi-geostrophic model using Monte Carlo methods to do forecast error statistics. *Journal Geophysical Research* 99(C5), 10143–10162.

Furrer, R. and T. Bengtsson (2006). Estimation of high-dimensional prior and posterior covariance matrices in Kalman filter variants. *Journal Multivariate Analysis* 98, 227–255.

Gaspari, G. and S.E. Cohn (1999). Construction of correlation functions in two and three dimensions. *Quarterly Journal Royal Meteorological Society* 125, 723–757.

Golub, G.H. and C.F. Van Loan (1996). *Matrix Computations.* Johns Hopkins University Press, Baltimore.

Houtekamer, P.L. and H.L. Mitchell (2001). A sequential ensemble Kalman filter for atmospheric data assimilation. *Monthly Weather Review*, 129, 123–137.

Hamill, T.M., J.S. Whitaker, and C. Snyder (2001). Distance-dependent filtering of background-error covariance estimates in an ensemble Kalman filter. *Monthly Weather Review*, 129, 2776–2790.

Harlim, J. and B.R. Hunt (2007). Four-dimensional local ensemble transform Kalman filter: Numerical experiments with a global circulation model. *Tellus A*, 59(5), 731–748, doi: 10.1111/j.1600-0870.2007.00255.x

Harvey, A., S.J. Koopman, and N. Shepard (2004). *State Space and Unobserved Component Models: Theory and Applications.* Cambridge Unversity Press, Cambridge, MA, 380 pp.

Jazwinski, A.H. (1970). *Stochastic Processes and Filtering Theory.* Academic Press, San Diego, 376 pp.

Kalman, R.E. (1960). A new approach to linear filtering and prediction problems. *Journal of Basic Engineering*, 82 (1): 35–45.

Kalman, R.E. and R.S. Bucy (1961). New results in linear filtering and prediction theory. *Transactions of the ASME—Journal of Basic Engineering*, 83, 95–107.

Kalnay, E. (2002). *Atmospheric Modeling, Data Assimilation and Predictability*. Cambridge University Press, Cambridge, MA.

Kistler, R., E. Kalnay, W. Collins et al. (2001). The NCEP-NCAR 50-year reanalysis: Monthly means CD-ROM and documentation. *Bulletin of the American Meteorological Society*, 82(2), 247–268.

Livings, D., S. Dance, and N. Nichols (2008). Unbiased ensemble square root filters. *Physica D*, 237, 1021–1028.

Szunyogh, I., E.J. Kostelich, G. Gyarmati, E. Kalnay, B.R. Hunt, E. Ott, E. Satterfield, and J.A. Yorke (2008). A local ensemble transform Kalman filter data assimilation system for the NCEP global model. *Tellus A*, 60(1), 113-130 doi:10.1111/j.1600-0870.2007.00274.x.

Le Dimet, F.-X. and O. Talagrand (1968). Variational algorithms for analysis and assimilation of meteorological observations: theoretical aspects. *Tellus. Series A, Dynamic Meteorology and Oceanography*, 38(2), 97–110.

Lewis, J.M. and J.C. Derber (1985). The use of adjoint equations to solve a variational adjustment problem with advective constraints. *Tellus. Series A, Dynamic Meteorology and Oceanography*, 37:44, 309–322.

Mitchell, H.L. and P.L. Houtekamer (2000). An adaptive ensemble Kalman filter. *Monthly Weather Review*, 128, 416–433.

Rabier, F., H. Jarvinen, E. Klinker, J.F. Mahfouf, and A. Simmons (2000). The ECMWF operational implementation of four-dimensional variational assimilation. I: experimental results with simplified physics. *Quarterly Journal Royal Meteorological Society*, 126, 1148–1170.

Tarantola, A. (1987). *Inverse Problem Theory: Methods for Data Fitting and Model Parameter Estimation*. Elsevier Science Publishers, New York.

Tippett, M.K., J.L. Anderson, C.H. Bishop, T.M. Hamill, and J.S. Whitaker (2003). Ensemble square root filters. *Monthly Weather Review* 131, 1485–1490.

West, M. and J. Harrison (1997). *Bayesian Forecasting and Dynamic Models.* Springer, New York, 680 pp.

Wikle, C. (2009). Hierarchical modeling with spatial data, Chapter 7, this volume.

Wikle, C.K. and L.M. Berliner (2006). A Bayesian tutorial for data assimilation. *Physica D: Nonlinear Phenomena*, 230, 1–16.

Whitaker, J.S. and T.M. Hamill (2002). Ensemble data assimilation without perturbed observations, *Monthly Weather Review*, 130, 1913–1924.

Zupanski, D. (1996). A general weak constraint applicable to operational 4DVAR data assimilation systems. *Monthly Weather Review*, 125, 2274–2283.

Part VI

Additional Topics

The objective of Part VI is to collect several important topics that build upon the material presented in the previous chapters, but in some sense seemed to us to be beyond these earlier ones. It is increasingly the case that multivariate data are being collected at locations, raising the issue of dependence between measurements at a location as well as dependence across measurements in space; hence, the investigation of multivariate spatial processes. Also, in many, arguably most, practical situations, we work with spatial data layers at different scales, e.g., at county level, post- or zipcode level, census units, and points (possibly different sets of point for different layers). The question of interpolation across scales within a layer and linking these layers to develop a regression model provides the need for approaches to handle misalignment and the modifiable areal unit problem. Somewhat connected to this is the issue of sensitivity of inference to spatial scale; dependence at one spatial scale can look and be modeled very differently from that at another scale. Furthermore, spatial scaling of inference must be undertaken with considerable care. Therefore, we consider the matter of spatial aggregation and disaggregation and its impact, usually referred to as ecological bias. Finally, in many situations there also may be interest in gradients associated with spatial surfaces, raising interesting analytical challenges when such surfaces are realizations of a stochastic process rather than explicit functions. A related problem arises from interest in taking a spatial data layer and imposing boundaries on it. Such boundaries are not viewed as administrative boundaries, but rather boundaries that reflect sharp changes or steep gradients in the spatial surface. This topic is known as wombling or boundary analysis and it is addressed in the last chapter of this handbook.

Chapter 28 takes up the general topic of multivariate spatial process models. Gelfand and Banerjee review the theoretical issues associated with this problem, particularly the development of valid cross-covariance functions. They discuss both theoretically motivated specifications as well as constructive approaches for these functions. In addition to multivariate response spatial regression models, an attractive application is to the setting of spatially varying coefficient models. Multivariate processes over time are also discussed and a challenging example is presented. In Chapter 29, Gelfand takes up the modifiable areal unit problem, reviewing the history in this area but focusing on model-based solutions along with the computational challenges. Chapter 30 addresses the ecological fallacy. This is a somewhat subtle matter and Wakefield and Lyons point out the variety of biases and potential misinterpretations that can be introduced if modeling is not carefully conceived. Finally, in Chapter 31, Banerjee deals with gradients and wombling, building the development of gradients to spatial surfaces from the idea of finite differences, formalizing gradient processes with associated covariance structure induced by the process model for the surface itself. Then, boundary analysis is discussed—for point referenced data through the use of polygonal curves to be employed as boundaries, for areal unit data through random entries in the proximity matrix.

28

Multivariate Spatial Process Models

Alan E. Gelfand and Sudipto Banerjee

CONTENTS

28.1 Introduction

Increasingly in spatial data settings there is need for analyzing multivariate measurements obtained at spatial locations. Such data settings arise when several spatially dependent response variables are recorded at each spatial location. A primary example is data taken at environmental monitoring stations where measurements on levels of several pollutants (e.g., ozone, $PM_{2.5}$, nitric oxide, carbon monoxide, etc.) would typically be measured. In atmospheric modeling, at a given site we may observe surface temperature, precipitation, and wind speed. In a study of ground level effects of nuclear explosives, soil and vegetation contamination in the form of plutonium and americium concentrations at sites have been collected. In examining commercial real estate markets, for an individual property at a given location, data include both selling price and total rental income. In forestry, investigators seek to produce spatially explicit predictions of multiple forest attributes using a multisource forest inventory approach. In each of these illustrations, we anticipate both dependence between measurements at a particular location, and association between measurements across locations.

In this chapter we focus on point-referenced spatial data.* To develop multivariate spatial process models for inference about parameters or for interpolation requires specification of either a valid cross-variogram or a valid cross-covariance function. Such specification with associated modeling detail and spatial interpolation is the primary focus for the sequel.

Cross-covariance functions are not routine to specify since they demand that for any number of locations and any choice of these locations the resulting covariance matrix for the associated data be positive definite. Various constructions are possible. As we shall see, separable forms are the easiest way to begin.

Another possibility is the moving average approach of Ver Hoef and Barry (1998). The technique is also called kernel convolution and is a well-known approach for creating rich classes of stationary and nonstationary spatial processes, as discussed in Higdon, Lee, and Holloman (2003) and Higdon, Swall, and Kern (1999). Yet another possibility would attempt a multivariate version of local stationarity, extending ideas in Fuentes and Smith (2001). Building upon ideas in Gaspari and Cohn (1999) and Majumdar and Gelfand (2007) use convolution of covariance functions to produce valid multivariate cross-covariance functions. An attractive, easily interpreted, flexible approach develops versions of the linear model of coregionalization (LMC) as in, e.g., Banerjee, Carlin, and Gelfand (2004); Gelfand, Schmidt, Banerjee, and Sirmans (2004); Grzebyk and Wackernagel (1994); Schmidt and Gelfand (2003); Wackernagel (2003).

Inference typically proceeds along one of three paths: somewhat informally, using, e.g., empirical covariogram/covariance estimates and least squares (with plug-in estimates for unknown parameters), using likelihood methods (with concerns regarding suitability of infill or increasing domain asymptotics), and within a fully Bayesian framework (requiring demanding computation).

In Section 28.2 we will review classical multivariate geostatistics. In Section 28.3 we develop some theory for cross-covariance functions. Section 28.4 focuses on separable cross-covariance functions while Section 28.5 takes up co-regionalization with an example. Section 28.9 elaborates alternative strategies for building valid cross-covariance functions. We conclude with a brief Section 28.10 discussing multivariate space–time data models.

28.2 Classical Multivariate Geostatistics

Classical multivariate geostatistics begins, as with much of geostatistics, with early work of Matheron (1973, 1979). The basic ideas here include cross-variograms and cross-covariance functions, intrinsic co-regionalization, and cokriging. The emphasis is on prediction. A thorough discussion of the work in this area is provided in Wackernagel (2003). See also Chiles and Delfiner (1999).

Consider $\mathbf{Y}(\mathbf{s})$, a $p \times 1$ vector where $\mathbf{s} \in \mathcal{D}$. We seek to capture the association both within components of $\mathbf{Y}(\mathbf{s})$ and across $\mathbf{s}$. The joint second-order (weak) stationarity hypothesis defines the cross-variogram as

$$\gamma_{ij}(\mathbf{h}) = \frac{1}{2} E(Y_i(\mathbf{s}+\mathbf{h}) - Y_i(\mathbf{s}))(Y_j(\mathbf{s}+\mathbf{h}) - Y_j(\mathbf{s})). \tag{28.1}$$

* With, say, regular lattice data or with areal unit data, we might instead consider multivariate random field models. For the latter, there exists recent literature on multivariate conditionally autoregressive models building on the work of Mardia (1988). See, e.g., Gelfand and Vounatsou (2003) and Jin, Banerjee, and Carlin (2007) for more current discussion. See also, Chapter 14.

Implicitly, we assume $E(Y(\mathbf{s}+\mathbf{h}) - Y(\mathbf{s})) = 0$ for all $\mathbf{s}$ and $\mathbf{s}+\mathbf{h} \in \mathcal{D}$. $\gamma_{ij}(\mathbf{h})$ is obviously an even function and, using the Cauchy–Schwarz inequality, satisfies $|\gamma_{ij}(\mathbf{h})|^2 \leq \gamma_{ii}(\mathbf{h})\gamma_{jj}(\mathbf{h})$.

The cross-covariance function is defined as

$$C_{ij}(\mathbf{h}) = E[(Y_i(\mathbf{s}+\mathbf{h}) - \mu_i)(Y_j(\mathbf{s}) - \mu_j)], \tag{28.2}$$

where we remark that, here, a constant mean μ_i is assumed for component $Y_i(\mathbf{s})$. Note that the cross-covariance function satisfies $|C_{ij}(\mathbf{h})|^2 \leq C_{ii}(\mathbf{0})C_{jj}(\mathbf{0})$, but $|C_{ij}(\mathbf{h})|$ need not be $\leq C_{ij}(\mathbf{0})$. In fact, $|C_{ij}(\mathbf{h})|$ need not be $\leq |C_{ij}(\mathbf{0})|$ because the maximum value of $C_{ij}(\mathbf{h})$ need not occur at $\mathbf{0}$.* Similarly, $|C_{ij}(\mathbf{h})|^2$ need not be $\leq C_{ii}(\mathbf{h})C_{jj}(\mathbf{h})$. The corresponding matrix $\mathbf{C}(\mathbf{h})$ of direct and cross-covariances (with $C_{ij}(\mathbf{h})$ as its (i, j)th element) need not be positive definite at any $\mathbf{h}$ though as $\mathbf{h} \to \mathbf{0}$, it converges to a positive definite matrix, the (local) covariance matrix associated with $\mathbf{Y}(\mathbf{s})$.

We can make the familiar connection between the cross-covariance and the cross-variogram. The former determines the latter and we can show that

$$\gamma_{ij}(\mathbf{h}) = C_{ij}(\mathbf{0}) - \frac{1}{2}(C_{ij}(\mathbf{h}) + C_{ij}(-\mathbf{h})). \tag{28.3}$$

Indeed, decomposing $C_{ij}(\mathbf{h})$ as $\frac{1}{2}(C_{ij}(\mathbf{h}) + C_{ij}(-\mathbf{h})) + \frac{1}{2}(C_{ij}(\mathbf{h}) - C_{ij}(-\mathbf{h}))$, we see that the cross-variogram only captures the even term of the cross-covariance function, suggesting that it may be inadequate in certain modeling situations. Such concerns led to the proposal of the pseudo cross-variogram (Clark, Basinger, and Harper, 1989; Cressie, 1993; Myers, 1991). In particular, Clark et al. (1989) suggested $\pi^c_{ij}(\mathbf{h}) = E(Y_i(\mathbf{s}+\mathbf{h}) - Y_j(\mathbf{h}))^2$ and Myers (1991) suggested a mean-corrected version, $\pi^m_{ij}(\mathbf{h}) = \text{var}(Y_i(\mathbf{s}+\mathbf{h}) - Y_j(\mathbf{h}))$. It is easy to show that $\pi^c_{ij}(\mathbf{h}) = \pi^m_{ij}(\mathbf{h}) + (\mu_i - \mu_j)^2$. The pseudo cross-variogram is not constrained to be an even function. However, the assumption of stationary cross-increments is unrealistic, certainly with variables measured on different scales and even with rescaling of the variables. A further limitation is the restriction of the pseudo cross-variogram to be positive. Despite the unattractiveness of "apples and oranges" comparison across components, Cressie and Wikle (1998) report successful cokriging using $\pi^m_{ij}(\mathbf{h})$.

28.2.1 Cokriging

Cokriging is spatial prediction at a new location that uses not only information from direct measurement of the spatial component process being considered, but also information from the measurements of the other component processes. Journel and Huijbregts (1978) and Matheron (1973) present early discussion, while Myers (1982) presents a general matrix development. Corsten (1989) and Stein and Corsten (1991) frame the development in the context of linear regression. Detailed reviews are presented in Chilès and Delfiner (1999) and Wackernagel (2003).

Myers (1982) points out the distinction between prediction of a single variable as above and joint prediction of several variables at a new location. In fact, suppose we start with the joint second order stationarity model in (28.1) above and we seek to predict, say, $Y_1(\mathbf{s}_0)$, i.e., the first component of $\mathbf{Y}(\mathbf{s})$ at a new location $\mathbf{s}_0$. An unbiased estimator based upon $\mathbf{Y} = (\mathbf{Y}(\mathbf{s}_1), \mathbf{Y}(\mathbf{s}_2), \ldots, \mathbf{Y}(\mathbf{s}_n))^T$ would take the form $\hat{Y}_1(\mathbf{s}_0) = \sum_{i=1}^n \sum_{l=1}^p \lambda_{il} Y_l(\mathbf{s}_i)$ where we have the constraints that $\sum_{i=1}^n \lambda_{il} = 0, l \neq 1, \sum_{i=1}^n \lambda_{i1} = 1$. On the other hand, if we sought

* We can illustrate this simply through the so-called delay effect (Wackernagel, 2003). Suppose, for instance, $p = 2$ and $Y_2(\mathbf{s}) = aY_1(\mathbf{s} + \mathbf{h}_0) + \epsilon(\mathbf{s})$ where $Y_1(\mathbf{s})$ is a spatial process with stationary covariance function $C(\mathbf{h})$, and $\epsilon(\mathbf{s})$ is a pure error process with variance τ^2. Then, the associated cross covariance function has $C_{11}(\mathbf{h}) = C(\mathbf{h})$, $C_{22}(\mathbf{h}) = a^2C(\mathbf{h})$ and $C_{12} = C(\mathbf{h} + \mathbf{h}_0)$. We note that delay effect process models find application in atmospheric science settings, such as the prevailing direction of weather fronts.

to predict $\mathbf{Y}(\mathbf{s}_0)$, we would now write $\hat{\mathbf{Y}}(\mathbf{s}_0) = \sum_{i=1}^n \Lambda_i \mathbf{Y}(\mathbf{s}_i)$. The unbiasedness condition becomes $\sum_{i=1}^n \Lambda_i = I$. Moreover, now, what should we take as the "optimality" condition? One choice is to choose the set $\{\Lambda_{0i}, 1 = 1, 2, \ldots, n\}$ with associated estimator $\hat{\mathbf{Y}}_0(\mathbf{s}_0)$ such that for any other unbiased estimator, $\tilde{\mathbf{Y}}(\mathbf{s}_0)$, $E(\tilde{\mathbf{Y}}(\mathbf{s}_0) - \mathbf{Y}(\mathbf{s}_0))(\tilde{\mathbf{Y}}(\mathbf{s}_0) - \mathbf{Y}(\mathbf{s}_0))^T - E(\hat{\mathbf{Y}}_0(\mathbf{s}_0) - \mathbf{Y}(\mathbf{s}_0))(\hat{\mathbf{Y}}_0(\mathbf{s}_0) - \mathbf{Y}(\mathbf{s}_0))^T$ is nonnegative definite (Ver Hoef and Cressie, 1993). Myers (1982) suggests minimizing $\text{tr}E(\hat{\mathbf{Y}}(\mathbf{s}_0) - \mathbf{Y}(\mathbf{s}_0))(\hat{\mathbf{Y}}(\mathbf{s}_0) - \mathbf{Y}(\mathbf{s}_0))^T = E(\hat{\mathbf{Y}}(\mathbf{s}_0) - \mathbf{Y}(\mathbf{s}_0))^T(\hat{\mathbf{Y}}(\mathbf{s}_0) - \mathbf{Y}(\mathbf{s}_0))$.

Returning to the individual prediction case, minimization of predictive mean square error, $E(Y_1(\mathbf{s}_0) - \hat{Y}_1(\mathbf{s}_0))^2$ amounts to a quadratic optimization subject to linear constraints and the solution can be obtained using Lagrange multipliers. As in the case of univariate kriging (see Chapter 3), the solution can be written as a function of a cross-variogram specification. In fact, Ver Hoef and Cress (1993) show that $\pi_{ij}(\mathbf{h})$ above emerges in computing predictive mean square error, suggesting that it is a natural cross-variogram for cokriging. But, altogether, given the concerns noted regarding $\gamma_{ij}(\mathbf{h})$ and $\pi_{ij}(\mathbf{h})$, it seems preferable (and most writers seem to agree) to assume the existence of second moments for the multivariate process, captured through a *valid* cross-covariance function, and to use it with regard to prediction. The next section discusses validity of cross-covariance functions in some detail. In this regard, the definition of a *valid* cross-variogram seems a bit murky. In Wackernagel (2003), it is induced by a valid cross-covariance function (as above). In Myers (1982) and Ver Hoef and Cressie (1993) the second-order stationarity above is assumed, but also a finite cross-covariance is assumed in order to bring $\gamma_{ij}(\mathbf{h})$ into the optimal cokriging equations. Rehman and Shapiro (1996) introduce the definition of a *permissible* cross-variogram requiring (i) the $\gamma(\mathbf{h})$ are continuous except possibly at the origin, (ii) $\gamma_{ij}(\mathbf{h}) \geq 0, \forall\, \mathbf{h} \in \mathcal{D}$, (iii) $\gamma_{ij}(\mathbf{h}) = \gamma(-\mathbf{h}), \forall\, \mathbf{h} \in \mathcal{D}$, and (iv) the functions, $-\gamma_{ij}(\mathbf{h})$, are conditionally nonnegative definite, the usual condition for individual variograms.

In fact, we can directly obtain the explicit solution to the individual cokriging problem if we assume a multivariate Gaussian spatial process. As we clarify below, such a process specification only requires supplying mean surfaces for each component of $\mathbf{Y}(\mathbf{s})$ and a valid cross-covariance function. For simplicity, assume $\mathbf{Y}(\mathbf{s})$ is centered to have mean $\mathbf{0}$. The cross-covariance function provides $\Sigma_{\mathbf{Y}}$, the $np \times np$ covariance matrix for the data $\mathbf{Y} = (\mathbf{Y}(\mathbf{s}_1)^T, \mathbf{Y}(\mathbf{s}_2)^T, \ldots, \mathbf{Y}(\mathbf{s}_n)^T)^T$. In addition, it provides the $np \times 1$ vector, $\mathbf{c}_0$, which is blocked as vectors $\mathbf{c}_{0j}$, $j = 1, 2, .., n$ with lth element $c_{0j,l} = \text{Cov}(Y_1(\mathbf{s}_0), Y_l(\mathbf{s}_j))$. Then, from the multivariate normal distribution of $\mathbf{Y}$, $Y_1(\mathbf{s}_0)$, we obtain the cokriging estimate,

$$E(Y_1(\mathbf{s}_0)|\mathbf{Y}) = \mathbf{c}_0^T \Sigma_{\mathbf{Y}}^{-1}\mathbf{Y}. \tag{28.4}$$

The associated variance, $\text{var}(Y_1(\mathbf{s}_0)|\mathbf{Y})$ is also immediately available, i.e., $\text{var}(Y_1(\mathbf{s}_0)|\mathbf{Y}) = C_{11}(\mathbf{0}) - \mathbf{c}_0^T \Sigma_{\mathbf{Y}}^{-1}\mathbf{c}_0$.

In particular, consider the special case of the $p \times p$ cross-covariance matrix, $\mathbf{C}(\mathbf{h}) = \rho(\mathbf{h})\mathbf{T}$, where $\rho(\cdot)$ is a valid correlation function and $\mathbf{T}$ is the local positive definite covariance matrix. Then, $\Sigma_{\mathbf{Y}} = \mathbf{R} \otimes \mathbf{T}$, where $\mathbf{R}$ is the $n \times n$ matrix with (i, j)th entry $\rho(\mathbf{s}_i - \mathbf{s}_j)$ and $\otimes$ denotes the Kronecker product. This specification also yields $\mathbf{c}_0 = \mathbf{r}_0 \otimes \mathbf{t}_{*1}$, where $\mathbf{r}_0$ is $n \times 1$ with entries $\rho(\mathbf{s}_0 - \mathbf{s}_j)$ and $\mathbf{t}_{*1}$ is the first column of $\mathbf{T}$. Then, Equation (28.4) becomes $t_{11}\mathbf{r}_0^T\mathbf{R}^{-1}\tilde{\mathbf{Y}}_1$ where t_{11} is the $(1, 1)$th element of $\mathbf{T}$ and $\tilde{\mathbf{Y}}_1$ is the vector of observations associated with the first component of the $\mathbf{Y}(\mathbf{s}_j)$s. This specification is known as the *intrinsic* multivariate correlation and is discussed in greater generality in Section 28.2.2. In other words, under an intrinsic specification, only observations on the first component are used to predict the first component at a new location. See Helterbrand and Cressie (1994), Wackernagel (1994), and Wackernagel, 2003, in this regard.

In all of the foregoing work, inference assumes the cross-covariance or the cross-variogram to be known. In practice, a parametric model is adopted and data-based estimates of the parameters are plugged in. A related issue here is whether the data are available for each variable at all sampling points (so-called *isotopy*—not to be confused with "isotropy"), some variables share some sample locations (partial *heterotopy*), or the variables have no sample locations in common (entirely heterotopic). (See Chapter 29 in this regard.) Similarly, in the context of prediction, if any of the $Y_l(\mathbf{s}_0)$ are available to help predict $Y_1(\mathbf{s}_0)$, we refer to this as "collocated cokriging." The challenge with heterotopy in classical work is that the empirical cross-variograms cannot be computed and empirical cross-covariances, though they can be computed, do not align with the sampling points used to compute the empirical direct covariances. Furthermore, the value of the cross-covariances at $\mathbf{0}$ cannot be computed.*

28.2.2 Intrinsic Multivariate Correlation and Nested Models

Recall that one way to develop a spatial model is through *structural analysis*. Such analysis usually suggests more than one variogram model, i.e., proposes a *nested* variogram model (Grzebyk and Wackernagel, 1994; Wackernagel, 2003), which we might write as $\gamma(\mathbf{h}) = \sum_{r=1}^{m} t_r \gamma_r(\mathbf{h})$. For instance, with three spatial scales, corresponding to a nugget, fine-scale dependence, and long-range dependence, respectively, we might write $\gamma(\mathbf{h}) = t_1\gamma_1(\mathbf{h}) + t_2\gamma_2(\mathbf{h}) + t_3\gamma_3(\mathbf{h})$, where $\gamma_1(\mathbf{h}) = 0$ if $|\mathbf{h}| = 0$, $= 1$ if $|\mathbf{h}| > 0$, while $\gamma_2(\cdot)$ reaches a sill equal to 1 very rapidly and $\gamma_3(\cdot)$ reaches a sill equal to 1 much more slowly.

Note that the nested variogram model corresponds to the spatial process $\sqrt{t_1}w_1(\mathbf{s}) + \sqrt{t_2}w_2(\mathbf{s}) + \sqrt{t_3}w_3(\mathbf{s})$—a linear combination of independent processes. Can this same idea be used to build a multivariate version of a nested variogram model? Journel and Huijbregts (1978) propose to do this using the specification $w_l(\mathbf{s}) = \sum_{r=1}^{m}\sum_{j=1}^{p} a_{rj}^{(l)} w_{rj}(\mathbf{s})$ for $l = 1, \ldots, p$. Here, the $w_{rj}(\mathbf{s})$ are such that they are independent process replicates across j and, for each r, the process has correlation function $\rho_r(\mathbf{h})$ and variogram $\gamma_r(\mathbf{h})$ (with sill 1). In the case of isotropic ρs, this implies that we have a different range for each r, but a common range for all components given r.

The representation in terms of independent processes can now be given in terms of the $p \times 1$ vector process $\mathbf{w}(\mathbf{s}) = [w_l(\mathbf{s})]_{l=1}^{p}$, formed by collecting the $w_l(\mathbf{s})$s into a column for $l = 1, \ldots, p$. We write the above linear specification as $\mathbf{w}(\mathbf{s}) = \sum_{r=1}^{m} \mathbf{A}_r \mathbf{w}_r(\mathbf{s})$, where each $\mathbf{A}_r$ is a $p \times p$ matrix with (l, j)th element $a_{rj}^{(l)}$ and $\mathbf{w}_r(\mathbf{s}) = (w_{r1}(\mathbf{s}), \ldots, w_{rp}(\mathbf{s}))^T$ are $p \times 1$ vectors that are independent replicates from a spatial process with correlation function $\rho_r(\mathbf{h})$ and variogram $\gamma_r(\mathbf{h})$ for $r = 1, 2, \ldots, p$.

Letting $\mathbf{C}_r(\mathbf{h})$ be the $p \times p$ cross-covariance matrix and $\Gamma_r(\mathbf{h})$ denote the $p \times p$ matrix of direct and cross-variograms associated with $\mathbf{w}(\mathbf{s})$, we have $\mathbf{C}_r(\mathbf{h}) = \rho_r(\mathbf{h})\mathbf{T}_r$ and $\Gamma_r(\mathbf{h}) = \gamma_r(\mathbf{h})\mathbf{T}_r$. Here, $\mathbf{T}_r$ is positive definite with $\mathbf{T}_r = \mathbf{A}_r\mathbf{A}_r^T = \sum_{j=1}^{p} \mathbf{a}_{rj}\mathbf{a}_{rj}^T$, where $\mathbf{a}_{rj}$ is the jth column vector of $\mathbf{A}_r$. Finally, the cross-covariance and cross-variogram nested model representations take the form $\mathbf{C}(\mathbf{h}) = \sum_{r=1}^{m} \rho_r(\mathbf{h})\mathbf{T}_r$ and $\Gamma(\mathbf{h}) = \sum_{r=1}^{m} \gamma_r(\mathbf{h})\mathbf{T}_r$.

The case $m = 1$ is called the intrinsic correlation model, the case $m > 1$ is called the intrinsic multivariate correlation model. Work (Vargas-Guzmán, Warrick, and Myers, 2002) allows the $w_{rj}(\mathbf{s})$ to be dependent.

Again, such modeling is natural when scaling is the issue, i.e., we want to introduce spatial effects to capture dependence at different scales (and, thus, m has nothing to do with p). If we have knowledge about these scales a priori, such modeling will be successful. However, to find datasets that inform about such scaling may be less successful. In different words, usually m will be small since, given m, mp process realizations are introduced.

* The empirical cross-variogram imitates the usual variogram (Chapter 3), creating bins and computing averages of cross products of differences within the bins. Similar words apply to the empirical cross-covariance.

28.3 Some Theory for Cross-Covariance Functions

In light of the critical role of cross-covariance functions, we provide some formal theory regarding the validity and properties of these functions. Let $\mathcal{D} \subset \Re^d$ be a connected subset of the d-dimensional Euclidean space and let $\mathbf{s} \in \mathcal{D}$ denote a generic point in $\mathcal{D}$. Consider a vector-valued spatial process $\{\mathbf{w}(\mathbf{s}) \in \Re^m : \mathbf{s} \in \mathcal{D}\}$, where $\mathbf{w}(\mathbf{s}) = [w_j(\mathbf{s})]_{j=1}^p$ is a $p \times 1$ vector. For convenience, assume that $E[\mathbf{w}(\mathbf{s})] = \mathbf{0}$. The *cross-covariance function* is a matrix-valued function, say $\mathbf{C}(\mathbf{s}, \mathbf{s}')$, defined for any pair of locations $(\mathbf{s}, \mathbf{s}') \in \mathcal{D} \times \mathcal{D}$ and yielding the $p \times p$ matrix whose (j, j')th element is $\text{Cov}(w_j(\mathbf{s}), w_{j'}(\mathbf{s}'))$:

$$\mathbf{C}(\mathbf{s}, \mathbf{s}') = \text{Cov}(\mathbf{w}(\mathbf{s}), \mathbf{w}(\mathbf{s}')) = [\text{Cov}(w_j(\mathbf{s}), w_{j'}(\mathbf{s}'))]_{j,j'=1}^p = E[\mathbf{w}(\mathbf{s})\mathbf{w}^T(\mathbf{s}')]. \tag{28.5}$$

The cross-covariance function completely determines the joint dispersion structure implied by the spatial process. To be precise, for any n and any arbitrary collection of sites $\mathcal{S} = \{\mathbf{s}_1, \ldots, \mathbf{s}_n\}$, the $np \times 1$ vector of realizations $\mathbf{w} = [\mathbf{w}(\mathbf{s}_j)]_{j=1}^n$ will have the variance–covariance matrix given by $\Sigma_{\mathbf{w}} = [\mathbf{C}(\mathbf{s}_i, \mathbf{s}_j)]_{i,j=1}^n$, where $\Sigma_{\mathbf{w}}$ is an $nm \times nm$ block matrix whose (i, j)th block is precisely the $p \times p$ cross-covariance function $\mathbf{C}(\mathbf{s}_i, \mathbf{s}_j)$. Since $\Sigma_{\mathbf{w}}$ must be symmetric and positive definite, it is immediate that the cross-covariance function must satisfy the following two conditions:

$$\mathbf{C}(\mathbf{s}, \mathbf{s}') = \mathbf{C}^T(\mathbf{s}', \mathbf{s}) \tag{28.6}$$

$$\sum_{i=1}^n \sum_{j=1}^n \mathbf{x}_i^T \mathbf{C}(\mathbf{s}_i, \mathbf{s}_j)\mathbf{x}_j > 0 \quad \forall \quad \mathbf{x}_i, \mathbf{x}_j \in \Re^p \backslash \{\mathbf{0}\}. \tag{28.7}$$

The first condition ensures that, while the cross-covariance function need not itself be symmetric, $\Sigma_{\mathbf{w}}$ is. The second condition ensures the positive definiteness of $\Sigma_{\mathbf{w}}$ and is, in fact, quite stringent; it must hold for all integers n and any arbitrary collection of sites $\mathcal{S} = \{\mathbf{s}_1, \ldots, \mathbf{s}_n\}$. Note that conditions (28.6) and (28.7) imply that $\mathbf{C}(\mathbf{s}, \mathbf{s})$ is a symmetric and positive definite function. In fact, it is precisely the variance–covariance matrix for the elements of $\mathbf{w}(\mathbf{s})$ within site $\mathbf{s}$.

We say that $\mathbf{w}(\mathbf{s})$ is *stationary* if $\mathbf{C}(\mathbf{s}, \mathbf{s}') = \mathbf{C}(\mathbf{s}' - \mathbf{s})$, i.e., the cross-covariance function depends only upon the separation of the sites, while we say that $\mathbf{w}(\mathbf{s})$ is *isotropic* if $\mathbf{C}(\mathbf{s}, \mathbf{s}') = \mathbf{C}(\|\mathbf{s}' - \mathbf{s}\|)$, i.e., the cross-covariance function depends only upon the distance between the sites. Note that for stationary processes we write the cross-covariance function as $\mathbf{C}(\mathbf{h}) = \mathbf{C}(\mathbf{s}, \mathbf{s} + \mathbf{h})$. From Equation (28.6) it immediately follows that

$$\mathbf{C}(-\mathbf{h}) = \mathbf{C}(\mathbf{s} + \mathbf{h}, \mathbf{s}) = \mathbf{C}^T(\mathbf{s}, \mathbf{s} + \mathbf{h}) = \mathbf{C}^T(\mathbf{h}).$$

Thus, for a stationary process, a symmetric cross-covariance function is equivalent to having $\mathbf{C}(-\mathbf{h}) = \mathbf{C}(\mathbf{h})$ (i.e., even function). For isotropic processes,

$$\mathbf{C}(\mathbf{h}) = \mathbf{C}(\|\mathbf{h}\|) = \mathbf{C}(\| - \mathbf{h}\|) = \mathbf{C}(-\mathbf{h}) = \mathbf{C}^T(\mathbf{h}),$$

hence, the cross-covariance function is even and the matrix is necessarily symmetric.

The primary characterization theorem for cross-covariance functions (Cramér, 1940; Yaglom, 1987) says that real-valued functions, say $C_{ij}(\mathbf{h})$, will form the elements of a valid cross-covariance matrix $\mathbf{C}(\mathbf{h}) = [C_{ij}(\mathbf{h})]_{i,j=1}^p$ if and only if each $C_{ij}(\mathbf{h})$ has the cross-spectral representation

$$C_{ij}(\mathbf{h}) = \int \exp(2\pi i \mathbf{t}^T \mathbf{h}) d(F_{ij}(\mathbf{t})), \tag{28.8}$$

with respect to a positive definite measure $F(\cdot)$, i.e., where the cross-spectral matrix $M(B) = [F_{ij}(B)]_{i,j=1}^{p}$ is positive definite for any Borel subset $B \subseteq \Re^d$. The representation in Equation (28.8) can be considered the most general representation theorem for cross-covariance functions. It is the analog of Bochner's theorem for covariance functions and has been employed by several authors to construct classes of cross-covariance functions. Essentially, one requires a choice of the $F_{ij}(\mathbf{t})$s. Matters simplify when $F_{ij}(\mathbf{h})$ is assumed to be square integrable ensuring that a spectral density function $f_{ij}(\mathbf{t})$ exists such that $d(F_{ij}(\mathbf{t})) = f_{ij}(\mathbf{t})d\mathbf{t}$. Now one simply needs to ensure that $[f_{ij}(\mathbf{t})]_{i,j=1}^{p}$ are positive definite for all $\mathbf{t} \in \Re^d$. Corollaries of the above representation lead to the approaches proposed in Gaspari and Cohn (1999) and Majumdar and Gelfand (2007) for constructing valid cross-covariance functions as convolutions of covariance functions of stationary random fields (see Section 28.9.3). For isotropic settings, we use the notation $||\mathbf{s}' - \mathbf{s}||$ for the distance between sites $\mathbf{s}$ and $\mathbf{s}'$. The representation in Equation (28.8) can be viewed more broadly in the sense that, working in the complex plane, if the matrix valued measure $M(\cdot)$ is Hermitian nonnegative definite, then we obtain a valid cross-covariance matrix in the complex plane. Rehman and Shapiro (1996) use this broader definition to obtain permissible cross variograms. Grzebyk and Wackernagel (1994) employ the induced complex covariance function to create a bilinear model of co-regionalization.

From a modeling perspective, it is often simpler to rewrite the cross-covariance matrix as $\mathbf{C}(\mathbf{s}, \mathbf{s}') = \mathbf{A}(\mathbf{s})\Theta(\mathbf{s}, \mathbf{s}')\mathbf{A}^T(\mathbf{s}')$. $\Theta(\mathbf{s}, \mathbf{s}')$ is called the *cross-correlation* function, which must not only satisfy Equation (28.6) and Equation (28.7), but in addition satisfies $\Theta(\mathbf{s}, \mathbf{s}) = I_p$. Therefore, $\mathbf{C}(\mathbf{s}, \mathbf{s}) = \mathbf{A}(\mathbf{s})\mathbf{A}^T(\mathbf{s})$ and $\mathbf{A}(\mathbf{s})$ identifies with the square root of the *within-site* dispersion matrix $\mathbf{C}(\mathbf{s}, \mathbf{s})$. Note that whenever $\Theta(\mathbf{s}, \mathbf{s}')$ is symmetric so is $\mathbf{C}(\mathbf{s}, \mathbf{s}')$ and also, if $\Theta(\mathbf{s}, \mathbf{s}')$ is positive, so is $\mathbf{C}(\mathbf{s}, \mathbf{s}')$.

For modeling $\mathbf{A}(\mathbf{s})$, without loss of generality, one can assume that $\mathbf{A}(\mathbf{s}) = \mathbf{C}^{1/2}(\mathbf{s}, \mathbf{s})$ is a lower-triangular square root; the one-to-one correspondence between the elements of $\mathbf{A}(\mathbf{s})$ and $\mathbf{C}(\mathbf{s}, \mathbf{s})$ is well known (see, e.g., Harville, 1997, p. 229).* Thus, $\mathbf{A}(\mathbf{s})$ determines the association between the elements of $\mathbf{w}(\mathbf{s})$ at location $\mathbf{s}$. If this association is assumed to not vary with $\mathbf{s}$, we have $\mathbf{A}(\mathbf{s}) = \mathbf{A}$, which results in a weakly stationary cross-covariance matrix with $\mathbf{A}\mathbf{A}^T = \mathbf{C}(\mathbf{0})$. In practice, the matrix $\mathbf{A}(\mathbf{s})$ will be unknown. Classical likelihood-based methods for estimating $\mathbf{A}(\mathbf{s})$ are usually difficult, although under stationarity an EM algorithm can be devised (see Zhang, 2006). For greater modeling flexibility, a Bayesian framework is often adopted. Here $\mathbf{A}(\mathbf{s})$ is assigned a prior specification and sampling-based methods are employed to obtain posterior samples for these parameters (see, e.g., Banerjee, Carlin, and Gelfand, 2004). A very general approach for such estimation has been laid out in Gelfand, Schmidt, Banerjee, and Sirmans (2004), where an inverse spatial-Wishart process for $\mathbf{A}(\mathbf{s})\mathbf{A}^T(\mathbf{s})$ is defined. Other approaches include using parametric association structures suggested by the design under consideration (e.g., in Banerjee and Johnson, 2006) or some simplifying assumptions.

Alternatively, we could adopt a spectral square root specification for $\mathbf{A}(\mathbf{s})$, setting $\mathbf{A}(\mathbf{s}) = \mathbf{P}(\mathbf{s})\Lambda^{1/2}(\mathbf{s})$, where $\mathbf{C}(\mathbf{s}, \mathbf{s}) = \mathbf{P}(\mathbf{s})\Lambda\mathbf{P}^T(\mathbf{s})$ is the spectral decomposition for $\mathbf{C}(\mathbf{s}, \mathbf{s})$. We can further parameterize the $p \times p$ orthogonal matrix function $\mathbf{P}(\mathbf{s})$ in terms of the $p(p-1)/2$ *Givens* angles $\theta_{ij}(\mathbf{s})$ for $i = 1, \ldots, p-1$ and $j = i+1, \ldots, p$ (following Cressie and Wikle, 1998). Specifically, $\mathbf{P}(\mathbf{s}) = \prod_{i=1}^{p-1}\prod_{j=i+1}^{p}\mathbf{G}_{ij}(\theta_{ij}(\mathbf{s}))$ where i and j are distinct and $\mathbf{G}_{ij}(\theta_{ij}(\mathbf{s}))$ is the $p \times p$ identity matrix with the ith and jth diagonal elements replaced by $\cos(\theta_{ij}(\mathbf{s}))$, and the (i, j)th and (j, i)th elements replaced by $\pm\sin(\theta_{ij}(\mathbf{s}))$, respectively. Given $\mathbf{P}(\mathbf{s})$ for any $\mathbf{s}$, the $\theta_{ij}(\mathbf{s})$s are unique within range $(-\pi/2, \pi/2)$. These may be further modeled by means of Gaussian processes on a suitably transformed function, say, $\tilde{\theta}_{ij}(\mathbf{s}) = \log(\frac{\pi/2+\theta_{ij}(\mathbf{s})}{\pi/2-\theta_{ij}(\mathbf{s})})$.

* Indeed, to ensure the one-to-one correspondence, we must insist that the diagonal elements of $\mathbf{A}(\mathbf{s})$ are greater than 0. This has obvious implications for prior specification in a Bayesian hierarchical modeling setting.

28.4 Separable Models

A widely used specification is the separable model

$$\mathbf{C}(\mathbf{s}, \mathbf{s}') = \rho(\mathbf{s}, \mathbf{s}')\mathbf{T}, \tag{28.9}$$

where $\rho(\cdot)$ is a valid (univariate) correlation function and $\mathbf{T}$ is a $p \times p$ positive definite matrix. Here, $\mathbf{T}$ is the nonspatial or "local" covariance matrix while ρ controls spatial association based upon proximity. In fact, under weak stationarity, $\mathbf{C}(\mathbf{s}, \mathbf{s}') = \rho(\mathbf{s} - \mathbf{s}'; \boldsymbol{\theta})\mathbf{C}(\mathbf{0})$. It is easy to verify that, for $\mathbf{Y} = (Y(\mathbf{s}_1), \ldots, Y(\mathbf{s}_n))^T$, $\Sigma_{\mathbf{Y}} = \mathbf{R} \otimes \mathbf{T}$, where $\mathbf{R}_{ij} = \rho(\mathbf{s}_i, \mathbf{s}_j)$ and $\otimes$ is the Kronecker product. Clearly, $\Sigma_{\mathbf{Y}}$ is positive definite since $\mathbf{R}$ and $\mathbf{T}$ are. In fact, $\Sigma_{\mathbf{Y}}$ is computationally convenient to work with since $|\Sigma_{\mathbf{Y}}| = |\mathbf{R}|^p\,|\mathbf{T}|^n$ and $\Sigma_{\mathbf{Y}}^{-1} = \mathbf{R}^{-1} \otimes \mathbf{T}^{-1}$. We note that Mardia and Goodall (1993) use separability for modeling multivariate spatio-temporal data so $\mathbf{T}$ arises from an autoregression (AR) model. We immediately see that this specification is the same as the intrinsic specification we discussed in Section 28.2.2. In the literature, the form (28.9) is called "separable" as it separates the component for spatial correlation, $\rho(\mathbf{s}, \mathbf{s}'; \boldsymbol{\theta})$, from that for association within a location, $\mathbf{C}(\mathbf{0})$. A limitation of (28.9) is that it is symmetric and, more critically, each component of $\mathbf{w}(\mathbf{s})$ shares the same spatial correlation structure.

The intrinsic or separable specification is the most basic co-regionalization model. Again, it arises as, say $\mathbf{Y}(\mathbf{s}) = \mathbf{A}\mathbf{w}(\mathbf{s})$, where, for our purposes, $\mathbf{A}$ is $p \times p$ full rank and the components of $\mathbf{w}(\mathbf{s})$ are iid spatial processes. If the $w_j(\mathbf{s})$ have mean 0 and are stationary with variance 1 and correlation function $\rho(\mathbf{h})$, then $E(\mathbf{Y}(\mathbf{s}))$ is $\mathbf{0}$ and the cross-covariance matrix, $\boldsymbol{\Sigma}_{\mathbf{Y}(\mathbf{s}),\mathbf{Y}(\mathbf{s}')} \equiv \mathbf{C}(\mathbf{s} - \mathbf{s}') = \rho(\mathbf{s} - \mathbf{s}')\mathbf{A}\mathbf{A}^T$, clarifying that $\mathbf{A}\mathbf{A}^T = \mathbf{T}$.

Inference based upon maximum likelihood is discussed in Mardia and Goodall (1993). Brown, Le, and Zidek (1994), working in the Bayesian framework, assign an inverse Wishart prior to $\Sigma_{\mathbf{Y}}$ centered around a separable specification. Hence, $\Sigma_{\mathbf{Y}}$ is immediately positive definite. It is also nonstationary since its entries are not even a function of the locations; we sacrifice connection with spatial separation vectors or distance. In fact, $\Sigma_{\mathbf{Y}}$ is not associated with a spatial process, but rather with a multivariate distribution.

The term "intrinsic" is usually taken to mean that the specification only requires the first and second moments of differences in measurement vectors and that the first moment difference is $\mathbf{0}$ and the second moments depend on the locations only through the separation vector $\mathbf{s} - \mathbf{s}'$. In fact, here $E[\mathbf{Y}(\mathbf{s}) - \mathbf{Y}(\mathbf{s}')] = \mathbf{0}$ and $\frac{1}{2}\boldsymbol{\Sigma}_{\mathbf{Y}(\mathbf{s})-\mathbf{Y}(\mathbf{s}')} = \Gamma(\mathbf{s} - \mathbf{s}')$, where $\Gamma(\mathbf{h}) = \mathbf{C}(\mathbf{0}) - \mathbf{C}(\mathbf{h}) = \mathbf{T} - \rho(\mathbf{s} - \mathbf{s}')\mathbf{T} = \gamma(\mathbf{s} - \mathbf{s}')\mathbf{T}$ with γ being a valid variogram. A possibly more insightful interpretation of "intrinsic" is that

$$\frac{cov(Y_i(\mathbf{s}), Y_j(\mathbf{s} + \mathbf{h}))}{\sqrt{\mathrm{Cov}(Y_i(\mathbf{s}), Y_i(\mathbf{s} + \mathbf{h}))\mathrm{Cov}(Y_j(\mathbf{s}), Y_j(\mathbf{s} + \mathbf{h}))}} = \frac{T_{ij}}{\sqrt{T_{ii}T_{jj}}}$$

regardless of $\mathbf{h}$, where $\mathbf{T} = [T_{ij}]_{i,j=1}^p$.

In the spirit of Stein and Corsten (1991), a bivariate spatial process model using separability becomes appropriate for regression with a single covariate $X(\mathbf{s})$ and a univariate response $Y(\mathbf{s})$. In fact, we treat this as a bivariate process to allow for missing Xs for some observed Ys and for inverse problems, inferring about $X(\mathbf{s}_0)$ for a given $Y(\mathbf{s}_0)$. We assume $\mathbf{Z}(\mathbf{s}) = (X(\mathbf{s}), Y(\mathbf{s}))^T$ is a bivariate Gaussian process with mean function $\boldsymbol{\mu}(\mathbf{s}) = (\mu_1(\mathbf{s}), \mu_2(\mathbf{s}))^T$ and a separable, or intrinsic, 2×2 cross-covariance function given by $\mathbf{C}(\mathbf{h}) = \rho(\mathbf{h})\mathbf{T}$, where $\mathbf{T} = [T_{ij}]_{i,j=1}^2$ is the 2×2 local dispersion matrix. The regression model arises by considering $Y(\mathbf{s})|X(\mathbf{s}) \sim N(\beta_0 + \beta_1 X(\mathbf{s}), \sigma^2)$, where $\beta_0 = \mu_2 - \frac{T_{12}}{T_{11}}\mu_1$, $\beta_1 = \frac{T_{12}}{T_{11}}$ and $\sigma^2 = T_{22} - \frac{T_{12}^2}{T_{11}}$. Banerjee and Gelfand (2002) (see also Banerjee et al., 2004) employ such models to analyze relationship between shrub density and dew duration for a dataset consisting of 1,129 locations in a west-facing watershed in the Negev Desert in Israel.

28.5 Co-Regionalization

In Section 28.2.2, we saw how linear combinations of independent processes can lead to richer modeling of cross-covariograms and cross-covariances. Such models, in general, are known as the *linear models of co-regionalization* (LMC) and can be employed to produce valid dispersion structures that are richer and more flexible than the separable or intrinsic specifications. A more general LMC arises if again $\mathbf{Y}(\mathbf{s}) = \mathbf{A}\mathbf{w}(\mathbf{s})$, where now the $w_j(\mathbf{s})$ are independent, but no longer identically distributed. In fact, let the $w_j(\mathbf{s})$ process have mean 0, variance 1, and stationary correlation function $\rho_j(\mathbf{h})$. Then $E(\mathbf{Y}(\mathbf{s})) = \mathbf{0}$, but the cross-covariance matrix associated with $\mathbf{Y}(\mathbf{s})$ is now

$$\Sigma_{\mathbf{Y}(\mathbf{s}),\mathbf{Y}(\mathbf{s}')} \equiv \mathbf{C}(\mathbf{s}-\mathbf{s}') = \sum_{j=1}^{p} \rho_j(\mathbf{s}-\mathbf{s}')\mathbf{T}_j, \tag{28.10}$$

where $\mathbf{T}_j = \mathbf{a}_j\mathbf{a}_j^T$ with $\mathbf{a}_j$ the jth column of $\mathbf{A}$. Note that the $\mathbf{T}_j$ have rank 1 and $\sum_j \mathbf{T}_j = \mathbf{T}$. Also, the cross-covariance function can be written as $\mathbf{C}(\mathbf{s}-\mathbf{s}') = \mathbf{A}\Theta(\mathbf{s}-\mathbf{s}')\mathbf{A}^T$, where $\mathbf{A}\mathbf{A}^T = \mathbf{T}$ and $\Theta(\mathbf{s}-\mathbf{s}')$ is a $p \times p$ diagonal matrix with $\rho_j(\mathbf{s}-\mathbf{s}')$ as its jth diagonal element. To connect with Section 28.3, we see that cross-covariance functions of LMCs arise as linear transformations of diagonal cross-correlation matrices.

More importantly, we note that such linear transformation maintains stationarity for the joint spatial process. With monotonic isotropic correlation functions, there will be a range associated with each component of the process, $Y_j(\mathbf{s})$, $j = 1, 2, \ldots, p$. This contrasts with the intrinsic case where, with only one correlation function, the $Y_j(\mathbf{s})$ processes share a common range. Again we can work with a covariogram representation, i.e., with $\Sigma_{\mathbf{Y}(\mathbf{s})-\mathbf{Y}(\mathbf{s}')} \equiv \Gamma(\mathbf{s}-\mathbf{s}')$, where $\Gamma(\mathbf{s}-\mathbf{s}') = \sum_j \gamma_j(\mathbf{s}-\mathbf{s}')\mathbf{T}_j$ where $\gamma_j(\mathbf{s}-\mathbf{s}') = \rho_j(\mathbf{0}) - \rho_j(\mathbf{s}-\mathbf{s}')$ (Geltner and Miller, 2001).

In applications, we would introduce Equation (28.10) as a component of a general multivariate spatial model for the data. That is, we assume

$$\mathbf{Y}(\mathbf{s}) = \boldsymbol{\mu}(\mathbf{s}) + \mathbf{v}(\mathbf{s}) + \boldsymbol{\epsilon}(\mathbf{s}), \tag{28.11}$$

where $\boldsymbol{\epsilon}(\mathbf{s})$ is a white noise vector, i.e., $\boldsymbol{\epsilon}(\mathbf{s}) \sim N(\mathbf{0}, \mathbf{D})$ where $\mathbf{D}$ is a $p \times p$ diagonal matrix with $(D)_{jj} = \tau_j^2$. In Equation (28.11), $\mathbf{v}(\mathbf{s}) = \mathbf{A}\mathbf{w}(\mathbf{s})$ following Equation (28.10) as above. In practice, we typically assume $\boldsymbol{\mu}(\mathbf{s})$ arises linearly in the covariates, i.e., from $\mu_j(\mathbf{s}) = \mathbf{X}_j^T(\mathbf{s})\boldsymbol{\beta}_j$. Each component can have its own set of covariates with its own coefficient vector.

Note that Equation (28.11) can be viewed as a hierarchical model. At the first stage, given $\{\boldsymbol{\beta}_j, j = 1, \cdots, p\}$ and $\{\mathbf{v}(\mathbf{s}_i)\}$, the $\mathbf{Y}(\mathbf{s}_i)$, $i = 1, \cdots, n$ are conditionally independent with $\mathbf{Y}(\mathbf{s}_i) \sim N(\boldsymbol{\mu}(\mathbf{s}_i) + \mathbf{v}(\mathbf{s}_i), \mathbf{D})$. At the second stage, the joint distribution of $\mathbf{v}$ (where $\mathbf{v} = (\mathbf{v}(\mathbf{s}_1), \cdots, \mathbf{v}(\mathbf{s}_n))$) is $N(\mathbf{0}, \sum_{j=1}^{p} \mathbf{R}_j \otimes \mathbf{T}_j)$, where $\mathbf{R}_j$ is $n \times n$ with $(R_j)_{ii'} = \rho_j(\mathbf{s}_i - \mathbf{s}_{i'})$. Concatenating the $\mathbf{Y}(\mathbf{s}_i)$ into an $np \times 1$ vector $\mathbf{Y}$, similarly $\boldsymbol{\mu}(\mathbf{s}_i)$ into $\boldsymbol{\mu}$, we can marginalize over $\mathbf{v}$ to obtain

$$f(\mathbf{Y}|\{\boldsymbol{\beta}_j\}, \mathbf{D}, \{\rho_j\}, \mathbf{T}) = N\left(\boldsymbol{\mu}, \sum_{j=1}^{p}(\mathbf{R}_j \otimes \mathbf{T}_j) + \mathbf{I}_{n\times n} \otimes \mathbf{D}\right). \tag{28.12}$$

Priors on $\{\boldsymbol{\beta}_j\}$, $\{\tau_j^2\}$, $\mathbf{T}$ and the parameters of the ρ_j complete a Bayesian hierarchical model specification.

28.5.1 Further Co-Regionalization Thoughts

Briefly, we return to the the nested modeling of Section 28.2.2. There, we obtained $c(\mathbf{h}) = \sum_{r=1}^{m} \rho_r(\mathbf{h})\mathbf{T}_r$ with $\mathbf{T}_r = \mathbf{A}_r\mathbf{A}_r^T$, positive definite. Here, with co-regionalization, $c(\mathbf{h}) = \sum_{j=1}^{p} \rho_j(\mathbf{h})\mathbf{T}_j$, where $\mathbf{T}_j = \mathbf{a}_j\mathbf{a}_j^T$ and rank $\mathbf{T}_j = 1$. Again, nesting is about multiple spatial scales with a common scale for each vector; co-regionalization is about multiple spatial scales with a different scale for each component of the vector.

We could imagine combining nesting and co-regionalization, asking in Section 28.2.2 that the components of $\mathbf{w}_r(\mathbf{s})$ are not replicates, but that each has its own correlation function. Then $c(\mathbf{h}) = \sum_{r=1}^{m}\sum_{j=1}^{p} A_{rj}A_{rj}^T\rho_{rj}(\mathbf{h})$. Though very flexible, such a model introduces mp correlation functions and it is likely that the data will not be able to inform about all of these functions. We are unaware of any work that has attempted to fit such models.

Also, co-regionalization may be considered in terms of dimension reduction, i.e., suppose that $\mathbf{A}$ is $p \times r, r < p$. That is, we are representing p processes through only r independent processes. Therefore $\mathbf{T} = \mathbf{A}\mathbf{A}^T$ has rank r and $\mathbf{Y}(\mathbf{s})$ lives in an r-dimensional subspace of R^p with probability 1. (This, of course, has nothing to do with whether $\mathbf{w}(\mathbf{s})$ is intrinsic.) Evidently, such a dimension reduction specification cannot be for a data model. However, it may prove adequate as a spatial random effects model.

28.6 Conditional Development of the LMC

For the process $\mathbf{v}(\mathbf{s}) = \mathbf{A}\mathbf{w}(\mathbf{s})$ as above, the LMC can be developed through a conditional approach rather than a joint modeling approach. This idea has been elaborated in, e.g., Matheron (1973) and in Berliner (2000) who refer to it as a hierarchical modeling approach for multivariate spatial modeling and prediction. It is proposed for cokriging and kriging with external drift.

We illuminate the equivalence of conditional and unconditional specifications in the special case where $\mathbf{v}(\mathbf{s}) = \mathbf{A}\mathbf{w}(\mathbf{s})$ with the $w_j(\mathbf{s})$ independent mean 0, variance 1 Gaussian processes. By taking $\mathbf{A}$ to be lower triangular, the equivalence and associated reparametrization will be easy to see. Upon permutation of the components of $\mathbf{v}(\mathbf{s})$ we can, without loss of generality, write $f(\mathbf{v}(\mathbf{s})) = f(v_1(\mathbf{s}))f(v_2(\mathbf{s})|v_1(\mathbf{s}))\cdots f(v_p(\mathbf{s})|v_1(\mathbf{s}),\cdots,v_{p-1}(\mathbf{s}))$. In the case of $p = 2$, $f(v_1(\mathbf{s}))$ is clearly $N(0, T_{11})$, i.e., $v_1(\mathbf{s}) = \sqrt{T_{11}}w_1(\mathbf{s}) = a_{11}w_1(\mathbf{s})$, $a_{11} > 0$. But $f(v_2(\mathbf{s})|v_1(\mathbf{s})) \sim N\left((T_{12}/T_{11})v_1(\mathbf{s}), T_{22} - T_{12}^2/T_{11}\right)$, i.e., $N\left((a_{21}/a_{11})v_1(\mathbf{s}), a_{22}^2\right)$. In fact, from the previous section we have $\boldsymbol{\Sigma}_{\mathbf{v}} = \sum_{j=1}^{p}\mathbf{R}_j \otimes \mathbf{T}_j$. If we permute the rows of $\mathbf{v}$ to $\tilde{\mathbf{v}} = (\mathbf{v}^{(1)T}, \mathbf{v}^{(2)T})^T$, where $\mathbf{v}^{(l)^T} = (v_l(\mathbf{s}_1),\cdots,v_l(\mathbf{s}_n))$, $l = 1, 2$, then $\boldsymbol{\Sigma}_{\tilde{\mathbf{v}}} = \sum_{j=1}^{p}\mathbf{T}_j \otimes \mathbf{R}_j$. Again with $p = 2$, we can calculate $E(\mathbf{v}^{(2)}|\mathbf{v}^{(1)}) = \frac{a_{21}}{a_{11}}\mathbf{v}^{(1)}$ and $\boldsymbol{\Sigma}_{\mathbf{v}^{(2)}|\mathbf{v}^{(1)}} = a_{22}^2\mathbf{R}_2$. But this is exactly the mean and covariance structure associated with variables $\{v_2(\mathbf{s}_i)\}$ given $\{v_1(\mathbf{s}_i)\}$, i.e., with $v_2(\mathbf{s}_i) = (a_{21}/a_{11})v_1(\mathbf{s}_i) + a_{22}w_2(\mathbf{s}_i)$. Note that there is no notion of a conditional process here, i.e., a process $v_2(\mathbf{s})|v_1(\mathbf{s})$ is not well defined (What would be the joint distribution of realizations from such a process?). There is only a joint distribution for $\mathbf{v}^{(1)}, \mathbf{v}^{(2)}$ given any n and any $\mathbf{s}_1,\cdots,\mathbf{s}_n$, thus, a conditional distribution for $\mathbf{v}^{(2)}$ given $\mathbf{v}^{(1)}$.

Suppose we write $v_1(\mathbf{s}) = \sigma_1 w_1(\mathbf{s})$ where $\sigma_1 > 0$ and $w_1(\mathbf{s})$ is a mean 0 spatial process with variance 1 and correlation function ρ_1 and we write $v_2(\mathbf{s})|v_1(\mathbf{s}) = \alpha v_1(\mathbf{s}) + \sigma_2 w_2(\mathbf{s})$ where $\sigma_2 > 0$ and $w_2(\mathbf{s})$ is a mean 0 spatial process with variance 1 and correlation function ρ_2. The parameterization $(\alpha, \sigma_1, \sigma_2)$ is obviously equivalent to (a_{11}, a_{12}, a_{22}), i.e., $a_{11} = \sigma_1, a_{21} = \alpha\sigma_1$, $a_{22} = \sigma_2$ and, hence, to $\mathbf{T}$, i.e., $T_{11} = \sigma_1^2$, $T_{12} = \alpha\sigma_1^2$, $T_{22} = \alpha^2\sigma_1^2 + \sigma_2^2$. For general p, we introduce the following notation. Let $v_1(\mathbf{s}) = \sigma_1 w_1(\mathbf{s})$ and given $v_1(\mathbf{s}), \ldots, v_{l-1}(\mathbf{s})$, $v_l(\mathbf{s}) = \sum_{j=1}^{l-1}\alpha_j^{(l)}v_j(\mathbf{s}) + \sigma_l w_l(\mathbf{s})$, $l = 2, \ldots, p$. Unconditionally, $\mathbf{T}$ introduces $p(p+1)/2$ parameters. Conditionally, we introduce $p(p-1)/2$ αs and p σs. Straightforward recursion shows the equivalence of the T parameterization and, in obvious notation, the $(\boldsymbol{\sigma}, \boldsymbol{\alpha})$ parametrization.

Gelfand et al. (2004) point out that if we want to introduce distinct, independent nugget processes for the components of $\mathbf{Y}(\mathbf{s})$ or if we want to use different covariates to explain the different components of $\mathbf{Y}(\mathbf{s})$, the equivalence between the conditional and unconditional approaches breaks down. Also, Schmidt and Gelfand (2003) present an example of a co-regionalization analysis for daily carbon monoxide, nitric oxide, and nitrogen dioxide monitoring station data for 68 monitoring sites in California. They illustrate both conditional and unconditional model-fitting and the benefits of the multivariate process modeling in interpolating exposures to new locations.

28.7 A Spatially Varying LMC

Replacing $\mathbf{A}$ by $\mathbf{A}(\mathbf{s})$, we can define

$$\mathbf{v}(\mathbf{s}) = \mathbf{A}(\mathbf{s})\mathbf{w}(\mathbf{s}) \tag{28.13}$$

for insertion into Equation (28.11). We refer to the model in Equation (28.13) as a spatially varying LMC (SVLMC). Let $\mathbf{T}(\mathbf{s}) = \mathbf{A}(\mathbf{s})\mathbf{A}(\mathbf{s})^T$. Again $\mathbf{A}(\mathbf{s})$ can be taken to be lower triangular for convenience. Now,

$$C(\mathbf{s}, \mathbf{s}') = \sum_j \rho_j(\mathbf{s} - \mathbf{s}')\mathbf{a}_j(\mathbf{s})\mathbf{a}_j^T(\mathbf{s}') \tag{28.14}$$

with $\mathbf{a}_j(\mathbf{s})$ the jth column of $\mathbf{A}(\mathbf{s})$. Letting $\mathbf{T}_j(\mathbf{s}) = \mathbf{a}_j(\mathbf{s})\mathbf{a}_j^T(\mathbf{s})$, again, $\sum_j \mathbf{T}_j(\mathbf{s}) = \mathbf{T}(\mathbf{s})$. From Equation (28.14), $\mathbf{v}(\mathbf{s})$ is no longer a stationary process. Letting $\mathbf{s} - \mathbf{s}' \to 0$, the covariance matrix for $\mathbf{v}(\mathbf{s}) = \mathbf{T}(\mathbf{s})$, which is a multivariate version of a spatial process with a spatially varying variance.

This suggests modeling $\mathbf{A}(\mathbf{s})$ through its one-to-one correspondence with $\mathbf{T}(\mathbf{s})$. In the univariate case, choices for $\sigma^2(\mathbf{s})$ include $\sigma^2(\mathbf{s}, \theta)$, i.e., a parametric function or trend surface in location; $\sigma^2(x(\mathbf{s})) = g(x(\mathbf{s}))\sigma^2$ where $x(\mathbf{s})$ is some covariate used to explain $\mathbf{Y}(\mathbf{s})$ and $g(.) > 0$ (then $g(x(\mathbf{s}))$ is typically $x(\mathbf{s})$ or $x^2(\mathbf{s})$); or $\sigma^2(\mathbf{s})$ is itself a spatial process (e.g., $\log \sigma^2(\mathbf{s})$ might be a Gaussian process). Extending the second possibility, we would take $\mathbf{T}(\mathbf{s}) = g(x(\mathbf{s}))\mathbf{T}$. In fact, in the example below we take $g(x(\mathbf{s})) = (x(\mathbf{s}))^{\psi}$ with $\psi \geq 0$, but unknown. This allows homogeneity of variance as a special case.

Extending the third possibility, we generalize to define $\mathbf{T}(\mathbf{s})$ to be a *matric-variate* spatial process. An elementary way to induce a spatial process for $\mathbf{T}(\mathbf{s})$ is to work with $\mathbf{A}(\mathbf{s})$, specifying independent mean 0 Gaussian processes for $b_{jj'}(\mathbf{s}), i \leq j' \leq j \leq p$ and setting $a_{jj'}(\mathbf{s}) = b_{jj'}(\mathbf{s}), j \neq j', a_{jj}(\mathbf{s}) = |b_{jj}(\mathbf{s})|$. However, such specification yields a nonstandard and computationally intractable distribution for $\mathbf{T}(\mathbf{s})$.

Instead, Gelfand et al. (2004) propose a matric-variate inverse Wishart spatial process for $\mathbf{T}(\mathbf{s})$ equivalently, a matric-variate Wishart spatial process for $\mathbf{T}^{-1}(\mathbf{s})$, where, marginally, $\mathbf{T}(\mathbf{s})$ has an inverse Wishart distribution. More detail on this process model is provided in Gelfand et al. (2004). However, if $\mathbf{T}(\mathbf{s})$ is random, $\mathbf{v}(\mathbf{s}) = \mathbf{A}(\mathbf{s})\mathbf{w}(\mathbf{s})$ is not only nonstationary, but non-Gaussian.

28.8 Bayesian Multivariate Spatial Regression Models

The multivariate spatial Gaussian process can be utilized in spatial regression models. Multivariate spatial regression envisions that each location $\mathbf{s}$ yields observations on q dependent variables given by a $q \times 1$ vector $\mathbf{Y}(\mathbf{s}) = [Y_l(\mathbf{s})]_{l=1}^q$. For each $Y_l(\mathbf{s})$, we also observe a $p_l \times 1$

vector of regressors $\mathbf{x}_l(\mathbf{s})$. Thus, for each location we have q univariate spatial regression equations, which we combine into a multivariate spatial regression model written as

$$\mathbf{Y}(\mathbf{s}) = \mathbf{X}^T(\mathbf{s})\boldsymbol{\beta} + \mathbf{Z}^T(\mathbf{s})\mathbf{w}(\mathbf{s}) + \boldsymbol{\epsilon}(\mathbf{s}), \tag{28.15}$$

where $\mathbf{X}^T(\mathbf{s})$ is a $q \times p$ matrix ($p = \sum_{l=1}^{q} p_l$) having a block diagonal structure with its lth diagonal being the $1 \times p_l$ vector $\mathbf{x}_l^T(\mathbf{s})$. Here $\boldsymbol{\beta} = (\boldsymbol{\beta}_1, \ldots, \boldsymbol{\beta}_p)^T$ is a $p \times 1$ vector of regression coefficients with $\boldsymbol{\beta}_l$ being the $p_l \times 1$ vector of regression coefficients for $\mathbf{x}_l^T(\mathbf{s})$, $\mathbf{w}(\mathbf{s}) \sim GP(\mathbf{0}, \mathbf{C}_{\mathbf{w}}(\cdot,\cdot))$ is an $m \times 1$ multivariate Gaussian process with cross-covariance function $\mathbf{C}_{\mathbf{w}}(\mathbf{s}, \mathbf{s}')$ and acts as a coefficient vector for the $q \times m$ design matrix $\mathbf{Z}^T(\mathbf{s})$ and $\boldsymbol{\epsilon}(\mathbf{s}) \sim MVN(\mathbf{0}, \Psi)$ is a $q \times 1$ vector modeling the residual error with dispersion matrix Ψ.

Model (28.15) acts as a general framework admitting several spatial models. For example, it accommodates the spatially varying coefficient models discussed in Gelfand et al. (2004). Letting $m = q$ and $\mathbf{Z}^T(\mathbf{s}) = \mathbf{I}_m$ leads to a multivariate spatial regression model with $\mathbf{w}(\mathbf{s})$ acting as a *spatially varying intercept*. On the other hand, we could envision all coefficients to be spatially varying and set $m = p$ with $\mathbf{Z}^T(\mathbf{s}) = \mathbf{X}^T(\mathbf{s})$. With multivariate spatial models involving a large number of locations, such computations can become burdensome. In such cases, the $\mathbf{w}(\mathbf{s})$ can be replaced by a lower-dimensional predictive process (Banerjee, Gelfand, Finley, and Sang, 2008) to alleviate the computational burden. Bayesian estimation of several spatial models that can be cast within Equation (28.15), as well as their reduced-rank predictive process counterparts, can now be accomplished using the `spBayes` package in `R` (see Banerjee et al., 2008 and Finley, Banerjee, and Carlin, 2007 for details).

28.8.1 An Illustration

The selling price of commercial real estate, e.g., an apartment property, is theoretically the expected income capitalized at some (risk-adjusted) discount rate. Since an individual property is fixed in location, upon transaction, both selling price and income (rent) are jointly determined at that location. The real estate economics literature has examined the (mean) variation in both selling price and rent. (See Benjamin and Sirmans, 1991; Geltner and Miller, 2001.)

We consider a dataset consisting of apartment buildings in three very distinct markets, Chicago, Dallas, and San Diego. Chicago is an older, traditional city where development expanded outward from a well-defined central business district. Dallas is a newer city where development tends to be polycentric with the central business district playing less of a role in spatial pattern. San Diego is a more physically constrained city with development more linear as opposed to the traditional "circular" pattern. We have 252 buildings in Chicago, 177 in Dallas, and 224 in San Diego. In each market, 20 additional transactions are held out for prediction of the selling price. The locations of the buildings in each market are shown in Figure 28.1. Note that the locations are very irregularly spaced across the respective markets. All of the models noted below were fitted using reprojected distance between locations in kilometers.

Our objective is to fit a joint model for selling price and net income and to obtain a spatial surface associated with the risk, which, for any building, is given by the ratio of net income and price. For each location $\mathbf{s}$, we observe log net income ($Y_1(\mathbf{s})$) and log selling price of the transaction ($Y_2(\mathbf{s})$). In addition, we have the following three regressors: average square feet of a unit within the building (sqft), the age of the building (age) and number of units within the building (unit). We fit a multivariate spatial regression model as in Equation (28.15) with $q = 2$, $\mathbf{X}^T(\mathbf{s}) = (\mathbf{I}_2 \otimes \mathbf{x}^T(\mathbf{s}))$, where $\mathbf{x}^T(\mathbf{s})$ is the 1×3 vector of regressors, $\mathbf{Z}(\mathbf{s}) = \mathbf{I}_2$, $\mathbf{w}(\mathbf{s})$ is a bivariate Gaussian process and $\Psi = Diag\,(\tau_1^2, \tau_2^2)$.

Within this framework, we investigate four model specifications that vary only in terms of the cross-covariance matrix. Model 1 is an intrinsic or separable specification, i.e., it

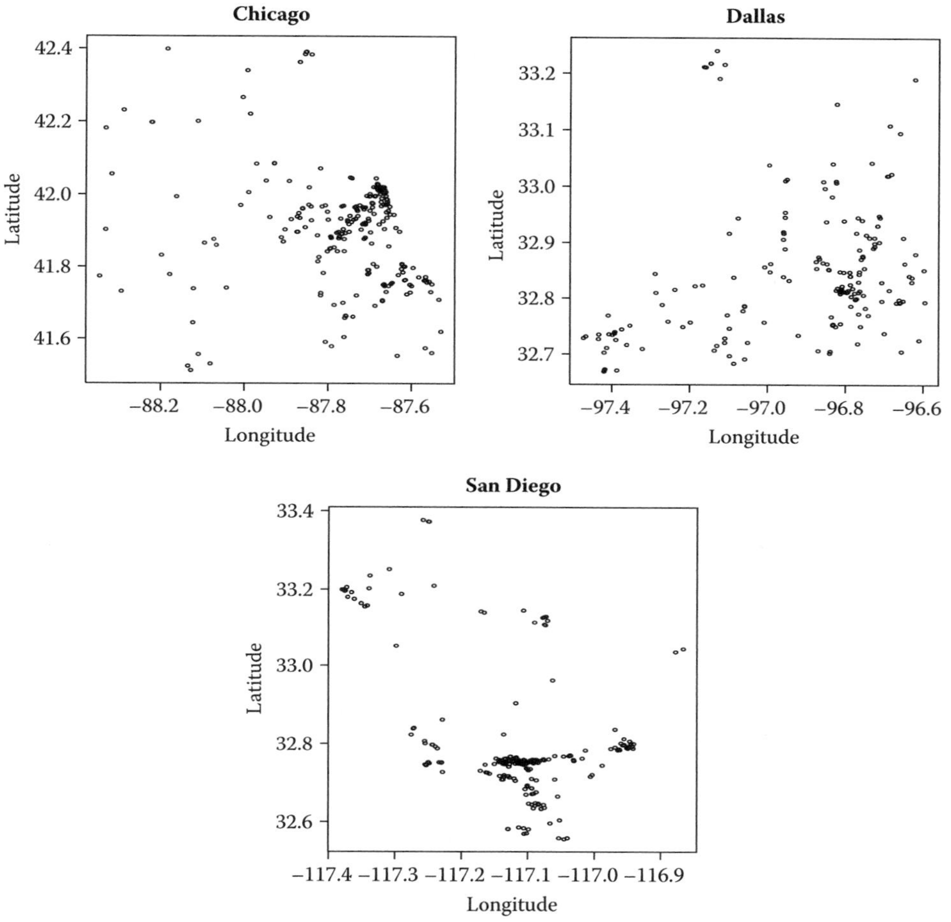

FIGURE 28.1
Sampling locations for the three markets in Chicago, Dallas, and San Diego.

assumes $\mathbf{C_w}(\mathbf{s}, \mathbf{s}') = \rho(\mathbf{s}, \mathbf{s}'; \phi)\mathbf{I}_2$. Model 2 assumes a stationary cross-covariance specification $\mathbf{C_w}(\mathbf{s}, \mathbf{s}') = \mathbf{A}Diag\,[\rho_k(\mathbf{s}, \mathbf{s}; \phi_k)]_{k=1}^2 \mathbf{A}^T$. Model 3 employs a spatially adaptive $\mathbf{C_w}(\mathbf{s}, \mathbf{s}) = (x(\mathbf{s}))^{\psi} \mathbf{A}\mathbf{A}^T$, where $x(\mathbf{s})$ is $unit(\mathbf{s})$. The supposition is that variability in $Y_1(\mathbf{s})$ and $Y_2(\mathbf{s})$ increases in building size. For Models 2 and 3, we modeled $\mathbf{A}\mathbf{A}^T$ as an inverse-Wishart distribution with $\mathbf{A}$ as the lower-triangular square root. The spectral square root approach mentioned at the end of Section 28.3 constitutes another feasible option here. Finally, Model 4 uses a matric-variate spatial Wishart process for $\mathbf{C_w}(\mathbf{s}, \mathbf{s}) = \mathbf{A}(\mathbf{s})\mathbf{A}^T(\mathbf{s})$ (see Gelfand et al., 2004, for details).

These models are fitted within the Bayesian framework where, for each, we use $\rho_k(\mathbf{s}, \mathbf{s}'; \phi_k) = e^{-\phi_k \|\mathbf{s}-\mathbf{s}'\|}$ and the decay parameters ϕ_k, $k = 1, 2$ have a gamma prior distribution arising from a mean range of one half the maximum interlocation distance, with infinite variance. Finally, τ_1^2 and τ_2^2 have inverse gamma priors with infinite variance. They are centered, respectively, at the ordinary least squares variance estimates obtained by fitting independent, nonspatial regression models to $Y_1(\mathbf{s})$ and $Y_2(\mathbf{s})$. Table 28.1 provides the model choice results for each of the markets using, for convenience, the posterior predictive criterion of Gelfand and Ghosh (1998) (see also Banerjee et al., 2004) with both the full

TABLE 28.1

Model Comparison Results for (a) the Full Dataset and (b) for the Holdout Sample; see Text for Details

(a)	Chicago			Dallas			San Diego		
Model	**G**	**P**	**D**	**G**	**P**	**D**	**G**	**P**	**D**
Model 1	0.1793	0.7299	0.9092	0.1126	0.5138	0.6264	0.0886	0.4842	0.5728
Model 2	0.1772	0.6416	0.8188	0.0709	0.4767	0.5476	0.0839	0.4478	0.5317
Model 3	0.1794	0.6368	0.8162	0.0715	0.4798	0.5513	0.0802	0.4513	0.5315
Model 4	0.1574	0.6923	0.8497	0.0436	0.4985	0.5421	0.0713	0.4588	0.5301
(b)	**Chicago**			**Dallas**			**San Diego**		
Model	**G**	**P**	**D**	**G**	**P**	**D**	**G**	**P**	**D**
Model 1	0.0219	0.0763	0.0982	0.0141	0.0631	0.0772	0.0091	0.0498	0.0589
Model 2	0.0221	0.0755	0.0976	0.0091	0.0598	0.0689	0.0095	0.0449	0.0544
Model 3	0.0191	0.0758	0.0949	0.0091	0.0610	0.0701	0.0087	0.0459	0.0546
Model 4	0.0178	0.0761	0.0939	0.0059	0.0631	0.0690	0.0074	0.0469	0.0543

dataset and also with holdout data.* In the table, G is the goodness of fit contribution, P is the penalty term, D is the sum, and small values of D are preferred. Evidently, the intrinsic model is the weakest. Models 2, 3, and 4 are quite close, but, since Model 4 is best in terms of G, we provide the results of the analysis for this model in Table 28.2.

In particular, Table 28.2 presents the posterior summaries of the parameters of the model for each market. Age receives a significant negative coefficient in Dallas and San Diego, but not in Chicago, perhaps because Chicago is an older city; a linear relationship for net income and selling price in age may not be adequate. Number of units receives a positive coefficient for both net income and price in all three markets. Square feet per unit is only significant in Chicago. The pure error variances (the τ^2s) are largest in Chicago, suggesting a bit more uncertainty in this market. The ϕs are very close in Dallas and San Diego, a bit less so in Chicago. The benefit of Model 4 lies more in the spatially varying $\mathbf{A}(\mathbf{s})$, equivalently $\mathbf{C}(\mathbf{s}, \mathbf{s})$, than in differing ranges for $w_1(\mathbf{s})$ and $w_2(\mathbf{s})$. Turning to Figure 28.2, we see the spatial surfaces associated with $\mathbf{C}_{11}(\mathbf{s})$, $\mathbf{C}_{22}(\mathbf{s})$, and $\mathbf{C}_{corr}(\mathbf{s}) = \mathbf{C}_{12}(\mathbf{s})/\sqrt{\mathbf{C}_{11}(\mathbf{s})\mathbf{C}_{22}(\mathbf{s})}$. Note that the $\mathbf{C}_{11}$ and $\mathbf{C}_{22}$ surfaces show considerable spatial variation and are quite different for all three markets. The correlations between $w_1(\mathbf{s})$ and $w_2(\mathbf{s})$ also show considerable spatial variation, ranging from .55 to .7 in Chicago, .3 to .85 in Dallas, .3 to .75 in San Diego. In Figure 28.3, we show the estimated residual spatial surfaces on the log scale (adjusted for the above regressors) for $Y_1(\mathbf{s})$, $Y_2(\mathbf{s})$ and $R(\mathbf{s})$. Most striking is the similarity between the $Y_1(\mathbf{s})$ and $Y_2(\mathbf{s})$ surfaces for each of the three markets. Also noteworthy is the spatial variation in each of the risk surfaces, suggesting that an aggregated market risk for each city is insufficient to make effective investment decisions.

28.9 Other Constructive Approaches

We review three other constructive approaches for building valid cross-covariance functions. The first is through moving average or kernel convolution of a process, the second extends local stationarity ideas in Fuentes and Smith (2001), the third describes a convolution of covariance functions as in Majumdar and Gelfand (2007).

* Twenty buildings were held out in each of the markets for cross-validatory purposes. In this regard, we can also consider validation of prediction using the holdout samples. For P, in Chicago 18 of 20 of the 95% predictive intervals contained the observed value, 20/20 in Dallas, and 20/20 in San Diego. For I, we have 19/20 in each market. It appears that Model 4 is providing claimed predictive performance.

TABLE 28.2

Posterior Median and Respective 2.5% and 97.5% Quantiles of the Parameters Involved in the Model for Net Income and Selling Price

	Chicago			Dallas			San Diego		
				Sales Price					
Parameter	50%	2.5%	97.5%	50%	2.5%	97.5%	50%	2.5%	97.5%
Intercept	2.63E+00	2.60E+00	2.65E+00	2.87E+00	2.84E+00	2.91E+00	2.61E+00	2.58E+00	2.65E+00
Age	8.64E−05	−1.07E−04	4.10E−04	−9.55E−04	−1.49E−03	−5.18E−04	−3.85E−04	−7.21E−04	−5.40E−05
No. Units	1.28E−03	1.01E−03	1.51E−03	4.64E−04	4.09E−04	5.36E−04	1.42E−03	1.22E−03	1.58E−03
Sqft/Unit	−2.83E−05	−5.93E−05	−1.10E−06	1.01E−04	−2.40E−06	2.21E−04	1.49E−05	−4.13E−05	7.82E−05
τ_1^2	7.08E−04	5.52E−04	8.86E−04	6.76E−04	5.05E−04	1.03E−03	5.45E−04	4.01E−04	7.25E−04
ϕ_1	1.34E−01	7.59E−02	4.42E−01	1.84E−01	7.28E−02	4.75E−01	1.18E−01	5.37E−02	4.66E−01
				Net Income					
Parameter	50%	2.5%	97.5%	50%	2.5%	97.5%	50%	2.5%	97.5%
Intercept	2.53E+00	2.51E+00	2.54E+00	2.45E+00	2.42E+00	2.49E+00	2.35E+00	2.32E+00	2.39E+00
Age	1.10E−04	−2.30E−04	3.69E−04	−1.15E−03	−1.67E−03	−5.98E−04	−4.55E−04	−8.57E−04	−1.29E−04
No. Units	1.56E−03	1.37E−03	1.79E−03	5.34E−04	4.60E−04	6.18E−04	1.69E−03	1.41E−03	1.87E−03
Sqft/Unit	−1.68E−05	−6.40E−05	1.19E−05	1.31E−04	−3.69E−05	3.26E−04	1.91E−05	−5.34E−05	8.22E−05
τ_2^2	9.93E−04	7.45E−04	1.25E−03	9.53E−04	7.17E−04	1.30E−03	6.71E−04	4.68E−04	9.69E−04
ϕ_2	1.79E−01	7.78E−02	4.79E−01	1.75E−01	8.56E−02	4.25E−01	1.22E−01	5.59E−02	4.54E−01

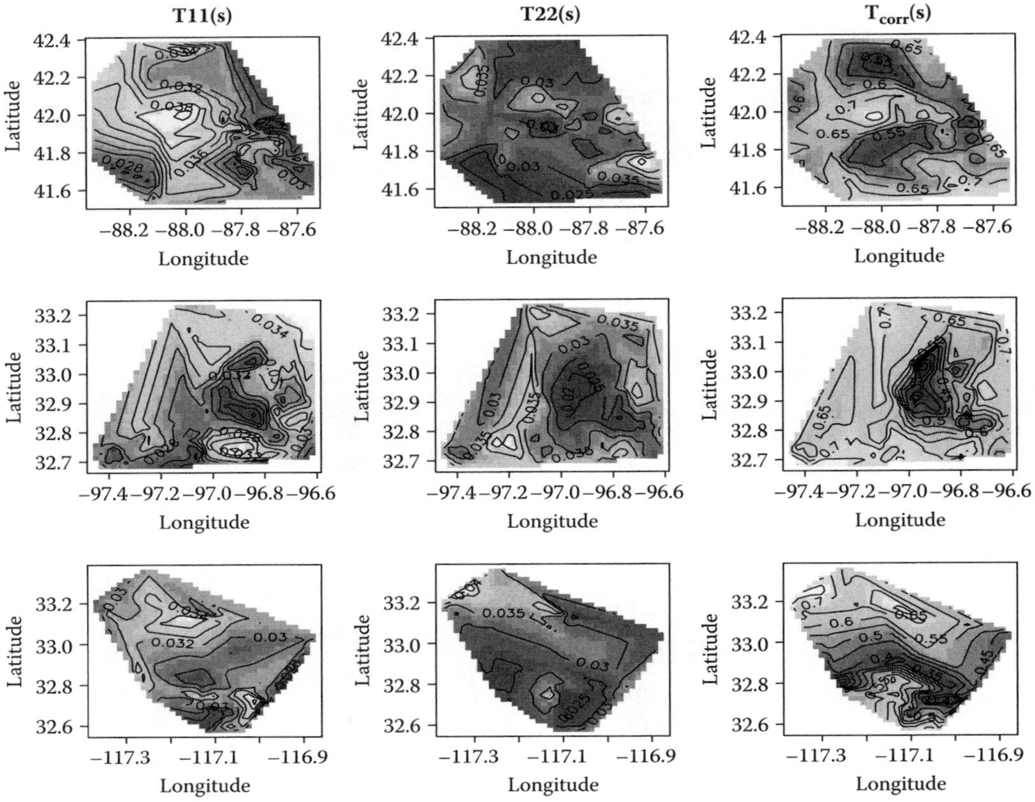

FIGURE 28.2
Spatial surfaces associated with the spatially varying $\mathbf{C}(\mathbf{s})$ for the three cities, Chicago (top row), Dallas (middle row), and San Diego (bottom row), with the columns corresponding to $C_{11}(\mathbf{s})$, $C_{22}(\mathbf{s})$, and $C_{corr}(\mathbf{s})$.

28.9.1 Kernel Convolution Methods

Ver Hoef and Barry (1998) describe what they refer to as a moving average approach for creating valid stationary cross-covariograms. The technique is also called kernel convolution and is a well-known approach for creating general classes of stationary processes. The one-dimensional case is discussed in Higdon et al. (1999, 2003). For the multivariate case, suppose $k_l(\cdot), l = 1, \ldots, p$ is a set of p square integrable kernel functions on $\mathbb{R}^2$ and, without loss of generality, assume $k_l(0) = 1$.

Let $w(\mathbf{s})$ be a mean 0, variance 1 Gaussian process with correlation function ρ. Define the p-variate spatial process $\mathbf{Y}(\mathbf{s})$ by

$$Y_l(\mathbf{s}) = \sigma_l \int k_l(\mathbf{s} - \mathbf{t}) w(\mathbf{t}) d\mathbf{t}, \quad l = 1, \cdots, p. \tag{28.16}$$

$\mathbf{Y}(\mathbf{s})$ is obviously a mean $\mathbf{0}$ Gaussian process with associated cross-covariance function $\mathbf{C}(\mathbf{s}, \mathbf{s}')$ having (l, l') entry

$$(\mathbf{C}(\mathbf{s}, \mathbf{s}'))_{ll'} = \sigma_l \sigma_{l'} \int\int k_l(\mathbf{s} - \mathbf{t}) k_{l'}(\mathbf{s}' - \mathbf{t}') \rho(\mathbf{t} - \mathbf{t}') d\mathbf{t} d\mathbf{t}'. \tag{28.17}$$

By construction, $\mathbf{C}(\mathbf{s}, \mathbf{s}')$ is valid. By transformation in Equation (28.17), we can see that $(\mathbf{C}(\mathbf{s}, \mathbf{s}'))_{ll'}$ depends only on $\mathbf{s} - \mathbf{s}'$, i.e., $\mathbf{Y}(\mathbf{s})$ is a stationary process. Note that $(\mathbf{C}(\mathbf{s} - \mathbf{s}'))_{ll'}$

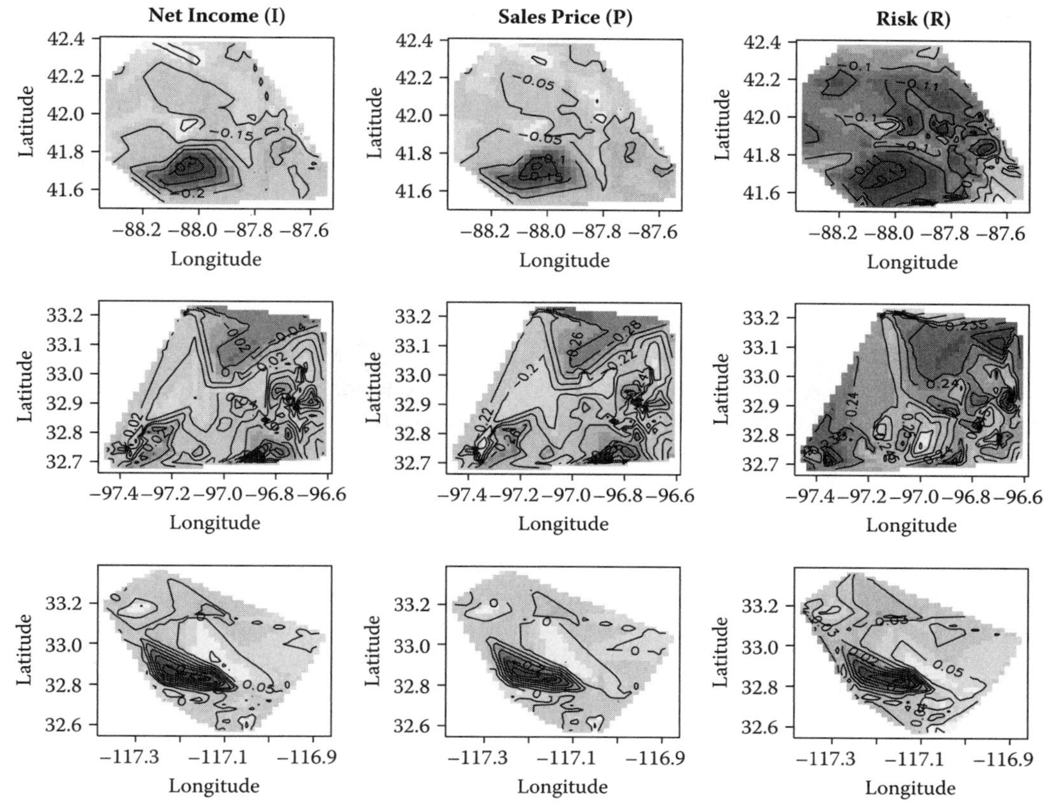

FIGURE 28.3
Residual spatial surfaces associated with the three processes, net income, sales price and risk, for the three cities, Chicago (top row), Dallas (middle row) and San Diego (bottom row), with the columns corresponding to net income (Y_1), sales price (Y_2) and risk (R).

need not equal $(\mathbf{C}(\mathbf{s}-\mathbf{s}'))_{l'l}$. If the k_l depend on $\mathbf{s}-\mathbf{s}'$ only through $||\mathbf{s}-\mathbf{s}'||$ and ρ is isotropic, then $C(\mathbf{s}-\mathbf{s}')$ is isotropic.

An objective in Ver Hoef and Barry (1998) is to be able to compute $\mathbf{C}(\mathbf{s}-\mathbf{s}')$ in Equation (28.17) explicitly. For instance, with kernels that are functions taking the form of a constant height over a bounded rectangle, zero outside, this is the case and an anisotropic form results. More recent work of Ver Hoef and colleagues (2004) no longer worries about this.

An alternative, as in Higdon et al. (1999), employs discrete approximation. Choosing a finite set of locations $\mathbf{t}_1, \ldots, \mathbf{t}_r$, we define

$$Y_l(\mathbf{s}) = \sigma_l \sum_{j=1}^{r} k_l(\mathbf{s}-\mathbf{t}_j) w(\mathbf{t}_j). \tag{28.18}$$

Now, $(\mathbf{C}(\mathbf{s},\mathbf{s}'))_{ll'}$ is such that

$$(\mathbf{C}(\mathbf{s},\mathbf{s}'))_{ll'} = \sigma_l \sigma_{l'} \sum_{j=1}^{r} \sum_{j'=1}^{r} k_l(\mathbf{s}-\mathbf{t}_j) k_{l'}(\mathbf{s}'-\mathbf{t}_{j'}) \rho(\mathbf{t}_j-\mathbf{t}_{j'}). \tag{28.19}$$

The form in Equation (28.19) is easy to work with, but note that the resulting process is no longer stationary.

Higdon et al. (1999) consider the univariate version of Equation (28.16), but with k now a spatially varying kernel, in particular, one that varies slowly in $\mathbf{s}$. This would replace $k(\mathbf{s}-\mathbf{t})$ with $k(\mathbf{s}-\mathbf{t};\mathbf{s})$. The multivariate analog would choose p square integrable (in the first argument) spatially varying kernel functions, $k_l(\mathbf{s}-\mathbf{t};\mathbf{s})$ and define $\mathbf{Y}(\mathbf{s})$ through

$$Y_l(\mathbf{s}) = \sigma_l \int k_l(\mathbf{s}-\mathbf{t};\mathbf{s})w(\mathbf{t})d\mathbf{t} \tag{28.20}$$

extending Equation (28.16). The cross-covariance matrix associated with Equation (28.20) has entries

$$(\mathbf{C}(\mathbf{s},\mathbf{s}'))_{ll'} = \sigma_l\sigma_{l'} \int k_l(\mathbf{s}-\mathbf{t};\mathbf{s})k_{l'}(\mathbf{s}'-\mathbf{t};\mathbf{s}')d\mathbf{t}. \tag{28.21}$$

Higdon et al. (1999) employ only Gaussian kernels, arguably, imparting too much smoothness to the $\mathbf{Y}(\mathbf{s})$ process. In recent work, Paciorek and Schervish (2004) suggest alternative kernels using, e.g., Matèrn forms to ameliorate this concern.

28.9.2 Locally Stationary Models

Fuentes and Smith (2001) introduce a class of univariate locally stationary models by defining $Y(\mathbf{s}) = \int b(\mathbf{s},\mathbf{t})w_{\theta(\mathbf{t})}(\mathbf{s})d\mathbf{t}$ where w_θ is a stationary spatial process having parameters θ with w_{θ_1} and w_{θ_2} independent if $\theta_1 \neq \theta_2$, and $b(\mathbf{s},\mathbf{t})$ is a choice of inverse distance function. Analogous to [28], the parameter $\theta(\mathbf{t})$ varies slowly in $\mathbf{t}$. In practice, the integral is discretized to a sum, i.e., $Y(\mathbf{s}) = \sum_{j=1}^{r} b(\mathbf{s},\mathbf{t}_j)w_j(\mathbf{s})$. This approach defines essentially locally stationary models in the sense that if $\mathbf{s}$ is close to $\mathbf{t}$, $Y(\mathbf{s}) \approx w_{\theta(\mathbf{t})}(\mathbf{s})$. The multivariate extension would introduce p inverse distance functions, $b_l(\mathbf{s},\mathbf{t}_j)$, $l = 1,\ldots,p$ and define

$$Y_l(\mathbf{s}) = \int b_l(\mathbf{s},\mathbf{t})w_{\theta(\mathbf{t})}(\mathbf{s})d\mathbf{t} \tag{28.22}$$

Straightforward calculation reveals that

$$(\mathbf{C}(\mathbf{s},\mathbf{s}'))_{ll'} = \int b_l(\mathbf{s},\mathbf{t})b_{l'}(\mathbf{s}',\mathbf{t})c(\mathbf{s}-\mathbf{s}';\theta(\mathbf{t}))d\mathbf{t}. \tag{28.23}$$

28.9.3 Convolution of Covariance Functions Approaches

Motivated by Gaspari and Cohn (1999), Majumdar and Gelfand (2007) discuss convolving p stationary one-dimensional covariance functions with each other to generate cross-covariance functions. Two remarks are appropriate. First, this approach convolves covariance functions as opposed to kernel convolution of processes as in the previous section. Second, the LMC also begins with p stationary one-dimensional covariance functions, but creates the cross-covariance function associated with an arbitrary linear transformation of p independent processes having these respective covariance functions.

Suppose that $C_1,\ldots,C_p$ are valid stationary covariance functions on R^d. Define $C_{ij}(\mathbf{s}) = \int C_i(\mathbf{s}-\mathbf{t})C_j(\mathbf{t})d\mathbf{t},\ i \neq j$ and $C_{ii}(\mathbf{s}) = \int C_i(\mathbf{s}-\mathbf{t})C_i(\mathbf{t})d\mathbf{t}\ i,\ j = 1,\cdots,k$. Majumdar and Gelfand (2007) show that, under fairly weak assumptions, the C_{ij} and C_{ii}s provide a valid cross-covariance structure for a p dimensional multivariate spatial process, i.e., $\text{Cov}(Y_i(\mathbf{s}),Y_j(\mathbf{s}')) = C_{ij}(\mathbf{s}-\mathbf{s}')$. Gaspari and Cohn (1999, p. 739) show that if C_i and C_j are isotropic functions, then so is C_{ij}.

If ρ_i are correlation functions, i.e., $\rho_i(\mathbf{0}) = 1$, then $\rho_{ii}(\mathbf{0}) = \int \rho_i(\mathbf{t})^2 d\mathbf{t}$ need not equal 1. In fact, if ρ_i is a parametric function, then $\text{var}(Y_i(\mathbf{s}))$ depends on these parameters. However, if one defines ρ_{ij} by the following relation

$$\rho_{ij}(\mathbf{s}) = \frac{C_{ij}(\mathbf{s})}{(C_{ii}(\mathbf{0})C_{jj}(\mathbf{0}))^{\frac{1}{2}}}, \tag{28.24}$$

then $\rho_{ii}(\mathbf{0}) = 1$. Let

$$\mathbf{D}_C = \begin{pmatrix} C_{11}(\mathbf{0}) & \dots & 0 \\ \vdots & \ddots & \vdots \\ 0 & \dots & C_{kk}(\mathbf{0}) \end{pmatrix} \tag{28.25}$$

and set $\mathbf{R}(\mathbf{s}) = \mathbf{D}_C{}^{-1/2}\mathbf{C}(\mathbf{s})\mathbf{D}_C{}^{-1/2}$, where $\mathbf{C}(\mathbf{s}) = [C_{ij}(\mathbf{s})]$. Then $\mathbf{R}(\mathbf{s})$ is a valid cross-correlation function and, in fact, if $\mathbf{D}_\sigma{}^{1/2} = diag\,(\sigma_1, \ldots, \sigma_k)$, we can take as a valid cross-covariance function $\mathbf{C}_\sigma = \mathbf{D}_\sigma{}^{1/2}\mathbf{R}(\mathbf{s})\mathbf{D}_\sigma{}^{1/2}$. Then $\text{var}(Y_i(\mathbf{s})) = \sigma_i{}^2$, but $\text{Cov}(Y_i(\mathbf{s}), Y_j(\mathbf{s})) = \sigma_i\sigma_j \frac{C_{ij}(\mathbf{0})}{\sqrt{C_{ii}(\mathbf{0})C_{jj}(\mathbf{0})}}$ and will still depend on the parameters in C_i and C_j. However, Majumdar and Gelfand (2007) show that $\rho_{ii}(\mathbf{s})$ may be looked upon as a "correlation function" and $\rho_{ij}(\mathbf{s})$ as a "cross-correlation function" since, under mild conditions, if the C_is are stationary, then $|\rho_{ij}(\mathbf{s})| \leq 1$ with equality if $i = j$ and $\mathbf{s} = \mathbf{0}$. Finally, in practice, $\rho_{l,j}(\mathbf{s}) = \int \rho_l(\mathbf{s} - \mathbf{t})\rho_j(\mathbf{t})d(\mathbf{t})$ will have no closed form. Mujamdar and Gelfand (2007) suggest Monte Carlo integration after transformation to polar coordinates.

28.10 Final Comments and Connections

We can introduce nonstationarity into the LMC by making the the $w_j(\mathbf{s})$ nonstationary. Arguably, this is more straightforward than utilizing spatially varying $\mathbf{A}(\mathbf{s})$ though the latter enables natural interpretation. Moreover, the kernel convolution model above used a common process $w(\mathbf{s})$ for each kernel. Instead, we could introduce a $p \times p$ kernel matrix and a $p \times 1$ $\mathbf{w}(\mathbf{s}) = \mathbf{A}\mathbf{v}(\mathbf{s})$. If the kernels decay rapidly, this model will behave like an LMC. Similar extension is possible for the model (Fuentes and Smith, 2001) with similar remarks. Details are given in the rejoinder in Gelfand et al. (2004). We can also introduce multivariate space–time processes through the LMC by setting $\mathbf{w}(\mathbf{s}, t) = \mathbf{A}\mathbf{v}(\mathbf{s}, t)$ where now the $\mathbf{v}(\mathbf{s}, t)$ are independent space–time processes (Chapter 23). Of course, we can also imagine spatially varying and/or temporally varying $\mathbf{A}$s. In this regard, Sansó, Schmidt, and Noble (2008) write the LMC indexed by time, i.e., $\mathbf{Y}_t = \mathbf{A}\mathbf{w}_t$ and explore a range of modeling specifications.

References

Banerjee, S., Carlin, B.P., and Gelfand, A.E. (2004). *Hierarchical Modeling and Analysis for Spatial Data.* Boca Raton, FL: Chapman & Hall/CRC Press.

Banerjee, S. and Gelfand, A.E. (2002). Prediction, interpolation and regression for spatially misaligned data. *Sankhya, Ser. A*, **64**, 227–245.

Banerjee, S., Gelfand, A.E., Finley, A.O., and Sang, H. (2008). Gaussian predictive process models for large spatial datasets. *Journal of the Royal Statistical Society Series B*, **70**, 825–848.

Banerjee, S. and Johnson, G.A. (2006). Coregionalized single- and multi-resolution spatially-varying growth curve modelling with applications to weed growth. *Biometrics* **61**, 617–625.

Benjamin, G. and Sirmans, G.S. (1991). Determinants of apartment rent. *Journal of Real Estate Research*, **6**, 357–379.

Berliner, L.M. (2000). Hierarchical Bayesian modeling in the environmental sciences, *Allgemeines Statistisches Archiv, Journal of the German Statistical Society*, **84**, 141–153.

Brown, P., Le, N. and Zidek, J. (1994). Multivariate spatial interpolation and exposure to air pollutants. *The Canadian Journal of Statistics*, **22**, 489–509.

Chilès, J.P. and Delfiner, P. (1999). *Geostatistics: Modeling Spatial Uncertainty*. New York: John Wiley & Sons.

Clark, I., Basinger, K.L., and Harper, W.V. (1989). MUCK: A novel approach to cokriging. In *Proceedings of the Conference on Geostatistical, Sensitivity, and Uncertainty Methods for Ground-Water Flow and Radionuclide Transport Modeling*, 473–493, ed. B.E. Buxton. Battelle Press, Columbus, OH.

Corsten, L.C.A. (1989). Interpolation and optimal linear prediction. *Statistica Neerlandica*, **43**, 69–84.

Cramér, H. (1940). On the theory of stationary random processes. *Annals of Mathematics*, **41**, 215–230.

Cressie, N.A.C. (1993). *Statistics for Spatial Data*, 2nd ed. New York: John Wiley & Sons.

Cressie, N.A.C. and Wikle, C.K. (1998). The variance-based cross-variogram: You can add apples and oranges. *Mathematical Geology*, **30**, 789–799.

Daniels, M.J. and Kass, R.E. (1999). Nonconjugate Bayesian estimation of covariance matrices and its use in hierarchical models. *Journal of the American Statistical Association*, **94**, 1254–1263.

Finley, A.O., Banerjee, S. and Carlin, B.P. (2007). **spBayes**: an R package for univariate and multivariate hierarchical point-referenced spatial models. *Journal of Statistical Software*, **19**, 4.

Fuentes, M. and Smith, R.L. (2001). Modeling nonstationary processes as a convolution of local stationary processes. Technical report, Department of Statistics, North Carolina State University, Ralligh.

Gaspari, G. and Cohn, S.E. (1999). Construction of correlation functions in two and three dimensions. *Quarterly Journal of the Royal Metereological Society*, **125** 723–757.

Gelfand, A.E. and Ghosh S.K. (1998). Model choice: A minimum posterior predictive loss approach. *Biometrika*, **85**, 1–11.

Gelfand, A.E., Kim, H.-J., Sirmans, C.F., and Banerjee, S. (2003). Spatial modeling with spatially varying coefficient processes. *J. Amer. Statist. Assoc.*, **98**, 387–396.

Gelfand, A.E., Schmidt, A.M., Banerjee, S. and C.F. Sirmans, C.F. (2004). Nonstationary multivariate process modelling through spatially varying coregionalization (with discussion), *Test* **13, 2,** 1–50.

Gelfand, A.E. and Vounatsou, P. (2003). Proper multivariate conditional autoregressive models for spatial data analysis. *Biostatistics*, **4**, 11–25.

Geltner, D. and Miller, N.G. (2001). *Commercial Real Estate Analysis and Investments*. Mason, OH: South-Western Publishing.

Goulard, M. and Voltz, M. (1992). Linear coregionalization model: Tools for estimation and choice of cross-variogram matrix. *Mathematical Geology*, **24**, 269–286.

Grzebyk, M. and Wackernagel, H. (1994). Multivariate analysis and spatial/temporal scales: Real and complex models. In *Proceedings of the XVIIth International Biometrics Conference*, Hamilton, Ontario, Canada: International Biometric Society, pp. 19–33.

Harville, D.A. (1997). *Matrix Algebra from a Statistician's Perspective*. New York: Springer-Verlag.

Helterbrand, J.D. and Cressie, N.A.C. (1994). Universal cokriging under intrinsic coregionalization. *Mathematical Geology*, **26**, 205–226.

Higdon, D., Lee, H., and Holloman, C. (2003). Markov chain Monte Carlo-based approaches for inference in computationally intensive inverse problems (with discussion). In *Bayesian Statistics 7*, eds. J.M. Bernardo, M.J. Bayarri, J.O. Berger, A.P. Dawid, D. Heckerman, A.F.M. Smith, and M. West. Oxford: Oxford University Press, pp. 181–197.

Higdon, D., Swall, J., and Kern, J. (1999). Non-stationary spatial modeling. In *Bayesian Statistics 6*, eds. J.M. Bernardo, J.O. Berger, A.P. Dawid, and A.F.M. Smith. Oxford: Oxford University Press, pp. 761–768.

Journel, A.G. and Huijbregts, C.J. (1978). *Mining Geostatistics*. London: Academic Press.

Jin, X., Banerjee, S., and Carlin, B.P. (2007). Order-free coregionalized lattice models with application to multiple disease mapping. *Journal of the Royal Statistical Society Series B*, **69**, 817–838.

Majumdar, A. and Gelfand, A.E. (2007). Multivariate spatial process modeling using convolved covariance functions, *Mathematical Geology*, **79**, 225–245.

Mardia, K.V. (1988). Multi-dimensional multivariate Gaussian Markov random fields with application to image processing. *Journal of Multivariate Analysis*, **24**, 265–284.

Mardia, K.V. and Goodall, C. (1993). Spatio-temporal analyses of multivariate environmental monitoring data. In *Multivariate Environmental Statistics*, eds. G.P. Patil and C.R. Rao. Amsterdam: Elsevier, pp. 347–386.

Matheron, G. (1973). The intrinsic random functions and their applications. *Advances in Applied Probability*, **5**, 437–468.

Matheron, G. (1979). Recherche de simplification dans un probleme de cokrigeage: Centre de Gostatistique, Fountainebleau, France. N-698.

Myers, D.E. (1982). Matrix formulation of cokriging. *Journal of the International Association for Mathematical Geology*, **15**, 633–637.

Myers, D.E. (1991). Pseudo-cross variograms, positive definiteness and cokriging. *Mathematical Geology*, **23**, 805–816.

Paciorek, C.J. and Schervish, M.J. (2004). Nonstationary covariance functions for Gaussian process regression. *Advances in Neural Information Processing Systems*, **16**, 273–280.

Rehman, S.U. and Shapiro, A. (1996). An integral transform approach to cross-variograms modeling. *Computational Statistics and Data Analysis*, **22**, 213–233.

Royle, J.A. and Berliner, L.M. (1999). A hierarchical approach to multivariate spatial modeling and prediction. *Journal of Agricultural, Biological and Environmental Statistics*, **4**, 29–56.

Sansó, B., Schmidt, A.M., and Noble, A. (2008). Bayesian spatio-temporal models based on discrete convolutions. *Canadian Journal of Statistics*, **36**, 239–258.

Schmidt, A.M. and Gelfand, A.E. (2003). A Bayesian coregionalization approach for multivariate pollutant data, *Journal of Geophysical Research—Atmosphere*, **108, D24**, 8783.

Stein, A. and Corsten, L.C.A. (1991). Universal kriging and cokriging as a regression procedure. *Biometrics*, **47**, 575–587.

Vargas-Guzmán, J.A., Warrick, A.W., and Myers, D.E. (2002). Coregionalization by linear combination of nonorthogonal components. *Mathematical Geology*, **34**, 405–419.

Ver Hoef, J.M. and Barry, R.P. (1998). Constructing and fitting models for cokriging and multivariable spatial prediction. *Journal of Statistical Planning and Inference*, **69**, 275–294.

Ver Hoef, J.M. and Cressie, N.A.C. (1993). Multivariable spatial prediction: *Mathematical Geology*, **25**, 219–240.

Ver Hoef, J.M., Cressie, N.A.C., and Barry, R.P. (2004). Flexible spatial models for kriging and cokriging using moving averages and the Fast Fourier Transform (FFT). *Journal of Computational and Graphical Statistics*, **13**, 265–282.

Wackernagel, H. (1994). Cokriging versus kriging in regionalized multivariate data analysis. *Geoderma*, **62**, 83–92.

Wackernagel, H. (2003). *Multivariate Geostatistics: An Introduction with Applications*, 3rd ed. New York: Springer-Verlag.

Yaglom, A.M. (1987). *Correlation Theory of Stationary and Related Random Functions*, vol. I. New York: Springer Verlag.

Zhang, H. (2006). Maximum-likelihood estimation for multivariate spatial linear coregionalization models. *Environmetrics*, **18**, 125–139.

29

Misaligned Spatial Data: The Change of Support Problem

Alan E. Gelfand

CONTENTS

29.1 Introduction

Handling spatial misalignment and the notion that statistical inference could change with spatial scale is a long-standing issue. By spatial misalignment we mean the summary or analysis of spatial data at a scale different from that at which it was originally collected. More generally, with the increased collection of spatial data layers, synthesis of such layers has moved to the forefront of spatial data analysis. In some cases the issue is merely interpolation—given data at one scale, inferring about data at a different scale, different locations, or different areal units. In other cases, the data layers are examining the same variable and the synthesis can be viewed as a fusion to better understand the behavior of the variable across space. In yet other settings, the data layers record different variables. Now the objective is regression, utilizing some data layers to explain others. In each of these settings, we can envision procedures that range from ad hoc, offering computational simplicity and convenience but limited in inference to fully model-based that require detailed stochastic specification and more demanding computation but allowing richer inference. In practice, valid inference requires assumptions that may not be possible to check from the available data.

As an example, we might wish to obtain the spatial distribution of some variable at the county level, even though it was originally collected at the census tract level. Or, we might have a very low-resolution global climate model for weather prediction, and seek to predict more locally (i.e., at higher resolution). For areal unit data, our purpose might be simply to understand the distribution of the variable at a new level of spatial aggregation (the so-called *modifiable areal unit problem*, or MAUP), or perhaps so we can relate it to another variable that is already available at this level (say, a demographic census variable collected over the tracts). For data modeled through a spatial process, we would envision *averaging* the process to capture it at different spatial scales, (the so-called *change of support problem*, or COSP), again possibly for connection with another variable observed at a particular scale.

A canonical illustration, which includes regression, arises in an environmental justice context. Ambient exposure to an environmental pollutant may be recorded at a network of monitoring stations. Adverse health outcomes that may be affected by exposure are provided at, say, zip or postcode level to ensure privacy. Finally, population at risk and demographic information may be available at census level units, e.g., at tract or block level. How can we use these layers to identify areas of elevated exposure? How can we assess whether a particular demographic group has disproportionately high exposure? How can we determine whether risk of an adverse outcome is increased by increased exposure and how much it is increased, adjusting for population at risk?

The format of this chapter is, first, to provide a historical perspective of the work on these problems. In particular, the origins of this work are not in the statistics community, but rather in the geography and social sciences worlds. Then, we will describe in some detail approaches for treating the foregoing problems. In this regard, we will build from naivé efforts to sophisticated hierarchical modeling efforts. In the process, we will make connections with aggregation/disaggregation issues and the ecological fallacy, which provide the subject matter for Chapter 30. The fallacy refers to the fact that relationships observed between variables measured at the ecological (aggregate) level may not accurately reflect (and may possibly overstate) the relationship between these same variables measured at the individual level. Reflecting the foregoing development on process modeling (Section 29.2) and discrete spatial variation (Section 29.3), we present our discussion according to whether the data are suitably modeled using a spatial process as opposed to a conditionally auto regressive (CAR) or simultaneous auto regressive (SAR) model. Here, the former assumption leads to more general modeling, since point-level data may be naturally aggregated to so-called *block* level, but the reverse procedure may or may not be possible; e.g., if the areal data are counts or proportions, what would the point-level variables be? However, since block-level summary data arise quite frequently in practice (often due to confidentiality restrictions), methods associated with such data are also of importance. We thus consider misalignment in the context of both point-level and block-level modeling.

To supplement the presentation here, we encourage the reader to look at the excellent review paper by Gotway and Young (2002). These authors give nice discussions of (as well as both traditional and Bayesian approaches for) the MAUP and COSP, spatial regression, and the *ecological fallacy*. Consideration of this last problem dates at least to Robinson (1950) (see Wakefield, 2003, 2004, 2008, for more modern treatments of this difficult subject).

We conclude this introduction with a more formal formulation of the problems. First, consider a single variable, a univariate setting and the COSP, areal interpolation, MAUP problems. We can envision four cases:

1. We observe point-referenced data $Y(\mathbf{s}_i)$ at locations $\mathbf{s}_i, i = 1, 2, \ldots, n$ and we seek to infer about the process at new locations $\mathbf{s}_j^*, j = 1, 2, \ldots m$. This problem is, in fact, spatial kriging as discussed in detail in Chapters 3, 4, and 7.
2. We observe point-referenced data $Y(\mathbf{s}_i)$ at locations $\{\mathbf{s}_i, i = 1, 2, \ldots, n\}$ and we seek to infer about the process at a collection of areal units, i.e., $Y(B_j)$ associated with $B_j, j = 1, 2, \ldots, m$. Here we are envisioning a suitable average for the process to assign to each of the B_j.
3. We have observations associated with areal units, i.e., $Y(B_i)$ at areal unit $B_i, i = 1, 2, \ldots, n$ and we seek to infer about the process at a collection of locations, say,

$\mathbf{s}_j^*, j = 1, 2, \ldots, m$. Here the caveat above applies; we may not be able to sensibly conceptualize the random variables $Y(\mathbf{s}_j^*)$.

4. We have observations associated with areal units, e.g., $Y(B_i)$ associated with $B_i, i = 1, 2, \ldots, n$ and we seek to infer about observations $Y(B_j^*)$ at $B_j^*, j = 1, 2, \ldots, m$. An issue here is whether or not the B_j^*s are nested within (a refinement of) the B_is or merely a different partition. In the latter case, it may be that $\cup_i B_i \neq \cup_j B_j^*$.

Second, we turn to the regression setting. We simplify to one explanatory variable and one response variable. Then, letting X denote the explanatory variable and Y denote the response variable, we seek to regress Y on X. Again, there are four cases:

1. We observe Y in the form $Y(B_i)$ at areal units $B_i, i = 1, 2, \ldots, n$ with X in the form $X(C_j)$ at $j = 1, 2, \ldots, m$.
2. We observe Y in the form $Y(B_i)$ at areal units $B_i, i = 1, 2, \ldots, n$ with X in the form $X(\mathbf{s}_j^*)$ at locations $\mathbf{s}_j^*, j = 1, 2, \ldots, m$.
3. We observe Y in the form $Y(\mathbf{s}_i)$ at locations $\mathbf{s}_i, i = 1, 2, \ldots, n$ with X in the form $X(\mathbf{s}_j^*)$ at locations $\mathbf{s}_j^*, j = 1, 2, \ldots, m$.
4. We observe Y in the form $Y(\mathbf{s}_i)$ at locations $\mathbf{s}_i, i = 1, 2, \ldots, n$ with X in the form $X(B_j)$ at areal units $B_j, j = 1, 2, \ldots, m$.

In general, we would seek to model at the highest resolution possible, at point-level if we can. Of course, such modeling will be the most computationally demanding. At the least, we would like to model the regression at the scale of the response variable but, for instance, case (4) might preclude that and we would instead have to average up the response variable. Also, in case (4), we note the potential for the *atomistic fallacy* (see Chapter 30), where invalid inference may arise if appropriate block-level contextual variables are not included in the model.

We shall see below that misaligned spatial regression is implemented as a COSP for X followed by a regression of Y on X (except perhaps, in case (4)). We shall also find that we implement the COSP itself through the use of covariate information. Here, we seek to improve upon naivé areally weighted interpolation. For instance, with Y_i observed at B_i, we infer about Y_{ij} at $B_{ij} \subset B_i$ using X_{ij} via a latent regression of Y_{ij} on X_{ij}.

Finally, we offer some discussion on a currently active misalignment issue, data fusion or assimilation, where, for a given variable, we have data layers at different scales that we seek to fuse in order to learn about the behavior of the variable at point-level everywhere in the region of interest.

29.2 Historical Perspective

It has long been recognized that the scale for study of spatial processes affects the interpretation of the process dynamics. Mechanisms operating at fine spatial scale may be irrelevant at large spatial scales and, conversely, mechanisms operating at large spatial scales may not even be seen at small scales. Such scaling issues are well appreciated in studying human, animal, and plant populations and have led to investigations by researchers in agriculture, geography, sociology, ecology, earth and environmental sciences, and, of course, statistics (see, e.g., Fairfield Smith, 1938; Jelinski and Wu, 1996; King, 1997; Levin, 1992; Richardson, 1992; Robinson, 1950). Indeed, with increasing research emphasis on process modeling and understanding, scaling concerns have become a much discussed problem, encompassing a broader range of problems than we consider in this chapter. Some aspects of this discussion will be taken up in Chapter 30. Here, our focus is on methodology for treating misaligned data situations.

Gotway and Young (2002) are careful to point out the COSP includes the MAUP, which in turn includes the ecological fallacy. That is, the ecological bias that is introduced when doing analysis based upon grouped individuals rather than on the individuals themselves is referred to as "ecological bias" (Greenland and Robins, 1994; Richardson, 1992). It is comprised of *aggregation bias* due to grouping of individuals and *specification bias* due to the fact that the distribution of confounding variables varies with grouping. In the MAUP literature, these two effects are referred to as the *scale* effect and the *zoning* effect (Openshaw, 1984; Openshaw and Taylor, 1979, 1981; Wong, 1996). The former acknowledges differences that emerge when the same dataset is grouped into increasingly larger areal units. The latter recognizes that, even at the same scale, different definitions / formations of areal units can lead to consequentially different levels of attributes that are associated with the units. Within the notion of the COSP, "support" can be viewed as the size or volume associated with a given data value so that this allows for geometrical size, shape, and orientation to be associated with measurements. Changing the support implies that a new random variable is created whose distribution may be developed from the original one but, in any event, has different statistical and spatial properties. Arbia (1989) uses the terminology *spatial data transformations* to refer to situations where the process of interest is inherently in one spatial form, but the data regarding the process are observed in another spatial form. Hence, a "transformation" is required from the observed form to the form of interest. In the sequel, we confine ourselves to points and to areal units (the latter stored as a shape file in the form of polygonal line segments).

Areal interpolation, which is the objective of the MAUP, is well discussed in the geography literature (Goodchild and Lam, 1980; Mrozinski and Cromley, 1999). Regions for which the data are available are referred to as source zones, regions for which we seek to interpolate are referred to as target zones. Effects of the MAUP have typically been studied by focusing on the variation in the correlation coefficient across choices of zones. Openshaw and Taylor (1979) studied this in the context of proportion of elderly voters by county in the state of Iowa. By using various groupings, they achieved spatial correlations between percent elderly and percent registered Republican ranging from −.81 to +.98. Similar results were found in earlier studies, e.g., Gehlke and Biehl (1934); Robinson (1950); Yule and Kendall (1950). With a single variable, these correlations are typically measured using Moran's I (see Chapter 14). One emergent suggestion from this work, as well as from follow-on papers, is the need to analyze areal unit data at multiple scales with multiple choices of zones. With widely available geographic information system (GIS) software, the diversity of outcomes associated with zone design can be implemented efficiently through repeated scaling and aggregation experiments (Longley, Goodchild, Maguire, and Rhind, 2001; Openshaw and Alvanides, 1999). The use of well-chosen grouping variables to adjust the area-level results can help to clarify individual-level relationships and, thus, assist in combating the ecological fallacy (see, e.g., Holt, Steel, and Tranmer, 1996). Sections 29.3 and 29.4 below formalize this approach in a fully model-based way.

Interesting debate in the geography literature has focused on whether the MAUP should be addressed solely with techniques independent of areal units. Tobler (1979) argued that all methods whose results depend upon areal units should be discarded a priori. Thus, improved spatial statistics tools are what is needed. (See also King, 1997, p. 250). In contrast, Openshaw and Taylor (1981) argued that "a context-free approach is contrary to geographic common sense, irrespective of what other virtues it might have." Thus, the right approach to the MAUP should be geographical rather than purely statistical or mathematical.

In the geography literature, there are a variety of algorithmic approaches to implement a solution to the MAUP, i.e., to implement an interpolation from a set of source zones to a set of target zones. We illustrate one such strategy here, due to Mrozinski and Cromley (1999), which relies heavily on GIS tools. In particular, overlays of polygonally defined data layers are used. Suppose one layer provides a set of areal units with their areas and populations (an aggregated variable), the source zones. We seek to assign population values to a new set of units, the target zones, knowing only their areas. Suppose we rasterize (i.e., partition) the entire region to small grid cells and assign to each grid cell a source zone and a target zone. Typically, edge effects are ignored, especially if the raster cells are small. (In any event, a majority rule with regard to area provides a unique assignment of cells to zones.) Next, suppose we areally allocate population to each of the raster cells using the source zone populations. That is, if there are, say, m_j cells in source zone B_j, we assign $1/m_j$ of the population to each. Next, we implement a smoothing or filtering function, assigning to each cell a new value that is the average of the current cell value and its four neighbors. This is done for all cells. Of course, now the sum of the cell populations does not agree with the total population for each of the source zones. So, a rescaling of the cell values is done to *preserve volume* relative to the source zones. This completes one iteration. This process is iterated until convergence is reached where convergence is measured using a suitable but arbitrarily selected criterion. Then, the cell values are summed appropriately to obtain values for the target cells.

The major weakness of this approach is the presumption that population is uniformly distributed over areas (we return to this point in Section 29.3). This concern encourages the use of additional covariate information that is connected to the variable being allocated and is not uniformly distributed (again, we will return to this). Working with GIS tools, rasterization to grid cells can be replaced with *resels*, resolution elements. These are the polygonal units that are created by overlaying the source and target areal units and can be used directly in the above iterative scheme. This avoids concern with edge issues. A variant of this problem, which has received little attention (again see Mrozinski and Cromley, 1999), is the case of two layers of source zones for the same attribute, where interest lies in assigning values to associated rasterized grid cells or resels.

Though spiritually similar to kriging (which is well discussed in the statistical literature), the areal interpolation problem has received little attention in the statistical literature. The first notable effort is by Tobler (1979). He offered a so-called *pycnophylactic* approach (volume preserving, as above). For what he referred to as *intensive* variables, i.e., variables that can be viewed as averaged to the areal units, such as rainfall, temperature, and environmental exposure, Tobler envisioned an intensity surface, $\lambda(\mathbf{s})$ over the study area. Then, for areal unit B, $E(Y(B)) = \int_B \lambda(\mathbf{s})d\mathbf{s}/|B|$ where $|B|$ denotes the area of B. In other words, the expected value for $Y(B)$ is the average of the intensity surface over B. Next, if $\hat{\lambda}(s)$ is an estimate of the intensity, $\hat{Y}(B) = \int_B \hat{\lambda}(\mathbf{s})d\mathbf{s}/|B|$. In particular, with observed $Y(B_i)$, $i = 1, 2, \ldots, n$, associated with exclusive and exhaustive sets B_i, he suggests method of moments estimation, equating $Y(B_i)$ to $E(Y(B_i))$, the pycnophylactic ("volume-preserving") property. Evidently, this fitting does not uniquely determine $\lambda(\mathbf{s})$; one choice is $\lambda(\mathbf{s}) = Y(B_i)$ if $\mathbf{s} \in B_i$. Tobler (1979) proposed smoothing λ by minimizing an integrated total squared differential form. A discretized implementation leads to a spline form solution. Kernel-based, locally weighted smoothing was suggested in Brillinger (1990). In the absence of smoothing, if $B = \bigcup A_i$, where the A_i are disjoint, it is easy to see that $\hat{Y}(B) = \sum_i w_i \hat{Y}(A_i)$ where $w_i = |A_i|/|B|$, i.e., an areal weighted average.

Flowerdew and Green (1989) introduced a more formal modeling approach. Working with so-called *extensive* variables, i.e., variables that cumulate over a region, such as counts, they introduced better covariates than area to implement the interpolation (in the spirit of Holt et al., 1996). Depending on the context, such areal unit variables could be land-use classifications, number of structures (perhaps of different types), topography information, such as elevation and aspect, etc. For counts at areal units, consider a classification covariate, with classes indexed by $\ell, \ell = 1, 2, \ldots, L$, let λ_ℓ denote the constant intensity associated with classification ℓ. That is, if classification ℓ operates exclusively over unit B, then $Y(B)$, the count associated with this unit, has a Poisson distribution, $\text{Po}(\lambda_\ell |B|)$. If the target zone, $|B|$, is a mix of classifications with areal proportions $|A_\ell|/|B| = p_\ell$, then $Y(B) \sim Po(\sum p_\ell \lambda_\ell |B| = \sum \lambda_\ell |A_\ell|)$. Finally, the λ_ℓ would be estimated from the source zone data using a Poisson regression.

Flowerdew and Green (1989) extended this idea using the expectation-maximization (EM) algorithm (Dempster, Laird, and Rubin, 1977; Tanner, 1996). Given a set of source zones $\{B_j\}$ and a set of target zones $\{C_k\}$, let $A_{jk} = B_j \bigcap C_k$. (Of course, many of the A_{jk} are empty.) The approach is to use the $Y(B_j)$ to impute $Y(A_{jk})$ and then obtain $Y(C_k)$ by summing over j. Again, with counts, it is assumed that $Y(A_{jk}) \sim Po(\lambda(A_{jk}))$ conditionally independent given the λs. The E step is implemented by computing $E(Y(A_{jk})|\lambda(A_{jk}), Y(B_j))$, under a conditional multinomial distribution. The M step is implemented by estimating $\lambda(A_{jk})$ again using a Poisson regression given the $\{Y(A_{jk})\}$. Apart from using $|A_{jk}|$, covariates specific to the target zone $X(C_k)$ can be employed.

Taking these ideas a bit further, Flowerdew and Green (1992, 1994) examined the intensive case. Viewing the continuous variables as, say, house prices, for B_j, $Y(B_j)$ is the average of, say, n_j observations in B_j. (The prices could be covariate adjusted or could be associated with a collection of similar houses.) Now, we seek to say something about average house prices associated with C_k using associated covariate information, $X(C_k)$. Though n_j is known, n_{jk} is not but a "cheap" interpolation, $n_{jk} = n_j \frac{|A_{jk}|}{|B_k|}$ is proposed. Then, a standard linear regression for $Y(A_{jk})$ is adopted, again assuming conditional independence. Since the conditional distribution of $Y(A_{jk})$ given $Y(B_j)$ is normal, the conditional mean provides the E step; estimating the regression model given $\{Y(A_{jk})\}$ provides the M step.

29.3 Misalignment through Point-Level Modeling

29.3.1 The Block Average

Consider a univariate variable that is modeled through a spatial process. It may be observed either at points in space, or over areal units (e.g., counties or zip codes), which we will refer to as *block* data. The *change of support problem* is concerned with inference about the values of the variable at points or blocks different from those at which it has been observed.

Let $Y(\mathbf{s})$ denote the spatial process (e.g., ozone level) measured at location $\mathbf{s}$, for $\mathbf{s}$ in some region of interest D. In our applications $D \subset \Re^2$, but our development works in arbitrary dimensions. A realization of the process is a surface over D. For point-referenced data the realization is observed at a finite set of sites, say, $\mathbf{s}_i, i = 1, 2, \ldots, I$. For block data we assume the observations arise as block averages. That is, for a block $B \subset D$,

$$Y(B) = |B|^{-1} \int_B Y(\mathbf{s}) d\mathbf{s}, \tag{29.1}$$

where $|B|$ denotes the area of B (see, e.g., Cressie, 1993). The integration in Equation (29.1) is an average of random variables, hence, a random or stochastic integral. (Though $Y(s)$ is a *function* over D, it has no analytical form that we can use.) Thus, the assumption of an underlying spatial process is only appropriate for block data that can be sensibly viewed as an averaging over point data; examples of this would include rainfall, pollutant level, temperature, and elevation. It would be inappropriate for, say, population, since there is no "population" at a particular point. It would also be inappropriate for most proportions. For instance, if $Y(B)$ is the proportion of college-educated persons in B, then $Y(B)$ is continuous, but even were we to conceptualize an individual at every point, $Y(\mathbf{s})$ would be binary.

Block averaging is distinct from a mean or density surface over D. That is, we could certainly envision a rainfall density over D, $\mu(\mathbf{s})$, such that the expected average rainfall in block B, $E(Y(B))$, is $\mu(B) = |B|^{-1} \int_B \mu(\mathbf{s})d\mathbf{s}$. We could also envision a population density over D such that the expected population in B cumulates via $\int_B \mu(\mathbf{s})d\mathbf{s}$. The latter provides an evident generalization of areal weighting (where $\mu(\mathbf{s})$ would be constant).

Inference about blocks through averages as in Equation (29.1) is not only formally attractive, but demonstrably preferable to ad hoc approaches. One such approach would be to average over the observed $Y(\mathbf{s}_i)$ in B. But this presumes there is at least one observation in any B and ignores the information about the spatial process in the observations outside of B. Another ad hoc approach would be to simply predict the value at some central point of B. But this value has larger variability than (and may be biased for) the block average. So, instead, we confine ourselves to block averages and take up a brief review of their properties.

Evidently, we cannot compute $Y(B)$ exactly. However, if we know the mean surface and covariance function associated with $Y(s)$, we can compute the mean and variance of $Y(B)$. In particular, if $E(Y(\mathbf{s})) = \mu(\mathbf{s}; \beta)$ and $\text{Cov}(Y(\mathbf{s}), Y(\mathbf{s}')) = C(\mathbf{s} - \mathbf{s}'; \phi)$,

$$E(Y(B)|\beta) \equiv \mu(B; \beta) = |B|^{-1} \int_B \mu(\mathbf{s}; \beta)d\mathbf{s} \tag{29.2}$$

and

$$\text{var}(Y(B)|\phi) \equiv v(B; \phi) = |B|^{-2} \int_B \int_B C(\mathbf{s} - \mathbf{s}'; \phi)d\mathbf{s}d\mathbf{s}'. \tag{29.3}$$

Moreover, if the process is Gaussian, then $Y(B) \sim N(\mu(B; \beta), v(B; \phi))$. In fact, for two sets B_1 and B_2, not necessarily disjoint, we have

$$\text{Cov}(Y(B_1), Y(B_2)|\phi) \equiv C(B_1, B_2; \phi) = |B_1|^{-1}|B_2|^{-1} \int_{B_1} \int_{B_2} C(\mathbf{s} - \mathbf{s}'; \phi)d\mathbf{s}d\mathbf{s}' \tag{29.4}$$

and

$$\text{Cov}(Y(B); Y(\mathbf{s})|\phi) \equiv C(B, \mathbf{s}; \phi) = |B|^{-1} \int_B C(\mathbf{s}', \mathbf{s}; \phi)d\mathbf{s}'. \tag{29.5}$$

Hence, in this situation, we can work with block averages without ever having to attempt to compute stochastic integrals.

In this regard, we summarize the ideas in Gelfand, Zhu, and Carlin (2001). If the Bs are regular in shape, the integrals in Equation (29.2) to Equation (29.5) can be well computed using Riemann approximation. However, such integration over irregularly shaped Bs may be awkward. Instead, noting that each such integration is an expectation with respect to a uniform distribution, we propose Monte Carlo integration. In particular, for each B we propose to draw a set of locations $\mathbf{s}_\ell$, $\ell = 1, 2, \ldots, L$, distributed independently and uniformly over B. Here, L can vary with B to allow for very unequal $|B|$s.

Thus, we replace $\mu(B; \beta)$, $v(B; \phi)$, $\text{Cov}(Y(B_1), Y(B_2))$, and $\text{Cov}(Y(B); Y(\mathbf{s}))$ with $\hat{\mu}(B; \beta) = L^{-1} \sum_\ell \mu(\mathbf{s}_\ell; \beta)$, $\hat{v}(B; \phi) = L^{-2} \sum_\ell \sum_{\ell'} C(\mathbf{s}_\ell - \mathbf{s}_{\ell'}; \phi)$, $\hat{C}(B_1, B_2; \phi) = L_1^{-1} L_2^{-1} \sum_\ell \sum_{\ell'} C(\mathbf{s}_{1\ell} - \mathbf{s}_{2\ell'}; \phi)$, and $\hat{C}(B, \mathbf{s}; \phi) = L^{-1} \sum_\ell C(\mathbf{s} - \mathbf{s}_\ell; \phi)$. In our notation, the "hat" denotes a Monte Carlo integration that can be made arbitrarily accurate and has nothing to do with the amount of data collected on the $Y(\mathbf{s})$ process. It is useful to note that if we define $\hat{Y}(B) = L^{-1} \sum_\ell Y(\mathbf{s}_\ell)$, then $\hat{Y}(B)$ is a Monte Carlo integration for $Y(B)$.

As a technical point, we might ask when $\hat{Y}(B) \xrightarrow{P} Y(B)$. An obvious sufficient condition is that realizations of the $Y(\mathbf{s})$ process are almost surely continuous. In the stationary case, Kent (1989) provides sufficient conditions on $C(\mathbf{s} - \mathbf{s}'; \phi)$ to ensure this. Alternatively, Stein

(1999) defines $Y(\mathbf{s})$ to be mean square continuous if $\lim_{\mathbf{h}\to 0} E(Y(\mathbf{s}+\mathbf{h}) - Y(\mathbf{s}))^2 = 0$ for all $\mathbf{s}$. But, then $Y(\mathbf{s}+\mathbf{h}) \xrightarrow{P} Y(\mathbf{s})$ as $\mathbf{h} \to 0$, which is sufficient to guarantee that $\hat{Y}(B) \xrightarrow{P} Y(B)$. Stein (1999) notes that if $Y(\mathbf{s})$ is stationary, we only require $C(\cdot\,; \phi)$ continuous at $\mathbf{0}$ for mean square continuity.

Beginning with point data observed at sites $\mathbf{s}_1, \ldots, \mathbf{s}_I$, let $\mathbf{Y}_s^T = (Y(\mathbf{s}_1), \ldots, Y(\mathbf{s}_I))$. Then, under a Gaussian process with mean and covariance function as above, with evident notation,

$$\mathbf{Y}_s \mid \beta, \phi \sim N(\boldsymbol{\mu}_s(\beta), C_s(\phi)). \tag{29.6}$$

Moreover, with $\mathbf{Y}_B^T = (Y(B_1), \ldots, Y(B_K))$, we have

$$f\left(\begin{pmatrix}\mathbf{Y}_s\\ \mathbf{Y}_B\end{pmatrix}\middle|\, \beta, \phi\right) = N\left(\begin{pmatrix}\boldsymbol{\mu}_s(\beta)\\ \boldsymbol{\mu}_B(\beta)\end{pmatrix}, \begin{pmatrix}C_s(\phi) & C_{s,B}(\phi)\\ C_{s,B}^T(\phi) & C_B(\phi)\end{pmatrix}\right). \tag{29.7}$$

From Equation (29.7), it is clear that we have the conditional distribution for any set of blocks given any set of points and vice versa. Hence, in terms of spatial prediction, we can implement any of the customary kriging approaches (as discussed in Part II). And, implicitly, we can do so for any set of points given another set of points and for any set of blocks given another set of blocks. So, with point referenced data, the misalignment problem for a single variable can be handled by available kriging methods.

For instance, following Journel and Huijbrects (1978) or Chiles and Delfiner (1999), we can implement universal block kriging to obtain $\hat{Y}(B) = \sum_{i=1}^{n} \lambda_i Y(\mathbf{s}_i)$ where the optimal weights, λ_i, are, in general, obtained via Lagrange multipliers, minimizing predictive mean square error subject to unbiasedness constraints as in usual kriging. Under normality, they arise from the conditional mean of $Y(B)|\{Y(\mathbf{s}_i), i = 1, 2, \ldots, \}$.

Within the Bayesian framework, with a prior on β and ϕ, to implement block kriging given point referenced data $\mathbf{Y}_s$, we would seek

$$f(\mathbf{Y}_B \mid \mathbf{Y}_s) = \int f(\mathbf{Y}_B \mid \mathbf{Y}_s; \beta, \phi) f(\beta, \phi \mid \mathbf{Y}_s) d\beta d\phi. \tag{29.8}$$

A Markov chain Monte Carlo (MCMC) algorithm would sample this distribution by composition, drawing (β, ϕ) from $f(\beta, \phi \mid \mathbf{Y}_s)$ and then the $Y(B)$s from $f(\mathbf{Y}_B \mid \mathbf{Y}_s; \beta, \phi)$. The Monte Carlo approximations above would replace $f\left((\mathbf{Y}_s, \mathbf{Y}_B)^T \mid \beta, \phi\right)$ with

$$\hat{f}((\mathbf{Y}_s, \mathbf{Y}_B)^T \mid \beta, \phi) = f((\mathbf{Y}_s, \hat{\mathbf{Y}}_B)^T \mid \beta, \phi), \tag{29.9}$$

where Equation (29.9) is interpreted to mean that the approximate joint distribution of $(\mathbf{Y}_s, \mathbf{Y}_B)$ is the exact joint distribution of $\mathbf{Y}_s, \hat{\mathbf{Y}}_B$. In practice, we would work with $\hat{f}$, converting to $\hat{f}(\mathbf{Y}_B \mid \mathbf{Y}_s, \beta, \phi)$ to sample $\mathbf{Y}_B$ rather than using Monte Carlo integrations to sample the $\hat{Y}(B_k)$s. But, evidently, we are sampling $\hat{\mathbf{Y}}_B$ rather than $\mathbf{Y}_B$.

For the modifiable areal unit problem (i.e., prediction at new blocks using data for a given set of blocks), suppose we take as our point estimate for a generic new set B_0 the posterior mean,

$$E(Y(B_0) \mid \mathbf{Y}_B) = E\{\mu(B_0; \beta) + C_{B,B_0}^T(\phi) C_B^{-1}(\phi)(\mathbf{Y}_B - \boldsymbol{\mu}_B(\beta)) \mid \mathbf{Y}_B\},$$

where $C_{B,B_0}(\phi)$ is $I \times 1$ with ith entry equal to $\mathrm{Cov}(Y(B_i), Y(B_0) \mid \phi)$. If $\mu(\mathbf{s}; \beta) = \mu_i$ for $\mathbf{s} \in B_i$, then $\mu(B_0; \beta) = |B_0|^{-1} \sum_i |B_i \cap B_0| \, \mu_i$. But $E(\mu_i \mid \mathbf{Y}_B) \approx Y(B_i)$ to a first-order approximation. So in this case, $E(Y(B_0) \mid \mathbf{Y}_B) \approx |B_0|^{-1} \sum_i |B_i \cap B_0| \, Y(B_i)$, the areally weighted estimate.

A noteworthy remark here is that, often, we are modeling a process, say $Z(\mathbf{s})$, which is necessarily positive, for instance, an intensity surface as in a Cox process (see Chapters 16 and 17). An obvious strategy is to model $Z(\mathbf{s})$ as a log Gaussian process, i.e., $Z(\mathbf{s}) = \exp(Y(\mathbf{s}))$

where $Y(\mathbf{s})$ is a Gaussian process. But now, block averaging as in Equation (29.1) becomes awkward; we require $Z(B) = |B|^{-1} \int_B exp(Y(\mathbf{s}))d\mathbf{s}$. Evidently, $Z(B)$ does not follow a lognormal distribution and its dependence structure is difficult to calculate. A frequent "cheat" is to use $\tilde{Z}(B) = \exp(|B|^{-1} \int_B Y(\mathbf{s})d\mathbf{s})$, which is lognormal. However, $\tilde{Z}(B) \leq Z(B)$ by Jensen's inequality. Moreover, the impact of such approximation is unclear, but can potentially create an ecologically fallacy in a regression setting (see Chapter 30). Similar concerns would arise using an alternative transformation from $R^1 \rightarrow R^+$. To avoid these problems, we will have to work with Monte Carlo or Riemann integrations for $Z(B)$, e.g., $\hat{Z}(B) = L^{-1} \sum_l exp(Y(\mathbf{s}_l))$. This basic issue arises with other nonlinear kriging situations. Consider, for example, indicator kriging where we look at the binary variable $I(Y(\mathbf{s}) \leq y) = 1$ if $Y(\mathbf{s}) \leq y$, 0 otherwise. If we are interested in indicator block kriging, $I(Y(B) \leq y)$, we again find that $I(Y(B) \leq y) \neq |B|^{-1} \int_B I(Y(\mathbf{s}) \leq y)d\mathbf{s}$. Again, we would employ numerical integration. Simulation methods have been used to handle such approximation strategies. Goovaerts (1997) proposes forward simulation conditional on estimation of the model parameters. The fully Bayesian approach would implement an *anchored* simulation, i.e., an MCMC algorithm, conditioned on the data to obtain posterior predictive samples (see, e.g., De Oliveira, Kedem, and Short, 1997, in this regard). The fully Bayesian approach is substantially more computationally demanding.

The foregoing development can easily be extended to space–time data settings. We briefly present the details. Suppose we envision a spatio-temporal Gaussian process with stationary covariance function $\text{Cov}(Y(\mathbf{s}, t), Y(\mathbf{s}', t')) = C(\mathbf{s} - \mathbf{s}', t - t'; \phi)$. We block average only in space, at particular points in time. The distribution theory is similar to that in Equation (29.2) to Equation (29.5) (with mean surface $\mu(\mathbf{s}, t; \beta)$), that is,

$$E(Y(B, t)|\beta) \equiv \mu(B, t; \beta) = |B|^{-1} \int_B \mu(\mathbf{s}, t; \beta)d\mathbf{s} \tag{29.10}$$

and

$$\text{var}(Y(B, t)|\phi) \equiv v(B, t; \phi) = |B|^{-2} \int_B \int_B C(\mathbf{s} - \mathbf{s}', 0; \phi)d\mathbf{s}d\mathbf{s}'. \tag{29.11}$$

Under stationarity in time, the variance is free of t. Moreover, if the process is Gaussian, then $Y(B) \sim N(\mu(B, t; \beta), v(B, t; \phi))$. In fact, for two sets B_1 and B_2, not necessarily disjoint, at times t_1 and t_2, we have

$$\text{Cov}(Y(B_1, t_1), Y(B_2, t_2)|\phi) \equiv C(B_1, B_2; \phi) = |B_1|^{-1}|B_2|^{-1} \int_{B_1} \int_{B_2} C(\mathbf{s} - \mathbf{s}', t_1 - t_2; \phi)d\mathbf{s}d\mathbf{s}' \tag{29.12}$$

and

$$\text{Cov}(Y(B, t'); Y(\mathbf{s}, t)|\phi) \equiv C(B, \mathbf{s}; \phi) = |B|^{-1} \int_B C(\mathbf{s}', \mathbf{s}, t' - t; \phi)d\mathbf{s}'. \tag{29.13}$$

Again, we use Riemann or Monte Carlo integration to approximate these integrals.

We confine ourselves to the case of predicting block averages given point data. Suppose we sample data $Y(\mathbf{s}_i, t_j)$ at n locations across T time points and we seek to predict $Y(B_0, t_0)$. The joint distributions are substantially simplified assuming a covariance function that is separable in space and time, i.e., $\text{Cov}(Y(\mathbf{s}, t), Y(\mathbf{s}', t')) = \sigma^2 \rho_1(\mathbf{s} - \mathbf{s}'; \phi_1)\rho_2(t - t'; \phi_2)$. Then, the distribution of $\mathbf{Y}$, the vector that concatenates the $Y(\mathbf{s}_i, t_j)$ by location and then time within location, is an nT-dimensional multivariate normal, $\mathbf{Y} \sim N(\boldsymbol{\mu}(\beta), \Sigma_{\mathbf{Y}} = \sigma^2 R_1(\phi_1) \otimes R_2(\phi_2))$ where R_1 is the $n \times n$ matrix with entries $\rho_1(\mathbf{s}_i - \mathbf{s}_j; \phi_1)$, R_2 is the $T \times T$ matrix with entries $\rho_2(t_i - t_j; \phi_2)$, and "$\otimes$" denotes the Kronecker product. Furthermore, the joint distribution of $\mathbf{Y}$, $Y(B_0, t_0)$ is an $nT + 1$–dimensional multivariate normal, which yields the conditional distribution

$$Y(B_0, t_0)|\mathbf{y}; \beta, \sigma^2, \phi_1, \phi_2 \sim N\big(\mu(B_0, t_0); \beta + \Sigma_{0,\mathbf{Y}}^T \Sigma_{\mathbf{Y}}^{-1}(\mathbf{Y} - \boldsymbol{\mu}(\beta)), v(B_0, t_0) - \Sigma_{0,\mathbf{Y}}^T \Sigma_{\mathbf{Y}}^{-1} \Sigma_{0,\mathbf{Y}}\big),$$

where $\Sigma_{0,Y}$ is an $nT \times 1$ vector with entries $\text{Cov}(Y(\mathbf{s}_i, t_j), Y(B_0, t_0))$. This is the requisite kriging distribution. Again, within the Bayesian framework, we would make predictive draws of $Y(B_0, \mathbf{s}_0)$ by composition.

29.4 Nested Block-Level Modeling

We now turn to the case of variables available only as block-level summaries. For example, it might be that data are known at the county level, but hypotheses of interest pertain to postcodes or census tracts. As above, we refer to regions on which data are available as source zones and regions for which data are needed as target zones.

In fact, we confine ourselves to variables that are aggregative, i.e., they can be viewed as sums over subblocks. We first focus on the *nested* misalignment setting where the target zonation of the spatial domain D is a refinement of the source zonation, following the work of Mugglin and Carlin (1998). In the setting below, the source zones are U.S. census tracts, while the target zones (and the zones on which covariate data are available) are U.S. census block groups.

Consider the diagram in Figure 29.1. Assume that a particular rectangular tract of land is divided into two regions (I and II), and spatial variables (say, disease counts) Y_1 and Y_2 are known for these regions (the source zones). But, suppose that the quantity of interest is Y_3, the unknown corresponding count in Region III (the target zone), which is comprised of subsections (IIIa and IIIb) of Regions I and II.

As already mentioned, a crude way to approach the problem is to assume that disease counts are distributed evenly throughout Regions I and II, and so the number of affected individuals in Region III is just

$$y_1 \left[\frac{area(IIIa)}{area(I)}\right] + y_2 \left[\frac{area(IIIb)}{area(II)}\right]. \tag{29.14}$$

This simple areal interpolation approach is available within many GIS packages, but seems unattractive due to the uniformity assumption. Moreover, it offers no associated estimate of uncertainty.

Suppose that each unit can be partitioned into smaller subunits, where on each subunit we can measure some other variable that is correlated with y. For instance, if we are looking at a

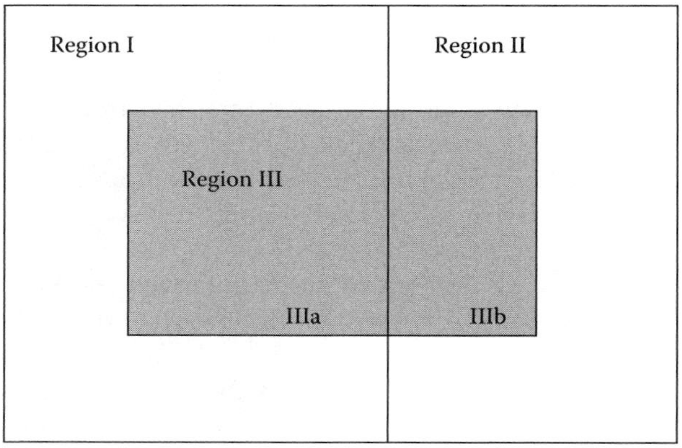

FIGURE 29.1
Regional map for motivating example.

particular tract of land, in each subunit we might record whether the land is predominantly rural or urban in character. Continuous covariates could also be used (say, the median household income in the unit). Suppose further, that the subunits arise as a refinement of the original scale of aggregation (e.g., if disease counts were available only by census tract, but covariate information arose at the census block group level), or as the result of overlaying a completely new set of boundaries (say, a zip code map) onto the original map. The statistical model is easier to formulate in the former case, but the latter case is, of course, more general.

To formalize this nested approach we model at the subunit level. Assuming counts are recorded for the units, we have $Y_{ij} \sim Po(n_{ij} p_{ij} = E_{ij} r_{ij})$ where n_{ij}; is the population at risk in unit i, j; p_{ij} is the incidence rate in unit i, j; E_{ij} is the expected count in unit i, j; and r_{ij} is the relative risk for unit i, j. If $\mathbf{X}_{ij}$ is covariate information for subunit i, j that is anticipated to help "explain" Y_{ij}, we can write, say, $\log r_{ij} = \mathbf{X}_{ij}^T \beta$. If the Y_{ij}s are assumed conditionally independent, then $Y_i \sim Po(\sum_j E_{ij} r_{ij})$.* Imposing the constraint that $\sum_j Y_{ij} = Y_i$, we have that $Y_{ij} | Y_i$ follows a binomial distribution with mean $\frac{E_{ij} r_{ij}}{\sum_j E_{ij} r_{ij}} Y_i$. We can now more clearly see the relationship between this approach and areal interpolation above. Areal interpolation sets $Y_{ij} = \frac{|B_{ij}|}{|B_i|} Y_i$. So, $|B_{ij}|$ plays the role of $E_{ij} r_{ij}$ and we ignore the uncertainty in Y_{ij}, just using its mean as a point estimate.

Our illustration envisions the Ys as disease counts. To be concrete, we suppose that they are observed at census tracts, i.e., Y_is, but we seek to impute them to census block groups contained within the census tracts, Y_{ij}s. Suppose the data record the block group-level population counts n_{ij} and, for example, covariate values u_{ij} and v_{ij} where $u_{ij} = 1$ if block group j within census tract i is classified as urban, 0 otherwise and $v_{ij} = 1$ if the block group centroid is within 2 km of a waste site, 0 if not. Hence, we recast the misalignment problem as a misaligned spatial regression problem. We can use a hierarchical model to incorporate the covariate information as well as to obtain variance estimates to accompany the block group-level point estimates. In this example, the unequal population totals in the block groups will play the weighting role that unequal areas would have played in Equation (29.14). Recalling the disease mapping discussion in Chapter 14, we introduce a first-stage Poisson model for the disease counts where we might specify the relative risk through $\log r_{ij} = \beta_0 + u_{ij}\beta_1 + v_{ij}\beta_2$.

In Mugglin and Carlin (1998), extending earlier work of Waller, Carlin, and Xia (1997), a Bayesian analysis using the above modeling is implemented for a leukemia dataset from upstate New York. Priors are added to complete a Bayesian hierarchical model and posterior inference is obtained for the r_{ij}.

29.5 Nonnested Block-Level Modeling

We next consider a framework for hierarchical Bayesian interpolation, estimation, and spatial smoothing over *nonnested* misaligned data grids. We confine our model development to the case of two misaligned spatial grids. Given this development, the extension to more than two grids will be conceptually apparent. The additional computational complexity and bookkeeping detail will also be evident.

Let the first grid have regions indexed by $i = 1, \ldots, I$, denoted by B_i, and let $S_B = \cup_i B_i$. Similarly, for the second grid we have regions C_j, $j = 1, \ldots, J$ with $S_C = \cup_j C_j$. In some

* Note the potential here for an ecological fallacy as we discussed in the previous section. If X_{ij} is a cumulative variable, $\log r_i$ is not linear in $\mathbf{X}_i$.

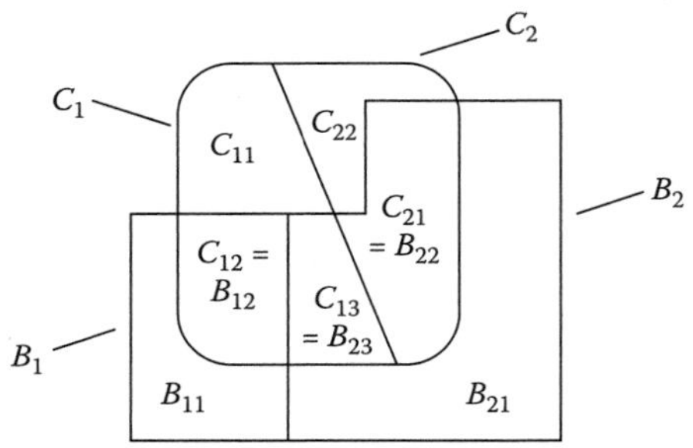

FIGURE 29.2
Illustrative representation of areal data misalignment.

applications $S_B = S_C$, i.e., the B cells and the C cells offer different partitions of a common region. Nested misalignment (e.g., where each C_j is contained entirely in one and only one B_i) is, evidently, a special case. Another possibility is that one data grid contains the other; say, $S_B \subset S_C$. In this case, there will exist some C cells for which a portion lies outside of S_B. In the most general case, illustrated by Figure 29.2, there is no containment and there will exist B cells for which a portion lies outside of S_C and C cells for which a portion lies outside of S_B.

Atoms are created by intersecting the two grids. For a given B_i, each C cell that intersects B_i creates an atom (which possibly could be a union of disjoint regions). There may also be a portion of B_i that does not intersect with any C_j. We refer to this portion as the *edge* atom associated with B_i, i.e., a B-edge atom. In Figure 29.2, atoms B_{11} and B_{21} are B-edge atoms. Similarly, for a given C_j, each B cell that intersects with C_j creates an atom, and we analogously determine C-edge atoms (atoms C_{11} and C_{22} in Figure 29.2). It is crucial to note that each nonedge atom can be referenced relative to an appropriate B cell, say B_i, and denoted as B_{ik}. It also can be referenced relative to an appropriate C cell, say C_j, and denoted by $C_{j\ell}$. Hence, there is a one-to-one mapping within $S_B \cap S_C$ between the set of iks and the set of $j\ell$s, as shown in Figure 29.2. Formally, we can define the function g on nonedge B atoms such that $g(B_{ik}) = C_{j\ell}$, and the *inverse* function b on C atoms such that $b(C_{j\ell}) = B_{ik}$. For computational purposes, we would create a "look-up" table to specify these functions.

Again, we refer to the first grid as the *source* grid, that is, at each B_i we observe a response Y_i. We seek to explain Y_i using covariates, some of which may be observed on the response grid; we denote the value of this vector for B_i by $\mathbf{W}_i$. But also, some covariates are observed on the second or *target* grid. We denote the value of this vector for C_j by $\mathbf{X}_j$. We seek to impute Ys to the C cells given Ys observed for the B cells. We can use covariates on the B areal partition (W_is) and also covariates on the C areal partition (X_js). We model at the atom level.

Again, we assume that the Ys are aggregated measurements so that Y_i can be envisioned as $\sum_k Y_{ik}$, where the Y_{ik} are unobserved or latent and the summation is over all atoms (including perhaps an edge atom) associated with B_i. To simplify, suppose that the Xs are also scalar aggregated measurements, i.e., $X_j = \sum_\ell X_{j\ell}$ where the summation is over all atoms associated with C_j. For the Ws, we assume that each component is either an aggregated measurement or an *inheritable* measurement. For component r, in the former case $W_i^{(r)} = \sum_k W_{ik}^{(r)}$ as with Y_i; in the latter case $W_{ik}^{(r)} = W_i^{(r)}$.

In addition to (or perhaps in place of) the W_is, we can introduce B cell random effects μ_i, $i = 1, \ldots, I$. These effects are employed to capture spatial association among the Y_is. The μ_i can be given a spatial prior specification, e.g., a CAR model (Barliner, 2000; Bernardinelli and Montomoli, 1992; Besag, 1974; Chapters 12 and 13), as described below, is convenient. Similarly, we will introduce C cell random effects ω_j, $j = 1, \ldots, J$ to capture spatial association among the X_js. It is assumed that the latent Y_{ik} inherit the effect μ_i and that the latent $X_{j\ell}$ inherit the effect ω_j.

With count data, we assume the latent variables are conditionally independent Poissons, as in the previous section. As a result, the observed measurements are Poissons and the conditional distribution of the latent variables given the observed is a product multinomial.* If we implement allocation proportional to area to the $X_{j\ell}$ in a stochastic fashion, we would obtain $X_{j\ell} \mid \omega_j \sim Po(e^{\omega_j}|C_{j\ell}|)$ assumed independent for $\ell = 1, 2, \ldots, L_j$. Then $X_j \mid \omega_j \sim Po(e^{\omega_j}|C_j|)$ and $(X_{j1}, X_{j2}, \ldots, X_{j,L_j} \mid X_j, \omega_j) \sim Mult(X_j; q_{j1}, \ldots, q_{j,L_j})$ where $q_{j\ell} = |C_{j\ell}|/|C_j|$.

Such strictly area-based modeling cannot be applied to the Y_{ik}s because it fails to connect the Ys with the Xs (as well as for the Ws). To do so, we again begin at the atom level. For nonedge atoms, we use the previously mentioned look-up table to find the $X_{j\ell}$ to associate with a given Y_{ik}. It is convenient to denote this $X_{j\ell}$ as X'_{ik}. Ignoring the $\mathbf{W}_i$ for the moment, we assume

$$Y_{ik} \mid \mu_i, \theta_{ik} \sim Po\left(e^{\mu_i}|B_{ik}|\, h(X'_{ik}/|B_{ik}|;\, \theta_{ik})\right), \tag{29.15}$$

independent for $k = 1, \ldots, K_i$. Here, h is a preselected parametric function that adjusts an expected proportional-to-area allocation according to X'_{ik}. See Mugglin, Carlin, and Gelfand (2000) for further discussion regarding choice of h.

If B_i has no associated edge atom, then

$$Y_i \mid \mu_i, \theta, \{X_{j\ell}\} \sim Po\left(e^{\mu_i} \sum_k |B_{ik}|\, h(X'_{ik}/|B_{ik}|;\, \theta_{ik})\right). \tag{29.16}$$

If B_i has an edge atom, say B_{iE}, since there is no corresponding $C_{j\ell}$, there is no corresponding X'_{iE}. Hence, we introduce a latent X'_{iE} whose distribution is determined by the nonedge atoms that are neighbors of B_{iE}, i.e., $X'_{iE} \mid \omega_i^* \sim Po(e^{\omega_i^*}|B_{iE}|)$, thus adding a new set of random effects $\{\omega_i^*\}$ to the existing set $\{\omega_j\}$. These two sets together are assumed to have a single CAR specification. An alternative is to model $X'_{iE} \sim Po(|B_{iE}|(\sum_{N(B_{iE})} X'_t / \sum_{N(B_{iE})} |B_t|))$, where $N(B_{iE})$ is the set of neighbors of B_{iE} and t indexes this set.

Now, with an X'_{ik} for all ik, Equation (29.15) is extended to all B atoms and the conditional distribution of Y_i is determined for all i as in Equation (29.16). But, also $Y_{i1}, \ldots, Y_{ik_i} | Y_i, \mu_i, \theta_{ik}$ is distributed multinomial $(Y_i; q_{i1}, \ldots, q_{ik_i})$, where

$$q_{ik} = |B_{ik}| h(X'_{ik}/|B_{ik}|; \theta_{ik}) / \sum_k |B_{ik}| h(X'_{ik}/|B_{ik}|; \theta_{ik}).$$

The entire specification can be given a representation as a graphical model, as in Figure 29.3. In this model, the arrow from $\{X_{j\ell}\} \to \{X'_{ik}\}$ indicates the inversion of the $\{X_{jl}\}$ to $\{X'_{ik}\}$, augmented by any required edge atom values X'_{iE}. The $\{\omega_i^*\}$ would be generated if the X'_{iE} are modeled using the Poisson specification above. Since the $\{Y_{ik}\}$ are not observed, but are distributed as multinomial given the fixed block group totals $\{Y_i\}$, this is a predictive step in our model, as indicated by the arrow from $\{Y_i\}$ to $\{Y_{ik}\}$ in the figure. In fact, as mentioned above, the further predictive step to impute Y'_j, the Y total associated with X_j in the jth target zone is of key interest. If there are edge atoms C_{jE}, this will require a model for the associated Y'_{jE}. Because there is no corresponding B atom for C_{jE}, a specification such as Equation (29.15) is not appropriate. Rather, we can imitate the above modeling for X'_{iE} by introducing $\{\mu_j^*\}$. The $\{\mu_j^*\}$ and $\{Y'_{jE}\}$ would add two consecutive nodes to the right side of Figure 29.3, connecting from λ_μ to $\{Y'_j\}$.

---

* It is not required that the Ys be count data. For instance, with aggregated measurements that are continuous, a convenient distributional assumption is conditionally independent gammas, in which case the latent variables would be rescaled to product Dirichlet. An alternative choice is the normal, whereupon the latent variables would have a distribution that is a product of conditional multivariate normals.

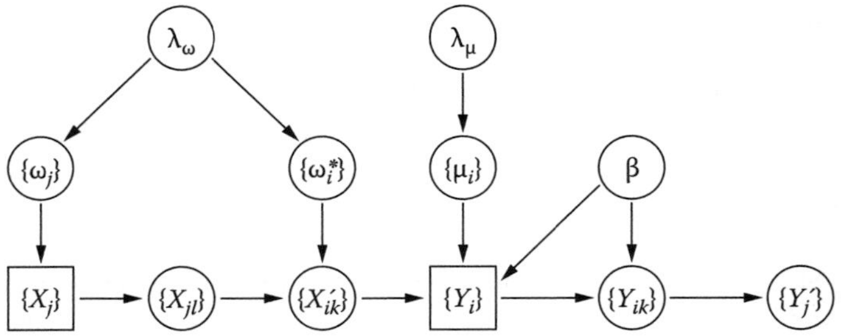

FIGURE 29.3
Graphical version of the model, with variables as described in the text. Boxes indicate data nodes, while circles indicate unknowns.

The distributional specification overlaid on this graphical model has been supplied in the foregoing discussion and (in the absence of C_{jE} edge atoms, as in Figure 29.3) takes the form

$$\prod_i f(Y_{i1}, \ldots, Y_{ik_i} \mid Y_i, \beta) \prod_i f(Y_i \mid \mu_i, \beta, \{X'_{ik}\}) \, f(\{X'_{ik}\} \mid \omega_i^*, \{X_{j\ell}\})$$
$$\times \prod_j f(X_{j1}, \ldots, X_{jL_j} \mid X_j) \prod_j f(X_j \mid \omega_j)$$
$$\times f(\{\mu_i\} \mid \lambda_\mu) \, f(\lambda_\mu) \, f(\{\omega_j\}, \{\omega_i^*\} \mid \lambda_\omega) \, f(\lambda_\omega) \, f(\beta) \,. \quad (29.17)$$

Bringing in the $\mathbf{W}_i$ merely revises the exponential term in Equation (29.15) from $\exp(\mu_i)$ to $\exp(\mu_i + \mathbf{W}_{ik}^T \beta)$.

29.6 Misaligned Regression Modeling

We first formalize the modeling details for each of the four potential regression misalignment settings. Again, we confine ourselves to a single explanatory variable. Consider first response data $Y(\mathbf{s}_i), i = 1, 2, \ldots, n$ with covariate data $X(\mathbf{s}_j^*), j = 1, 2, \ldots, m$. Here, the set of $\mathbf{s}_i$s may overlap with the set of $\mathbf{s}_j^*$s, but need not. (So, we might call this a missing data problem rather than a misalignment problem.) The naïve solution might be to regress $Y(\mathbf{s}_i)$ on the nearest $X(\mathbf{s}_j^*)$ or perhaps on an average of neighboring $X(\mathbf{s}_j)$s. (See Royle, Berliner, Wikle, and Mitliff, 1997, in this regard.) A fully model-based version, which interpolates to $X(\mathbf{s}_i)$, will require a process model for $X(\mathbf{s})$. In fact, it will require a bivariate process model for $(X(\mathbf{s}), Y(\mathbf{s}))$ (see Chapter 28). If θ denotes the parameters of this bivariate process, then the misaligned regression model takes the form

$$f(\{Y(\mathbf{s}_i), i = 1, 2, \ldots, n\} | \{X(\mathbf{s}_i), 1 = 1, 2, \ldots, n\}, \theta)$$
$$\times f(\{X(\mathbf{s}_i), i = 1, 2, \ldots, n\} | \{X(\mathbf{s}_j^*), j = 1, 2, \ldots, m\}, \theta) f(\{X(\mathbf{s}_j^*), j = 1, 2, \ldots, m\}, \theta). \quad (29.18)$$

With a prior on θ, we obtain a fully specified Bayesian model and we can implement posterior inference for θ. In the Bayesian framework, using MCMC model fitting, prediction of $Y(\mathbf{s}_0)$ at a new $X(\mathbf{s}_0)$ follows by sampling the predictive distribution, $f(Y(\mathbf{s}_0)|X(\mathbf{s}_0), \{Y(\mathbf{s}_i), i = 1, 2, \ldots, n\}, \{X(\mathbf{s}_j^*), j = 1, 2, \ldots, m\})$. This would be done by composition; under a Gaussian process, we would take a posterior draw of θ and insert this posterior draw in the univariate normal conditional distribution, $f(Y(\mathbf{s}_0)|X(\mathbf{s}_0), \{Y(\mathbf{s}_i), i = 1, 2, \ldots, n\}, \{X(\mathbf{s}_j^*), j = 1, 2, \ldots, m\}, \theta)$, to draw $Y(\mathbf{s}_0)$. (For an illustrative example, see Banerjee and Gelfand

2002.) Note that the version that kriges the $X(\mathbf{s}_j^*)$s to the $X(\mathbf{s}_i)$s and then just plugs the kriged values into a regular regression model converts the COSP into a spatial errors-in-variables problem. However, if we ignore the uncertainty in the kriged values, we underestimate the uncertainty in the regression model. We also remind the reader that such prediction as above is generally referred to as cokriging (see Chapter 28 and the fuller discussion in Chiles and Delfiner, 1999).

The remaining three cases proceed similarly. Suppose we have $Y(B_i), i = 1, 2, \ldots, n$ with explanatory $X(\mathbf{s}_j), j = 1, 2, \ldots, m$. Now the misaligned regression model takes the form

$$\begin{aligned} &f(\{Y(B_i), i = 1, 2, \ldots, n\} | \{X(B_i), 1 = 1, 2, \ldots, n\}, \theta) \\ &\times f(\{X(B_i), i = 1, 2, \ldots, n\} | \{X(\mathbf{s}_j), j = 1, 2, \ldots, m\}, \theta) f(\{X(\mathbf{s}_j), j = 1, 2, \ldots, m\}, \theta). \end{aligned} \tag{29.19}$$

Again, under a bivariate process model, we can imitate block average calculations as in Equation (29.2) to Equation (29.5) with corresponding Monte Carlo integrations to approximate the multivariate normal distributions in Equation (29.19). Posterior inference follows as in the point–point case. If instead, we have $Y(\mathbf{s}_i), i = 1, 2, .., n$ with $X(B_j), j = 1, 2, \ldots, m$, as noted earlier, we must decide whether it is sensible to imagine $X(\mathbf{s})$s. If so, we can follow Equation (29.19), switching points and blocks for the Xs. If not, we can adopt a naíve version, assigning the value $X(\mathbf{s}_i) = X(B_j)$ if $\mathbf{s}_i \in B_j$. The preferred way, in order to fully capture the uncertainty in the regression, is to regress $Y(B_j)$ on $X(B_j)$, averaging up the $Y(\mathbf{s})$ process. In this regard, we condition on the $X(B_j)$s, viewing them as fixed. Last, in the case where we have $Y(B_i), i = 1, 2, \ldots, n$ with $X(C_j), j = 1, 2, \ldots, m$, we have the MAUP for the $X(B)$s as in the previous section. We obtain the MAUP regression model as

$$\begin{aligned} &f(\{Y(B_i), i = 1, 2, \ldots, n\} | \{X(B_i), 1 = 1, 2, \ldots, n\}, \theta) \\ &\times f(\{X(B_i), i = 1, 2, \ldots, n\} | \{X(Cj), j = 1, 2, \ldots, m\}, \theta) f(\{X(C_j), j = 1, 2, \ldots, m\}, \theta). \end{aligned} \tag{29.20}$$

Occasionally the Cs will be nested within the Bs, but, more typically, we will have to deal with the nonnested case. We next describe a version of the MAUP regression.

29.6.1 A Misaligned Regression Modeling Example

In Agarwal, Gelfand, and Silander (2002), the authors apply the ideas above in a rasterized datasetting. Such data are common in remote sensing, where satellites can collect data (say, land use) over a pixelized surface, which is often fine enough so that town or other geopolitical boundaries can be (approximately) taken as the union of a collection of pixels.

The focal area for the study in Agarwal et al. (2002) is the tropical rainforest biome within Toamasina (or Tamatave) Province of Madagascar. This province is located along the east coast of Madagascar, and includes the greatest extent of tropical rainforest in the island nation. The aerial extent of Toamasina Province is roughly 75,000 square km. Four geo-referenced GIS coverages were constructed for the province: town boundaries with associated 1993 population census data, elevation, slope, and land cover. Ultimately, the total number of towns was 159, and the total number of pixels was 74,607. Below, the above 1 km raster layers are aggregated into 4 km pixels.

Figure 29.4 shows the town-level map for the 159 towns in the Madagascar study region. In fact, there is an escarpment in the western portion where the climate differs from the rest of the region. It is a seasonally dry grassland/savanna mosaic. Also, the northern part is expected to differ from the southern part, because the north has fewer population areas with large forest patches, while the south has more villages with many smaller forest patches

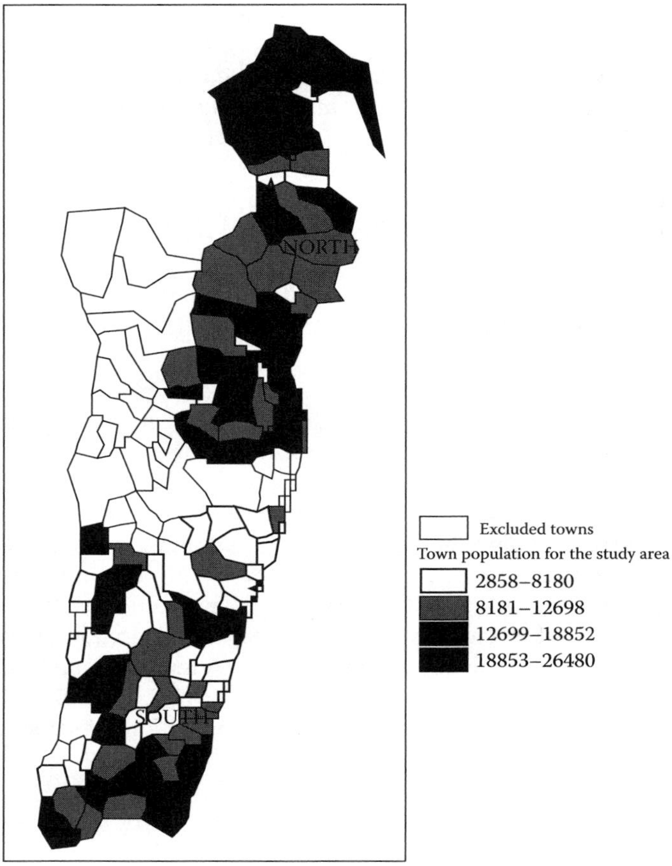

FIGURE 29.4
Northern and southern regions within the Madagascar study region, with population overlaid.

and more extensive road development, including commercial routes to the national capital west of the study region. The north and south regions with a transition zone were created, as shown in Figure 29.4.

The joint distribution of land use and population count is modeled at the pixel level. Let L_{ij} denote the land use value for the jth pixel in the ith town and let P_{ij} denote the population count for the jth pixel in the ith town. Again, the L_{ij} are observed, but only $P_{i.} = \sum_j P_{ij}$ are observed at the town level. Collect the L_{ij} and P_{ij} into town-level vectors $\mathbf{L}_i$ and $\mathbf{P}_i$, and overall vectors $\mathbf{L}$ and $\mathbf{P}$.

Covariates observed at each pixel include an assigned elevation, E_{ij}, and an assigned slope, S_{ij}. To capture spatial association between the L_{ij}, pixel-level spatial effects φ_{ij} are introduced; to capture spatial association between the $P_{i.}$, town-level spatial effects δ_i are introduced. That is, the spatial process governing land use may differ from that for population.

The joint distribution, $p(\mathbf{L}, \mathbf{P}|E_{ij}, S_{ij}, \varphi_{ij}, \delta_i)$ is specified by factoring it as

$$p(\mathbf{P} \mid E_{ij}, S_{ij}, \delta_i)\; p(\mathbf{L} \mid \mathbf{P}, E_{ij}, S_{ij}, \varphi_{ij}). \tag{29.21}$$

Conditioning is done in this fashion in order to explain the effect of population on land use. Causality is *not* asserted; the conditioning could be reversed. (Also, implicit in (29.21) is a marginal specification for L and a conditional specification for $\mathbf{P}|\mathbf{L}$.)

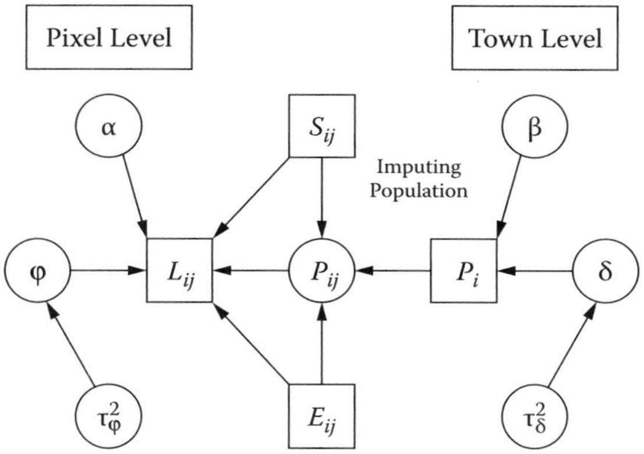

FIGURE 29.5
Graphical representation of the land use population model.

Turning to the first term in (29.21), the P_{ij} are assumed conditionally independent given the Es, Ss, and δs. In fact, we assume $P_{ij} \sim \text{Poisson}(\lambda_{ij})$, where

$$\log \lambda_{ij} = \beta_0 + \beta_1 E_{ij} + \beta_2 S_{ij} + \delta_i. \tag{29.22}$$

Thus, $P_{i.} \sim \text{Poisson}(\lambda_{i.})$, where $\log \lambda_{i.} = \log \sum_j \lambda_{ij} = \log \sum_j \exp(\beta_0 + \beta_1 E_{ij} + \beta_2 S_{ij} + \delta_i)$. In other words, the P_{ij} inherit the spatial effect associated with $P_{i.}$. Also, $\{P_{ij}\} \,|\, P_{i.} \sim \text{Multinomial}(P_{i.}; \{\gamma_{ij}\})$, where $\gamma_{ij} = \lambda_{ij}/\lambda_{i\cdot}$.

In the second term in (29.21), conditional independence of the L_{ij} given the Ps, Es, Ss, and φs is assumed. To facilitate computation, we aggregate to 4 km × 4 km resolution. Since L_{ij} lies between 0 and 16, it is assumed that $L_{ij} \sim \text{Binomial}\ (16, q_{ij})$, i.e., that the 16 1 km × 1 km pixels that comprise a given 4 km × 4 km pixel are iid Bernoulli random variables with q_{ij}, such that

$$\log\left(\frac{q_{ij}}{1 - q_{ij}}\right) = \alpha_0 + \alpha_1 E_{ij} + \alpha_2 S_{ij} + \alpha_3 P_{ij} + \varphi_{ij}. \tag{29.23}$$

For the town-level spatial effects, a CAR prior is assumed (see Chapters 12 and 13), using only the adjacent towns for the mean structure, with variance τ_δ^2, and similarly for the pixel effects using only adjacent pixels, with variance τ_φ^2. To complete the hierarchical model specification, priors for α, β, τ_δ^2, and τ_φ^2 (when the φ_{ij} are included) are required. Under a binomial, with proper priors for τ_δ^2 and τ_φ^2, a flat prior for α and β will yield a proper posterior. For τ_δ^2 and τ_φ^2, inverse gamma priors may be adopted. Figure 29.5 offers a graphical representation of the full model.

At the 4 km × 4 km pixel scale, two versions of the model in Equation (29.23) were fit, one with the φ_{ij} (Model 2) and one without them (Model 1). Models 1 and 2 were fitted separately for the northern and southern regions. The results are summarized in Table 29.1, point (posterior median) and interval (95% equal tail) estimate. The population-count model results are little affected by the inclusion of the φ_{ij}. For the land-use model, this is not the case. Interval estimates for the fixed effects coefficients are much wider when the φ_{ij} are included. This is not surprising from the form in Equation (29.23). Though the P_{ij} are modeled and are constrained by summation over j and though the ϕ_{ij} are modeled dependently through the CAR specification, since neither is observed, strong collinearity between the P_{ij} and ϕ_{ij} is expected, inflating the variability of the αs.

Specifically, for the population count model in Equation (29.22), in all cases, the elevation coefficient is significantly negative; higher elevation yields smaller expected population.

TABLE 29.1

Parameter Estimation (Point and Interval Estimates) for Models 1 and 2 for the Northern and Southern Regions

Model:	M_1		M_2	
Region:	North	South	North	South
Population model parameters:				
β_1	−.577	−.245	−.592	−.176
(elev)	(−.663, −.498)	(−.419, −.061)	(−.679, −.500)	(−.341, .019)
β_2	.125	−.061	.127	−.096
(slope)	(.027, .209)	(−.212, .095)	(.014, .220)	(−.270, .050)
τ_{δ^2}	1.32	1.67	1.33	1.71
	(.910, 2.04)	(1.23, 2.36)	(.906, 1.94)	(1.22, 2.41)
Land use model parameters:				
α_1	.406	−.081	.490	.130
(elev)	(.373, .440)	(−.109, −.053)	(.160, .857)	(−.327, .610)
α_2	.015	.157	.040	−.011
(slope)	(−.013, .047)	(.129, .187)	(−.085, .178)	(−.152, .117)
α_3	−5.10	−3.60	−4.12	−8.11
$(\times 10^{-4})$	(−5.76, −4.43)	(−4.27, −2.80)	(−7.90, −.329)	(−14.2, −3.69)
τ_{φ^2}	—	—	6.84	5.85
			(6.15, 7.65)	(5.23, 6.54)

Interestingly, the elevation coefficient is more negative in the north. The slope variable is intended to provide a measure of the differential in elevation between a pixel and its neighbors. However, a crude algorithm is used within the ARC/INFO software for its calculation, diminishing its value as a covariate. Indeed, higher slope would typically encourage lower expected population. While this is roughly true for the south under either model, the opposite emerges for the north. The inference for the town-level spatial variance component τ_δ^2 is consistent across all models. Homogeneity of spatial variance for the population model is acceptable.

Turning to Equation (29.23), in all cases the coefficient for population is significantly negative. There is a strong relationship between land use and population size; increased population increases the chance of deforestation, in support of the primary hypothesis for this analysis. The elevation coefficients are mixed with regard to significance. However, for both Models 1 and 2, the coefficient is always at least .46 larger in the north. Elevation more strongly encourages forest cover in the north than in the south. This is consistent with the discussion of the preceding paragraph but, apparently, the effect is weaker in the presence of the population effect. Again, the slope covariate provides inconsistent results, but is insignificant in the presence of spatial effects. Inference for the pixel-level spatial variance component does not criticize homogeneity across regions. Note that τ_φ^2 is significantly larger than τ_δ^2. Again, this is expected. With a model having four population parameters to explain 3186 $q_{ij}'s$ as opposed to a model having three population parameters to explain 115 $\lambda_i's$, we would expect much more variability in the $\varphi_{ij}'s$ than in the $\delta_i's$. Finally, Figure 29.6 shows the imputed population (on the square root scale) at the 4 km $\times$ 4 km pixel level.

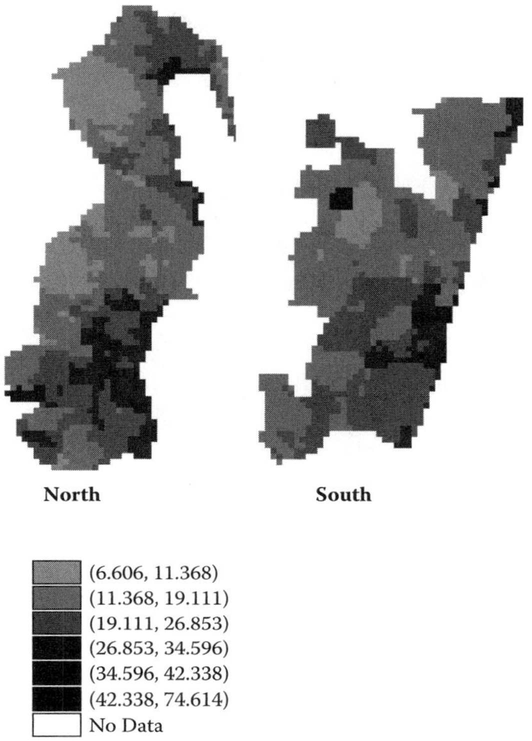

FIGURE 29.6
Imputed population (on the square root scale) at the pixel level for north and south regions.

The fully nonnested approach will be difficult to implement with more than two mutually misaligned areal data layers, due mostly to the multiple labeling of atoms and the needed higher-way look-up table. However, the approach of this section suggests a simpler strategy for handling this situation. First, rasterize all data layers to a common scale of resolution. Then, build a suitable latent regression model at that scale, with conditional distributions for the response and explanatory variables constrained by the observed aggregated measurements for the respective layers.

We conclude by noting that Zhu, Carlin, and Gelfand (2003) consider regression in the point-block misalignment setting, illustrating with the Atlanta ozone data pictured in Figure 29.7. In this setting, the problem is to relate several air quality indicators (ozone, particulate matter, nitrogen oxides, etc.) and a range of sociodemographic variables (age, gender, race, and a socioeconomic status surrogate) to the response, pediatric emergency room (ER) visit counts for asthma in Atlanta, Georgia. Here, the air quality data is collected at fixed monitoring stations (point locations) while the sociodemographic covariates and response variable are collected by zip code (areal summaries).* In fact, the air quality data is available as daily averages at each monitoring station, and the response is available as daily counts of visits in each zip code. In Zhu et al. (2003), they use the foregoing methods to realign the data and then fit a Poisson regression model at the zip code scale. In fact, they do this dynamically, across days.

* Again, the possibility of ecological arises here. (See Wakefield and Shaddick, 2006, in this regard.)

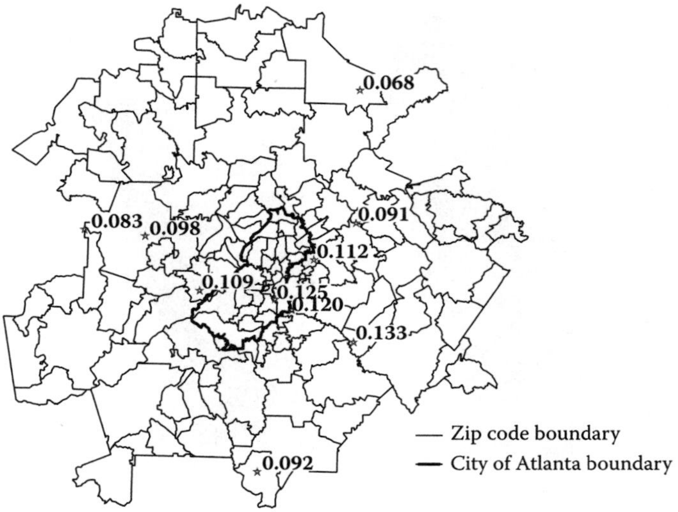

FIGURE 29.7
Atlanta, Georgia, ozone data.

29.7 Fusion of Spatial Data

From Section 29.1, recall the misalignment problem in the context of data assimilation or data fusion (see, Chapter 27 and, e.g., Kalnay, 2002, in the context of atmospheric modeling). Here the objective is to combine data on a particular variable that is available at different scales. For instance, in considering environmental exposure to acid deposition, Cowles and Zimmerman (2003) fuse two misaligned sets of station data. Here, we consider particulate matter, where we can obtain monitoring station level data (which is obviously point referenced and is relatively sparse) along with modeled output data (such as CMAQ—Community Multiscale Air Quality data), which is available at grid cell resolutions, such as 36, 12, or even 4 km^2 squares, and satellite data (e.g., MODIS—Moderate Resolution Imaging Spectroradiometer) obtained in swaths of 5 km or less. The overall objective would be improved prediction of exposure at arbitrary locations in a study region.

To illustrate how such fusion might be implemented in the static case, we present two model-based strategies, one due to Fuentes and Raftery (2005), the other from McMillan, Holland, Morara, and Feng (2008). In fact, McMillan et al. (2008) also consider the dynamic case. Earlier references in this regard include Best, Ickstadt, and Wolpert (2000); Davis, Nychka, and Bailey (2000); and Gelfand et al. (2001). Suppose we consider just two data layers, $Y(\mathbf{s}_i), i = 1, 2, \ldots, n$ and $\tilde{Y}(B_j), j = 1, 2, \ldots, m$ where the B_j are exclusive and exhaustive for a region D and the $\mathbf{s}_i$ are all in D.

The approach of Fuentes and Raftery (2005) assumes that both data layers inform about the "true" values. In other words, a latent true process $Y_{true}(\mathbf{s})$ is imagined. The station data and modeled output data are viewed as observations of the true process subject to measurement error and calibration error, respectively. More precisely, the latent $Y_{true}(\mathbf{s})$ is modeled as a Gaussian process using a spatial regression. Then, $Y(\mathbf{s}_i) = Y_{true}(\mathbf{s}_i) + \epsilon(\mathbf{s}_i)$ while $\tilde{Y}(B_j) = |B_j|^{-1} \int_{B_j} (a(\mathbf{s}) + b(\mathbf{s}) Y_{true}(\mathbf{s}) + \eta(\mathbf{s})) d\mathbf{s}$. Here, $\epsilon(\mathbf{s})$ is a pure error process, while $a(\mathbf{s})$ and $b(\mathbf{s})$ provide spatially varying calibration for the areal unit data. The $\eta(\mathbf{s})$ are again a pure error process yielding independent $\eta(B_j)$ with variance inversely proportional to $|B_j|$. Fuentes and Raftery (2005) note that it is difficult to identify both $a(\cdot)$ and $b(\cdot)$ and, thus, they introduce a trend surface for $a(\mathbf{s})$ and set $b(\cdot)$ to a constant. The resultant integration over B_j above produces a block average of the true process. The hierarchical model is fitted using MCMC, but fitting will be very slow if the number of areal units is large (e.g., with a high-resolution computer output model) due to the large number of block integrations.

The approach of McMillan, Holland, Morara, and Feng (2008) avoids block averaging by modeling the process at the areal unit level. With sparse monitoring data and model output data available at all grid cells, such modeling is easier and faster. Again, a latent true level for the variable is presumed, but associated with B_j, say, $Y_{true}(B_j)$. Now, $\tilde{Y}(B_j) = g(Y_{true}(B_j)) + \epsilon(B_j)$ where the ϵs are iid. Since $|B_j|$ is constant, $g(\cdot)$ takes the form of Y_{true} plus a B-spline deviation. And, if $Y(\mathbf{s}_i) \in B_j$, $Y(\mathbf{s}_i)$ has a measurement error model, i.e., $Y(\mathbf{s}_i) = Y_{true}(B_j) + \delta(\mathbf{s}_i)$ where the δ's are iid. Obviously, any B_j can contain 0, 1, or more monitoring sites. Finally, the $Y_{true}(B_j)$ are modeled with a spatial regression, using spatial random effects with a CAR specification to capture spatial dependence. In fact, McMillan et al. (2008) present their model dynamically with AR(1) temporal updating of the CAR's.

In summary, we have attempted to illuminate the range of issues that arise in attempting to combine spatially incompatible data. As Gotway and Young (2002) note, the increasing use of GIS software makes the change of support problem increasingly relevant and emphasizes the need for suitable statistical methodology to address it.

References

Agarwal, D.K., Gelfand, A.E., and Silander, J.A., Jr. (2002). Investigating tropical deforestation using two stage spatially misaligned regression models. *Journal of Agricultural, Biological and Environmental Statistics*, **7**, 420–439.

Arbia, G. (1989) Statistical effect of data transformations: a proposed general framework. In *The Accuracy of Spatial Data Bases*, eds. M. Goodchild and S. Gopal, London: Taylor & Francis, 249–259.

Banerjee, S., Carlin, B.P., and Gelfand, A.E. (2004). *Hierarchical Modeling and the Analysis of Spatial Data*. Boca Raton, FL: Chapman & Hall.

Banerjee, S. and Gelfand, A.E. (2002). Prediction, interpolation and regression for spatially misaligned data. *Sankhya, Ser. A*, **64**, 227–245.

Berliner, L.M. (2000). Hierarchical Bayesian modeling in the environmental sciences. *Allgemeines Statistisches Archiv (Journal of the German Statistical Society)*, **84**, 141–153.

Bernardinelli, L. and Montomoli, C. (1992). Empirical Bayes versus fully Bayesian analysis of geographical variation in disease risk. *Statistics in Medicine*, **11**, 983–1007.

Besag, J. (1974). Spatial interaction and the statistical analysis of lattice systems (with discussion). *Journal of the Royal Statistical Society*, **36**, 192–236.

Best, N.G., Ickstadt, K., and Wolpert, R.L. (2000). Spatial Poisson regression for health and exposure data measured at disparate resolutions. *Journal of the American Statistical Association*, **95**, 1076–1088.

Brillinger, D.R. (1990). Spatial-temporal modeling of spatially aggregate birth data. *Survey Methodology*, **16**, 255–269.

Chiles, J.P. and Delfiner, P. (1999). *Geostatistics: Modeling Spatial Uncertainty*. New York: John Wiley & Sons.

Cowles, M.K. and Zimmerman, D.L. (2003). A Bayesian space-time analysis of acid deposition data combined for two monitoring networks. *Journal of Geophysical Research*, **108**, 90–106.

Cressie, N.A.C. (1993). *Statistics for Spatial Data*, 2nd ed. New York: John Wiley & Sons.

Cressie, N.A.C. (1996). Change of support and the modifiable areal unit problem. *Geographical Systems*, **3**, 159–180.

Davis, J.M., Nychka, D., and Bailey, B. (2000). A comparison of regional oxidant model (ROM) output with observed ozone data. *Atmospheric Environment*, **34**, 2413–2423.

De Oliveira, V., Kedem, B., and Short, D.A. (1997). Bayesian prediction of transformed Gaussian random fields. *Journal of the American Statistical Association*, **92**, 1422–1433.

Dempster, A.P., Laird, N.M., and Rubin, D.B. (1977). Maximum likelihood estimation from incomplete data via the EM algorithm (with discussion). *Journal of the Royal Statistical Society, Ser. B*, **39**, 1–38.

Fairfield Smith, H. (1938). An empirical law describing heterogeneity in the yields of agricultural crops. *Journal of Agricultural Science*, **28**, 1–23.

Flowerdew, R. and Green, M. (1989). Statistical methods for inference between incompatible zonal systems. In *Accuracy of Spatial Databases*, eds. M. Goodchild and S. Gopal, London: Taylor & Francis, pp. 239–247.

Flowerdew, R. and Green, M. (1992). Developments in areal interpolating methods and GIS. *Annals of Regional Science*, **26**, 67–78.

Flowerdew, R. and Green, M. (1994). Areal interpolation and types of data. In *Spatial Analysis and GIS*, eds. S. Fotheringham and P. Rogerson, London: Taylor & Francis, pp. 121–145.

Fuentes, M. and Raftery, A. (2005). Model evaluation and spatial interpolation by Bayesian combination of observations with outputs from numerical models. *Biometrics*, **61**, 36–45.

Gehlke, C. and Biehl, K. (1934). Certain effects of grouping upon the size of the correlation coefficient in census tract material. *Journal of American Statistical Association*, **29**, 169–170.

Gelfand, A.E., Zhu, L., and Carlin, B.P. (2001). On the change of support problem for spatio-temporal data. *Biostatistics*, **2**, 31–45.

Goodchild, M.F. and Lam, N. S-N. (1980). Areal interpolation: A variant of the traditional spatial problem. *Geo-Processing*, **1**, 297–312.

Goovaerts, P. (1997) *Geostatistics for Natural Resources Evaluation*, New York: Oxford University Press.

Gotway, C.A. and Young, L.J. (2002). Combining incompatible spatial data. *Journal of the American Statistical Association*, **97**, 632–648.

Greenland, S. and Robins, J. (1994). Ecologic studies—biases, misconceptions, counterexamples. *American Journal of Epidemiology*, **139**, 747–760.

Holt, D., Steel, D., and Tranmer, M. (1996). Areal homogeneity and the modifiable areal unit problem. *Geographical Systems*, **3**, 181–200.

Jelinski, D.E. and Wu, J. (1996). The modifiable areal unit problem and implications for landscape ecology, *Landscape Ecology*, **11**, 129–140.

Journel, A.G. and Huijbregts, C.J. (1978). *Mining Geostatistics*. New York: Academic Press.

Kalnay, E. (2002). *Atmospheric Modeling, Data Assimilation and Predictability*. Cambridge, U.K.: Cambridge University Press.

Kent, J.T. (1989). Continuity properties for random fields. *Annals of Probability*, **17**, 1432–1440.

King, G. (1997). *A Solution to the Ecological Inference Problem*. Princeton, NJ: Princeton University Press.

Levin, S. (1992). The problem of pattern and scale in ecology. *Ecology*, **73**, 1943–1967.

Longley, P.A., Goodchild, D.J. Maguire, D.J., and Rhind, D.W. (2001). *Geographic Information Systems and Science*. Chichester, U.K.: John Wiley & Sons.

McMillan, N., Holland, D., Morara, M., and Feng, J. (2008). Combining numerical model output and particulate data using Bayesian space-time modeling. *Environmentrics*, DOI 10.1002/env. 984.

Mrozinski, R. and Cromley, R. (1999). Singly and doubly constrained methods of areal interpolation for vector-based GIS, *Transactions in GIS*, **3**, 285–301.

Mugglin, A.S. and Carlin, B.P. (1998). Hierarchical modeling in Geographic Information Systems: Population interpolation over incompatible zones. *Journal of Agricultural, Biological and Environmental Statistics*, **3**, 111–130.

Mugglin, A.S., Carlin, B.P., and Gelfand, A.E. (2000). Fully model based approaches for spatially misaligned data. *Journal of the American Statistical Association*, **95**, 877–887.

Openshaw, S. (1984). *The Modifiable Areal Unit Problem*. Norwich, U.K.: Geobooks.

Openshaw, S. and Alvanides, S. (1999). Applying geo-computation to the analysis of spatial distributions. In *Geographical Information Systems: Principles, Techniques, Management, and Applications*, Eds. P.A. Longley, M.F. Goodchild, D.J. Maguire, and D.W. Rhind, New York: John Wiley & Sons.

Openshaw, S. and Taylor, P. (1979). A million or so correlation coefficients. In *Statistical Methods in the Social Sciences*, ed. N. Wrigley, London: Pion, 127–144.

Openshaw, S. and Taylor, P. (1981). The modifiable areal unit problem. In *Quantitative Geography: A British View*, eds. N. Wrigley and R. Bennett, London: Routledge and Kegan Paul, 60–69.

Richardson, S. (1992). Statistical methods for geographic correlation studies. In *Geographical and Environmental Epidemiology: Methods for Small Area Studies*, Eds. P. Elliott, J. Cuzick, D. English, and R. Stern, New York: Oxford University Press, 181–204.

Robinson, W.S. (1950). Ecological correlations and the behavior of individuals. *American Sociological Review*, **15**, 351–357.

Royle, A., Berliner, L.M., Wikle, C.K., and Mitliff, R. (1997). A hierarchical spatial model for constructing wind fields from scatterometer data in the Labrador Sea. Technical Report 97–30, the Ohio State University, Columbus.

Stein, M.L. (1999). *Interpolation of Spatial Data: Some Theory for Kriging*. New York: Springer-Verlag.

Tanner, M.A. (1996). *Tools for Statistical Inference: Methods for the Exploration of Posterior Distributions and Likelihood Functions*, 3rd ed. New York: Springer-Verlag.

Tobler, W.R. (1979). Smooth pycnophylactic interpolation for geographical regions (with discussion). *Journal of the American Statistical Association*, **74**, 519–536.

Wakefield, J. (2003). Sensitivity analyses for ecological regression. *Biometrics*, **59**, 9–17.

Wakefield, J. (2004). Ecological inference for 2×2 tables (with discussion). *Journal of the Royal Statistical Society, Ser. A*. **167**, 385–445.

Wakefield, J. (2008). Ecological studies revisited. *Annual Review of Public Health*, **29**, 75–90.

Wakefield, J. and Shaddick, G. (2006). Health-exposure modelling and the ecological fallacy. *Biostatistics*, **7**, 438–455.

Waller, L.A., Carlin, B.P., and Xia, H. (1997). Structuring correlation within hierarchical spatio-temporal models for disease rates. In *Modelling Longitudinal and Spatially Correlated Data*, eds. T.G. Gregoire, D.R. Brillinger, P.J. Diggle, E. Russek-Cohen, W.G. Warren, and R.D. Wolfinger, New York: Springer-Verlag, pp. 308–319.

Waller, L.A., Carlin, B.P., Xia, H., and Gelfand, A.E. (1997). Hierarchical spatio-temporal mapping of disease rates. *Journal of the American Statistical Association*, **92**, 607–617.

Wong, D.W.S. (1996). Aggregation effects in geo-referenced data. In *Advanced Spatial Statistics*, ed. D. Griffiths, Boca Raton, FL: CRC Press, 83–106.

Yule, G.U. and Kendall, M.G. (1950). *An Introduction to the Theory of Statistics*, London: Griffin.

Zhu, L., Carlin, B.P., and Gelfand, A.E. (2003). Hierarchical regression with misaligned spatial data: Relating ambient ozone and pediatric asthma ER visits in Atlanta. *Environmetrics*, **14**, 537–557.

30

Spatial Aggregation and the Ecological Fallacy

Jonathan Wakefield and Hilary Lyons

CONTENTS

30.1 Introduction

In general, ecological studies are characterized by being based on grouped data, with the groups often corresponding to geographical areas, so that spatial aggregation has been carried out. Such studies have a long history in many disciplines including political science (King, 1997), geography (Openshaw, 1984), sociology (Robinson, 1950), and epidemiology and public health (Morgenstern, 1998). Our terminology will reflect the latter application; however, the ideas generalize across disciplines. Ecological studies are prone to unique drawbacks, in particular the potential for *ecological bias*, which describes the difference between estimated associations based on ecological- and individual-level data. Ecological data are a special case of spatially misaligned data, a discussion of which is the subject of Chapter 29.

Ecological data may be used for a variety of purposes including mapping (the geographical summarization of an outcome, see Chapter 14), and cluster detection (in which anomalous areas are flagged); here, we focus on spatial regression in which the aim is to investigate associations between an outcome and covariates, which we will refer to as exposures. In mapping, ecological bias is not a problem as prediction of area-level outcome summaries is the objective rather than the estimation of associations. Although ecological covariates may be used within a regression model to improve predictions, the coefficients are not of direct interest. Interesting, within-area, features may be masked by the process of aggregation; however, Wakefield (2007) provides more discussion. Cluster detection is also not concerned with regression analysis and, again, though small area anomalies may be "washed away" when data are aggregated, ecological bias as defined above is not an issue.

There are a number of reasons for the popularity of ecological studies, the obvious one being the wide and increasing availability of aggregated data. Improved ease of analysis also contributes to the widespread use of ecological data. For example, a geographic information system (GIS) allows the effective storage and combination of datasets from different sources and with differing geographies, and recent advances in statistical methodology allow a more refined analysis of ecological data Elliott, Wakefield, Best, and Briggs (2000) and Waller and Gotway (2004) contain reviews in a spatial epidemiological setting).

The fundamental problem of ecological analyses is the loss of information due to aggregation — the mean function, upon which regression is often based, is usually not identifiable from ecological data alone. This lack of identifiability can lead to the *ecological fallacy* in which individual and ecological associations between the outcome and an explanatory variable differ, and may even reverse direction. There are two key issues that we wish to emphasize throughout this chapter. First, hierarchical models cannot account for the loss of information, and the use of spatial models in particular will not resolve the ecological fallacy. Second, the only solution to the ecological fallacy, and thereby to provide reliable inference, is to supplement the ecological-level data with individual-level data.

30.2 Motivating Example

To motivate the discussion that follows, we introduce an example. Of interest is an investigation of the association between asthma hospitalization and air pollution, specifically $PM_{2.5}$ (particulate matter less than 2.5 microns in diameter) in California. This example is typical of many studies performed in environmental epidemiology. Ideally we would have access to individual-level outcomes, along with individual-level predictors and some measure of exposure. Such data are costly and logistically difficult to collect, however, and often unavailable for reasons of patient confidentiality. Instead, we consider the analysis of county-level asthma hospitalization data collected in 58 counties in California over the period 1998 to 2000. We wish to combine these data with point level pollution monitor data, so strictly speaking the exposure data are not aggregate in nature. We have data from 86 monitor sites.

We let Y_i represent the total disease counts in county i over the 3-year period, and x_{ik} the average log exposure in the last year of that period ($PM_{2.5}$ measured at monitor k in county i, $i = 1, \ldots, m$, $k = 1, \ldots, m_i$). The m_is range between 0 and 9, with $\sum_{i=1}^{58} m_i = 86$. We also have population counts N_{ic} in area i and for confounder stratum c, $c = 1, \ldots, C$. In our case, we have two age categories ($\leq 14/>15$) and four race categories (non-Hispanic white/black/Asian or Pacific Islander/Hispanic), so that $C = 8$. We also have the elevation of the centroid of block groups (elevation has been shown to have an association with asthma incidence) within the area, which may be population-averaged to create a single number per county, z_i. There are 22,133 block groups in California. A common descriptive measure for data of this type is the standardized morbidity ratio (SMR) that is given by Y_i/E_i where the expected numbers $E_i = \sum_{c=1}^{C} N_{ic}\widehat{q}_c$ control for differences in outcome by stratum. The SMR is a summary (across confounder stratum) of the area level relative risk. Here $\widehat{q}_j$ are reference risks; one must be wary in the manner by which these are calculated in a regression setting, so as to not bias the regression estimates (Wakefield, 2007). Here we use reference rates calculated from data for California from a previous period. Figure 30.1 maps the county-level SMRs and we see a relatively large range of variability across California with minimum of 0.07 and maximum of 2.22. The variability associated with the SMR in

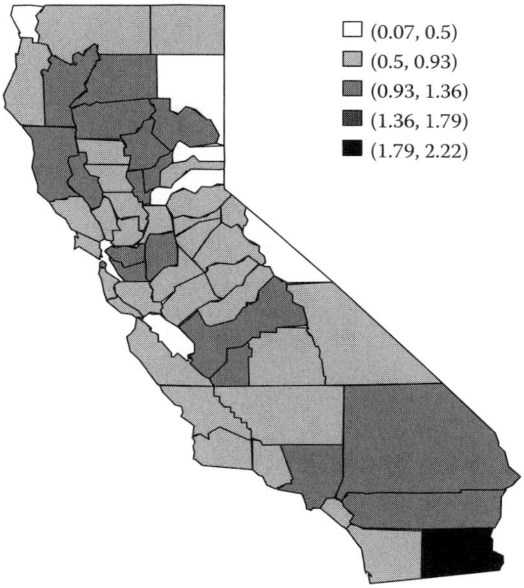

FIGURE 30.1
SMRs for asthma hospitalization in California counties.

area i is proportional to E_i^{-1}, however, so it is unclear as to the extent the map is displaying true differences, as compared to sampling variability.

Figure 30.2 plots logs of the SMRs versus the mean of the monitors in the 42 counties containing monitors. There is no clear pattern, though a slight suggestion of increased county-level relative risks in those counties with higher average log $PM_{2.5}$.

A simple model is provided by the county-level regression:

$$E[Y_i|\overline{x}_i, z_i] = \mu_i = E_i \exp(\beta_0 + \beta_1 \overline{x}_i + \beta_2 z_i), \tag{30.1}$$

where $\overline{x}_i$ is the (log) exposure within county i, $i = 1, \ldots, m$. These exposures were obtained as kriged values at the county centroid (Section 30.7 contains details on this procedure). In this model, $\exp(\beta_1)$ is the ecological county-level relative risk associated with

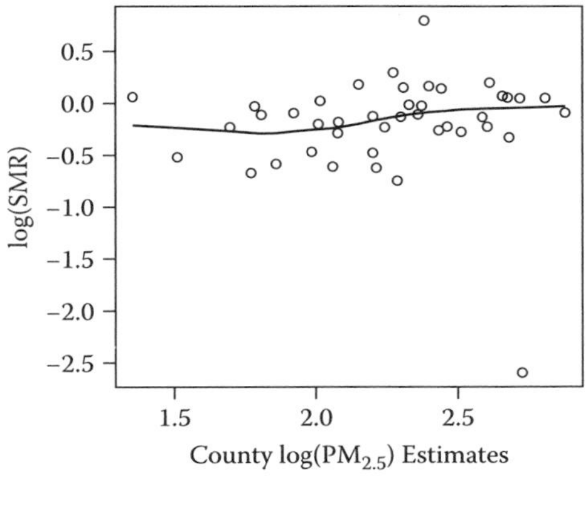

FIGURE 30.2
Log SMR versus mean log $PM_{2.5}$ for counties with monitors, with a local smoother imposed.

TABLE 30.1

Association between Asthma Hospitalization and Log $PM_{2.5}$ ($\widehat{\beta}_1$) and Elevation ($\widehat{\beta}_2$)

Mean Model	Estimation Model	$\widehat{\beta}_1$	Std. Err.	$\widehat{\beta}_2$	Std. Err.
Log-linear model	Poisson	0.306	0.013	−0.017	0.007
Log-linear model	Quasi-likelihood	0.306	0.128	−0.017	0.064
Log-linear model	Negative binomial	0.227	0.171	−0.143	0.045
Aggregate Exposure Model	Quasi-likelihood	0.261	0.092	0.011	0.058
Log-linear model	Hierarchical nonspatial	0.240	0.177	−0.146	0.047
Log-linear model	Hierarchical convolution	0.230	0.217	−0.146	0.048

Note: The four models are fitted using maximum and quasi-maximum likelihood estimation and the bottom two are Bayesian hierarchical models (and assume a Poisson likelihood). The log-linear model refers to model (30.1) while the aggregate exposure model refers to model (30.15) using modeled exposures. Convolution refers to the model with nonspatial and spatial random effects modeled via an ICAR model.

a unit increase in log $PM_{2.5}$; similarly, $\exp(\beta_2)$ is the ecological county-level relative risk associated with a unit increase in elevation, in each case with the other variable held constant. Model (30.1) may be fitted via likelihood-based methods under the assumption that $Y_i \sim \text{Poisson}(\mu_i)$. However, for data such as these there is usually excess-Poisson variability due to unmeasured variables, measurement error in the exposures, problems with the population/confounder data, or other sources of model misspecification (Wakefield and Elliott, 1999). A simple fix is to utilize quasi-likelihood (McCullagh and Nelder, 1989) to allow for overdispersion in a semiparametric way via the second moment assumption $\text{var}(Y_i) = \kappa \times E[Y_i]$. Alternatively, we may assume a negative binomial model so that overdispersion is incorporated via a parametric specification.

The first three rows of Table 30.1 give the maximum likelihood estimates (MLEs) along with their asymptotic standard errors for the Poisson, quasi-likelihood and negative binomial models. Under a Wald test, the Poisson model gives significant (at the 5% level) associations for both exposure and elevation, with $PM_{2.5}$ harmful. When moving from the Poisson to the quasi-likelihood model, we see a very large increase in the standard error, reflecting the huge excess-Poisson variability in these data. The standard errors are multiplied by $\widehat{\kappa}^{1/2} = 9.8$. We see a further increase in the standard error with the negative binomial model, and a substantial decrease in the point estimate. On examination of residuals versus fitted values (not shown), we are led to prefer the negative binomial model (under this model, we have a quadratic mean variance model, as compared to a linear relationship under the quasi-Poisson model). Neither the quasi-Poisson or negative binomial models suggest a significant association.

These results are all subject to ecological bias because we have used aggregate risk and exposure measures averaged over monitors within counties. We now discuss sources of ecological bias.

30.3 Ecological Bias: Introduction

There is a vast literature describing sources of ecological bias (Greenland, 1992; Greenland and Morgenstern, 1989; Greenland and Robins, 1994; Künzli and Tager, 1997; Morgenstern, 1998; Piantadosi, Byar, and Green, 1988; Plummer and Clayton, 1996; Richardson and Montfort, 2000; Richardson, Stucker, and Hémon, 1987; Steel and Holt, 1996; Wakefield, 2008; Wakefield, 2003; Wakefield, 2004; Wakefield, 2007). The fundamental problem with ecological inference is that the process of aggregation reduces information, and this information loss usually prevents identification of parameters of interest in the underlying individual-level model. When trying to understand ecological bias, it is often beneficial to specify an individual-level model, and aggregate to determine the consequences (Sheppard, 2003; Wakefield, 2004; Wakefield and Salway, 2001). The majority of the literature on ecological bias is less specific about the model, however. For example, in Robinson's famous 1950 paper, the correlation between literacy and race was calculated at various levels of geographic aggregation, and compared with the individual-level correlation, without reference to an explicit model.

If there is no within-area variability in exposures and confounders, then there will be no ecological bias; hence, ecological bias occurs due to within-area variability in exposures and confounders. There are a number of distinct consequences of this variability. Throughout, unless stated otherwise, we assume that at the individual level the outcome, y, is a 0/1 disease indicator, though ecological bias can occur for any type of outcome.

30.4 Ecological Bias: Pure Specification Bias

So-called pure specification bias (Greenland, 2002) (also referred to as model specification bias (Sheppard, 2003)) arises because a nonlinear risk model changes its form under aggregation. We initially assume a single exposure x and the linear individual-level model

$$E[Y_{ij}|x_{ij}] = \beta_0 + \beta_1 x_{ij}, \tag{30.2}$$

where Y_{ij} and x_{ij} are the outcome and exposure for individual j within area $i, i = 1, \ldots, m$, $j = 1, \ldots, n_i$. The aggregate data are assumed to correspond to the average risk $\overline{y}_i = \frac{1}{m_i}\sum_{j=1}^{m_i} y_{ij}$ and average exposure $\overline{x}_i = \frac{1}{m_i}\sum_{j=1}^{m_i} x_{ij}$. On aggregation of Equation (30.2), we obtain

$$E[\overline{Y}_i|\overline{x}_i] = \beta_0 + \beta_1 \overline{x}_i, \tag{30.3}$$

so that in this very specific scenario of a linear model we have not lost anything by aggregation (and this is clearly true regardless of whether Y is discrete or continuous).

Unfortunately, a linear model is often inappropriate for the modeling of risk and for rare diseases the individual-level model:

$$E[Y_{ij}|x_{ij}] = \mathrm{e}^{\beta_0 + \beta_1 x_{ij}} \tag{30.4}$$

is often more appropriate. In this model, e^{β_0} is the risk associated with $x = 0$ (baseline risk) and e^{β_1} is the relative risk corresponding to an increase in x of one unit. The logistic model, which is often used for nonrare outcomes, is unfortunately not amenable to analytical study and so the effects of aggregation are difficult to discern (Salaway and Wakefield, 2005). Aggregation of Equation (30.4) yields:

$$E[\overline{Y}_i|x_{ij}, j = 1, \ldots, n_i] = \frac{1}{n_i}\sum_{j=1}^{n_i} \mathrm{e}^{\beta_0 + \beta_1 x_{ij}}, \tag{30.5}$$

so that the ecological risk is the average of the risks of the constituent individuals. A naivé ecological model would assume

$$E[\overline{Y}_i|\overline{x}_i] = \mathrm{e}^{\beta_0^e + \beta_1^e \overline{x}_i}, \tag{30.6}$$

where the ecological parameters, β_0^e, β_1^e, have been superscripted with "e" to distinguish them from the individual-level parameters in Equation (30.4). Model (30.6) is actually a so-called *contextual effects* model since risk depends on the average exposure in the area (contextual variables are summaries of a shared environment). Interpreting $e_1^{\beta^e}$ as an individual association would correspond to a belief that it is average exposure that is causative, and that individual exposure is irrelevant. As an aside, we mention the atomistic fallacy that occurs when inference is required at the level of the group, but is incorrectly estimated using individual-level data (Diez-Roux, 1998).

The difference between (30.5) and (30.6) is clear; while the former averages the risks across all exposures, the latter is the risk corresponding to the average exposure. Without further assumptions on the moments of the within-area exposure distributions, we can guarantee no ecological bias, i.e., $e^{\beta_1} = e^{\beta_1^e}$, only when there is no within-area variability in exposure so that $x_{ij} = \overline{x}_i$ for all $j = 1, \ldots, n_i$ individuals in area i and for all areas, $i = 1, \ldots, m$. Therefore, pure specification bias is reduced in size as homogeneity of exposures within areas increases — small areas are advantageous in this respect. Unfortunately data aggregation is usually carried out according to administration groupings and not in order to obtain areas with constant exposure. As we will shortly describe, there are other specific circumstances when pure specification is likely to be small and these depend on the moments of the exposure distributions.

Binary exposures are the simplest to study analytically. Such exposures may correspond to, for example, an individual being below or above a pollutant threshold. For a binary exposure, Equation (30.4) can be written

$$e^{\beta_0+\beta_1 x_{ij}} = (1 - x_{ij})e^{\beta_0} + x_{ij}e^{\beta_0+\beta_1},$$

which is linear in x_{ij}. This form yields the aggregate form:

$$E[\overline{Y}_i|\overline{x}_i] = (1 - \overline{x}_i)e^{\beta_0} + \overline{x}_i e^{\beta_0+\beta_1}, \qquad (30.7)$$

where $\overline{x}_i$ is the proportion exposed in area i. Hence, with a linear risk model, there is no pure specification bias so long as model (30.7) is fitted using the binary proportion, $\overline{x}_i$, and not model (30.6). If model (30.6) is fitted, there will be no correspondence between e^{β_1} and $e^{\beta_1^e}$ because they are associated with completely different comparisons.

The extension to general categorical exposures is straightforward, and the parameters of the disease model are identifiable so long as we have observed the aggregate proportions in each category. We now demonstrate that for a continuous exposure pure specification bias is dominated by the within-area mean-variance relationship. In an ecological regression context, a normal within-area exposure distribution $N(x|\overline{x}_i, s_i^2)$, and the log-linear model (30.4), has been considered by a number of authors (Plummer and Clayton, 1996; Richardson, Stucker, and Hémon, 1987; Wakefield and Salway, 2001). We assume that n_i is large so that the summation in (30.5) can be approximated by an integral. For a normally distributed exposure, this integral is available as

$$E[\overline{Y}_i|\overline{x}_i] = \exp\left(\beta_0 + \beta_1\overline{x}_i + \beta_1^2 s_i^2/2\right), \qquad (30.8)$$

which may be compared with the naivé ecological model $e^{\beta_0^e+\beta_1^e\overline{x}_i}$. To gain intuition as to the extent of the bias, we observe that in Equation (30.8) the within-area variance s_i^2 is acting like a confounder and, consequently, there is no pure specification bias if the exposure is constant within each area or if the variance is independent of the mean exposure in the area. The expression (30.8) also allows us to characterize the direction of bias. For example, suppose that $s_i^2 = a + b\overline{x}_i$ with $b > 0$ so that the variance increases with the mean (as is often observed with environmental exposures, for example). In this case, the parameter we are estimating from the ecological data is

$$\beta_1^e = \beta_1 + \beta_1^2 b/2.$$

If $\beta_1 > 0$, then overestimation will occur using the ecological model, and, if $\beta_1 < 0$, the ecological association, β_1^e, may reverse sign when compared to β_1.

In general, there is no pure specification bias if the disease model is linear in x, or if all the moments of the within-area distribution of exposure are independent of the mean. If β_1 is close to zero, pure specification bias is also likely to be small (since then the exponential model will be approximately linear for which there is no bias), though in this case confounding is likely to be a serious worry (Section 30.5). Unfortunately the mean-variance relationship is impossible to assess without individual-level data on the exposure. If the exposure is heterogeneous within areas, we need information on the variability within each area in order to control the bias. Such information may come from a sample of individuals within each area; how to use this individual-level data (beyond assessing the within-area exposure mean-variance relationship) is the subject of Section 30.6.

30.5 Ecological Bias: Confounding

We assume a single exposure x_{ij}, a single confounder z_{ij}, and the individual-level model

$$E[Y_{ij}|x_{ij}, z_{ij}] = e^{\beta_0+\beta_1 x_{ij}+\beta_2 z_{ij}}. \tag{30.9}$$

As with pure specification bias, the key to understanding sources of, and correction for, ecological bias is to aggregate the individual-level model to give

$$E[\overline{Y}_i|x_{ij}, z_{ij}, j = 1, \ldots, n_i] = \frac{1}{n_i}\sum_{j=1}^{n_i} e^{\beta_0+\beta_1 x_{ij}+\beta_2 z_{ij}}. \tag{30.10}$$

To understand why controlling for confounding is in general impossible with ecological data, we consider the simplest case of a binary exposure (unexposed/exposed) and a binary confounder, which for ease of explanation we assume is gender. Table 30.2 shows the distribution of the exposure and confounder within area i. The complete within-area distribution of exposure and confounder can be described by three frequencies, but the ecologic data usually consist of two quantities only, the proportion exposed, $\overline{x}_i$, and the proportion male, $\overline{z}_i$. From Equation (30.10), the aggregate form is

$$\begin{aligned} E[\overline{Y}_i|p_{i00}, p_{i01}, p_{i10}, p_{i11}] &= (1 - \overline{x}_i - \overline{z}_i + p_{i11})e^{\beta_0} \\ &\quad + (\overline{x}_i - p_{i11})e^{\beta_0+\beta_1} + (\overline{z}_i - p_{i11})e^{\beta_0+\beta_2} + p_{i11}e^{\beta_0+\beta_1+\beta_2} \end{aligned}$$

showing that the marginal prevalences, $\overline{x}_i, \overline{z}_i$, alone, are not sufficient to characterize the joint distribution unless x and z are independent, in which case, z is not a within-area

TABLE 30.2

Exposure and Gender Distribution in Area i, $\overline{x}_i$ Is the Proportion Exposed and $\overline{z}_i$ Is the Proportion Male; $p_{i00}, p_{i01}, p_{i10}, p_{i11}$ Are the Within-Area Cross-Classification Frequencies

	Female	Male	
Unexposed	p_{i00}	p_{i01}	$1 - \overline{x}_i$
Exposed	p_{i10}	p_{i11}	$\overline{x}_i$
	$1 - \overline{z}_i$	$\overline{z}_i$	1.0

confounder. This scenario has been considered in detail elsewhere (Lasserre, Guihenneuc-Jouyaux, and Richardson, 2000), where it was argued that if the proportion of exposed males (p_{i11}) is missing, it should be estimated by the marginal prevalences ($\overline{x}_i \times \overline{z}_i$). It is not possible to determine the accuracy of this approximation without individual-level data, however. This is a recurring theme in the analysis of ecological data, bias can be reduced under model assumptions, but estimation is crucially dependent on the appropriateness of these assumptions, which are uncheckable without individual-level data.

We now examine the situation in which we have a binary exposure and a continuous confounder. Let the confounders in the unexposed be denoted, z_{ij}, $j = 1, \ldots, n_{i0}$, and the confounders in the exposed, z_{ij}, $j = n_{i0} + 1, \ldots, n_{i0} + n_{i1}$, with $n_{i0} + n_{i1} = n_i$. In this case, the ecological form corresponding to Equation (30.9) is

$$E[\overline{Y}_i | q_{i0}, q_{i1}] = q_{i0} \times r_{i0} + q_{i1} \times r_{i1},$$

where $q_{i0} = n_{i0}/n_i$ and $q_{i1} = n_{i1}/n_i$ are the probabilities of being unexposed and exposed, and

$$r_{i0} = \frac{e^{\beta_0}}{n_{i0}} \sum_{j=1}^{n_{i0}} e^{\beta_2 z_{ij}}, \quad r_{i1} = \frac{e^{\beta_0+\beta_1}}{n_{i1}} \sum_{j=n_{i0}+1}^{n_{i0}+n_{i1}} e^{\beta_2 z_{ij}}$$

are the (aggregated) risks in the unexposed and exposed. The important message here is that we need the confounder distribution within each exposure category, unless z is not a within-area confounder. Again it is clear that if we fit the model:

$$E[\overline{Y}_i | \overline{x}_i, \overline{z}_i] = e^{\beta_0^e + \beta_1^e \overline{x}_i + \beta_2^e \overline{z}_i},$$

where $\overline{z}_i = \frac{1}{n_i} \sum_{j=1}^{n_i} z_{ij}$, then it is not possible to equate the ecological coefficient β_1^e with the individual-level parameter of interest β_1.

We now extend our discussion to multiple strata and show the link with the use of expected numbers (as defined in Section 30.2). Consider the continuous exposure x_{icj} for the jth individual in stratum c and area i, and suppose the individual level model is given by

$$E[Y_{icj} | x_{icj}, \text{ strata } c] = e^{\beta_0 + \beta_1 x_{icj} + \beta_{2c}},$$

for $c = 1, \ldots, C$ stratum levels with relative risks $e^{\beta_{2c}}$ (with $e^{\beta_{21}} = 1$ for identifiability), and with $j = 1, \ldots, n_{ic}$. For ease of exposition, suppose c indexes age categories. Let $\overline{Y}_{ic} = \frac{1}{n_{ic}} \sum_{j=1}^{n_c} Y_{icj}$ be the proportion with the disease in area i, stratum c. Then

$$E[\overline{Y}_{ic} | x_{icj}, j = 1, \ldots, n_{ic}] = \frac{e^{\beta_0+\beta_{2c}}}{n_{ic}} \sum_{j=1}^{n_{ic}} e^{\beta_1 x_{icj}}.$$

Summing over stratum and letting $\overline{Y}_i$ be the proportion with disease in area i:

$$E[\overline{Y}_i | x_{icj}, j = 1, \ldots, n_{ic}, c = 1, \ldots, C] = \frac{1}{n_i} \sum_{c=1}^{C} n_{ic} \left\{ \frac{e^{\beta_0+\beta_{2c}}}{n_{ic}} \sum_{j=1}^{n_{ic}} e^{\beta_1 x_{icj}} \right\}. \tag{30.11}$$

If we assume a common exposure distribution across stratum and let x_{ij}, $j = 1, \ldots, m_i$ be a representative exposure sample, then we could fit the model

$$\begin{aligned} E[Y_i | x_{ij}, j = 1, \ldots, m_i] &= \sum_{c=1}^{C} n_{ic} e^{\beta_{2c}} \times \frac{e^{\beta_0}}{m_i} \sum_{j=1}^{m_i} e^{\beta_1 x_{ij}} \\ &= E_i \times \frac{e^{\beta_0}}{m_i} \sum_{j=1}^{m_i} e^{\beta_1 x_{ij}}, \end{aligned} \tag{30.12}$$

where $E_i = \sum_{c=1}^{C} n_{ic} e^{\beta_{2c}}$ are the expected numbers. Model (30.12) attempts to correct for pure specification bias, but assumes common exposure variability across areas. Therefore, we see that in this model (which has been previously used (Guthrie, Sheppard, and Wakefield, 2002)), we have standardized for age (via indirect standardization), but for this to be valid we need to assume that the exposure is constant across age groups (so that age is not a within-area confounder). This can be compared with the model that is frequently fitted:

$$E[Y_i|\overline{x}_i] = E_i \times e^{\beta_0+\beta_1\overline{x}_i}.$$

Validity of this model goes beyond a constant exposure distribution across stratum within each area. We also require no within-area variability in exposure (or, recalling our earlier discussion, the exposure variance being independent of the mean, in addition to constant distributions across stratum).

This discussion is closely related to the idea of mutual standardization in which, if the response is standardized by age, say, the exposure variable must also be standardized for this variable (Rosenbaum and Rubin, 1984). The correct model is given by Equation (30.11), and requires the exposure distribution by age group, or at least a representative sample of exposures from each age group. The above discussion makes it clear that we need *individual-level data* to characterize the within-area distribution of confounders and exposures.

The extension to general exposure and confounder scenarios is obvious from the above. If we have true confounders that are constant within areas (for example, access to healthcare), then they are analogous to conventional confounders because the area is the unit of analysis, and so the implications are relatively easy to understand and adjustment is straightforward.

Without an interaction between exposure and confounder, the parameters of a linear model are estimable from marginal information only, though, if an interaction is present, within-area information is required (Wakefield, 2003).

30.6 Combining Ecological and Individual Data

As we saw in Section 30.3, the only solution to the ecologic inference problem that does not require uncheckable assumptions is to add individual-level data to the ecological data. Here, we briefly review some of the proposals for such an endeavor. Another perspective is that ecological data can supplement already-available individual data in order to increase power.

Table 30.3 summarizes four distinct scenarios in terms of data availability (Künzli and Tager, 1997; Sheppard, 2003). All entries but the individual–individual cell concern change of support situations (Chapter 29). The obvious approach to adding individual-level data is to collect a random sample of individuals within areas. For a continuous outcome, Raghunathan, Diehr, and Cheadle (2003) show that moment and maximum likelihood

TABLE 30.3

Study Designs by Level of Outcome and Exposure Data

		Exposure	
		Individual	**Ecological**
Outcome	Individual	Individual	Semiecological
	Ecological	Aggregate	Ecological

estimates of a common within-group correlation coefficient will improve when aggregate data are combined with individual data within groups, and Glynn, Wakefield, Handcock, and Richardson (2008) derive optimal design strategies for the collection of individual-level data when the model is linear. With a binary nonrare outcome, the benefits have also been illustrated (Steele, Beh, and Chambers, 2004; Wakefield, 2004).

For a rare disease, few cases will be present in the individuals within the sample, and so only information on the distribution of exposures and confounders will be obtained via a random sampling strategy (which is, therefore, equivalent to using a survey sample of covariates only). This prompted the derivation of the so-called aggregate data method of (Prentice and Sheppard 1995; Sheppard, Prentice, and Rossing, 1996; Sheppard and Prentice, 1995), which is the bottom left entry in Table 30.3. Inference proceeds by constructing an estimating function based on the sample of $m_i \leq n_i$ individuals in each area. For example, with samples for two variables, $\{x_{ij}, z_{ij}, j = 1, \ldots, m_i\}$ we have the mean function:

$$E\left[\overline{Y}_i | x_{ij}, z_{ij}, j = 1, \ldots, m_i\right] = \frac{1}{n_i} \frac{n_i}{m_i} \sum_{j=1}^{m_i} \mathrm{e}^{\beta_0 + \beta_1 x_{ij} + \beta_2 z_{ij}}.$$

There is bias involved in the resultant estimator since the complete set of exposures are not available, but Prentice and Sheppard give a finite sample correction to the estimating function based on survey sampling methods. This is an extremely powerful design since estimation is not based on any assumptions with respect to the within-area distribution of exposures and confounders (though this distribution may not be well characterized for small samples, Salway and Wakefield, 2008). An alternative approach is to assume a particular distribution for the within-area variability in exposure, and fit the implied model (Best, Cockings, Bennett, Wakefield, and Elliott, 2001; Jackson, Best, and Richardson, 2008; Jackson, Best, and Richardson, 2006; Richardson, Stucker, and Hémon, 1987; Wakefield and Salway, 2001). The normal model is usually assumed, in which case (for a single exposure) the mean model is (30.8). This method implicitly assumes that a sample of within-area exposures is available since the within-area moments need to be available. More recently an approach has been suggested that takes the mean as a combination of the Prentice and Sheppard and the parametric approaches, with the latter dominating for small samples (when the aggregate data method can provide unstable inference) (Salway and Wakefield, 2008).

In the same spirit as Prentice and Sheppard, Wakefield and Shaddick (2006) described a likelihood-based method for alleviating ecological bias. We describe the method in the situation in which a single exposure is available. If data on all individual exposures were available, then we would fit the model

$$E[\overline{Y}_i | x_{ij}, j = 1, \ldots, n_i] = \frac{\mathrm{e}^{\beta_0}}{n_i} \sum_{j=1}^{n_i} \mathrm{e}^{\beta_1 x_{ij}}, \tag{30.13}$$

which will remove ecological bias, but will result in a loss of power relative to an individual-level study because we have not used the linked individual disease-exposure data. Usually we will not have access to all of the individual exposures, but instead we may have access to data in subareas at a lower level of aggregation, e.g., block groups within counties. Suppose that we have m_i sub-groups within area i with an exposure measure x_{ij} for $j = 1, \ldots, m_i$. Suppose also that N_{ij} is the number of individuals in subarea j so that $n_i = \sum_{j=1}^{m_i} N_{ij}$. For example, x_{ij} may represent a measure of exposure at the centroids of subarea j within area i. We then alter model (30.13) slightly, and act as if there were N_{ij} individuals each with exposure x_{ij} (so that we are effectively ignoring within subarea variability in exposure):

$$E[\overline{Y}_i | x_{ij}, j = 1, \ldots, m_i] = \frac{e^{\beta_0}}{n_i} \sum_{j=1}^{m_i} N_{ij} e^{\beta_1 x_{ij}}, \tag{30.14}$$

where j again indexes the number of subareas within area i. If we have population information at the subarea level, e.g., age and gender, then we may calculate expected numbers, $E_{ij} = \sum_{c=1}^{C} N_{icj} e^{\beta_{2c}}$, and these will then replace the N_{ij} within Equation (30.14). We can also add in an area-level covariate z_i to give

$$E[Y_i | x_{ij}, j = 1, \ldots, m_i, z_i] = e^{\beta_0} \sum_{j=1}^{m_i} E_{ij} e^{\beta_1 x_{ij}}. \tag{30.15}$$

We call this latter form the *aggregate exposure* model. Valid estimates from this model require that population subgroups have the same exposure within each subarea (and these are constant across the stratum over which the expected numbers were calculated), but if the heterogeneity in exposure is small within these subareas, little bias will result. Often the collection $x_{i1}, \ldots, x_{im_i}$ will be obtained via exposure modeling and the validity of estimation requires that these exposure measures are accurate, which may be difficult to achieve unless the monitoring network is dense (relative to the variability of the exposure).

A different approach to adding individual data in the context of a rare response is outcome-dependent sampling, which avoids the problems of zero cases encountered in random sampling. For the situation in which ecologic data are supplemented with individual case-control information gathered within the constituent areas, inferential approaches have been developed (Haneuse and Wakefield, 2007, 2008a, 2008b). The case-control data remove ecological bias, while the ecological data provide increased power and constraints on the sampling distribution of the case-control data, which improves the precision of estimates.

Two-phase methods have a long history in statistics and epidemiology (Breslow and Cain, 1988; Breslow and Chatterjee, 1999; Walker, 1982; White, 1982) and are based on an initial cross-classification by outcome and confounders and exposures; this classification providing a sampling frame within which additional covariates may be gathered via the sampling of individuals. Such a design may be used in an ecological setting, where the initial classification is based on one or more of area, confounder stratum, and possibly error-prone measures of exposure (Wakefield and haneuse, 2008).

In all of these approaches, it is clearly vital to avoid response bias in the survey samples or selection bias in outcome-dependent sampling, and establishing a relevant sampling frame is essential.

30.7 Example Revisited

For the California data we have information on census block groups within counties. Recall there are 58 counties and 22,133 census block groups. We assume that the logged $PM_{2.5}$ monitor data are a realization from a Gaussian random field, and we fit this model to the monitor data using restricted maximum likelihood (REML) and a mean linear in population density. Given the fitted model we impute exposures at each block group centroid. Figure 30.3 displays exposure maps at both the county and the block group level. We now treat these exposures as known and examine the association between asthma hospitalization and exposure to $PM_{2.5}$. We note that because we have not jointly modeled the health and exposure variables, the uncertainty in the exposure predictions is not propagated through

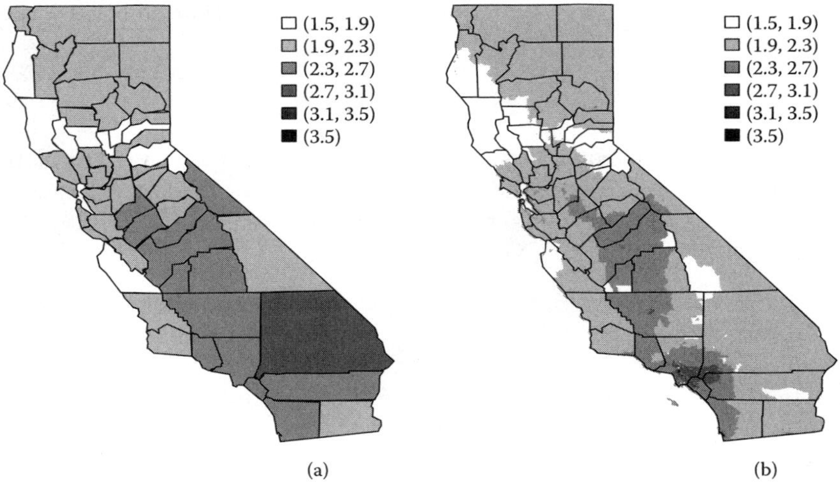

FIGURE 30.3
Predicted exposures (log $PM_{2.5}$) for (a) counties and (b) census block groups in California.

the estimation, and there is no feedback in the model. This can be advantageous, however, since misspecification of either the health or the exposure component can cause problems for the other component. Probably the biggest problem with this approach, though, is that we need sufficient monitor data to impute accurate exposures for unmonitored locations.

To utilize the census block group information, we used a quasi-likelihood model with $E[Y_i] = \mu_i, \text{var}(Y_i) = \kappa \times E[Y_i]$ and where μ_i is given by (30.15), and obtained the estimates, via quasi-likelihood, in Table 30.1, line 4. We see that the association is attenuated when compared with the previous quasi-likelihood estimate, as we might expect if the within-area variability in exposure increases with the mean, all other things being equal (Section 30.4). The standard error is increased also.

The biggest advantage of the above approach is that ecological bias can be overcome. An alternative approach that was followed by Zhu, Carlin, and Gelfand (2003) is to obtain county *average* exposures via kriging (Chapter 3) and then use model (30.6). However, this will produce ecologically biased estimates since within-area variability in exposure has not been acknowledged. We reiterate that it is the average of the risk functions that is required for individual-level inference, and not the risk function evaluated at the average exposure.

30.8 Spatial Dependence and Hierarchical Modeling

When data are available as counts from a set of contiguous areas, we might expect residual dependence in the counts, particularly for small-area studies, due to the presence of unmeasured variables with spatial structure. The use of the word "residual" here acknowledges that variables known to influence the outcome have already been adjusted for in the mean model. Analysis methods that ignore the dependence are strictly not applicable, with inappropriate standard errors being the most obvious manifestation. A great deal of work has focused on models for spatial dependence (Besag, York, and Mollié, 1991; Best, Ickstadt, and Wolpert, 2000; Christensen and Waagepetersen, 2002; Clayton, Bernardinelli, and Montomoli, 1993; Cressie and Chan, 1989; Diggle, Tawn, and Moyeed, 1998; Kelsall and Wakefield, 2002; Leroux, Lei, and Breslow, 1999). Richardson (2003) provides an excellent review of this literature (see also Chapter 14). With respect to ecological bias, however, the most important message is that unless the mean model is correct, adjustment for spatial dependence is a pointless exercise (Wakefield, 2007).

Spatial smoothing models have also been proposed to control for "confounding by location" (Clayton et al., 1993). A subtle but extremely important point is that such an endeavor is fraught with pitfalls since the exposure of interest usually has spatial structure and so one must choose an appropriate spatial scale for smoothing. If the scale is chosen to be too small, the exposure effect may be attenuated, while if too large a scale is chosen, the signal that is due to confounding may be absorbed into the exposure association estimate. Practically, one can obtain estimates from models with and without spatial smoothing, and with a variety of spatial models, to address the sensitivity of inference concerning parameters of interest. (See Reich, Hodges, and Zadnik, 2006, for further discussion.) Similar issues arise in time series analysis when one must control trends by selecting an appropriate level of temporal smoothing (Dominici, Sheppard, and Clyde, 2003). Such analyses are more straightforward since time is one-dimensional, the data are generally collected at regular intervals (often daily), and the data are also abundant, perhaps being collected over many years.

In a much-cited book (King, 1997), a hierarchical model was proposed for the analysis of ecologic data in a political science context, as "a solution to the ecological inference problem." Identifiability in this model is imposed through the random effects prior, however, and it is not possible to check the appropriateness of this prior from the ecological data alone Freedman, Klein, Ostland, and Roberts, 1998; Wakefield, 2004).

We have concentrated on Bayesian hierarchical spatial models, but a number of frequentist approaches are possible, though currently they have not been investigated in detail. Thurston, Wand, and Wiencke (2000) describe a negative binomial additive model that could provide a useful alternative to the models described here. The negative binomial aspect allows for overdispersion, while a generalized additive model would allow flexible modeling of latitude and longitude to model nonsmall scale spatial variability. More generally, recent work on generalized linear models with splines may be applicable in the setting described here (see, for example, Lin and zhang, 1999; Gu and Ma, 2005). Allowing for small-scale residual spatial dependence in these models would also be desirable, however. It would also be desirable to perform sandwich estimation in a spatial regression setting, but unfortunately the nonlattice nature of the data does not easily allow any concept of replication across space (as has been used for lattice data, (Heagerty and Lumley, 2000)).

30.9 Example Revisited

The first hierarchical model we fit adds a single nonspatial random effect, $V_i \sim_{iid} N(0, \sigma_v^2)$ to the linear predictor; this gives the same (quadratic) marginal mean-variance structure as the negative binomial model, and we would expect to see similar inference under the two models. This is confirmed in Table 30.1 for the naivé models.

To acknowledge spatial dependence, we now fit the log-linear model $Y_i|x_i \sim \text{Poisson}(E_i e^{\beta_1 x_i + U_i + V_i})$ where $V_i \sim_{iid} N(0, \sigma_v^2)$ are independent random effects and the U_i have spatial structure, in particular, we choose an intrinsic conditional autoregressive (ICAR) model, as suggested elsewhere (Besag et al., 1991). For inference we utilized, in addition to Markov chain Monte Carlo (MCMC), the integrated nested Laplace approximation scheme described in Rue, Martino, and Chopin (2009). MCMC displayed very poor convergence for the spatial log-linear model for these data. We attempted to fit the aggregate data model using MCMC, but the chain was extremely poorly behaved.

We place a prior on the total residual variance and upon the proportion of the variance that is spatial, approximately scaling the conditional variance in the ICAR model so that it is comparable with σ_v^2 (Wakefield, 2007). From Table 30.1, we see that when the ICAR component is included in the model we see a very similar estimate of the association with $PM_{2.5}$ as with the nonspatial hierarchical model, but with an increased standard error.

For these ecological data we would conclude that there is no evidence of an association. The reliability of the aggregate exposure model here is questionable, however, since validation of the exposure model has not been carried out. Unmeasured confounding is a serious worry here, and in particular ecological bias due to within-area confounding.

30.10 Semiecological Studies

In a semiecological study, sometimes more optimistically referred to as a "semiindividual study" (Künzli and Tager, 1997), individual-level data are collected on outcome and confounders, with exposure information arising from another source. The Harvard six-cities study (Dockery, Pope, Xiping, Spengler et al., 1993) provides an example in which the exposure was city-specific and was an average exposure from pollution monitors over the follow-up of the study.

We consider the risk for individual j in confounder stratum c and area i, $c = 1, \ldots, C$, $j = 1, \ldots, n_{ic}$, $i = 1, \ldots, m$. Let x_{icj} be the exposures of the individuals within stratum c and area i, and β_{2c} the baseline risk in stratum c. Under exposure aggregation, we have

$$\begin{aligned} E[Y_{icj} | x_{ic1}, \ldots, x_{icn_c}] &= E_{x|x_{ic1},\ldots,x_{icn_c}} \{E[Y_{icj}|x]\} \\ &= \frac{e^{\beta_0+\beta_{2c}}}{n_{ic}} \sum_{j=1}^{n_c} e^{\beta_1 x_{icj}} \end{aligned}$$

since the distribution of $x | x_{ic1}, \ldots, x_{icn_c}$ is discrete over the n_{ic} exposures, $x_{ic1}, \ldots, x_{icn_{ic}}$. A naivé semiecologic model is

$$E[Y_{icj} | \overline{x}_i] = e^{\beta_0^e + \beta_{2c}^e + \beta_1^e \overline{x}_i}, \tag{30.16}$$

where $\overline{x}_i$ is an exposure summary in area i. Künzli and Tager (1997) argue that semiecological studies are free of ecological bias, but this is incorrect because there are two possible sources of bias in model (30.16); the first is that we have pure specification bias because we have not acknowledged within-area variability in exposure, and the second is that we have not allowed the exposure to vary by confounder stratum, so we have not controlled for within-area confounding. In an air pollution study in multiple cities, x may correspond to a monitor average or an average over several monitors. In this case, Equation (30.16) will provide an approximately unbiased estimate of β_1 if there is small exposure variability in cities and if this variability is similar across confounder stratum.

Semiecological studies frequently have survival as an endpoint, but there has been less focus on the implications of aggregation in the context of survival models, with few exceptions (Abrahamowicz, du Berger, Krewski, Burnett et al., 2004; Haneuse, Wakefield, and Sheppard, 2007).

30.11 Concluding Remarks

The use of ecological data is ubiquitous, and so is the potential for ecological bias. A skeptic might conclude from the litany of potential biases described in Section 30.3 that ecological inference should never be attempted, but this would be too pessimistic a view.

A useful starting point for all ecological analyses is to write down an individual-level model for the outcome-exposure association of interest, including known confounders. Ecological bias will be small when within-area variability in exposures and confounders is small. A serious source of bias is that due to confounding, since ecological data on exposure are rarely stratified by confounder strata within areas. For a well-conducted ecological study, estimate associations may add to the totality of evidence for a hypothesis. When comparing ecological and semiecological estimates with individual-level estimates, it is clearly crucial to have a common effect measure (e.g., a relative risk or a hazard ratio). So, for example, it will be difficult to compare an ecological correlation coefficient, which is a measure that is often reported, with an effect estimate from an individual-level study.

Less well-designed ecological studies can be suggestive of hypotheses to investigate if strong ecological associations are observed. An alternative to the pessimistic outlook expressed above is that when a strong ecological association is observed an attempt should be made to explain how such a relationship could have arisen.

There are a number of issues that we have not discussed. Care should be taken in determining the effects of measurement error in an ecological study since the directions of bias may not be predictable. For example, in the absence of pure specification and confounder bias for linear and log-linear models, if there is nondifferential measurement error in a binary exposure, there will be overestimation of the effect parameter, in contrast to individual-level studies (Brenner, Savitz, Jockel, and Greenland, 1992). We refer interested readers to alternative sources, (Elliott and Wakefield, 1999, Wakefield and Elliott, 1999), for other issues, such as consideration of migration, latency periods, and the likely impacts of inaccuracies in population and health data.

Studies that investigate the acute effects of air pollution are another common situation in which ecological exposures are used. For example, daily disease counts in a city are often regressed against daily and/or lagged concentration measurements taken from a monitor, or the average of a collection of monitors to estimate the acute effects of air pollution. If day-to-day exposure variability is greater than within-city variability, then we would expect ecological bias to be relatively small. We have not considered ecological bias in a space–time context, little work has been done in this area (see Wakefield, 2004) for a brief development).

With respect to data availability, exposure information is generally not aggregate in nature (unless the "exposure" is a demographic or socio-economic variable), and in an environmental epidemiological setting the modeling of pollutant concentration surfaces will undoubtedly grow in popularity. However, an important insight is that in a health-exposure modeling context it may be better to use measurements from the nearest monitor rather than model the concentration surface because the latter approach may be susceptible to large biases, particularly when, as is usually the case, the monitoring network is sparse (Wakefield and Shaddick, 2006). A remaining challenge is to diagnose when the available data are of sufficient abundance and quality to support the use of complex models.

In Section 30.6 we described a number of proposals for the combination of ecological and individual data. Such endeavors will no doubt increase and will hopefully allow the reliable exploitation of ecological information.

Acknowledgments

This work was supported by grant R01 CA095994 from the National Institutes of Health.

References

M. Abrahamowicz, R. du Berger, D. Krewski, R. Burnett, G. Bartlett, R.M. Tamblyn, and K. Leffondré. Bias due to aggregation of individual covariates in the Cox regression model. *American Journal of Epidemiology*, 160:696–706, 2004.

J. Besag, J. York, and A. Mollié. Bayesian image restoration with two applications in spatial statistics. *Annals of the Institute of Statistics and Mathematics*, 43:1–59, 1991.

N. Best, S. Cockings, J. Bennett, J. Wakefield, and P. Elliott. Ecological regression analysis of environmental benzene exposure and childhood leukaemia: Sensitivity to data inaccuracies, geographical scale and ecological bias. *Journal of the Royal Statistical Society, Series A*, 164:155–174, 2001.

N.G. Best, K. Ickstadt, and R.L. Wolpert. Ecological modelling of health and exposure data measured at disparate spatial scales. *Journal of the American Statistical Association*, 95:1076–1088, 2000.

H. Brenner, D. Savitz, K.-H. Jockel, and S. Greenland. Effects of non-differential exposure misclassification in ecologic studies. *American Journal of Epidemiology*, 135:85–95, 1992.

N.E. Breslow and K.C. Cain. Logistic regression for two-stage case-control data. *Biometrika*, 75:11–20, 1988.

N.E. Breslow and N. Chatterjee. Design and analysis of two-phase studies with binary outcome applied to Wilms tumour prognosis. *Applied Statistics*, 48:457–468, 1999.

O.F. Christensen and R. Waagepetersen. Bayesian prediction of spatial count data using generalised linear mixed models. *Biometrics*, 58:280–286, 2002.

D. Clayton, L. Bernardinelli, and C. Montomoli. Spatial correlation in ecological analysis. *International Journal of Epidemiology*, 22:1193–1202, 1993.

N. Cressie and N.H. Chan. Spatial modelling of regional variables. *Journal of the American Statistical Association*, 84:393–401, 1989.

A.V. Diez-Roux. Bringing context back into epidemiology: Variables and fallacies in multilevel analysis. *American Journal of Public Health*, 88:216–222, 1998.

P.J. Diggle, J.A. Tawn, and R.A. Moyeed. Model-based geostatistics (with discussion). *Applied Statistics*, 47:299–350, 1998.

D. Dockery, C.A. Pope III, X. Xiping, J. Spengler, J. Ware, M. Fay, B. Ferris, and F. Speizer. An association between air pollution and mortality in six U.S. cities. *New England Journal of Medicine*, 329:1753–1759, 1993.

F. Dominici, L. Sheppard, and M. Clyde. Health effects of air pollution: A statistical review. *International Statistical Review*, 71:243–276, 2003.

P. Elliott, J.C. Wakefield, N.G. Best, and D.J. Briggs. *Spatial Epidemiology: Methods and Applications*. Oxford University Press, Oxford, U.K., 2000.

P. Elliott and J.C. Wakefield. Small-area studies of environment and health. In V. Barnett, A. Stein, and K.F. Turkman, Eds., *Statistics for the Environment 4: Health and the Environment*, pages 3–27. John Wiley & Sons, New York, 1999.

D.A. Freedman, S.P. Klein, M. Ostland, and M.R. Roberts. A solution to the ecological inference problem (book review). *Journal of the American Statistical Association*, 93:1518–1522, 1998.

A. Glynn, J. Wakefield, M. Handcock, and T. Richardson. Alleviating linear ecological bias and optimal design with subsample data. *Journal of the Royal Statistical Society, Series A*, 171:179–202, 2008.

S. Greenland. Divergent biases in ecologic and individual level studies. *Statistics in Medicine*, 11:1209–1223, 1992.

S. Greenland. A review of multilevel theory for ecologic analyses. *Statistics in Medicine*, 21:389–395, 2002.

S. Greenland and H. Morgenstern. Ecological bias, confounding and effect modification. *International Journal of Epidemiology*, 18:269–274, 1989.

S. Greenland and J. Robins. Ecological studies: Biases, misconceptions and counterexamples. *American Journal of Epidemiology*, 139:747–760, 1994.

C. Gu and P. Ma. Generalized non-parametric mixed-effects model: Computation and smoothing parameter selection. *Journal of Computational and Graphical Statistics*,14:485–504, 2005.

K. Guthrie, L. Sheppard, and J. Wakefield. A hierarchical aggregate data model with spatially correlated disease rates. *Biometrics*, 58:898–905, 2002.

S. Haneuse and J. Wakefield. Hierarchical models for combining ecological and case control data. *Biometrics*, 63:128–136, 2007.

S. Haneuse and J. Wakefield. The combination of ecological and case-control data. *Journal of the Royal Statistical Society, Series B*, 70:73–93, 2008a.

S. Haneuse and J. Wakefield. Geographic-based ecological correlation studies using supplemental case-control data. *Statistics in Medicine*, 27:864–887, 2008b.

S. Haneuse, J. Wakefield, and L. Sheppard. The interpretation of exposure effect estimates in chronic air pollution studies. *Statistics in Medicine*, 26:3172–3187, 2007.

P.J. Heagerty and T. Lumley. Window subsampling of estimating functions with application to regression models. *Journal of the American Statistical Association*, 95:197–211, 2000.

C. Jackson, N. Best, and S. Richardson. Hierarchical related regression for combining aggregate and individual data in studies of socio-economic disease risk factors. *Journal of the Royal Statistical Society, Series A*, 171:159–178, 2008.

C.H. Jackson, N.G. Best, and S. Richardson. Improving ecological inference using individual-level data. *Statistics in Medicine*, 25:2136–2159, 2006.

J.E. Kelsall and J.C. Wakefield. Modeling spatial variation in disease risk: A geostatistical approach. *Journal of the American Statistical Association*, 97:692–701, 2002.

G. King. *A Solution to the Ecological Inference Problem*. Princeton University Press, Princeton, NJ, 1997.

N. Künzli and I.B. Tager. The semi-individual study in air pollution epidemiology: A valid design as compared to ecologic studies. *Environmental Health Perspectives*, 10:1078–1083, NJ, 1997.

V. Lasserre, C. Guihenneuc-Jouyaux, and S. Richardson. Biases in ecological studies: Utility of including within-area distribution of confounders. *Statistics in Medicine*,19:45–59, 2000.

B.G. Leroux, X. Lei, and N. Breslow. Estimation of disease rates in small areas: A new mixed model for spatial dependence. In M.E. Halloran and D.A Berry, Eds., *Statistical Models in Epidemiology. The Environment and Clinical Trials*, pages 179–192. Springer, New York, 1999.

X. Lin and D. Zhang. Inference in generalized additive mixed models by smoothing splines. *Journal of the Royal Statistical Society, Series B*, 61:381–400, 1999.

P. McCullagh and J.A. Nelder. *Generalized Linear Models*, 2nd ed. Chapman & Hall, London, 1989.

H. Morgenstern. Ecologic study. In P. Armitage and T. Colton, Eds., *Encyclopedia of Biostatistics*, vol. 2, pp. 1255–1276. John Wiley & Sons, New York, 1998.

S. Openshaw. *The Modifiable Areal Unit Problem*. CATMOG No. 38, Geo Books, Norwich, U.K., 1984.

S. Piantadosi, D.P. Byar, and S.B. Green. The ecological fallacy. *American Journal of Epidemiology*, 127:893–904, 1988.

M. Plummer and D. Clayton. Estimation of population exposure. *Journal of the Royal Statistical Society, Series B*, 58:113–126, 1996.

R.L. Prentice and L. Sheppard. Aggregate data studies of disease risk factors. *Biometrika*, 82:113–125, 1995.

T.E. Raghunathan, P.K. Diehr, and A.D. Cheadle. Combining aggregate and individual level data to estimate an individual level correlation coefficient. *Journal of Educational and Behavioral Statistics*, 28:1–19, 2003.

B.J. Reich, J.S. Hodges, and V. Zadnik. Effects of residual smoothing on the posterior of the fixed effects in disease-mapping models. *Biometrics*, 62:1197–1206, 2006.

S. Richardson. Spatial models in epidemiological applications. In P.J. Green, N.L. Hjort, and S. Richardson, Eds., *Highly Structured Stochastic Systems*, pp. 237–259. Oxford Statistical Science Series, Oxford, U.K., 2003.

S. Richardson and C. Montfort. Ecological correlation studies. In P. Elliott, J.C. Wakefield, N.G. Best, and D. Briggs, Eds., *Spatial Epidemiology: Methods and Applications*, pages 205–220. Oxford University Press, Oxford, U.K., 2000.

S. Richardson, I. Stucker, and D. Hémon. Comparison of relative risks obtained in ecological and individual studies: Some methodological considerations. *International Journal of Epidemiology*, 16:111–20, 1987.

W.S. Robinson. Ecological correlations and the behavior of individuals. *American Sociological Review*, 15:351–357, 1950.

P.R. Rosenbaum and D.B. Rubin. Difficulties with regression analyses of age-adjusted rates. *Biometrics*, 40:437–443, 1984.

H. Rue, S. Martino, and N. Chopin. Approximate Bayesian inference for latent Gaussian models using integrated nested Laplace approximations (with discussion). *Journal of the Royal Statistical Society, Series B*, 71:319–392, 2009.

R. Salway and J. Wakefield. A hybrid model for reducing ecological bias. *Biostatistics*, 9:1–17, 2008.

R.A. Salway and J.C. Wakefield. Sources of bias in ecological studies of non-rare events. *Environmental and Ecological Statistics*, 12:321–347, 2005.

L. Sheppard. Insights on bias and information in group-level studies. *Biostatistics*, 4:265–278, 2003.

L. Sheppard, R.L. Prentice, and M.A. Rossing. Design considerations for estimation of exposure effects on disease risk, using aggregate data studies. *Statistics in Medicine*,15:1849–1858, 1996.

L. Sheppard and R.L. Prentice. On the reliability and precision of within- and between-population estimates of relative rate parameters. *Biometrics*, 51:853–863, 1995.

D.G. Steel and D. Holt. Analysing and adjusting aggregation effects: The ecological fallacy revisited. *International Statistical Review*, 64:39–60, 1996.

D.G. Steele, E.J. Beh, and R.L. Chambers. The information in aggregate data. In G. King, O. Rosen, and M. Tanner, Eds., *Ecological Inference: New Methodological Strategies*. Cambridge University Press, Cambridge, U.K., 2004.

S.W. Thurston, M.P. Wand, and J.K. Wiencke. Negative binomial additive models. *Biometrics*, 56:139–144, 2000.

J. Wakefield. Ecologic studies revisited. *Annual Review of Public Health*, 29:75–90, 2008.

J. Wakefield and P. Elliott. Issues in the statistical analysis of small area health data. *Statistics in Medicine*, 18:2377–2399, 1999.

J. Wakefield and S. Haneuse. Overcoming ecological bias using the two-phase study design. *American Journal of Epidemiology*, 167:908–916, 2008.

J. Wakefield and G. Shaddick. Health-exposure modelling and the ecological fallacy. *Biostatistics*, 7:438–455, 2006.

J.C. Wakefield. Sensitivity analyses for ecological regression. *Biometrics*, 59:9–17, 2003.

J.C. Wakefield. A critique of statistical aspects of ecological studies in spatial epidemiology. *Environmental and Ecological Statistics*, 11:31–54, 2004.

J.C. Wakefield. Ecological inference for 2 × 2 tables (with discussion). *Journal of the Royal Statistical Society, Series A*, 167:385–445, 2004.

J.C. Wakefield and R.E. Salway. A statistical framework for ecological and aggregate studies. *Journal of the Royal Statistical Society, Series A*, 164:119–137, 2001.

J.C. Wakefield. Disease mapping and spatial regression with count data. *Biostatistics*, 8:158–183, 2007.

A.M. Walker. Anamorphic analysis: Sampling and estimation for covariate effects when both exposure and disease are known. *Biometrics*, 38:1025–1032, 1982.

L.A. Waller and C.A. Gotway. *Applied Spatial Statistics for Public Health Data*. John Wiley & Sons, New York, 2004.

J.E. White. A two stage design for the study of the relationship between a rare exposure and a rare disease. *American Journal of Epidemiology*, 115:119–128, 1982.

L. Zhu, B.P. Carlin, and A.E. Gelfand. Hierarchical regression with misaligned spatial data: Relating ambient ozone and pediatric asthma ER visits in Atlanta. *Environmetrics*, 14:537–557, 2003.

31

Spatial Gradients and Wombling

Sudipto Banerjee

CONTENTS

31.1 Introduction

Spatial data are widely modeled using spatial processes that assume, for a study region D, a collection of random variables $\{Y(\mathbf{s}) : \mathbf{s} \in D\}$ where $\mathbf{s}$ indexes the points in D. This set is viewed as a randomly realized surface over D, which, in practice, is only observed at a finite set of locations in $\mathcal{S} = \{\mathbf{s}_1, \mathbf{s}_2, \ldots, \mathbf{s}_n\}$. Once such an interpolated surface has been obtained, investigation of rapid change on the surface may be of interest. Here, interest often lies in the rate of change of the surface at a given location in a given direction. Examples include temperature or rainfall gradients in meteorology, pollution gradients for environmental data, and surface roughness assessment for digital elevation models. Since the spatial surface is viewed as a random realization, all such rates of change are random as well.

Such local assessments of spatial surfaces are not restricted to points, but are often desired for curves and boundaries. For instance, environmental scientists are interested in ascertaining whether natural boundaries (e.g., mountains, forest edges, etc.) represent a zone of rapid change in weather, ecologists are interested in determining curves that delineate differing zones of species abundance, while public health officials want to identify change in healthcare delivery across municipal boundaries, counties, or states. The above objectives require the notion of gradients and, in particular, assigning gradients to curves (*curvilinear gradients*) in order to identify curves that track a path through the region where the surface is rapidly changing. Such boundaries are commonly referred to as difference boundaries or *wombling boundaries*, named after Womble (1951), who discussed their importance in understanding scientific phenomena (also see Fagan, Fortin, and Soykan, 2003).

Visual assessment of the surface over D often proceeds using contour and image plots of the surface fitted from the data using surface interpolators. Surface representation and contouring methods range from tensor-product interpolators for gridded data (e.g., Cohen, Riesenfeld, and Elber, 2001) to more elaborate adaptive control-lattice or tessellation-based interpolators for scattered data (e.g., Akima, 1996; Lee, Wolberg, and Shin, 1997). Mitas and Mitasova, (1999), provide a review of several such methods available in geographic information system (GIS) software (e.g., GRASS: http://grass.itc.it/). These methods are often fast and simple to implement and produce contour maps that reveal topographic features. However, they do not account for association and uncertainty in the data. Contrary to being competitive with statistical methods, they play a complementary role, creating descriptive plots from the raw data in the premodeling stage and providing visual displays of estimated response or residual surfaces in the postmodeling stage. It is worth pointing out that while contours often provide an idea about the local topography, they are not the same as wombling boundaries. Contour lines connect points with the same spatial elevation and may or may not track large gradients, so they may or may not correspond to wombling boundaries.

As a concept, wombling is useful because it attempts to quantify spatial information in objects, such as curves and paths, that are not easy to model as regressors. Existing wombling methods for point referenced data concentrate upon finding points having large gradients and attempt to connect them in an algorithmic fashion, which then defines a "boundary." These have been employed widely in computational ecology, anthropology, and geography. For example, Barbujani, Jacquez, and Ligi, (1990) and Barbujani, Oden, and Sokal, (1989) used wombling on red blood cell markers to identify genetic boundaries in Eurasian human populations by different processes restricting gene flow; Bocquet-Appel, and Bacro (1994) investigated genetic, morphometric, and physiologic boundaries; Fortin (1994, 1997) delineated boundaries related to specific vegetation zones, and Fortin, and Drapeau, (1995) applied wombling on real environmental data.

Building upon an inferential theory for spatial gradients in Banerjee, Gelfand, and Sirmans (2003), Banerjee and Gelfand (2006) formulated a Bayesian framework for point-referenced curvilinear gradients or boundary analysis, a conceptually harder problem due to the lack of definitive candidate boundaries. Spatial process models help in estimating not only response surfaces, but residual surfaces after covariate and systematic trends have been accounted. Depending on the scientific application, boundary analysis may be desirable on either. Algorithmic methods treat statistical estimates of the surface as "data" and apply interpolation-based wombling to obtain boundaries. Although such methods produce useful descriptive surface plots, they preclude formal statistical inference. Indeed, boundary assessment using such reconstructed surfaces suffers from inaccurate estimation of uncertainty. Evidently, gradients are central to wombling and the concerned spatial surfaces must be sufficiently smooth. This precludes methods, such as wavelet analysis, that have been employed in detecting image discontinuities (e.g., Csillag, and Kabos, 2002), but do not admit gradients.

While this chapter will primarily focus upon gradient-based wombling, it is worth pointing out that spatial data do not always arise from *fixed* locations. They are often observed only as summaries over geographical regions (say, counties or zip codes). Such data, known as *areal data*, are more commonly encountered in public health research to protect patient confidentiality. Boundary analysis, or "areal wombling," is concerned with identifying edges across which areal units are significantly different. In public health, this is useful for detecting regions of significantly different disease mortality or incidence, thus improving decision making regarding disease prevention and control, allocation of resources, and so on. Models for such data (see, e.g., Chapters 13 and 14; Banerjee, Carlin, and Gelfand,

2004; Cressie, 1993) do not yield smooth spatial surfaces, thereby precluding gradient-based wombling. We will briefly discuss wombling in these contexts.

31.2 Directional Finite Difference and Derivative Processes

Derivatives (more generally, linear functionals) of random fields have been discussed in Adler (1981), Banerjee et al. (2003) and Mardia, Kent, Goodall, and Little (1996). Let $Y(\mathbf{s})$ be a real-valued stationary spatial process with covariance function $Cov(Y(\mathbf{s}), Y(\mathbf{s}')) = K(\mathbf{s}-\mathbf{s}')$ where K is a positive definite function on $\Re^d$. Stationarity is not strictly required, but simplifies forms for the induced covariance function. The process $\{Y(\mathbf{s}) : \mathbf{s} \in \Re^d\}$ is L_2 (or mean square) continuous at $\mathbf{s}_0$ if $\lim_{\mathbf{s}\to\mathbf{s}_0} E(|Y(\mathbf{s}) - Y(\mathbf{s}_0)|)^2 = 0$. Under stationarity, we have $E(|Y(\mathbf{s}) - Y(\mathbf{s}_0)|)^2 = 2(K(\mathbf{0}) - K(\mathbf{s}-\mathbf{s}_0))$, thus, the process $Y(\mathbf{s})$ is mean square continuous at all sites $\mathbf{s}$ if K is continuous at $\mathbf{0}$.

The notion of a mean square differentiable process can be formalized using the analogous definition of total differentiability of a function in $\Re^d$ in a nonstochastic setting. To be precise, we say that $Y(\mathbf{s})$ is mean square differentiable at $\mathbf{s}_0$ if it admits a first-order linear expansion for any scalar h and any unit vector (direction) $\mathbf{u} \in \Re^d$,*

$$Y(\mathbf{s}_0 + h\mathbf{u}) = Y(\mathbf{s}_0) + h\langle \nabla Y(\mathbf{s}_0), \mathbf{u}\rangle + o(h) \tag{31.1a}$$

in the L_2 sense as $h \to 0$, where $\nabla Y(\mathbf{s}_0)$ is a $d \times 1$ vector called the gradient vector and $\langle \cdot, \cdot \rangle$ is the usual Euclidean inner-product on $\Re^d$.* That is, for any unit vector $\mathbf{u}$, we require

$$\lim_{h\to 0} E\left(\frac{Y(\mathbf{s}_0 + h\mathbf{u}) - Y(\mathbf{s}_0)}{h} - \langle \nabla Y(\mathbf{s}_0), \mathbf{u}\rangle\right)^2 = 0. \tag{31.1b}$$

This linearity ensures that mean square differentiable processes are mean square continuous.

Spatial gradients can be developed from *finite difference processes*. For any *parent process* $Y(\mathbf{s})$ and given direction $\mathbf{u}$ and any scale h, we have

$$Y_{\mathbf{u},h}(\mathbf{s}) = \frac{Y(\mathbf{s} + h\mathbf{u}) - Y(\mathbf{s})}{h}. \tag{31.2}$$

Clearly, for a fixed $\mathbf{u}$ and h, $Y_{\mathbf{u},h}(\mathbf{s})$ is a well-defined process on $\Re^d$ — in fact, with $\boldsymbol{\delta} = \mathbf{s}' - \mathbf{s}$, its covariance function is given by

$$C_{\mathbf{u}}^{(h)}(\mathbf{s}, \mathbf{s}') = \frac{(2K(\boldsymbol{\delta}) - K(\boldsymbol{\delta} + h\mathbf{u}) - K(\boldsymbol{\delta} - h\mathbf{u}))}{h^2}, \tag{31.3}$$

where $\mathrm{var}\big(Y_{\mathbf{u},h}(\mathbf{s})\big) = 2(K(\mathbf{0}) - K(h\mathbf{u}))/h^2$. The *directional derivative process* or *directional gradient process* is defined as $D_{\mathbf{u}}Y(\mathbf{s}) = \lim_{h\to 0} Y_{\mathbf{u},h}(\mathbf{s})$ when this mean square limit exists. Indeed when the parent process is mean square differentiable, i.e., Equation (31.1b) holds for every $\mathbf{s}_0$, then it immediately follows that, for each $\mathbf{u}$, $D_{\mathbf{u}}Y(\mathbf{s}) = \langle \nabla Y(\mathbf{s}), \mathbf{u}\rangle$ exists with equality again in the L_2 sense. In fact, under stationarity of the parent process, whenever the second-order partial and mixed derivatives of K exist and are continuous, $D_{\mathbf{u}}Y(\mathbf{s})$ is a well-defined process whose covariance function is obtained from the limit of Equation (31.3)

* Note that, unlike the nonstochastic setting, $Y(\mathbf{s}_0)$ is a random realization at $\mathbf{s}_0$. As well, $\nabla Y(\mathbf{s}_0)$ is not a function, but a random d-dimensional vector.

as $h \to 0$, yielding $C_{\mathbf{u}}(\mathbf{s}, \mathbf{s}') = -\mathbf{u}^T H_K(\boldsymbol{\delta})\mathbf{u}$, where $H_K(\boldsymbol{\delta}) = ((\partial^2 K(\boldsymbol{\delta})/\partial\delta_i\partial\delta_j))$ is the $d \times d$ Hessian matrix of $K(\boldsymbol{\delta})$.

More generally, collecting a set of p directions in $\Re^d$ into the $d \times p$ matrix, $U = [\mathbf{u}_1, \ldots, \mathbf{u}_p]$, we can write $D_U Y(\mathbf{s})$ as the $p \times 1$ vector, $D_U Y(\mathbf{s}) = (D_{\mathbf{u}_1} Y(\mathbf{s}), \ldots, D_{\mathbf{u}_p} Y(\mathbf{s}))^T$, so that $D_U Y(\mathbf{s}) = U^T \nabla Y(\mathbf{s})$. In particular, setting $p = d$ and taking U as the $d \times d$ identity matrix (i.e., taking the canonical basis, $\{\mathbf{e}_1, \ldots, \mathbf{e}_d\}$, as our directions), we have $D_I Y(\mathbf{s}) = \nabla Y(\mathbf{s})$, which gives a representation of $\nabla Y(\mathbf{s})$ in terms of the *partial derivatives* of the components of $Y(\mathbf{s})$. Explicitly, $\nabla Y(\mathbf{s}) = \left(\frac{\partial Y(\mathbf{s})}{\partial s_1}, \ldots, \frac{\partial Y(\mathbf{s})}{\partial s_d}\right)^T$, where $\mathbf{s} = \sum_{i=1}^d \mathbf{s}_i \mathbf{e}_i$, so $\mathbf{s}_i$s are the coordinates of $\mathbf{s}$ with respect to the canonical basis, and $D_U Y(\mathbf{s}) = U^T D_I Y(\mathbf{s})$. Thus, the derivative process in a set of arbitrary directions is a linear transformation of the partial derivatives in the canonical directions and all inference about the directional derivatives can be built from this relationship.

Formally, finite difference processes require less assumption for their existence. To compute differences, we need not worry about a numerical degree of smoothness for the realized spatial surface. However, issues of numerical stability can arise if h is too small. Also, with directional derivatives in, say, two-dimensional space, from the previous paragraph, we only need work with north and east directional derivatives processes in order to study directional derivatives in arbitrary directions. The nature of the data collection and the scientific questions of interest would often determine the choice of directional finite difference processes versus directional derivative processes. Differences, viewed as discrete approximations to gradients, may initially seem less attractive. However, in applications involving spatial data, scale is usually a critical question (e.g., in environmental, ecological, or demographic settings). Infinitesimal local rates of change may be of less interest than finite differences at the scale of a map of interpoint distances. On the other hand, gradients are of fundamental importance in geometry and physics and researchers in the physical sciences (e.g., geophysics, meteorology, oceanography) often formulate relationships in terms of gradients. Data arising from such phenomena may require inference through derivative processes.

31.3 Inference for Finite Differences and Gradients

Let $Y(\mathbf{s}) \sim GP(\mu(\mathbf{s}), K(\cdot))$ be a second-order stationary Gaussian process with mean $\mu(\mathbf{s})$ and covariance function $K(\boldsymbol{\delta})$. Both $\mu(\mathbf{s})$ and $K(\boldsymbol{\delta})$ typically depend upon unknown model parameters that we suppress for now. Consider a collection of p fixed direction vectors $U = [\mathbf{u}_1, \ldots, \mathbf{u}_p]$ and a fixed scale h and let $\mathbf{Y}_{U,h}(\mathbf{s}) = (Y_{\mathbf{u}_1,h}(\mathbf{s}), \ldots, Y_{\mathbf{u}_p,h}(\mathbf{s}))^T$. The $(p+1)\times 1$ process $\mathbf{Z}_{U,h}(\mathbf{s}) = (Y(\mathbf{s}), \mathbf{Y}_{U,h}^T(\mathbf{s}))^T$ is a well-defined multivariate Gaussian process (see, e.g., Banerjee et al., 2003) enabling inference for finite differences in a predictive fashion making use of the distribution theory arising from the multivariate process $\mathbf{Z}_{U,h}(\mathbf{s})$.

Let $\mathcal{S} = \{\mathbf{s}_1, \ldots, \mathbf{s}_n\}$ be a set of n points at which the outcome $Y(\mathbf{s})$ has been observed. The realizations of $\mathbf{Z}_{U,h}(\mathbf{s})$ over $\mathcal{S}$ is the $n(p+1)\times 1$ vector $\mathbf{Z}_{U,h} = (\mathbf{Z}_{U,h}^T(\mathbf{s}_1), \ldots, \mathbf{Z}_{U,h}^T(\mathbf{s}_n))^T$, which follows a multivariate normal distribution. The mean vector and dispersion matrix (more specifically, the cross-covariance matrices that specify the dispersion) are easily obtained from the linearity of the finite difference operation (see Banerjee et al., 2003 for details).

We seek the predictive distribution $p(Y_{U,h}(\mathbf{s}) \mid \mathbf{Y}) = \int p(Y_{U,h}(\mathbf{s}) \mid \boldsymbol{\theta}, \mathbf{Y}) p(\boldsymbol{\theta} \mid \mathbf{Y}) d\boldsymbol{\theta}$, where $\mathbf{Y} = (Y(\mathbf{s}_1), \ldots, Y(\mathbf{s}_n))^T$ and $\boldsymbol{\theta}$ is the collection of model parameters. Drawing samples from this distribution is routine using Bayesian composition sampling; given L posterior samples $\{\boldsymbol{\theta}^{(l)}\}_{l=1}^L$ from $p(\boldsymbol{\theta} \mid \mathbf{Y})$, we draw, one-for-one, $Y_{U,h}^{(l)}$ from $p(Y_{U,h}(\mathbf{s}) \mid \boldsymbol{\theta}^{(l)}, \mathbf{Y})$ — itself a normal distribution (see Banerjee et al., 2003 for details).

For spatial gradients, the process $\mathbf{Z}(\mathbf{s}) = (Y(\mathbf{s}), \nabla Y(\mathbf{s}))$ is a multivariate Gaussian process with a stationary cross-covariance function

$$\begin{pmatrix} Cov(Y(\mathbf{s}), Y(\mathbf{s}+\boldsymbol{\delta})) & Cov(Y(\mathbf{s}), \nabla Y(\mathbf{s}+\boldsymbol{\delta})) \\ Cov(\nabla Y(\mathbf{s}), Y(\mathbf{s}+\boldsymbol{\delta})) & Cov(\nabla Y(\mathbf{s}), \nabla Y(\mathbf{s}+\boldsymbol{\delta})) \end{pmatrix} = \begin{pmatrix} K(\boldsymbol{\delta}) & -(\nabla K(\boldsymbol{\delta}))^T \\ \nabla K(\boldsymbol{\delta}) & -H_K(\boldsymbol{\delta}) \end{pmatrix}.$$

This can be derived from the cross-covariance expressions for the finite differences $\mathbf{Z}_{U,h}(\mathbf{s})$ by setting U to be the $d \times d$ identity matrix and letting $h \to 0$. Indeed, $\mathbf{Z}(\mathbf{s}) = \lim_{h \to 0} \mathbf{Z}_{I,h}(\mathbf{s})$. The cross-covariance matrix enables the joint distribution $p(\mathbf{Y}, \nabla Y(\mathbf{s}) \mid \boldsymbol{\theta})$, allowing predictive inference for not only the gradient at arbitrary points, say $\mathbf{s}_0$, but also for functions thereof, including the direction of the maximal gradient $(\nabla Y(\mathbf{s}_0)/||\nabla Y(\mathbf{s}_0)||)$ and the size of the maximal gradient $(||\nabla Y(\mathbf{s}_0)||)$. Simplifications arise when the mean surface, $\mu(\mathbf{s})$, admits a gradient $\nabla \mu(\mathbf{s})$. Let $\boldsymbol{\mu} = (\mu(\mathbf{s}_1), \ldots, \mu(\mathbf{s}_n))$, let Σ_Y denote the $n \times n$ dispersion matrix for the data $\mathbf{Y}$, and let $\gamma^T = (\nabla K(\boldsymbol{\delta}_{01}), \ldots, \nabla K(\boldsymbol{\delta}_{0n}))$ be the $d \times n$ matrix with $\boldsymbol{\delta}_{0j} = \mathbf{s}_0 - \mathbf{s}_j$. Then, $p(\mathbf{Y}, \nabla Y(\mathbf{s}_0) \mid \boldsymbol{\theta})$ is distributed as the $d+n$ dimensional normal distribution

$$N_{d+n}\left(\begin{pmatrix} \boldsymbol{\mu} \\ \nabla \mu(\mathbf{s}_0) \end{pmatrix}, \begin{pmatrix} \Sigma_Y & \gamma \\ \gamma^T & -H_K(\mathbf{0}) \end{pmatrix}\right). \tag{31.4}$$

and the conditional predictive distribution for the gradient, $p(\nabla Y(\mathbf{s}_0) \mid \mathbf{Y}, \boldsymbol{\theta})$, is the d dimensional normal distribution $N_d\left(\nabla \mu(\mathbf{s}_0) - \gamma^T \Sigma_Y^{-1}(\mathbf{Y} - \boldsymbol{\mu}), -H_K(\mathbf{0}) - \gamma^T \Sigma_Y^{-1} \gamma\right)$. Simulation from the posterior predictive distribution $p(\nabla Y(\mathbf{s}_0) \mid \mathbf{Y})$ is again routine; for each $\boldsymbol{\theta}_{(l)}$ obtained from $p(\boldsymbol{\theta} \mid \mathbf{Y})$, we draw $\nabla Y^{(l)}(\mathbf{s}_0)$ from the $p\left(\nabla Y(\mathbf{s}_0) \mid \mathbf{Y}, \boldsymbol{\theta}^{(l)}\right)$. As long as $\nabla \mu(\mathbf{s}_0)$ is computable, obtaining samples from the above distribution is routine. In practice, we could have $\mu(\mathbf{s}, \boldsymbol{\beta}) = \mu$, a constant, in which case $\nabla \mu(\mathbf{s}) = \mathbf{0}$. More generally, we would have $\mu(\mathbf{s}, \boldsymbol{\beta}) = \mathbf{f}(\mathbf{s})^T \boldsymbol{\beta}$, where $\mathbf{f}(\mathbf{s})$ may represent some spatial regressors. If $\mathbf{f}(\mathbf{s})^T \boldsymbol{\beta}$ describes a trend surface, then explicit calculation of $\nabla \mu(\mathbf{s})$ will be possible. For a continuous regressor, such as elevation, we can interpolate a surface and approximate $\nabla \mu(\mathbf{s})$ at any location $\mathbf{s}$.

Note that it might suffice to consider gradients of a residual (or intercept) spatial process, say, $w(\mathbf{s})$, where $Y(\mathbf{s}) = \mu(\mathbf{s}; \boldsymbol{\beta}) + w(\mathbf{s}) + \epsilon(\mathbf{s})$, where $w(\mathbf{s}) \sim GP(0, K(\cdot; \phi))$ is a zero-mean Gaussian process, and $\epsilon(\mathbf{s}) \sim N(0, \tau^2)$ is a nugget or white-noise process. Inference will then proceed from the posterior distribution of $p(\nabla w(\mathbf{s}) \mid \mathbf{Y})$. Based on the above distribution theory, formal statistical inference on gradients can be performed. For instance, given any direction $\mathbf{u}$ and any location $\mathbf{s}_0$, a statistically "significant" directional gradient would mean that a 95% posterior credible interval for $D_{\mathbf{u}} w(\mathbf{s}_0)$ would not include 0. Since $D_{-\mathbf{u}} w(\mathbf{s}_0) = -D_{\mathbf{u}} w(\mathbf{s}_0)$, inference for $\mathbf{u}$ is the same as for $-\mathbf{u}$. Also, assessing significance of the spatial residual process, $w(\mathbf{s})$ is more general than for the parent process $Y(\mathbf{s})$. Indeed, when $\nabla \mu(\mathbf{s}_0)$ exists and there is no nugget ($\tau^2 = 0$), then the former is equivalent to testing the significance of $D_{\mathbf{u}} Y(\mathbf{s}_0)$ as a departure from the trend surface gradient $D_{\mathbf{u}} \mu(\mathbf{s}_0)$ (the *null value*). But even when $\nabla \mu(\mathbf{s}_0)$ is inaccessible, or there is a nugget $\tau^2 > 0$, assessment of spatial gradients for $w(\mathbf{s}_0)$ is still legitimate.

31.4 Illustration: Inference for Differences and Gradients

A simulation example (also see Banerjee et al., 2003) is provided to illustrate inference on finite differences and directional gradients. We generate data from a Gaussian random field with constant mean μ and a covariance structure specified through the Matèrn ($\nu = 3/2$) covariance function, $\sigma^2(1+\phi d)\exp(-\phi d)$. The field is observed on a randomly sampled set of points within a 10×10 square. We set $\mu = 0$, $\sigma^2 = 1.0$, and $\phi = 1.05$. In the subsequent

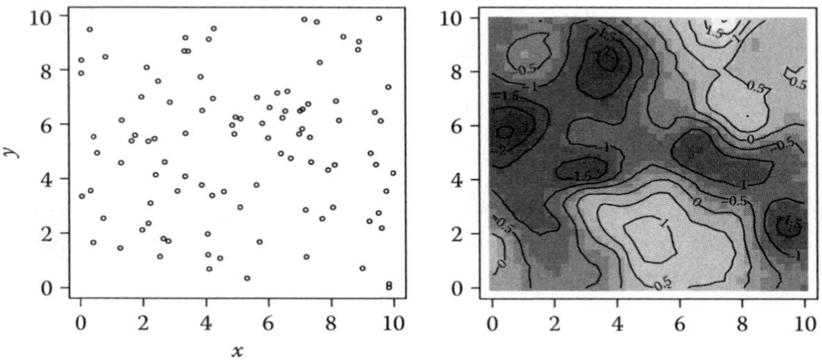

FIGURE 31.1
Left: Location of the 100 sites where the random field has been observed. Right: A grayscale plot with contour lines showing the topography of the random field in the simulation example.

illustration, our data consists of $n = 100$ observations at the randomly selected sites shown in the left panel of Figure 31.1. The maximum observed distance in our generated field is approximately 13.25 units. The value of $\phi = 1.05$ provides an effective isotropic range of about 4.5 units. We also perform a Bayesian kriging on the data to develop a predicted field. The right panel of Figure 31.1 shows a grayscale plot with contour lines displaying the topography of the "kriged" field. We will see below that our predictions of the spatial gradients at selected points are consistent with the topography around those points, as depicted in the right panel of Figure 31.1. Adopting a flat prior for μ, an IG(2, 0.1) (mean = 10, infinite variance) prior for σ^2, a $G(2, 0.1)$ prior (mean = 20, variance = 200) for ϕ, and a uniform on (1, 2) for ν, we obtain the posterior estimates for our parameters shown in Table 31.1.

We next predict the directional derivatives and directional finite differences for the unit vectors corresponding to angles of 0, 45, 90, 135, 180, 225, 270 and 315 degrees with the horizontal axis in a counterclockwise direction at the point. For the finite differences we consider $h = 1.0, 0.1$ and 0.01. Recall that $D_{-\mathbf{u}}Y(\mathbf{s}) = -D_{\mathbf{u}}Y(\mathbf{s})$. Table 31.2 presents the resulting posterior predictive inference for the point (3.5, 3.5) in Figure 31.1. We see that (3.5, 3.5) seems to be in a rather interesting portion of the surface, with many contour lines nearby. It is clear from the contour lines that there is a negative northern gradient (downhill) and a positive southern gradient (uphill) around this point. On the other hand, there does not seem to be any significant east–west gradient around that point as seen from the contour lines through that point running east–west. This is brought out very clearly in column 1 of Table 31.2. The angles of 0 and 180 degrees that correspond to the east–west gradients are not at all significant. The north–south gradients are indeed pronounced as seen by the 90 and 270 degree gradients. The directional derivatives along the diagonals also indicate presence of a gradient. There is a significant downhill gradient toward the northeast and (therefore) a significant uphill gradient toward the southwest. Hence, the directional derivative process provides inference consistent with features captured descriptively and visually in Figure 31.1.

TABLE 31.1
Posterior Estimates for Model Parameters

Parameter	50% (2.5%, 97.5%)
μ	−0.39 (−0.91, 0.10)
σ^2	0.74 (0.50, 1.46)
ϕ	1.12 (0.85, 1.41)
ν	1.50 (1.24, 1.77)

TABLE 31.2

Posterior Medians and (2.5%, 97.5%) Predictive Intervals for Directional Derivatives and Finite Differences at Point (3.5,3.5)

	(1)	(2)	(3)	(4)
Angle	$D_{\mathbf{u}}Y(\mathbf{s})\ (h=0)$	$h = 1.0$	$h = 0.1$	$h = 0.01$
0	−0.06 (−1.12, 1.09)	0.51 (−0.82, 1.81)	−0.08 (−1.23, 1.20)	−0.07 (−1.11, 1.10)
45	−1.49 (−2.81, −0.34)	−0.01 (−1.29, 1.32)	−1.55 (−2.93, −0.56)	−1.53 (−2.89, −0.49)
90	−2.07 (−3.44, −0.66)	−0.46 (−1.71, 0.84)	−2.13 (−3.40, −0.70)	−2.11 (−3.41, −0.69)
135	−1.42 (−2.68, −0.23)	−0.43 (−1.69, 0.82)	−1.44 (−2.64, −0.23)	−1.43 (−2.70, −0.23)
180	0.06 (−1.09, 1.12)	−0.48 (−1.74, 0.80)	0.08 (−1.19, 1.23)	0.06 (−1.10, 1.12)
225	1.49 (0.34, 2.81)	0.16 (−1.05, 1.41)	1.61 (0.52, 3.03)	1.52 (0.48, 2.90)
270	2.07 (0.66, 3.44)	0.48 (−0.91, 1.73)	2.12 (0.68, 3.43)	2.10 (0.68, 3.42)
315	1.42 (0.23, 2.68)	1.12 (−0.09, 2.41)	1.44 (0.24, 2.68)	1.42 (0.23, 2.70)

For the directional finite differences in columns 2, 3, and 4 of Table 31.2, note, for instance, the difference between column 2 and columns 3 and 4. In the former, none of the directional finite differences are significant. The low resolution (large h) fails to capture local topographic properties. On the other hand, the latter very much resemble column 1. As expected, at high resolution, the directional finite difference process results match those of the directional derivative process. Computational simplicity and stability (difficulties may arise with very small h in the denominator of (2)) encourage the use of the latter (see Banerjee et al., 2003 for details).

31.5 Curvilinear Gradients and Wombling

31.5.1 Gradients along Parametric Curves

Banerjee and Gelfand (2006), extend the concept of wombling with gradients from points to curves. The conceptual challenge here is to formulate a measure to associate with a curve, in order to assess whether it can be declared a wombling boundary. In applications, such curves might be proposed, for instance, as topographic or legislated boundaries or perhaps as level curves arising from a contouring routines.

Let C be an open curve in $\Re^2$ and we want to ascertain whether C is a wombling boundary with regard to $Y(\mathbf{s})$. To do so, we seek to associate an average gradient with C. In particular, for each point $\mathbf{s}$ lying on C, we let $D_{n(\mathbf{s})}Y(\mathbf{s})$ be the directional derivative in the direction of the unit normal $n(\mathbf{s})$. The rationale behind the direction normal to the curve is that, for a curve tracking rapid change in the spatial surface, lines orthogonal to the curve should reveal sharp gradients. We can define the *wombling measure* of the curve either as the total gradient along C,

$$\int_C D_{n(\mathbf{s})}Y(\mathbf{s})\,d\nu = \int_C \langle \nabla Y(\mathbf{s}), n(\mathbf{s})\rangle d\nu, \tag{31.5a}$$

or perhaps as the average gradient along C,

$$\frac{1}{\nu(C)}\int_C D_{n(\mathbf{s})}Y(\mathbf{s})\,d\nu = \frac{1}{\nu(C)}\int_C \langle \nabla Y(\mathbf{s}), n(\mathbf{s})\rangle d\nu, \tag{31.5b}$$

where $\nu(\cdot)$ is an appropriate measure. For Equation (31.5a) and Equation (31.5b), ambiguity arises with respect to the choice of measure. For example, $\nu(C) = 0$ if we take ν as two-dimensional Lebesgue measure and, indeed, this is true for any ν that is mutually absolutely continuous with respect to Lebesgue measure. Upon reflection, an appropriate choice for ν turns out to be arc-length. This can be made clear by a parametric treatment of the curve C.

In particular, a curve C in $\Re^2$ is a set parametrized by a single parameter $t \in \Re^1$ where $C = \{\mathbf{s}(t) : t \in \mathcal{T}\}$, with $\mathcal{T} \subset \Re^1$. We call $\mathbf{s}(t) = (s_{(1)}(t), s_{(2)}(t)) \in \Re^2$ the position vector of the curve – $\mathbf{s}(t)$ traces out C as t spans its domain. Then, assuming a differentiable curve with nonvanishing derivative $\mathbf{s}'(t) \neq 0$ (such a curve is often called *regular*), we obtain the (component-wise) derivative $\mathbf{s}'(t)$ as the "velocity" vector, with unit velocity (or tangent) vector $\mathbf{s}'(t)/||\mathbf{s}'(t)||$. Letting $n(\mathbf{s}(t))$ be the parametric unit normal vector to C, again if C is sufficiently smooth, then $\langle \mathbf{s}'(t), n(\mathbf{s}(t))\rangle = 0, a.e.\mathcal{T}$. In $\Re^2$, we see that $n(\mathbf{s}(t)) = \frac{(s'_{(2)}(t), -s'_{(1)}(t))}{||\mathbf{s}'(t)||}$.

Now the arc-length measure ν can be defined as $\nu(\mathcal{T}) = \int_{\mathcal{T}} ||\mathbf{s}'(t)|| \, dt$. In fact, $||\mathbf{s}'(t)||$ is analogous to the "speed" (the norm of the velocity) at "time" t, so the above integral is interpretable as the distance traversed or, equivalently, the arc-length $\nu(C)$ or $\nu(\mathcal{T})$. In particular, if $\mathcal{T}$ is an interval, say $[t_0, t_1]$, we can write $\nu(\mathcal{T}) = \nu_{t_0}(t_1) = \int_{t_0}^{t_1} ||\mathbf{s}'(t)|| \, dt$. Thus, we have $d\nu_{t_0}(t) = ||\mathbf{s}'(t)|| \, dt$ and, taking ν as the arc-length measure for C, we have the wombling measures in Equation (31.5a) (total gradient) and Equation (31.5b) (average gradient) respectively as

$$\Gamma_{Y(\mathbf{s})}(\mathcal{T}) = \int_{\mathcal{T}} \langle \nabla Y(\mathbf{s}(t)), n(\mathbf{s}(t))\rangle \, ||\mathbf{s}'(t)|| \, dt \text{ and } \bar{\Gamma}_{Y(\mathbf{s})}(\mathcal{T}) = \frac{1}{\nu(\mathcal{T})} \Gamma_{Y(\mathbf{s})}(\mathcal{T}). \qquad (31.6)$$

This result is important because we want to take ν as the arc-length measure, but it will be easier to use the parametric representation and work in t space. Also, it is a consequence of the implicit mapping theorem in mathematical analysis (see, e.g., Rudin, 1976) that any other parametrization $\mathbf{s}^*(t)$ of the curve C is related to $\mathbf{s}(t)$ through a differentiable mapping g such that $\mathbf{s}^*(t) = \mathbf{s}(g(t))$. This immediately implies (using Equation (31.6)) that our proposed wombling measure is invariant to the parametrization of C and, as desired, a feature of the curve itself.

For some simple curves, the wombling measure can be evaluated quite easily. For instance, when C is a segment of length 1 of the straight line through the point $\mathbf{s}_0$ in the direction $\mathbf{u} = (u_{(1)}, u_{(2)})$, then we have $C = \{\mathbf{s}_0 + t\mathbf{u} : t \in [0, 1]\}$. Under this parametrization, $\mathbf{s}'(t)^T = (u_{(1)}, u_{(2)})$, $||\mathbf{s}'(t)|| = 1$, and $\nu_{t_0}(t) = t$. Clearly, $n(\mathbf{s}(t)) = (u_{(2)}, -u_{(1)})$, (independent of t), which we write as $\mathbf{u}^{\perp}$ – the normal direction to $\mathbf{u}$. Therefore $\Gamma_{Y(\mathbf{s})}(\mathcal{T})$ in Equation (31.6) becomes

$$\int_0^1 \langle \nabla Y(\mathbf{s}(t)), n(\mathbf{s}(t))\rangle dt = \int_0^1 D_{\mathbf{u}^{\perp}} Y(\mathbf{s}(t)) \, dt.$$

Banerjee and Gelfand (2006), also consider the "flux" of a region bounded by a closed curve C (e.g., C might be the boundary of a county, census unit, or school district). The integral over a closed curve is denoted by $\oint_C$ and the average gradient in the normal direction to the curve C is written as

$$\frac{1}{\nu(C)} \oint_C \langle \nabla Y(\mathbf{s}), n(\mathbf{s})\rangle d\mathbf{s} = \frac{1}{\nu(C)} \oint_C \langle \nabla Y(\mathbf{s}(t)), n(\mathbf{s}(t))\rangle \, ||\mathbf{s}'(t)|| \, dt.$$

For a twice mean square differentiable surface $Y(\mathbf{s})$, the closed line integral can be written down as a double integral over the domain of $\mathbf{s}$ and no explicit parametrization by t is required. This offers computational advantages because the righthand integral can be computed by sampling within the region, which, in general, is simpler than along a curve. Working exclusively with line segments as described below, we confine ourselves to (4.5′).

Using the above formulation, we can give a formal definition of a curvilinear *wombling boundary*:

Definition: A curvilinear wombling boundary is a curve C that reveals a large wombling measure, $\Gamma_{Y(\mathbf{s})}(\mathcal{T})$ or $\bar{\Gamma}_{Y(\mathbf{s})}(\mathcal{T})$ (as given in Equation (31.6)) in the direction normal to the curve.

Were the surface fixed, we would have to set a threshold to determine what "large," say, in absolute value, means. Since the surface is a random realization, $\Gamma_{Y(\mathbf{s})}(\mathcal{T})$ and $\bar{\Gamma}_{Y(\mathbf{s})}(\mathcal{T})$ are random. Therefore, we declare a curve to be a wombling boundary if, say, a 95% credible set for $\bar{\Gamma}_{Y(\mathbf{s})}(\mathcal{T})$ does not contain 0. It is worth pointing out that while one normal direction is used in Equation (31.6), $-n(\mathbf{s}(t))$ would also have been a valid choice. Since $D_{-n(\mathbf{s}(t))}Y(\mathbf{s}(t)) = -D_{n(\mathbf{s}(t))}Y(\mathbf{s}(t))$, we note that the wombling measure with respect to one is simply the negative of the other. Thus, in the above definition, large positive as well as large negative values of the integral in Equation (31.6) would signify a wombling boundary. Being a local concept, an uphill gradient is equivalent to a downhill gradient across a curve, as are the fluxes radiating outward or inward for a closed region.

We also point out that, being a continuous average (or sum) of the directional gradients along a curve, the wombling measure may "cancel" the overall gradient effect. For instance, imagine a curve C that exhibits a large positive gradient in the $n(\mathbf{s})$ direction for the first half of its length and a large negative gradient for the second half, thereby canceling the total or average gradient effect. A remedy is to redefine the wombling measure using absolute gradients, $|D_{n(\mathbf{s})}Y(\mathbf{s})|$, in Equation (31.5a) and Equation (31.5b). The corresponding development does not entail any substantially new ideas, but would sacrifice the attractive distribution theory below. It will also make calibrating the resulting measure with regard to significance much more difficult. Moreover, in practice, a descriptive contour representation is usually available where sharp gradients will usually reflect themselves and one could instead compute the wombling measure for appropriate subcurves of C (such subcurves are usually unambiguous). More fundamentally, in certain applications, a signed measure may actually be desirable; one might want to classify a curve as a wombling boundary if it reflects either an overall "large positive" or a "large negative" gradient effect across it.

As for directional gradients at points, inference for curvilinear gradients will also proceed in posterior predictive fashion. Let us suppose that $\mathcal{T}$ is an interval, $[0, T]$, which generates the curve $C = \{\mathbf{s}(t) : t \in [0, T]\}$. For any $t^* \in [0, T]$ let $\nu(t^*)$ denote the arc-length of the associated curve C_{t^*}. The line integrals for total gradient and average gradient along C_{t^*} are given by $\Gamma_{Y(\mathbf{s})}(t^*)$ and $\bar{\Gamma}_{Y(\mathbf{s})}(t^*)$, respectively, as

$$\Gamma_{Y(\mathbf{s})}(t^*) = \int_0^{t^*} D_{n(\mathbf{s}(t))}Y(\mathbf{s}(t))\|\mathbf{s}'(t)\|dt \quad \text{and} \quad \bar{\Gamma}_{Y(\mathbf{s})}(t^*) = \frac{1}{\nu(t^*)}\Gamma_{Y(\mathbf{s})}(t^*). \tag{31.7}$$

We seek to infer about $\Gamma_{Y(\mathbf{s})}(t^*)$ based on data $Y = (Y(\mathbf{s}_1), \ldots, Y(\mathbf{s}_n))$. Although $D_{n(\mathbf{s})}Y(\mathbf{s})$ is a process on $\Re^2$, our parametrization of the coordinates by $t \in \mathcal{T} \subseteq \Re^1$ induces a valid process on $\mathcal{T}$. In fact, it is shown in Banerjee and Gelfand (2006) that $\Gamma_{Y(\mathbf{s})}(t^*)$ is a mean square continuous, but nonstationary (even if $Y(\mathbf{s})$ is stationary), Gaussian process. This enables us to carry out predictive inference as detailed in Banerjee and Gelfand (2006).

Returning to the model $Y(\mathbf{s}) = \mathbf{x}^T(\mathbf{s})\beta + w(\mathbf{s}) + \epsilon(\mathbf{s})$ with $\mathbf{x}(\mathbf{s})$ a general covariate vector, $w(\mathbf{s}) \sim GP(0, \sigma^2\rho(\cdot, \phi))$ and $\epsilon(\mathbf{s})$ a zero-centered, white-noise process with variance τ^2, consider boundary analysis for the residual surface $w(\mathbf{s})$. In fact, boundary analysis on the spatial residual surface is feasible in generalized linear modeling contexts with exponential families, where $w(\mathbf{s})$ may be looked upon as a nonparametric latent structure in the mean of the parent process as originally proposed in Diggle, Tawn, and Moyeed (1998). (See Section 31.6 for an example.)

Denoting by $\Gamma_{w(\mathbf{s})}(t)$ and $\bar{\Gamma}_{w(\mathbf{s})}(t)$ as the total and average gradient processes (as defined in (31.6)) for $w(\mathbf{s})$, we seek the posterior distributions $p(\Gamma_{w(\mathbf{s})}(t^*)|\mathbf{Y})$ and $p(\bar{\Gamma}_{w(\mathbf{s})}(t^*)|\mathbf{Y})$. Thus,

$$p(\Gamma_{w(\mathbf{s})}(t^*)|\mathbf{Y}) = \int p(\Gamma_{w(\mathbf{s})}(t^*)|\mathbf{w}, \boldsymbol{\theta})p(\mathbf{w}|\boldsymbol{\theta}, \mathbf{Y})p(\boldsymbol{\theta}|\mathbf{Y})d\boldsymbol{\theta}d\mathbf{w}, \tag{31.8}$$

where $\mathbf{w} = (w(s_1), \ldots, w(s_n))$ and $\boldsymbol{\theta} = (\boldsymbol{\beta}, \sigma^2, \phi, \tau^2)$. Posterior predictive sampling proceeds using posterior samples of $\boldsymbol{\theta}$, and is expedited in a Gaussian setting since $p(\mathbf{w}|\boldsymbol{\theta}, \mathbf{Y})$ and $p(\Gamma_{w(\mathbf{s})}(t^*)|\mathbf{w}, \boldsymbol{\theta})$ are both Gaussian distributions.

Formal inference for a wombling boundary is done more naturally on the residual surface $w(\mathbf{s})$, i.e., for $\Gamma_{w(\mathbf{s})}(t^*)$ and $\bar{\Gamma}_{w(\mathbf{s})}(t^*)$, because $w(\mathbf{s})$ is the surface containing any nonsystematic spatial information on the parent process $Y(\mathbf{s})$. Since $w(\mathbf{s})$ is a zero-mean process, one needs to check for the inclusion of this null value in the resulting 95% credible intervals for $\Gamma_{w(\mathbf{s})}(t^*)$ or, equivalently, for $\bar{\Gamma}_{w(\mathbf{s})}(t^*)$. Again, this clarifies the issue of the normal direction mentioned earlier; significance using $n(\mathbf{s}(t))$ is equivalent to significance using $-n(\mathbf{s}(t))$. One only needs to select and maintain a particular orthogonal direction relative to the curve.

31.6 Illustration: Spatial Boundaries for Invasive Plant Species

We present, briefly, an analysis considered in Banerjee and Gelfand (2006). The data were collected from 603 locations in Connecticut with presence/absence and abundance scores for some individual invasive plant species, along with some environmental predictors. The outcome variable $Y(\mathbf{s})$ is a presence–absence binary indicator (0 for absence) for one species *Celastrus orbiculatus* (Oriental bittersweet) at location $\mathbf{s}$. There are three categorical predictors: (1) habitat class (representing the current state of the habitat) of four different types; (2) land use and land cover (LULC) types (land use/cover history of the location, e.g., always forest, formerly pasture now forest, etc.) at five levels; and (3) a 1970 category number (LULC at one point in the past: 1970, e.g., forest, pasture, residential, etc.) with six levels. In addition, we have an ordinal covariate, canopy closure percentage (percent of the sky that is blocked by "canopy" of leaves of trees), a binary predictor for heavily managed points (0 if "no"; "heavy management" implies active landscaping or lawn mowing) and a continuous variable measuring the distance from the forest edge in the logarithm scale. A location under mature forest would have close to 100% canopy closure while a forest edge would have closer to 25% with four levels in increasing order. Figure 31.2 is a digital terrain image of the study domain, with the labeled curves indicating forest edges extracted using the GIS software ArcView (http://www.esri.com/). Ecologists are interested in evaluating spatial gradients along these 10 natural curves and identifying them as wombling boundaries.

We fit a logistic regression model with spatial random effects,

$$\log\left(\frac{P(Y(\mathbf{s}) = 1)}{P(Y(\mathbf{s}) = 0)}\right) = \mathbf{x}^T(\mathbf{s})\boldsymbol{\beta} + w(\mathbf{s}),$$

where $\mathbf{x}(\mathbf{s})$ is the vector of covariates observed at location $\mathbf{s}$ and $w(\mathbf{s}) \sim GP(0, \sigma^2\rho(\cdot; \phi, \nu))$ is a Gaussian process with $\rho(\cdot; \phi, \nu)$ as a Matérn correlation function. While $Y(\mathbf{s})$ is a binary surface that does not admit gradients, conducting boundary analysis on $w(\mathbf{s})$ may be of interest. The residual spatial surface reflects unmeasured or unobservable environmental features in the mean surface. Detecting significant curvilinear gradients to the residual surface tracks rapid change in the departure from the mean surface.

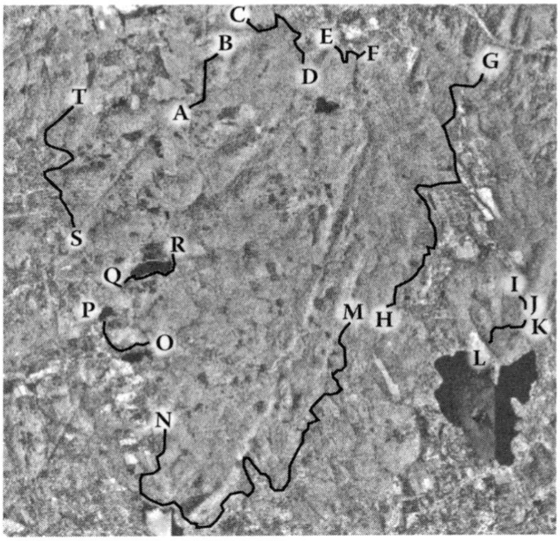

FIGURE 31.2
A digital image of the study domain in Connecticut indicating the forest edges as marked curves. These are assessed for significant gradients. Note: Eastings range from 699148 to 708961; Northings range from 4604089 to 4615875 for the image.

It was reported in Banerjee and Gelfand (2006) that the inference was fairly robust to the choice of priors. The results presented here result from a completely noninformative flat prior for β, an inverted-gamma $IG(2, 0.001)$ prior for σ^2 and the Matérn correlation function with a gamma prior for the correlation decay parameter, ϕ, specified so that the prior spatial range has a mean of about half of the observed maximum intersite distance (the maximum distance is 11,887 meters based on a UTM projection), and a $U(1, 2)$ prior for the smoothness parameter ν. (Further details may be found in Banerjee and Gelfand, 2006.)

Table 31.3 presents the posterior estimates of the model parameters. Most of the categorical variables reveal significance: Types 2 and 4 for habitat class have significantly different effects from Type 1; all the four types of LULC show significant departure from the baseline Type 1; for the 1970 category number, category 2 shows a significant negative effect, while categories 4 and 6 show significant positive effects compared to category 1. Canopy closure is significantly positive, implying higher presence probabilities of *Celastrus orbiculatus* with higher canopy blockage, while points that are more heavily managed appear to have a significantly lower probability of species presence as does the distance from the nearest forest edge.

Finally, Table 31.4 presents the formal curvilinear gradient analysis for the 10 forest edges in Figure 31.2. Six out of the 10 edge curves (with the exception of CD, EF, KL, and MN) are formally found to be wombling boundaries. The methodology proves useful here because some of these edge curves meander along the terrain for substantially long distances. Indeed, it is difficult to make visual assessments on the size (and significance) of average gradients for longer curves. Furthermore, with nongridded data as here, the surface interpolators (in this case, Akima, 1996) often find it difficult to extrapolate beyond a convex hull of the site locations. Consequently, parts of the curve (e.g., endpoints C, G, and (almost) T) lie outside the fitted surface, making local visual assessment quite impossible.

TABLE 31.3

Parameter Estimates for the Logistic Spatial Regression Example

Parameters	50% (2.5%, 97.5%)
Intercept	0.983 (−2.619, 4.482)
Habitat Class (Baseline: Type 1)	
Type 2	−0.660 (−1.044, −0.409)
Type 3	−0.553 (−1.254, 0.751)
Type 4	−0.400 (−0.804, −0.145)
Land Use, Land Cover Types (Baseline: Level 1)	
Type 2	0.591 (0.094, 1.305)
Type 3	1.434 (0.946, 2.269)
Type 4	1.425 (0.982, 1.974)
Type 5	1.692 (0.934, 2.384)
1970 Category Types (Baseline: Category 1)	
Category 2	−4.394 (−6.169, −3.090)
Category 3	−0.104 (−0.504, 0.226)
Category 4	1.217 (0.864, 1.588)
Category 5	−0.039 (−0.316, 0.154)
Category 6	0.613 (0.123, 1.006)
Canopy Closure	0.337 (0.174, 0.459)
Heavily Managed Points (Baseline: No)	
Yes	−1.545 (−2.027, −0.975)
Log Edge Distance	−1.501 (−1.891, −1.194)
σ^2	8.629 (7.005, 18.401)
ϕ	1.75E-3 (1.14E-3, 3.03E-3)
ν	1.496 (1.102, 1.839)
Range (in meters)	1109.3 (632.8, 1741.7)

Quickly and reliably identifying forest edges could be useful in determining boundaries between areas of substantial anthropogenic activity and minimally managed forest habitats. Such boundaries are important because locations at which forest blocks have not been invaded by exotic plant species may be subject to significant seed rain from these species. It is worth pointing out that the spatial residual surface can greatly assist scientists in finding important missing predictors. Identifying significant wombling boundaries on such surfaces can throw light upon unobserved or "lurking" predictors that may be causing local disparities on this residual surface.

TABLE 31.4

Curvilinear Gradient Assessment for the 10 Forest Edges Labeled in Figure 31.2 for the Logistic Regression Example

Curve	Average Gradient	Curve	Average Gradient
AB	1.021 (0.912, 1.116)	KL	0.036 (−0.154, 0.202)
CD	0.131 (-0.031, 0.273)	MN	0.005 (−0.021, 0.028)
EF	0.037 (-0.157, 0.207)	OP	0.227 (0.087, 0.349)
GH	1.538 (1.343, 1.707)	QR	0.282 (0.118, 0.424)
IJ	0.586 (0.136, 0.978)	ST	0.070 (0.017, 0.117)

31.7 A Stochastic Algorithm for Constructing Boundaries

The foregoing strategy can be used to *test* whether a given boundary is a difference boundary. However, in practice it will often be of interest to develop a stochastic algorithm to *construct* boundaries. It is both natural and easiest to construct boundaries using polygonal curves, hence, to consider piecewise linear boundaries. One option is a posterior predictive construction that only requires a starting point, say $\mathbf{s}_0^{(1)}$, and then travels an unknown distance $t_{(1)}^*$ in the direction $\mathbf{u}^{(1)}$ that is perpendicular to the direction of maximal spatial gradient at $\mathbf{s}_0^{(1)}$ (i.e., $\nabla Y(\mathbf{s}_0^{(1)})/\|\nabla Y(\mathbf{s}_0^{(1)})\|$). (We call this direction the *optimal* direction.) The starting point $\mathbf{s}_0^{(1)}$ will be a point showing a significant maximal spatial gradient, perhaps suggested by a contour plot of the surface, but formally tested as discussed in Section 31.2. Of course, if the surface is smooth and there is at least one location with a significant maximal gradient, there will be an uncountable collection of such points, revealing that curvilinear wombling boundaries offer a simplified representation for a *zone of transition*. Moreover, typically, the contour plot will suggest starting points in several different parts of the region. Thus, we may use the algorithm below to construct boundaries in a local fashion. Posterior sampling, obtained as described below, will enable us to estimate the distance $t_{(1)}^*$. Evidently, this distance along with the direction and $\mathbf{s}_0^{(1)}$ determine a candidate, $\mathbf{s}_0^{(2)}$, to enable the algorithm to proceed. Note, that the direction vector $\mathbf{u}^{(1)}$ is determined only up to *sign*. The level curves of the contour plot will be needed to enable us to resolve any ambiguities in direction as the algorithm proceeds.

With regard to choosing T, without loss of generality (see below), the $\mathbf{s}(t)$ can always be linear segments of at most unit length, so we simply write $\bar{\Gamma}(t)$ for $\bar{\Gamma}_{Y(\mathbf{s})}(t)$. Let $\bar{\Gamma}_{[0,1]} = \{\bar{\Gamma}(t) : t \in [0, 1]\}$ be a complete evaluation of the average gradient process over its domain. Clearly, for a given starting point $\mathbf{s}_0^{(1)}$ and the corresponding optimal direction $\mathbf{u}^{(1)}$ (as described above), we want to move a distance $\tilde{t}$ so that the average gradient is maximized. We, therefore, define $\tilde{t}$ as arg sup $\bar{\Gamma}_{[0,1]}$ and seek the posterior predictive distribution $P(\tilde{t}|\mathbf{Y})$. While intractable in closed form, our sampling-based framework enables routine computation of $P(\tilde{t}|\mathbf{Y})$ as follows. We consider a set of points $t_0 = 0 < t_1 < \cdots < t_{M-1} < 1 = t_M$ and obtain samples from the predictive distribution $p(\bar{\Gamma}(t)|\mathbf{Y})$ for each of the points t_k, $k = 0, \ldots, M$, say, $\{\bar{\Gamma}^l(t_k)\}$. This immediately renders samples from $P(\tilde{t}|\mathbf{Y})$ by computing, for each l, $\{\tilde{t}^l\} = \arg\max_k \bar{\Gamma}^l(t_k)$.

Having obtained the full posterior distribution for $\tilde{t}$, we compute $\tilde{t}^{(1)}$ as a summary measure (posterior mean or median) and set $\mathbf{s}_0^{(2)} = \mathbf{s}_0^{(1)} + \tilde{t}^{(1)}\mathbf{u}^{(1)}$. We then check whether $\mathbf{s}_0^{(2)}$ has a significant spatial gradient $\nabla Y(\mathbf{s}_0^{(2)})$. If not, then we stop and the boundary ends there. If the gradient is significant, we repeat the above steps moving in a direction $\mathbf{u}^{(2)}$ perpendicular to $\nabla Y(\mathbf{s}_0^{(2)})/\|\nabla Y(\mathbf{s}_0^{(2)})\|$ and evaluating a $\tilde{t}^{(2)}$. The procedure is continued, stopping when we arrive at a point with insignificant gradient. Note that wombling boundaries constructed in this manner will be closed with probability zero. In fact, it is possible that the algorithm never stops, for instance, if we are moving around under, say, a roughly conical portion of the surface. Of course, in such a case, essentially every point under the cone is a boundary element and the notion of an associated boundary is not well defined. To avoid endless circling, ad hoc intervention can be imposed, e.g., stopping at the first segment J such that $\sum_{j=1}^{J} \angle_j \geq 2\pi$, where $\angle_j$ is the angle between $\mathbf{u}^{(j)}$ and $\mathbf{u}^{(j+1)}$.

Finally, note that there is no loss of generality in restricting attention to t in [0, 1]. Should the $(j-1)$st line segment *want* to be greater than 1, we will expect the estimated $\tilde{t}$ to be close to 1, but the $\mathbf{u}^{(j)}$ that we compute for the next segment will still "point" us in essentially the same direction (as $\mathbf{u}^{(j-1)}$). Also, the boundary drawn is the "mean" or "median" boundary in the sense that it uses the posterior mean or median of $\tilde{t}$. Uncertainty in these drawn boundaries can be displayed by choosing, at each segment, a random $\tilde{t}$ from $p(\tilde{t}|\mathbf{Y})$ (rather than the mean or median) to create a random posterior boundary associated with the particular starting point. Overlaying a sample of, say, 100 such boundaries on the region would reveal posterior variation.

31.8 Other Forms of Wombling

31.8.1 Areal Wombling: A Brief Overview

As noted in the Introduction, "areal wombling" refers to the exercise of ascertaining boundaries on *areally* referenced data. In the absence of smooth realizations of spatial surfaces, areal wombling, therefore, cannot employ spatial gradients. The gradient is not explicitly modeled; boundary effects are looked upon as edge effects and modeled using Markov random field specifications. Boundaries in areal wombling are just a collection of segments (or *arcs*, in GIS parlance) dually indexed by ij, corresponding to the two adjacent regions i and j the segment separates. In the fields of image analysis and pattern recognition, there has been much research in using statistical models for capturing "edge" and "line" effects (see, e.g., Besag, 1986; Geman and Geman, 1984; Geman and McClure, 1985; Geman and McClure, 1987; Helterbrand, Cressie, and Davidson, 1994). See also Cressie (1993) (Section 7.4) and references therein. Such models are based on probability distributions, such as Gibbs distributions or Markov random fields (MRFs) (see Chapters 12 and 13 and Rue and Held, 2006) that model pixel intensities as conditional dependencies using the neighborhood structure (see, e.g., Chellappa and Jain, 1993). Modeling objectives include identification of edges based on distinctly different image intensities in adjacent pixels.

In these models, local spatial dependence between the observed image characteristics is captured by a *neighborhood structure*, where a pixel is independent of the rest given the values of its neighbors. Various neighborhood structures are possible (see, e.g., Banerjee et al., 2004, pp. 70–71), but all propose stronger statistical dependence between data values from areas that are spatially closer, thus inducing local smoothing. However, this leads to a new problem: When real discontinuities (boundaries) exist between neighboring pixels, MRF models often lead to oversmoothing, blurring these edges.

Although the boundary analysis problem for public health data resemble the edge-detection problem in image processing, significant differences exist. Unlike image pixels, geographical maps that form the domain of most public health data are not regularly spaced, but still have a well-defined neighborhood structure (a topological graph). Furthermore, there are usually far fewer of these areas than the number of pixels that would arise in a typical image restoration problem, so we have far less data. Finally, the areal units (polygons) are often quite different in size, shape, and number of neighbors, leading, for example, to different degrees of smoothing in urban and rural regions as well as near the external boundary of the study region.

Deterministic methods for areal wombling do exist, and include the polygonal wombling algorithm implemented in the `BoundarySeer` software (http://www.terraseer.com). For instance, Jacquez and Greiling (2003) employed polygonal wombling to estimate boundaries for breast, lung, and colorectal cancer rates in males and females in Nassau, Suffolk, and Queens counties in New York. Lu and Carlin (2005) proposed a fully model-based hierarchical Bayesian wombling model and showed its advantages over the algorithmic version implemented in `BoundarySeer`. This approach uses MRF methods to account for spatial structure, but suffers from the oversmoothing problems mentioned above. It also generally fails to produce the long series of connected boundary segments often desired by practitioners. This work was extended further by Lu, Reilly, Banerjee, and Carlin (2007), who developed a hierarchical framework that models the adjacency matrix underlying these maps, while Ma and Carlin (2007) have extended this work to multiple disease maps. More recently, Ma, Carlin, and Banerjee (2008) proposed a compound Gaussian–Markov random field model using the Ising distribution as priors on edges.

31.8.2 Wombling with Point Process Data

In disease mapping and public health, a spatial layer for which boundary analysis would be of considerable interest is the pattern of disease incidence. In particular, we would seek to identify transition from areas with low incidence to areas with elevated incidence. For cases aggregated to counts for areal units, e.g., census blocks or zip codes (in order to protect confidentiality), this would require obtaining standardized incidence rates for the units. Wombling for such data is discussed in Jacquez and Greiling (2003).

Note that the data format here is similar to that of areal wombling. However, a fundamental conceptual difference is that here we model an assumed *point-level process* that is being aggregated to produce the areal summary. Whether we analyze the random locations or the aggregated counts, we assume that the point pattern is driven by an intensity surface, $\lambda(\mathbf{s})$. (See Part IV of this book.) Wombling for the observed point pattern would be achieved by wombling the estimated $\lambda(\mathbf{s})$. A bit more generally, Liang, Banerjee, and Carlin (2008) and Liang, Carlin, and Gelfand (2007) propose viewing the data, including locations and nuisance covariates, as a random realization from some nonhomogeneous Poisson process with intensity function $\lambda(\mathbf{s}, \mathbf{v})$ defined over the product space $D \times \mathcal{V}$, where $\mathcal{V}$ is the nuisance covariate space. These authors further let $\lambda(\mathbf{s}, \mathbf{v}, k) = r(\mathbf{s})\pi(\mathbf{s}, \mathbf{v}, k)$, where $r(\mathbf{s})$ is the population density surface at location $\mathbf{s}$, where $r(\mathbf{s})$ serves as an offset and $\pi(\mathbf{s}, \mathbf{v}, k)$ is interpreted as a population-adjusted (or *relative*) intensity surface. We then set $\pi(\mathbf{s}, \mathbf{v}, k) = \exp\{\beta_{0k} + \mathbf{z}(\mathbf{s})^T \boldsymbol{\beta}_k + \mathbf{v}^T \boldsymbol{\alpha}_k + w_k(\mathbf{s})\}$, where $w_k(\mathbf{s})$ is a zero-centered stochastic process, and β_{0k}, $\boldsymbol{\beta}_k$, and $\boldsymbol{\alpha}_k$ are unknown regression coefficients. With $w_k(\mathbf{s})$ a Gaussian process and no nonspatial covariates, the original point process becomes a log *Gaussian–Cox process*. Wombling (point and curves) can now proceed by applying the gradient-based wombling methods to $\mathbf{w}(\mathbf{s})$. Details, including an extension to *marked point processes* is discussed in Liang et al. (2008).

31.9 Concluding Remarks

In this chapter, we have discussed recently proposed theories concerning the use of spatial gradients to detect points and curves that represent rapid change. Here we have confined ourselves to curves that track zones of rapid change. However, as we have alluded to above, zones of rapid change are areal notions; description by a curve may be an unsatisfying simplification. As the simplest illustration, consider an entirely flat landscape apart from a cone somewhere in the interior. Should the wombling boundary be a contour of the cone? Which one, since all the contours of the cone have the same average wombling measure? Is the entire footprint of the cone the more appropriate boundary? Thus, describing zones as areal quantities, i.e., as sets of nonzero Lebesgue measure in $\Re^2$ is an alternative. We need to build suitable models for random sets. To proceed, the crucial issue is to formalize shape-oriented definitions of a wombling boundary. While much work has been done on statistical shape analysis, its use in the point-referenced spatial data context we have set out is unexplored. There are many possibilities, using formal differential geometry and calculus of variations, providing directions for future research. In summary, current approaches are built entirely upon the specification of a point-referenced spatial process model. One might examine the boundary analysis problem from an alternative modeling perspective, where curves or zones arise as random processes. Possibilities include line processes, random tessellations, and random areal units.

References

Adler, R.J. (1981). *The Geometry of Random Fields*, Chichester, U.K.: John Wiley & Sons.

Akima, H. (1996). Algorithm 761: Scattered-data surface fitting that has the accuracy of a cubic polynomial. *ACM Transactions on Mathematical Software*, **22**, 362–371.

Allard, D., Gabriel, E., and Bacro, J.N. (2005). Estimating and testing zones of abrupt changes for spatial data. Research report, Unit de Biomtrie, INRA (Institut National de la Recherche Agronomique), Avignon, France.

Banerjee, S., Carlin, B.P., and Gelfand, A.E. (2004). *Hierarchical Modeling and Analysis for Spatial Data*. Boca Raton, FL: Chapman & Hall.

Banerjee, S. and Gelfand, A.E. (2006). Bayesian wombling: Curvilinear gradient assessment under spatial process models. *Journal of the American Statistical Association*, **101**, 1487–1501.

Banerjee, S., Gelfand, A.E., and Sirmans, C.F. (2003). Directional rates of change under spatial process models. *Journal of the American Statistical Association*, **98**, 946–954.

Barbujani, G., Jacquez, G.M., and Ligi, L. (1990). Diversity of some gene frequencies in European and Asian populations. V. Steep multilocus clines. *American Journal of Human Genetics* **47**, 867–875.

Barbujani, G., Oden, N.L., and Sokal, R.R. (1989). Detecting areas of abrupt change in maps of biological variables. *Systematic Zoology*, **38**, 376–389.

Besag, J. (1986). On the statistical analysis of dirty pictures. *Journal of the Royal Statistical Society, Ser. B*, **48**, 259–302.

Bocquet-Appel, J.P. and Bacro, J.N. (1994). Generalized wombling. *Systematic Biology*, **43**, 442–448.

Chellappa, R. and Jain, A.K. (1993), *Markov Random Fields*. New York: Academic Press.

Cohen, E., Riesenfeld, R.F., and Elber, G. (2001). *Geometric Modeling with Splines: An Introduction*. Natick, MA: A.K. Peters.

Cressie, N.A.C. (1993), *Statistics for Spatial Data*, 2nd ed. New York: John Wiley & Sons.

Csillag, F. and Kabos, S. (2002). Wavelets, boundaries and the analysis of landscape pattern. *Ecoscience*, **9**, 177–190.

Dass, S.C. and Nair, V.N. (2003). Edge detection, spatial smoothing, and image reconstruction with partially observed multivariate data. *Journal of the American Statistical Association*, **98**, 77–89.

Diggle, P.J., Tawn, J.A., and Moyeed, R.A. (1998). Model-based geostatistics (with discussion). *Applied Statistics*, **47**, 299–350.

Fagan, W.F., Fortin, M.J., and Soykan, C. (2003). Integrating edge detection and dynamic modeling in quantitative analyses of ecological boundaries. *BioScience*, **53**, 730–738.

Fortin, M.J. (1994). Edge detection algorithms for two-dimensional ecological data. *Ecology*, **75**, 956–965.

Fortin, M.J. (1997). Effects of data types on vegetation boundary delineation. *Canadian Journal of Forest Research*, **27**, 1851–1858.

Fortin, M.J. and Drapeau, P. (1995). Delineation of ecological boundaries: Comparisons of approaches and significance tests. *Oikos*, **72**, 323–332.

Geman, S. and Geman, D. (1984). Stochastic relaxation, Gibbs distributions and the Bayesian restoration of images. *IEEE Transactions on Pattern Analysis and Machine Intelligence*, **6**, 721–742.

Geman, S. and McClure, D.E. (1985). Bayesian image analysis: An application to single photon emission tomography. *Proceedings of the Statistical Computing Section, American Statistical Association*. Washington, D.C., 12–18.

Geman, S. and McClure, D.E. (1987). Statistical methods for tomographic image reconstruction. *Bulletin of the International Statistical Institute*, **52**, 5–21.

Helterbrand, J.D., Cressie, N., and Davidson, J.L. (1994). A statistical approach to identifying closed object boundaries in images. *Advances in Applied Probability*, **26**, 831–854.

Jacquez, G.M. and Greiling, D.A. (2003). Geographic boundaries in breast, lung and colorectal cancers in relation to exposure to air toxins in Long Island, New York. *International Journal of Health Geographics*, **2**, 4.

Lee, S., Wolberg, G., and Shin, S.Y. (1997). Scattered data interpolation with multilevel B-splines. *IEEE Transactions on Visualization and Computer Graphics*, **3**, 228–244.

Liang, S., Banerjee, S., and Carlin, B.P. (2008). Bayesian wombling for spatial point processes. Forthcoming. *Biometrics*.

Liang, S., Carlin, B.P., and Gelfand, A.E. (2007). Analysis of marked point patterns with spatial and non-spatial covariate information. Research Report 2007–019, Division of Biostatistics, University of Minnesota, Minneapolis.

Lu, H. and Carlin, B.P. (2005). Bayesian areal wombling for geographical boundary analysis. *Geographical Analysis*, **37**, 265–285.

Lu, H., Reilly, C., Banerjee, S., and Carlin, B.P. (2007). Bayesian areal wombling via adjacency modeling. *Environmental and Ecological Statistics*, **14**, 433–452.

Ma, H. and Carlin, B.P. (2007). Bayesian multivariate areal wombling for multiple disease boundary analysis. *Bayesian Analysis*, **2**, 281–302.

Ma, H., Carlin, B.P., and Banerjee, S. (Forthcoming). Hierarchical and joint site-edge methods for medicare hospice service region boundary analysis. *Biometrics*.

Mardia, K.V., Kent, J.T., Goodall, C.R., and Little, J.A. (1996). Kriging and splines with derivative information. *Biometrika*, **83**, 207–221.

Mitas, L. and Mitasova, H. (1999). Spatial interpolation. In P. Longley, M.F. Goodchild, D.J. Maguire, D.W. Rhind (eds.) *Geographical Information Systems: Principles, Techniques, Management and Applications*, GeoInformation International. New York: John Wiley & Sons, 481–492.

Møller, J. (ed.) (2003). *Spatial Statistics and Computational Methods*. New York: Springer.

Rudin, W. (1976). *Principles of Mathematical Analysis*. New York: McGraw-Hill Book Co.

Rue, H. and Held, L. (2006). *Gaussian Markov Random Fields: Theory and Applications*. Boca Raton, FL: Chapman & Hall.

Stein, M.L. (1999). *Interpolation of Spatial Data: Some Theory of Kriging*. New York: Springer.

Womble, W.H. (1951). Differential systematics. *Science*, **114**, 315–322.

Index

D

H

I

J

K

L

M

O

T